PROCEEDINGS SERIES

POPULATION DOSE EVALUATION AND STANDARDS FOR MAN AND HIS ENVIRONMENT

PROCEEDINGS OF THE SEMINAR
ON RADIOLOGICAL SAFETY EVALUATION
OF POPULATION DOSES AND
APPLICATION OF RADIOLOGICAL SAFETY STANDARDS
TO MAN AND THE ENVIRONMENT
ORGANIZED BY THE
INTERNATIONAL ATOMIC ENERGY AGENCY
AND THE WORLD HEALTH ORGANIZATION
WITH THE SUPPORT OF THE
UNITED NATIONS ENVIRONMENT PROGRAMME
AND HELD IN PORTOROŽ, 20-24 MAY 1974

INTERNATIONAL ATOMIC ENERGY AGENCY
VIENNA, 1974

POPULATION DOSE EVALUATION
AND STANDARDS FOR MAN AND HIS ENVIRONMENT
IAEA, VIENNA, 1974
STI/PUB/375

ISBN 92-0-020374-4

Printed by the IAEA in Austria
September 1974

FOREWORD

Radiation protection practice for wastes produced during the uranium fuel cycle requires that almost all the wastes be contained. However, after careful assessment of the risks of exposure of man, or of accident, small amounts of low radioactive effluents are occasionally released to the environment from nuclear installations.

The basic principles and criteria involved in such releases to the environment can be summarized as follows: (1) to avoid all unnecessary and unjustifiable exposure; (2) to justify any necessary exposure and to keep it as low as reasonably possible in the light of the economic and social factors, ensuring at the same time that the exposure in no case exceeds the dose limits prescribed by the ICRP and other international and national competent authorities.

Radioactive material released to the biosphere can reach man through a variety of pathways; consequently the justifiable exposure, in accordance with principle (2) above, should be kept as low as is readily achievable. This requirement can be met by the process of optimization of exposure. Optimization of exposure can be rationally achieved by the process of differential cost-benefit analysis, and this was the main subject of a Seminar on Population Dose Evaluation and Standards for Man and his Environment organized by the International Atomic Energy Agency and the World Health Organization with the support of the United Nations Environment Programme and held at Portorož, Yugoslavia, from 20 to 24 May 1974. The Seminar also examined some of the possible ecological effects of the operation of a nuclear installation on the balance of nature in the environment of man. The Seminar provided a valuable forum for the exchange of experience in Member States relating to the exposure of populations or individuals as well as to environmental studies.

The Agency gratefully acknowledges the invitation for this Seminar from the Government of the Socialist Federal Republic of Yugoslavia, and the assistance and co-operation of the staffs of the Jožef Stefan Institute, Ljubljana, and the Zavod za Turizem, Portorož.

EDITORIAL NOTE

The papers and discussions incorporated in the proceedings published by the International Atomic Energy Agency are edited by the Agency's editorial staff to the extent considered necessary for the reader's assistance. The views expressed and the general style adopted remain, however, the responsibility of the named authors or participants.

For the sake of speed of publication the present Proceedings have been printed by composition typing and photo-offset lithography. Within the limitations imposed by this method, every effort has been made to maintain a high editorial standard; in particular, the units and symbols employed are to the fullest practicable extent those standardized or recommended by the competent international scientific bodies.

The affiliations of authors are those given at the time of nomination.

The use in these Proceedings of particular designations of countries or territories does not imply any judgement by the Agency as to the legal status of such countries or territories, of their authorities and institutions or of the delimitation of their boundaries.

The mention of specific companies or of their products or brand-names does not imply any endorsement or recommendation on the part of the International Atomic Energy Agency.

CONTENTS

BASIC CONCEPTS AND METHODOLOGY

BASIC CONCEPTS AND PRINCIPLES OF ASSESSMENT (Sessions I and II)

APPLICATION OF BASIC CONCEPTS IN RADIATION PROTECTION REGARDING POPULATION EXPOSURE (Sessions III and IV)

PRACTICAL EXPERIENCE AND MONITORING

POPULATION DOSES RESULTING FROM RADIONUCLIDES OF WORLDWIDE DISTRIBUTION (Session IV continued)

MONITORING PROGRAMMES FOR THE DETERMINATION OF POPULATION DOSES (Session V)

POPULATION DOSES (Session VI)

POTENTIAL HARM AND ECOLOGICAL ASPECTS AND CONTROL

POTENTIAL HARM TO POPULATIONS FROM EXPOSURES (Session VII)

TRANSFER OF RADIONUCLIDES TO MAN THROUGH ENVIRONMENTAL PATHWAYS (Session VII continued and Session VIII)

BASIC CONCEPTS AND METHODOLOGY

Basic concepts and principles of assessment

(Sessions I and II)

Chairmen: M. ČOPIČ (Yugoslavia)
D. BENINSON (Argentina)

Invited Review Paper

BASIC CONCEPTS AND PRINCIPLES OF ASSESSMENT
A review of the Advisory Committee Report on the Biological Effects of Ionizing Radiation

J. C. VILLFORTH
Bureau of Radiological Health,
US Department of Health, Education, and Welfare,
Rockville, Md.,
United States of America

Abstract

BASIC CONCEPTS AND PRINCIPLES OF ASSESSMENT: A REVIEW OF THE ADVISORY COMMITTEE REPORT ON THE BIOLOGICAL EFFECTS OF IONIZING RADIATION.

The present concepts of a single upper limit for individual and population doses, with the understanding that the risks should be kept as low as practicable, may not be adequate for the future uses of nuclear radiation. This is because of the potential for exposing large populations from the uses of nuclear power and medical radiation. There is a need to compare the biological risks and benefits of radiation applications and its alternatives. The recommendations of the US National Academy of Sciences — National Research Council's Advisory Committee on the Biological Effects of Ionizing Radiation (BEIR) are presented. The Committee estimates of the genetic, somatic and ill health risks from population exposures of 170 mrem are included.

INTRODUCTION

The potential effects of ionizing radiation on human populations have been a concern of the scientific community for several decades. The oldest of the scientific bodies now having responsibility in this area is the International Commission on Radiological Protection (ICRP), formed in 1928. The ICRP has maintained continuing studies of radiation protection problems that are of special relevance to the radiation control programs of many nations.

In the 1940's with the establishment of the U.S. Atomic Energy Commission and its program, there was recognition of possible radiation problems and large-scale animal experiments were initiated. In the early 1950's, as a result of the testing of nuclear weapons, public concern arose about the potential effects of ionizing radiation on human populations. In 1955, as a response to this concern, the President of the National Academy of Sciences (NAS) appointed a group of scientists to conduct a continuing appraisal of the effects of atomic radiation on living organisms. That study, entitled "Biological Effects of Atomic Radiation," led to a series of reports by six committees issued from 1956-1963 and which are generally referred to as the BEAR reports.

The BEAR reports led to a basis for public understanding of the expected effects of the testing of nuclear devices that had occurred to that date and introduced the important concept of regulation of average population doses on the basis of genetic risk to future generations. These reports also emphasized medical-dental x rays as the greatest source of man-made radiation exposure of the population.

Also, in 1955, the General Assembly of the United Nations established the UN Scientific Committee on the Effects of Atomic Radiation (UNSCEAR), which, among other tasks associated with monitoring and assembling reports of radiation exposure throughout the world, was "to make yearly progress reports and to develop a summary of reports received on radiation levels and radiation effects on man and his environment..." (UNSCEAR 1969). The periodic reports issued by UNSCEAR (the latest in 1972), in accordance with its objective, have served as a review of worldwide scientific information and opinion concerning human exposure to atomic radiation [1].

In the United States in 1959, the Federal Radiation Council (FRC) was formed to provide a Federal policy on human radiation exposure. A major function of the FRC was to "advise the President with respect to radiation matters, directly or indirectly affecting health, including guidance for all Federal agencies in the formulation of radiation standards and in the establishment and execution of programs of cooperation with States..."

In the late 1960's, concern arose that developing peacetime applications of nuclear energy, particularly the growth of a nuclear power industry for production of electricity, could cause serious exposure of the human population to radiation. Thus, in February 1970, the FRC asked the NAS-NRC Advisory Committee[1] to consider a complete review and re-evaluation of the existing scientific knowledge concerning radiation exposure to human populations. This request from the FRC came about because of: (1) a naturally developing sequence of the Advisory Committee's concern that there had been no detailed overall review since the BEAR reports; (2) new factors that might need to be considered, such as optional methods of producing electrical energy and types of environmental contamination different from those previously encountered; and (3) a growing number of allegations made in the public media and before Congressional committees that the existing radiation protection guides were inadequate and could lead to serious hazard to the health of the general population. The following sections summarize the results of the NAS-NRC Advisory Committee review [2].

QUANTIFICATION OF RISK

Deleterious effects in individuals and populations of living organisms cannot be attributed to exposure to ionizing radiation at levels near that of average natural background except by inference. Such effects are not directly observable. It has been taken for granted by many that exposure to additional radiation near background levels, and especially within variations of natural background, represents a risk so small compared with other hazards of life that any associated nontrivial benefit would far offset any harm caused. The effects of such radiation exposures have been variously regarded as insiginificant, negligible, tolerable, permissible, acceptable. But if in fact any level of radiation will cause some harm (no threshold), and if in fact entire populations of nations or of the world are exposed to additional man-made radiation, then for decisions about radiation protection, it becomes necessary to quantify the risks; that is, to estimate the probabilities or frequencies of effects.

Such estimates are fraught with uncertainty. However, they are needed as a basis for logical decision-making and may serve to stimulate the gaining of data

[1] The NAS-NRC Advisory Committee, on March 25, 1970, accepted the task proposed by the FRC, as a part of the contract agreement between NAS and the Department of Health, Education, and Welfare, signed September 1, 1970. On December 2, 1970, the activities and functions of the FRC were transferred to the Environmental Protection Agency because the FRC had ceased to exist as a specific body.

for assessment of comparative hazards from technological options and development, at the same time promoting better public understanding of the issues.

The present U.S. Radiation Protection Guide for the general population was based on genetic considerations and conforms to the BEAR Committee recommendations that the average individual exposure be less than 10 R (Roentgens) before the mean age of reproduction (30 years) [3]. The FRC did not include medical radiation in its limits and set 5 rem as the 30-year limit (0.17 rem per year).

Present estimates of genetic risk are expressed in four ways: (a) *Risk Relative to Natural Background Radiation.* Exposure to man-made radiation below the level of background radiation will produce additional effects that are less in quantity and no different in kind from those that man has experienced and has been able to tolerate throughout his history. (b) *Risk Estimates for Specific Genetic Conditions.* The expected effect of radiation can be compared with current incidence of genetic effects by use of the concept of doubling dose (the dose required to produce a number of mutations equal to those that occur naturally). Based mainly on experimental studies in the mouse and *Drosophila* and with some support from observations of human populations in Hiroshima and Nagasaki, the doubling dose for chronic radiation in man is estimated to fall in the range of 20-200 rem. It is calculated that the effect of 170 mrem per year (or 5 rem per 30-year reproduction generation) would cause in the U.S. in the first generation between 100 and 1800 cases of serious, dominant or X-linked diseases and defects per year (assuming 3.6 million births annually in the U.S.). This is an incidence of 0.05 percent. At equilibrium (approached after several generations) these numbers would be about fivefold larger. Added to these would be a smaller number caused by chromosomal defects and recessive diseases. (c) *Risk Relative to Current Prevalence of Serious Disabilities.* In addition to those in (b) caused by single-gene defects and chromosome aberrations are congenital abnormalities and constitutional diseases which are partly genetic. It is estimated that the *total* incidence from all these including those in (b) above, would be between 1100 and 17,000 per year at equilibrium (again, based on 3.6 million births). This would be about 0.75 percent at equilibrium or 0.1 percent in the first generation. (d) *The Risk in Terms of Overall Ill-Health.* The most tangible measure of total genetic damage is probably "ill-health" which includes but is not limited to the above categories. It is thought that between 5 percent and 50 percent of ill-health is proportional to the mutation rate. Using a value of 20 percent and a doubling dose of 20 rem, we can calculate that 5 rem per generation would eventually lead to an increase of 5 percent in the ill-health of the population. Using estimates of the financial costs of ill-health, such effects can be measured in monetary units if this is needed for cost-benefit analysis.

Until recently, it has been taken for granted that genetic risks from exposure of populations to ionizing radiation near background levels were of much greater import than were somatic risks. However, this assumption can no longer be made if linear nonthreshold relationships are accepted as a basis for estimating cancer risks. Based on knowledge of mechanisms (admittedly incomplete), it must be stated that tumor induction as a result of radiation injury to one or a few cells of the body cannot be excluded. Risk estimates have been made based on this premise and using linear extrapolation from the data from the A-bomb survivors of Hiroshima and Nagasaki, from certain groups of patients irradiated therapeutically, and from groups occupationally exposed. Such calculations based on these data from irradiated humans lead to the prediction that additional exposure of the U.S. population of 5 rem per 30 years could cause from roughly 3,000 to 15,000 cancer deaths annually, depending on the assumptions used in the calculations. The Committee considers the most likely estimate to be approximately 6,000 cancer deaths annually, an increase of about 2 percent in the spontaneous cancer death rate which is an increase of about 0.3 percent in the overall death rate from all causes.

Given the estimates for genetic and somatic risk, the question arises as to how this information can be used as a basis for radiation protection guidance.

Logically the guidance or standards should be related to risk. Whether we regard a risk as acceptable or not depends on how avoidable it is, and, to the extent not avoidable, how it compares with the risks of alternative options and those normally accepted by society.

COST-BENEFIT ANALYSIS

When the risk from radiation exposure from a given technological development has been estimated, it is then logical for the decision-making process that comparisons be made and consideration given to (a) benefits to be attained, (b) costs of reducing the risks, or (c) risks of the alternative options including abandonment of the development. The concept of always balancing the risk of radiation exposure against the expected benefit has been well-recognized and accepted, but it was not until the publication of ICRP Publication 22 that an attempt has been made to evaluate both sides of the equation in any way that could lead to operational guidance. Offical recommendations call for radiation exposure to be kept at a level "as low as practicable," a policy that emphasizes and encourages sound practice. However, risk-estimates and cost-benefit analysis are needed for decision-making. An additional important point, often overlooked, is that even if the benefit outweighs the biological cost, it is in the public interest that the latter must still be reduced to the extent possible providing the health gains achieved per unit of expenditure are compatible with the cost-effectiveness of other societal efforts.

It appears logical to attempt to express both risks and benefits in comparable terms - monetary units. To a limited degree, risks can be estimated in such terms. For example, the statement of risk can be expressed in terms of cost to an individual or to his family and society since there are specific expenses attributable to an effect. ICRP Publication 22 summarizes a number of published estimates of the monetary value of avoiding the detriment possibly associated with a population or collective dose of 1 man-rem [4]. In spite of the intuitive nature of these estimates, they all fall within the range of $10 to $250 per man-rad. Similarly, estimates can be made of expenses required to effect given reductions of exposure to harmful agents. In some instances, it may not be necessary to use absolute monetary costs: that is, one can compare the cost of different ways of producing the same desired objective. Given the need for additional electrical power, one might compare nuclear plants and fossil fuel plants directly in terms of total biological and environmental costs per unit of electricity produced. Often however, there will be need for information on absolute costs. This will occur when decisions have to be made on whether the public interest is better served by spending our limited resources on health gains from reducing contamination or by spending for other societal needs.

Cyril Comar, Chairman of the Advisory Committee on the Biological Effects of Ionizing Radiation, stated in his analysis of the implications of the BEIR Report that it is obvious that any risk can be decreased at an increased financial cost [5]. In a resource-limited society the allocations must be made where they will do the most good. It is a misuse of resources and a disservice to society to add costs for the purpose of decreasing the risks of any one system greatly below acceptable levels, when other societal activities with unacceptable risks are not being attended to. For examples of some choices that could be made: a national program to persuade people to use seat belts is estimated to cost less than $100 for each death averted; a program of early cancer detection and treatment is estimated to cost up to about $40,000 for each death averted. At the height of fallout, it was calculated that the removal of ^{90}Sr from milk, while costing 2 to 3 cents per quart, would cost about $20 million for each case of cancer averted. It has been estimated that money spent on improved collimation of x-ray machines would be 1,000 to 10,000 times more effective in reducing radiation dosages than money spent on improving present reactor waste systems.

It must be emphasized that there are many inherent problems in cost-benefit analysis that will prevent rigorous application in the very complex systems of present concern to society. These include the implication of assigning a monetary value to human life, suffering or productivity; the difficulty in assessment of factors related to the quality of life such as recreational water and land resources; the fact that the costs and benefits may not accrue to the same members of the population, or even to the same generation; and the virtual impossibility of establishing a single cost system that would be socially acceptable and still take into account differences in individual willingness to accept various types of risks. An illustration of the latter points is the observation that health and environmental effects from power plants would be reduced by their location in relatively unpopulated areas. Yet the people in such areas generally are not the ones who need the additional electrical energy.

Despite these uncertainties, there are important advantages in attempting cost-benefit analyses. There is a focus on the biological and environmental cost from technological developments and the need for specific information becomes apparent. Thus, for example, we find relatively little data available on the health risks of effluents from the combustion of fossil fuels. Furthermore, it is becoming increasingly important that society not expend enormously large resources to reduce very small risks still further, at the expense of greater risks that go unattended; such imbalances may pass unnoticed unless a cost-benefit analysis is attempted. If these matters are not explored, the decisions will still be made and the complex issues resolved either arbitrarily or by default since the setting and implementation of standards represent such a resolution.

STANDARDS

The present radiation standards used by the U.S. Federal Government are based on the recommendations of the Federal Radiation Council (FRC). The FRC developed the Radiation Protection Guide that is defined as "the radiation dose which should not be exceeded without careful consideration of the reasons for doing so, every effort should be made to encourage the maintenance of radiation doses as far below this guide as practicable." The FRC also indicated that "there should not be any man-made radiation exposure without the expectation of benefit resulting from such exposure."

The present status of Radiation Protection Guides for the U.S. general population is presented from FRC Report No. 1 [3]:

> "5.2 We believe that the current population exposure resulting from background radiation is a most important starting point in the establishment of Radiation Protection Guides for the general population. This exposure has been present throughout the history of mankind, and the human race has demonstrated an ability to survive in spite of any deleterious effects that may result. Radiation exposures received by different individuals as a result of natural background are subject to appreciable variation. Yet, any differences in effects that may result have not been sufficiently great to lead to attempts to control background radiation or to select our environment with background radiation in mind.
>
> "5.3 On this basis, and after giving due consideration to the other bases for the establishment of Radiation Protection Guides, it is our basic recommendation that the *yearly radiation exposure to the whole body of individuals in the general population (exclusive of natural background and the deliberate exposure of patients by practitioners of the healing arts) should not exceed 0.5 rem*. We note the essential agreement between this value and current recommendations of the ICRP and NCRP. It is not reasonable to establish Radiation Protection Guides for the population

which take into account all possible combinations of circumstances. Every reasonable effort should be made to keep exposures as far below this level as practicable. Similarly, it is obviously appropriate to exceed this level if a careful study indicates that the probable benefits will outweigh the potential risk. Thus, the degree of control effort does not depend solely on whether or not this Guide is being exceeded. Rather, any exposure of the population may call for some control effort, the magnitude of which increases with the dose.

"5.4 Under certain conditions, such as widespread radioactive contamination of the environment, the only data available may be related to average contamination or exposure levels. Under these circumstances, it is necessary to make assumptions concerning the relationship between average and maximum doses. The Federal Radiation Council suggests the use of the arbitrary assumption that the majority of individuals do not vary from the average by a factor greater than three. *Thus, we recommend the use of 0.17 rem for yearly whole-body exposure of average population groups.* (It is noted that this guide is also in essential agreement with current recommendations of the NCRP and the ICRP.) It is critical that this guide be applied with reason and judgment. Especially, it is noted that the use of the average figure as a substitute for evidence concerning the dose to individuals, is permissible only when there is a probability of appreciable homogeneity concerning the distribution of the dose within the population included in the average. Particular care should be taken to assure that a disproportionate fraction of the average dose is not received by the most sensitive population elements. Specifically, it would be inappropriate to average the dose between children and adults, especially if it is believed that there are selective factors making the dose to children generally higher than that for adults.

"5.5 When the size of the population group under consideration is sufficiently large, consideration must be given to the contribution to the genetically significant population dose. The Federal Radiation Council endorses in principle the recommendations of such groups as the NAS-NRC, the NCRP, and the ICRP concerning population genetic dose, and recommends the use of the Radiation Protection Guide of 5 rem in 30 years (exclusive of natural background and the purposeful exposure of patients by practitioners of the healing arts) for limiting the average genetically significant exposure of the total U.S. population. The use of 0.17 rem per capita per year, as described in paragraph 5.4 as a technique for assuring that the basic Guide for individual whole-body dose is not exceeded, is likely in the immediate future to assure that the gonadal exposure Guide is not exceeded. The data ..... indicates that allocation of this population dose among various sources is not needed now or in the immediate future."

A major difficulty has been the misinterpretation of these standards, particularly in the public mind. The intent as stated is that no individual in the general population should receive whole-body exposure of more than 0.5 rem/year and that the average exposure of population groups should not exceed 0.17 rem/year. What is often not realized is that one or the other of these limits may be governing depending on the nature of exposure. For example, if the exposure were to arise from specific locations such as nuclear power plants or reprocessing plants and it were assured that no individual at the boundaries of the installations could be exposed to more than 0.5 rem/year, it would be physically impossible for the U.S. population averages to approach anywhere near the level of 0.17 rem/year from such sources. Accordingly, the Committee felt that both individual and average population guidelines should be maintained but that clarification should be included as an integral part of the regulatory statement.

In addition to individual and average population guidelines, the BEIR Committee recommended that an additional limitation be formulated (not as a basic standard but for generating guidance) that takes into account the product of the radiation exposure and the number of persons exposed; this might be expressed in terms of person-rems. This need arises from acceptance of the nonthreshold approach in risk estimates which implies that absolute harm in the population will be related to such a product. Operationally, for example, there would be advantage in assessment of trade-offs in connection with the siting of nuclear installations as related to the population densities of areas under consideration.

The above recommendations could be implemented with present knowledge. We now come to an important area that requires newer approaches. The BEIR Committee suggested that numerical radiation standards be considered for each major type of radiation exposure based upon the results of cost-benefit analysis. As a start, consideration should be given to exposures from medical practice because of present relatively high levels of exposure and from nuclear power development because of future problems of energy production and the need for public understanding.

With the development of modern health care programs in the Western world, there has been a marked increase in the use of radiation in the healing arts -- medical diagnostic radiology, clinical nuclear medicine, and radiotherapy. This has resulted in the recognition that medical radiation now contributes the largest fraction, by one or two orders of magnitude, of the dose from man-made radiation to the United States public. In 1970 it is estimated that 129 million persons, or 63 percent of the population in the United States, received 210 million diagnostic radiological examinations; that is, a rate of 68.5 examinations per 100 persons, and an increase of about 1 percent per year since 1964. The exposure rate is further increased by the estimated 8 million pregnant females at risk during the year 1970. At present, the estimated dose which is genetically significant to the population is of the order of 20 mrem per person per year (the actual value will be reported in the spring of 1975). The significance of this lies in the absolute reduction of exposure that could be brought about at relatively low cost with no reduction in medical benefit and in addressing four important issues which center on the continued growth of health care delivery in this country. (a) At present, it does not appear feasible that the large number of variables involved in the use of radiation in health care to the public permit valid efficacious guidelines for medical practice. However, there is convincing evidence that certain nonselective mass screening radiographic procedures do not provide sufficient diagnostic health rewards for costs incurred; for example, mass chest radiography for carcinoma of the bronchus and possibly for pulmonary tuberculosis, mass gastric radiography, routine pre-employment radiography for insurance purposes of some foodhandlers, and possibly screening mammography. (b) Attention must be directed toward the reduction of medical radiation dose to the pregnant or potentially pregnant female, in view of the evidence for significantly greater radiation sensitivity of the developing ovum and fetus. (c) Significant reduction of mean genetically significant dose can be brought about through programs of education, improvement of equipment, and certification of all persons who use radiation for diagnosis and therapy. Special attention should be given to testis shielding. On the basis of mouse data, we would expect the human male to be much more susceptible to radiation-induced mutation than the female. Also, the gentically significant dose of medical radiation is about twice as great in males as in females. For these two reasons, testis shielding, which is relatively simple, could reduce the number of radiation-induced mutations to a small fraction of the present number. (d) Control and regulation of present and future technological equipment responsible for medical exposure may be among the most feasible avenues to effect a continued reduction of dose due to medical radiation exposure.

The difficulties in attaining a useful cost-benefit analysis for nuclear power are formidable and will require interdisciplinary approaches well beyond those that have yet been attempted. Areas that require evaluation include: (a) projection of energy demands; (b) availability of fuel resources; (c) technological developments (clean combustion techniques, coal gasification, breeder reactors, fusion processes, magnetohydrodynamics, etc.); (d) public health and environmental costs of electrical energy production from both nuclear and fossil fuel, including aspects of fuel extraction, conversion to electrical energy, and transmission and distribution.

It might be interesting to examine how the concerns for the protection of the environment, biosphere, and health and welfare of man have been manifest in the United States. On January 1, 1970, the Congress passed the National Environmental Policy Act (P.L. 91-190) which will, among other things, assure for all Americans safe, healthful surroundings; and will attain the widest range of beneficial uses of the environment without degradation, risk to health or safety, or other undesirable and unintended consequences. As a result of the Act, the Council on Environmental Quality was established, and on August 1, 1973, a regulation applicable to all Federal departments, agencies, and establishments was published [6]. As a result, these organizations are now required to prepare statements in connection with their proposals for legislation and other major Federal actions significantly affecting the quality of the human environment. Contained in these statements is the requirement for an analysis of alternatives and their environmental benefits, costs, and risks as well as the assessment of the positive and negative effects of the proposed action as it affects both the national and international environment.

For example, the U.S. Food and Drug Administration's Bureau of Radiological Health has recently completed environmental assessment reports for two significant regulatory performance standards for x-ray equipment. One concerns diagnostic x-ray equipment and the other cabinet x-ray equipment used in industry and for airport inspection of carry-on luggage. In both these analyses the cost of implementing the standard was compared to the benefit from the reduced person-rems resulting from the standard. The implications of the requirements of the Council on Environmental Quality should do much to regain the public's confidence in radiation standard setting agencies.

PRINCIPLES

It is apparent that sound decisions require technical, economic, and sociological considerations of a complex nature. However, we can state some general principles, many of which are well-recognized and in use, and some of which may represent a departure from present practice.

(a) No exposure to ionizing radiation should be permitted without the expectation of a commensurate benefit.

(b) The public must be protected from radiation but not to the extent that the degree of protection provided results in the substitution of a worse hazard for the radiation avoided. Additionally, there should not be attempted the reduction of small risks even further at the cost of large sums of money that spent otherwise would clearly produce greater benefit.

(c) There should be an upper limit of man-made nonmedical exposure for individuals in the general population such that the risk of serious injury from somatic effects in such individuals is very small relative to risks that are normally accepted. Exceptions to this limit in specific cases should be allowable only if it can be demonstrated that meeting it would cause individuals to be exposed to other risks greater than those from the radiation avoided.

(d) There should be an upper limit of man-made nonmedical exposure for the general population. The average exposure permitted for the population should be considerably lower than the upper limit permitted for individuals.

(e) Medical radiation exposure can and should be reduced considerably by limiting its use to clinically indicated procedures utilizing efficient exposure techniques and optimal operation of radiation equipment.

(f) Guidance for the nuclear power industry should be established on the basis of cost-benefit analysis, particularly taking into account the total biological and environmental risks of the various options available and the cost-effectiveness of reducing these risks. The quantifying of the "as low as practicable" concept and consideration of the net effect on the welfare of society should be encouraged.

(g) In addition to normal operating conditions in the nuclear power industry, careful consideration should be given to the probabilities and estimated effects of uncontrolled releases. It has been estimated that a catastrophic accident leading to melting of the core of a large nuclear reactor could result in mortality comparable to that of a severe natural disaster. Hence, extraordinary efforts to minimize this risk are clearly called for.

(h) Occupational and emergency exposure limits should be based on results relating to somatic risk to the individual.

(i) In regard to possible effects of radiation on the environment, it is felt that if the guidelines and standards are accepted as adequate for man, then it is highly unlikely that populations of other living organisms would be perceptibly harmed. Nevertheless, ecological studies should be improved and strengthened and programs put in force to answer the following questions about release of radioactivity to the environment: (1) how much, where, and what type of radioactivity is released; (2) how are these materials moved through the environment; (3) where are they concentrated in natural systems; (4) how long might it take for them to move through these systems to a position of contact with man; (5) what is their effect on the environment itself; (6) how can this information be used as an early warning system to prevent potential problems from developing?

Every effort should be made to assure accurate estimates and predictions of radiation equivalent dosages from all existing and planned sources. This requires use of present knowledge on transport in the environment, on metabolism, and on relative biological efficiencies of radiation as well as further research on many aspects.

CONCLUSION

In anticipation of the widespread increased use of nuclear energy, it is time to think anew about radiation protection. We need standards for the major categories of radiation exposure, based insofar as possible on risk estimates and on cost-benefit analyses which compare the activity involving radiation with the alternative options. Such analyses, crude though they must be at this time, are needed to provide a better public understanding of the issues and a sound basis for decision. These analyses should seek to clarify such matters as: (a) the environmental and biological risks of given developments, (b) a comparison of these risks with the benefits to be gained, (c) the feasibility and worth of reducing these environmental and biological risks, (d) the net benefit to society of a given development as compared to the alternative options.

In the foreseeable future, the major contributors to radiation exposure of the population will continue to be natural background with an average whole-body dose of about 100 mrem/year, and medical applications which now contribute comparable exposures to various tissues of the body. Medical exposures are not under control or guidance by regulation or law at present. The use of ionizing radiation in medicine is of tremendous value but it is essential to reduce exposures since this can be accomplished without loss of benefit and at relatively low cost. The aim is not only to reduce the radiation exposure to the individual but also to have procedures carried out with maximum efficiency so that there can be a continuing increase in medical benefits accompanied by a minimum radiation exposure.

REFERENCES

[1] UNITED NATIONS SCIENTIFIC COMMITTEE ON THE EFFECTS OF ATOMIC RADIATION, Ionizing Radiation: Levels and Effects - A report of the UNSCEAR to the General Assembly, UN Publication E.72.IX.17 and 18 (1972).

[2] NATIONAL ACADEMY OF SCIENCES - NATIONAL RESEARCH COUNCIL, The Effects on Populations of Exposure to Low Levels of Ionizing Radiation: Report of the Advisory Committee on the Biological Effects of Ionizing Radiation, Superintendent of Documents, U.S. Government Printing Office, Washington, D. C. 20402 (1972).

[3] FEDERAL RADIATION COUNCIL, Background Material for the Development of Radiation Protection Standards (Report No. 1 of the Federal Radiation Council), Superintendent of Documents, U.S. Government Printing Office, Washington, D. C. 20402 (1960).

[4] INTERNATIONAL COMMISSION ON RADIOLOGICAL PROTECTION, Implications of Commission Recommendations that Dose Be Kept as Low as Readily Achievable (ICRP Publication 22), Pergamon Press (1973).

[5] COMAR, C. L., "Implications of the BEIR report," 5th Annual National Conference on Radiation Control, DHEW Publication No. (FDA) 74-8008, Superintendent of Documents, U.S. Government Printing Office, Washington, D. C. 20402, GPO Stock No. 1715-0006, $3.25 (1973).

[6] FEDERAL REGISTER, Preparation of Environmental Impact Statements: Guidelines, Council on Environmental Quality, Federal Register 38 147, U.S. Government Printing Office, Washington, D. C. 20402 (August 1, 1973) 20550.

DISCUSSION

M.J.A. DELPLA: Prompted by the paper just presented, I feel it necessary to stress, at the start of our Seminar, that the effect of a given dose varies with the conditions of the irradiation. One cannot extrapolate, say, a dose of 500 rems in one minute to a dose of 5 rems spread over a year. I do not believe that nuclear power stations involve a risk of cancer or a genetic risk and, in spite of the good intentions of international commissions in their efforts to establish proportional relations between dose and effect, they only introduce an element of fear by so doing; this in turn leads to heavy expenditure on unnecessary safety measures.

J.C. VILLFORTH: I agree that the biological effect will depend to some extent on the rate at which the dose is delivered. According to the UNSCEAR report, animal data show that protracted continuous irradiation and fractionated irradiation both have a smaller carcinogenic effect than a single administration of the same total dose; this suggests that the same difference might also be observed in man, although it was admitted that "considerably more data are required".

For planning purposes and for cost-benefit analysis I still feel that a linear non-threshold model is the best, even though it is probably the most conservative. Far from producing fear in the population, the proper application and interpretation of the results of this model should in fact restore people's confidence in their radiation protection authority. In the United States of America this loss of confidence has been one of the primary causes of criticism of the nuclear power programme. I believe that the intelligent use of cost-benefit analysis based on non-threshold models will help both the nuclear power programmes and the medical radiation programmes.

H. T. DAW: I should just like to add a comment on the concept of cost-benefit analysis mentioned by Mr. Villforth. In assuming linearity of the dose-effect relationship, it is important to keep in mind that the estimated risk is then an upper limit. The actual risk may be anything from zero up to that limit, depending on the true form of the relation between dose and effect. In other words, if the relation has some other form, the money and effort spent in order to reduce the risk assuming linearity may not be justified by the actual benefit.

POPULATION EXPOSURE AND THE INTERPRETATION OF ITS SIGNIFICANCE

A. MARTIN, H. ApSIMON
Associated Nuclear Services,
London,
United Kingdom

Abstract

POPULATION EXPOSURE AND THE INTERPRETATION OF ITS SIGNIFICANCE.

The paper illustrates the dose commitment approach in assessing the significance of radioactivity released to the environment. The detriment resulting from population exposure is discussed in terms of risk, loss of life and in monetary terms.

Estimates of the dose commitments for unit releases of tritium, carbon-14 and krypton-85 are given. The commitments are expressed in three ways, namely to individual members of a standardized critical group, to the whole UK population and to the global population. The results for tritium release are very dependent on assumptions as to washout and deposition of atmospheric tritium. The estimates of dose commitment from tritium are higher than previous estimates and suggest that, taking account of their respective yields in nuclear reactors, tritium is of greater radiological significance than krypton-85. Attention is drawn to the much greater significance of carbon-14 as a source of population exposure.

In a discussion of the detriment resulting from population exposure, it is suggested that loss of life, expressed in man·years, is a more meaningful unit than risk. The problems of assessing the detriment in monetary terms are discussed and some illustrative costs are presented, expressed in terms of 'V', the average valuation of one year of life. The question as to whether it is appropriate to apply economic discounting techniques is examined.

1. INTRODUCTION

Releases of radioactivity to the environment result in the exposure of the population to ionising radiation. The doses received by individuals within the population are subject to a wide variation depending on their location, age, dietary and other habits, and other statistical factors. More generally, this can be expressed in the terms that a given release of radioactivity results in a dose distribution in the population. The area under the distribution curve represents the total population dose, which may be assessed in relation to the risk both to individuals and to the population as a whole. In the former case the individuals at greatest risk are the members of the so-called critical exposure group at the upper extremity of the dose distribution. In the latter case, if a linear dose-risk relationship is assumed, the total risk within the population is proportional to the area under the distribution curve.

In this paper, various approaches to the assessment of the significance of population dose are examined quantitatively by reference to releases of tritium, carbon-14 and krypton-85. The problems of expressing the consequences of population exposure in monetary terms are also discussed briefly.

2. POPULATION DOSE COMMITMENT

In common with developments in the philosophy of internal radiation dosimetry, the concept of dose commitment is now recognised as a useful basis

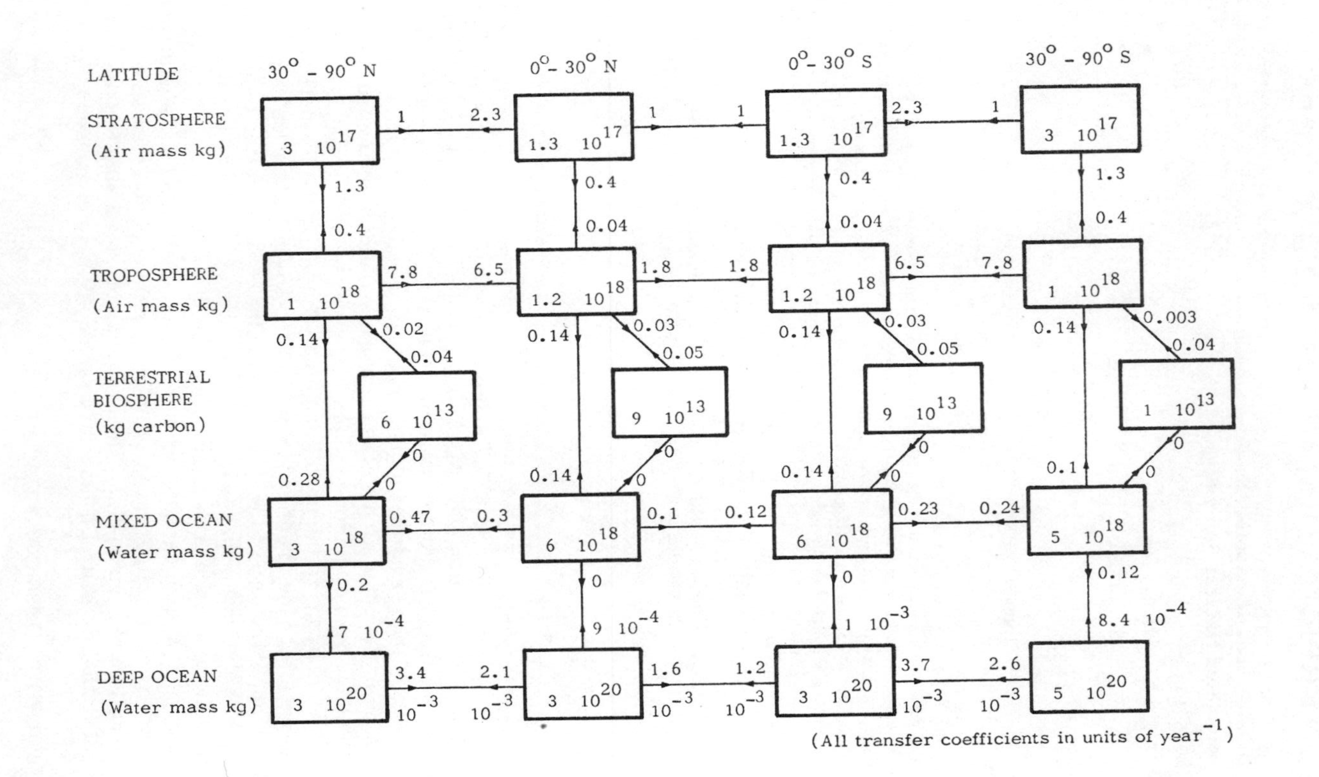

FIG. 1. Global transport model for carbon-14.

on which to establish release limits. The dose commitments per unit release from a typical UK site to atmosphere of three important nuclides, H-3, C-14 and Kr-85, have been computed. The dose commitments are expressed in three ways, namely, to individual members of a standardised critical group, to the UK population and to the global population.

The doses to individual members of the critical exposure group from releases of the nuclides to atmosphere have been estimated on the basis of a standard dilution factor of 10^{-7} sec m^{-3}, corresponding to a distance of a few hundred metres to a few km from the point of release, depending on the site characteristics and stack height.

In order to calculate the UK population dose, the radioactivity was assumed to be released from a 100-m stack at Windscale. Dispersion calculations were based on Pasquill's method, using typical distributions of wind direction and atmospheric stability. The population distribution was defined in terms of eight 45°-sectors each divided into six segments with outer radii increasing geometrically up to 1095 km.

Estimates of the global population dose have been made using a twenty-compartment global model developed for this purpose. Four latitude bands are considered, 0° to 30° and 30° to 90°, north and south. Each latitude band is divided into five compartments, namely, stratosphere, troposphere, mixed ocean, deep ocean and a fifth compartment which may represent either the biosphere or surface water. The rates of exchange between atmosphere and ocean or atmosphere and biosphere depend on the physical and chemical form of the contaminant. Within the atmosphere or the ocean, the dispersion of radioactivity depends largely on the natural circulation cycles. The transfer coefficients between the various compartments in the model have been estimated from a review of a number of authors, including Craig [1], Lal and Rama [2], Nydal [3], Krey and Krajewski [4] and Pritchard et al. [5].

In the atmosphere, mixing within each latitude band is quite rapid with time scales of weeks to months. The troposphere is rather shallower at the higher latitudes than in equatorial regions and mixes more rapidly with the stratosphere. Mixing between the latitudes depends on prevailing wind systems and on Hadley cell circulation. Transfer between the hemispheres takes place both by the mean Hadley cell circulation and by eddies in the upper troposphere.

In the oceanic circulation system, the surface layers down to a depth of 50 to 100 m are well mixed by the influence of winds and surface currents. Residence times in these surface layers are estimated to be a few years. Transfer rates within and from the deep ocean are much slower. During winter in the arctic and antarctic, cold water sinks and a deep water current is induced. The deep water from the arctic flows south at rates as low as one mile per day and resurfaces in the antarctic region. Cold water sinking in the antarctic is even denser and flows underneath the arctic current. Radioactivity penetrating to these deeper layers is likely to remain there for hundreds of years.

The global transport model is illustrated in Fig. 1, in which the coefficients used to estimate radiocarbon dispersion are shown.

3 TRITIUM

Estimates of the dose commitments from tritium have been based on a re-assessment of the metabolic, dosimetric and environmental factors involved.

These are to be discussed in a paper by Martin and Fry [6], but the main factors used in our calculations are summarised below.

The daily water intake is assumed to be 2.8 kg via ingestion and 0.2 kg absorption from the atmosphere, the latter being equally divided between inhalation and transpiration. This intake corresponds to a 10-day biological half-life for 43 kg of body water and to a dose commitment of 1.0×10^{-4} rem for a 1 μCi intake via any route. On the assumptions of evaporation factors of 85% for rainfall and 95% for 'dry' deposition, an effective washout coefficient during rainfall of 3×10^{-5} sec^{-1} and an effective deposition velocity of 10^{-3} m sec^{-1} have been used. The two modes of deposition, which must be calculated separately in order to obtain the concentration of tritium in ground water, are thus assumed to contribute equally to a net annual deposition of 0.3 m.

In estimating the critical group dose from tritium, consideration must be given to the location of drinking water catchment areas. Our estimates of critical group dose are based on the critical group consuming water having a tritium concentration equivalent to the average ground water value up to a distance of 40 km. On this basis, ingestion and exposure to atmospheric tritium contribute equally to the dose commitment. If drinking water is derived from local sources the critical group dose could be higher by up to an order of magnitude.

Tritium does not follow quite the same pattern of circulation in the atmosphere as krypton and carbon. The relative proportion of water vapour in the atmosphere decreases quite sharply with increasing height and precipitation from rising and cooling masses of air reduces the transfer rate up to the stratosphere from the troposphere. Tritium is lost from the atmosphere to the ocean and land by rain and by dry deposition at the surface. A large proportion of the tritium lost to the land is almost immediately returned by evaporation. The rest runs off to join lakes and mix with underground water, some of it eventually reaching the ocean.

4. CARBON-14

The intake of carbon into the body is almost entirely by ingestion in foodstuffs, with inhalation contributing only about 0.3%. The dose from C-14 is therefore dependent on distribution and consumption patterns of foodstuffs which are assumed to be in equilibrium with the atmosphere at the point of production. In calculating the critical group and UK population dose commitments, it has been assumed that 10% and 50% respectively of food consumed is locally produced, the remainder being from areas of much lower specific activity. In the global situation, the dose estimates are based on the specific activity of the biosphere. In assessing the UK population dose, no effective plume depletion was assumed, since the rate of exchange of carbon in much of the biosphere is quite high.

The average dose to the body was taken to be 0.17 mrem y^{-1} for a specific activity of 1pCi g^{-1} of carbon, derived from UNSCEAR [7]. The population dose commitment from C-14 is spread over many generations and so it was thought to be useful to show the first-generation dose commitment, as well as the value integrated to infinity.

Carbon mixes evenly through the atmosphere and follows the general pattern of atmospheric circulation. It passes from the atmosphere into the biosphere and into the ocean. The total carbon reservoir in the biosphere is of the same order as in the atmosphere. About 3% of the atmospheric carbon exchanges annually with carbon in the biosphere. Part of this, probably about a half, is returned quite quickly on a time scale of a year or so, but the rest is retained far longer in the more durable part of the biomass and dead matter. Transfer rates from the atmosphere to ocean can be estimated from the Suess effect, the change in the relative abundance of C-12 and C-14 isotopes in the atmosphere resulting from the burning of fossil fuels. This leads to a time scale of 5 to 10 years for transfer from troposphere to the mixed surface layers of the ocean, a faster transfer rate than to the biosphere. Ultimately most of the carbon reaches the deep ocean, though this takes hundreds of years.

5. KRYPTON-85

The dose rates resulting from continuous exposure to a concentration of 1 μCi m^{-3} of krypton-85 are assumed to be 1.4×10^{-2} rem y^{-1} whole-body average [7] and 1 rem y^{-1} to the skin at a depth of 7 mg cm^{-2} [8]. The male gonad dose is some 25% greater than that to the whole body.

A reduction factor of 0.4 has been applied to the gamma component to allow for time spent indoors.

Since krypton is virtually insoluble, it circulates almost completely in the atmosphere; less than 3% of the stable krypton is estimated to be in the ocean. Within the atmosphere, it follows the same global circulation pattern as the air masses in the atmospheric circulation system.

6. RESULTS

The estimates of the dose commitments per Ci released are shown in Table I. Also shown in the case of tritium is the global dose commitment resulting from discharge of the nuclide to the surface mixed ocean. No estimates of the local dose commitments have been made in this case, but it is clear that a substantial reduction in population dose commitment is achieved by discharging the tritium to sea rather than to atmosphere. Table I shows that, in terms of whole body dose, the dose commitments per Ci release, both individual and collective, are much higher for C-14 than for the other nuclides.

In order to judge the relative importance of the three nuclides, account must be taken of their production rates in power reactors. This is illustrated in Table II, which shows that, in terms of the world population dose, C-14 is 20 times more significant than Kr-85 if the first generation dose only is considered. However, if the dose is integrated to infinity, C-14 appears to be 400 times more important. According to our estimates, tritium is of greater significance than Kr-85 in terms of the UK population dose, but of comparable significance when viewed on a global basis.

Feeding into our global model an extended version of the world nuclear power programme as projected by Spinrad [9] and our own estimates of C-14 release, the dose rate from man-made C-14 up to the year 2020 has been derived. This is plotted in Fig. 2, which shows that the dose rate from this source will be about 0.7 milli-rem per year at the end of this period. The environmental persistence of C-14 is illustrated by the relatively slow decline of the dose rate if no further releases were made from 2020 onwards.

The estimated release rate at 2020 is about 0.2 MCi y^{-1} compared to a natural C-14 production rate of 0.03 MCi y^{-1}.

TABLE I. DOSE COMMITMENTS FROM UNIT RELEASES OF H-3, C-14 AND Kr-85

Nuclide	Mode of exposure	Dose commitment per Ci released		
		Standardised critical group (a) rem	UK (b) population man-rem	Rest of world population man-rem
H-3 release to - atmosphere - sea	whole body	8×10^{-9} -	1.1×10^{-2} -	3×10^{-3} 2×10^{-4}
C-14 (c)	whole body	5×10^{-7} -	1 (i) 2 (ii)	14 (i) 300 (ii)
Kr-85	whole body skin	1.8×10^{-11} 3×10^{-9}	7×10^{-6} 1.3×10^{-3}	8×10^{-5} 1.5×10^{-2}

Notes
a. Based on dilution factor of 10^{-7} sec m^{-3}
b. Including first pass and subsequent exposure
c. Dose commitment to i) 30y, ii) infinity

TABLE II. RELATIVE IMPORTANCE OF H-3, C-14 AND Kr-85 AS CONTRIBUTORS TO POPULATION WHOLE-BODY DOSE

Nuclide	Typical production rate Ci/GWy(e)	Importance relative to Kr-85	
		UK population dose	World population dose (incl. UK)
H-3 (a)	2×10^{4}	10	1.2
C-14 (b)	3×10^{1}	15 (i) 30 (ii)	20 (i) 400 (ii)
Kr-85	3×10^{5}	1	1

Notes
a. Assuming 90% to sea, 10% to atmosphere
b. Considering dose commitment to i) 30y, ii) infinity

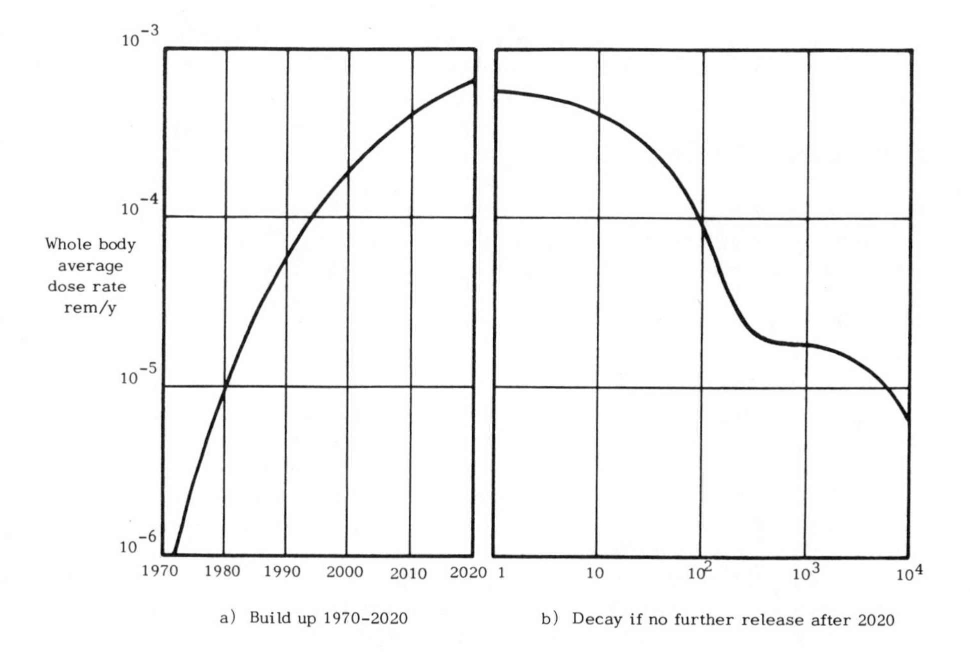

FIG. 2. Dose rate from ^{14}C from world nuclear power programme.

7. THE COST OF POPULATION EXPOSURE

An obvious extension of the concept of dose commitment is that of risk commitment. This may be defined as the product of the dose commitment and the appropriate risk coefficient. If n people are committed to a dose X, the population dose commitment is nX and the risk commitment R is cnX, where c is the appropriate risk coefficient. Now c may depend on age at exposure and so the risk commitment can be expressed

$$R = \sum_a c_a n_a X$$

where n_a is the population in age group a.

A more meaningful way of expressing the detriment is in terms of the man-years of life lost. Thus, considering c_a as the coefficient for the risk of death from radiation-induced cancer, if e is the average remaining expectation of life and p the mean induction period for appearance of the cancer, all for irradiation at age a, the loss of life for each casualty is $(e-p)_a$. The total loss of life L in the case considered above is then

$$L = \sum_a c_a n_a X (e-p)_a$$

Apart from the uncertainties inherent in any assessment of the detriment, the expression of this detriment in monetary terms poses further problems.

First, it would be necessary to consider what value should be placed on one year of life, how it would vary with age and how opinions as to this value would vary from country to country. A second problem is, if economic techniques are to be used, to what extent is it appropriate to discount costs incurred in the future to bring them to present day values?

Considering first the question of the value to be placed on one year of life, one approach is to assume that this value is independent of age. If an average value can be taken, which we shall express as 'V', the cost 'W' of the dose commitment nX becomes

$$W = \sum_a c_a n_a X (e-p)_a V$$

This cost is not incurred immediately and an accountant's approach might be to apply a discounting procedure in order to obtain the present worth of the lost years of life. Two separate factors are involved in this discounting process. The first of these factors corrects to present worth the value of a future loss of life from year p to year e resulting from dose received now, and the second factor makes allowance for the delay in receiving some of the dose when the nuclide is long lived.

If such a procedure is to be used, what would be the appropriate discount rate and would the resulting difference in cost estimates be significant in view of the other uncertainties? Since the valuation of human life is likely to increase, the economic discount rate would be partly offset. For illustrative purposes we have performed an analysis on the basis of an effective discounting rate of 3%. In the analysis, a population dose of 1 man-rem was assumed to be evenly spread over a population of typical age distribution. Risk coefficients varying from 10^{-3} rem^{-1} in the foetus to 10^{-4} rem^{-1} in the adult were taken, with a weighted mean of 2.4×10^{-4} rem^{-1}. On this basis and with assumptions about the delay in appearance of the damage based on induction periods for cancer, the loss of life per man-rem of population exposure was estimated to be 7×10^{-3} man-year. If no discounting is performed, the associated cost could then be said to be 7×10^{-3}V, whereas, if discounting at an effective rate of 3% is applied, the cost reduces to a present worth of about 3×10^{-3}V. This difference is small in relation to uncertainties in the risk coefficient and in the selection of the value V.

The second factor which might be taken into account when considering releases of long-lived radioactivity into the environment is that much of the dose may not be received until many years after the release has occurred. Again, it would be possible to apply a discounting factor and this would lead to the concept of a discounted population dose commitment. The latter would include the first-pass dose, undiscounted in most cases since the dose is delivered relatively quickly, and a discounted contribution from the globally dispersed radioactivity.

In the case of tritium, the first pass population dose commitment comprises about 80% of the total and so the effect of discounting is not significant. For Kr-85, the first pass dose is relatively much less important and the effect of discounting the dose from the globally dispersed krypton would be a reduction of about 30%. Again, this difference is rather small compared to the other uncertainties.

The concept of discounting assumes greater importance when considering long-lived, environmentally persistent nuclides. For C-14, the discounted population dose commitment would be only 8% of the undiscounted figure.

TABLE III. ESTIMATES OF COST OF POPULATION EXPOSURE

Cost estimate for	Cost in units of V (a)	
	Undiscounted	Discounted at 3% p.a.
1 Ci release of H-3 (b)	1 10^{-4}	4 10^{-5}
1 Ci release of C-14	2	7 10^{-2}
1 Ci release of Kr-85	6 10^{-7}	2 10^{-7}
1 man-rem (c)	7 10^{-3}	3 10^{-3}

Notes a. V is average valuation of 1 year of life
b. To atmosphere
c. Short-term dose

On the basis of the results discussed above, it is possible to express the consequences of 1 Ci releases of the three nuclides in monetary terms, with or without allowance for discounting. These estimates are shown in Table III, where we have again preferred to express the results in terms of the value V. Also shown are the cost estimates for a 1 man-rem population dose commitment from a short-lived nuclide.

8. DISCUSSION

Estimates have been presented of the dose commitments resulting from releases to the environment of H-3, C-14 and Kr-85.

A criticism which is often made of this method of integration of population dose is that much of the dose arises from the exposure of large populations in regions remote from the point of release to dose rates which may be orders of magnitude below natural radiation background. On the other hand, the application of a threshold dose rate in the integration poses a number of difficulties. First, an arbitrary and possibly controversial decision is required as to the threshold level. Typically, the level is set at 1 mrem y^{-1}, representing about 1% of the dose rate from natural background. Secondly, the population dose estimated in this way is not proportional to the rate of release of a nuclide since, for example, doubling the release rate not only doubles the dose to the population previously considered, but also increases the numbers affected. A third difficulty is that, if the threshold level is to be applied rigorously, the contribution of every nuclide from every contributory source must be considered. In some cases, such as the three nuclides previously discussed, virtually every nuclear power plant and fuel reprocessing plant in the world could contribute. From the data in Table I, it can be seen that the first pass critical group dose from a typical release of 100 Ci y^{-1} of C-14 would be only 5 x 10^{-5} rem y^{-1}. Thus the significance, assessed on the basis of a threshold for the individual site, would be zero. Yet, as shown in Fig 2, the contribution of all sources probably makes C-14 the most significant nuclide released from nuclear installations.

Ultimately, the absolute significance of releases of these and other radionuclides depends on the nature of the dose-risk relationship and this is the area of major uncertainty. We are inclined towards the view that the relationship includes a linear term which dominates the response to low dose rates. Whether or not this is the case, it will probably be necessary for the foreseeable future to assume that it is true. The estimation of the risk coefficient is a matter of importance. When considering the cancer risk from the whole-body exposure of a typical population, we feel that it would be difficult to justify the use of a value of less than 10^{-4} rem^{-1}, that is 100 cases per million per rem.

On the basis of a linear dose-risk assumption, it is possible to make estimates of the detriment resulting from population exposure. As a first step, it is suggested that the loss of life, in man-years, is a more meaningful expression of detriment than the risk of death. The former method has the advantage of giving greater weighting to the exposure of younger age groups.

Attention is drawn to the controversial question as to whether or not economic discounting techniques should be used in assessing the consequences of population exposure in monetary terms. This is a matter of great importance when considering the longer lived radionuclides. The idea of discounting the future consequences may be repugnant to many people. On the other hand, if cost-benefit analysis is to become a useful tool in judging the best allocation of resources, it is important that it is used in a rigorous and consistent manner.

9. ACKNOWLEDGEMENTS

Much of the work reported in this paper has been performed under contract to the Nuclear Installations Inspectorate of the Department of Energy, to whom we are indebted for permission to publish.

We are very pleased to acknowledge the advice and assistance of Messrs. T.M. Fry, H.A. Shotter and G.H. Firmin.

REFERENCES

[1] CRAIG, H. The natural distribution of radiocarbon and the exchange time of carbon dioxide between atmosphere and sea. Tellus, 9, 1 (1957).

[2] LAL, D. and RAMA. Characteristics of global tropospheric mixing based on man-made C-14, H-3 and Sr-90. J. Geophys. Res. 71, 2865 (1966).

[3] NYDAL, R. Further investigations on the transfer of radiocarbon in nature. J. Geophys. Res. 73, 3617 (1968).

[4] KREY, P.W. and KRAJEWSKI, B.T. Comparison of atmospheric transport model calculations with observations of radioactive debris. J. Geophys. Res. 75, 2901 (1970).

[5] PRITCHARD, D.W. et al. Physical processes of water movement and mixing. "Radioactivity in the Marine Environment". National Academy of Sciences (1971).

[6] MARTIN, A. and FRY, T.M. Population dose commitments from releases of tritium, carbon-14 and krypton-85. International Symposium on Radiation Protection - Philosophy and Implementation. Aviemore, Scotland (June, 1974).

[7] UNSCEAR. "Ionising Radiation: Levels and Effects". A report of the United Nations Scientific Committee on the Effects of Atomic Radiation (1972).

[8] HENDRICKSON, M.M. The dose from Kr-85 released to the earth's atmosphere. Paper no. SM 146/12 Proceedings of IAEA Symposium, New York (1970).

[9] SPINRAD, B.I. The role of nuclear power in meeting world energy needs. Paper no. IAEA-SM-146/2. Proceedings of IAEA Symposium, New York (1970).

DISCUSSION

D. BENINSON: In assessing the collective dose commitment per curie of carbon-14 to infinite time, what population increase was assumed? An exponential population increase would make the integration diverge.

A. MARTIN: We assumed that the global population would increase to 10^{10} and then remain static.

D. BENINSON: What production modes were considered for carbon-14 and in what type of reactors?

A. MARTIN: In light-water reactors carbon-14 arises from $^{17}O(n,\alpha)$ reactions on the oxygen in fuel oxide and in water, and from $^{14}N(n,p)$ reactions on dissolved nitrogen. In high-temperature reactors it depends on the particular design, on the method of reprocessing, and on the disposal of the graphite. The production of ^{14}C in fast reactors appears to be rather low.

K.-J. VOGT: It is usually assumed that the ^{131}I released in nuclear industry makes an important contribution to the population dose. Could you give any figures for the relative detriment from unit releases of ^{131}I?

A. MARTIN: We estimate that the UK population dose per Ci of ^{131}I release to atmosphere is about 200 man·rems. This value could vary by a factor of 2 or more depending on location and height of release.

A. BAYER: Table II of your preprint indicates the importance of tritium relative to ^{85}Kr, and the value given obviously refers only to tritium originating from nuclear facilities. Have you also calculated the relative importance of tritium from bomb fall-out for the period considered?

A. MARTIN: No, I'am afraid not, but you will find these estimates in the UNSCEAR literature.

N. T. MITCHELL: Table I of your paper gives no estimates for exposure of UK populations via disposal to sea, although you assume that 90% of the tritium produced will be released by this route. Could you suggest what these actual values would be?

A. MARTIN: We have not estimated local dose from tritium release to sea, but we feel that it will be orders of magnitude less than the dose from release to atmosphere.

IAEA-SM-184/25

RADIOPROTECTION: RISQUES ET CONSEQUENCES

M. DELPLA, J. HEBERT
Electricité de France,
Paris, France

Abstract–Résumé

RADIATION PROTECTION: RISKS AND CONSEQUENCES.

As a contribution to better understanding between the public health specialist and the lawyer, the authors define "risk" and "damage" – first as they are used in French law and then as they are used in radiation protection. By way of example they employ the mathematical model of the public health specialist in calculating theoretical collective damage (somatic and genetic). The lawyer is looking for a system of presumptions to simplify the task of proving a causal relationship and he may be misled by abuse of the "damage" concept (particularly if it is not made clear that the damage in question is theoretical).

RADIOPROTECTION: RISQUES ET CONSEQUENCES.

A titre de contribution à une meilleure compréhension entre l'hygiéniste et le juriste, les auteurs définissent risque et dommage, d'abord d'après le droit français, puis en radioprotection. A titre d'exemple ils appliquent le modèle mathématique de l'hygiéniste au calcul de dommages collectifs théoriques (somatique et génétique). Le juriste recherche un système de présomptions pour simplifier la preuve du lien de causalité. L'abus de la notion de dommage (surtout s'il n'est pas qualifié de théorique) risque d'entraîner le juriste dans une direction erronée.

1. INTRODUCTION

En raison de la nature particulière des effets des rayonnements ionisants sur l'homme, l'hygiéniste, pour la première fois, a voulu prévoir, prévoir pour prévenir. Pour cela, pour la première fois il a défini un risque théorique. Se donnant ce risque, il calcule facilement le dommage qui en découlerait. Or, pour tout dommage, l'opinion publique désire qu'on lui désigne un responsable, afin qu'il soit contraint à réparation. Ainsi, au-delà de l'hygiéniste, le législateur se sent concerné et, partant, le juriste. Tous doivent bien se comprendre, donc, d'abord, adopter un langage commun.

Avec l'intention de leur ouvrir la voie, nous reprendrons les conceptions du risque et leurs conséquences, d'abord en droit. En ce qui concerne la radioprotection, nous n'aborderons la notion de risque qu'après avoir rappelé, pour les préciser au besoin, différentes grandeurs dosimétriques.

2. ENDROIT

Le droit, constituant un moyen essentiel d'organisation et d'action des Etats, il s'ensuit, comme le montre le droit comparé, que les concepts utilisés par les divers droits nationaux ne sont qu'exceptionnellement synonymes, quant au sens, de la délimitation de l'objet défini par le concept considéré ou de l'articulation de celui-ci avec la construction juridique d'ensemble.

Il ne nous est pas possible, dans le cadre de ce rapport, d'effectuer une étude comparative. Aussi nous nous bornerons à nous référer au droit français, tout en étant conscients qu'il ne s'agit que d'une première «approche» d'une meilleure communication entre hygiénistes et juristes. Nous rechercherons tout d'abord les sens que revêt le concept de risque. Nous examinerons ensuite la signification et les caractères du dommage.

2.1. Le risque

Le concept de risque est utilisé dans diverses branches du droit français pour désigner un événement dont la survenance dans l'avenir est incertaine au moment où les sujets de droit prévoient, ou doivent prévoir, sa réalisation comme possible. Ce concept de risque diffère donc de celui de fait générateur, qui, en matière de responsabilité, désigne l'action particulière d'un homme d'où est résulté un dommage constaté. Il diffère également du concept de force majeure, qui désigne un événement présentant en particulier le caractère d'être normalement imprévisible.

2.1.1. Droit des assurances

Dans le droit des assurances, on trouve la définition suivante du risque : «un événement incertain et qui ne dépend pas uniquement de la volonté des parties (au contrat d'assurance), spécialement de celle de l'assuré» [1].

Le risque, élément fondamental du contrat d'assurance, implique donc l'idée d'éventualité préjudiciable à l'une ou l'autre des parties. Celle-ci concerne soit la réalisation même de l'événement envisagé dans le contrat (par exemple, l'incendie d'une maison), soit le moment de réalisation d'un événement qui se produira nécessairement (par exemple, décès, dans les polices d'assurance sur la vie). L'événement incertain ne peut faire l'objet d'un contrat d'assurance que dans la mesure où sa réalisation suppose l'intervention, au moins partielle, du hasard.

2.1.2. Droit civil

Le concept et parfois le mot même du risque se retrouve dans diverses dispositions du droit civil pour désigner la perte de la chose objet du contrat (vente, louage, prêt, société). Mais lorsqu'elle survient pour une cause fortuite, elle est supportée par l'acheteur; si elle survient du fait de la mauvaise qualité de la chose, de ses vices, elle incombe au vendeur[2]. On voit que dans ce dernier cas l'incertitude qui plane sur la nature du risque peut être assez réduite.

Enfin, le mot risque apparaît dans le titre d'une théorie, dite «du risque» [3], selon laquelle, dans sa forme la plus absolue tout au moins, on serait responsable par le seul fait qu'en agissant on a causé un dommage, autrement dit, sans qu'il soit besoin d'établir que l'action (ou inaction) constitue une faute en s'écartant d'un modèle théorique de conduite. Les justifications et formulations de cette théorie, ou plutôt des théories de la même famille, sont nombreuses, si bien que le mot «risque» n'est guère qu'un titre générique évoquant pour le juriste français la non-exigence de la preuve d'une faute. Repoussée comme principe général de responsabilité, la théorie du risque n'en a pas moins exercé une influence indéniable, souvent sous des formes atténuées, sur le développement du droit de la responsabilité civile.

Ce droit connaît une autre acception du mot «risque», s'agissant d'exonérer l'auteur d'un dommage en cas de faute de la victime ou de partager entre eux la responsabilité, il recherche dans certains cas si la victime a accepté le risque (par exemple, de blessures dans l'exercice de certains sports). Dans ces situations, la réalisation du dommage n'est pas voulue par la victime mais devait être sérieusement envisagée par elle comme possible.

2.1.3. Droit administratif

Le droit administratif (qui en France constitue un corpus juris largement autonome et réglant les activités de l'Etat) reconnaît la responsabilité de la puissance publique, sans exiger, dans de nombreuses hypothèses, la preuve qu'on peut lui imputer une faute.

Certaines de ces hypothèses sont rattachées à l'idée de risque. Il en est ainsi lorsque la responsabilité de l'Etat est engagée en raison de dommages causés à des tiers (personnes privées) par l'usage d'une chose ou par l'exercice d'une activité comportant un certain danger (cas de passants blessés par un tir de la police au cours de la poursuite de malfaiteurs). Plus généralement l'idée de risque exprime la corrélation entre les avantages escomptés pour l'Etat (par exemple du fonctionnement d'un ouvrage public)et les charges qui en découlent pour certains citoyens [4]. Il est intéressant de relever cette dérive de la notion de «risque» vers une analyse coût-bénéfice, celle-ci étant cependant largement implicite et présumée,s'agissant de l'avantage escompté par l'Etat.

2.2. Le dommage

Le dommage, en droit français, a un sens beaucoup plus univoque que le terme risque. Son emploi est cantonné au droit de la responsabilité dont il constitue un élément essentiel, avec le fait générateur, le lien de causalité et, en principe, la faute de l'auteur du fait générateur [5]. A la différence du risque, événement dont la réalisation est incertaine, le dommage est un fait déjà réalisé ou qui se réalisera nécessairement (dommage futur).

2.2.1. Droit civil

Le dommage (synonyme de préjudice) est un élément constitutif de la responsabilité civile, délictuelle ou contractuelle, et de la responsabilité de la puissance publique. Cependant, «le Code Civil français ne contient aucune définition du dommage réparable, c'est une question de pur fait, le dommage étant aussi varié dans ses aspects que le fait dommageable qui l'engendre» [6].

En droit civil français, le dommage doit présenter les caractères suivants: susceptible d'être prouvé et évalué; n'avoir pas déjà été réparé; porter atteinte à un intérêt certain, et légitime; être direct. Sauf exception il s'agit d'un dommage individuel.

2.2.2. Droit administratif

Au moins dans les hypothèses où, par application de la théorie du risque ou de l'idée d'égalité des citoyens devant les charges publiques, la responsabilité de la puissance publique, sans faute prouvée, est reconnue, la jurisprudence administrative exige en outre que le préjudice soit spécial, c'est-à-dire qu'il n'atteigne que certains citoyens et non la généralité de ceux-ci et parfois aussi qu'il soit anormal, c'est-à-dire qu'il excède les sujétions courantes qu'entraîne, par exemple, l'existence d'ouvrages publics.

3. GRANDEURS DOSIMETRIQUES

Parce que l'hygiéniste raisonne en termes de dommage collectif, le praticien en radioprotection tend de plus en plus à interpréter les résultats de la dosimétrie individuelle sous forme de dosimétrie collective.

3.1. Dosimétrie individuelle

Sans reprendre les grandeurs et unités dosimétriques, bien définies par la CIUMR (Commission internationale des unités et mesures radiologiques [7], nous donnerons quelques indications sur la façon de résoudre les difficultés relatives à la dispersion de l'irradiation dans l'espace, et à son étalement dans le temps. Bien connues des radiologues, ces difficultés sont, trop souvent, perdues de vue en radioprotection.

3.1.1. Dispersion dans l'espace

Les rayonnements ionisants, en raison de leur nature, agissent dans la matière au niveau de l'atome, et au hasard des rencontres favorables. La définition physique de la dose fait abstraction de la localisation ponctuelle des processus de transfert d'énergie entre les rayons et les atomes, mais elle tient compte de la nature stochastique de cette localisation[7]. La définition biologique de la dose tient compte, dans une certaine mesure, à l'aide de facteurs de modification, de l'influence de la répartition des événements primaires ponctuels dans le milieu; elle se traduit numériquement par l'équivalent de dose, exprimé en rems. Pour simplifier l'expression, suivant l'habitude, nous appelons cette grandeur "dose" et nous l'exprimons en rems.

Par suite de phénomènes physiques de divergence (à partir de la source), d'absorption et de diffusion (par la matière), à l'intérieur d'un organe irradié la dose varie de façon mesurable. Aussi calcule-t-on la dose moyenne dans l'organe (ou dans l'organisme). Pour cela, on découpe l'organe (ou l'organisme) en n petits volumes, de sorte qu'à l'intérieur du volume j, de masse Δm_j, la dose conserve sensiblement la même valeur D_j ; puis on fait la somme des produits ($\Delta m_j \times D_j$) obtenue en donnant à j toutes les valeurs entières allant de 1 à n. La dose moyenne D_o est égale au rapport de cette somme par la masse m de l'organe (ou de l'organisme) :

$$D_o = \Sigma(\Delta m_j \times D_j)/m$$

3.1.2. Dispersion dans le temps

Pour les travailleurs qui utilisent des sources de rayonnements ionisants, pour les malades soumis à la radiothérapie, l'irradiation est fractionnée, c'est-à-dire scindée par des intervalles libres. On calcule la dose moyenne cumulée, D_c, durant la période considérée, et dans une unité physiologique (l'organisme entier, un organe, ou, pour un malade, le siège d'une lésion); pour cela, on fait la somme des doses qui correspondent aux intervalles de temps durant lesquels, pour la période d'irradiation considérée, le débit de dose n'a pas été nul :

$$\underline{D}_c = \Sigma \underline{D}_o$$

3.2. Dosimétrie collective

On définit une dose (cumulée) pour la population ou pour des groupes particuliers de personnes [8], et une dose moyenne (cumulée).

3.2.1. Dose collective, dose-population

Dans une population, la dose moyenne cumulée par irradiation professionnelle varie grandement entre individus, aussi la CIRP (Commission internationale de protection radiologique) a-t-elle considéré des groupes particuliers de personnes placées dans des conditions semblables [9]. Cela l'a incitée à dégager les deux grandeurs de dose collective et de dose-population [8]. L'une et l'autre s'obtiennent par la somme des doses moyennes cumulées : la première, $\underline{D}_g$, dans un groupe particulier; la seconde $\underline{D}_p$ pour l'ensemble de la population. Elles s'expriment en hommerems [10]. On écrit :

$$\underline{D}_g \text{, ou } \underline{D}_p = \Sigma \underline{D}_c$$

On doit remarquer qu'à l'intérieur même d'un groupe et, a fortiori, dans la population tout entière, la dose moyenne varie grandement d'un individu à un autre; elle varie, aussi, dans l'organisme d'un même individu. Pour tenir compte de telles variations, s'il y a lieu, on considère la dose collective au niveau d'un organe particulier, pour un groupe particulier : en cas de contamination du milieu par de l'iode radioactif, par exemple, on exprime sa valeur en nourrissons - thyroïderems [11].

3.2.2. Dose collective moyenne[1]

La dose collective moyenne cumulée dans un groupe (ou dans la population tout entière) $\underline{D}_m$, s'obtient par le quotient de la dose collective, $\underline{D}_g$ (ou $\underline{D}_p$) par l'effectif $\underline{N}$ du groupe (ou de la population). On écrit :

$$\underline{D}_m = \underline{D}_g / \underline{N}$$

4. EN RADIOPROTECTION

Voulant prévoir à longue échéance, l'hygiéniste ne retient que l'éventualité d'effets différés, voire très longuement différés. Voulant prévoir en vue de prévenir, il ne retient que des effets délétères : des dommages.

4.1 Effets, dommages et risques

Il lui faut traduire quantitativement les effets dommageables en risques.

---

[1] Il nous paraît utile de conserver l'épithète collective pour désigner cette grandeur, pour éviter toute confusion avec la dose moyenne particulière à chaque individu (§§ 1.1.1., 1.1.2.). Toutes deux s'expriment en rems.

4.1.1. Effet et dommage observés

Considérons, pour fixer les idées, l'effet différé le plus souvent pris en compte : la leucémie. Dommage individuel, elle atteint l'individu ou le laisse indemne, suivant la loi du tout ou rien. Soit un groupe de N individus qui ont été irradiés et, durant la période d'observation, de durée t, soit n le nombre des leucémies diagnostiquées. Ce nombre exprime le dommage collectif observé, grandeur mesurable (et variable, en particulier avec t).

4.1.2. Dommages observés et risques

Au début, l'observateur ne saurait distinguer les individus qu'il va voir devenir leucémiques; à la fin, il sait que, pour chacun, la probabilité moyenne de le devenir était : $p = n/N$.

Cette probabilité est un risque moyen.

En procédant de même pour d'autres affections différées, on trouve autant de valeurs de risques que de dommages possibles. Un même individu peut en subir plusieurs (leucémie et cataracte, par exemple) : indépendants, ils ne s'excluent pas l'un l'autre. Soit p_i et p_j les risques moyens respectifs pour les dommages i et j, et p_{ij} le risque moyen d'apparition de i et j sur le même individu ; la probabilité moyenne (p) pour qu'un individu subisse au moins l'un des deux dommages peut s'écrire[2]:

$$p = p_i + p_j - p_{ij}$$

En pratique l'observation se réduit aux dommages les plus graves; en petit nombre, ils demeurent, pour la plupart, assez exceptionnels. Ainsi, l'expression précédente prend la forme simplifiée :

$$p \approx p_i + p_j$$

Lorsque l'observateur dénombre, par exemple, les morts attribuées à différentes causes, les causes sont exclusives; alors :

$$p = p_i + p_j$$

On peut donc, le plus souvent, calculer le risque moyen pour qu'un individu subisse au moins l'un des n dommages considérés par la formule :

$$p = \Sigma p_i$$

où i prend toutes les valeurs entières allant de 1 à n.

Aucun des effets possibles n'étant spécifique de l'irradiation, l'observateur doit comparer le groupe qui a subi une irradiation particulière à un groupe témoin. Dans la mesure où les deux groupes sont représentatifs d'une même population, le risque moyen ajouté par cette irradiation s'obtient par la différence entre les deux valeurs observées, pour les irradiés (p) et pour

---

[2] Pour 3 effets (i, j et k), avec des notations semblables, l'expression de la probabilité pour qu'un individu subisse au moins l'un des trois devient:

$$p = p_i + p_j + p_k - (p_{ij} + p_{jk} + p_{ki}) + p_{ijk}$$

si l'on tient compte des victimes de plusieurs effets, elle se compliquent, rapidement, d'autant plus qu'ils sont plus nombreux.

les témoins ($\underline{p}_o$) . Ce risque ($\underline{p} - \underline{p}_o$) peut ne pas différer de zéro de façon statistiquement significative ($\underline{P} > 0,05$) ; dans le cas contraire, il peut être positif; il peut, aussi, être négatif, et cette dernière éventualité, généralement négligée, devrait, au contraire, en radioprotection, retenir toute l'attention.

Pour une période d'observation de longue durée, la mort (pour toutes causes) réduit notablement les effectifs; en outre, dans les groupes de personnes, certaines sont perdues de vue [3]. Il est alors indiqué de rapporter les dommages collectifs observés, non à $\underline{N}$ individus, mais à la somme des temps d'observation de tous les individus qui ont fait partie du groupe; en représentant la valeur de cette somme par $\underline{Nt}$, le risque moyen observé par unité de temps s'écrit, pour le dommage $\underline{i}$ (§ 4.1.2. ci-dessus) :

$$\underline{p}_i = \underline{n}_i / \underline{Nt}$$

Dans les mêmes conditions d'observation, la dose collective moyenne (§ 3.2.2. ci-dessus) s'exprime aussi par unité de temps; elle prend la forme :

$$\underline{D}_m = \underline{D}_g / \underline{Nt}$$

4.2. La méthode prévisionnelle de l'hygiéniste

Nous avons vu que l'observation de groupes représentatifs d'une même population permet — du moins en principe, pour cette population — de déduire, de dommages observés, le risque ajouté par une irradiation. Inversement, la connaissance du risque permet de prévoir le dommage avant que l'irradiation n'ait eu lieu. Pour cela, à l'aide d'hypothèses, l'hygiéniste traduit la situation en un modèle mathématique.

4.2.1. Les observations qu'il retient

Le risque moyen ajouté par l'irradiation ne peut différer de zéro de façon significative ($\underline{P} < 0,05$) que si l'on observe longuement des groupes d'effectif considérable, à moins qu'une très forte dose n'ait été administrée en peu de temps. L'hygiéniste ne retient que des observations consécutives à de telles doses administrées ainsi.

4.2.2. Ses hypothèses

L'ensemble des hypothèses de l'hygiéniste est exposé dans une communication présentée durant ces journées [12] , aussi nous bornerons-nous à rappeler brièvement qu'elles admettent que le risque ajouté par une irradiation, indépendant du débit de dose (ou du fractionnement de l'irradiation) est proportionnel à la dose moyenne cumulée (§ 3.1.2. ci-dessus).

4.2.3. Son modèle mathématique prévisionnel

Soit un pays dont la population, stable, est de 50 millions de personnes, âgées en moyenne de 40 ans. Le développement des utilisations industrielles de l'énergie atomique va augmenter l'irradiation de cette population. Avant de fixer une limite supérieure pour cette augmentation (par exemple, suivant la CIPR, cinq trentièmes de rem par an, en moyenne) les autorités sanitaires consultent l'hygiéniste .

[3] Lorsqu'il le peut, l'observateur compense les sorties par des entrées.

Voyons, par exemple, comment raisonne cet expert en ce qui concerne les dommages, en leucémies (effet somatique), et en mutations létales récessives (effet génétique).

A Nagasaki, suivant la dose reçue par l'explosion nucléaire, la population a été répartie en classes; la plus exposée (à plus de 200 rems) et la moins exposée (à moins de 10 rems) comptent, respectivement 25 000 et 210 000 personnes-ans qui, de 1950 à 1970, ont montré, respectivement, 15 et 11 leucémies [13]. En prenant, faute de mieux, la classe la moins exposée comme témoin, on voit que le risque moyen ajouté (§ 4.1.2. ci-dessus) dans la classe la plus irradiée :

$$(p - p_o) = 550 \text{ leucémies par million de personnes-ans.}$$

Suivant ses hypothèses, l'hygiéniste définit un risque moyen théorique (RMT), pour le dommage i (la leucémie), tel que :

$$(RMT)_i = k_i \; D_c$$

où k_i représente la constante de proportionnalité, et D_c, la dose moyenne cumulée dans la moelle osseuse rouge. En fait, il ne connaît que la dose collective moyenne, D_m, qui, mesurée dans l'air, est proche de 400 rems; il écrit (avec k'_i, mesuré dans l'air, inférieur à k_i, mesuré *in situ*) :

$$(RMT)_i = k'_i \; D_m$$

attribuant à $(RMT)_i$ la valeur de $(p - p_o)$ de la classe la plus exposée, il a:

$$k'_i = 1,5 \text{ leucémie par million de personnes-ans et par rem}$$

ce résultat ne diffère pas trop de celui de Court-Brown et Doll [14] : ils avaient trouvé, en 1957, sur un groupe de rhumatisants qui avaient été traités par irradiation de la colonne rachidienne, que la valeur de k_i pouvait être de 1 à 1,5 .

Entre (RMT) et dommage collectif théorique (DCT), on peut écrire (§ 4.1.2. ci-dessus) :

$$(DCT)_i = (RMT)_i \; Nt$$

d'où, en combinant cette égalité à la précédente

$$(DCT)_i = k'_i \; Nt \; D_m$$

ou encore (§ 4.1.2. ci-dessus), en fonction de D_p, dose-population exprimée en hommerems :

$$\begin{aligned}(DCT)_i &= k'_i \; D_p \\ &= 1,5 \times 10^{-6} \times 50 \times 10^{6} \times \frac{5}{30} \times 40 \\ &= 500 \text{ leucémies par an.}\end{aligned}$$

En ce qui concerne les prévisions génétiques, l'hygiéniste ne peut que recourir à des résultats expérimentaux. Par exemple, à ceux de Lyon, Phillips et Searle [15] relatifs aux mutations létales récessives : avec des souris dont les testicules avaient reçu 1200 rems, accouplées avec leurs petits femelles, la proportion des ovules non arrivés à terme a diminué (par rapport aux témoins) de 3,2 %; cela correspond à une proportion de gamètes mâles porteurs de mutation létale récessive voisine de 25 %; d'où :

$$k_i \approx 200 \times 10^{-6} \text{ mutation par gamète mâle et par rem}$$

En admettant, avec W.L.Russell [16] que l'ovaire restaure intégralement les lésions génétiques, le risque moyen de transmission d'une telle mutation s'écrit, pour la population considérée :

$$\mu_i = \frac{1}{2} \times 200 \times 10^{-6} \times 5 = 500 \times 10^{-6}$$ mutation par zygote et par génération,

les générations humaines ayant un rythme trentenaire. On démontre qu'au cours des générations successives le risque (de mort) dû à une mutation récessive tend vers μ_i(atteint vers la centième génération) ; d'où :

$$(DCT)_i = 500 \times 10^{6} \times 50 \times 10^{6} = 25\ 000$$ morts par génération.

Ainsi, l'alourdissement du fardeau génétique [17] de la population par les applications industrielles de l'énergie atomique causerait un nombre croissant de morts qui,à l'équilibre (dans 3 millénaires)atteindrait un millier par an.

4.2.4. Discussion de ses prévisions

Bien sûr, on discute les résultats des calculs de DCT (donnés ici à titre d'exemples) : ils dépendent des valeurs — fort incertaines — attribuées à la constante de proportionnalité ($\underline{k}_i$). Mais peu importe : ils n'ont rien de commun, ni avec les observations cliniques, ni avec les résultats expérimentaux, qu'il s'agisse d'effets somatiques ou d'effets génétiques. C'est le modèle mathématique prévisionnel lui-même qui est en cause, et qu'il faut rejeter : grossièrement erroné [12] , il est dangereux. En effet, l'angoisse des masses qu'il réveille ou, pour le moins, qu'il polarise sur l'industrie nucléaire, conduit les responsables à prendre des mesures draconiennes de plus en plus coûteuses. Sans gain de quiétude pour quiconque, sans intérêt pour la santé des personnes. Pur gaspillage.

Et l'industriel — présumé responsable, malgré tous ses efforts — ne se voit-il pas menacé d'être contraint d'indemniser les victimes, directes ou indirectes, de tous les dommages individuels observés (non ajoutés par l'irradiation) dans ses installations, voire alentour (cancers, et même malformations congénitales) ? Le RMT risque de lui rendre la défense difficile.

Que penser de l'hygiéniste qui propose ce RMT pour guide au législateur [8, p.16, appendix III] ?

5. CONCLUSION

Les ressources d'un pays — aussi riche soit-il — étant limitées, les dépenses consenties pour la santé — toujours insuffisantes — devraient être réparties entre la prévention des risques effectifs les plus préoccupants, l'amélioration des soins, et la prise en charge des incurables. Et cela devrait suffire.

Or, voici, brièvement résumée, la conclusion d'un observateur américain [18]:«en cette époque d'égalitarisme», des inégalités aussi «monstrueuses» que celles que l'on peut relever (par exemple en ce qui concerne l'indemnisation d'accidents du travail ou de maladies professionnelles) paraissent «anachroniques».

REFERENCES

[1] PICARD, M. et BESSON, A., Les assurances terrestres en droit français, n°22, Librairie générale de droit et jurisprudence, Paris (1950).

[2] Code Civil, art. 1647 .

[3] MAZEAUD, H. et TUNC, A., Traité théorique et pratique de la responsabilité civile délictuelle et contractuelle, Montchrétien, Paris (1965) 66-95.

[4] VEDEL, G., Droit administratif, collection THEMIS, P.U.F. Paris (1964) 679 pages, cite EISENMANN (p.273).

[5] Code Civil, art. 1382.

[6] GRANDCOURT (de), J.L. et SUSSEL-YACEF, Jurisclasseur-Responsabilité Civile et des Assurances, Editions Techniques, fasc. III (1971).

[7] WYCKOFF, H.O. (Chairman), Radiation quantities and units, ICRU, Washington (1971), 18 pages .

[8] ICRP, Implications of Commission Recommendations that doses be kept as low as readily achievable (adopted April 1973), Publication 22, Pergamon, Oxford (1973), 17 pages.

[9] CIPR, Recommandations (adoptées le 17 sept. 1965), Publication 9, (traduct. H. JAMMET et A. DUCHENE) Serv. central doc., CEN de Saclay (non datée), 62 pages.

[10] DELPLA, M., L'hommerem en milieu de travail en centrale nucléaire, Pollut. atmosph., 15 (1973), 17-26.

[11] USAEC, Draft environmental statement concerning proposed rule making action (janvier 1973), 1.18.

[12] DELPLA, M. et VIGNES, S., Seuils d'action et radioprotection, Ces comptes rendus, IAEA-SM-184/21.

[13] MAYS, C.W. and MARSHALL, J.H., Malignancy risk to humans from total body γ - ray irradiation, Troisième congrès internat. AIRP, Washington (9-14 sept. 1973), à paraître.

[14] COURT BROWN, W.M. and DOLL, R., Leukaemia and aplastic anaemia in patients irradiated for ankylosing spondylitis, HMSO , London (1957) 135 .

[15] LYON, M.F., PHILLIPS, J.S. and SEARLE, A.G., The overall rates of dominant and recessive lethal and visible mutation induced by spermatogonial X - irradiation of mice, Genet. Res., 5 (1964) 448-467.

[16] RUSSELL, W.L., The genetic effects of radiation, 4e Conf. int. util. énergie atom. fins pacif. (Actes Conf. Genève, 1971) 13, ONU, New York, AIEA, Vienne (1972) 487-500.

[17] MULLER, H.J., Our load of mutations, Am. J. Human Genetics, 2 (1950) 111-176.

[18] SAGAN, L.A., Human costs of nuclear power, Science, 177 (1972) 487-493.

SEUILS D'ACTION ET RADIOPROTECTION

M. DELPLA, Suzanne VIGNES
Electricité de France,
Paris, France

Abstract–Résumé

ACTION THRESHOLDS AND RADIATION PROTECTION.

The hypothesis on which the calculations of the public health specialist rest are set forth. The main one – the hypothesis that the added risk deriving from irradiation is proportional to the dose – is then disproved by a number of examples of threshold dose rate values obtained with animals even for such very radiosensitive processes as lymphopoiesis and spermatogenesis. There follow a number of other examples – for animals and for man – of dose thresholds in relation to somatic and genetic effects, and even to beneficial effects observed after the receipt of doses below the threshold value. Such thresholds are related to the phenomenon of active and possibly complete recovery, accompanied by a high degree of selection – especially genetic selection. Unfortunately, the proportionality hypothesis takes no account of this; it ignores the existence of dose and dose rate thresholds since it is applied to cumulative doses through extrapolation towards zero of the results obtained with high doses. It is therefore unrealistic. Efforts should be made to determine precisely what doses produce no observable effect or even produce beneficial effects. In that way one would put an end to a growing psychosis and to the wasting of money that could be used elsewhere.

SEUILS D'ACTION ET RADIOPROTECTION.

L'ensemble des hypothèses sur lesquelles s'appuient les calculs de l'hygiéniste sont énoncées. La principale, l'hypothèse de proportionnalité entre risque ajouté par l'irradiation et dose, est ensuite réfutée par quelques exemples de valeurs de seuils de débits de dose, obtenues chez l'animal, et cela pour les effets pourtant les plus radiosensibles, tels la lymphopoïèse et la spermatogénèse; puis quelques autres exemples sont cités, chez l'animal et chez l'homme, de seuils de dose vis-à-vis d'effets somatiques et génétiques, et même d'effets bénéfiques relevés pour des valeurs inférieures aux seuils. Ces seuils existent par suite d'une restauration active, qui peut être totale, et à laquelle s'ajoute une forte sélection, en particulier en génétique. Or l'hypothèse de proportionnalité ne tient aucun compte de ces faits. Elle méconnaît l'existence de seuils de débits de dose et de dose, puisqu'on l'applique à des doses cumulées, en extrapolant vers zéro les résultats obtenus à fortes doses. Elle n'est donc pas réaliste. Des efforts devraient être entrepris en vue de préciser la zone de doses sans effet observable, voire celle où les effets sont bénéfiques. Ainsi mettrait-on un terme à une psychose grandissante et à un gaspillage intempestif de deniers qui seraient si utiles ailleurs.

1. INTRODUCTION

Grâce aux précautions prises d'emblée, l'irradiation des personnes par les utilisateurs de l'énergie atomique dans l'industrie a toujours été très faible ; néanmoins les industriels reçoivent des directives pour mettre en œuvre des moyens supplémentaires afin de la diminuer encore. Cette action continue, dans le sens d'une rigueur accrue, repose sur la recommandation de la C.I.P.R. (commission internationale de protection radiologique) de maintenir toutes les doses aux valeurs les plus faibles auxquelles l'on peut parvenir sans difficulté, compte tenu des aspects sociaux et économiques. Cette recommandation, traduite récemment en français [1], a revêtu plusieurs formes.[1] Dès l'abord, elle paraît raisonnablement prudente, cependant les textes où elle figure sont émaillés de phrases

---

[1] As low as practicable et as low as readily achievable [2] et, plus récemment : as low as is reasonably achievable [3].

telles que : certains effets des rayonnements ionisants sont irréversibles et cumulatifs (et l'on sait que de tels effets comprendraient, en particulier, des cancers et des anomalies à la naissance). On comprend,dès lors, l'empressement mis par l'industriel, impressionné, se sentant incompétent, pour obtempérer. On comprend aussi, malgré les précautions prises, la répugnance manifestée par les personnes du public qui voient approcher de leur foyer des constructeurs de centrales nucléaires.

Nous allons examiner quelles sont les hypothèses qui ont servi de base au raisonnement de l'hygiéniste, soucieux de prévoir , raisonnement repris à leur compte, sans esprit critique, par des personnalités tout aussi bien intentionnées ; nous essaierons d'en montrer les failles et l'inexactitude, et nous suggèrerons une mise au point sur le plan international. Elle serait libératrice.

2. LES HYPOTHESES

Après les avoir énoncées, nous verrons leur intérêt, leurs justifications, leurs inconvénients.

2.1. Enoncés et commentaires

Les hypothèses relatives à l'irradiation collective découlent de celles qui concernent l'irradiation individuelle.[2] Commençons donc par énoncer ces dernières.

2.1.1. Irradiation individuelle

Nous trouvons trois hypothèses ; la troisième, la principale, souvent la seule invoquée, ne suffit pas, sauf exception.

Première hypothèse : On admet que, même en cas d'irradiation partielle, l'individu court le même risque que si la dose avait, en chaque point de l'organe (ou de l'organisme), même valeur numérique que la dose moyenne.

Ainsi sont éludées les difficultés relatives à la dispersion géométrique de l'irradiation à l'intérieur de l'organe considéré (ou de l'organisme entier).

Deuxième hypothèse : On admet que l'individu court le même risque, pour une dose donnée, quelle qu'ait été sa distribution dans le temps ; en d'autres termes, l'effet produit serait indépendant du débit de dose ; ou encore, le risque inhérent à l'irradiation ne dépendrait que de la dose moyenne cumulée dans l'organe considéré (ou dans l'organisme entier).

Ainsi sont éludées les difficultés relatives à la dispersion chronologique de l'irradiation.

Troisième hypothèse : On admet que le risque moyen ajouté par une irradiation est proportionnel à la dose moyenne cumulée dans l'organe considéré (ou dans l'organisme entier).

---

[2] Les définitions des grandeurs utilisées ici ne sont pas rappelées lorsqu'elles ont été précisées dans une autre communication présentée à ces mêmes journées [4].

Cette hypothèse élude l'incertitude quantitative.

En vue d'éviter de confondre un risque ainsi défini avec un risque effectif, nous avons proposé de lui adjoindre l'épithète de théorique : c'est le risque moyen théoriquement ajouté par l'irradiation — le R.M.T. — qui est proportionnel à la dose moyenne cumulée [5] .

2.1.2. Irradiation collective

Les hypothèses précédentes étant admises, pour prévoir les conséquences de l'irradiation d'une collectivité, l'hygiéniste en a imaginé trois autres ; la dernière n'est, d'ailleurs, qu'un corollaire de la troisième.

Quatrième hypothèse : On admet que l'on peut assimiler un groupe composé d'un certain nombre d'individualités dissemblables à un même nombre d'individualités fictives identiques entre elles.

Dans la mesure où le groupe est représentatif d'une population, pour cette population, cette hypothèse élude les difficultés dues à la variabilité des réactions individuelles.

Cinquième hypothèse : On admet que le risque moyen ajouté par l'irradiation est, pour le groupe, indépendant des variations que la dose moyenne cumulée présente entre individus.

Cette hypothèse élude les difficultés dues à la variabilité des irradiations individuelles.

Sixième hypothèse : Le dommage collectif provoqué par l'irradiation d'un groupe (ou d'une population) est proportionnel à la dose collective (ou à la dose-population).

En vue d'éviter de confondre un dommage ainsi défini avec un dommage effectif, nous avons proposé de lui adjoindre l'épithète de théorique : c'est le dommage collectif théoriquement ajouté par l'irradiation — le D.C.T. — qui est proportionnel à la dose collective (ou à la dose-population) [5].

2.2. Leur intérêt : prévoir

C'est la première fois que l'hygiéniste a voulu prévoir les dommages susceptibles d'être causés par l'irradiation des personnes par les applications industrielles de sources de rayonnements ionisants. Voyons donc pourquoi, et comment.

2.2.1 - Prévoir pour prévenir

En matière d'hygiène industrielle, l'employeur est responsable des conditions de travail. Sa responsabilité est précisée, du moins en France, par les tableaux relatifs aux maladies professionnelles. Les risques n'y sont mentionnés que lorsqu'ils se sont traduits par des dommages effectifs sur un nombre jugé suffisant de travailleurs. L'un de ces tableaux a été établi ainsi pour les rayonnements ionisants.

En effet, si les médecins ont su tirer profit des rayons X et des rayons du radium aussitôt après leurs découvertes, ces rayons, d'abord mystérieux, bientôt merveilleux (bénéfiques pour certains malades), ne tardèrent pas à s'avérer puissamment délétères (sur des malades excessivement irradiés;et sur des médecins , accidentellement) et à donner lieu à des procédures de contentieux [6].

Mais les médecins ont mis longtemps pour s'apercevoir, à leur dépens, que ces rayons étaient, en outre, très insidieusement délétères. Aussi, en raison de la gravité de certains effets, et de leur latence prolongée, l'hygiéniste a innové : d'abord, sans même attendre la leçon de l'expérience, il a réduit progressivement les doses maximales admissibles (D.M.A), et étendu ses recommandations jusqu'à la population tout entière ; ensuite, il a voulu prévoir quelles seraient les conséquences, à échéance lointaine, de l'irradiation de groupes de personnes jusqu'aux limites recommandées, pour cela, il a calculé, pour chaque groupe, le D.C.T. correspondant aux affections les plus redoutées, somatiques et génétiques.

2.2.2. Modèle mathématique

Pour faire ses calculs, grâce à ses hypothèses, il a traduit chaque situation en un modèle mathématique [4].

A l'usage, ces hypothèses, qui ont leurs défenseurs, présentent des inconvénients.

2.3. Leurs justifications

Certains ont cherché à les justifier. Vanité.

2.3.1. Justification théorique :

Selon BLAIR [7, 8]la restauration des dommages radio-induits ne serait jamais totale. Il s'appuie sur la théorie de la cible et en tire les assertions suivantes :

- il existe toujours une fraction réversible et une autre irréversible d'un dommage radio-induit ;
- la part réversible du dommage diminue exponentiellement avec la dose.

Nous critiquerons ces vues (§§ 2.3.3, 3.4).

2.3.2. Justifications cliniques ou expérimentales

Des observations ont montré que le risque de cancer (en particulier, de leucémie), croît avec la dose. On peut admettre une croissance linéaire, mais seulement pour de fortes doses (de 250 à 2500 rads à la moelle osseuse pour les leucémies des rhumatisants britanniques [9], de 100 à plus de 400 rads pour les leucémies de Nagasaki, de 10 à plus de 400 rads pour celles d'Hiroshima, où ont coexisté des rayons gamma et des neutrons [10]). Par ailleurs, STEWART, suivie de McMAHON, a insisté sur l'augmentation des cancers durant les dix premières années de vie des enfants irradiés *in utero* [11-14] .HEMPELMANN, et aussi d'autres auteurs, ont accusé l'irradiation de jeunes enfants sur le thymus de déterminer des tumeurs de la thyroïde, et des leucémies [15-18]. Citons aussi une enquête américaine sur la longévité des radiologistes, elle a conduit à dénoncer l'irradiation chronique comme devant raccourcir la longévité [19-21]. Enfin, en génétique, les premiers résultats provenant de l'observation clinique [22-24] semblaient révéler l'influence de l'irradiation parentale sur le taux de masculinité des enfants. L'expérimentation sur des souris, mâles et femelles [25, 26], a montré l'augmentation du taux de mutations en fonction de la dose.

2.3.3. Carence des justifications

La théorie de la cible confond matière inerte et matière vivante ; cette dernière, à toute agression, répond par une réaction, réparatrice, bénéfique.

Les observations cliniques précédentes n'ont jamais été confirmées par la comparaison de groupes représentatifs d'une même population [27-30]. Les conclusions génétiques, un peu hâtives, sur la variation radio-induite du taux de masculinité ont été reniées par les plus notoires de leurs auteurs [31].

De toute façon, les observations qui constituent le fondement de l'hypothèse de proportionnalité concernent des irradiations à fortes doses et à forts débits de dose. Or la valeur de ce débit de dose a été trouvée déterminante sur l'importance de l'effet [32-37] et nous verrons, au chapitre suivant, que des seuils existent.

2.4. Leurs inconvénients

Elles inquiètent ; elles coûtent cher, inutilement.

2.4.1. Impact psychologique

L'hypothèse selon laquelle toute irradiation, aussi faible soit-elle, se solde par un dommage, ne laisse pas d'inquiéter. Même le millirem peut, par le calcul, conduire à un D.C.T. non négligeable, s'il est appliqué à une vaste population. Le public n'admet pas un tel risque, d'autant moins qu'il lui est imposé.

2.4.2. Incidence économique

Les dispositions à prendre pour épargner un hommerem à l'environnement sont d'autant plus coûteuses que le nombre en a déjà été plus réduit.

2.4.3. Incidence sur la santé publique

Les sommes dépensées ainsi pour préserver — théoriquement — la santé publique, font cruellement défaut dans certains secteurs.

3. MISE EN DEFAUT DES HYPOTHESES

Examinons quelques faits qui, tirés de l'expérimentation animale, et aussi de l'observation de groupes humains, contredisent les hypothèses de l'hygiéniste. En effet, des seuils de débit de dose peuvent être avancés et, lorsque ceux-ci sont franchis, nous verrons qu'il faut encore atteindre un seuil de dose pour obtenir un effet délétère ; de plus, des effets bénéfiques sont observables en deçà de ces seuils.

3.1. Existence de seuils de débit de dose

L'étude systématique de toute une gamme de valeurs de débit de dose par irradiation continue de l'animal conduit à trouver des seuils, même pour les effets réputés être les indicateurs les plus précoces et les plus sensibles d'une irradiation : spermatogénèse, ovogénèse, hématopoïèse, embryogénèse, mitoses intestinales, longévité, et génétique.

3.1.1. Spermatogénèse

Chez le chien, espèce particulièrement radiosensible, irradié 10 minutes par jour, 5 jours par semaine, la vie durant, CASARETT a trouvé un seuil de débit de dose vers 0,6 rem/j, à partir duquel une diminution nette et progressive de la spermatogénèse apparaît [38]. La stérilité complète nécessite 3 rem/j, et le cumul de 475 rems. Pour les souris et les rats mâles une valeur de seuil vers 3 rem/j peut être avancée, puisque, respectivement, des débits de dose de 2 rem/j, et de 2,6 rem/j, toute la vie et sur plus de dix générations, ne réussissent pas à affecter leur fertilité [39,40]; avec 3 rem/j, d'après CASARETT, on provoquerait une inhibition de la spermatogénèse, laquelle, en sept semaines, atteint un palier à 50 % de la normale [38].

3.1.2. Ovogénèse

LORENTZ et coll. [41] ont dû irradier des souris femelles 18 h/j, 6 j/semaine, à 8,8 rem/j, pour les rendre stériles. Quant à BROWN, étudiant les rates [39], il n'a décelé un effet de stérilité qu'à partir de 10 rem/j. Ainsi, en irradiation continuellement répétée, l'ovogénèse serait-elle plus radiorésistante que la spermatogénèse.

3.1.3. Hématopoïèse

Un effet dépressif sur l'hématopoïèse apparaît à partir de débits de dose variables selon l'espèce animale et selon la lignée cellulaire. CASARETT [38] a réuni l'ensemble des observations, dont l'essentiel est reproduit au tableau I. Les valeurs les plus faibles de seuils de débit de dose s'observent pour le cobaye, espèce très radiosensible ; pour tous les cas, c'est la lymphopoïèse qui accuse le maximum de sensibilité.

Notons que, chez le rat, LAMERTON et coll. [42] ne trouvent aucun trouble pathologique consécutif à une exposition journalière à 16 rems, et cela jusqu'à 330 jours (dose cumulée : 5000 rems) ; à 50 rem/j, la dépression des érythrocytes se limite à quelques semaines, puis est suivie d'un retour à la normale qui se maintient durant l'observation (plusieurs mois) [43,44].

TABLEAU I. SEUILS DE DEBITS DE DOSE POUR QUELQUES ESPECES ANIMALES

Espèce	Seuil de débit de dose (rem/j, mesure dans l'air)			
animale	Lymphocytes	Neutrophiles	Plaquettes	Erythrocytes
Cobaye	0,11	2,2	2,2	2,2
Chien	0,5	3,0	3,0	6,0
Souris	2,2	4,4	8,8	8,8
Rat	> 1,0	10,0	10,0	10,0
Lapin	2,2	> 10,0	10,0	> 10,0

3.1.4. Embryogénèse

Aucune anomalie n'a pu être constatée, ni sur la taille des portées, ni sur le nombre des nouveau-nés vivants, chez des rats, exposés durant toute leur vie intrautérine, à des débits de dose journaliers allant de 2 à 10 rems. Une légère diminution de poids des petits a été trouvée seulement à partir de 20 rem/j , qui constituerait donc une valeur de seuil pour cette espèce animale [39].

3.1.5. Mitoses dans les cryptes intestinales

L'irradiation continue de rats à raison de 16 rem/j , et même de 50 rem/j , conduit à observer un état stationnaire non pathologique [43]. Bien plus, un débit de dose de 84 rem/j détermine une restauration tellement rapide que les rats demeurent en bonne santé [45] ; et même à 350 rem/j , débit de dose incompatible avec une survie prolongée du rat, il apparaît au niveau des cryptes un effacement rapide des lésions [46]. En effet, on assiste à une accélération des mitoses, d'autant plus nette que la dose est plus élevée, et aussi à une extension de la zone proliférante de cellules-souches.

3.1.6. Longévité

Aucun abaissement de longévité n'a pu être observé chez le chien, aux débits de dose inférieurs à 0,6 rem/j , chez la souris à moins de 1 rem/j , et chez le rat à moins de 4 rem/j [38,41,47,48] . En dessous de ces valeurs, on note, au contraire, une élévation de la durée de la vie.

3.1.7. Génétique

RUSSELL a montré qu'à 9 mrem/min 400 rems ne produisent aucun effet délétère sur la descendance : la restauration de l'ovaire est totale pour ce débit de dose [49].

Quant au testicule de la souris, aucun seuil n'a pu être trouvé par le même auteur, qui est descendu jusqu'à 1 mrem/min pour obtenir une dose de 600 rems (l'irradiation a duré plus d'un an) [50].

3.2. Existence de seuils de dose

De même que des seuils de débit de dose peuvent être mis en évidence en irradiation continue, des seuils de dose existent, en irradiation unique ou fractionnée.

3.2.1. Spermatogénèse

Il faut, par une irradiation unique, plus de 2000 rems, à 70 rem/min, pour stériliser définitivement le chien (en fait, 90 % du groupe étudié) [38]. Chez l'homme, des cas d'irradiés accidentels, tel celui de Los Alamos (21 mai 1946) ou les pêcheurs japonais dans les eaux du Pacifique, près de Bikini (20 mars 1954), ont manifesté une azoospermie se prolongeant de deux à quatre ans, selon les cas. L'accidenté de Los Alamos avait reçu 400 rems, il eut deux nouveaux enfants, le premier cinq ans après l'accident [51]; quant aux marins japonais, les doses ont varié de 250 à 550 rems ; jeunes, pour la plupart, ils avaient 17 enfants avant l'irradiation, ils en ont eu 45 de plus dans les dix ans qui suivirent, tous normaux [52].

Des hommes relativement jeunes, (âgés d'une trentaine d'années en moyenne)après orchiectomie pour séminome, ont reçu par radiothérapie locale, à la peau, selon les cas de 2500 à 4500 rems (soit environ le dixième aux testicules) fractionnés en plusieurs séances, sur 20 à 30 jours. Tous ont, après une période de stérilité prolongée de 2 à 4 ans, retrouvé une spermatogénèse normale ; ils ont pu avoir de nouveaux enfants, normaux [53]. Ainsi la stérilisation définitive de l'homme ne peut-elle pas être facilement provoquée par l'irradiation.

Notons d'ailleurs que la gonade mâle restaure beaucoup mieux que la gonade femelle, après une irradiation unique, ou fractionnée, alors que l'inverse est constaté après une irradiation continue (§ 3.12).

3.2.2. Ovogénèse

Les doses susceptibles de stériliser les femelles diffèrent beaucoup d'une espèce animale à l'autre [54]. Chez la femme, les doses dans l'ovaire varient de 450 rems en 1 séance, vers la quarantaine, à 1500 rems en 4 séances quotidiennes, lorsqu'elle est jeune [55].

3.2.3. Cataracte

L'induction d'opacités cristalliniennes, puis de cataracte, exige des doses (en rayons gamma) supérieures, respectivement, à 200 et 1000 rems, en irradiation unique [56]. Si l'irradiation est fractionnée, étalée sur des semaines, voire des mois, les doses requises sont trois ou quatre fois plus élevées [57].

3.2.4. Cancers cutanés

ROWELL et MAISIN [58, 59] n'ont pas observé de cancer, après radiothérapie, sur des peaux exposées à moins de 1500 rems. Les doses cancérogènes, pour les praticiens négligents qui ont introduit leurs mains dans le faisceau de rayons X, seraient de l'ordre de 3000 rems [59].

3.2.5. Tumeurs osseuses

Toutes les expérimentations animales (sur le chien ou le rat) et toutes les observations humaines, concernant des contaminations par le radium-226 ou le strontium-90, permettent de définir un seuil de dose cancérogène dans le squelette. Pour le strontium-90 les charges inoffensives sont de 0,8 μCi/g de poids de la souris, 0,4 μCi/g de rat et 0,15 μCi/g de chien [60]. L'homme contaminé par du radium-226 doit cumuler plus de 1000 rads dans le squelette pour voir apparaître un cancer [61, 62].

3.2.6. Longévité

Chez l'animal, la mise en évidence d'un abaissement de longévité sous l'effet d'une irradiation inférieure à 100 rems est rare et difficile.

Citons toutefois UPTON, sur la souris femelle, qui a réussi à observer cet effet par l'administration de 30 rems en une fois [63]. Par contre, après avoir irradié le rat, HURSH et coll. [64], par 20 séances quotidiennes de 30 rems, n'ont pas provoqué de décalage entre les courbes de survie des irradiés et des témoins. La population, à Hiroshima et à Nagasaki, suivie plus de vingt ans, n'a manifesté, pour le groupe exposé à moins de 100 rads, aucune élévation du risque de mortalité par rapport à celui que faisaient prévoir les statistiques nationales japonaises de l'époque considérée [65] .

3.2.7. Génétique

Une dose de 275 rems administrée en 1 séance, avec 75 rem/min, augmente le risque de mutation létale, pour les caractères dominants et les caractères récessifs. Par contre, la même dose, fractionnée en 55 séances quotidiennes, n'augmente pas ce risque [76]. Ainsi, contrairement à la conclusion de RUSSELL, un seuil de dose existerait, même chez le mâle. SHERIDAN a même obtenu, en fractionnant les 275 rems, un risque de mutation létale récessive inférieur à celui des témoins ; donc, un effet bénéfique.

3.3. Existence d'effets bénéfiques

Si des seuils peuvent être définis pour la plupart des effets délétères radio-induits, en deçà des seuils de débit de dose ou de dose, il existe des effets bénéfiques. En voici d'autres exemples.

3.3.1. Longévité

Les souris et les rats mâles ont manifesté une longévité supérieure à la moyenne, à la suite d'une irradiation externe continue, respectivement à 0,11 rem/j et 2,6 rem/j [47,48]. UPTON a trouvé un résultat similaire sur des souris femelles irradiées à raison de 1 rem/j [63]. Selon les conclusions de cet auteur, la qualité de l'effet qui, de délétère devient bénéfique, s'inverse pour la souris entre 30 et 15 rems, en irradiation aiguë.

Les expérimentations de contamination interne sur souris ont conduit à des observations analogues. C'est ainsi qu'après avoir abreuvé des souris, leur vie durant, et cela sur dix générations, à l'aide d'eau renfermant du césium-137 (4 nCi/cm^3) et du strontium-90 (1nCi/cm^3), NISCHIO et coll. [66] constatèrent une élévation statistiquement significative ($P< 0,01$), de la longévité moyenne des générations successives (442 j chez les irradiées contre 346 j chez les témoins). De plus, par injection intra-veineuse, FINKEL [67] a aussi obtenu, sur des souris femelles, une élévation variable avec la dose (ramenée, comme en thérapeutique, à l'unité de poids corporel) et avec le radionucléide. Par exemple, pour le plus redouté, le plutonium-239, des doses de 0,08 et de 0,4 μCi/kg ont provoqué des élévations respectives de 14 et de 10 % (la première est statistiquement significative).

3.3.2. Leucémies

En irradiant une seule fois à l'aide de neutrons des souris mâles et femelles, MEWISSEN a trouvé une diminution de la fréquence de lymphomes thymiques, pour des doses respectivement de 5 rads chez les mâles et de 7 rads chez les femelles [68]. Les différences avec les témoins non irradiés ne sont toutefois pas statistiquement significatives.

L'analyse des leucémies, déclarées en vingt ans chez les 20 000 survivants de Nagasaki, apporte aussi des éléments nouveaux [69]. MAYS et MARSHALL ont, faute de mieux, proposé le groupe le plus faiblement irradié comme témoin ; la valeur du risque observé, pour ce groupe, est de 50 par million de personnes-ans. On trouve, pour les groupes suivants, aux irradiations allant, respectivement, de 10 à 50, et de 50 à 100 rems, des risques observés de 30 et de 0 par million de personnes-ans. Toutefois, les effectifs sont faibles et les variations observées non significatives statistiquement.

3.3.3 Immunité

BURCH a suggéré que l'allongement de la longévité moyenne observé sur souris et rats serait imputable à une stimulation des réactions de défense des animaux contre les affections nombreuses qui les menacent surtout dans leur jeune âge [70]. L'irradiation X accroît l'activité bactéricide des phagocytes du hamster [71]. Il est regrettable que la recherche effectuée sur l'effet du débit de dose par GENGOZIAN et coll. [72] n'ait porté que sur des doses au moins égales à 200 rems, seuil pour la dépression immunitaire.

3.4. Interprétation

L'existence de seuils est due à des processus de restauration ou de sélection ; des doses inférieures à un seuil peuvent provoquer des effets bénéfiques par stimulation des processus métaboliques.

3.4.1. Restauration

Les phénomènes de restauration se constatent chez tous les êtres vivants : mis d'abord en évidence sur les bactéries, ils l'ont, depuis, été également sur les tissus de mammifères. Un équipement enzymatique spécifique permet de remédier aux ruptures de chaînes dans l'édifice, complexe et fragile, de l'acide désoxyribonucléique (ADN) ainsi qu'à diverses autres altérations de structure [73, 74]. Ainsi est effacée la quasi-totalité des mutations. Néanmoins, si un débit de dose trop élevé provoque la mort de cellules, d'autres processus compensateurs interviennent pour accélérer les phénomènes de multiplication des cellules-souches ; ces phénomènes ont été particulièrement étudiés dans la peau, la moelle osseuse et la muqueuse de l'intestin grêle (§ 3.1.5.).

3.4.2. Sélection

Aux processus de restauration s'ajoutent ceux de sélection. Par exemple, à tous les stades des lignées germinales (mâle et femelle), des processus d'élimination des gamètes anormaux entrent en jeu, réalisant ainsi en permanence une sélection efficace. C'est ainsi que, chez la souris mâle, selon PRESTON [75] le sixième seulement des translocations induites sur les spermatogonies subsiste chez les spermatocytes et que, selon SHERIDAN et FORD [76, 77], une diminution appréciable apparaît encore lors de la maturation des spermatocytes en spermatozoïdes. Les processus sélectifs se comprennent lorsque l'on considère l'importance des pertes cellulaires spontanées dans les lignées germinales: 30 % chez la souris et 50 % chez le rat mâles. Chez la femme, seulement 0,2‰ des follicules primordiaux présents à la naissance évolue jusqu'à la maturité. Ces phénomènes sont encore intensifiés par une irradiation, ce qui permet d'expliquer les écarts existants entre les prévisions théoriques et les résultats de l'expérience [78, 79].

3.5. Conséquences

L'existence de seuils et *a fortiori* celle d'effets bénéfiques, pour des doses ou des débits de dose qui ne sont pas faibles, ne justifie pas les méthodes actuelles de contrôle de l'efficacité de la radioprotection (comptabilisation des doses cumulées par des travailleurs, des doses génétiquement significatives). Dans les conditions normales de travail, seules sont à considérer, comme limites supérieures, les valeurs les plus faibles des seuils.

La position des organisations internationales, confirmée récemment [80, 81, 3], s'oppose délibérément à une telle conversion. Elles ne devraient pourtant pas tarder à se rendre à la raison.

4. CONCLUSION

Dans le but de rationaliser les choix budgétaires, plus encore que l'hygiéniste, l'économiste recourt à des modèles mathématiques prévisionnels. Tout modèle extrapole : il ne saurait prévoir loin.

Au fur et à mesure que le temps passe, l'économiste, qui observe, rectifie ses hypothèses : il a bien accrédité sa méthode.

En matière de radioprotection, l'hygiéniste, au contraire, s'enferre dans ses hypothèses, extrapolant sans aucune retenue, escamotant les seuils d'action et, en deçà de chaque seuil, toute la zone où le risque a changé de signe. Or, c'est vers chacune de ces zones qu'il faut orienter le chercheur, pour bien la délimiter pour chaque risque, en étendue et en profondeur. C'est dans la sous-zone commune à l'ensemble qu'il faut introduire les personnes : en toute quiétude, elles y seront mieux qu'en toute sécurité. Au contraire, cantonnées de plus en plus étroitement auprès du zéro par un coûteux déploiement de moyens de radioprotection, elles entendent dire qu'elles ne courent qu'un risque "acceptable". Et cela les inquiète. Oui, un risque imposé peut-il, vraiment, paraître acceptable ?

REFERENCES

[1] C.I.P.R, Recommandations, Publication 9 (Adoptées le 17 Sept. 1965), Service Central Doc., Saclay (sans date).

[2] I.C.R.P., Recommendations, Publication 9 (Adopted 17 September 1965), Pergamon Press, Oxford (1966).

[3] I.C.R.P., Implications of Commission Recommendations that Doses be kept as Low as Readily Achievable, Publication 22 (Adopted in April - 1973), Pergamon Press, Oxford (1973) 18 pages.

[4] DELPLA, M. et HEBERT, J., Radioprotection : risques et conséquences, Ces comptes rendus, IAEA-SM-184/25.

[5] DELPLA, M. et VIGNES, S., Tendances nouvelles en radioprotection, SFRP, Montrouge (1972) 19-43.

[6] HEBERT, J. et LAUDE, F., Cahiers Droit En. Nucl. (1969) 31-63, dactyl.

[7] BLAIR, H. in Conf. int. util. énergie atom. fins pacif. (Actes Conf. Genève, 1955) 11, ONU, New York (1956) 118-20.

[8] BLAIR, H. in "Some aspects internal irradiation", Pergamon Press, Oxford (1962) 233.

[9] COURT BROWN W.M. et DOLL R., HMSO, London (1957).

[10] ISHIMARU T, HOSHIMO T. et OKADA H., ABCC, Technical Report, 25-69.

[11] STEWART A., WEBB J. and HEWITT D., Brit. Med. J. (1958), 1495 - 1508.

[12] STEWART A., Brit. Med. J. (1961) 452-460.

[13] STEWART A. and KNEALE G.W., Lancet, 6, (1970) 1185-1188.

[14] Mc MAHON B., J. Nat. Cancer Inst., 28, (1962) 1173-1191.

[15] SIMPSON L. and HEMPELMANN L.H., Cancer, 10, (1957) 42-56.

[16] PIFER J.W., TOYOOKA E.T., MURRAY R.W., AMES W.R. and HEMPELMANN L.H., J. Nat. Cancer Inst., 31, (1963) 1333-1336.

[17] HEMPELMANN L.H., PIFER J.W., BURKE G.W., TERRY R. and AMES W.R., J. Nat. Cancer Inst., 38, (1967) 317-341.

[18] HEMPELMANN L.H., Science, 160, (1968) 159-163

[19] WARREN S., JAMA, 162, (1956) 464-468.

[20] SELTZER R. and SARTWELL P.E., Am. J. Epidemiol, 81, (1965) 2-22.

[21] WARREN S., Arch. Environ. Health, 13, (1966) 415-421.

[22] NEEL J.V., THOMAS Ch., Changing perspectives on the genetic effects of radiation, EAST LAWRENCE AVE SPRINGFIELD, ILLINOIS, (1963), 21-36.

[23] SCHULL J.W., Nucleonics, 21, (1963) 54-57.

[24] LEJEUNE J., TURPIN R. et RETHORE M.O., Thérapie, 16, (1961) 521-529.

[25] RUSSELL W.L., Cold Spring Harbor Sym. Quant. Biol., 16, (1951) 327-336.

[26] RUSSEL, W.L., in 4e Conf. int. util. énergie atom. fins pacif. (Actes Conf. Genève, 1971) 13, ONU, New York, AIEA, Vienne (1972) 487.

[27] SAENGER E.L., SILVERMAN F.N., STERLING T.D. and TURNER M.E., Radiology, 74, (1960) 889-904.

[28] GRIEM M.L., MEIER P. and DOBBEN G.D., Radiology, 88, (1967) 347-349.

[29] CONTI E.A., PATTON G.D., CONTI J.E. and HEMPELMANN, Radiology, 74, (1960) 386-391.

[30] JABLON S. and KATO H., Lancet, 14, (1970) 1000-1003.

[31] SCHULL W.J. and NEEL J.V., Amer. J.Human Gen., 18, (1966) 328-338.

[32] MICHAELSON M., WOODEVARD K.T., OLDLAND L.T. et HOLLAND J.W. Conf. 680 410 - Symposium on "Dose rate in mammalian radiation biology" Oak-Ridge, (1968) 7.1-7.21.

[33] BROWN D.G. et CRAGLE R.G.[id. 32], 5.1-5.13.

[34] TAYLOR J.F., PAGE N.P., STILL E.T., LEONG G.F. et AINSWORTH E.J., N.R.D.L.T.R. 69-96 San-Francisco Californie, (1969) 271.

[35] PAGE N.P., AINSWORTH E.J., LEONG G.F., et TAYLOR J.F., Rad.res., 31, (1967), 532-533 et Rad. res. 33, (1968) 94-106.

[36] LANGHAM W., BROOKS Ph. et GRAHN D., Aerosp.med. 36, (1965) 15.

[37] SPALDING J.F. and HOLLAND L.M. "The Species recovery from radiation injury", CONF. 700909, AEC (1971) 245-258.

[38] CASARETT C.W. in the Proceedings of a Colloquium on "Late Effects of Radiation", Chicago, (1969) 85-100.

[39] BROWN S.O., Genetics, 50, (1964) 1101-1113.

[40] STADLER J. et GOWEN J.W. "Effects of Ionizing radiations on the reproductive system", Pergamon Press (1964) 45-58.

[41] LORENTZ E., HESTON W.E., ESCHENBRENNER A.B. et DERINGER M.K, Radiology, 49, (1947) 274-285.

[42] LAMERTON L.F., PONTIFEX A.H, BLACKETT N.M. et ADAMS K., Br. J. Radiol, 33, (1960) 287.

[43] LAMERTON L.F. et COURTENAY V.D., in The Proceedings of a Symposium on "Dose rate in mammalian radiation biology", Oak-Ridge, (1968) USAEC, DTI, 3-1-3-11.

[44] BLACKETT N.M., Br.J.Haemat., 13, (1967) 915.

[45] QUASTLER H., BENSTED J.P.M., LAMERTON L.F. et SIMPSON S.M., Br.J.Radiol, 38, (1959) 501.

[46] CAIRNIE A.B., Rad. Res, 32, (1967) 240.

[47] LORENTZ E., HOLLCROFT J.W., MILLER E., CONGDON C.C. et SCHWEISTHAL R., J. Nat. Canc. Inst., 15, (1955) 1049-1058.

[48] CARLSON L.D. et JACKSON B.H., Rad.Res., 11, (1959) 509-519.

[49] RUSSELL W.L., in 4e Conf. int. util. énergie atom. fins pacif. (Actes Conf. Genève, 1971) 13, ONU, New York, AIEA, Vienne (1972) 487.

[50] RUSSELL W.L., Pediatrics, 41, (1968), 223-230.

[51] HEMPELMANN L.H., OMS, Genève, (1961) 51.

[52] KUMATORI T. et coll, A report after 10 years (1965)

[53] AMELAR R.D., DUBIN L. et HOTCHKISS R.S., The Jl. of Urol., 106, (1971), 714-718.

[54] BAKER T.G., Amer.J. Obst.Gyn., 110, (1971) 745.

[55] ALDERSON M.R. et JACKSON S.M. Brit.J.Radiol., 44, (1971) 295-298.

[56] MERRIAM G.R. et FOCHT E.F. The Am.J.Roentg.Rad.Th.and Nucl.Med., 77, (1957) 759-785.

[57] BACLESE F. et DOLFUS A. J.Radiol. ; 39, (1958) 832-840.

[58] MAISIN J., Symposium Euratom, Bruxelles, (1961) 285-296.

[59] ROWELL N.R., Special Report n°6, Brit.J.Radiol., 45, (1972) 610-620.

[60] NELSON A. in the Proceedings of a Colloquium on "Late Effects of Radiation", Chicago, (mai 1969) 204.

[61] MAYS C.W., DOUGHERTY T.F., TAYLOR G.N. et al.Com. Congrès Evian sur Delayed effects of bone-seeking radionuclides, MAYS-JEE-LLOYD. - , Utah Press, (1969) 387-408.

[62] ROWLAND R.E."Medical Radionuclides : radiation dose and effects", AEC Symposium Series 20 Athènes (1970) 381-383.

[63] UPTON A.C. J. Gerontology, 12, (1957) 306-313.

[64] HURSH J.B., NOONAN T.R., LASARETT G. et VAN SLYKE F., Am.J.Roent , 74, (1955) 130-134.

[65] JABLON S. et KATO H., ABCC. Report n°6, T.R. 10-71, (1971).

[66] NISCHIO K., MEGUMI T., et YONEZAWA M., An.Report Rad. Center Osaka Pref., 8, (1967) 123-128.

[67] FINKEL M.P. Rad.Res.suppl.1, (1959) 265-279.

[68] MEWISSEN D.J. in "Medical radionuclides : radiation dose and effects", AEC Symposium series 20, Athènes (1970) 414-424.

[69] MAYS C.W. et LLOYD R.D., Malignancy risk to humans from total body γ-ray irradiation, in Proceedings of III Int.Cong.IRPA Washington, (9-14 Sept.1973) à paraître.

[70] BURCH R.J. Nuclear Safety 10, (1969) 161-169.

[71] MURHERJEE A.K.B. et STRAUSS P.R., J.Ret.Soc, 5, (1968) 529-537.

[72] GENGOZIAN N., CARLSON D.E. and GOTTLIEB C.F. Radiation exposure rates : effects ont the immune system ,in proceedings of Symposium on "Dose rate in mammalian radiation biology", Oak-Ridge, USAEC, DTI, (1968) 16-1-16-22.

[73] DEVORET R., Sc. Prog. Découv., (oct.1971) 8-14.

[74] LATARJET R., Current Topics Rad.Res.Quaterly, 8, (1972) 1-38.

[75] PRESTON R.J., BREWEN J.G. and GENEROSO W.M., Abst. Rad.Res. 51 (1972) 514.

[76] SHERIDAN W. and WARDELL I., Mut.Res., 5 (1968) 313-321.

[77] FORD C.E., SEARLE A.G., EVANS E.P. et WEST B.J., Cytogenetics, 8 (1969) 447-470.

[78] LEONARD A. IV Cong.Int.Gen.Hum., Conf. série 233, n° 19, (1971) Paris.

[79] NEWCOMBE H.B., IV Cong.Int.Gen.Hum., Conf. série 233, n° 19, (1971) Paris.

[80] The effects on populations of exposure to low levels of ionizing radiation, Nat. Acad. SC., Washington (1972) 217 p.

[81] Ionizing radiation : levels and effects, N.U.E.72 IX. 18, I et II (1972) 447 p.

DISCUSSION

ON PAPERS IAEA-SM-184/25 AND 21

A. MARTIN: I should like to say that I particularly welcome these papers because the philosophies of radiation protection have been constrained for too long by the linear dose-risk relationship. The threshold hypothesis suffers from the same drawback of being too specific. We require a more general approach, such as a dose plus dose squared plus dose cubed relationship, with appropriate time-dependent factors. Effort could then be expended on estimating the coefficients of the different terms rather than on determining this meaningless linear coefficient.

F.D. SOWBY: On this same point I should like to explain that the linear dose-risk relationship has been assumed to describe the probable upper boundary for stochastic effects such as cancer or genetic damage, and it is appropriate for setting dose limits. For these stochastic effects the true relationship may actually be other than linear; if the shape of the curve were known, it would theoretically be possible to use it in radiological protection, although this would complicate the addition of doses. It is recognized that non-stochastic effects, such as the development of cataract, probably require a threshold dose, and this fact has to be taken into account in the setting of dose limits.

M.J.A. DELPLA: The straight line, the quadratic or cubic curves, etc., all come from infinity and pass through the origin, the zero point. These curves are all above the x-axis, whereas the true curve — whatever risk is considered — goes through zero at a negative value; in other words the average risk at doses which are not too high is negative and the effect is actually beneficial, even as regards leukaemia or genetic lethality (mice). The risk is thus less in the irradiated group than in the control group.

S.M. MITROVIČ: I should like to make a more general kind of comment. In conventional industry there is a well-established allocation of responsibility between industrialists, authorities and employees as regards the prevention of any given risks or dangers, but this is not so clear in the case of nuclear industry. We have already heard about the linear dose-risk relationship and the various other relations based on thresholds, but there is no general agreement as regards their validity. If the possibility exists that an increase in the background of natural radiation can increase the number of cancer deaths, we must make every effort to safeguard the environment and provide adequate protection.

O. ILARI: The discussion seems to be taking the form of a debate between Mr. Delpla and his colleagues on the one hand and the various radiation protection experts on the other. This is all very interesting but at present we have to keep on scientific ground and cannot consider the practical attitudes of regulatory and public-health authorities. Actually, it is the duty of these authorities to adopt the most appropriate procedures for reviewing and licensing nuclear plants, and this amounts to taking the conservative approach. This means that for the moment the authorities are obliged to accept the ICRP policy of assuming a linear dose-effect relationship, without threshold, until other unequivocal, confirmed data become available to justify a modification in their attitude.

J. M. STOLZ: Could Mr. Ilari explain what exactly he means by "appropriate".

O. ILARI: I mean conservative without going to extremes. Thus, one should adopt a conservative approach in making measurements and then apply common sense in their interpretation.

ON THE USE OF THE RISK CONCEPT AND COST-BENEFIT ANALYSIS IN THE SAFETY ASSESSMENT OF NUCLEAR INSTALLATIONS

D.M. SIMPSON, B.C. WINKLER, J.O. TATTERSALL
Licensing Branch,
Atomic Energy Board,
Pretoria,
South Africa

Abstract

ON THE USE OF THE RISK CONCEPT AND COST-BENEFIT ANALYSIS IN THE SAFETY ASSESSMENT OF NUCLEAR INSTALLATIONS.

The risk concept adopted by the Licensing Branch of the South African Atomic Energy Board for use in the safety assessment of nuclear installations is described. Standards for the population average risk resulting from the presence of a nuclear installation are defined and from this a radioactive release criterion is specified that relates the accidental release magnitude to permissible frequency of occurrence, Individual and population average dose limits are also derived for the normal operational release condition.

Some problem areas in the use of such a risk concept are identified and discussed. An inconsistency in the use of a linear dose-effect relationship in relation to a defined population at risk is discussed. The use of the concept in defining the measures of population control around the site and the extent to which emergency procedures are implemented are also described.

Although the application of the risk concept is a step towards a more rational approach to reactor siting and safety, in the opinion of the authors a cost-benefit approach is the technique which should be actively pursued. Some possibilities for the application of such a technique are discussed.

1. INTRODUCTION

In the realm of human endeavour there is inevitably some finite risk associated with all major constructions. In the case of a nuclear installation there is a considerable potential risk, and it is the duty of the authority concerned with assessing such a plant on grounds of public health and safety, to ensure that it does not present an excessive hazard to society. Tattersall et al [1] have indicated that it might perhaps be generally argued that a new phenomenon is acceptably safe if the summation of the benefits consequent upon its introduction, suitably weighted to allow for their time distribution, exceeds by some positive margin the summation of similarly weighted costs, risks and other detractions, and further that the margin at least equals that obtainable from any available alternative proposal. Such a contention leads to the important conclusions that acceptable standards are conditional upon features of the phenomenon under consideration and that these standards may change with time.

Whilst the above may be satisfactory as a statement of principle, much of a cost-benefit balance of this nature can at present only be described in broad qualitative terms. In the meantime a simpler, more readily applicable and rational approach is required.

For this reason the Licensing Branch of the South African Atomic Energy Board is making use of a risk philosophy in evaluating the safety of nuclear installations. The basis of this philosophy has been indicated by Tattersall et al [loc. cit.]. It is implicit in this approach that the risks involved are quantified as far as possible in order to provide the best possible evidence for the final judgement as to whether a plant is acceptably safe. This paper describes how the concept of risk can be used in the generation of criteria for radioactive releases of discharges from a nuclear installation, both during normal operation and under accident conditions.

The basis for decisions on subjects such as the control of population development in areas around nuclear sites and the characteristics of, and some problems associated with, the philosophy are discussed. Finally the introduction of limited cost-benefit analyses is proposed to improve the knowledge of the situation in areas where information is lacking.

The hazard to the community as a result of the presence of a nuclear installation is due mainly to the effects on the human population of the discharge of radioactive material. The effects are complex and range from statistical variation in the long-term probability of an individual contracting, at the low-dose end, some radiation-induced disease such as cancer and genetic effects, to the possibility of death as a result of a very high radiation dose.

Here a general distinction must be made between

1) the actual risk resulting from the normal operating discharges of radioactivity which will be released during the reactor lifetime, and
2) the potential risk which will be realized only if active material is released from the installation as a result of accidents within the plant.

Methods for generating radioactivity release criteria related to both of these risks are described below. In any assessment on the basis of risk, the problem arises as to which risk parameter or measure of risk should be used. If the ratio of the risk of all effects to the risk of mortality in respect of conventional hazards is not greatly different from this ratio for the nuclear hazard, the mortality risk can be used as the basis for comparison between the nuclear industry and other industries.

In the presence of only limited data, it is assumed that this ratio is broadly similar; it is therefore felt that the use of the mortality risk for the development of release criteria is justified.

2. ACCIDENTAL RADIOACTIVITY RELEASE CRITERIA

In order to develop a standard on the basis of the population average mortality risk, the average annual risks to members of society resulting from a range of conventional hazards were studied. It was assumed that, on the whole, the existing risk levels in society were tolerated by that society and that the imposition of an additional risk which did not significantly alter the population average risk would also be acceptable. It was apparent from the study that society is concerned not only with the average risk levels, but is also less tolerant of a few major disasters compared with a large number of minor ones presenting the same risk. Consequently, the magnitude-frequency relationship for the occurrence of several types of disaster was examined. It furthermore appears [2] that there is a difference of as much as three orders of magnitude between risks accepted voluntarily (e.g. private flying) and those imposed on society by, for example, natural phenomena. Table I gives values of some population average mortality risk levels. As a result of this investigation it was concluded that an additional mortality risk of 10^{-6} per person per

TABLE I

POPULATION RISKS

Type of risk	Average annual mortality risk per individual
Motor Vehicle, USA	$2,7 \times 10^{-4}$
Motor Vehicle, Japan	2×10^{-4}
Motor Vehicle, SA	3×10^{-4}
Abnormal births, Japan	5×10^{-2}
Railway, Japan	2×10^{-5}
Earthquakes, Japan	3×10^{-5}
Water Transport, Japan	9×10^{-6}
Water Transport, USA	8×10^{-6}
Aircraft, USA	$7,5 \times 10^{-6}$
Medical and Surgical, USA	$5,5 \times 10^{-6}$ x
Electric Current, USA	5×10^{-6}
Lightning, USA	$5,5 \times 10^{-7}$
Railway, SA	1×10^{-5}
Drowning, SA	6×10^{-5}
Lightning, SA	2×10^{-6}
Venomous Insects, SA	2×10^{-6}

year would be considered acceptable to society. Furthermore, Tattersall et al [1] suggested that the risk standard should be more conservative by between one and two orders of magnitude in order to allow for the uncertainties in predicting hazards and the suspected current trends for society to be less tolerant of risks the more affluent it becomes.

Consequently, the population average risk standard was fixed at an average of 10^{-8} deaths per person per year. This can be compared with an individual risk of 10^{-7}, leading to a population average risk of $2,5 \times 10^{-9}$ given by Farmer [3], and an individual risk of 10^{-7} used by Otway et al [4]. Figures 1 and 2 give some examples of the magnitude-frequency relationship for a variety of hazards. If the nuclear hazard magnitude-frequency relationship is to be similar to these curves, it can be expressed by a frequency distribution of the form

$$dP(N) = AN^{-1,5}dN$$

where dP(N) represents the probable frequency per year of all events in which the number of casualties lie between N and N+dN. Hence the probable frequency for all events between N_1 and N_2 would be

$$P(N_2) - P(N_1) = \int_{N_1}^{N_2} AN^{-1,5}dN$$

The value of the constant A depends on the characteristics of a particular site.

For any nuclear installation in a particular environment there may well be one critical nuclide which will contribute the major share to the total mortality risk. Therefore, in order to provide a more readily useful standard in a particular situation, the above frequency distribution can be converted to an activity release criterion which is coupled to the critical nuclide and which can be generated in the following manner.

On the assumption that, for accidental releases from a nuclear power plant, the critical nuclide is ^{131}I, the permitted frequency distribution for varying magnitudes of ^{131}I release is defined as

$$dP(C) = g(C)dC$$

so that, analogous to the above procedure,

$$P(C_2) - P(C_1) = \int_{C_1}^{C_2} g(C)dC$$

where C is the radioactivity release magnitude in curies and P(C) is the probable frequency of all events having release magnitudes between 0 and C curies.

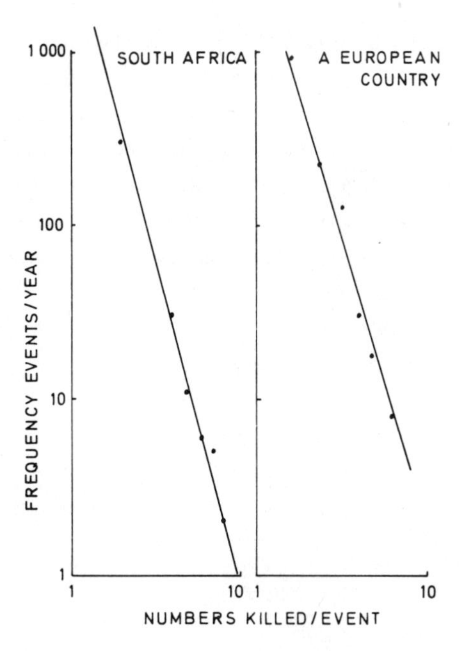

FIG.1. Motor vehicle accidents: frequency-magnitude diagram.

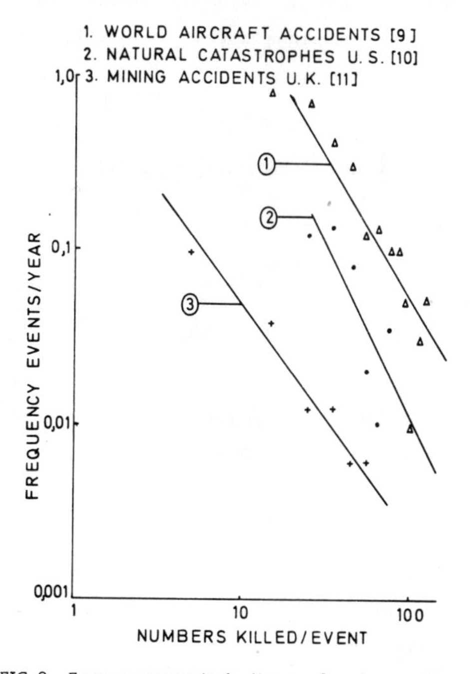

FIG.2. Frequency-magnitude diagram for other accidents.

For a particular magnitude of ^{131}I release, the population average thyroid dose, and hence risk, will depend on the atmospheric dispersion characterictics existing when the release occurs and on the population distribution downwind at the time. For this reason several different population thyroid doses could result from the release of one curie of ^{131}I. Let these doses be denoted by D_i. The corresponding probabilities of occurrence of these doses, given that the release has occurred, are given by Q_i where $1 \leqslant i \leqslant m$, and m is the number of different population doses. For example, if it is assumed that only wind direction and weather stability can influence the population dose, and if the wind direction is divided into j sectors and the weather stability into k groups, then

$$m = j \times k$$

A given number of casualties N can therefore be caused by m different release magnitudes, C_i,

$$N = C_i D_i R$$

where R is the mortality risk coefficient for delayed deaths from thyroid cancer as a result of ^{131}I inhalation and a linear dose-effect relationship is assumed. The frequency of all events resulting in up to N casualties is equal to the sum for all i of the total frequency for all release magnitudes up to C_i multiplied by the probability Q_i.

That is

$$\int_0^N AN^{-1,5}\,dN = \sum_i Q_i \int_0^{C_i} g(C)\,dC$$

It can be shown that, under these conditions,

$$g(C) = GC^{-1,5}$$

where G is a constant which is again site specific.
The potential number of casualties expected from a release of C curies of ^{131}I is

$$\sum_i Q_i . D_i . R . C.$$

No account has been taken of the fact that the person in question may die from other causes during the induction period. The potential number of casualties per year is therefore given by the expression

$$\sum_i Q_i D_i R \int_O^{C_{max}} g(C).CdC$$

where C_{max} is the maximum possible release of ^{131}I under any circumstance.

That is

$$\sum_i Q_i D_i R \int_O^{C_{max}} GC^{-0,5} dC$$

But, if the total population deemed to be at risk comprises P individuals, the number of casualties per year permitted in terms of the primary standard is $10^{-8} \times P$.

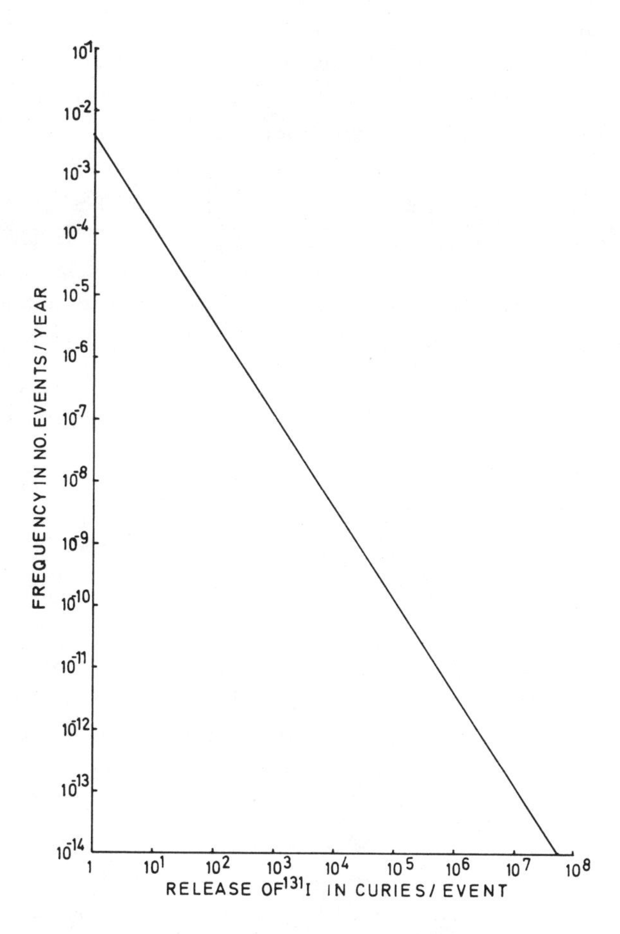

FIG.3. Permitted frequency density function for ^{131}I.

Hence, in the limit

$$G = \frac{10^{-8} \times P}{\sum_i 2Q_i D_i R\, C_{max}^{0,5}}$$

A typical example of a frequency-magnitude relationship $dP(C) = GC^{-1,5}\,dC$ versus C is depicted in Figure 3. In order to facilitate the use of this criterion for the purpose of a safety assessment of a nuclear installation, the frequency distribution given above, viz.

$$dP(C) = GC^{-1,5}\,dC$$

can be integrated over release magnitudes of a defined interval (for example decade intervals) to provide the total permitted frequency of accidental releases within each decade. The summation of the expected frequencies of all events that would give rise to an accidental release of ^{131}I with a magnitude within a given interval must be compared with the permitted frequency thus derived.

For situations where ^{131}I is not the dominant nuclide, 'equivalent' ^{131}I releases would be calculated for the other contributing nuclides on the basis of their risk potential. In calculating the potential risk no account has been taken of the time required to reinstate the installation after a major accident. Indeed, in some cases the plant would not merit reinstatement at all. The potential population risk could thus in fact be less than the calculated value.

3. OPERATIONAL DISCHARGES

For operational releases of radioactive material from an installation, the actual risk to the population can be estimated and hence suitable release limits can be determined on the basis of the population average risk level. Winkler et al [5] have described typical limits for members of the general public for exposure resulting from normal operational releases which were derived in such a manner. It should be noted that these figures are based on a population average risk level of 10^{-7} casualties/person/year which is a relaxation of a factor of 10 on the accidental release value. This factor is introduced because the releases are made in a controlled fashion and there is a consequent reduction in the uncertainties in the risk to the public compared with the potential risk in the case of accidental releases. The risk coefficient for whole-body irradiation is taken [5] to be 200 cases of cancer per 10^6 man rem, or an average of 2×10^{-4} cases/man rem. Each case is equated to a fatality and the resulting population average whole-body dose limit would therefore be 5×10^{-4} rem/year.

The risk is, however, not uniformly distributed throughout the population, since those living close to the installation can, by the very nature of things, be expected to receive a greater radiation exposure. Thus, although the population average risk might be acceptable, the peak individual risk might be excessive; it should therefore likewise be limited by the imposition of an individual dose limit. We have used the term peak-to-average risk ratio to identify this characteristic. Variations in risk level between different members of society do exist for most other hazards, and investigations were made to establish values of the risk ratio thought to be acceptable. Earthquakes and dam disasters provide two examples of cases similar to that of nuclear power stations, where the risk reduces with increasing distance from the focus of the hazard and where, although the consequences of the occurrence of an accident could be severe, the associated probabilities are very low. Unfortunately data of this nature are very scarce. From the data that are available it is thought that the risk ratio should not exceed a value of 50 for the population defined to be at risk. By using the whole-body dose risk coefficient referred to above, the maximum individual annual whole-body dose becomes 25 mrem/year which is 1/200 th of the occupational dose limit recommended by the ICRP. In the case of sites with typical population distributions, the individual dose limit will be more restrictive.

For exposure other than whole-body exposure, we have specified [5] that the dose to any individual shall not exceed 1/200 th of the occupational dose limits recommended by the ICRP. This is based on the assumption that these occupational dose limits for the various categories of exposure reflect the same mortality risk to the individual exposed.

4. DISCUSSION ON THE USE OF THE RISK CONCEPT

Although the adoption of the risk concept as a basis for safety assessment provides one with a more rational means of deciding whether an installation is acceptably safe, there are a number of problem areas and conflicting requirements which are discussed below.

First and foremost is the paucity of data. Although data on accidents are generally available in large quantities, information on the number of casualties per event and risk distributions cannot be readily extracted. Risk coefficients are derived from data on acute exposure and, in this method, have to be used for chronic exposure levels as well. The effect of dose-rate dependence, which could involve a relaxation of the release criteria, has to be disregarded on account of the lack of quantified data, and risk coefficients for single-organ exposure, except in respect of a few instances, have to be inferred. If more data were available, the factors of conservatism used in deriving the average risk levels of 10^{-7} and 10^{-8} casualties/person/year mentioned above could be reviewed.

Furthermore, the linear dose-effect relationship is assumed in our calculations, as is the common practice, in view of the absence of justifiable alternatives. Logically, this would require one to consider the radiological impact on the entire global population when assessing the effects of a nuclear installation [6]. In general we sympathize with this approach. However, since at large distances from an installation the risk of radiation-induced diseases and of consequent deaths represents but a small fraction of the variation in the natural incidence of such diseases, we consider it justifiable to limit the population at risk for decision-making purposes by excluding that part of the global population for whom the additional risk represents less than a certain fraction of that which is the result of the natural incidence of the particular disease. We suggest that 0,1% would represent a sufficiently conservative fraction. It should be realized that it can be argued that the incidence of diseases due to radioactivity releases from the installation cannot be proven to exist at these levels.

We recognize, however, that the risk to these people outside the defined population as a result of nuclear power does, in theory, exist and should be studied as a separate exercise so long as the linear dose-effect relationship is not superseded.

If an area within which the population at risk resides is defined, the effect of changing the population distribution within that area is to alter the average risk and hence the release criteria. On the other hand, if the distribution remains essentially the same but the total population is increased, the population average risk and the release criteria remain the same. It can be seen that changing the release criteria as a result of changes in the population distribution could result in expensive backfitting of safety equipment to the installation. It is therefore necessary to set the release criteria on the basis of the projected population for the lifetime of the plant, and also to apply some measure of population control. This control should be based on a guide which indicates the size of development, at any given distance from the site, that would cause a significant increase in the population average risk. As an example, for a typical site a general increase in the population of 20%, with no developments within the 10 km radius from the site, would increase the overall risk by 5%. The addition of a block of 30% of the total population between 10 and 15 km from the site would increase the overall risk by 45%. Table II indicates the size of population increases in relation to their distance from the site, each of which would involve an increase of about 0,1% in the population average risk. Such a Table suitably adjusted to take account of environmental features could be used in compiling a guide to the control of developments around the site. Population growths larger than those given in the Table would be referred

TABLE II

GUIDE TO A SUGGESTED FRAMEWORK FOR POPULATION CONTROL AROUND NUCLEAR SITES

Distance	Additional people permissible within $22\frac{1}{2}^{\circ}$ segments without reference (see text)
Within 5 km	All developments referred
5 - 7,5 km	100
7,5 - 10 km	200
10 - 15 km	400
15 - 20 km	1 000
Above 20 km	4 500

to the local authority for comment. However, it should be noted that if simultaneous developments were to take place at levels just below those given in the Table, then the population risk figure would alter significantly. Therefore, in practice, information about the actual changes in population during the life of the installation would be required to avoid the violation of the various risk limits.

It might in many cases be true that particular nuclear installations meet the specified criteria, and that the risk imposed on the affected population as a result of accidental releases - even the maximum individual risk - would be acceptable and there would be no necessity for invoking emergency procedures from considerations of the risk alone. However, even if the risk standards are complied with, in the event of an accident the risk to the public has to be minimized; here the risk concept indicates the extent to which emergency procedures need to be implemented. Minimizing the risk would involve among other things a balance between the reduction in risk resulting from the limitation of radiation exposure, and the imposition of additional risk as a result of the implementation of certain emergency actions such as, for instance, evacuation of members of the public from the vicinity of the site.

5. COST-BENEFIT ANALYSIS

From the above it can be seen that a risk philosophy can provide quantitative information for a more rational decision on the safety of a nuclear installation. However, as mentioned in the introduction, the philosophy described above was developed as an interim measure for use until some technique for the assessment of nuclear plant safety along cost-benefit lines could be developed.

It can be argued that all major technological developments should be aimed at improving the 'quality of life' and hence the decision on whether a particular installation would be desirable for the sociaty should be based on the overall requirement that it should represent a significant improvement to this quality of life. In an ideal situation it is felt that in any community some special organization should be responsible for carrying out a complex analysis to determine the overall plan for maximum benefit to society. As a part of such an analysis the need for and location of power generation plant would be determined, and the choice between different types of power plant using different fuels would be made.

However, in carrying out such an analysis, all the benefits and costs in the fullest sense relating to all the members of the society bearing those costs or reaping the benefits, should be considered and not just, as we suspect is the common practice for this type of analysis, the economic argument alone.

It is clear that a risk philosophy focusses attention only on the detrimental characteristics of an installation; moreover, in the method outlined in this paper, only one risk parameter (the mortality risk) is used as a measure of the whole spectrum of detractions. It therefore represents only a limited - albeit most important - part of the cost-benefit analysis.

Some of the factors that would be involved in such an analysis are given in Table III. In order that the final cost-benefit balance can be carried out with minimal use of subjective judgement, as many of these factors as possible must be quantified and then reduced to some common base. The question of what quantity to use for this base then arises. We believe that some effort should be devoted to this critical question of what the right unit for use in cost-benefit analysis is. However, in the meantime we accept that money, or some directly proportional equivalent such as a fraction of a society's gross national product, seems to be the best measure at the present time.

Many of the basic factors themselves have a changing value to society. Whereas an industrial development may be acceptable at a given point in time, even though costs are incurred as a result of, for example pollution of the atmosphere, the same society may at a later date, if faced with a similar decision, reject the proposal. In addition, the apparent views of a community, as expressed by the majority of its members, may not be in the best interest of that community since, if these members were in possession of all the facts and had the necessary expertise themselves to make a realistic valuation of those parameters that are at present difficult to quantify, they might hold alternative views.

The problem of quantifying costs and benefits associated with parameters that are currently difficult to quantify - in fact the whole problem of catering for

TABLE III

SOME FACTORS TO BE CONSIDERED IN A COST-BENEFIT ANALYSIS

Costs	Benefits
1. Cost of design, construction, operation	1. Worth of the product(s)
2. Risk of death and injury to humans	2. Reduction of some existing risks
3. Damage to other forms of life	3. Ecological benefits
4. Reduction of existing amenities	4. New amenities
5. Reduction in aesthetics	5. Gain in aesthetic value of the environment
6. Other forms of pollution	6. Improvement in employment proposals

the spectrum of societal views on matters such as these - together with the multitude of other costs and benefits arising from a technology such as electric power generation, indicates that cost-benefit analysis in its broadest sense presents a formidable task.

The important point is, however, to realize that no solution to any equation relating risks to benefits and costs to benefits will ever be perfect. It is right to identify the areas of uncertainty and important perhaps to stress how large some of these uncertainties are or how considerable they may appear to be at the present time. Nevertheless we must not be so dismayed by the confusion and uncertainty as to abandon the attempt. If the concept is right, and we have yet to find a better one, then any application, however limited, is likely to improve our perspective. Where, for some parameters, quantification has, through lack of information or understanding at the present time, perforce to be so imprecise as to have no significance in the mathematical balance of the equation, then the direction of the effect of these parameters should be expressed in qualitative terms to form a qualifying comment on the overall result obtained purely from the balance of those terms that can be quantified.

Clearly then, although we have a long way to go, every step in this direction will present an improvement. We believe that the use of cost-benefit techniques in a limited sense presents a way in which such improvements can be made. For instance, once a decision has been taken, by whatever means, to provide a certain electrical generation capacity within a given broadly defined area, alternative means of achieving this aim can be analyzed with a cost-benefit analysis, i.e. 'the comparative cost-benefit analysis'. Many of the imponderables, e.g. the use to which the electricity is put, may be common to all proposals and can be eliminated. The USAEC [7] for example has recently moved in this direction.

Cost-benefit techniques can also be employed in a limited sense for the analysis of particular subsystems within a plant; if an alternative system (e.g. emergency core cooling and shutdown systems) is available, the analysis can give guidance on the choice of the best system to be used. The imposition of fuel temperature reduction, with its associated costs in terms of reduced output, can be compared with the reduction in risk achieved. An example of the limited type of analysis is contained in the recently published work on numerical guides for design objectives and limiting conditions to meet the criterion 'as low as practicable' for radioactive material in light-water-cooled nuclear power reactor effluents (WASH 1258) [8]. Further work of this nature by Beattie et al [12] details an analysis of the cost of large accidental releases.

Proper use of even these limited techniques involves some understanding of the value of a human life and cost of human injury and although we recognize that it is a difficult problem to establish such valuations, it cannot be avoided.

6. CONCLUSION

We have outlined the use of the risk concept in the assessment of nuclear installations and have identified some associated problem areas. The application of a true cost-benefit analysis may at present seem to present a formidable task, but we believe that it should none the less be pursued at every opportunity, even if only in a limited sense, since it is only in this way that progress towards our understanding of this technique will be maintained and its eventual general use in the fields of safety and technological assessment will come about.

REFERENCES

[1] TATTERSALL, J.O., SIMPSON, D.M., REYNOLDS, R.A. 4th Int. Conf. Peaceful Uses At. Energy (Proc. Conf. Geneva, 1971) 11 UN, IAEA, (1972) 487.

[2] STARR, C. Social benefit versus technological risk, Science 165 (1969) 1232.

[3] FARMER, F.R. Containment and Siting of Nuclear Power Plants (Proc. Symposium Vienna, 1967), IAEA, Vienna (1967) 303.

[4] OTWAY, H.J., ERDMANN, R.C. Reactor Siting and Design from a Risk Viewpoint, Nuclear Engineering and Design 13 2 (1970) 365.

[5] WINKLER, B.C., SIMPSON, D.M. Regional Conference on Radiation Protection (Proc. Conf. Jerusalem, 1973).

[6] HEDGRAN, A., LINDELL, B. On the Swedish policy with regard to the Limitations of Radioactive Discharges from Nuclear Power Stations. An Interpretation of Current International Recommendations. Statens Strålskyddinstitut, Verksamheten (1970) 75.

[7] USAEC, Preparation of Environmental Reports for Nuclear Power Plants, Regulatory Guide 4.2, US Atomic Energy Commission Directorate of Regulatory Standards (March 1973).

[8] USAEC, Final Environmental Statement concerning Proposed Rule-making Action: Numerical Guides for Design Objectives and Limiting Conditions for Operation to meet the Criterion 'as low as practicable' for Radioactive Material in Light-water-cooled Nuclear Power Reactor Effluents. WASH-1258 (1973).

[9] World Airline accident Summary, Air Registration Board, UK.

[10] Statistical Bulletin, Metropolitan Life Insurance Company, USA.

[11] UK Ministry of Technology, Private Communication.

[12] BEATTIE, J.R., BELL, G.D. A Possible Standard of Risk for Large Accidental Releases, SRS Report SRS/GR/16 (1973).

DISCUSSION

M. J. A. DELPLA: Your paper seems to take the opposite line to that of papers 25 and 21 just presented.

D. M. SIMPSON: That is correct. The main difference is in our use of a linear dose-effect relationship without a threshold. We have used this because it is the best documented evidence available to us at present. However, in the paper we do refer to a general lack of data in this area.

M. J. A. DELPLA: The linear relation may be the only one used in practice but this is no justification for it. The linear relation is in fact disastrous and we should pay very serious attention to other types of relationship, which should of course be prudent but not excessively so, as is the linear relation.

THE PREDICTION OF POPULATION DOSES DUE TO RADIOACTIVE EFFLUENTS FROM NUCLEAR INSTALLATIONS

R.H. CLARKE
Central Electricity Generating Board,
Berkeley Nuclear Laboratories,
Berkeley, Glos

A.J.H. GODDARD, J. FITZPATRICK
Department of Mechanical Engineering,
Imperial College of Science and Technology,
London, United Kingdom

Abstract

THE PREDICTION OF POPULATION DOSES DUE TO RADIOACTIVE EFFLUENTS FROM NUCLEAR INSTALLATIONS.

Studies on the siting and safety aspects of nuclear installations have been made more comprehensive recently by the development of fast and economical computer programs describing the complex pathways by which radioactivity finds its way from irradiated fuel to man. Within the CEGB the WEERIE program has been developed to evaluate inhaled doses to adult human body organs, cloud β and γ doses and the ground deposition from discharges of airborne effluents under normal operational or accident conditions. WEERIE may thus be used to generate spatial distributions of dose to a variety of organs and the possibility arises of integrating the dose isopleths with real population distributions to assess population doses, which may then be combined with UNSCEAR figures for risks per 10^6 man·rads. In the paper the methodology used to set up the evaluation of population dose using economical methods developed for the storage, retrieval and modification of population data is described. As illustrative examples of the work, population dose evaluations are made for ^{85}Kr discharges from a reprocessing plant. The major contributions to population dose are identified, the effects of cutting off dose contributions at 1% of background doses are investigated and the results presented as contours of equal population dose.

1. INTRODUCTION

The computer program WEERIE, which has been described elsewhere [1, 2], calculates the estimated release rates for a wide range of nuclides into the atmosphere and after allowing for meteorological dispersion, the inhaled doses, cloud β-dose and ground deposition of activity may be found, while integration over the volume of the plume leads to estimates of cloud γ exposure. The program has been widely used for investigations of accident conditions at Magnox power stations [3], to provide guidance for improved emergency monitoring procedures together with data for assessing the dosimetric consequences. More recently WEERIE has been used to consider the radiological safety following a fuel pin failure in an advanced gas cooled reactor including representation of the free pin volume activity [4].

WEERIE enables predictions to be made of spatial distributions of dose under normal operational or accident conditions and in the present paper these calculations have been extended to the next logical stage of development by integrating them with real population distributions within the U.K. This enables the assessment to be made of population doses from single and multiple sources so that the effects of a complete nuclear power programme may be estimated. The model is illustrated by examples of continuous releases of nuclides: firstly the population dose from the ^{85}Kr discharged from a notional reprocessing plant serving a 5 GW(E) power programme is estimated, and secondly assuming the plant to be operating to the I.C.R.P. criteria for the critical group of the population, the population dose is evaluated as a function of the effective height of release and distance to the critical group. The effects of dose cut-off at 1% of background dose are investigated and the spatial distribution of population dose is presented graphically by means of population dose isopleths.

2. METHODS OF CALCULATION

2.1 The WEERIE program

The ^{85}Kr inventories for the example cases here were generated by the FISP program [5] which is a module of the WEERIE code [6] and the ^{85}Kr was assumed to be emitted directly to the atmosphere, without filtration. The Gaussian plume model forms the basis of the meteorological calculation, with allowance for ground deposition of activity, and the dispersion parameters σ_y and σ_z are usually those of Pasquill [7] for releases of duration up to a few hours. There are also instantaneous release parameters for short "puffs" of activity. Finally, there is a continuous meteorology option in which the concentrations at any downwind distance are averaged over the six Pasquill categories weighted with their mean frequency of occurrence in the U.K. using data presented by Bryant [8].

In the final stage of the WEERIE program the complete radioactive environment is used to assess inhalation doses to organs of exposed individuals by use of tabulated values of rems per Curie inhaled which were calculated for all the fission product nuclides in WEERIE using I.C.R.P. 2 metabolic data [9]. The cloud β dose is evaluated from point concentrations as are the inhaled doses, but the cloud-γ exposure requires integration over a large volume of the plume because of the long mean free path of γ-rays in air. Special techniques in WEERIE have enabled the code to be extremely fast in a multi-energy group cloud-γ evaluation utilizing Gaussian quadrature numerical integration over the real plume geometry and not making approximations to the semi-infinite cloud model [2].

2.2 Storage and display of population data

A detailed study of the local population distribution have always formed a necessary part of the assessment of a nuclear site [10]. In this work direct access by a computer program to census data has been developed, influenced partly by present trends in the presentation of census information. For example, the United Kingdom 1971 Census, soon to be generally available, has been planned for compatibility of information, use in computations which involve locations and in the display of data by graphical means [11]. The standard data to be made available will be on a 10, 1 or 0.1 km grid.

The earlier 1966 Sample Census data is not so readily adaptable to computer use, but a reduction to a 1.59 km x 2.64 km grid covering England and Wales has been published [12]. This was derived from population totals of local authority areas by condensing the population upon the centroid of each of these areas and assigning a population class to a particular grid cell if a centroid with population in a certain range lay within the cell. In the present work some adjustments were necessary to this largely experimental presentation in order to obtain the correct total populations for England and Wales and that for the Greater London area.

In working with stored population data computing costs must be considered. An acceptable compromise between definition of cell population and the description of overall variations may be achieved by using population classes identified by characters assigned to cells. A single character may be stored more economically than the actual population figure. A second advantage of character representation is that grey-level (line printer) maps [13] of population density may be generated directly and this technique has been used to generate the population density maps in Section 3.

Depending upon the computer used, different strategies may be adopted. The present work has been carried out using the Imperial College CDC 6400 computer which has a 60 bit word. 20 characters have been packed into each word and up to 8 different characters may be used; this range of characters could be increased by using more bits per character. The complete population map of England and Wales may be stored in 4 kilo-words. In work in progress using the C.E.G.B. IBM 370 computer, characters are assigned to 8 bit "bytes", 4 to a word. The complete range of approximately 50 characters may be used, with a penalty of increased core requirement.

2.3 Calculation and display of population doses

A program has been written which makes use of a library of data generated by WEERIE and of population data in "packed store" as described in 2.2. This work is concerned with continuous releases and for the two cases considered WEERIE was used to generate dose patterns as a function of downwind distance for a range of release heights.

When the site of the release has been defined the program scans the population cells, extracts the population density from packed store and interpolates in the dose pattern to yield man-rads for that cell. Since for realistic sites the wind does not distribute activity uniformly into each sector, the values of man-rads estimated are weighted by the probability of the wind blowing in the direction of interest. For inhaled or cloud β doses it is permissible to separate the "wind rose" from the WEERIE calculation, but for cloud-γ doses which require spatial integration over the plume, slight errors may be introduced within a few hundred meters of the stack (the mean free path of 1 MeV γs is $\sim$130 m in air). Since in this work the spatial resolution of population density is at minimum 1.59 km, the wind rose is again assumed separable from the WEERIE calculation. The same frequency of occurrence of weather categories has been assumed to apply within 200 km of each site; it is known that in large urban areas this frequency will be modified and doses in these areas may thus be overestimated.

Contour maps of man-rads may be generated by the program, on a flat-bed or drum plotter, to the same scale as the population grey - level map. In order to apply some degree of smoothing and to reduce storage requirements for the automatic contouring routine, a condensation to man-rads per year per square kilometre averaged over 16 cells was carried out before plotting.

3. ILLUSTRATIVE EXAMPLES OF POPULATION DOSES

3.1 Continuous gaseous release from a reprocessing plant

The first example deals with population doses resulting from a notional reprocessing plant situated on the N.E. coast of England and discharging ^{85}Kr and ^{3}H into the atmosphere from a 5000 MW(E) nuclear power programme. Assuming a mean thermal efficiency of 30% for the power reactors and a

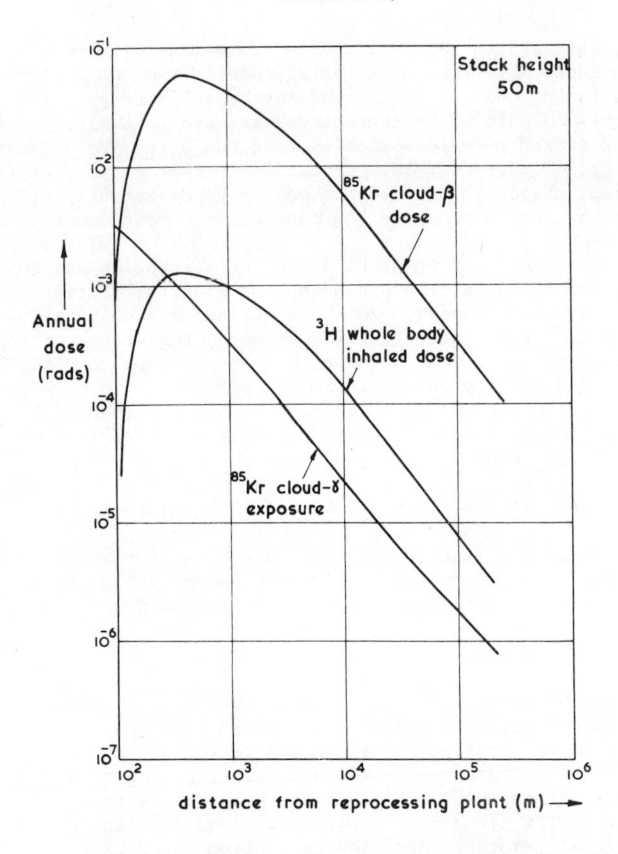

FIG. 1. Annual doses as a function of distance from a reprocessing plant serving a 5000-MW(e) nuclear programme assuming ^{85}Kr and ^{3}H are discharged to atmosphere from a 50-m stack.

burn up of 20,000 MWD.T^{-1}, a 5000 MW(E) program requires 300 tonnes of fuel to be reprocessed per year giving rise to about 2MCi of ^{85}Kr released to atmosphere per year. This figure compares with 500 Ci $(MW(E)yr)^{-1}$ quoted by U.N.S.C.E.A.R. [14]. The corresponding ^{3}H inventory is 10^{5}Ci.yr^{-1} which is here assumed to reach the atmosphere, although in practice it may be in liquid effluents. The downwind doses are shown in Fig. 1; where the cloud β and cloud γ dose are due to ^{85}Kr and the whole body dose results from ^{3}H inhalation, a 50 m effective stack height having been assumed.

Dunster and Warner [15] and the U.S. Environmental Protection Agency [16] give comparable data for ^{85}Kr arisings and subsequent doses when allowance is made for variations in stack height and the reduction from cloud β dose to skin dose. The ICRP criterion for limiting the discharge would be that the peak skin dose should not exceed 3 rem.yr^{-1} and the results of Fig. 1 may be scaled to this level by multiplying by a factor of 52.5. Implicit in this assumption is equality of cloud-β dose in air and skin dose. This is not quite true, but for the purposes of this paper cloud-β dose is assumed to effectively be skin dose. Thus under the present assumption of a 50 m effective stack height the processing plant

TABLE I. CALCULATED ANNUAL POPULATION DOSES WITHIN 200 KM RADIUS OF NOTIONAL PROCESSING PLANT SERVING 5 GW(e) PROGRAMME

Population	^{85}Kr cloud β dose (man-rad year^{-1})	^{85}Kr cloud γ dose (man-rad year^{-1})	^{3}H inhalation dose (man-rad year^{-1})
1.706×10^{7}	3,152	13.8	70.2

TABLE II. CALCULATED ANNUAL POPULATION DOSES FROM NOTIONAL REPROCESSING PLANT SERVING 250 GW(e) PROGRAMME AND MEETING ICRP CRITERION FOR CLOUD β DOSE TO SKIN

	Population	^{85}Kr cloud β dose (man-rad year^{-1})	^{85}Kr cloud γ dose (man-rad year^{-1})	^{3}H inhalation dose (man-rad year^{-1})
(a)	1.23×10^{3}	523	2.31	12.3
(b)	1.706×10^{7}	1.60×10^{5}	726	3.69×10^{3}

(a) - for regions where cloud γ dose exceeds 1% of natural background.

(b) - within a 200 km radius.

(with present technology and filtration techniques) could deal with a 250 GW(E) nuclear power programme. Tables I, II and Fig. 2 demonstrate the population doses from the 5 GW(E) and 250 GW(E) power programmes and in the latter case the effect of a 1% of natural background cut-off on the cloud-γ contribution to whole body irradiation. Even for the 250 GW(E) programme only a small fraction of the population receives more than 1 mrad.yr^{-1} whole body dose. If the 200 km range is used for the 250 GW(E) power programme, the ^{85}Kr cloud β (i.e. skin), population dose is approaching 10% of the natural background population skin dose. This is seen in Fig. 1 to be mainly due to a cloud-β dose averaging about 10 mrad yr^{-1} to the population at distances of 100-200 km from the site. In Fig. 2 the contour levels on the grey area map are at 10^{-1} man rads. yr^{-1} km^{-2} for cloud-β ^{85}Kr dose from the 5 GW(E) programme.

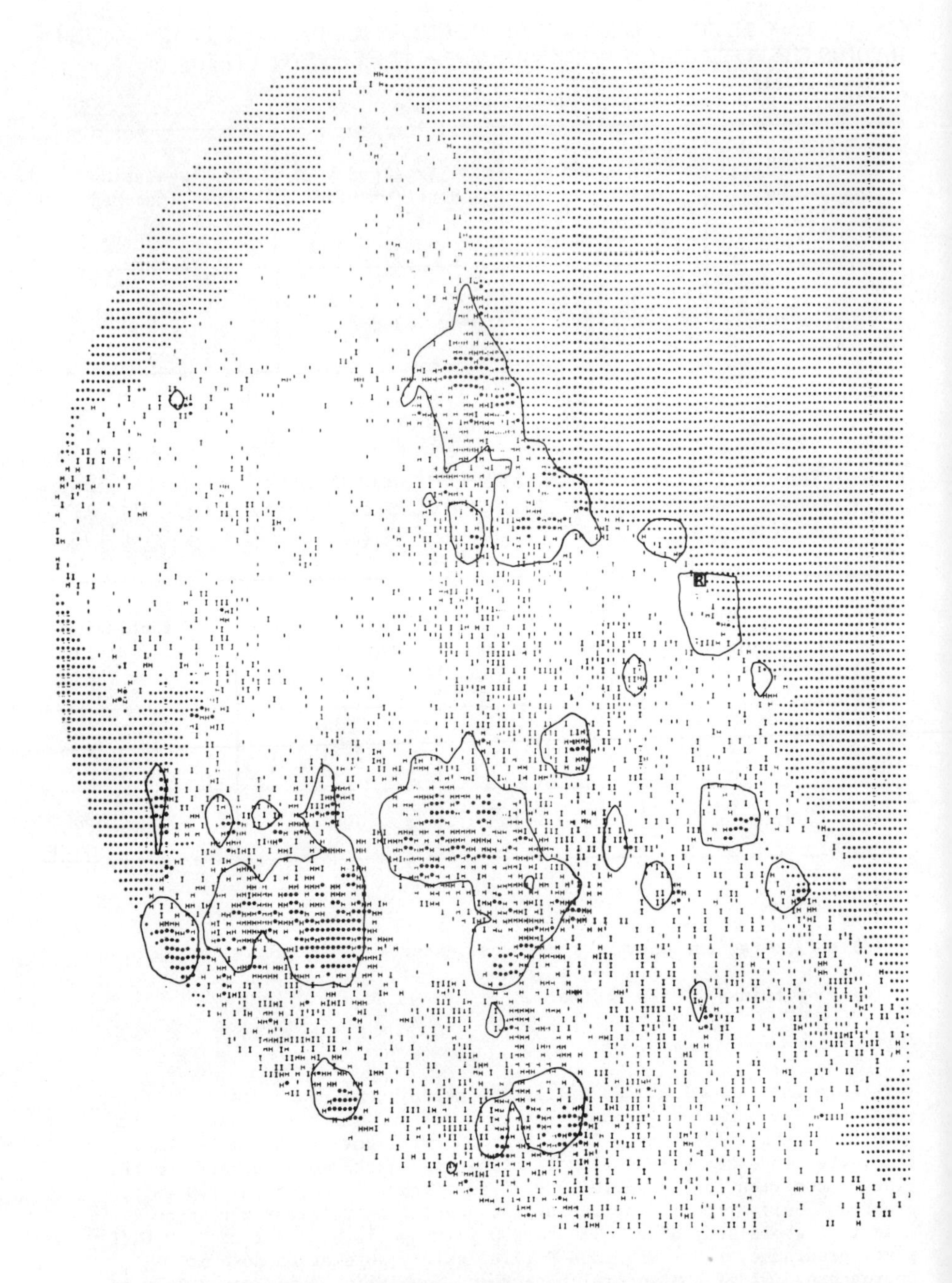

FIG. 2. Grey scale population map of NE England with contour plots of cloud-β population doses at 10^{-1} man·rad·a^{-1}·km^{-2} (———) for the release of ^{85}Kr from a 5-GW(e) nuclear programme from a reprocessing plant marked [R]. (Grey scale map Crown copyright reserved.)

3.2 Effects of height of release on population dose

In this second example, the same hypothetical reprocessing plant as in the last example is assumed to emit ^{85}Kr and the population doses are evaluated for varying effective heights of release. In Fig. 3 the relative cloud-β doses are shown as a function of distance and the I.C.R.P. suggested maximum skin dose to an individual member of the public is 3 rads. yr^{-1} which is normally taken to be the maximum dose to which the 'critical group' may be exposed. Thus the annual doses in Fig. 3 can be scaled so that the maximum dose to any individual for a specific height of release and distance of closest approach meets the 3 rad. yr^{-1} limit.

Fig. 4 shows the resulting population cloud-β doses out to 200 km from the plant for effective release heights up to 200 m with the distance to the critical group being up to 2 km. For a tall stack, if the nearest population is closer than the peak in the close-distance curve, the release rate is set as if that population were at the peak dose point, so that population dose will be constant until the critical group is farther away than the peak in the dose-distance curve.

One of the interesting features of Fig. 4 is that although the population dose is increasing at a faster linear rate with critical group distance for taller stacks, all four curves must meet at large distances due to the convergence of the dose-distance curves, as shown in Fig. 3.

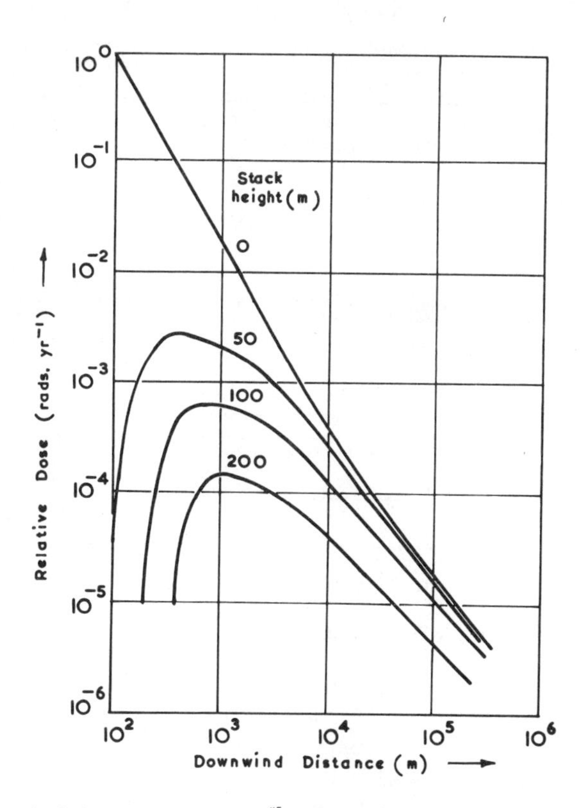

FIG. 3. Annual relative cloud-β doses from ^{85}Kr discharged from stacks of differing heights.

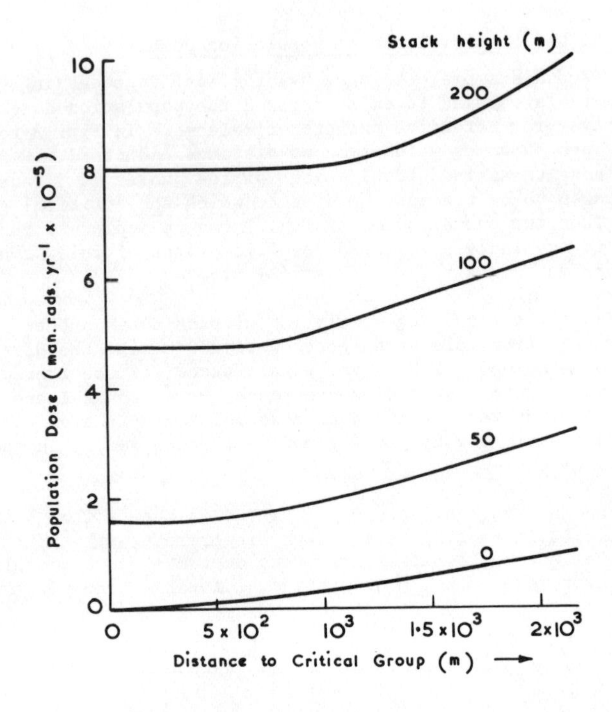

FIG. 4. Annual population cloud-β dose calculated out to 200 km from a reprocessing plant emitting ^{85}Kr at a rate that meets the ICRP suggestion of 3 rad·a^{-1} to the 'critical group' as a function of height of release and distance to that 'critical group'.

TABLE III. POPULATION CLOUD-β DOSES FROM ^{85}Kr ADDED UP TO 200 KM FROM A REPROCESSING PLANT AS A FUNCTION OF EFFECTIVE HEIGHT OF RELEASE AND DISTANCE TO THE CRITICAL GROUP (population dose units are man·rad·a^{-1})

Population	Distance to 'critical group' (m)	Stack height (m)			
		0	50	100	200
1.706×10^7	200	1.78×10^3	-	-	-
	400	5.81×10^3	1.60×10^5	-	-
	800	1.95×10^4	1.88×10^5	4.80×10^5	-
	1000	2.89×10^4	2.04×10^5	5.11×10^5	8.00×10^5
	2000	9.62×10^4	3.07×10^5	6.35×10^5	9.60×10^5

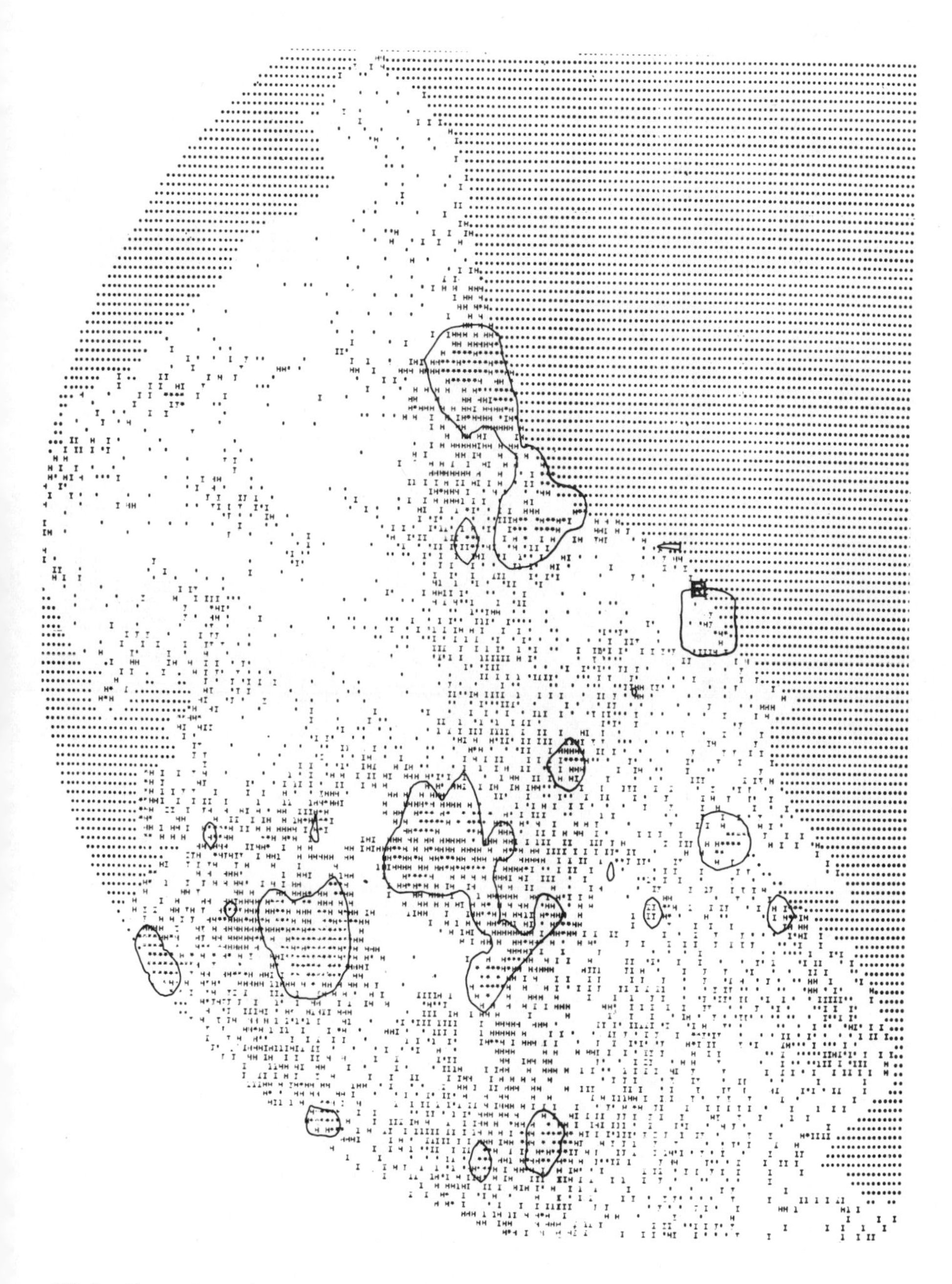

FIG. 5. Grey scale population map of NE England with plots of cloud-ε population doses at 10^1 man·rad·a^{-1}·km^{-2} (———) for ^{85}Kr released from a plant (marked [R]) to meet the ICRP criterion of 3 rad·a^{-1} at 400 m from a 50-m stack. (Grey scale map Crown copyright reserved.)

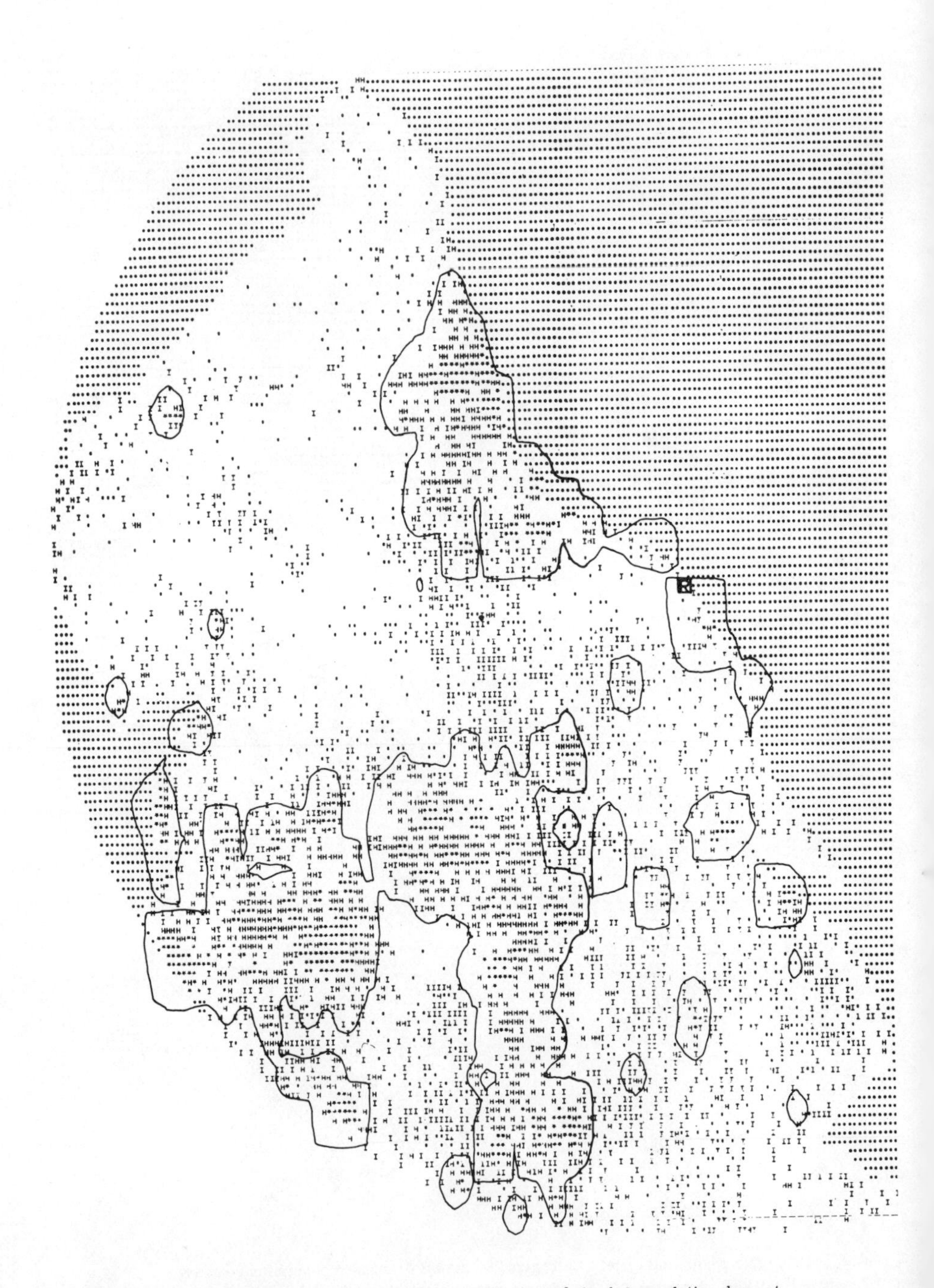

FIG. 6. Grey scale population map of NE England with plots of cloud-β population doses at 10^1 man·rad·a^{-1}·km^{-2} (———) for ^{85}Kr released from a plant (marked R) to meet the ICRP criterion of 3 rad·a^{-1}·km^{-2} at 1 km from a 200-m stack.

The total population within 200 km of the plant in this example is 1.706 x 10^7 and the population dose varies from 1800 man-rads. yr^{-1} for a ground level release and 200m fence to 10^6 man-rads.yr^{-1} for a 200m stack and 2 km fence (both cases requiring 3 rads. yr^{-1} at the fence) as shown in Table III. If background radiation is taken as giving 100 mrad. yr^{-1} to the skin, the population dose would be 1.7 x 10^6 man-rads. yr^{-1} from this source and the contribution from a reprocessing plant could be as high as 59% or as low as 0.1% of the skin dose due to natural background. It is to be expected that these figures would vary with the individual site and if cloud-γ rather than cloud-β doses were used as the limiting criteria.

Finally Fig. 5 and 6 show the population dose contours for 10 man-rads $yr.^{-1}$ km^{-2} superimposed on grey scale maps for the cases of a 50m stack with a critical group distance of 400m and a 200m stack with a 1 km critical group distance respectively. It is clear that the population doses increase and the number of people above a given dose limit increases with stack height. Also the population dose contributions are significant at large distances and clearly would be increased by the siting of subsequent nuclear installations.

4. CONCLUSIONS

The principal conclusion to be drawn from this work is that it has been shown feasible to estimate collective doses for realistic population distributions around actual nuclear sites by the development of computing techniques which are economical both in fast core requirements and execution time.

Illustrative examples have been given for notional nuclear sites which demonstrate that the relationship between population dose and site characteristics is extremely complex for a nuclear installation operating to the I.C.R.P. criteria for individual members of the public.

The methods outlined in the paper provide a basis for siting calculations in that the effect of a number of sites on a population group may be considered and changes in population distribution with time can be investigated.

The immediate future development of the work is towards representation of long range (>10^3 km) meteorological dispersion and the consideration of the accident situation in terms of population dose commitment.

REFERENCES

[1] CLARKE, R.H., Physical Aspects of the Effects of Nuclear Reactors in Working and Public Environments, C.E.G.B., Berkeley Nuclear Laboratories, Berkeley, Gloucestershire (1973).

[2] CLARKE, R.H., The WEERIE program for assessing the radiological consequences of airborne effluents from nuclear installations, Health Physics, 25, 3 (1973) 267.

[3] MACDONALD, H.F., CLARKE, R.H., The prediction of environmental releases from nuclear reactors under normal and accident conditions, VIth International Congress of the Société Francais de Radioprotection, Bordeaux, VI SFRP/15 (1972).

[4] MACDONALD, H.F., DARLEY, P.J., CLARKE, R.H., Recent developments in the prediction of environmental consequences of radioactive releases from nuclear power reactors, IAEA Symposium on the Physical Behaviour of Radioactive Contaminants in the Atmosphere, Vienna, IAEA-SM-181/33 (1973).

[5] CLARKE, R.H., FISP, A comprehensive computer program for generating fission product inventories, Health Physics, 23, 4 (1972), 565.

[6] CLARKE, R.H., A Users' Guide to WEERIE, C.E.G.B. Berkeley Nuclear Laboratories, Berkeley, Gloucestershire, RD/B/N2407 (1973).

[7] PASQUILL, F., The Meteorological Magazine, 90, 1063, (1961), 33.

[8] BRYANT, P.M., UKAEA Rep. AHSB (RP) R42 (1964).

[9] CLARKE, R.H., UTTING, R.E., Initial Estimates of Doses in Human Body Organs Following Acute Exposure to Radioactive Fission Products, C.E.G.B. Berkeley Nuclear Laboratories, Berkeley, Gloucestershire, RD/B/N1762 (1970).

[10] GRONOW, W.S., GAUSDEN, R., Licensing and regulatory control of thermal power reactors in the U.K., IAEA Symposium on Principles and Standards of Reactor Safety, Julich,IAEA-SM-169/23 (1973).

[11] OFFICE OF POPULATION CENSUSES AND SURVEYS, 1971 Census, Information Papers 1-5, Census Customer Service, Titchfield, Fareham, Hampshire (1973).

[12] MINISTRY OF HOUSING AND LOCAL GOVERNMENT (now Department of the Environment), Population Density 1966, LINMAP 147, (1969).

[13] HOWARTH, R.J., FORTRAN IV program for grey-level mapping of spatial data, Journal of the International Association for Mathematical Geology, 3, 2, (1971), 95.

[14] U.N.S.C.E.A.R., Ionising Radiation: Levels and Effects, A report of the United Nations Scientific Committee on the Effects of Atomic Radiation, Table 71, 1 (1972), 106.

[15] DUNSTER, H.J., WARNER, B.F., UKAEA rep. AHSB (RP) R101, (1970).

[16] KLEMENT, A.W., MILLER, C.R., MINX, R.P., SHLEINEN, B., Estimates of ionising radiation doses in the U.S.A., 1960-2000, OPR/CSD/72-1 (1972), 36.

INFLUENCE DE LA PUISSANCE ET DE LA DISTANCE SUR LES RISQUES PRESENTES PAR UN REACTEUR NUCLEAIRE — FACTEUR ATMOSPHERIQUE DE SITE

R. LE QUINIO
CEA, Centre d'études nucléaires de Saclay,
Département de sûreté nucléaire,
Gif-sur-Yvette, France

Abstract–Résumé

HAZARDS OF NUCLEAR REACTORS AS A FUNCTION OF REACTOR POWER AND DISTANCE — THE SITE ATMOSPHERIC FACTOR.

The paper discusses the characterization of a reactor site by a relative value representing hazards transmitted via the atmosphere. To do this the author proposes a general method which, with a few supplementary hypotheses, can be applied to particular cases. First, on the assumption that there is an upper damage threshold, one determines — as a function of reactor power — a maximum distance from the reactor at which damage will occur by taking effluent emission together with atmospheric diffusion coefficients. In evaluating the risk for each distance, account is taken of atmospheric diffusion and the probability of an accident of a certain severity. The population inside the circle of maximum radius is then weighted, both as a function of the previously obtained distance factor and as a function of the prevailing wind directions. The practical application of the method is discussed in the light of the accuracy of the currently available data. A numerical example is given.

INFLUENCE DE LA PUISSANCE ET DE LA DISTANCE SUR LES RISQUES PRESENTES PAR UN REACTEUR NUCLEAIRE — FACTEUR ATMOSPHERIQUE DE SITE.

L'objectif de cet exposé est de caractériser un site par une valeur relative concernant les risques transmis par voie aérienne. Pour ce faire on propose une méthode générale que l'on applique avec quelques hypothèses supplémentaires à des cas concrets. Tout d'abord une distance maximale d'atteinte en fonction de la puissance est déterminée par la combinaison des émissions d'effluents et des coefficients de diffusion atmosphérique, en supposant une nuisance limitée par un seuil. Pour chaque distance le risque est évalué en tenant compte de la diffusion atmosphérique et de la probabilité de voir survenir un accident d'une certaine intensité. Enfin la population à l'intérieur du rayon maximal est pondérée, d'une part en fonction du facteur distance obtenu précédemment et d'autre part en fonction de la rose des vents. Les modalités pratiques de l'application de cette méthode sont discutées en regard de la précision des données actuellement disponibles. Une application numérique est citée en exemple.

1. INTRODUCTION

Le développement de l'énergie nucléaire va poser de plus en plus le problème du choix des sites.

On peut espérer que le développement rapide de la sûreté des réacteurs conduise un jour à la banalisation de ces installations, c'est-à-dire à une liberté d'implantation au moins aussi grande que celle d'autres établissements industriels; dans l'état actuel de nos connaissances les pays privilégiés qui ont encore la liberté de choix du site se doivent d'essayer de trouver, en matière de sûreté, des critères objectifs.

Ces critères doivent permettre, au minimum, de comparer les risques relatifs de deux installations sur un même site et surtout, d'une même installation sur deux sites différents.

La vitesse relativement grande du transfert de la contamination par voie atmosphérique, les difficultés,dans ce cas,de prévention et de lutte, imposent de choisir ce mode de transfert comme critère principal.

2. DEFINITIONS ET HYPOTHESES

2.1. Limite d'influence

Tout d'abord la notion d'environnement d'une centrale a besoin d'être précisée. La question se pose en effet immédiatement lorsque l'on veut définir le domaine d'intégration des doses qui conduit à la notion d'hommes-rems. L'expression mathématique associée étant généralement divergente,il y a lieu de donner une borne supérieure aux distances considérées, à moins d'envisager comme Machta [1] des panaches faisant plusieurs fois le tour de la terre, les doses au-delà de 100 km étant alors du même ordre que les doses en-deçà pour l'installation étudiée.

Cette borne supérieure peut être fixe (50 miles par exemple) ou mieux être associée à un niveau individuel d'exposition. En effet dans ce cas elle est plus facilement liée à un rejet potentiel déterminé. D'autre part, il est possible de définir ces seuils soit par comparaison avec les fluctuations de la radioactivité naturelle, soit par une limite au-dessous de laquelle il sera pratiquement impossible de déterminer un effet (10 rems à la thyroïde par exemple).

Théoriquement ceci entraîne une borne différente pour chaque condition météorologique, chaque nucléide et chaque seuil considéré.

2.2. Diffusion atmosphérique

Il n'est pas dans notre propos de revenir sur la définition des classes de stabilité par exemple. Dans les pays construits et habités que sont ou seront les alentours des centres nucléaires, il nous paraît assez vain d'imaginer un gradient thermique "basses couches" constant au-dessus de sols de nature différente sur des dizaines de kilomètres. Les essais de corrélation menés avec nos résultats expérimentaux n'ont pas abouti : des conditions de diffusion moyenne nous paraissent suffisants;mais nous pourrions améliorer la méthode en tenant compte d'une distribution probabiliste globale des concentrations telle que celle que nous avons proposée en [2] .

Le seul paramètre que nous retiendrons momentanément,la direction du vent,peut lui-même être sujet à caution si nous considérons certains sites et les distances de quelques dizaines de kilomètres : ce sont les trajectoires, souvent sinueuses, qu'il faudrait envisager.

Quoi qu'il en soit,à chaque direction de vent nous supposerons un panache à l'intérieur duquel les concentrations au sol sont réparties uniformément. La largeur angulaire de ce panache décroît en

$$x^{-\gamma}$$

La concentration moyenne (ou le dépôt qui lui est proportionnel) décroît avec la distance selon également une loi puissance dans les modèles les plus répandus. L'exposant de cette loi puissance varie selon ces modèles, selon les conditions météorologiques, voire avec la distance elle-même mais nous pensons pouvoir rendre compte de l'ensemble des phénomènes dans le domaine considéré (quelques heures, quelques dizaines de kilomètres) avec un exposant unique.

2.3. Dommages retenus

En principe la nuisance potentielle de l'installation sera la somme de toutes les conséquences possibles de tous les rejets possibles. Ceci implique de regarder l'évolution de tous les nucléides relâchés et d'évaluer le coût sous forme de vies humaines ou seulement de soins, de mesures de décontamination ou d'évacuation de populations, d'immobilisation de machines coûteuses etc. Le but principal de cette étude étant d'obtenir un coefficient relatif, nous ne retiendrons que les premières et même,dans le cas d'applications numériques,que les problèmes associés aux iodes et à la thyroïde. Il n'est pas nécessaire pour cela que -ce qui est admis généralement- ce risque soit le principal mais cela suppose implicitement que les autres dommages lui soient proportionnels (doses globales proportionnelles à la dose thyroïde, activité industrielle proportionnelle à la population etc...).

En toute rigueur il faudrait que les seuils retenus pour chacun des phénomènes soient également compatibles, par exemple si la dose d'irradiation globale est en tous points le 1/10 de la dose thyroïde, le seuil à retenir serait également le 1/10, sous peine de devoir dans un cas considérer des populations qui ne seraient pas comptées dans le second.

Il n'est pas permis de trop entrer dans le détail car il est probable que,dans le cas d'accident important,le seuil d'action ou d'indemnisation sera fixé en fonction de considérations économiques ou politiques qui échappent à notre analyse.

Nous ne considérons que les rejets accidentels, les rejets normaux étant considérés comme acceptés et dans tous les cas donnant des doses bien inférieures à celle retenue comme limite.

2.4. Définition du détriment

Enfin nous caractériserons l'agression potentielle d'une installation sur son environnement par le "détriment",c'est-à-dire par l'"espérance mathématique" ou la "moyenne stochastique", c'est-à-dire la somme des produits des conséquences des accidents par leur probabilité, et plus particulièrement dans ces conséquences un nombre de décès proportionnel aux irradiations.

3. METHODE

3.1. Calcul du détriment

La quantité Q de nucléides potentiellement rejetable est proportionnelle à la puissance W.

$$Q_i = a\, b_i\, W$$

L'indice i va être caractéristique d'un accident. Les termes a , quantité de nucléide contenu par unité de puissance, et W, puissance, peuvent varier (a en particulier avec la nature et surtout l'âge du combustible) mais nous éliminons ces variations en nous plaçant à la puissance nominale et en supposant les quantités maximales (fin de vie).

La gravité de l'accident sera essentiellement caractérisée par le terme b_i fraction de la quantité de nucléide contenue dans le coeur rejetée dans l'environnement; b_i variera de 0 à 1. On pourrait imaginer des $b_i > 1$ dans le cas d' "emballement" du réacteur. Mais d'une part ces accidents de criticité ont des probabilités deux ordres de grandeur en dessous des précédentes [3] , d'autre part ils concerneront surtout des nucléides à période courte qui n'intéresseront que les alentours immédiats.

Toute nuisance individuelle D_{xi} (dose associée à une concentration ou un dépôt à une distance x) est proportionnelle à Q et décroît avec la distance selon la loi puissance considérée plus haut.

On peut donc écrire :

$D_{xi} = a\, b_i\, c\, W\, x^{-\alpha}$ où α est positif et c représente la dose à 1 km associée à une émission de 1 curie.

Ceci suppose que l'on néglige la décroissance de la radioactivité durant le transport, du réacteur au point considéré.

Si l'on s'est fixé un seuil D_1 au-dessous duquel l'effet sera nul (ou indiscernable) on peut ainsi lui associer -pour chaque accident- une distance maximale d'atteinte x_1 telle que

$$x_{1i} = \left(\frac{a\, b_i\, c\, W}{D_1}\right)^{1/\alpha}$$

On pourrait aussi définir une borne inférieure (x_{2i}) de distance correspondant à une dose D_2 au-delà de laquelle les "rems" n'ont plus d'effet (il est indifférent d'irradier un cadavre) mais dans les cas pratiques cette distance est inférieure aux dimensions du site.

On pourra, pour l'accident maximal envisageable, définir une distance au-delà de laquelle le réacteur n'a plus d'influence détectable :

$$x_m = \left(\frac{a\, b_m\, c\, W}{D_1}\right)^{1/\alpha}$$

b_m pouvant être pris égal à 1 ou une fraction du même ordre si l'on admet que la volatilisation complète d'une installation est physiquement impossible.

Le nombre d'habitants soumis au panache à une distance x est :

$p\,(x, \theta)\quad \Delta\theta\quad x\, dx$

où $p\,(x, \theta)$ est la densité de population

$\Delta\theta\; x\; dx$ l'aire du secteur balayé par les effluents, $\Delta\theta$ exprimé en radians.

Nous pouvons écrire

$$\Delta\theta = \Delta\theta_1 x^{-\gamma}$$

$$(\Delta\theta_1 = \text{largeur à 1 km})$$

Donc, en cas d'un accident unique, le détriment subi par une population située à une distance x dans une direction θ est

$$N_{i\theta} = P(i) \;\; P(\theta) \; a \, b_i c \, g \, W \;\; x^{-\alpha} \; p(x,\theta) \;\; x \;\; \Delta\theta_1 \, x^{-\gamma} dx$$

P (i) représentant la probabilité de survenue de l'accident

P (θ) probabilité que le vent souffle vers le secteur de direction θ et de largeur $\Delta\theta$

$a \, b_i \, c \, W \, x^{-\alpha}$ la dose individuelle reçue à cette distance x, g le facteur de proportionnalité (supposé unique) entre la dose individuelle et le risque de décès. En fait, ce facteur a aussi une distribution probabiliste qui dépendra de la composition de la population et des contre-mesures qui pourraient être prises (surveillance médicale, inhalation d'iode stable ...)

Ces calculs sous cette forme, ou arrêtés à la notion d'hommes-rems, sont souvent effectués pour un accident baptisé "maximal" ou de "référence" ou de "dimensionnement".

La densité de population n'y est pondérée en fonction de la distance que par le facteur $x^{1-\alpha-\gamma}$ qui dépend exclusivement de la dilution des effluents.

En réalité nous avons vu que le détriment est la somme de tous les $N_{i\theta}$. A une distance proche de la distance x_m précédemment définie seul l'accident apocalyptique, pratiquement impossible, pourra intervenir.

A des distances plus faibles, le même accident aura des conséquences plus élevées auxquelles s'ajouteront les conséquences faibles des accidents modérés, plus probables.

Pour connaître le détriment total, il suffirait en principe, pour chaque secteur ($\Delta\theta \;\; \Delta x$) de sommer les résultats de chaque accident sur le spectre total de ces derniers, mais un tel travail exhaustif n'est pas d'une approche facile.

Toujours dans le cadre de valeurs relatives de site, il est plus simple de supposer une distribution des accidents en fonction de leur probabilité. Il est fréquent de voir présenter sur un diagramme rejets-probabilité un tel accident par un point.

En fait cet accident, défini par une séquence d'événements malheureux ou arbre de défaillances, conduit à un rejet qui ne devrait être lui-même donné que dans un intervalle donné avec une distribution assez large venue des imprécisions sur les taux de fuite, l'efficacité des filtres, le piégeage de parois etc... De plus les points correspondant à divers "arbres" considérés n'ont aucune raison apparente de s'aligner sur une courbe. Mais quelques accidents ou groupes d'accidents ont généralement des détriments nettement supérieurs aux autres et conditionnent l'allure d'une courbe qui donnera la densité de probabilité en fonction du rejet R = a b W sous une forme

$$\Delta P_i = k \, R_i^{-\beta} \, \Delta R = k_1 \, b_i^{-\beta} \, \Delta b \;\; W^{1-\beta} \quad (\beta \text{ positif})$$

où ΔP_i est la somme des probabilités des accidents entraînant des rejets compris entre R_i et $R_i + \Delta R$. Les constantes k et k_1 qui en dérivent sont propres à l'installation et permettent en principe de déterminer le risque et de le comparer à un critère quelconque.

Remarquons en passant que le critère proposé est souvent indépendant de l'installation, ce qui implique une valeur de k_1 variant en sens inverse de la puissance et va à l'encontre d'analyses type coût-avantage.

Pour sommer sur le spectre d'accidents (supposé alors quasi continu), les détriments intéressant la population du secteur (x, θ), la borne supérieure de b sera b_m qu'on prendra égale à 1 dans ces premières applications.

La borne inférieure b_f correspondra à l'accident qui donne la dose seuil individuelle à la distance considérée c'est-à-dire telle que

$$b_f = \frac{x^{\alpha} D_1}{acW}$$

les accidents de rejets inférieurs à b_f n'entraînant aucune conséquence à la distance x considérée.

Le détriment $N(x, \theta)$ subi par la population considérée dans le secteur x, θ est donc de :

$$N(x, \theta) = P(\theta)\, a\, c\, g\, W^{2-\beta}\, x^{-\alpha}\, p(x, \theta)\, x\, \Delta\theta_1\, x^{-\gamma} dx\, k_1 \int_{\frac{x^{\alpha} D_1}{a\, c\, W}}^{1} b_i\, b_i^{-\beta} \Delta b_i$$

$$= \frac{K_1}{2-\beta} \left[x^{1-\alpha-\gamma} - x^{1-\gamma+\alpha(1-\beta)} D \right]$$

où D représente l'expression $\left(\frac{D_1}{a\, c\, W}\right)^{2-\beta}$

cas général

et, si $\beta = 2$,

$$N(x, \theta) = K_2 \left(-\log\left(\frac{x^{\alpha} D_1}{a\, c\, W}\right) x^{1-\alpha-\gamma}\right)$$

Pour obtenir le détriment total du site, il faut alors intégrer par rapport à θ (sur tout le cercle) et par rapport à x à partir de x_2 distance qu'on peut confondre avec la limite du site jusqu'à x_1 distance maximale où un effet peut être perceptible (x_1 et x_2 définis plus haut).

On peut obtenir des fonctions analytiques qui définissent la population en fonction de l'azimut et de la distance (la densité est souvent une fonction croissante de cette dernière dans les premiers kilomètres).

Il faut toutefois prendre garde que mathématiquement on ne peut parler de population en un point donné comme on le fait couramment en localisant les habitants d'une ville à la mairie. On peut parler de populations cumulées jusqu'à une distance ou du taux d'accroissement par unité de distance : les difficultés sont un peu analogues à celles rencontrées dans les "courbes" de probabilités. On les évite en utilisant la densité classique. Mais pratiquement on disposera de valeurs finies dont la précision pourra être de moins en moins bonne quand on s'éloignera de l'installation et la somme pourra se faire soit par ordinateur, soit souvent à la main, les approximations et les extrapolations ne nécessitant pas un pas d'intégration très fin.

Un simple paramètre, proportionnel à la densité de population, pondéré par la fréquence des vents et l'expression précédente pour la distance, et sommé jusqu'à la distance x_m suffira donc à comparer des sites entre eux pour une installation donnée. Pour deux installations différentes il faudra faire intervenir k_1.

Pour un indice absolu de nocivité, il faudra tenir compte de plusieurs nucléides et de plusieurs effets. Seul ce dernier indice pourrait être comparé à la nocivité des rejets normaux, ou aux avantages obtenus du fonctionnement de l'installation.

3.2. Influence de la puissance

En ne regardant qu'un seul nucléide on pourrait encore simplement déterminer l'influence de la puissance et donc comparer deux installations entre elles. Il est relativement aisé de tenir compte de la répartition de la population. Par contre il semble que le facteur k_1 ou plus prosaïquement l'importance et la probabilité des accidents envisageables n'est pas une fonction simple de la taille. Au contraire l'établissement de critères indépendants de la puissance et surtout la redondance des sécurités dans les nouvelles constructions (sans parler de l'expérience acquise tous les jours) fait qu'un PWR de 1000 MW(th) ne serait pas plus dangereux qu'un réacteur de recherches de 8 MW(th) [3]. En postulant une densité de population sensiblement uniforme avec la distance, et un coefficient k_1 identique on obtient d'abord le dommage provoqué dans l'environnement à la distance x.

$$N(x, \theta) = K_3 \, W^{2-\beta}\left[x^{1-\alpha-\gamma} - x^{1-\gamma+\alpha(1-\beta)}\left(\frac{D\,1}{a\,c\,W}\right)^{2-\beta}\right] dx$$

Le détriment dû au réacteur peut dans ce cas être intégré analytiquement selon x. La borne supérieure sera la distance maximale considérée.

$$x_m = \left(\frac{a\,c\,W}{D\,1}\right)^{1/\alpha}$$

La borne inférieure devrait être celle définie au début distance en deçà de laquelle les conséquences ne sont plus fonction de la dose individuelle. Plus aisément on peut prendre $x = 1$ km, limite habituelle du site, ou plus aisément encore $x = 0$ l'intégration étant facilitée si

$$2 - \alpha - \gamma > 0, \quad \text{et, } 2 - \gamma + \alpha(1-\beta) > 0$$

l'on trouve dans ce cas :

$$N(\theta) = K_4 \, W^{\frac{2-\gamma}{\alpha} + 1 - \beta}$$

formule qui suppose donc :

-une nuisance individuelle limitée par un seuil
-des rejets proportionnels à la puissance
-une densité de population constante

-des concentations d'effluents et la largeur du panache décrites par des lois puissance
-enfin, puisque nous n'avons pas fait intervenir la décroissance radioactive, des nucléides dont la période est supérieure à une dizaine d'heures.

REMARQUES :

Si l'on admet que les probabilités ne dépendent pas du rejet en curies, mais uniquement de la fraction de la quantité contenue dans le coeur, $\Delta P_i = k_2 b_i^{-\beta} \Delta b$ et on obtient une formule particulièrement simple pour l'influence de la puissance.

$$N(\theta) = K_5 W^{\frac{2-\gamma}{\alpha}}$$

Cette hypothèse apparaît assez logique puisque, à un arbre de défaillance identique dans deux installations, elle associe des rejets proportionnels à la puissance.

4. APPLICATION

Nous prendrons :

a = 50 Ci/kW d'iodes contenus
c = 5.10^{-4} rem à la thyroïde par curie émis ce qui suppose un coefficient moyen de diffusion de :

$$5.10^{-6} \ s.m^{-3}$$

D_1 sera pris arbitrairement égal à 30 rems à la thyroïde, dose au-dessus de laquelle une surveillance médicale est obligatoire en France.

$\alpha = 1,5$
$\gamma = 0,2$ (la largeur angulaire du panache est divisée par 2 à quelques dizaines de kilomètres)
$\beta = 1$ les accidents participant alors également au détriment, les graves conséquences étant compensées par une densité de probabilités inversement proportionnelle.

On peut d'abord définir la distance x_m d'atteinte pour un réacteur de puissance W. On obtient, en simplifiant légèrement

$$x_m = W^{2/3}$$

x_m exprimé en kilomètres
W en mégawatts
ce qui implique pour un réacteur de l'ordre du gigawatt thermique une étude d'environnement jusqu'à une centaine de kilomètres.

Mais le terme de pondération de la densité de population devient alors

$$\frac{W}{x^{0,7}} - x^{0,8}$$

Ce facteur, approché par $Wx^{-0,7}$ jusqu'à une trentaine de kilomètres pour ce même réacteur, décroît après très rapidement de sorte qu'une étude fine n'est plus nécessaire.

Si l'on prenait $\beta = 3$ tel qu'on pourrait (avec un peu de bonne volonté) le déduire de [3] on obtiendrait, en gardant les mêmes valeurs des autres paramètres, un facteur de pondération égal à

$$\left(x^{-2,2} - \frac{x}{W}^{-0,7}\right)$$

c'est à dire une décroissance beaucoup plus rapide.

En particulier, pour 1000 MW, le facteur de pondération est déjà égal à 1/1000 à 20 km ce qui rend négligeable les distributions de population au-delà, sauf évidemment si la densité est nulle en deçà de cette distance.

Nous prendrons $\beta = 2$, tel que ce coefficient pourrait être tiré de la courbe critère proposée [4] ceci ne change rien à la valeur x_m ; par contre on obtient alors un terme de pondération proportionnel à

$$x^{-0,7} \log \frac{W}{x^{1,5}}$$

que nous retiendrons pour l'application numérique.

Il faut souligner que ce terme $\beta = 2$ suppose une "courbe" critère classique de pente - 1. En effet la "courbe" critère relie des conséquences d'accidents individuels. Pour une exploitation quantifiée d'une telle courbe il faut une hypothèse supplémentaire sur le "nombre d'accidents" dans chaque intervalle de rejets.

La définition des secteurs (x, θ) est liée à l'imprécision sur les autres facteurs.

Lorsqu'on passe d'une probabilité de 5% à une probabilité de 95% dans l'évaluation des conséquences d'un rejet donné dans l'atmosphère, les concentrations à une distance donnée varient d'un facteur supérieur à 20 [2]. Dans le même intervalle une même concentration peut être obtenue à des distances de rapport voisin de 10. On voit donc qu'une échelle telle que 1 à 2, 2 à 5, 5 à 10 km etc est largement suffisante.

L'emploi d'un modèle de diffusion plus élaboré donnerait une précision illusoire car personne n'a encore chiffré la dispersion des résultats à l'intérieur d'une classe donnée. Il est envisageable par contre de rajouter une telle distribution globale à l'analyse, mais aux dépens de la simplicité.

La définition de $\Delta \theta$ pose aussi quelques problèmes. C'est par abus de langage que l'on donne une fréquence de vent de direction θ alors que cette fréquence est mathématiquement nulle, la seule donnée ayant un sens étant la fréquence dans un secteur donné.

Dans tous les cas lorsque le vent moyen souffle vers un secteur donné, la largeur du panache est telle que son atteinte dépasse le plus souvent ce secteur. Il faudrait donc pondérer la probabilité d'un secteur par celles des secteurs voisins. Pratiquement la continuité générale des roses des vents permet d'estimer que ce terme correctif n'est pas très important. Si cette discontinuité entre deux directions existe, c'est qu'il existe un accident topographique mais alors il est prudent de considérer les trajectoires, les déviations du panache sur une centaine de kilomètres étant probables. De plus dans les vallées étroites, exemple de ces accidents

TABLEAU I. FACTEUR DE PONDERATION DES DENSITES

Puissance thermique (MW) / distance (km)	100	1000	3000
1 à 2	150	200	250
2 à 5	50	100	150
5 à 10	20	50	70
10 à 25	4	20	25
25 à 50		8	10
50 à 100		1	3
100 à 200			1

topographiques, la largeur angulaire va décroître beaucoup plus rapidement qu'indiqué ici. Enfin, ces fréquences de vent serviront à pondérer des populations qui seront supposées groupées à la mairie alors que l'habitat dépassera souvent les limites des secteurs.

Il faut déterminer une rose des vents totale. La largeur des secteurs importe peu, pourvu bien entendu qu'elle soit la même sur les sites que l'on veut comparer. Pour la déterminer on n'oubliera pas de se raccorder à une station climatologique voisine (pour tenir compte d'une trentaine d'années) et surtout d'attribuer une direction aux vents dits "calmes", de façon à ce que la somme des fréquences (assimilées ici à $P(\theta)$) soit égale à 1.

La population sera répartie dans des secteurs ayant comme largeur angulaire celle de la rose des vents et comme longueur les intervalles définis précédemment. La densité de population dans ces secteurs sera déterminée. Pour les mêmes intervalles de distance, la valeur moyenne de la fonction $x^{-0,7} \log \frac{W}{x^{1,5}}$ sera calculée avec x en kilomètres et W en mégawatts.
Chaque densité sera multipliée par la fréquence du vent correspondante et la valeur moyenne trouvée de l'expression précédente. La somme de ces produits caractérisera le site au point de vue atmosphérique pour l'installation de puissance W. Par exemple, on aura le tableau approximatif I.

5. DISCUSSION

Les termes α et γ peuvent varier légèrement selon les auteurs ou le progrès de nos connaissances. En tous cas ils restent physiquement déterminés et l'on peut estimer que la distance "maximale" d'atteinte ne dépendra que de la dose D_1 prise comme seuil ou du rejet b_i W qui sera jugé "impossible".

Le terme β par contre restera toujours assez arbitraire. En effet,ou bien nos connaissances resteront incomplètes et ce coefficient restera basé sur des impressions subjectives ("les gros accidents" doivent être psychologiquement- ou politiquement- évités) ou bien nos connaissances dans le spectre des accidents aura progressé et il sera alors relativement facile (avec des moyens de calcul plus puissants) de faire la somme discrète des détriments en un point dus à chaque accident, sans se soucier de l'importance relative des "détriments".

Il faut prendre garde à ce que les distances considérables envisagées ne préjugent en rien du "détriment" absolu de l'installation. Au contraire à la distance maximale d'atteinte x_m, ce dernier est mathématiquement nul. En fait toutes les études ont montré que les installations nucléaires ont atteint un niveau de sûreté largement supérieur à celui de beaucoup d'autres industries [3]. Ce facteur intrinsèque est même si bon que le paramètre β défini principalement à partir de la probabilité des grands accidents dépendra de plus en plus des facteurs d'agression de l'environnement contre le réacteur : séismes, inondations, sans doute, mais aussi explosions industrielles et,surtout,négligence ou malveillance humaines, difficilement chiffrables.

Sur un site à densité constante le tableau permet de voir que la puissance a relativement peu d'importance pour le détriment. Ceci tient essentiellement à l'hypothèse -généralisée- que les probabilités seront liées aux rejets, indépendamment de l'installation. La puissance n'intervient que par la possibilité de rejets plus élevés, reculant les bornes d'intégration. Si l'on avait supposé une distribution des probabilités fonction de la fraction des nucléides contenus dans le coeur, on aurait obtenu :

pour $\beta = 1$: le même facteur $\frac{W}{x^{0,7}} - x^{0,8}$

pour $\beta = 3$: $W^2 x^{-2,2} - W x^{-0,7}$

et pour $\beta = 2$: enfin $W x^{-0,7} \log \frac{W}{x^{1,5}}$

c'est-à-dire, comme on pourrait s'y attendre, une importance beaucoup plus grande de la puissance.

6. CONCLUSION

En faisant des hypothèses relativement simples, mais réalistes, on a pu donner une méthode de calcul permettant de déterminer un facteur atmosphérique caractérisant un site en fonction d'une installation donnée.

Ce facteur est particulièrement sensible aux probabilités associées aux rejets et particulièrement à l'importance relative des graves accidents et des incidents plus bénins.

Les distributions de ces probabilités conduisent à des influences très diverses de la puissance.

Les études ultérieures devront porter sur cet aspect ainsi que sur l'influence des distributions probabilistes des coefficients de diffusion et des décès dus à une irradiation donnée.

D'ores et déjà des applications numériques objectives peuvent être obtenues mais les facteurs de pondération de la densité de population vont varier au fur et à mesure du progrès de nos connaissances.

REFERENCES

[1] MACHTA, L., FERBER, G.J., HEFFTER, J.L., «Regional and global scale dispersion of ^{85}Kr for population-dose calculations», Physical Behaviour of Radioactive Contaminants in the Atmosphere (C.R. Coll. Vienne, 1973), AIEA, Vienne (1973) 411.

[2] LE QUINIO, R., «Concentrations sur une heure de polluants dues à des émissions ponctuelles près du sol: Présentation probabiliste», Principles and Standards of Reactor Safety (C.R. Coll. Juliers, 1973), AIEA, Vienne (1973) 215.

[3] USAEC, The Safety of Power Reactors (Light Water Cooled) and Related Facilities, Rep. WASH-1250 (1973) p. 6.31.

[4] BEATTIE, J.R., BELL, G.D., EDWARDS, J.E., Methods for the Evaluation of Risk, UKAEA Rep. AHSB-S-R-159 (1969).

DISCUSSION

D. M. SIMPSON: If one considers a very low-probability accident, say 10^{-6} events per year, there is a possibility of one such event in 1000 years for a 1000 reactor programme. In other words, such an event may never occur within the nuclear energy programme as we know it today. Do you think that the probability theory, which estimates the number of events that are likely to happen in a long time-period, can be extended to short time-periods when the event is not likely to occur at all?

R. LE QUINIO: It is normal for very low-probability events to be regarded as practically impossible, and this is why most of the present national regulatory bodies restrict their studies to minor "reference" accidents or "dimensioning" accidents, which have no serious consequences. Nevertheless, much more serious accidents are always possible, and the responsible authorities clearly realize this because nuclear power stations are always located at a distance from populous areas in spite of the clear economic interest of having them close by.

L'EVALUATION DES DOSES ENGAGEES POUR LES POPULATIONS LORS DE L'EXAMEN DES RAPPORTS DE SURETE DES CENTRALES NUCLEAIRES FRANÇAISES

P. CANDES
Ministère de l'industrie, du commerce et de l'artisanat,
Service central de sûreté des installations nucléaires,
Paris

P. SLIZEWICZ
CEA, Centre d'études nucléaires de Saclay,
Département de sûreté nucléaire,
Gif-sur-Yvette,
France

Abstract–Résumé

ESTIMATION OF POPULATION DOSE COMMITMENTS DURING THE EXAMINATION OF SAFETY REPORTS ON FRENCH NUCLEAR POWER STATIONS.

After describing briefly the French legislation relating to the safety of nuclear facilities, the authors examine in greater detail the way in which population dose commitments – or potential population doses – are estimated. The administrative procedures are of fairly recent origin and their formulation coincided with a considerable growth in the number of major facilities. One purpose of a safety assessment is to make the prospective operator ask himself about the effects which the facility will have on the environment and in particular about the constraints placed on future industrial development over an area which may extend for tens of kilometres around the site. It appears to be increasingly difficult at present to assess the safety of a nuclear facility without taking into account the other facilities (nuclear or otherwise) located within a considerable radius. It is in the general context of the development of the nuclear industry in a given country – or a group of countries – that nuclear safety must be assessed.

L'EVALUATION DES DOSES ENGAGEES POUR LES POPULATIONS LORS DE L'EXAMEN DES RAPPORTS DE SURETE DES CENTRALES NUCLEAIRES FRANÇAISES.

Après avoir décrit sommairement la législation française concernant la sûreté des installations nucléaires, les auteurs examinent plus en détail la façon dont sont évaluées les doses engagées – ou potentielles – pour les populations. Les procédures administratives sont relativement récentes et leur mise en place coïncide avec un développement considérable du nombre des installations importantes. Un des buts de l'examen de la sûreté est d'amener le futur exploitant à s'interroger sur l'influence que son installation aura sur l'environnement et par exemple sur les contraintes qu'il impose au développement futur des industries dans une zone qui peut s'étendre sur plusieurs dizaines de kilomètres autour du site. Il paraît actuellement de plus en plus difficile de juger la sûreté d'une installation nucléaire sans tenir compte des autres installations – nucléaires ou non – implantées dans un rayon assez large. C'est dans le cadre général du développement de l'industrie nucléaire dans un pays donné – et même dans un ensemble de pays – que la sûreté nucléaire doit être évaluée.

Avant même que les programmes nucléaires atteignent le niveau qu'on leur connaît actuellement, la sûreté des installations atomiques a été une des préoccupations essentielles des exploitants. On peut d'ailleurs se demander si, dans une certaine mesure, les précautions prises ne se sont pas retournées contre l'utilisation de cette nouvelle source d'énergie en laissant croire au public qu'elle cachait des dangers importants mal connus ou mal maîtrisés.

Dans ce domaine chaque pays a progressivement mis sur pied une législation ou une règlementation qui tenait compte de ses structures juridiques propres. En fait tous les textes publiés sont fondés sur les Recommandations de la Commission Internationale de Protection Radiologique (CIPR) ainsi que sur les nombreuses publications de l'Agence Internationale de l'Energie Atomique. Cette Organisation Internationale a suscité de très nombreux échanges et grâce à elle il existe une assez grande uniformité de vue dans l'évaluation des conséquences pour les populations du développement de l'énergie nucléaire.

Nous nous proposons dans cette Communication de présenter la façon dont, en FRANCE, sont examinés sur le plan de la sûreté nucléaire, les projets de centrales, et comment, à ce stade, sont estimées les doses engagées pour les populations. Ces estimations sont faites à la fois en considérant le fonctionnement normal et les accidents de dimensionnement. Il faut bien remarquer que les procédures ne sont pas encore figées et que les renseignements que doit fournir le futur exploitant sont encore discutés et précisés au cours de l'examen des rapports de sûreté.

Notons également que nous ne parlerons pas des informations qui doivent être fournies à la Commission des Communautés Européennes (CCE) au sens de l'article 37 du traité d'EURATOM. Cet article stipule que chaque Etat Membre est tenu de communiquer les données générales de tout projet de rejet d'effluents radioactifs, à la Commission, qui lui permettent d'émettre un avis sur les conséquences radiologiques pouvant en résulter pour un autre Etat Membre.

Après avoir donné un aperçu de la règlementation existante ou en préparation, nous préciserons les informations que doit fournir l'étude du site pour permettre d'évaluer les conséquences radiologiques pour le public du rejet des effluents. Nous préciserons ensuite la méthode utilisée pour calculer ces conséquences aussi bien en fonctionnement normal qu'en cas d'accident. Nous décrirons également les conditions du contrôle de l'environnement.

1. LA REGLEMENTATION

La sûreté des installations nucléaires, et donc celle des réacteurs de puissance est, en FRANCE, sous la responsabilité du Ministère de l'Industrie, du Commerce et de l'Artisanat et plus particulièrement d'un Service de ce Ministère: le Service Central de Sûreté des Installations Nucléaires (SCSIN) [1] [2] . Ce Service est "notamment chargé de préparer et de mettre en oeuvre toutes actions techniques (...) relatives à la sûreté des installations nucléaires et en particulier: Elaborer la règlementation technique concernant la sûreté des installations nucléaires et suivre son application".

Par ailleurs un décret [3] définit le cadre juridique et administratif dans lequel la construction et l'exploitation des installations dites "Installations Nucléaires de Base" (INB) doivent se dérouler. Bien entendu les réacteurs nucléaires constituent des Installations Nucléaires de Base. C'est ainsi que "les installations (...) ne peuvent être créées qu'après autorisation", cette autorisation étant délivrée par décret du Premier Ministre après avis d'l Commission Interministérielle des Installations Nucléaires de Base (CIINB).

Des Instructions ministérielles précisent par ailleurs la procédure exacte que doit suivre le futur exploitant qui demande une autorisation de création. Sans entrer dans le détail des cheminements administratifs, disons simplement que le SCSIN joue le rôle de "plaque tournante" dans la circulation des dossiers soit vers le Commissariat à l'Energie Atomique dont nous préciserons le rôle plus loin, pour l'étude technique des demandes d'autorisation, soit vers la CIINB pour solliciter son avis, soit vers le Ministre et le Premi Ministre, pour proposer une décision.

Le document de base sur lequel est jugée la sûreté de l'installation est le "Rapport de Sûreté". Il n'est bien entendu pas possible qu'un document définitif concernant une installation aussi complexe qu'un réacteur nucléaire de puissance puisse être établi dès la demande d'autorisation de construction. Aussi la règlementation a-t-elle prévu trois stades successifs pour arriver à une rédaction finale:

- Un rapport préliminaire de sûreté qui joue le rôle de notice descriptive et définit avec précision les principales options techniques concernant la sûreté pour la demande d'autorisation de création.

- Un rapport provisoire de sûreté qui doit être remis six mois avant le premier chargement du réacteur.

- Un rapport définitif de sûreté remis après les essais de mise en service.

L'examen de ces trois rapports de sûreté est effectué par un"Groupe Permanent chargé des réacteurs" qui assiste le SCSIN et est composé de cinq représentants du Ministère de l'Industrie, du Commerce et de l'Artisanat, de quatre experts du Commissariat à l'Energie Atomique et de quatre experts de l'exploitant (Electricité de France). Le Commissariat à l'Energie Atomique joue un rôle important aussi bien dans le développement des études techniques de sûreté, que dans l'analyse de sûreté des installations. D'une part il est prévu dans ses missions de "proposer les mesures propres à assurer la protection des personnes et des biens contre les effets de l'énergie atomique et contribuer à leur mise en oeuvre" et à ce titre, il dispose de moyens d'études techniques et expérimentales importants dans les domaines de la sûreté et de la protection, d'autre part la décision ministérielle relative à la création des groupes permanents lui confie la mission de rapporteur auprès de ces groupes pour l'analyse de sûreté.

Dans le cadre de l'Instruction ministérielle relative à l'application du décret sur les installations nucléaires de base, le plan indicatif des rapports de sûreté est précisé. Chacun des trois rapports signalés plus haut comprend trois volumes:

- Volume I - Introduction et Généralités.
- Volume II - Equipements de la centrale et fonctionnement.
- Volume III - Analyse de la sûreté.

Chaque volume comprend plusieurs chapitres et les chapitres qui traitent directement ou indirectement de l'exposition des populations aux rayonnements sont:

- Les chapitres 2 (Site), 5 (Stockage, controle et evacuation des dechets et effluents radioactifs), 6 (Résumé de l'analyse de sûreté. Conséquences radiologiques pour la population) et 7 (Organisation au stade de la construction et de l'exploitation - Protection du personnel) du volume I.

- Le chapitre 4 (Fonctionnement normal et accidentel - Conséquences radiologiques de l'exploitation pour la population) et le chapitre 5 (Radioprotection) du volume III.

Il convient de signaler le rôle important joué par le Ministère de la Santé Publique. Ce Ministère est saisi comme les autres Ministères intéressés par le Ministre de l'Industrie, du Commerce et de l'Artisanat des demandes d'autorisation de création d'installations nucléaires et il est représenté à la CIINB participant ainsi à la mise au point de l'avis de cette Commission. Mais de plus l'autorisation de construction ne peut être délivrée qu'avec l'accord explicite (avis conforme) de ce Ministère.

Enfin ce Ministère interviendra conjointement avec le Ministère chargé de la Protection de la Nature et de l'Environnement dans l'examen des projets d'arrêtés prévus dans le cadre d'une règlementation en préparation pour les rejets d'effluents liquides ou gazeux. Nous y reviendrons plus loin.

Pendant plusieurs années, alors que la législation était moins structurée - et les installations moins nombreuses et moins importantes - l'examen des conséquences du fonctionnement d'une installation sur l'environnement se limitait à étudier les conséquences d'un ou plusieurs accidents graves pris en compte pour le dimensionnement de l'installation. Le fonctionnement normal dont l'influence sur l'environnement était négligeable n'était pas pris en compte [4 La multiplication des installations et la concentration de plusieurs installations sur un même site ont conduit à envisager également les conséquences de ce fonctionnement normal sans incident grave tout en maintenant l'étude du ou des accidents dits de "dimensionnement". De toute façon, il était de pratique courante de présenter une étude du site; les conditions actuelles n'ont fait qu'accroître l'intérêt porté à cet aspect de l'évaluation de la sûreté.

2. L'ETUDE DU SITE

Nous avons vu que le chapitre 2 du volume I des rapports de sûreté devait être consacré au site. Le contenu de ce chapitre doit, dès le rapport préliminaire, permettre de connaître au moins dans leurs grandes lignes, les condition de dispersion dans l'environnement des effluents liquides et gazeux. Un canevas pour une étude de ce type a été proposé par l'Agence Internationale de l'Energie Atomique [5] et le futur exploitant a tout intérêt à s'en inspirer. Parmi les renseignements qui doivent être fournis notons:

- La distribution des habitants dans la zone entourant le site en fonction de la distance et de la direction.
- La prévision du développement escomptée dans l'avenir de cette zone en tenant compte de l'influence que l'installation nucléaire projetée aura sur ce développement.
- La description détaillée de l'utilisation des terrains en portant son attention sur l'utilisation agricole, notamment les pâturages.
- L'implantation des industries.
- La configuration des voies de communications et d'accès.

Par ailleurs la connaissance du transfert des effluents gazeux suppose une étude météorologique qui, dans une première étape, pourra se référer aux mesures de la station météorologique la plus proche. Cependant sur le site lui-même, des mesures sont indispensables au moins pendant quelques années afin de préciser les conditions locales.

La circulation des eaux souterraines et des eaux de surface sera, bien entendu, décrite en mettant l'accent sur l'utilisation présente et future en aval et en amont du site. En effet, les eaux de surface, souvent un cours d'eau recevront les effluents liquides de l'installation, et les eaux souterraines risquent d'être contaminées par les fuites d'un stockage ou d'une piscine. Le point important est de demander au futur exploitant de s'interroger sur l'avenir de son installation - y compris son arrêt définitif - et également sur les contraintes qu'il risque d'imposer aux activités installées en aval.

Enfin le chapitre "Ecologie" doit traiter des processus de transfert de la contamination par les organismes vivants, c'est-à-dire principalement les produits agricoles et les élevages. C'est sans doute à ce niveau que l'évaluation de la dose engagée est la plus délicate car elle implique la connaissance de nombreux paramètres qui ne sont pas nécessairement connus ou facilement obtenus au moment du dépôt de la demande d'autorisation de création.

A ce stade, il faut demander au futur exploitant de commencer les investigations nécessaires en soulignant que, par ailleurs, le contrôle de l'environnement qu'il sera amené à faire dans le futur ne prendra toute sa signification que si les transferts de contamination à partir du site sont correctement appréciés.

Pratiquement, donc le rapport préliminaire de sûreté pourra ne contenir qu'une évaluation approchée de la dose engagée. Cependant il peut fournir une évaluation assez précise des effluents radioactifs prévus en fonctionnement normal.

3. LES EFFLUENTS EN FONCTIONNEMENT NORMAL

La législation française doit dans un très proche avenir être complétée par deux décrets fixant les conditions de rejet d'effluents radioactifs que tout exploitant devra respecter. En fait il devra fournir au Ministère de l'Industrie, du Commerce et de l'Artisanat une étude préliminaire précisant les équivalents de dose qui résulteront pour le public. Cette évaluation doit s'appliquer au fonctionnement normal, y compris les incidents mineurs. Cette étude préliminaire doit précéder le dépôt d'une demande d'autorisation de rejets. On peut penser que dans ses grandes lignes cette étude reprendra les chapitres correspondants des rapports de sûreté dans lesquels on trouve déjà les évaluations des conséquences radiologiques des rejets des effluents gazeux et liquides.

3.1. Evaluation des conséquences radiologiques des rejets des effluents gazeux

Cette évaluation passe nécessairement par la connaissance des produits de fission présents dans le combustible. Des programmes de calcul appropriés ont été mis au point à cet effet [6] . Ensuite il faut faire un certain nombre d'hypothèses sur le pourcentage de gaines fissurées, sur le taux de fuite des différentes parties du circuit primaire pour finalement aboutir à chiffrer les activités des produits de fission gazeux émis dans l'atmosphère. Pour un réacteur à eau pressurisée de 1 000 MWe on peut ainsi calculer un rejet annuel de l'ordre de 25 000 Ci de mélange de gaz rares. En supposant alors un rejet permanent et continu, et en utilisant la rose des vents fournie par l'étude de site, on peut calculer le transfert de cette activité dans les différentes directions. On divisera ainsi le site en différents secteurs, ou mailles géographiques, déterminés par l'azimut et la distance. On relèvera le nombre d'habitants résidant à l'intérieur de chaque maille.

Une question délicate se pose: c'est celle du choix de la distance maximale à partir du site jusqu'où on doit faire ce découpage. Selon des critères tel celui développé par R. LE QUINIO [7] , on se limite généralement à une distance comprise entre 60 et 100 km: la présence d'une grande agglomération étant une raison de dépasser 60 km, mais il paraît inutile d'aller au-delà de 100 km. On peut alors calculer dans chaque maille géographique, pour deux catégories de stabilité des masses d'air les concentrations volumiques moyennes (en Ci/m^3) correspondant au rejet considéré. On distingue deux catégories de conditions de diffusion: celles dites bonnes sont associées aux vents supérieurs à 2 m/s et celles dites mauvaises sont associées à des vents inférieurs à 1 m/s.

Les concentrations moyennes ainsi obtenues sont transformées en concentrations intégrées annuelles partielles par multiplication par une certaine durée, déduite de la probabilité de persistance, elle-même déduite de la fréquence tirée des statistiques.

On passe ensuite à la concentration intégrée individuelle totale en faisant la somme de tous les résultats partiels provenant des différentes plages de vitesse de vent pour les deux régimes de diffusion.

Puis à partir de la concentration intégrée annuelle totale et des Concentrations Maximales Admissibles, on calcule en utilisant les correspondances proposées par la CIPR, les équivalents de dose individuelle en rems. Une telle investigation permettra de déterminer en premier lieu la dose moyenne à laquelle seront exposés les habitants de chaque maille géographique. A l'avenir les résultats pourraient éventuellement être exprimés en homme-rems. Une telle présentation aurait pour effet de focaliser l'attention sur l'exposition des grandes agglomérations. Elle peut constituer un moyen pratique pour guider l'exploitant pour les rejets concertés en l'incitant à profiter des situations météorologiques favorables à une irradiation minimale des zones à forte population. Cependant l'utilisation de cette unité suscite encore des réticences et on doit se montrer réservé sur son utilisation. Cette utilisation ne serait envisagée qu'après accord des autorités intéressées.

On comprend également que les trajectoires réelles des effluents gazeux doivent être connues avec une bonne précision et dans le cas d'un relief accidenté la direction du vent sur le site peut ne pas suffire à prévoir les objectifs atteints à quelques dizaines de km. Il sera alors indispensable de procéder à une étude plus rigoureuse en utilisant des lâchers de traceurs atmosphériques et en étudiant leur dispersion sur le terrain.

Bien entendu, si les rejets gazeux contiennent des radionucléides solides, susceptibles de se déposer sur les cultures et par là de contaminer l'alimentation d'une fraction de la population, il faudra en tenir compte.

On comprend également que, pour que cette estimation ait quelque signification sur le plan de la sécurité radiologique, il faut la faire simultanément, non seulement pour les installations groupées sur un même site, mais également pour les installations implantées sur des sites différents dont les rejets peuvent cependant atteindre les mêmes populations.

Ces questions seront naturellement évoquées au moment de l'examen des différents rapports de sûreté et on comprend d'autant mieux la nécessité d'un examen centralisé de la sûreté des installations nucléaires.

3.2. Evaluation des conséquences radiologiques des rejets des effluents liquides

De même que pour les rejets d'effluents gazeux, le futur exploitant devra fournir une étude évaluant les conséquences des rejets liquides qu'il pense avoir à effectuer. Au cours de l'examen de cette étude, une attention particulière sera apportée au "réalisme" des évaluations. De toute bonne foi, le futur exploitant peut sous-estimer les quantités à rejeter. Il lui est peut-être difficile de prévoir les conséquences pour les rejets des incidents inévitables qui se produiront durant les vingt ou trente années de fonctionnement de la centrale. Dans cette optique, un examen attentif des comptes rendus d'exploitation des centrales en service fournit des informations très précieuses.

Si la "dispersion" des rejets liquides n'est pas soumise aux fluctuations directionnelles que connaît celle des rejets gazeux, l'évaluation des doses engagées pour les populations n'en est guère simplifiée. Il est certes relativement facile de calculer les concentrations volumiques en Ci/m^3 atteintes dans le cours d'eau récepteur. Notons cependant que le mélange de l'eau rejetée, généralement plus chaude que l'eau réceptrice, peut ne pas se faire rapidement surtout s'il s'agit d'un fleuve au débit calme et tranquille.

Mais de toute façon comment passer à l'équivalent de dose reçu par une personne du public?

Une première approche simpliste consiste à calculer la dose annuelle reçue par un individu qui utiliserait à longueur d'année cette eau comme eau de boisson. Mais cette eau peut être utilisée aussi comme eau d'irrigation et donc contaminer des aliments. On sait également que des organismes aquatiques peuvent concentrer certains radionucléides et devenir ainsi une source de contamination pour l'homme. De très nombreuses études ont été consacrées à ces évaluations et présentées dans différents Congrès [8] [9] .

Une méthode rationnelle pour aborder ce problème a, par ailleurs été décrite [10] mais sa mise en application nécessite de nombreuses études particulières pour définir par exemple la voie critique, le groupe de population critique, etc. Aussi ne semble-t-il pas possible qu'elle puisse être conduite à son terme au moment où sont étudiés les rapports de sûreté. On voit au contraire apparaître la nécessité de contrôles adaptés au contexte écologique, nous développerons ce point dans un prochain paragraphe.

Signalons enfin pour en terminer avec les effluents liquides le problème particulier posé par les grands fleuves internationaux dont la surveillance exige à l'évidence une collaboration internationale efficace et confiante.

4. LES EFFLUENTS APRES UN ACCIDENT

Les rapports de sûreté doivent analyser les conséquences radiologiques d'un certain nombre d'évènements fortuits considérés comme des accidents au sens commun du mot, c'est-à-dire qu'ils entraînent l'indisponibilité de l'installation pour une durée non négligeable, parfois même l'indisponibilité définitive. Classiquement on considère parmi ces accidents:

- La rupture circonférentielle et totale d'une canalisation du circuit primaire (branche froide pour les pressurisés, boucle de recirculation en amont de la pompe pour les bouillants).
- La chute d'un élément combustible au cours du déchargement.
- L'éjection d'une barre ou d'une grappe de commande.

Nous allons tenter dans ce qui suit de donner quelques indications sur les paramètres pris en compte en discutant ou commentant les hypothèses ou les faits qui les justifient.

Notons tout d'abord que la connaissance précise de la puissance à laquelle a fonctionné le réacteur n'est pas un facteur essentiel à connaître car une erreur de quelques pourcents sur la puissance introduit une incertitude négligeable sur la quantité des produits de fission disponibles. Par contre dans le cas de l'accident le plus grave qui est la rupture du circuit primaire il est essentiel de préciser le taux de ruptures de gaines. Il est admis que pour les accidents de dimensionnement, la température des gaines ne dépasse pas 815° C pour les bouillants et 1 200° C pour les pressurisés: l'US AEC prend 100% de gaines rompues dans ces deux cas.

On pourrait cependant penser que la température maximale atteinte dans les bouillants, étant plus faible, devrait conduire à un nombre également plus faible de gaines rompues.

L'autre donnée indispensable à connaître concerne les produits de fission qui sont émis par les ruptures de gaine. On peut l'envisager en deux étapes: premièrement, émission par le combustible dans l'espace gaine-combustible (le "gap" selon la terminologie américaine), deuxièmement, l'émission proprement dite à travers la rupture de gaine. La méthode américaine a tendance à confondre ces deux étapes en se plaçant d'ailleurs systématiquement du côté pessimiste c'est-à-dire en prenant une émission de 100% pour les gaz et les iodes contenus dans le combustible concerné. Ce taux d'émission est en fait celui que l'on observerait pour la fusion du coeur. Dans ce cas, il conviendrait d'y ajouter d'autres radionucléides relativement volatils. Il nous semblerait plus logique de prendre 10% pour l'émission des gaz rares, sauf pour le Krypton-85 pour lequel on prendrait 30% [11] . En fait la dose due à l'irradiation par ce radioélément est faible si on la compare à celle due à la totalité des gaz rares [12]

En ce qui concerne les radioiodes, une question importante est de connaître l'espèce chimique sous laquelle on les retrouve dans l'environnement. Certains travaux conduisent à penser que l'iode peut se trouver soit sous forme d'iode moléculaire, soit sous forme d'iodure de méthyle, soit sous forme d'acide hypoiodeux: ces deux dernières formes constituant ce que l'on appellerait l'iode pénétrant [13] . Par ailleurs, si l'ensemble des gaz rares est disponible pour le rejet, on doit considérer que différents phénomènes réduisent la quantité des iodes disponibles, à savoir: le dépôt sur les structures (parois de l'enceinte, parois des condenseurs à glace), la précipitation avec l'eau de la douche de l'enceinte, etc.

Finalement la proportion d'iode "disponible" pour les rejets est de l'ordre de 1 à 10% suivant l'importance attribuée aux différents facteurs de réduction. De plus, la nature de l'espèce chimique est importante à connaître. En effet, on considère les conséquences radiologiques du rejet sous deux aspects. Tout d'abord l'inhalation pendant le passage du nuage qui entraîne une irradiation de la thyroïde due à l'iode qui s'y dépose: cet effet est considéré comme indépendant de l'espèce chimique.

Mais on doit également considérer le dépôt sur les pâturages et la contamination du lait des vaches se nourrissant de l'herbe ainsi contaminée. L'expérience a montré que ce dépôt, fonction bien entendu de la concentration dans l'air, était en fait la voie qui entraînait la contamination interne la plus importante surtout si l'on considère que des nourrissons boivent le lait concerné [14] [15] . Mais alors des contre-mesures efficaces existent: limitation de la consommation du lait et transformation de ce lait en fromage ou en lait en poudre pour lesquels un stockage d'une durée suffisante assurera la disparition de l'iode contaminant. Cependant l'iodure de méthyle qui est souvent considéré comme la forme la plus fréquente d'iode difficile à piéger, se dépose beaucoup moins que l'iode moléculaire [16] . Finalement les précipitations qui entraînent un dépôt maximal sont, dans un certain sens, donc favorables par le "lavage" des feuilles qu'elles entraînent à l'élimination de l'iode de la chaîne alimentaire.

Un autre facteur qui intervient dans la contamination de l'environnement est le taux de fuite de l'enceinte: celui-ci est défini par le constructeur et tient compte de la surpression qui se produit au moment de l'accident. On prend en général un taux de fuite de 0,1% par jour: mais la pression à l'intérieur de l'enceinte doit assez rapidement revenir à la pression normale et les fuites non contrôlées cesser.

Le dernier facteur à prendre en compte est lié aux conditions météorologiques existant au moment de l'accident. Les caractéristiques météorologiques du site sont, nous l'avons dit, étudiées pour évaluer les conséquences radiologiques des rejets normaux. On les connaît donc bien et il s'agit de choisir parmi les conditions possibles les conditions réalistes. Il peut sembler logique de choisir des concentrations au sol de transfert ayant une probabilité de 95% de ne pas être dépassées à une distance déterminée [17] .

En ce qui concerne la durée de l'accident on peut soit choisir une valeur arbitraire, deux heures par exemple pendant lesquelles seront mises sur pied des actions de protection, soit essayer d'évaluer la valeur exacte de la durée de la surpression dans l'enceinte pendant laquelle des rejets non contrôlables ont lieu.

Finalement pour les conséquences radiologiques les calculs sont conduits de la même façon que pour les rejets normaux.

Le deuxième accident qui peut avoir des conséquences significatives pour l'environnement est l'accident de manutention: chute d'un assemblage irradié dans la piscine de stockage par exemple. On peut admettre que tous les crayons sont alors rompus mais on bénéficie d'un temps de refroidissement qui n'est pas inférieur à vingt-quatre heures et qui serait même plutôt de l'ordre de quatre jours. Parmi les facteurs à prendre en compte pour évaluer les conséquences radiologiques, on trouve:

- Taux d'émission des produits de fission dans l'espace gaine-combustible.
- Forme chimique des iodes émis.
- Taux de renouvellement horaire de l'atmosphère du bâtiment.
- Rétention par l'eau de la piscine.
- Efficacité des pièges à iode.
- Conditions de diffusion atmosphérique.

Dans ce cas encore c'est l'iode qui paraît être le radioélément le plus "important", et le problème est traité de la même façon que dans le cas précédent.

5. LE CONTROLE DE L'ENVIRONNEMENT

L'étude de site ayant mis en évidence les points où risquent de se retrouver les produits radioactifs relâchés par l'installation, une surveillance adéquate doit y être poursuivie. Mais bien entendu il semble logique - et même impératif car toute surveillance est incomplète - de surveiller les rejets eux-mêmes que ce soit les effluents gazeux et les effluents liquides. C'est ainsi qu'en ce qui concerne ces derniers des capacités en nombre suffisant doivent être prévues pour stocker les effluents qui ne pourraient être rejetés immédiatement et devraient subir un traitement de décontamination.

Il paraît également utile d'établir une carte de l'irradiation naturelle autour du site. On évaluera ainsi la dose d'irradiation que le public reçoit en dehors de toute activité nucléaire. Par la suite, il sera ainsi possible de comparer les doses dues au fonctionnement des installations avec celles dues à la radioactivité naturelle. A la limite d'ailleurs, ainsi que cela a été suggéré [18] un site dont la radioactivité naturelle serait trop élevée pourrait être éliminé.

En évitant de développer d'une façon inconsidérée les contrôles extérieurs [19] on se limitera à la mesure de la radioactivité atmosphérique (gaz et poussières), de la radioactivité des eaux recevant les rejets liquides et d'un certain nombre de produits agricoles (le lait notamment). Ce problème a été abondamment développé dans de nombreuses publications [20] [21] [22] [23]

L'ensemble des contrôles effectués sur le territoire seront centralisés par le Ministère de la Santé Publique qui sera ainsi en mesure de calculer les doses dues au fonctionnement des différentes installations et de porter une appréciation sur leur impact sur les populations.

Un juste équilibre doit être trouvé entre des contrôles trop espacés qui laisseraient échapper des contaminations temporaires et des contrôles trop nombreux qui prendraient beaucoup de temps, et risqueraient finalement de soulever des craintes injustifiées. Pendant un court laps de temps, il peut être intéressant d'effectuer des campagnes de mesures très précises telles que celles proposées par H.L. BECK [24] [25] . C'est bien finalement à partir de ces contrôles que devraient être non plus évalués mais mesurés les équivalents de dose reçus effectivement par le public. Un point nous paraît important à souligner: la nécessité que les méthodes utilisées par l'exploitant responsable de la surveillance autour du site soient parfaitement définies. En effet les comparaisons de ces faibles niveaux de contamination n'ont de valeur que dans la mesure où les résultats sont effectivement comparables entre eux. Des comparaisons ont d'ailleurs été effectuées sous les auspices de l'Agence par exemple dans le domaine de la contamination du milieu marin [26] .

CONCLUSIONS

Nous avons rapidement décrit la façon dont, en FRANCE, sont examinés les rapports de sûreté des réacteurs de puissance et notamment comment sont évaluée les doses engagées pour les populations.

Une règlementation complète se met progressivement en place qui s'inspire de ce qui est réalisé dans d'autres pays et des Recommandations de l'AIEA.

Nous pensons que l'un des rôles d'une "Commission" de sûreté est d'incliner le futur exploitant à réfléchir davantage sur les problèmes de sécurité notamment ceux concernant la contamination de l'environnement. Il faut également attirer son attention sur les problèmes posés par la présence de plusieurs réacteurs sur le même site, sur les incidents qui se produiront certainement tout au long de la vie des réacteurs. Les installations de décontamination doi être capables de traiter ces incidents et de les rendre inoffensifs pour le pub tout en évitant une diminution excessive de la production d'électricité.

Il nous semble également très utile d'attirer l'attention sur la nécessi et l'utilité d'une surveillance raisonnable de l'environnement, sur l'intérêt de parfaire la connaissance des caractéristiques géophysiques et écologiques du site notamment si des installations nouvelles risquent d'apporter de nouvelles sources de rayonnements ou de modifier certains paramètres: circulation des eau souterraines par exemple ou modification du régime des eaux superficielles par suite de la construction d'un barrage.

Bien entendu les problèmes de sûreté ne sont pas terminés avec la rédaction du rapport définitif de sûreté, au contraire le fonctionnement de l'installation va permettre de juger la valeur des études et apporter l'expérience qui servira à mieux définir les solutions adéquates pour une sûreté toujours plus grande.

REFERENCES

[1] SERVANT, J., La Sûreté Nucléaire au Ministère du Développement Industriel et Scientifique, Revue Française de l'Energie, 24, 254 (1973) 355-360.

[2] Décret n° 73-278 du 13 mars 1973 portant création d'un Conseil Supérieur de la Sûreté Nucléaire et d'un Service Central de Sûreté des Installations Nucléaires. Journal Officiel de la République Française, n° 63 (15 mars 1973) 2 808.

[3] Décret n° 63-1 228 du 11 décembre 1963 relatif aux installations nucléaires. Journal Officiel de la République Française du 14 décembre 1963 page 11 092, modifié par Décret 73-405 du 27 mars 1973, Journal Officiel de la République Française, n° 80 (4 avril 1973) 3 798.

[4] DUVERGER DE CUY, G. et al., L'examen des projets d'installations sous l'angle de la sûreté radiologique, Bull. Inf. Sci. Techn., n° 158 (1971) 21.

[5] Anonyme., Présentation et contenu des rapports de sûreté sur les centrales nucléaires fixes, VIENNE, Agence Internationale de l'Energie Atomique (1971) "Collection Sécurité n° 34".

[6] BESSIS, J., VIREFLEAU, R., Evaluation des produits de fission dans un combustible irradié dans des conditions variables, Programme PROFIS, Rapport CEA R 3987 (1970).

[7] LE QUINIO, R., Influence de la puissance et de la distance sur les risques présentés par un réacteur nucléaire - Facteur atmosphérique du site, ces comptes rendus, IAEA-SM-184/23.

[8] COWSER, K.E., et al. Evaluation of radiation dose to man from radionuclides released to the Clinch river, "Disposal of radioactive wastes into seas, oceans and surface waters, VIENNA 16-20 mai 1966 "VIENNE,AIEA (1966) 639-671.

[9] STRAUCH, S. et al., A study of the potential radiological impact of an expanding nuclear power industry on the Tennessee valley region, "Environmental behaviour of radionuclides released in the nuclear industry, AIX EN PROVENCE, 14-18 mai 73" VIENNE, AIEA (1973) 431-447.

[10] LACOURLY, G., La capacité radiologique de l'environnement, base des études prospectives de protection, "Environmental behaviour of radionuclides released in the nuclear industry, AIX EN PROVENCE, 14-18 mai 1973", VIENNE, AIEA (1973) 603-612.

[11] JAPAVAIRE, R., ROUSSEAU, L., Emission de produits de fission à partir du combustible en cas d'accident, Réunion de spécialistes sur la sécurité des éléments combustibles pour réacteurs à eau, AEN-CSIN, SACLAY 22-24 octobre 1973.

[12] HENDRICKSON, M.M., The dose from Kr-85 released to the earth's atmosphere, "Environmental aspects of nuclear power stations, NEW YORK 10-14 août 1970", VIENNE, AIEA (1971) 237-246.

[13] KELLER, J.H., DUCE, F.A., and MAECK, W.J., A selective adsorbent sampling system for differentiating airborne iodine species, " 11th AEC Air cleaning Conference, RICHLAND Wash. 31 août-3 septembre 1970", CONF 700816, Vol. 2 (1970) 621-634.

[14] BURNETT, Th.J., A derivation of the "factor of 700" for I-131, Health Physics Vol. 18 (1970) 73-75.

[15] HOFFMAN, F. O., Environmental variables involved with the estimation of the amount of I-131 in milk and subsequent dose to the thyroid, IRS W 6 (juin 1973).

[16] ATKINS, D. H. F., CHADWICK, R. C., CHAMBERLAIN, A. C., Deposition of radioactive methyl iodide to vegetation, Health Physics, 13 (1967) 91.

[17] LE QUINIO, R., Concentrations sur une heure de polluants dues à des émissions ponctuelles près du sol - Présentation probabiliste, Atmospheric Environment, Vol. 7 (1973) 423-428.

[18] Anonyme, Basic protection criteria, NCRP report N° 39, WASHINGTON D.C., National Council on radiation protection and measurements (15 janvier 1971) 5.

[19] PELLERIN, P., Surveillance de l'environnement des installations nucléaires - L'heure du réalisme, "AIEA Colloque sur la surveillance de l'environnement auprès des installations nucléaires, VARSOVIE, 5-9 novembre 1973", AIEA SM-180/76.

[20] Anonyme, Manuel sur le contrôle radiologique du milieu (en période normale) VIENNE AIEA (1967), "Collection Sécurité n° 16".

[21] Anonyme, Le contrôle de la pollution de l'air due aux installations nucléaires, VIENNE, AIEA (1967), "Collection Sécurité n° 17".

[22] Anonyme, Environmental radioactivity surveillance guide, U.S. Environmental Protection Agency, ORP/SID 72 - 2 (1972).

[23] GRANDIN, M., ROUX, R., et al., Douze ans d'expériences de surveillance de l'environnement auprès du Centre d'Etudes Nucléaires de CADARACHE," AIEA Colloque sur la surveillance de l'environnement auprès des installations nucléaires, VARSOVIE, 5-9 novembre 1973", AIEA SM-180/48.

[24] BECK, H. L., DE CAMPO, J. A., et al., New perspectives on low level environmental radiation monitoring around nuclear facilities, Nuclear Technology, Vol. 14 (1972) 232.

[25] McLAUGHLIN, J. E., and BECK, H. L., Environmental radiation dosimetry and problems, IEEE Transactions on Nuclear Science, Vol. NS 20, n° 1 (Febr. 1973) 36.

[26] FUKAI, R., BALLESTRA, S., and MURRAY, C. N., Intercalibration of methods for measuring fission products in sea water samples, "Radioactive contamination of the marine environment, SEATTLE, 10-14 juillet 1972", VIENNE, AIEA (1973) 3-27.

Application of basic concepts in radiation protection regarding population exposure

(Sessions III and IV)

Chairmen: O. ILARI (Italy)
N. T. MITCHELL (United Kingdom)

THE UNCHANGING ASPECTS OF RADIATION EXPOSURE LIMITS

C.B. MEINHOLD
Health Physics and Safety Division,
Brookhaven National Laboratory,
Upton, New York,
United States of America

Abstract

THE UNCHANGING ASPECTS OF RADIATION EXPOSURE LIMITS.

Beginning with the introduction of the first formal recommendation on radiation protection practices, a series of revisions in recommended standards has occurred. A review of the influences and pressures on the scientific community as a whole, and therefore, on the standards setters may provide some valuable insight into the standard-setting process.

The introduction of the hot cathode X-ray tube coupled with the public outcry over the battlefield X-ray casualties during World War I provided the needed impetus for the formal adoption of that first 'standard'. In the mid-1920s Mutscheller developed the recommendation of $1/100^{th}$ of the erythema dose in 30 days. This led directly to the first internationally accepted limit of 0.2 R/d and later to 0.1 R/d. Even the 1950 limit of 300 mrem/week for organs at 5 cm in tissue and 600 mrem/week limit for skin had a strong basis in the original Mutscheller work. By the late 1950s the public and scientific debate over the fall-out controversy ultimately led to a review of the available radiobiological data by the National Academy of Sciences in the USA and the Medical Research Council in Great Britain. The reports of these committees led the ICRP and the NCRP to the present limits of 5 (age - 18) rem for certain critical tissues and 15 rem/a for others. Since 15 rem/a is 300 mrem/week, the tie to 1925 is intact.

Today the scientific and public debate is over nuclear power plant effluents. Again, as in the 1950s the National Academy of Sciences initiated a review of the available radiobiological data. Concurrently the United Nations Scientific Committee on the Effects of Atomic Radiation performed a similar study. Both reviews were published in 1972. By early 1974 the NCRP had not changed its basic recommendations. The ICRP, after an extensive review issued a press release in 1972 stating that no changes in the radiation protection standards are warranted.

There is a popular notion that over the past several decades radiation exposure limits have steadily and dramatically decreased in response to an increased appreciation of radiobiological effects, and that they will therefore continue to decrease. The purpose of this paper is to show that, in the first place, very little change has taken place since the mid-1920's and in the second place, those changes which have come about have only rarely been a direct result of the introduction of new radiobiological data.

1920 - Radiation protection practices adopted

Although physicists experimenting on gas discharge effects had been exposed to low energy x rays since the 1850's and evidence of biological damage was clearly evident in the first ten years of the twentieth century following Roentgen's discovery of the x ray in 1895, it was not until 1921 in Great Britain[1] and 1922 in the United States[2] that a formal endorsement of radiation protection practices came about. The introduction of the hot cathode tube which caused great concern among the radiologists due to the much larger currents available, and a series of sensational headlines and horror stories describing the anemia induced as a result of battlefield x-ray practices during World War I apparently created the pressure for this change.

1925 - 1/100th of the erythema dose in 30 days

When the studies done by Mutscheller[3] in Germany in the 1920's are examined, it becomes apparent that he provided a basic foundation for setting radiation protection exposure standards. This is not only to imply that his methodology was instructive, but also that his recommended exposure limit is still relevent.

Mutscheller visited a number of well run x-ray departments and observed working practices. He discerned that such operations were not hindered when the medical personnel involved were restricted to less than 1/100th of the acute erythema dose in 30 days. He further noted that employees working under these guidelines were in good health. The context in which these recommendations were made can be seen from Mutscheller's words, "... for in order to be able to calculate the thickness of the protective shield there must be known the dose which an operator can, for a prolonged period of time, tolerate without ultimately suffering injury." This "tolerance" dose was based on the assumption of a threshold in the dose effect curve. Such a relationship is common to toxic agents when tested for their acute effects. There were strong reasons to believe that this held true for chronic radiation exposure as well.

1934, 1936 - .2 R/day, .1 R/day

In 1934 the International Commission on Radiological Protection (ICRP) adopted a tolerance dose of .2 Roentgen/day[4]. One can show that this is the exposure, in Roentgens, essentially equivalent to Mutscheller's 1/100th of the acute erythema dose in 30 days. The acute exposure in Roentgens required to produce an erythema with the equipment then in use is approximately 600 R (Roentgens). 1/100th of this exposure is 6 R, which, when distributed over 30 days is .2 R/day. Taylor[5] indicates some rather more sophisticated approaches would lead to the same number. The equivalence to Mutscheller's number, however, is clear.

In the U.S. a limit of .1 R/day was adopted[6]. It is instructive to understand the rationale behind the difference between the ICRP and the U.S. limit. Failla testified before the Joint Committee on Atomic Energy[7] that he played a major role in the adoption of .1 R/day. He stated that he had been recording the radiation exposure and blood count results from three technicians who had been handling a radium source at Columbia Presbyterian Hospital. His records indicated that two of them had received .1 R/day over a three-year period and had showed no abnormal blood counts. He further testified that on this basis he recommended .1 R/day.

The National Council on Radiation Protection and Measurements' (NCRP) Report No.17[8] indicates that the difference between .2 R/day and .1 R/day was due to a difference in measuring technique. In the U.S. the measurements were in free air while in Great Britain and Europe the measurements were made at the surface of the body.

1954 - 300 mrem/week and 600 mrem/week

The next period during which a change in the limits came about was in the early 1950's. This was prompted by the changed situation regarding the energy of the available radiations. Both NCRP[9] and the ICRP[10] recommended limits of 600 mrem/wk to the skin and 300 mrem/wk to the blood forming organs, the gonads and the lens of the eye. Extensive animal studies had been performed during the Second World War to investigate dose rate effects, energy dependence, fractionation and repair and the relative effectiveness of high LET radiation. Leukemia had been identified as a late effect and there was a growing acceptance that the total dose over a working lifetime should be limited.

As will be demonstrated, there were still some very strong ties to previous recommendations. 600 mrem/wk to the skin is the same as the .1 R/day for the six-day workweek which prevailed in the 1920's. Further, the energies encountered prior to the Manhattan Project (the U.S. nuclear weapons program) were roughly 200 kvp with a half value layer of 5 cm. Since the blood forming organs are considered to be at 5 cm, the limit of 300 mrem/wk to the blood forming organs was essentially in effect in the late 20's, the 30's and the 40's! The weekly dose to the bone marrow, when the skin is irradiated with 200 kvp x rays to a limit of .1 R/day, is 300mrem. With the advent of new high energy radiations, specific bone marrow limits were required. For example, a daily skin exposure to .1 R/day of cobalt-60 radiation with an energy in excess of 1 MeV would result in very nearly 600 mrem/wk to bone marrow. Genetic effects were also considered during this time. Indeed, Failla reports[11] that he had asked H. J. Muller and Donald R. Charles, both prominent geneticists, to participate on his NCRP Committee. The NCRP Report No. 17, however, indicates that the gonad dose limit came as a result of suspected fertility impairment and a judgment that the gonads were nominally at or below 5 cm in tissue.

Both the NCRP and ICRP recommended 1/10th of the occupational limit for individuals in the general population. Again from NCRP Report No. 17 the reason does not seem to be based on population genetics considerations, but rather on reducing the total lifetime dose of potential occupational workers.

1957 - 5 (Age -18) rem

By the mid-1950's very significant changes were made in the recommendations of the NCRP[12] and the ICRP[13]. These changes came about for two interrelated reasons. The first was the scientific and public concern over weapons testing fallout. The second was the emergence of genetic damage as the controlling effect in low dose situations.

An extensive review of the available radiobiological knowledge was undertaken in the U.S. by the National Academy of Sciences(NAS)--National Research Council, Committee on the Biological Effects of Atomic Radiation (BEAR)[14]. The subcommittee on genetic effects provided the primary input. On the basis of the radiobiological data then available, they adopted a linear no-threshold dose effect relationship. In addition, they attempted to estimate the fraction of the ambient genetic burden that could be ascribed to the natural radiation environment. On this basis they recommended that the general population be limited to 10 rem in 30 years and that individuals be limited to 50 rem to age 30. In Great Britain the Medical Research Council (MRC)[15] did a similar review at the same time and produced very similar recommendations. Although they did not arrive at a definite population limit they did suggest it should be no more than twice the natural background (approximately 6 rem in 30 years).

On the somatic side, the MRC recommended a 200 rem lifetime dose. They felt there was good evidence to indicate that leukemia incidence showed a statistical increase after acute exposure to that dose, and that the effect from protracted exposure was less and within the acceptable range of risks of other kinds commonly incurred in industry and home life. They recommended a continuance of the .3 rem/week limit with the admonition that these exposures be kept as low as practicable.

Following closely on the issurance of these reports the NCRP and the ICRP developed the first new exposure limits since Mutsheller's recommendations in the 1920's.

In order to comply with the suggestions for occupational exposure limits, the two Commissions adopted a limit of 5(Age - 18) rem for the critical organs. For the ICRP, these were the gonads, blood forming organs and the lens of the eye. The NCRP considered these organs to be the whole body, the head and trunk, and active blood forming organs. The NCRP opted to limit the skin to 10 rem/year which is twice the critical organ limit to conform with prior practice of having the skin limit equal to twice the bone marrow limit. The ICRP decided on 8 rem/13 weeks and 30 rem/year since there was no radiobiological data to warrant reducing the skin limit. All other organs were limited to 15 rem/year. The changes at this time were stimulated primarily by genetic considerations.

With regard to non-occupational exposures, the NCRP recommended a limit of .5 rem/year for any individual. The ICRP used a rather complicated set of categories of non-occupational exposure but recommended a limit of .5 rem/year to individual members of the public. This limit was based on the consideration that children formed part of such a population.

The ICRP took the matter one step further in making recommendations concerning the exposure of populations. They adopted a maximum genetic dose to the whole population of 5 rem/30 years in addition to natural background and medical exposure. This is in agreement with the MRC and NAS recommendations of 6 and 10 rem/30 years, respectively, since the addition of medical exposure to the 5 rem/30 years will result in 6 - 10 rem/30 years.

In 1960 the U.S. Federal Radiation Council (FRC)[16] adopted a value of .5 rem/year for individuals in the general population. They further recommended that the average population exposure not exceed .170 rem/year because there might be a variability of a factor of three between the average population exposure and the maximum an individual in the population might receive. It is interesting to note that 5 rem in 30 years (ICRP genetic dose) is .170 rem/year. Indeed, the FRC acknowledges in its Report #1 that adherence to the .170 rem/year will assure that 5 rem/30 years is not exceeded.

During the recent and continuing power reactor controversy in the U.S. much attention has been focused on low level radiation effects, particularly on those related to low level releases resulting from reactor operations. In this situation, much like the fallout controversy of the 1950's, the NAS was once again asked to review the radiobiological data and make recommendations. The Academy appointed a committee on the Biological Effects of Ionizing Radiation (BEIR). At approximately the same time the United Nations Scientific Committee on the Effects of Atomic Radiation (UNSCEAR) undertook a similar study. Both committee reports were released in late 1972.

Both reports suggest the estimates of genetic risks made in the mid-1950's were probably high. The UNSCEAR Report[17] suggests a doubling dose of approximately 200 rem might be a reasonable estimate for low dose-rate exposure, whereas the BEIR Committee[18] uses a range of 20 - 200 rem. These estimates can be compared to the 1956 values of approximately 30 - 80 rem as the doubling dose mentioned in both the BEAR Committee and MRC reports.

From a genetic point alone then, the radiobiological basis for the standard would dictate either no change or an increase (less stringent).

The somatic case is less clear. The BEIR Committee, using only human data, decided that a linear dose effect curve drawn through data at high doses and dose rates and extrapolated to the origin, would be the basis of their risk estimates. It should be noted that the suggestion for reducing the population dose limit by a factor of 100 was based on their assumption that the power reactor program could operate within this lower guideline.

The UNSCEAR Committee, using animal data, concluded as follows:

> "The data from gamma- or x-irradiated animals suggest that for low-LET radiation, while both linear and curvilinear dose-response curves are seen, linear curves in the dose range of less than 100 rads occur principally when the target tissue is highly susceptible to induction of neoplasms by radiation. In man, target tissues which show such high sensitivity to tumour induction by radiation have not been identified except possibly in the foetus (see annex H). If, as the data suggests, most human tumours induced by radiation arise from relatively resistant tissues, then it could be predicted in the light of experimental animal data that the dose-response curves for such neoplasms will be non-linear in the low-dose ranges.
>
> "Another important consideration is the reduced effect of protracted irradiation as compared to an equal dose administered in a short period of time. While considerably more data are required, the animal data available indicate that both protracted continuous irradiation and fractionated irradiation produce less carcinogenic effect than a single administration of the same total dose, suggesting that such an effect might be expected to occur in man as well."

This can be interpreted to mean that one should not extrapolate in a linear fashion from effects seen at high doses and dose rates to estimate effects that might occur at low doses and dose rates.

In fact, although the weight of evidence seems to be swinging in favor of the UNSCEAR conclusion, a definitive estimate of the "recovery factor" is not now available. However, there is no crisis since large populations simply are not going to be exposed to more than 1% of the population limit in the foreseeable future.

The ICRP completed a review of this same material in mid-1972 with a view toward the impact of any new data on its recommendation. At its November 1972 meeting, the Commission issued the following statement:

> "As a result of this review the Commission does not see grounds for making any reduction in its dose limits for exposures of the whole body or of individual organs, either for workers or for members of the general public. In fact, it recognizes that the limits for red bone-marrow and gonads might appropriately be raised to a moderate extent to become consistent with the levels adopted for the whole body and for other tissues. It does not however regard it as necessary to modify the existing system of dose limits by recommending an immediate increase in these limits."[19]

Today's skin limit is 30 rem/year. This is equivalent to 600 mrem/week (1949), .1 R/day (1934) and 1/100th erythema dose in 30 days (1925). Other organs limit is 15 rem/year. This is equivalent to 300 mrem/week, one half of .1 R/day (1934) and one half of 1/100th of the erythema dose in 30 days

TABLE I. HIGHLIGHTS IN THE EVOLUTION OF RADIATION PROTECTION STANDARDS

Date	Standard	Background	Relationship to Previous Recommendations
1920	Protection practices	Hot cathode Battlefield x ray casualties	-
1925	1/100th of erythema in 30 days	Search for a standard	-
1934	.2 R/day and .1 R/day	Roentgen adopted in 1928	.2 R and .1 R is the measurement of 1/100th of the erythema in 30 days
1954	300 mrem/wk; 600 mrem/wk	High energy radiation and radiobiological data from U.S. "atomic bomb" research	For 1934 radiations .1 R/day delivers 600 mrem to the skin and 300 mrem at 5 cm
1957	5(Age -18) rem 15 rem/year	Fallout controversy Genetic considerations	Organ limit of 15 rem/year is equivalent to 300 mrem/week
1975	?	Power reactor controversy	?

at 200 kvp (1925). Bone marrow and gonads limits today are 5 rem/year based on genetic considerations of 1956. It is currently estimated that the genetic effect may be overestimated by a factor of six. Using a conservative factor of three, we might be all the way back to 1925 (Table I).

REFERENCES

[1] ———— J. Roentgen Soc. 17, 100 (1921).

[2] TAYLOR, L.S., Radiation Protection Standards, p. 11, CRC Press, The Chemical Rubber Co., Cleveland, Ohio (1971).

[3] MUTSCHELLER, A.M., Am. J. Roentenol. Radium Therapy Nucl. Med. 13, 65 (1925).

[4] ———— International Recommendations for X-ray and Radium Protection, Radiology 23, 682 (1934).

[5] TAYLOR, L.S., Radiation Protection Standards, p. 14, CRC Press, The Chemical Rubber Co., Cleveland, Ohio (1971).

[6] ———— "Radium Protection for Amounts Up to 300 Milligrams, Natl. Bur. Std. (U.S.) Handbook 18, [published for Advisory Committee on X-ray and Radium (now NCRP)] (1934).

[7] JOINT COMMITTEE ON ATOMIC ENERGY, Congress of the United States, Selected Materials on Radiation Protection Criteria and Standards: Their Basis and Use, p. 202 (May 1960).

[8] ———— Permissible Dose from External Sources of Ionizing Radiation, Handbook 59, National Bureau of Standards (1954).

[9] Addendum to National Bureau of Standards Handbook 59 (revised 1957).

[10] INTERNATIONAL RECOMMENDATIONS ON RADIOLOGICAL PROTECTION (1950) Radiology 56, 431 (1951).

[11] JOINT COMMITTEE ON ATOMIC ENERGY, Congress of the United States, Selected Materials on Radiation Protection Criteria and Standards: Their Basis and Use, p. 203 (May 1960).

[12] NATIONAL COMMITTEE ON RADIATION PROTECTION AND MEASUREMENTS, Maximum Permissible Radiation Exposure to Man--A Preliminary Statement of the NCRP (January 8, 1957) Amer. J. Roentgen. 77, 910 (1957) and Radiology 68, 260 (1957).

[13] INTERNATIONAL COMMISSION ON RADIOLOGICAL PROTECTION, Recommendations of the ICRP (adopted September 9, 1958) ICRP Publ. 1, Pergamon Press, London (1959).

[14] NATIONAL ACADEMY OF SCIENCES--NATIONAL RESEARCH COUNCIL, The Biological Effects of Atomic Radiation--Summary Reports (1956).

[15] MEDICAL RESEARCH COUNCIL, The Hazards to Man of Nuclear and Allied Radiations, Cmd 9780, H.M.S. Office (1956).

[16] FEDERAL RADIATION COUNCIL, Background Material for the Development of Radiation Protection Standards, p. 26 (1960).

[17] UNITED NATIONS SCIENTIFIC COMMITTEE ON THE EFFECTS OF ATOMIC RADIATION, Ionizing Radiation: Levels and Effects, Vol. II: Effects. United Nations, New York (1972).

[18] ADVISORY COMMITTEE ON THE BIOLOGICAL EFFECTS OF IONIZING RADIATIONS, The Effects on Populations of Exposure to Low Levels of Ionizing Radiation. National Academy of Sciences--National Research Council, Washington, D.C. (1972).

[19] ——— Statement issued following ICRP meeting in November 1972, Health Physics 24, 360 (1973).

BASIC CONCEPTS FOR ENVIRONMENTAL RADIATION STANDARDS

W.D. ROWE, A.C.B. RICHARDSON
US Environmental Protection Agency,
Washington, D.C.,
United States of America

Abstract

BASIC CONCEPTS FOR ENVIRONMENTAL RADIATION STANDARDS.

The elements required for informed decisions in the establishment of standards to protect the public from the effects of environmental releases of radioactive materials are reviewed. These include explicit estimation of all potential health effects and an analysis of the costs and capabilities of effluent controls, as well as consideration of the overall benefit/cost balance for each type of activity releasing radioactive materials. Subsidiary issues are the distribution in time and over populations of potential health impacts and the possibility of different value judgements by different nations in situations where radioactive materials may cross international boundaries. It is concluded 1) that standards based on health considerations alone are not appropriate in the present age of widespread use of radioactive materials, 2) that consideration of benefit/cost tradeoffs and the cost-effectiveness of health risk reduction must be explicitly made in deriving quantitative radiation standards, and 3) that international agreement is required on the basis for control of global radioactive pollutants.

1. INTRODUCTION

For a number of years we have all been exposed to the general socio-economic concepts that are central to this meeting--concepts such as "cost-benefit balancing," "cost-effectiveness of risk reduction," and "as low as practicable" (or practical, depending upon the thickness of your dictionary). For the purpose of developing guidance for radiation protection these have been coupled with our own strictly radiological health concepts such as the nonthreshold, linear dose-effect relationship, and dose commitments both to individuals and to populations measured in "person-rems" (or "man-rads"). Each of these concepts has different uses which depend upon the scope and nature of the problem at hand. I would like to describe how, in the United States, we have attempted to sort out and use these in the development of radiation standards for protection of the general population--we call these standards "environmental radiation standards."

2. THE ASSESSMENT AND IMPLICATIONS OF HEALTH EFFECTS

A prerequisite for any standards-setting rationale is an assumed relationship between cause and effect--in this case for the effects of exposure of human beings to ionizing radiation. Figure 1 shows some of the more basic forms this relationship could take in the low dose region of interest to us in establishing standards for protection of the general population.

Curve 1 represents the linear, nonthreshold relationship usually prudently assumed for purposes of radiation protection. Curve 2 represents a threshold assumption for radiation effects. Curve 3 is the type of relationship that could result if repair mechanisms are

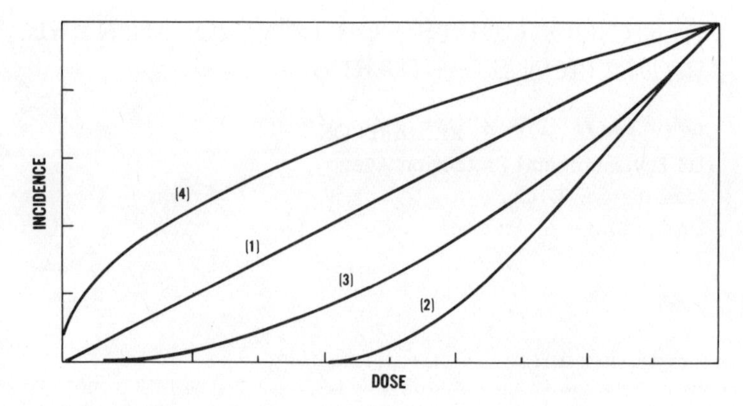

FIG. 1. Dose-effect relationships for radiation exposure of human populations. See text for description.

important at low dose rates. Finally, curve 4 could result for the average response of a population if particularly radiation-sensitive individuals exist that are preferentially singled out by low doses uniformly applied to a population. Although there appear to be data available to support each of these viewpoints, we continue to base our judgments for standards on curve 1, the linear nonthreshold dose-effect relationship. This decision stems from the need for reasonable prudence in matters affecting public health protection, especially in the absence of definitive scientific information, and reflects, as well, a consensus of the collective value judgments of experts in the field.

The rejection of a threshold relationship, such as that shown by curve 2, has basic significance for standards-setting, of course, and that is that there is no acceptable non-zero dose level based on elimination of health risk alone. Thus, at any level of exposure we must examine the benefits associated with an activity producing public radiation exposure and the cost-effectiveness of risk-reduction through effluent control and through other measures. In carrying forward the process of developing standards based on this examination, we must proceed to make a series of decisions on judgmental issues. These include such matters as the appropriate limiting level of spending for measures to reduce exposure, the equity of both the absolute and relative distributions over the population of benefits and radiation risks attributable to an activity producing public radiation exposures, and finally, the implications of the distribution in time of these benefits and risks. The consideration of time is required because many uses of radioactive materials involve, in at least some degree, incurring long-term risks in order to acquire short-term benefits.

Before discussing each of these factors and developing a perspective on their various roles in the development of a framework for setting environmental standards, I would like to emphasize that two objectives should be of overriding importance in choosing a methodology to be used for standards-setting. The first is that as complete an assessment of the impact on public health must be made as is possible, an assessment that reflects an up-to-date consensus of available knowledge. The second is that, in addition to explicitly assessing public health impact, the cost and effectiveness of measures available to reduce or eliminate radioactive effluents must be carefully examined. It would be irresponsible to set standards that impose unnecessary health risks on the public (unnecessary in the sense that exposures permitted can be avoided at a small or reasonable cost), and it would equally be irresponsible to set standards that impose unreasonable costs (unreasonable in the sense that control costs imposed by standards provide little or no health benefit to the public). Thus, the necessity to examine the economics of effluent control proceeds directly from the nature of the relationship assumed to exist between radiation dose and its effects on health. As pointed out above, it is true that if a threshold level were to exist it might then be possible to establish a standard based on health considerations alone, by setting it below that level so as to avoid any public health impact. However, such a threshold cannot be assumed for the vast majority of radiation effects, and a standard set at any level other than zero must be justified on the basis that the activity provides offsetting benefits, and that all reasonable measures to minimize the risk due to radioactive effluents are required by the standards.

There are several elements to a complete assessment of the potential health impact of radioactive effluents. The first is an accurate determination of effluent source terms as a function of the level of possible effluent control. Next, the radionuclides released must be followed, using appropriate environmental modeling techniques, through the biosphere over as wide an area and for as long a period as they may expose human populations. The totality of potential human dose must next be calculated, and the probabilities of incurring the various somatic and genetic health effects attributable to these doses must be estimated and applied.

The health assessments made by EPA for deriving environmental standards depart in two respects from practice common in the past. The first of these is the use of the concept of total population dose commitment in assessing the impact of an environmental release. We have defined the term "environmental dose commitment" for this total population dose commitment due to an environmental release. It differs from the UNSCEAR definition of dose commitment, which is, strictly speaking, a calculation of the limiting potential individual dose from an environmental level of radioactivity, and is not the total population dose due to a specific environmental release. Previous assessments of the impact of radioactive effluents from specific facilities have usually focused upon the calculation of radiation dose commitments to the hypothetical critical individual, who is generally found in the local population and whose exposure is usually incurred immediately following the release of an effluent. For short-lived radionuclides this may suffice to limit population impact, but when long-lived materials are involved this practice can lead to gross underestimates of the total impact of an environmental release. Instead of just local annual dose, the totality of doses to all populations over the lifetime of the radionuclide in the biosphere must be considered. The underlying assumption justifying the practice of assessing only the annual dose to local populations around nuclear facilities has usually been that maximum individual doses are of paramount concern and that doses to other than local populations and at times after the "first pass" of an effluent are so small as to be indistinguishable from those due to natural background radiation and are, therefore, ignorable. This point of view is not acceptable for use in deriving environmental standards because it not only neglects the implications of a nonthreshold hypothesis for radiation effects, but also the consideration that the radiation doses involved are due to avoidable man-made releases of materials and are not doses due to natural phenomena.

The second departure from usual practice is our use of explicit estimates of health effects rather than the use of exposure as the endpoint to be minimized by standards. It is perhaps obvious, in retrospect, that the proper focus for determination of the appropriate level for a standard to protect public health impact of that level, but in the past minimization of exposure has often served as a useful surrogate for this impact because of uncertainties about or a reluctance to accept the consequences of assumptions concerning the form of the relationship between exposure and effect. In this regard, the protection of biota other than man must also be considered. Although, on the basis of existing data for a wide variety of life forms, none appear to be sufficiently more sensitive than man to require separate protection, we must continue to exercise vigilance to insure that standards provide adequate protection for other life forms as well.

3. THE ECONOMICS OF RISK REDUCTION

Before discussing in detail economic aspects such as cost-benefit balancing and cost-effectiveness of risk reduction, it is useful to consider, in the context of a specific example such as the nuclear power industry, the various perspectives in which consideration of standards for radioactive effluents from the nuclear power industry can be placed. At least the following are possible:

1. The public health impact of each effluent stream of radioactive materials from each type of facility in the fuel cycle;

2. the combined impact of the various components of the fuel cycle required to support the production of a given quantity of electrical power; and

3. the integrated impact of the entire fuel cycle due to the projected future growth of the industry through some future year, for example, the year 2000.

The first of these perspectives is required for assessing the effectiveness of control of particular effluent streams from specific types of facilities. It is particularly useful for regulatory purposes, such as the development of technical guides and regulations specifying "as low as practicable" design and operation of facilities.

The second viewpoint, which can be expressed as an assessment of the total impact of the industry for each unit of the beneficial end-product (electrical power) as a function of the level of effluent control, provides the perspective required for an assessment of the relationship of this benefit to the related environmental cost - in this case the potential public health impact. We find this perspective most useful for development of environmental standards for specific activities.

Finally, although each of these perspectives can assist the forming of judgments as to the proper level of control and the acceptable impact of typical facilities or for a unit of output from the entire fuel cycle, only the third perspective provides an assessment of the potential overall impact of the entire industry. The magnitude of this impact can be either considerable or relatively small, depending upon the level of effluent control required by environmental standards. This third viewpoint will be most useful at the political level, where decisions must be made concerning the social acceptability of major alternative national courses of action for future energy supply. Of course, the overall potential radiological impact is only one small part of that consideration. This third perspective can also be useful to us as standards-setters in the establishment of priorities for action in our own area of responsibility.

The nuclear industry also provides a useful example for the examination of the explicit use of cost/benefit balancing and cost-effectiveness of risk reduction in deriving standards. Figure 2 displays the general form that the cost-effectiveness of risk reduction function for a complex activity such as the entire uranium fuel cycle will take. Each point on the curve represents the addition of a new control over environmental releases. These have been arranged in decreasing order of effectiveness. In the case of a large industry involving many different types of facilities, each with many types of effluents and many options for degree of control, this curve will actually approach the idealized smooth curve shown. The horizontal axis represents accumulated costs. These costs must be expressed as the present worth of all costs over the life expectancy of the controls, including both capital and operating expenses, and must also be

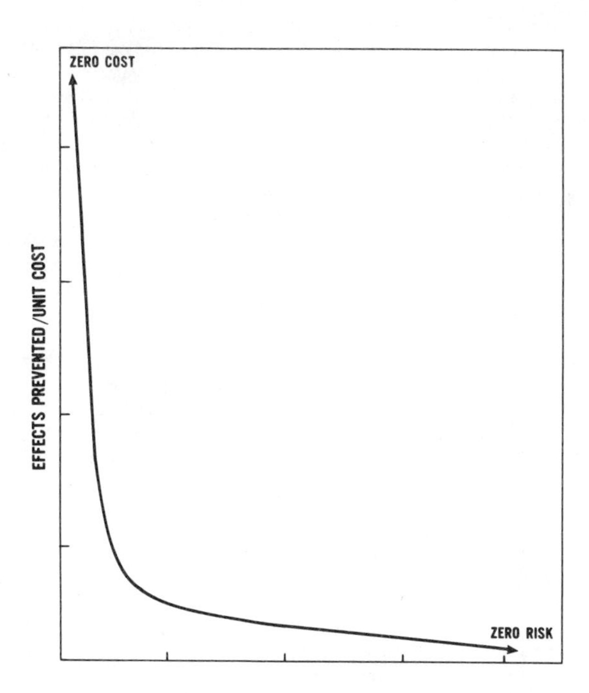

FIG.2. Cost-effectiveness of risk reduction for a complex activity.

normalized to a unit of benefit (in this case, a gigawatt year of electricity), since different types of facilities have different quantitative relationships to the end product. The vertical axis is cost-effectiveness, expressed as the number of projected effects prevented per unit of present worth of expenditure. This number of projected effects must be carefully estimated -- it should include all of the potential impact of each release over its entire projected time of residence in the biosphere, and also include all anticipated releases over the assumed lifetime of the control system concerned.

This sort of display (and its integral, which yields the total number of health effects prevented by any particular level of expenditure) gives us a rather complete assessment of the total potential radiological impact of environmental releases from an activity. Its main deficiencies are 1) a failure to display the distribution of that impact across the population and over time and 2) a failure to relate the total impact of the activity to its benefits. We will deal with these separately later.

What does a display such as that in figure 2 tell us about the choices for standards setting? There are two simple cases. These are no control, represented by the extreme upper left-hand end of the curve, and no risk, represented by the extreme lower right-hand end. In general, neither of these are reasonable possibilities. There are at least three more realistic choices. First, if one is given (from other considerations) an acceptable level of impact, one can require the imposition of controls, starting at the left of the curve, until the impact is reduced to the desired level. A second alternative arises if the amount of expenditure available for control is predetermined. Then one simply imposes all controls to the left of that amount on the horizontal axis. Usually neither of these value judgments are available (or even appropriate) and a third alternative must be used - either explicitly or implicitly. This is to determine a maximum acceptable rate of spending for risk avoidance and require the imposition of all controls above that rate on the vertical axis. Although this procedure may seem distasteful, in actual practice it usually turns out to be the least so of the choices available. We have also found that in real situations a discrete break-point often occurs within the range of acceptable rates of expenditure that facilitates such decision-making. Based on recent U.S. experience this acceptable range appears to lie in the neighborhood of a few hundred thousand to about half a million dollars per health effect averted.

Now, what relation does benefit/risk balancing bear to this discussion? The first and most obvious use is expressed as a necessary, but not a sufficient, condition. This is the condition that the total benefits of a particular activity must exceed its total costs, including its public health and environmental impact, in order that its pursuit be a societally acceptable endeavor. Such an evaluation is difficult, if not impossible, to achieve in a quantitative way because of the diverse characteristics of the various benefits and costs involved. We are generally reduced to making qualitative judgments, once the various categories of benefit and cost have been reduced to quantitative terms (usually in different units), to the extent possible.

Often, however, such tradeoffs may be subdivided and made on a category by category basis, such as comparing the health benefit of an x-ray diagnosis to the radiation risk (as in tubercular screening) and then considering the dollar benefits and costs to society of the process as a separate matter. This concept of categorization is useful when we attempt to make further use of benefit/cost considerations in setting standards for radiation protection. Suppose we are considering, as before, the nuclear production of electricity via the uranium fuel cycle, and we focus our attention on health benefits. Although the health benefits of electrical power, at the margin, are extremely difficult to quantify and will be quite dependent on the time frame considered, some generalizations are possible. If we consider the health benefit of electrical power lost due to investments in control costs, it is clear that if we make no expenditure for controls the investment in power capacity can be maximized, and the benefit foregone will be zero. Now, as we divert investment in construction of power production capacity to investment in construction of environmental controls, benefit foregone will increase, and probably at an increasing rate, as the lack of power capacity becomes more acute. Figure 3 shows, superimposed on the integral of the cost-effective curve of figure 2, two such curves which are concave upward and represent, in a general way, what may be conjectured to be reasonable limits of uncertainty on such a benefits foregone function. The point of interest derives from the observation that, in general, the cost-effectiveness function by its very nature asymptotically approaches both axes, and has a rather small region of sharp inflection near the origin. The region of intersection of the limiting benefit foregone functions with this curve represents the appropriate region for choice of standards whose objective is the

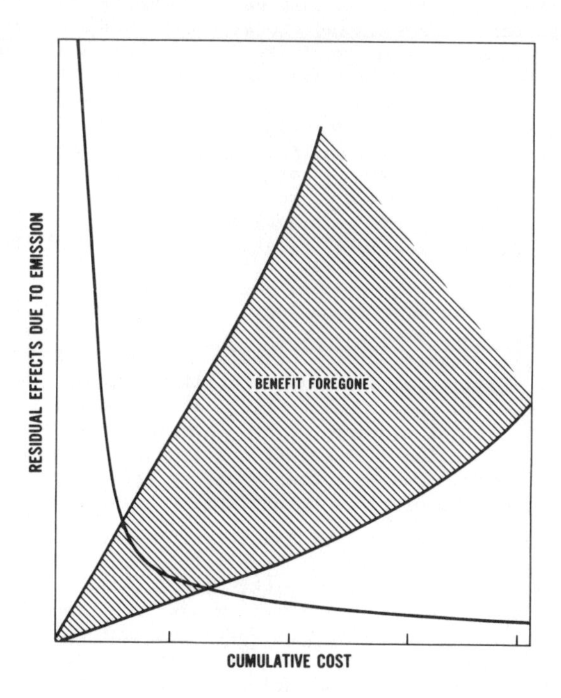

FIG. 3. A conjecture on the functional form of health benefits and its relation to health costs due to effluents for nuclear power.

balancing of benefits and costs. As you can see, it can easily turn out that the region of uncertainty for benefit/cost balancing will encompass most of the region of rapidly changing cost-effectiveness - the region where we have found that the acceptable choices based on cost-effectiveness alone usually lie. Thus, cost-effective standards will usually also satisfy, within the present limits of uncertainty, criteria for an acceptable benefit/cost balance - at least as far as health benefits are concerned. The other types of benefits and costs really fall outside of the area of consideration for standards-setting to protect health, except for the need to satisfy the necessary but not sufficient criterion that benefits exceed total costs mentioned earlier.

In summary, one may identify, out of a large variety of standards-setting rationales, at least the following choices:

A. Zero Emission - Zero Risk: The underlying premise is that in order to have zero environmental or human health risk due to an activity it is required that there be zero emission of pollutants to the environment. This approach eliminates the environmental health risks associated with emissions from the activity. The cost of implementation is usually prohibitive; in many cases the only way to accomplish such an objective is to suspend the activity, otherwise this approach can result in a requirement to expend more resources than are available to sustain the activity.

B. Best Available Technology: This approach is based on the philosophy that the best methods available for reducing pollution or emissions should be utilized. Factors favoring this approach are that additional health risks due to emissions from the particular activity are held to a minimum. On the other hand, the cost may be prohibitive. As technology improves the activity may have to upgrade through frequent installation of new emission control equipment. In some instances this approach may require installation of emission control equipment that is not cost-effective. On the other hand it can also permit pursuit of activities with unacceptably high impact, if the best available technology is insufficient to provide an acceptable relationship of total benefits to total costs.

C. "Lowest Practicable" Technology: This approach has as its objective the maintenance of emissions at a level that is as low as practicable, giving principal consideration to the costs and availability of controls. Utilization of this approach will usually keep health risk at a low level; the cost of emission control equipment and availability of alternative courses of action will affect the amount of emissions allowed. This approach, however, does not specifically require that either benefits or risks be explicitly considered.

D. Cost-effectiveness of Risk Reduction: In this approach the cost of reducing an emission is explicitly balanced against the environmental and health risks associated with that emission. Public health and environmental risks are kept low through consideration of the practical parameters of cost-effectiveness of risk reduction. Use of this approach, however, requires a difficult quantitative balancing of costs against known or assumed risks. In common with the preceding alternatives, it does not require explicit consideration of the relationship of total benefits to total costs.

E. Benefit/Cost balancing: This approach is based on the philosophy that the benefits created by an activity should be cost-effectively balanced against the associated risks. This approach keeps total health and environmental risk at a minimum level, since it requires that residual environmental or health risks of activities have offsetting benefits. Allowable environmental risks, however, may be greater than for the first two approaches, and an extremely difficult assessment and balancing of benefits and costs is required.

We have found, as a practical matter, that approach D, the cost-effective reduction of risk to a societally acceptable level of expenditure, coupled with a requirement for a subjective judgment that total benefits exceed total costs, is the most satisfactory approach available for standards-setting at the present level of development of these types of analyses. We have recently applied this methodology to the uranium fuel cycle. The principal finding has been that there is an urgent need for some new requirements to limit long-lived materials such as krypton-85, iodine-129, and the long-lived actinides. We were pleased to find that the levels of control now being instituted in the United States for most other radioactive effluents, particularly from reactors, satisfy our cost-effectiveness criteria, although in a few minor instances the level of control may have surpassed the point of easily justified return.

4. FUTURE RISK, RISK DISTRIBUTION, AND DIFFERING SOCIETAL VALUES

Another aspect of standards-setting that has not in the past received sufficient consideration is that of the role of time. Most of the potential health impact of radiation exposure is expressed many years in the future and, in the case of long-lived radioactive materials, the exposure itself can occur a long time after their release to the environment.

In assessing the impact of an action to reduce exposure, such as installation of an effluent control, it is customary to express the train of future costs implied by that action in terms of their present worth. Economic theory provides us the means for calculating this, and the essential prerequisites are two: 1) the future cost must be capable of being met in one of two alternative ways: either by obligating the future cost at the time it will be incurred, or by making an investment now that will grow in size so as to cover the future cost when it arises, and 2) agreement on an appropriate discount rate which represents the interest that will be earned and the depreciation that will accrue on a present investment up to the time of the future cost. The method is well-established for use in dealing with future monetary costs. There is, however, much less guidance available that is useful for more general kinds of future costs, such as potential health effects. In particular, when one is dealing with future costs whose remission cannot be purchased with a monetary expenditure in the future, but are irrevocably committed in the present, the problem is far more difficult. Health effects of radiation fall into this category. In our consideration of this program we have not been able to progress much beyond identification of some of the more important factors that are involved.

First, the concept of "discounting" future health effects is most obviously applied to the monetary value required to compensate for an effect after it has occurred. However, the monetary value of an effect which has occurred and the amount society (or an individual) is willing to invest to avoid occurrence of an effect are two quite different concepts, which must be distinguished.

Secondly, the monetary value of most of the health effects considered in radiation standards-setting, which involve either potential loss of life or extremely serious disability, is not easily assessed now, and is even more uncertain for the future. Leaving aside the moral implications of assigning a monetary value to compensate for such effects and considering only the experience we can draw upon for what society is willing to spend to prevent their occurrence, we can distinguish several characteristics. The amount depends heavily upon whether the risk of incurring the effect is imposed voluntarily or involuntarily (the latter case carrying a much greater willingness to spend) and how far into the future it is anticipated to occur. The amount also depends upon who is supplying it and upon how the burden of payment is distributed. In addition, the historical trend is for steadily increasing amounts and there is no reason to believe that this trend will not continue.

Finally, the time spans involved fall into three quite different categories: those for somatic effects, those for genetic effects, and those for delayed exposure due to releases of long-lived radionuclides. Somatic effects, with some exceptions, typically begin to appear ten to fifteen years following exposure and continue for a period ranging from several decades to the lifespan of the exposed individual. Most, but not all, genetic effects occur with decreasing frequency in successive generations, a majority having been expressed by the end of the first five or so generations. The time distribution of dose commitments due to environmental pollution by long-lived nuclides depends heavily upon the half-life of the nuclide and the details of its pathways through the biosphere, and ranges from decades through time frames typical of recorded history to geological time frames.

We have not found it feasible to use the concept of present worth of a monetary value assigned to specific health effects for two reasons. First, victims of the effects of radiation exposure are not identifiable. Thus the usefulness of the concept of compensation for real loss is only as a surrogate for the purpose of estimating the worth of a life, since the specific victims cannot be identified. However, the concept "worth of a life" is not, in our opinion, the relevant one. Indeed, in some instances the loss of a life may be perceived, in purely economic terms, to have negligible or even positive monetary value to society. The relevant consideration is what amount is society willing to invest to prevent such effects from occurring, with all of the psychological, economic, and ethical implications that become factored into such a societal value. Secondly, we find it significant that one can only prevent such effects through a decision to spend now, and the alternative of expending an amount at a later time to avoid the impact of an earlier release of radioactive materials is simply not available. This would appear to rule out use of the concept of discounting the value of future health effects. The problem posed by the possibility of commitments for health effects thousands of years hence remains unsolved - its solution must ultimately depend upon society expressing in some manner an ethic concerning its responsibility to future generations.

Once one has dealt with the question of determining what value society places on the prevention of health effects now, the further question arises, will this value change in the future? We can examine the current range of values, but we cannot pretend to be able to make a prediction for the future, except to point to the historical trend, which has been a steady increase, dependent principally upon society's growing ability to pay for such investments in the future quality of life. A subsidiary question is the possibility of the emergence of other means to deal with the effects of radiation after they have occurred, such as the development of an effective treatment for cancer or leukemia, or new developments resulting in future societal changes in view on the acceptability of the use of such measures as abortion or euthanasia to avoid genetically deficient individuals. The only supportable position appears to be to predicate today's decisions on today's values and capabilities and to make future changes only when and if these values or capabilities change.

Another issue that must be recognized is that different cultures and societies may place different values on risk, life, and benefits. While it may be possible for all to agree on a health-effect to dose relationship, standards can still vary from country to country since risk/benefit judgments may differ. While this may be acceptable for effects of radioactivity confined within a country's borders, it will not suffice for situations in which one country's effluents affect the populations of other nations. The need for international agreement in this area on mutually acceptable risk/benefit judgments is obvious, but the achievement of such agreement may be a difficult process. Perhaps a useful place to start would be the matter of krypton-85 releases associated with the production of nuclear power. Here we have at least the advantage of clearly defined models for environmental behavior.

A second major consideration in standards-setting, beyond a look at total impact, is the matter of the distribution of benefits and detrimental impact across the population. This problem is the one traditionally emphasized by radiation standards, which are usually expressed as limits on annual individual exposure. It is not as straightforward a problem to deal with as total impact, partly because the level of individual risks are usually very small, especially in comparison with other more everyday forms of individual risk-taking. We have found that individual exposures from various nuclides and pathways are usually relatively uniformly low as a result of measures taken to control total population impact. In those few instances where this is not the case (radioiodine in milk consumed by a dairy farmer's children comes to mind) we have used the argument that maximum individual risk associated with such pathways should be arbitrarily reduced to levels comensurate with that for other nuclides and pathways using the most cost-effective means available. To date, this modus operandi for handling gross maldistribution of dose appears to be satisfactory. We have not attempted to tackle the rather esoteric further question of the relative distribution of benefit to risk.

5. CONCLUSIONS

I would like to summarize our major conclusions for radiation standards. First, standards for acceptable radiation exposure based on health alone, and without specific consideration of the sources concerned, are not appropriate in an age where use of radioactive materials is ubiquitous. This conclusion applies to most of existing numerical recommendations for limits on radiation exposure. We recognize that the shortcomings of such universally applicable numerical guidance are usually mitigated by the additional guidance that exposures should be maintained as far below such guides as practical.

Second, United States policy for environmental radiation standards-setting is to use analyses of risks, costs, and benefits for each major category of radiation sources and to develop separate standards for each of these major categories. This process must include an assessment of the long term as well as of the short term implications of the activity concerned. In establishing these standards, especially when only small doses are concerned, the primary function of standards is to limit total potential societal impact, and a secondary but still essential function is to provide a limitation on the distribution of that impact through limits on maximum individual exposure. This implies that measures for standards must be expanded beyond the usual units for individual doses, millirems or rads, to also encompass units for total impact or potential impact, person-rems (or man-rads) and, in the case of long-lived materials, curies released to the environment.

Finally, we realize that value judgments appropriate to the United States may not be appropriate for control of radiation within the borders of other countries. Even if there is international agreement on the relationship between radiation dose and health effects, differences in other value judgments required for standards-setting are still not precluded. However, in cases where radioactive effluents do not honor national boundaries, there is a need, which is yearly growing more urgent, for international agreement on a mutually acceptable basis for control of these materials.

DISCUSSION

H. T. DAW: I should like to ask a question relating to Fig. 1, which shows four possible curves for the dose-effect relationship. Curve 4 has a higher slope at low doses, which can be interpreted to mean that radiation causes more damage at low doses. This is in addition to your explanation that there is selective action on certain more sensitive individuals in the population exposed. Would you care to expand a little on this?

W. D. ROWE: Yes. Instead of intersecting Curve 1 at the extreme right of the figure, Curve 4 could very well intersect at a lower value so that the total effects under Curve 4 would average the same as for Curve 1, or even less if taken over the whole range. We did not try to indicate the magnitude of the effects, but only the various processes that might exist.

O. ILARI (Chairman): You referred to the need for a public health authority to perform a complete assessment of the total environmental impact associated with any given activity. In particular, you mentioned the overall assessment of the environmental impact of the entire nuclear fuel cycle. Such overall assessments can lead to surprising results as far as concerns the relative contributions due to the various parts of the activity under review. For instance, in several Environmental Statements published by the USAEC for nuclear power stations it is surprising to find that the principal contribution to the population collective dose is not due to operation of the plants but to transportation of fuel and wastes. Owing to the particular assumptions made for this assessment, however, it is possible that the detriment evaluation relating to the transport phase of the cycle has a health meaning different from that of the other parts of the cycle. In other words, the results of an overall assessment might be misleading if you are not careful with the assumptions made and the interpretation of results.

W.D. ROWE: This is always possible when viewing the total cycle. However, if the assessment is not only concerned with the magnitude of the health effect, but also considers the cost-effectiveness of health-effect reduction on an equitable basis, there is little danger of taking the 'wrong' action.

P. CANDÈS: I am interested in this matter of evaluation of population doses due to the transport of radioactive materials, and I understand that you are suggesting the possibility of establishing nuclear parks to reduce this dose. Have you evaluated the risks associated with the transport of all the small amounts of radioisotopes which would then be involved?

W.D. ROWE: We have not yet considered these risks in detail, but they will be fully evaluated when the proponents of nuclear parks put the concept forward as a serious alternative.

S. AMARANTOS: Will the EPA (Environmental Protection Agency) standards be concerned with accidental conditions?

W.D. ROWE: Yes, but in exactly what manner has not yet been decided. The Atomic Energy Commission is responsible for ensuring that equipment and facilities meet specifications. The specifications may be such that an acceptable level of risk is met on a predictable basis. The role of EPA might be to fix the acceptance risk and to determine the validity of assumptions for predictability.

H.P. JAMMET: I just want to say a few words about the confusion that can arise through the various interpretations of the word 'standards'. A distinction should be made between recommendations or regulations based on radiological principles and those based on economic, administrative and social considerations. The former may be universal in character, but the latter are only of national or local significance.

P. RECHT: I should like to join Mr. Jammet in drawing attention to the importance of terminology. Standards such as the "primary protection standard" used in the USA and the norms employed in Europe are limits in the regulatory sense. We should take a lead from the definitions used at the Conference on the Environment in Stockholm, 1972, where a distinction was made between scientific recommendations and standards having a regulatory, administrative, or political character.

W.D. ROWE: Value judgements or 'scientific standards' certainly can and should be made without regard for national boundaries, while standards laid down in a regulatory sense must be addressed to the political entity for which they are established. Of course, regulatory standards might well be international in scope. However, I agree entirely that we should know exactly what is meant by the different types of standard.

O. ILARI: In the context of the present discussion I should like to add that frequently safety and protection standards are exported together with technology. In fact, the public health authorities of countries importing nuclear technology from abroad often have great difficulty in trying to set national standards which are different from those proposed by the technologists. This situation often arises in the setting of acceptable reference levels for accident conditions, exclusion radius, etc.

W.D. ROWE: This is true, and there is the additional complication that a major accident will have serious repercussions around the world.

For example, an accident of this nature outside the United States of America in a plant using United States technology and standards would provide significant ammunition to those who are opposed to nuclear power in any form. Thus the problem is extremely complex. I did not claim in my paper that I would provide any answers; I am not even sure that we've asked all the questions yet!

G. BRESSON: In your presentation you explained what you understand by such concepts as 'undertaking', 'activity', etc., which have an economic, commercial or technical significance, but what do you mean by "society", a concept which has no legal, economic or sociological basis?

W.D. ROWE: By society, I mean that collective population which is involuntarily subjected to risk without directly participating in the benefits of an undertaking. One must differentiate between direct benefits — e.g. profit to an industry, energy supply — and indirect benefits, such as arise for instance through increases in the gross national product.

S.M. MITROVIĆ: If man is biologically the same all over the world, he is everywhere affected in the same way by the same dangers and risks. I therefore fail to see how different limits can be applied in different countries. It must also be borne in mind that these dose limits also correspond to limits of responsibility.

W.D. ROWE: It is nevertheless understandable that each country or population will have its own criteria as to the importance of the risks involved.

D. BENINSON: A system of dose limitation has two components. First, there are the dose limits, related ideally to values of acceptable risks and, secondly, there are the levels derived for each source, based on the justification and the 'as low as reasonably achievable' concepts. When discussing standards, a very clear distinction should also be made between these two components.

W.D. ROWE: If one assumes a true linear relationship, and can define the levels and benefits of radiation from all sources, then the dose limits you mentioned first are meaningless, at least in theory.

D. BENINSON: Even if every source is justified and 'optimized', the possibility of multiplication of sources calls for some sort of limit for the total average dose.

W.D. ROWE: This is not true in theory, but could occur in practice, for instance with a total dose of 1000 rem/a for individuals, all would cease to exist. Fortunately, our experience with radiation is quite the reverse, i.e. source standards based upon practicality are so low that the sum from all sources will be significantly less than existing population exposure limits, e.g. 170 mrem/a.

EXPERIENCE GAINED IN APPLYING THE ICRP CRITICAL GROUP CONCEPT TO THE ASSESSMENT OF PUBLIC RADIATION EXPOSURE IN CONTROL OF LIQUID RADIOACTIVE WASTE DISPOSAL

A. PRESTON
Ministry of Agriculture, Fisheries and Food,
Fisheries Laboratory, Lowestoft

N.T. MITCHELL, D.F. JEFFERIES
Ministry of Agriculture, Fisheries and Food,
Fisheries Radiobiological Laboratory, Lowestoft,
Suffolk, United Kingdom

Abstract

EXPERIENCE GAINED IN APPLYING THE ICRP CRITICAL GROUP CONCEPT TO THE ASSESSMENT OF PUBLIC RADIATION EXPOSURE IN CONTROL OF LIQUID RADIOACTIVE WASTE DISPOSAL.

The control of liquid radioactive waste disposal to the aquatic environment of the United Kingdom is based on the limitation of public radiation exposure. Experience has shown that in general somatic dose has been of greater importance than genetically significant exposure. UK waste disposal policy is briefly reviewed, showing the importance attached to the estimation of radiation exposure of the individual or of small groups of the population in a somatic context.

Control of such exposure has been based on critical pathway analysis, and originally on the most highly exposed individual — isolated from a representative sample of the exposed population; latterly the critical group concept has been used as outlined in the current recommendations of the ICRP. The problems of applying this concept, together with its advantages and disadvantages, are discussed. The paper goes on to review experience gained in attempting its application. It concludes with a comparison, using the two most important internal exposure pathways for Windscale sea discharges (Porphyra/ laverbread and fish), of different methods of critical group identification which have been examined.

1. INTRODUCTION

The objectives of United Kingdom policy for the control of radioactive waste disposal include specific limits for human radiation exposure, both for individual members of the public and collectively for the population of the country as a whole. These limits, published in 1959 in the White Paper [1] "The Control of Radioactive Waste", may be summarized as requiring:

(a) that exposure of individual members of the public is within the dose limits recommended by the International Commission on Radiological Protection, and

(b) that the collective dose to the population of the country as a whole does not, on average, exceed 1 rem per person in 30 years.

It is normal practice in setting controls on liquid radioactive wastes from major nuclear sites to consider each site individually. This is regarded as a practical necessity because of the many variables, particularly environmental, which have to be taken into account. It is also closely in accord with the overall waste disposal policy, since in almost all cases individual exposure from a discharge is much more significant than the contribution made to population exposure, and assessment of this category of exposure requires specific investigation. The population exposure of the country as a whole from controlled disposal of liquid radioactive wastes is only in the region of a few hundred man-rem per year and is thus of the order of 10^{-2}% of the limit specified as national policy, with individual contribution from disposals from many sites being less than 10^{-4}%. The lowest levels of individual exposure can only be quantified on a 'not greater than' basis where they are lower than about 0.1% of the ICRP-recommended dose limit. Maximum individual values in a few cases reach a few per cent or so.

These findings have made the estimation and assessment of exposure of individual members of the public of paramount importance, although choice of a particular value for an exposure factor such as food consumption rate will, of course, have a consequential effect on population exposure. The computational model used in assessing exposure has been reviewed extensively[2, 3, 4] and is based on critical pathway techniques utilizing parameters identified and quantified by means of surveys of the exposed population. The aim has always been to protect those who are most highly exposed and, in the early years of application of the control policy, this was achieved on the assumption that the individual in the sample, with, for example, the highest observed consumption rate, would be the most highly exposed. Because of the innate variability of the metabolic factors which also help to determine the dose received, this is an assumption which may not always be valid. Furthermore, the acute dependence on one single individual places a very great onus on the conduct of the surveys. Since waste disposal controls are based on these data and are thus dependent on the habits of a single, often very exceptional and sometimes even eccentric, individual, the scale of the permissible discharges may fluctuate widely between successive surveys. Nevertheless, despite its disadvantages and weaknesses, there was at that time no acceptable alternative procedure to using the single highest observed value in a representative sample of the exposed population, until the concept of a critical group was developed by the ICRP[5] in its 1966 recommendations. Following endorsement by the British Medical Research Council's Committee on Protection against Ionizing Radiation, these recommendations were adopted for use in the UK and this paper is an attempt to review experience in the use of the concept by one of the authorizing departments in the UK.

2. THE ICRP CRITICAL GROUP CONCEPT

2.1. Basis of the concept

The ICRP's recommendations for exposure of members of the general public take the form of a system of dose limitations to body organs or the whole body. In the case of internal radiation, following ingestion of contaminated food or drink, or the inhalation of contaminated air, the specific dose limits are related to derived organ or body burdens of individual radionuclides. For practical purposes in the application of the dose limits, the organ body burdens have to be related to derived rates of intake of specific radionuclides sufficient to generate and maintain these burdens. In turn these rates of intake of radioactivity are equated to concentration limits in food and water, the so-called derived working limits, which require a knowledge of food and water intake rates. To establish such a model, assumptions have to be made regarding the anatomy and physiology of exposed individuals and the relevant metabolic turnover of radionuclides in the body. The ICRP[6] has defined the important characteristics of a 'standard man'; in relating rates of intake to organ and body burdens, mean values of metabolic turnover are taken for this 'typical' person and individual variations are not taken into account, neither for these factors nor for size, sex, physical conditions and so on. 'Standard man' so defined was taken by the ICRP to represent a typical occupationally exposed adult selected as to medical fitness, and we may therefore conclude that the range of metabolic parameters in the public at large might be even wider than in standard man.

It is quite clear that variations do exist, so that even within a group of people which appears to be homogeneous with respect to factors such as size, sex and other physical parameters, the dose received from a standard rate of intake will vary. For example, an estimate of the range of doses to the lower large intestine of 54 individuals, using an unabsorbed radionuclide, lanthanum-140, as a tracer, has been studied[7]. The individuals ranged in age from 7 to 76 years and had no apparent abnormalities in GI tract function. The authors concluded that a sizable proportion of the population may experience doses many times in excess of that assumed for the average or 'standard man'. Their measurements indicated that about 15% of the general population may experience a dose 3 times, and 6% as much as 5 times, that for 'standard man'. In a report prepared for Committee II of the ICRP in relation to GI tract exposure, Eve[8] gave data on the mean weight of faecal samples in a group of male radiation workers at AERE, Harwell. The data suggested that the distribution might be of the log-Gaussian type. The estimated median value was 141 $g \cdot d^{-1}$ and 90% of the population lay within the range of 70-275 $g \cdot d^{-1}$. Tipton and Cook[9] measured the concentrations of a number of trace elements in human tissue. The data were obtained from 150 adult subjects who were considered to be a representative sample of the normal adult population of the United States. The authors concluded that for most trace elements, in most tissues, the distributions of concentration values were not normally distributed. Table I shows a selection of their data for five elements in liver and kidney. The median and the 80% range (10th-90th percentiles) are shown. The values of the concentrations for 'standard man'[6], adjusted to $\mu g \cdot g^{-1}$ (ash), are shown for comparison. It can be seen that the observations are skewed toward low values and the 90th percentiles for Al, Cu, Mn and Sr in liver and for Al and Sr in kidney are more than twice the median.

Because of the innate variability of these factors, the person with, for instance, the maximum consumption rate of a critical foodstuff may not necessarily be the most highly exposed member of the population. The nearest approach to characterizing the highest degree of exposure which can reasonably be achieved is therefore probably that of taking the average behaviour of a group of those who are likely to be the most highly exposed.

This is consistent with the current recommendation of the ICRP, which is that for control and assessment of a specific waste disposal operation, the appropriate dose limit(s) shall be applied to the mean dose of a small group of the population; this group should be representative of those expected to receive the highest dose and as homogeneous as practicable with respect to those factors which determine radiation dose.

TABLE I. STABLE ELEMENTS IN HUMAN TISSUE

Element	Concentration of element ($\mu g.\,g^{-1}$ of tissue ash)							
	Liver				Kidney			
	Median	Percentiles		Standard man	Median	Percentiles		Standard man
		10th	90th			10th	90th	
Al	47	17	120	55	25	14	82	37
Cu	510	320	1300	660	260	190	340	260
Mn	100	55	220	100	82	45	150	77
Sr	1.6	0.7	3.5	2.3	5.2	2.8	12	7.3
Zn	3500	2200	5900	3500	4500	3200	7400	4400

From Tipton and Cook, Health Physics 9 (1963) 103.

3. APPLICATION OF THE CONCEPT

It is thus necessary to ensure that the critical group is representative of that particular segment of the exposed population which is at highest risk, not only due to a combination of characteristics such as age, sex, location, etc., but also with respect to their dietary, occupational or recreational habits, which may be of prime importance in determining their exposure. The procedure adopted has become one of estimating the relevant parameter, which may be the consumption rate of a foodstuff or the time spent in an external exposure regime, by means of a 'habits/consumption survey'. This has a number of functions, the first of which is simply to identify the critical pathway(s), though its ultimate and most important purpose is to collect data on a sample of the population exposed via this critical pathway, on which waste disposal control procedures may be based. Whilst the qualitative selection of the critical group may be fairly straightforward, the precise identification of the range of consumption or occupancy rate for the group is a much more difficult proposition.

It is neither practicable nor necessary to survey the whole of a population exposed via a particular pathway; it is sufficient for a sample of it to be examined and steps taken to ensure that this sample is as representative as possible, with particular emphasis on those most likely to be highly exposed. The critical group can be identified either directly from the raw sample data or, more correctly, from a mathematical distribution of the whole of the exposed population constructed from the sample data. A typical distribution, taking consumption rate of a foodstuff as an example, is shown in Fig. 1. The skewness of the distribution is a characteristic of this type of survey. The practical problem is one of deciding the point at which to set the lower limit to consumption rate (G_{min}) within the critical group, such that the dose to individuals in this end of the distribution remains reasonably homogeneous. It is our view that the ratio between the maximum for the group (G_{max}) and the mean ($\overline{G}$) should be no greater than the unavoidable variability which exists in metabolic factors, and preferably less. The type of mean value ($\overline{G}$) used is open to question; the ICRP, for instance, is not specific on this point. Whilst the arithmetic or geometric means might be used - or even the modal value - we have chosen to use the geometric mean or median as the standpoint from which to estimate the dose to individuals.

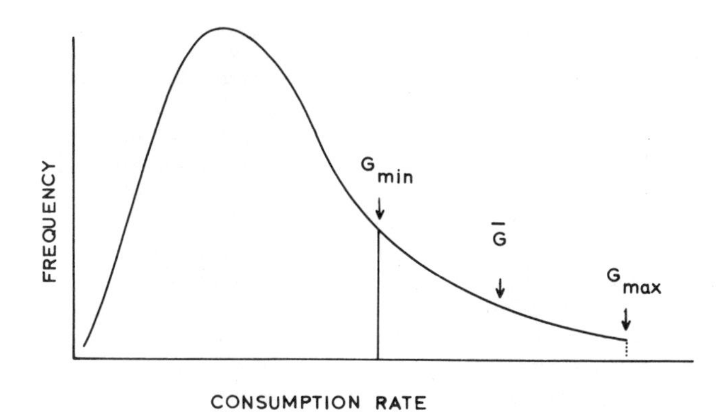

FIG. 1. Typical distribution found from habit surveys for consumption of a foodstuff.

Ideally, the acceptable limit to the quantity $G_{max}/\overline{G}$, which may be termed the homogeneity or 'H' ratio, should be decided against a searching analysis of metabolic variability; failing this it has to be set on a more arbitrary basis at 2 or, at the most, 3, which should ensure that the implied variation in dose, due to differing consumption rates, is less than that caused by metabolic factors. This might constitute a reasonable basis on which to judge whether a particular system of critical group identification is acceptable and it has been applied here to the situations reviewed in section 4, using some of the methods discussed below.

The ICRP gives only general guidance on the considerations needed to define the critical group, leaving national authorities to decide on a more precise mathematical or statistical approach to the problem. To date, there is an almost complete lack of published attempts at its solution, the only example being the well-known Porphyra/laverbread consumption pathway at Windscale[10]. One of the first problems which has to be faced is the often small size of both the exposed population and the sample of it for which data can be obtained. Some of the more refined statistical methods are therefore of academic interest only, since they can be applied in only a small minority of cases where a large amount of data is available. In addition, the refinements they introduce may be invalidated by inaccuracies in the estimates of consumption rate or occupancy factor due to the method by which these data are produced. Whilst it may be possible to check the information by, for instance, having people record their use of a foodstuff, this will always be retrospective, and the data in the survey proper will often depend on the subject's memory and judgement at interview. A great deal depends on the design of the survey[11] and the skill of the interviewer, but some uncertainty in the data must always remain. The ideal solution would be fairly simple and applicable to the results of any survey, large or small, whilst at the same time having as objective a basis as possible.

Attempts at a solution may be considered - according to whether they are wholely arbitrary or have some mathematical/statistical basis. In each case the results of a particular method can be judged in terms of the 'H' ratio as a test of relative acceptability.

3.1. Arbitrary methods

3.1.1. If simplicity were the only consideration then the most obvious of arbitrary methods, selecting data at the upper end of the distribution which stand out from the rest as exceptional, would be an automatic choice. This is most easily achieved by examining the data displayed in histogram form, and one way would be to choose the lowest interval above the mode in which there are no observations, as the lower cut-off value appropriate to G_{min} of Fig. 1; alternatively, this point may be selected on the basis of a minimum in the smooth curve drawn on the data. These methods have the advantage that they can be applied to any set of data; their chief disadvantage is their subjective nature, particularly with the former approach, because the position of G_{min} can be manipulated to some extent by varying the size of interval chosen for a histogram. Nevertheless, it is a useful method and is clearly consistent with the ICRP recommendation that the critical group shall be representative of those who may be expected to receive the highest dose, for it selects those whose habits are exceptional.

3.1.2. 'An alternative method is to select an arbitrary fraction of the upper end of the distribution and use the arithmetic mean or median value within it. The upper quartile, and the top 10, 5, 2 or even 1% have been suggested. Whilst their choice may meet the broad ICRP requirements it does not provide a particularly systematic basis for a selection procedure, there being no logic behind selection of any of these values.

3.2. Mathematical and statistical analysis

3.2.1. Although all the distributions of habits which we have studied turn out to be skewed, frequently with a long tail towards the higher values, the data often fit a log-Gaussian distribution. The degree of fit can itself provide a method of critical group identification in cases where there is a significant deviation from the log-Gaussian plot at the upper end of the range. This then forms a sub-group with significantly higher consumption or occupancy rates than those indicated by the main group. As illustrated with data from the Windscale/laverbread pathway discussed in section 4, the data may be insufficient to identify precisely the point at which separation of the sub-group occurs, but a reasonable approximation may be made by extrapolation.

3.2.2. The derivation of a Gaussian-type distribution by taking the logarithm of the variable provides a further opportunity for analysis of survey data. One such method is to take those values in the upper end of the range which lie outside some statistically defined limits. This can be done in terms of standard deviation of the log-Gaussian distribution and, according to choice and the degree to which the result meets the test of homogeneity, the position of G_{min} may be based on 1σ or 2σ, giving associated probabilities of 84 and 97.5% respectively that all observations obtained will be below the critical group values.

3.2.3. Another method which has recently been proposed[12] uses the statistical extreme value theory of Gumbel[13], the situation to which it was applied being the identification of a critical group for the *Porphyra*/laverbread pathway. The survey data were sub-sampled by random, Monte-Carlo, techniques, taking groups of 10, 20, 30, 40 and 50 observations of laverbread consumption rate. The highest value out of each sub-sample was plotted and the extreme-value frequency distributions produced in this way were examined, separately for adults and children, male and female. A large number of sub-samples were generated by an iterative technique, observations from each pool of consumption rate data being replaced for further use, so that the product of the number and size of the sub-samples far exceeded the number of observations in the consumption rate survey. The median values for each distribution were examined, from which it was found that men had the highest consumption rates and that the lower values found for children were not such as to be likely to lead to a higher degree of exposure after allowance has been made for other factors related to their age. As for women and children the median of the extreme-value distribution for men increased with sample size, but the rate of increase became small for a sample size of more than about 30. For this reason it was proposed that the median value for the extreme-value distribution of groups of 30 men should be taken as the mean value of consumption rate for the critical group. This particular value of sample size was regarded as being consistent with the intentions of the ICRP when referring to critical groups[5].

The appropriate choice of size of sample necessarily involves judgement but with this reservation the method would have been one of the most objective proposed so far, for it is based on a statistical technique which it is claimed is widely used. Its chief practical disadvantage lies in the large amount of survey data that is required. The situation to which it was applied by Beach is ideal in this respect, for there were more than 2500 observations (1206 men, 1067 women and some children). This is a very much larger amount of data than that available from most surveys, so that it is unlikely that the method will often be applicable in practice. In fact it has only been possible to apply it to one other pathway so far, that involving fish consumption in relation to Windscale discharges (section 4.1.2).

4. EXPERIENCE OF APPLICATION OF THE CONCEPT

Although the critical group concept can be applied to the evaluation of any waste disposal operation, the need for the greater refinement and sophistication it offers over the previous simpler approach of selecting an individual varies. In general, the larger the potential dose to individuals in the population the greater will be the need to secure an accurate assessment, whereas when exposure is unusually low a relatively crude estimate will suffice. Since a relatively large body of data is an almost essential prerequisite for successful application of the critical group concept, it is an expensive system to apply, because it incurs a lot of time and effort to secure and collate the data; this additional expense is more easily justified at higher levels of exposure.

In consequence, and though a number of surveys have now been examined in the context of critical group identification, strenuous efforts to identify such groups have only been made for the more important pathways - for Windscale (the laverbread and fish consumption) and for Trawsfynydd (fish consumption). Control of the majority of waste disposals, for which exposure is extremely low, often with contamination below limits of detection, has continued along simple lines involving the selection of an exceptional individual from a representative sample.

4.1. Windscale

Discharges from this site have offered unusual opportunities for critical group identification, because two of the critical pathways involve large groups of people. It has been estimated that at least 26 000 people eat laverbread, whilst the British fish-eating population who receive supplies from the north-east Irish Sea must number as many as 500 000. The third pathway, involving external exposure, concerns so very few people in any significant way that it has not been possible to apply the critical group concept, and control has had to continue on the basis of the individual spending the most time in the more highly contaminated areas.

4.1.1. Laverbread

1967 survey This was our first opportunity to evaluate the practicability and value of the critical group concept, more than 1500 observations being secured for adults as well as nearly 300 for children. The first step was to establish the radiological status of children as opposed to adults. After being satisfied that the consumption rate range observed for children did not place them in a specially

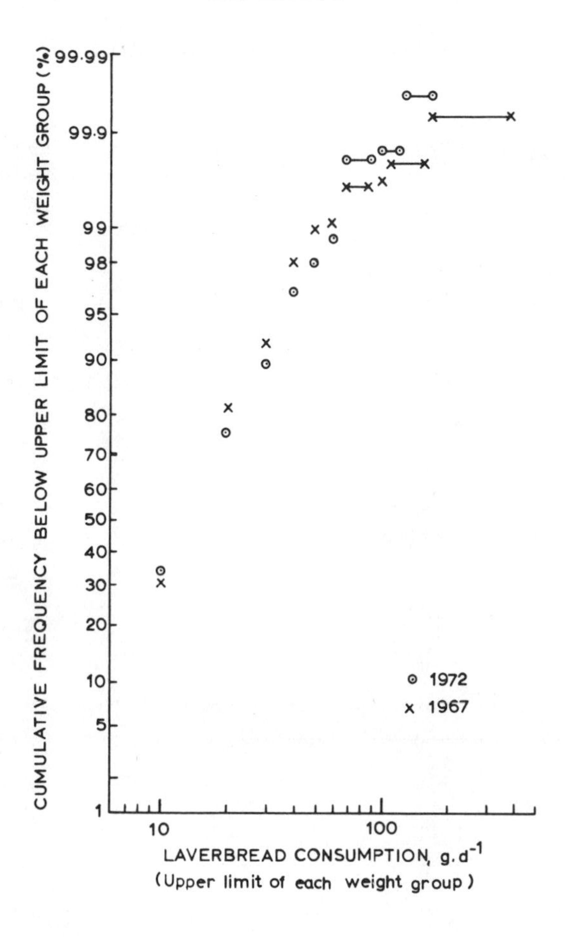

FIG. 2. The distribution of laverbread consumption rates for adults, 1967 and 1972.

important category, and that their exposure was likely to be of lower significance than that of adults, attention was concentrated on adults. Laverbread was found to be eaten by almost as many women as men but at a lower rate. Since there was no evidence to suggest that the radiation sensitivity of the sexes differed significantly in this particular context, the whole of the data for adults was combined for analysis.

The results showed a typically skewed distribution but one in which almost all of the data fitted a log-Gaussian curve, as demonstrated on a log-probability plot and illustrated in Fig. 2. However, a small number of observations, the whole of the upper end of the range and equal to only about 0.5% of the observations, lay well off the main distribution at higher values of consumption than would have been expected from the line fitted to the rest of the data. Precise separation of this sub-group was not possible but the deviation was sufficient for a decision to be made; the observations beyond the maximum value at which log-normality still

TABLE II. DISTRIBUTION OF ADULT CONSUMPTION RATE OF LAVERBREAD, 1972

Mean per capita consumption rate, g. d^{-1}	Frequency, number of observations	Cumulative frequency	Cumulative relative frequency as % of total
0.1- 10	777	777	34.18
10.1- 20	944	1721	75.71
20.1- 30	280	2001	88.03
30.1- 40	191	2192	96.44
40.1- 50	33	2225	97.89
50.1- 60	18	2243	98.68
60.1- 70	25	2268	99.78
70.1- 80	0	2268	99.78
80.1- 90	0	2268	99.78
90.1-100	1	2269	99.82
100.1-110	0	2269	99.82
110.1-120	0	2269	99.82
120.1-130	3	2272	99.96
130.1-140	0	2272	99.96
140.1-150	0	2272	99.96
150.1-160	0	2272	99.96
160.1-170	0	2272	99.96
170.1-180	1	2273	100.0

held were taken to constitute the critical group. The median value within this subgroup was 161.9 g/day, and a rounded-off value of 160 g/day was adopted as the mean consumption rate of the critical group. This treatment of the data gives an 'H' ratio of 2.4 between the maximum and median values, meeting the postulated requirement of reasonable homogeneity.

1972 survey In the repeat survey in 1972 further efforts were made to increase the number of observations, and data were obtained for 2273 adults (1206 men and 1067 women) and 250 children. Adults were confirmed as being more highly exposed than children and their combined distribution is summarized in Table II. Unlike the data from the previous survey a good fit to a log-normal distribution was maintained throughout the sample (Fig. 2), even at the higher consumption rates. This is in part attributed to an apparent increase in the proportion of the population with consumption rates in the 20 to 40 g/day range, and there is no basis for identifying a critical group of exceptional consumers on the basis of a deviation from log-normality.

Visual examination of the data shows frequent observations up to 60-70 g/day, the continuum being broken above this band, with only five observations of unequivocally high consumption rates, though the dividing line between these high-rate consumers and the remainder is somewhat arbitrary and cannot be expressed on a mathematical or statistical basis. They represent only about 0.2% and the range of consumption rates within this sub-group is quite narrow, the 'H' value being 1.3. On a provisional basis while further methods of evaluation were being studied, this group was adopted as the critical group with a median value of 130 g/day.

Most of the other methods discussed in section 3 have now been tested on the data of this survey, and the results are summarized in Table III with their 'H' values alongside. The result of using a straightforward method of examining the log-normal distribution, as described in section 3.2.2, is clearly unacceptable if selection is based on 1σ, because the 'G' value is 4.9. Each of the other systems gives more readily acceptable 'H' values, though that produced from application of the extreme value is rather higher than the value of 3 arbitrarily posed as an acceptable limit.

4.1.2. Fish

The other major internal exposure pathway for liquid discharges from Windscale involves a relatively large number of people and results from the consumption of fish. The north-east Irish Sea to which discharges are made is a productive fishery for many species, notably plaice, cod, whiting, herring and, to a lesser extent, for shellfish. Common experience in situations such as this is that though fish may be eaten by many people, most of these consumers live far from the source of supply and their consumption rates are much lower than those of the nearby coastal communities, and particularly the fishermen.

National dietary statistics show that the average per capita intake of fish is 21 g/day[14]. Against this norm, fish consumption was surveyed along the Cumberland and north Lancashire coasts, with special emphasis on the fishing communities at Whitehaven and Morecambe Bay. The results of this survey are summarized in Table IV.

The methods of analysis used for the laverbread survey were applied to these data and revealed a deviation from a log-Gaussian distribution in which the exceptional consumption rate values are lower than would have been predicted from extrapolation, the reverse of the situation for the 1967 laverbread survey. The results are shown in Table III alongside the analysis of the 1972 laverbread survey.

4.2. Trawsfynydd

The critical pathway for discharges from this power station to the lake of the same name is provided by fish consumption - principally trout though some perch are eaten. The results of a recent consumption survey amongst anglers and

TABLE III. COMPARISON OF CRITICAL GROUP IDENTIFICATION PROCEDURES

Model*	Windscale				Trawsfynydd	
	Laverbread consumption, 1972		Fish consumption		Fish consumption	
	Median consumption rate, $g.d^{-1}$	Homogeneity ratio	Median consumption rate, $g.d^{-1}$	Homogeneity ratio	Median consumption rate, $g.d^{-1}$	Homogeneity ratio
3.1.1.	130	1.3:1	270	3.1:1	115	2.0:1
3.1.2. (e.g. 5%)	48	3.6:1	275	3.0:1	65	3.6:1
3.2.1	No deviation detectable		205	4.1:1	No deviation detectable	
3.2.2., data outside 1σ	35	4.9:1	220	3.8:1	45	5.2:1
data outside 2σ	52	3.3:1	400	2.1:1	74	3.2:1
3.2.3.	55	3.1:1	310	2.7:1	-	-

*The numbers refer to sections within the text.

TABLE IV. DISTRIBUTION OF FISH FLESH CONSUMPTION RATES; NORTH-EAST IRISH SEA COASTAL COMMUNITIES

Mean *per capita* consumption rate, $g.d^{-1}$	Frequency, number of observations	Cumulative frequency	Cumulative relative frequency as % of total
0.1- 10	52	52	10.4
10.1- 20	59	111	22.2
20.1- 30	49	160	32.0
30.1- 40	37	197	39.4
40.1- 50	41	238	47.6
50.1- 60	37	275	55.0
60.1- 70	41	316	63.2
70.1- 80	19	335	67.0
80.1- 90	14	349	69.8
90.1-100	33	382	76.4
100.1-110	17	399	79.8
110.1-120	16	415	83.0
120.1-130	9	424	84.8
130.1-140	11	435	87.0
140.1-150	9	444	88.8
150.1-160	4	448	89.6
160.1-170	9	457	91.4
170.1-180	5	462	92.4
180.1-190	2	464	92.8
190.1-200	3	467	93.4
200.1-210	7	474	94.8
210.1-220	2	476	95.2
220.1-230	1	477	95.4
230.1-240	2	479	95.8
240.1-250	1	480	96.0
250.1-260	4	484	96.8
260.1-270	1	485	97.0
270.1-280	5	490	98.0
280.1-290	0	490	98.0
290.1-300	1	491	98.2
300.1-310	2	493	98.6
310.1-320	0	493	98.6
320.1-330	1	494	98.6
370.1-380	1	495	99.0
410.1-420	1	496	99.2
450.1-460	1	497	99.4
480.1-490	1	498	99.6
500.1-510	1	499	99.8
830.1-840	1	500	100.0

TABLE V. DISTRIBUTION OF TROUT AND PERCH CONSUMPTION RATES; TRAWSFYNYDD, 1973

Mean per capita consumption rate, $g.d^{-1}$	Frequency, number of observations	Cumulative frequency	Cumulative relative frequency as % of total
0.1- 10	93	93	41.33
10.1- 20	52	145	64.44
20.1- 30	44	189	84.00
30.1- 40	14	203	90.22
40.1- 50	5	208	92.44
50.1- 60	2	210	92.33
60.1- 70	12	222	98.67
70.1- 90	0	222	98.67
90.1-100	1	223	99.11
100.1-110	0	223	99.11
110.1-120	1	224	99.56
120.1-230	0	224	99.56
230.1-240	1	225	100.0

their families and friends are summarized in Table V. Because perch are more highly contaminated with the critical radionuclide (caesium-137) than trout, a weighting factor for perch consumption rates has been applied before adding them to data on trout. This was done so that a true total intake of radioactivity could be computed more easily from levels of radioactivity in trout alone; the sum total of 'consumption' so calculated is termed a trout-equivalent value, though, in practice, it does not make a great deal of difference to the result.

The data from 224 observations fit a log-Gaussian distribution and several methods of analysis have been attempted (Table III). Though there are not as many observations for the Windscale pathways just discussed, it is, nevertheless, an unusually high number in our experience, matched only by one other recent set of data, that for the Hartlepools survey for fish/shellfish consumption. The latter survey is particularly interesting in that the data again deviate from a log-Gaussian distribution over the upper end of the range, in a manner similar to the Windscale fish-survey data, i.e. the high values are less than would be predicted by extrapolation.

5. DISCUSSION AND CONCLUSIONS

In practice, the ideal solution to the problem of identifying a critical group, i.e. that it should be completely objective, is difficult to approach. Chief among the reasons for this is the nature of so many of the exposed population groups and particularly their small size, though the basic requirement of homogeneity with respect to those factors which affect the dose received makes for difficulty right from the start.

Our experience of conducting habits/consumption surveys in the UK is that opportunities to amass very large numbers of observations by survey are rare, the examples referred to in the context of Windscale being unique. Even to produce a hundred or two observations is unusual, the number generated in most cases being only a few tens. In such circumstances it is not surprising that the means of selecting a critical group are severely restricted. Application of complex methods such as the extreme-value analysis are ruled out and whilst it may be possible to attempt approaches which require a display of the data in a log-normal fashion, the degree of fit to such a distribution is likely to be poor, making methods dependent on it of poorer precision than where a large body of data is available. The existence of a deviation from log-normality is in any case subject to chance and if the degree of fit is poor it will be difficult to decide the point at which such deviation begins to occur. As seen from Table II, the application of a 1σ limit usually gives an 'H' value which is barely acceptable. Application of 2σ produces a narrower range to the values within the critical group and often gives an 'H' value within the chosen arbitrarily-posed limit. However, its application to surveys where only a few tens of observations are available will be hardly logical, for unless there are at least 80 observations the result will be based on less than two known individuals' habits. In the majority of exposure pathways resulting from the discharge of liquid radioactive wastes in the UK, recourse has therefore had to be made to the use of the exceptional individual in a representative sample of the exposed population.

6. ACKNOWLEDGEMENTS

We are indebted to many of the staff of the Fisheries Radiobiological Laboratory for their support in the surveys which have been referred to and to Mr C. J. Hewett in particular. Our thanks are also due to Mr L. E. Woolner for help in the mathematical and statistical aspects of the paper.

REFERENCES

[1] The Disposal of Radioactive Wastes, Cmnd 884, HMSO, London (1959).

[2] FOSTER, R. F., OPHEL, I. L., PRESTON, A., "Evaluation of human radiation exposure", Ch. 10, Radioactivity in the Marine Environment, National Academy of Sciences, Washington, DC (1971).

[3] PRESTON, A., The United Kingdom approach to the application of ICRP standards to the controlled disposal of radioactive waste resulting from nuclear power programmes, Environmental Aspects of Nuclear Power Stations (Proc. Symp. New York, 1970), IAEA, Vienna (1971) 147.

[4] PRESTON, A., MITCHELL, N. T., The evaluation of public radiation exposure from the controlled marine disposal of radioactive waste (with special reference to UK), Interaction of Radioactive Contaminants with Constituents of the Marine Environment (Proc. Symp. Seattle, 1972), IAEA, Vienna (1973) 575.

[5] ICRP, Recommendations of the International Commission on Radiological Protection (Adopted September 17, 1965), ICRP Publication 9, Pergamon Press, Oxford (1966).

[6] ICRP, Report of Committee II on Permissible Dose for Internal Radiation (1959), Recommendations of the International Commission on Radiological Protection, ICRP Publication 2, Pergamon Press, Oxford (1960).

[7] HAYES, R. L., CARLTON, J. E., BUTLER, W. R. Jr., Radiation dose to the human intestinal tract from internal emitters, Hlth Phys. 9 (1963) 915.

[8] EVE, I. S., A review of the physiology of the gastro-intestinal tract in relation to radiation doses from radioactive materials, Hlth Phys. 12 (1966) 131.

[9] TIPTON, I. H., COOK, M. J., Trace elements in human tissue, Part II Adult subjects from the United States, Hlth Phys. 9 (1963) 103.

[10] PRESTON, A., JEFFERIES, D. F., The ICRP critical group concept in relation to the Windscale sea discharges, Hlth Phys. 16 (1969) 33.

[11] MOSER, C. A., Survey Methods in Social Investigation, Heinemann (1958).

[12] BEACH, S. A., The identification of an homogeneous critical group using statistical extreme-value: Application to laverbread consumers and the Windscale liquid effluent discharges, Nat. Rad. Prot. Board Rep. NRPB R-19 (1974), HMSO, London.

[13] GUMBEL, E. J., Statistics of Extremes, Columbia University Press, New York (1960).

[14] ANON., Household Food Consumption and Expenditure, Annual Rep., National Food Survey Committee, HMSO, London (1973).

IAEA-SM-184/30

DOSE INDIVIDUELLE ET DOSE COLLECTIVE A LA LUMIERE DES RECOMMANDATIONS DE LA CIPR

H. JAMMET, D. MECHALI
CEA, Centre d'études nucléaires
de Fontenay-aux-Roses,
Département de protection,
Fontenay-aux-Roses,
France

Abstract–Résumé

INDIVIDUAL AND COLLECTIVE DOSES IN THE LIGHT OF THE RECOMMENDATIONS OF ICRP.

In activities involving exposure to radiation, protective measures must take into account both the individual and the population as a whole if their objectives are to be achieved. Hence the 'individual dose' and 'collective dose' concepts, the first of them being applied when the risks that might be incurred by individuals have to be kept within acceptable limits and the second when the burden that such activities might impose on the population as a whole is being assessed. The methods for ensuring that dose limits are respected have reached a high degree of perfection. Studies relating to the practical application of the 'collective dose' concept should be pursued and elaborated.

DOSE INDIVIDUELLE ET DOSE COLLECTIVE A LA LUMIERE DES RECOMMANDATIONS DE LA CIPR.

Pour remplir ses objectifs, la protection mise en œuvre dans les activités impliquant une exposition aux rayonnements doit tenir compte de considérations relatives aux individus et de considérations relatives à la société. Les notions de dose individuelle et de dose collective répondent à ces deux aspects, la première permettant de limiter à une valeur acceptable les risques que pourraient courir les individus, la seconde permettant d'apprécier la charge que ces activités pourraient faire peser sur la société. Les méthodes permettant d'assurer le respect des limites de dose sont maintenant bien au point. Les études relatives à l'application pratique du concept de dose collective doivent être poursuivies et développées.

On assiste depuis quelques années à une évolution dans l'importance relative à accorder aux différents critères permettant de définir la protection à mettre en oeuvre dans les activités comportant une exposition aux rayonnements. Alors que pendant longtemps l'accent avait été mis sur les conséquences que l'exposition pouvait avoir sur le plan individuel, on s'est de plus en plus attaché, au cours de ces dernières années, à l'aspect collectif du problème et la CIPR a développé ces considérations dans ses dernières publications. Après avoir rappelé, à la lumière de nos connaissances sur les effets pathologiques des rayonnements, les objectifs que devait se proposer la protection, on verra comment, dans ce contexte, tenir compte des considérations relatives aux individus et de celles relatives à la collectivité et la façon dont ces deux ordres de considérations se combinent et se complètent pour atteindre les objectifs visés.

Lorsque l'on considère les conditions dans lesquelles se manifestent les effets pathologiques des rayonnements ionisants, on peut classer ceux-ci en deux grands types.

Les effets du premier type se manifestent à coup sûr lorsque la dose reçue a atteint ou dépassé une certaine valeur et jamais dans le cas contraire. La valeur de ce seuil d'action varie, pour un même effet, d'un individu à l'autre mais dans des limites relativement étroites et on peut donc, lorsque l'on considère une population irradiée, déterminer pour un effet donné, deux valeurs de la dose, l'une au-dessous de laquelle aucun individu ne sera atteint, l'autre au-delà de laquelle tous les individus de la population subiront l'effet considéré. C'est le cas, par exemple, des lésions de la peau, de la cataracte ou des différentes manifestations du syndrome aigu d'irradiation. Il faut noter également que lorsque l'on considère les effets sur un organe ou un tissu, ils sont d'autant plus sévères que la dose reçue a été plus élevée.

Les conditions dans lesquelles apparaissent les effets du second type sont tout à fait différentes. Lorsqu'une population est irradiée, ces effets n'apparaissent que chez quelques individus et ceci au hasard, d'où leur nom d'effets stochastiques. C'est en particulier le cas des affections malignes, cancers et leucémies et des effets génétiques. Lorsque de tels effets se manifestent leur gravité est totalement indépendante de la dose reçue et ils ne se distinguent en rien des affections de même nature qui apparaissent spontanément. Leur mise en évidence ne peut donc reposer que sur la comparaison statistique des fréquences avec lesquelles ils apparaissent dans deux populations comparables mais dont l'une seulement a subi une irradiation. Les données dont on dispose en ce domaine proviennent d'une part des nombreuses expérimentations effectuées sur l'animal et d'autre part de l'observation de populations humaines soumises à une irradiation, qu'il s'agisse d'irradiation à des fins médicales, d'irradiation professionnelle ou de l'irradiation subie par les populations d'Hiroshima et de Nagasaki. Ces données permettent d'affirmer que l'irradiation à dose élevée peut entraîner l'apparition d'affections malignes et de tares génétiques et montrent que d'une façon générale la fréquence d'apparition de ces affections croît lorsque la dose s'élève. Mais, dans la majeure partie des cas, ces observations concernent des irradiations à dose élevée, souvent délivrées en un temps court, et lorsqu'on veut évaluer les effets résultant de l'irradiation à des doses beaucoup plus faibles telles qu'on les rencontre dans le domaine de la radioprotection, deux problèmes se posent :

- le premier concerne l'existence ou l'absence d'un seuil d'action, analogue à celui mis en évidence pour les effets non stochastiques. En fait, il est pratiquement impossible de résoudre ce problème par l'expérimentation animale ou l'observation humaine, car pour pouvoir juger sur le plan statistique de l'existence ou de l'absence de très faibles modifications de la fréquence d'apparition de ces affections il faudrait faire porter l'étude sur des populations extrêmement grandes ;
- le second concerne l'évaluation quantitative des effets éventuels à faible dose à partir des résultats des observations effectuées chez l'homme à dose élevée et donc la forme de la relation entre la dose reçue et l'augmentation de fréquence de l'affection considérée. Les données dont on dispose ne permettent pas de résoudre ce problème de façon certaine, tant en raison des incertitudes statistiques sur les fréquences observées, les populations étudiées étant limitées, que des incertitudes sur les doses effectivement reçues. Comme l'a indiqué le Comité Scientifique des Nations Unies dans son dernier rapport, les expérimentations animales permettent cependant de penser qu'au moins pour les rayonnements à faible TLE, les relations dose-effet pour les affections observées chez l'homme seraient de forme curvilinéaire. Des considérations théoriques, telles celles développées par Rossi et son école, conduiraient également à adopter des relations polynomiales. On ne peut actuellement faire état de certitude en ce domaine, mais il apparaît que de toutes les formes de relation compatibles avec les

données disponibles la relation linéaire est certainement celle qui conduit aux évaluations les plus prudentes.

En l'absence de certitudes sur ces deux points, la CIPR, considérant que pour fixer des règles de protection il pouvait suffire d'évaluer la limite supérieure du risque, a décidé de fonder ses recommandations sur les hypothèses les plus prudentes, absence de seuil et proportionnalité entre la dose et l'effet.

Si l'on essaie, après ce bref rappel sur les effets pathologiques des rayonnements, de définir les objectifs que doit se fixer la protection, on voit que la question se pose en termes totalement différents selon que l'on considère les effets stochastiques ou les effets non stochastiques. Pour ces derniers, il existe de façon certaine un seuil au-dessous duquel l'irradiation ne provoque aucun effet nocif et il suffit donc, pour assurer une protection absolue, de fixer des règles de protection propres à maintenir l'exposition au-dessous de ce seuil. Pour les effets stochastiques, dès lors que l'on admet qu'il n'existe pas de seuil à l'action des rayonnements et que toute irradiation même très faible peut entraîner une augmentation, si faible soit-elle, de la probabilité d'apparition de certaines affections, il n'est plus possible de viser une protection absolue et on doit donc accepter un certain risque ; le problème consiste alors à limiter ce risque à un niveau acceptable. Le concept de risque acceptable qui apparaissait déjà dans la Publication CIPR 1 a été longuement developpé dans les Publications CIPR 9 et CIPR 22. C'est ainsi que le paragraphe (34) de la Publication CIPR 9 indique que l'on "doit limiter la dose de rayonnement à un niveau tel que le risque encouru puisse être jugé acceptable par l'individu et par la société en raison des avantages qui découlent des activités impliquant une exposition aux rayonnements ionisants". D'autre part, puisque l'on admet que toute exposition peut comporter un certain risque, on doit, comme le recommande la CIPR, réduire toutes les irradiations autant qu'on peut raisonnablement le faire compte tenu des aspects économiques et sociaux. C'est là d'ailleurs une philosophie qui n'est pas particulière au domaine des rayonnements ionisants et qui, explicitement ou implicitement, a servi de guide dans bien d'autres activités humaines, qu'il s'agisse d'activités professionnelles, de la vie courante ou même d'activités de loisirs comme la plupart des sports.

Avant d'examiner les critères sur lesquels juger de l'acceptabilité du risque, il convient de préciser la signification du terme "risque", qui, dans la Publication CIPR 9, est utilisé dans son sens général. Dans la Publication 22, la CIPR a clairement distingué les deux concepts que recouvre ce terme :

- celui de risque qui correspond à la probabilité qu'un individu subisse un effet nocif du fait d'une irradiation. Cette probabilité globale est, en première approximation, égale à la somme des probabilités partielles correspondant aux différents effets possibles ;
- celui de détriment qui tient compte non seulement de la probabilité d'apparition de chacun des effets nocifs possibles mais également de leur gravité. Il est évident que l'apparition d'une tumeur bénigne, par exemple, ne peut être mise sur le même pied que celle d'une tumeur maligne et que même pour les affections malignes leur gravité dépend largement de l'efficacité des thérapeutiques.

Pour juger de l'acceptabilité d'une exposition aux rayonnements, selon la philosophie développée par la CIPR on doit donc comparer, tant du point de vue des individus que de la société les avantages qui découlent de l'activité comportant cette exposition au détriment qui pourrait en résulter.

Dans certains cas, le bilan des avantages et du détriment se pose en termes sensiblement identiques sur le plan individuel ou sur le plan de la société. Il en est ainsi, par exemple, des irradiations à des fins médicales, qu'il s'agisse d'examens diagnostiques ou d'actes thérapeutiques : les avantages retirés de ces actes médicaux et le détriment qui pourrait résulter de l'exposition aux rayonnements concernent les mêmes individus et sont donc distribués de façon identique dans la population.

Mais le plus souvent, et c'est en particulier le cas des activités qui comportent des rejets de substances radioactives dans l'environnement, la distribution du détriment ne coïncide pas avec la distribution des avantages. C'est ainsi, par exemple, que les individus soumis à l'exposition la plus importante du fait de rejets d'effluents ne sont pas forcément ceux qui retirent le plus d'avantages de l'activité considérée. Aussi, d'une façon générale, la balance entre les avantages et le détriment n'a-t-elle vraiment un sens et ne peut-elle se faire que sur le plan collectif. Les termes qui entrent dans le bilan concernent alors les avantages qu'apporte à la société l'activité considérée et le détriment total qui en résulte.

Il en est de même lorsqu'il s'agit non de vérifier par une analyse globale coût-avantages que l'activité envisagée est justifiée mais de s'assurer qu'elle est exercée dans les meilleures conditions de protection, c'est-à-dire à un niveau tel que l'effort que nécessiterait une réduction supplémentaire de l'exposition ne serait pas justifié par la diminution du détriment qui en résulterait : c'est là encore le détriment collectif que l'on considère.

C'est donc par des analyses globales ou différentielles dont l'un des termes concerne le détriment pour la collectivité que l'on pourra, dans chaque cas, déterminer le niveau optimal de la protection à mettre en oeuvre et juger de la justification d'une activité comportant une exposition aux rayonnements.

Mais on ne peut négliger pour autant l'aspect individuel du problème et il n'est raisonnable de se fonder sur le détriment total qu'à condition que le détriment individuel soit, pour tout individu, limité à une valeur que la société juge acceptable. C'est là encore un problème qui n'est pas particulier à la protection contre les rayonnements et qui se pose en des termes analogues dans la plupart des activités humaines. On doit donc définir le niveau jusqu'auquel le détriment individuel peut être considéré comme acceptable. Les doses correspondantes constitueront des limites individuelles, de portée générale, qui ne devront en aucun cas être dépassées.

Comme l'a indiqué la CIPR dans sa publication 9, à propos de la fixation des limites de dose pour les travailleurs et pour les personnes du public, les dangers entraînés par l'exposition ne devraient pas dépasser pour les travailleurs ceux qui sont acceptés dans la plupart des activités industrielles ou scientifiques présentant un niveau de sécurité élevé et pour les personnes du public ceux qui sont habituellement acceptés dans la vie courante. En raison des hypothèses adoptées pour l'évaluation du détriment, les limites de dose ainsi déterminées comportent d'ailleurs très probablement un facteur de sécurité.

Le problème qui se pose en pratique dans chaque cas particulier consiste à définir des mesures de protection propres à garantir que les limites de dose ne seront pas dépassées. Pour évaluer l'exposition des personnes du public résultant du rejet dans l'environnement de substances radioactives, on a pendant un temps fait appel à la notion de concentration maximale admissible dans l'air ou dans l'eau en admettant implicitement que l'exposition des individus était proportionnelle à la concentration des radionucléides dans le milieu où s'effectuait la dispersion initiale et on se contentait de s'assurer que cette concentration maximale était respectée. Mais, en raison de la complexité des voies qui peuvent conduire à une exposition de l'homme, cette approximation est souvent très insuffisante pour évaluer l'irradiation des individus du public. Une méthodologie s'est développée, qui mettant en évidence dans chaque cas les voies qui peuvent conduire à une exposition de la population permet d'évaluer de façon plus sûre l'irradiation correspondante. Elle tient compte de très nombreux facteurs, les uns caractérisant l'environnement les autres caractérisant les populations. Il n'est bien entendu pas possible d'évaluer ainsi l'irradiation de chaque individu de la population, mais on peut généralement définir au sein de celle-ci des groupes de population composés d'individus ayant en ce qui concerne les facteurs qui conditionnent l'exposition (situation géographique, âge, habitudes alimentaires, habitudes professionnelles, mode de vie...) des caractéristiques semblables. A condition que les groupes de population ainsi définis soient suffisamment homogènes, il est raisonnable de considérer l'exposition moyenne à l'intérieur de chaque groupe comme représentative de l'exposition des individus qui le composent et c'est cette exposition moyenne que l'on comparera aux limites de dose pour les personnes du public. En pratique, les études menées dans chaque cas permettent d'établir l'existence d'un ou deux groupes de population dont les caractéristiques sont telles qu'ils seront soumis à une exposition plus élevée que le reste de la population et il suffira, pour garantir le respect des limites de dose, de s'assurer que l'exposition du ou des groupes critiques est inférieure à ces limites.

Il ne faut cependant pas oublier, dans cette analyse, que les limites de dose ne s'appliquent pas à l'irradiation résultant d'une source donnée, mais à l'irradiation totale subie par les personnes du public du fait de l'ensemble des sources, à l'exclusion toutefois de l'irradiation naturelle et de l'irradiation à des fins médicales. Dans le cas où plusieurs sources peuvent contribuer de façon significative à l'exposition et ceci sera probablement de plus en plus fréquent dans l'avenir, la définition du groupe critique et l'évaluation de son irradiation devront tenir compte de ces différentes sources.

Comme nous l'avons vu précédemment, pour juger de la charge sociale que l'exposition aux rayonnements peut faire peser sur une population, il est nécessaire d'évaluer le détriment total c'est-à-dire la somme des détriments individuels. Dans l'hypothèse où les différents risques sont liés à la dose reçue par des relations linéaires, le détriment individuel est directement proportionnel à la dose. On peut alors évaluer le détriment total dans une population en multipliant la somme des doses reçues par les individus qui composent cette population par un coefficient de proportionnalité. La somme des doses individuelles ou dose collective peut également s'exprimer par le produit de la dose individuelle moyenne par le nombre d'individus de la collectivité étudiée. D'une façon plus générale la dose collective s'exprime par l'intégrale $\int H\, N(H)\, dH$ où H représente la dose et N(H) dH le nombre de personnes exposées à des doses comprises entre H et H + dH. La dose collective correspondant au produit d'une dose par un nombre de personnes aura pour unité l'homme-rem.

Comme les doses individuelles qui la composent, la dose collective peut être évaluée pour l'organisme entier ou pour un organe particulier ; elle concerne la dose effectivement reçue ou la dose qui résultera de l'incorporation de substances radioactives.

Lorsque des radionucléides à vie longue sont rejetés dans l'environnement, l'exposition des populations se poursuivra pendant de nombreuses années à un niveau qui variera avec le temps. Pour apprécier le détriment total qui, dans une population, résultera d'un rejet il faudra donc évaluer la dose collective accumulée au cours des années ; mathématiquement celle-ci correspond à l'intégrale par rapport au temps du débit de dose collective dans la population considérée. Ce concept a été développé et utilisé par le Comité Scientifique des Nations Unies sous le nom de "dose engagée". Dans le cas où une installation rejette chaque année dans l'environnement les mêmes quantités de substances radioactives, la dose collective engagée par les rejets effectués en une année est égale à la dose collective annuelle lorsque l'équilibre est atteint dans l'environnement. La même philosophie s'applique d'ailleurs lorsque l'on veut s'assurer du respect des limites de dose et c'est l'exposition annuelle du groupe critique après équilibre dans l'environnement ou la dose engagée résultant du rejet annuel que l'on compare à ces limites.

La dose collective peut être évaluée pour un groupe de population plus ou moins large. Ce peut être, par exemple, dans le cas de l'exposition professionnelle, les travailleurs d'une installation ; dans le cas de l'exposition de la population, il peut être utile d'évaluer la dose collective pour le groupe critique, pour la population d'une région ou celle d'un pays. Quand on applique ce concept à la population mondiale on lui donne le nom de dose population.

Il faut noter à ce propos que la notion de détriment total dans une population est appliquée depuis longtemps dans le domaine des effets génétiques et qu'elle est à la base des recommandations de la CIPR sur la limite de dose génétique.

La validité de l'utilisation du concept de dose collective, tel qu'il vient d'être défini, pour évaluer le détriment pour une population exposée repose sur la validité de l'hypothèse de la linéarité des relations entre la dose et les risques. Dans le cas contraire, en effet, le coefficient de proportionnalité entre la dose et le détriment ne serait pas constant et varierait avec la valeur de la dose et on devrait alors pour apprécier le détriment pour la population multiplier la fraction de la dose collective correspondant à chaque niveau de dose par un coefficient approprié. L'utilisation d'un coefficient de proportionnalité constant, déduit d'observations à dose élevée, conduirait alors à une surestimation du détriment, et celle-ci serait d'autant plus importante que les niveaux de dose seraient plus faibles.

La CIPR a insisté sur les conséquences qui pourraient, dans certains cas, résulter d'une surestimation du détriment. Elle a indiqué que ceci pouvait conduire à remplacer des techniques comportant une exposition aux rayonnements par des techniques en fait plus dangereuses et que, lorsqu'un tel problème se posait, l'estimation du détriment fondée sur des hypothèses délibérément prudentes ne devait être utilisée qu'avec beaucoup de prudence et en ayant conscience de la possibilité que le détriment réel soit bien inférieur à cette estimation. Il en est de même dans les analyses visant à l'optimisation de la protection où la surestimation du détriment différentiel résultant d'une réduction de l'exposition pourrait conduire à consentir un effort qui ne serait pas compensé par le gain social correspondant.

Sous ces réserves, la dose collective traduit, au moins de façon grossière, le détriment qui, pour une population, résulte d'une activité comportant une exposition aux rayonnements ou d'une source d'exposition déterminée. Deux points cependant méritent attention lorsqu'on l'utilise en pratique.

Le premier, qui concerne surtout l'optimisation de la protection, a trait à la façon dont on doit combiner, pour la prise de décisions, les considérations relatives aux doses reçues par les individus et celles relatives à la dose collective. Lorsque les doses individuelles sont proches des limites de dose, on ne peut négliger l'aspect individuel du problème et une réduction des doses permet à la fois de s'assurer que les limites ne seront pas dépassées et de réduire le détriment individuel à un niveau plus faible. Lorsque les doses individuelles sont très inférieures aux limites de dose et faibles devant les variations locales de l'irradiation naturelle, l'aspect individuel du détriment perd toute signification et seul le détriment collectif garde un sens. Il est donc légitime de consentir, pour une même réduction de la dose collective, un effort plus grand dans le premier cas que dans le second et la CIPR a estimé qu'un facteur de l'ordre de 10 pouvait être convenable.

Le second problème qui se pose en pratique lorsque l'on évalue la dose collective ou la dose population concerne les limites de l'intégration. Il est certain qu'en principe, dès lors que l'on admet que le détriment est lié à la dose par une relation linéaire sans seuil, la dose collective doit comprendre toutes les doses individuelles, si faibles soient-elles, et la borne inférieure de l'intégrale doit être égale à zéro. Mais souvent les doses reçues par une fraction de la population sont si faibles que leur contribution à la dose totale est peu significative malgré le grand nombre d'individus concernés. Aussi, en pratique, lorsque les doses individuelles sont très inférieures aux limites de dose, peut-on arrêter l'intégration lorsqu'on a la certitude que cela ne modifierait pas l'évaluation de la dose collective d'un facteur supérieur à deux ou trois, ou encore comme l'indique également la CIPR dans la Publication 22 lorsqu'il est évident que le détriment résiduel est insignifiant par rapport aux avantages apportés par les activités qui sont à l'origine de l'exposition.

Doses individuelles et dose collective constituent donc les deux critères permettant de s'assurer que les activités comportant une exposition aux rayonnements sont exercées dans les meilleures conditions de protection.

Sur le plan pratique, les méthodes permettant de garantir le respect des limites de dose individuelles, en particulier dans le cas de rejets de substances radioactives dans l'environnement, sont maintenant bien au point et une longue expérience a permis de s'assurer de la validité des modèles qui ont été développés pour l'évaluation de l'exposition des groupes critiques.

En ce qui concerne les doses collectives et en particulier la dose population, si le concept est bien clair, son application dans la pratique soulève encore des problèmes tant du point de vue des méthodes d'évaluation car les modèles nécessaires sont bien plus complexes que dans le cas précédent que du point de vue de son utilisation. Aussi, bien que de nombreux travaux aient déjà vu le jour en ce domaine au cours de ces dernières années, les études doivent-elles se poursuivre et se développer.

DISCUSSION

P. RECHT: In paper IAEA-SM-184/10 presented by Mr. N. T. Mitchell we find an example of the application of ICRP Publication 7 to a critical group of laverbread eaters. How can the concepts of collective dose and cost-benefit balance be applied in this case, where the ICRP dose limit is used, and what would be the detriment? If this detriment had been calculated, I wonder whether the British authorities would have fixed release levels at their present values.

D. MECHALI: Publication 7 gives a practical method of ensuring that, in the case of releases to the environment, nobody will be subjected to a dose higher than the limits set for members of the public, limits which are of general application. Cost-benefit or cost-effectiveness analyses, used to determine that the activity is justified and to ensure optimal protection conditions, can be applied not to the critical group but to the population as a whole and must be performed in each case. They are of significance not for individuals but for the collectivity. Although complementary, these two aspects, dose limits and cost-benefit analyses, must be clearly distinguished.

H.P. JAMMET: As a co-author of the paper perhaps I can add that basically the purpose of this new methodology is to avoid confusing discrepancies in the estimation of benefits and disadvantages resulting from the use of radiations and to rationalize these estimations and provide solutions proved by comparable methods.

P. RECHT: Another thing I should like to point out is that there is no ICRP population dose limit for somatic effects, which complicates the assessment of detriment except for genetic effects.

D. BENINSON: I should just like to comment that protection assessments are of two types. First, there are those related to people, in which case dose limits ensure that the risks incurred are acceptable. Secondly, there are those related to the source; here it is necessary to assess the total (hypothetical) impact of the source, for example to compare it with that of alternative sources or procedures. These assessments require the use of collective doses. Of course the collective dose concept is not new; for many years use has been made of population-average doses, obtained through simply dividing the collective dose by the number of people involved.

RADIOLOGICAL SAFETY GUIDES FOR POPULATION IN POLAND: THE PRINCIPLES FOR SETTING AND APPLICATION BY DESIGNERS OF NUCLEAR INSTALLATIONS

J. PEŃSKO
Institute for Nuclear Research,
Radiation Protection Department,
Świerk/Otwock,
Poland

Abstract

RADIOLOGICAL SAFETY GUIDES FOR POPULATION IN POLAND: THE PRINCIPLES FOR SETTING AND APPLICATION BY DESIGNERS OF NUCLEAR INSTALLATIONS.

The paper discusses the bases for setting the radiological safety guides for the population in Poland. The discussion covers the individual and collective dose limits for the population in normal operation of nuclear installations as well as emergency guides for accidental situations. The principles of the application of these guides by designers of nuclear installations in this country are also defined.

1. PROPOSED GUIDES FOR POPULATION EXPOSURE IN POLAND FROM NORMAL OPERATION OF NUCLEAR REACTORS

According to the recommendations of the International Commission for Radiological Protection [1], it is necessary in the event of the radiation exposure of the whole or a large part of the population to consider the resulting consequences not only for an individual but also for all exposed persons. Table I presents the maximum permissible doses for adult individual members of the population in Poland, which have been based on our recent domestic regulations [2].

The corresponding data from international recommendations are given in Table II. One can see immediately that there are slight differences between the Polish standards and the ICRP recommendations, which lie mainly in the age group considered for our population, a somewhat different determination of critical organs and the absence of differentiation of thyroid dose for children and adults in our regulations. The dose values in Tables I and II are based mainly on possible somatic effects in human body due to ionizing radiation.

In the international recommendations [1, 3] a guide for evaluation of the whole population exposure is postulated as well. This dose limit, with possible genetic effects considered, was set at 5 rem for the first 30 years of human life. The following population exposure limits result from these recommendations:

Maximum dose limit for an individual member of the population: 500 mrem/a

Average dose limit for an individual member of the population: 170 mrem/a.

TABLE I. MAXIMUM PERMISSIBLE DOSES OF IONIZING RADIATION FOR ADULT PERSONS WITHIN AND IN THE VICINITY OF NUCLEAR FACILITIES IN POLAND [2]

Organ	Annual dose (rem)
Gonads and bone marrow	0,5
Muscles, fat tissue, liver, spleen, kidneys, G-I tract, lungs, eye lenses and other organs of human body not mentioned in the remaining groups	1,5
Bones, thyroid and skin of the whole body	3,0
Hands, forearms and feet	7,5

TABLE II. THE ANNUAL DOSE LIMITS FOR INDIVIDUAL MEMBERS OF THE PUBLIC ACCORDING TO THE INTERNATIONAL RECOMMENDATIONS [1, 3]

Organ	Annual dose limit (rem)
Whole body, gonads, red bone marrow	0,5
Any single organ excluding the red bone marrow, gonads, bone, thyroid and skin	1,5
Bone, thyroid[a], skin of the whole body (excluding skin of hand, forearms, feet and ankles)	3,0
Hands, forearms, feet and ankles	7,5

[a] The exposure of the thyroid of children below the age of 16 shall be limited to 1,5 rem/a.

In setting these values the International Commission for Radiological Protection has paid special attention to limiting the possibility of genetic damage due to ionizing radiation. Subsequent research has proved, however, that for small doses the genetic risk is somewhat less, while the somatic risk is slightly bigger than had been thought [4]. The considerable quantity of tumours observed among children whose mothers had received radiation doses between 1 and 3 rem before delivery suggests that an individual dose of 0,5 rem/a may not be within the acceptable margin of safety.

TABLE III. PROPOSED BREAKDOWN OF THE AVERAGE INDIVIDUAL DOSE FOR THE POPULATION OF THE USA [5]

Kind of exposure	Dose rate equivalent (mrem/a)
Medical diagnoses	5
Medical therapy	10
Nuclear industry, excl. occupational exposure	10
Other radiation sources available for the population	5
Occupational exposure including employees of nuclear industry	10
Sum	40
Reserve	130
Total	170

With this in mind some authors [5] consider that as far as the somatic effects are concerned the recently set maximum permissible doses may be accepted as long as radiation protection measures in nuclear facilities can ensure real radiation levels much below the maximum permissible values. Morgan, whose view on this issue has been mentioned above, has proposed the breakdown of the average individual dose for the members of population in the USA [5], as quoted in Table III.

Exposures for medical purposes were also included in the average dose breakdown, which was largely because of the need to limit the big contribution of this irradiation to the total exposure of the population in the USA. It is worth noting that a similar situation may exist in many other countries [6]. Presumably medical exposures contribute to the population exposure in Poland, too. As yet, however, no representative evaluation of this has been made in this country.

The plans for electricity production by nuclear power plants in Poland and their realization should proceed together with the necessary breakdown of the average dose limit for individual members of the population in the country. The International Commission for Radiological Protection has stated in its Publication 9 (§ 92) that it is very unlikely that the peaceful use of radioisotope technology and nuclear energy will cause greater genetic exposure of the whole population than the fifth part of the genetic dose limit, which is equivalent to 1 rem per 30 years.

Nuclear power production will probably provide the biggest share of this exposure in the foreseeable future. Selection of the genetic dose limit for the population due to all sources of radiation, excluding medical exposures and natural background radiation, must be carried out separately and, to a certain degree, will be arbitrary for each country. Taking into account the ICRP recommendations [1] it would seem that in our country a dose of 1 rem/30 years is acceptable. This will represent half of the

TABLE IV. DOSE LIMITS FOR 30 YEARS FOR THE WHOLE POPULATION FROM THE LITERATURE

Country or organization	Dose limit[a] (rem/30a)	Reference
ICRP	2,0	[7]
Sweden	1,0	[8]
United Kingdom	1,0	[9]
Czechoslovakia	1,0	[10]
USSR	2,0	[11]

[a] Excluding medical exposure and natural background radiation.

TABLE V. PROPOSED BREAKDOWN OF THE AVERAGE INDIVIDUAL DOSE FOR POLAND

Type of exposure	Dose rate equivalent (mrem/a)
Whole population, excluding groups of occupational exposure The expected breakdown is as follows: 15 mrem/a – reserved for possible accidents in nuclear facilities 8 mrem/a – exposure of population from normal operation of nuclear facilities 5 mrem/a – exposure of population from auxiliary activities for reactor safety (handling and fuel reprocessing, decontamination of effluents etc.) 5 mrem/a – other sources available to population	33
Occupational exposure including employees of nuclear industry	10
Sum	43
Reserve	127
Total	170

exemplary value given by the ICRP in the years 1958-1964 for the breakdown of an average dose limit for the population, which was at the level of 5 rem/30 years, among the different uses of nuclear energy [7] and is consistent with the value accepted so far by various countries (see Table IV).

The proposed value of 1 rem/30 a is equivalent to 33 mrem/a. It seems reasonable to reserve about half of this value for unexpected exposures resulting from uncontrolled accidental releases, which might occur during the realization of a nuclear energy development programme.

Therefore the upper limit of the average exposure of population due to normal operation of nuclear power plant may reach about 15 mrem/a. About half of this value, i.e. 8 mrem/a, should be reserved for the realization of the appropriate programme for the discharge of radioactive wastes into the environment during normal operation of nuclear facilities. A smaller value of about 5 mrem/a should, on the other hand, be reserved for limiting radionuclides released by auxiliary activities such as handling and reprocessing spent fuel, decontamination of radioactive effluents etc. Finally, 5 mrem/a should cover the effects of the application of other radioactive sources, which might be both used in different branches of the national economy and available to the population and, hence, might represent a certain hazard.

According to the foregoing considerations the final breakdown of the average individual dose for members of the whole population in Poland may be accepted. The results are summarized in Table V.

On the basis of the nuclear power development programme in Poland and demographic forecasts an approximate average factor of the installed nuclear power for this country was calculated as 0,0012 MW(e)/man. Bearing this and the above-mentioned average dose limit of 8 mrem/a (see Table V) in mind, an average annual population dose for 1 MW of installed electric power can be calculated to be about 7 man·rem/a per MW(e) and may be recommended as a factor in the design of stationary power reactors in Poland. It should be anticipated, however, that the possible growth of population around nuclear facilities, greater electric energy production from nuclear power plants than predicted and the possible growth of electric energy production per capita after AD 2000 may lead to an increase in the reserved value. For this reason it is recommended to adopt an average population dose of 5 man·rem/a per 1 MW(e) of installed power or a better value of 1,6 man·rem/a per 1 MW of thermal power. The latter is due to the possible development of nuclear heat generating plants, which, as in other countries, is predicted for Poland.

Using the available data, a comparison of postulated population dose burdens from the predicted generation of electricity in nuclear power plants in AD 2000 and the appropriate demographic forecasts for various countries was carried out. The results are given in Table VI and prove that the hitherto proposed exposure limits for the population of Poland under normal operation of nuclear power plants are, projected to AD 2000 and beyond, comparatively well within the safety margin.

2. DESIGN CRITERIA FOR THE ESTABLISHMENT OF CRITICAL AREAS AROUND NUCLEAR FACILITIES IN POLAND

2.1. Design basis for the establishment of critical areas

When designing a nuclear reactor it is necessary to consider the possibility of uncontrolled exposure of the population in its vicinity during an emergency situation. For this reason the Design Basis Accident concept should be adopted, as well as the evaluation of the doses that might

TABLE VI. COMPARISON OF POPULATION DOSE BURDENS FROM THE PREDICTED GENERATION OF ELECTRICITY IN NUCLEAR POWER PLANTS AND DEMOGRAPHIC FORECASTS IN VARIOUS COUNTRIES FOR AD 2000

Country	Predicted number of population ($\times 10^6$)	Power installed in nuclear power plants (MW(e)/person)	Average dose limit (mrem/a)	Population dose limit per 1 MW(e) (man·rem/a)	Radiation burden factor of 10^6 people (man·rem/a·MW(e) $\times 10^6$)	Reference
Czechoslovakia	17	0,002	10	5	0,29	[10]
Poland	38,6	0,0012	8	5[a]	0,13	
Sweden	9	0,01	10	1	0,11	[12]
USA	270	0,003	10	3	0,01	[5, 13]

[a] This value results from the reduction of the calculated original value by a factor of 1,4 for the formation of a margin of safety for future development of the nuclear energy programme.

be received by the population. The possible effects of the doses received should be estimated from both the somatic and the genetic points of view.

In the case of an uncontrolled population exposure during a reactor accident it is not possible to apply the risk-benefit approach, which is of use for normal operation with ionizing radiation. Thus the only possibility is to plan the appropriate emergency measures to prevent and limit the exposure to the population. In planning an emergency action the following should be taken into account:

The range of emergency measures

Cost-benefit analyses.

Because of their impact on construction costs and operation, these problems should be considered at an early stage of the design and site selection of a nuclear facility.

For a postulated accidental release of radionuclides from a nuclear reactor to the environment it is not possible, as for occupational exposure of employees or for the population, to establish the universal values of doses that would cover all the emergency situations and the presumed exceeding of which would justify special precautions or the application of extremely costly engineered safeguards techniques. This is largely because the probability of an accident may in different ways depend on particular design features and random circumstances. On the other hand, the risk associated with special preventive action during an accident and the effectiveness of this action will largely depend on the particular situation during an emergency.

Although it is in practice not possible to set these values, it is clear that the reactor should be designed in the manner that would fulfil the risk-benefit condition. It is also certain that the person in charge of emergency action will have to undertake rapid decisions ranging from the exposure prevention of employees within the plant to radiological protection of the population in its vicinity. These duties might be accomplished with more precision and accuracy if the safety guides for classifying the accidental situation into the appropriate emergency category are set. For this reason Emergency Reference Levels (ERL), discussed later, have been set in our country.

The ERL have been divided into primary and secondary ones. The primary ERL comprise exposure dose and doses absorbed by the whole body or by particular organs. The secondary ERL, which are in practice easier to deal with, comprise cloud dosage, concentration of critical radionuclides in milk and contamination levels of pastures.

2.2. Emergency Reference Levels — primary guides

As a reference point for the institution of critical areas around a nuclear reactor it is necessary to set the reference doses received by critical organs in both external and internal exposure. The proposed values for these primary guides are given in Tables VII and VIII.

If an exposure dose of 15 R is applied as a reference level for the institution of critical areas due to external gamma radiation, it may be assumed that within the radiation energy range of interest the exposure dose of 15 R as measured in air approximately corresponds to that of 10 rads absorbed in bone marrow.

TABLE VII. EMERGENCY REFERENCE LEVELS FOR EXTERNAL EXPOSURE

Radiation	Critical organ	ERL
Gamma	Whole body	15 R in air

TABLE VIII. DOSES TO CRITICAL ORGANS AS EMERGENCY REFERENCE LEVELS FOR INTERNAL EXPOSURE

Radionuclide	Half-life	Critical organ	ERL (rad)
Iodine isotopes	-	Thyroid	25
^{89}Sr	51 d	Bone marrow	10
^{90}Sr	28 a	Bone marrow	10
^{137}Cs	27 a	Whole body	10

TABLE IX. CLOUD DOSAGE AS AN EMERGENCY REFERENCE LEVEL IN INCORPORATION OF CRITICAL RADIONUCLIDES BY THE RESPIRATORY TRACT

Radioisotope	^{131}I	^{137}Cs	^{89}Sr	^{90}Sr
Critical group of population	Children [17]	Adults [17]	Newborn children [17]	Newborn children [17]
Critical organ	Thyroid	Whole body	Bone marrow	Bone marrow
Dose on critical organ as an ERL (rad)	25	10	10	10
Cloud dosage as an ERL in incorporation of critical organ by respiratory tract (Ci·s/m^3	0,03[a]	0,93	0,85	0,05

[a] If the dose to thyroid may be caused also by other radioisotopes of iodine and/or ^{132}Te after their accidental release, the appropriate ERL value for ^{131}I should be reduced by a factor of 2.

TABLE X. ERL FOR CRITICAL RADIONUCLIDES REPRESENTING POPULATION HAZARD VIA THE GASTRIC TRACT

Radioisotope	^{131}I	^{137}Cs	^{89}Sr	^{90}Sr
Critical group of population	Children	Children [17]	Newborn children	Newborn children
Critical organ	Thyroid	Whole body	Bone marrow	Bone marrow
Dose on critical organ as an ERL (rad)	25	10	10	10
Concentration of radioactivity in milk as an ERL ($\mu Ci/m^3$)	250	6700	3500	150
Contamination of pasture as an ERL ($\mu Ci/m^2$)	1,5	22	175	7,5

2.3. Emergency reference levels — secondary guides

The emergency reference levels (secondary guides) are applied in the internal exposure of both the respiratory and gastric tracts. The values were set for absorbed doses for critical organs (see Table II) using the calculation methods given in a previous paper [14] and were based on recent British data [15] calculated for ^{90}Sr and ^{89}Sr. The latter refers to recent data for the retention of ^{90}Sr in human bones plotted against age and to its impact on radiation dose [16]. The proposed cloud dosage, recommended in Poland as ERL for internal exposure due to inhalation of ^{131}I, ^{132}Te, ^{137}Cs, ^{89}Sr and ^{90}Sr, is given in Table IX.

Table X gives the proposed values of critical radionuclide concentrations in milk and the contamination of pastures as the ERL in internal exposure due to consumption of milk contaminated by ^{131}I, ^{137}Cs, ^{89}Sr and ^{90}Sr. Radioisotopes with short half-lives do not affect the proposed values.

For the transition of post-accidentally released ^{131}I, ^{137}Cs, ^{89}Sr and ^{90}Sr from food to the human body the following chain comprising milk consumption has been adopted for our country:

Atmosphere → grass → cow → milk → man.

Within a few days after the post-accidental release of these critical radionuclides to atmosphere a rapid increase in the contamination level of milk from cows pastured in the surrounding area will occur. The contamination level will slowly fall within a period of weeks or months, largely depending on the physical and chemical properties of the contaminating material, as well as on some environmental and biological factors.

3. PRACTICAL APPLICATION OF THE ERL IN REACTOR DESIGN AND INSTITUTION OF CRITICAL AREAS AROUND NUCLEAR INSTALLATIONS

The designers of nuclear facilities in Poland should apply the recommended ERL in the following manner. After assuming the type and range of the design basis accident (DBA) for a given nuclear facility and adopting the most unfavourable meteorological, demographic and agricultural conditions, the cloud dosage and contamination of soil and milk in a given area should be calculated.

As a next step the average individual dose absorbed within contaminated area by the critical organs of an individual of the critical age should be calculated. This may be accomplished by simple proportional relation between secondary and primary ERL. Finally, the critical areas for which the dose and/or contamination levels are within the following ranges should be established:

First Critical Area: 10 ERL ≤ doses and/or contaminations
Second Critical Area: 1 ERL < dose and/or contaminations < 10 ERL

These calculations should be revised for the same postulated reactor accident using available typical data for the actual meteorological and other conditions at the given site.

If necessary, the First Critical Area thus calculated may be regarded as an exclusion area for a given site.

REFERENCES

[1] INTERNATIONAL COMMISSION ON RADIOLOGICAL PROTECTION, Publication 9, Pergamon Press, London (1966).

[2] Zarządzenie Ministerstwa Zdrowia i Opieki Społecznej oraz Pełnomocnika Rządu d/s WEJ z dnia 15 grudnia 1969 r. w sprawie największych dopuszczalnych dawek promieniowania oraz innych wskaźników z zakresu ochrony przed promieniowaniem. Monitor Polski Nr 1 z dnia 13 stycznia 1970 r., poz.7, §9.

[3] INTERNATIONAL ATOMIC ENERGY AGENCY, Basic Safety Standards for Radiation Protection, Safety Series No 9, IAEA, Vienna (1967).

[4] RUSSELL, W.L., Studies in mammalian radiation genetics, Nucleonics 23 1 (1965) 53.

[5] MORGAN, K.Z., "Health physics and the environment", Rapid Methods for Measuring Radioactivity in the Environment (Proc. Symp. Neuherberg, 1971), IAEA, Vienna (1971) 3-21.

[6] UNSCEAR, Report to the UN General Assembly, New York (1972)

[7] INTERNATIONAL COMMISSION ON RADIOLOGICAL PROTECTION, Publication 6, Pergamon Press, London (1964) 31.

[8] BERGSTROM, S.W.O., "Environmental consequences from the normal operation of an urban nuclear power plant", Proc. Fifth Ann. Health Physics Society Mid-Year Topical Symp. (VOILLEQUE, R., Ed.), Eastern Idaho Chapter HPS (1971).

[9] KENNY, A.W., MITCHELL, N.T., "United Kingdom waste-management policy", Management of Low and Intermediate-Level Radioactive Wastes, (Proc. Symp. Aix-en-Provence, 1970), IAEA, Vienna (1970) 69-90.

[10] SEVC, J., KUNZ, E., NAMESTEK, L., "Kriteria radiacni ochrany pri vystavbe jadernych teplaren", Konf. Jadernem Teplarenstvi, Brno, 1973.

[11] Atomn. Energ. 28 6 (1970) 463-67.

[12] HEDGRAN, A., LINDELL, B., On the Swedish Policy with Regard to the Limitation of Radioactive Discharges from Nuclear Power Stations: An Interpretation of Current International Recommendations, Rep. SSI: 1970-027, National Institute of Radiation Protection, Stockholm (1970).

[13] BEIR Report, National Academy of Sciences, Washington (1972) 15.
[14] PEŃSKO, J., NOWICKI, K., Wskaźniki dla oceny radiologicznego zagrożenia otoczenia w wypadkach awarii obiektów jądrowych. PTJ, Seria: Ochrona przed promieniowaniem Nr.70 (547) (1973).
[15] BRYANT, Pamela M., "Developments in the United Kingdom in the derivation of emergency reference levels in environmental materials", Environmental Surveillance Around Nuclear Installations (Proc. Symp. Warsaw, 1973), IAEA, Vienna (in press).
[16] PAPWORTH, D.G., VENNART, J., Retention of ^{90}Sr in human bone at different ages and the resulting radiation doses, Phys. Med. Biol. 18 (1973) 169-86.
[17] BRYANT, Pamela, M., Date for assessment concerning controlled and accidental release of ^{131}I and ^{137}Cs to the atmosphere, Health Phys. 17 (1969) 51.

DISCUSSION

F.D. SOWBY: The current recommendations of ICRP no longer include a breakdown of the genetic dose limit into components corresponding to different sources. The Commission now recommends that each component of the genetic dose should be treated on the basis of risk-benefit analysis. The genetic dose limit should not be regarded as a cake that can be cut up into rations.

J. PEŃSKO: Although we do not yet have a big nuclear industry in Poland, we are already trying to establish general criteria for the designers of nuclear installations to help them with the problems of safety analysis in respect of populations living beyond the fence. At the present time we have no possibility of setting more exact values, based on risk-benefit analysis, but we do need some figures — if only to answer the questions on environmental safety that could be raised by members of the public living in the vicinity of the selected site. For these reasons we think it is justified and useful in our particular case to regard the ICRP genetic dose limit as a cake, part of which could serve as a rough indication for our nuclear designers.

E. KUNZ: I should like to make it clear that there are no official population dose limits for the nuclear industry in Czechoslovakia. The figures, quoted as relating to Czechoslovakia, for genetic dose limit apportionment given in Table IV of your paper and for collective dose given in Table VI are only proposals which have been put forward.

O. ILARI (Chairman): The apportionment of the genetic dose limit in Table V apparently takes no account of the need for establishing population exposure limits that are 'as low as readily achievable', but is derived from the ICRP upper limits. Actually, the doses indicated in Table V are quite high if intended as average values over the whole Polish population, and present technology permits much lower exposure levels to be achieved. The apportionment presented may be considered appropriate as a first approximation to the problem, but a second approximation will undoubtedly be needed.

J. PEŃSKO: You are quite right. The partitioning of the genetic dose limit presented in my paper is only a first approximation of the problem corresponding to Polish conditions because at the moment we have no better data. We are fully conscious of both the present and the future population burden due to nuclear reactors in Poland; this burden will be far below the proposed value of 8 mrem/a, which was adopted only for calculation purposes.

CRITERIOS Y NORMATIVA PARA LA EVALUACION DE LA DOSIS RECIBIDA POR LA POBLACION EN LAS ZONAS DE INFLUENCIA DE LAS INSTALACIONES NUCLEARES

E. IRANZO, F. DIAZ DE LA CRUZ
Junta de Energía Nuclear,
Madrid, Spain

Abstract–Resumen

CRITERIA AND STANDARDS FOR EVALUATING THE DOSE RECEIVED BY THE POPULATION IN AREAS AFFECTED BY NUCLEAR FACILITIES.

The paper sets forth the criteria that should be taken into consideration in determining the exposure levels of the population living in the area affected by a nuclear facility during its normal operation and in deducing the total dose received from external exposure, inhalation of radioactive nuclides released with the gaseous effluents and ingestion of food and water contaminated by gaseous and liquid releases of both types of effluent. Following these criteria, the authors indicate what studies are necessary and what calculation programmes are to be used in order to determine the zones that can be considered critical in the light of the meteorological and topographic characteristics of the area where the nuclear facility is situated, so as to define the critical population groups from the standpoint of inhalation and of individual external exposure, and to determine the total population that would be involved as far as genetic effects are concerned. The paper also gives the standards to be used in determining the critical areas connected with radioactive contamination of foodstuffs produced in the area affected by the facility and its incidence on the internal contamination of the population as a consequence of consumption of such food and of water affected by the direct release of liquid effluents and by the indirect effects of liquid and gaseous effluents. On the basis of these studies and appropriate calculation programmes, the authors show how to determine the most suitable locations for external-exposure measurement apparatus and for samplers of gases, aerosols, foodstuffs and water. A description is also given of the calculation programmes to be used in establishing the total radiation dose received by the critical population groups and all others liable to be affected, to a greater or lesser degree, by the operation of the nuclear facility.

CRITERIOS Y NORMATIVA PARA LA EVALUACION DE LA DOSIS RECIBIDA POR LA POBLACION EN LAS ZONAS DE INFLUENCIA DE LAS INSTALACIONES NUCLEARES.

En la memoria se exponen los criterios que se considera deben tenerse en cuenta para la determinación de los niveles de exposición del público que habita en el área de influencia de una instalación nuclear, como consecuencia de su operación normal, y la deducción de la dosis total recibida a causa de su exposición externa, de la inhalación de los núclidos radiactivos evacuados con los efluentes gaseosos y de la ingestión de los alimentos y agua que resultan contaminados radiactivamente por las descargas gaseosas y líquidas de ambos tipos de efluentes.

De acuerdo con dichos criterios se exponen los estudios a realizar y los programas de cálculo a utilizar para llegar a la deducción de las zonas que podemos considerar críticas, según las características meteorológicas y topográficas de la zona donde se encuentra el emplazamiento de la instalación nuclear, con el fin de deducir los grupos críticos de población desde el punto de vista de la inhalación y de la exposición externa individual y la población total que podría verse implicada en la consideración de los efectos genéticos. Se expone también la normativa para la deducción de las zonas críticas relacionadas con la contaminación radiactiva de los productos alimenticios producidos en el área de influencia de la instalación y su incidencia sobre la contaminación interna de la población como consecuencia de su ingestión, y de la ingestión del agua afectada por la evacuación directa de los efluentes líquidos y la indirecta derivada de éstos y de los gaseosos.

De dichos estudios y con la aplicación de programas de cálculo apropiados se indica cómo se deducen los lugares más idóneos para la colocación de los aparatos que han de medir los valores de exposición externa y los de muestreo de gases, aerosoles, alimentos y aguas, así como los programas de cálculo que se emplean para deducir la dosis total de radiación recibida por los grupos críticos de población y todos los restantes que en mayor o menor grado pueden verse afectados por el funcionamiento de la instalación nuclear.

1. INTRODUCCION

En España, el Ministerio de Industria en colaboración con la Junta de Energía Nuclear y la industria eléctrica nacional, ha definido un plan eléctrico que establece el desarrollo de ésta para un periodo de 10 años. Este plan se revisa cada 2 años de forma que las previsiones para los años siguientes estén siempre actualizadas.

Como consecuencia de dicho plan eléctrico nacional y a partir de las extrapolaciones consideradas válidas en 1.972, se ha previsto la potencia y el porcentaje correspondiente de energía eléctrica de origen nuclear hasta el año 1985, tabla I.

De acuerdo con dicho plan energético y dado el papel preponderante que en él se ha concedido a las centrales nucleares, las diversas compañías eléctricas españolas han iniciado sus previsiones, estableciendo la localización de las que les corresponden en base a la distribución del mercado y de acuerdo con las previsiones establecidas en la planificación industrial prevista en los planes de desarrollo.

Hasta la fecha se han cursado los expedientes de solicitud para la autorización de un total de 24 unidades, ubicadas en 18 emplazamientos, con una potencia total del orden de los 22000 MW(e).

En la tabla II se han incluido las características principales, localización y fecha de explotación prevista de dichas centrales; en la figura 1 se muestra un plano de la localización de las mismas.

La construcción de las primeras centrales nucleares estuvo acompañada por una gran escasez de experiencia propia en relación con las posibles consecuencias derivadas de la evacuación al medio ambiente de los desechos radiactivos producidos durante la explotación. Tras la construcción de la primera central nuclear española, se introdujo el concepto de central de referencia, con las excepciones que impusieran las características propias del emplazamiento. Así, los criterios de los límites de evacuación de desechos radiactivos al medio ambiente de la C.N. de Sta Mª de Garoña son similares a los de la central Dresden II; del mismo modo la C.N. de Vandellós puede ser considerada como una reproducción de la central francesa de Saint-Laurent-des-Eaux (S.L.2.).

Hoy en día se está confeccionando un proyecto de reglamento de acuerdo con la legislación nacional, para que el explotador de cada central nuclear vigile el entorno de la misma al objeto de asegurarse que la población no recibe dosis superiores a unos límites fijados que, además de ser aceptables desde el punto de vista de la protección radiológica, no supongan un porcentaje importante sobre el fondo radiactivo natural.

2. INFLUENCIA DEL EMPLAZAMIENTO EN LAS DOSIS

La experiencia adquirida con la explotación de las tres centrales nucleares actualmente en servicio ha hecho que se preste gran atención a los problemas que la Demografía, Hidrología y Meteorología plantean en cuanto a la estimación de las dosis que la población recibe como consecuencia de la explotación de una central.

TABLA I. POTENCIA ELECTRICA INSTALADA DE ORIGEN NUCLEAR (Revision provisional correspondiente a 1974)

Año	1975	1979	1982	1985
MW(e)	10^3	$7,6 \times 10^3$	$13,7 \times 10^3$	$23,5 \times 10^3$
Porcentaje	4	20	25	33

FIG. 1. Emplazamiento de las centrales nucleares.

2.1. Demografía

La elección de los emplazamientos, ha sido fijada por las necesidades de energía eléctrica previstas para las zonas más industrializadas en la actualidad y las que en un futuro próximo han de alcanzar índices importantes de industrialización; esto implica que, si bien puede presentarse una densidad de población casi nula en su inmediata proximidad, resulte prácticamente imposible situar los emplazamientos en zonas de muy baja densidad de población. Si se observa el mapa expuesto en la figura 2, se puede deducir que la mayor parte de las centrales nucleares se van a situar en zonas que poseen una densidad de población superior a los 30 habitantes por km^2, a excepción

TABLA II. CARACTERISTICAS GENERALES DE LAS CENTRALES NUCLEARES DEL PLAN ELECTRICO NACIONAL

Central nuclear	Tipo de reactor	Potencia [MW(e)]	Fecha explotación	Localización	Agua de refrigeración	Sistema de refrigeración	Zona de exclusión, radio (m)	Toneladas UO_2
José Cabrera	PWR	153	1968	Guadalajara	Río Tajo	Circuito abierto	230	21
Garoña	BWR/4	440	1970	Burgos	Río Ebro	Circuito abierto	320	89
Vandellós	Grafito-gas	480	1972	Tarragona	Mediterráneo	Circuito abierto	250	500[a]
Almaraz I	PWR	930	1977	Cáceres	Río Tajo	Embalse	1 000	80
Lemoniz I	PWR	930	1977	Vizcaya	Cantábrico	Circuito abierto	750	80
Almaraz II	PWR	930	1978	Cáceres	Río Tajo	Embalse	1 000	80
Lemoniz II	PWR	930	1978	Vizcaya	Cantábrico	Circuito abierto	750	80
Ascó I	PWR	930	1978	Tarragona	Mediterráneo	Circuito abierto	750	80
Ascó II	PWR	930	1979	Tarragona	Mediterráneo	Circuito abierto	750	80
Cofrentes	BWR/6	970	1979	Valencia	Río Júcar	Torres		126
Santilán I		1 000	1980	Santander	Cantábrico	Circuito abierto		
Tarifa I		1 000	1981	Cádiz	Atlántico	Circuito abierto		
Aragón		1 000	1981	Zaragoza	Río Ebro			
Cabo Cope		1 000	1981	Murcia	Mediterráneo			
Trillo I		1 000	1982	Guadalajara	Río Tajo			

TABLA II (continuación)

Punta Endala I		1 000	1982	Guipúzcoa	Cantábrico			
Regodola		1 000	1982	Lugo	Cantábrico			
Punta Endala II		1 000	1983	Guipúzcoa	Cantábrico			
Sayago I		1 000	1984	Zamora	Río Duero			
Tarifa II		1 000	1984	Cádiz	Atlántico			
Vergara		1 000	1985	Navarra	Río Ebro			
Asperillo I		1 000	1986	Huelva	Atlántico			
Oguella I		1 000	1987	Vizcaya	Cantábrico			
Oguella II		1 000	1988	Vizcaya	Cantábrico			
Asperillo II		1 000	1988	Huelva	Atlántico			

[a] Aleación U-Mo.

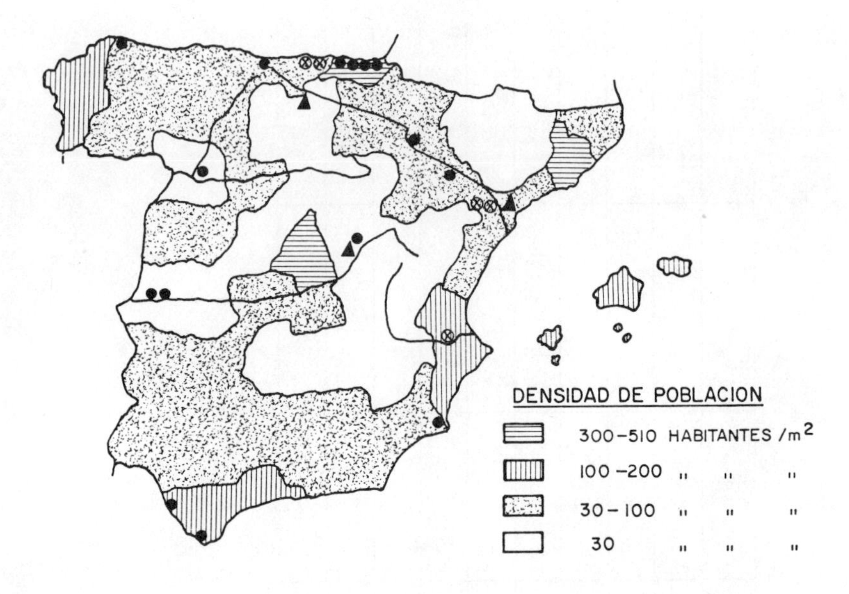

FIG. 2. Distribución demográfica y emplazamiento de las centrales nucleares.

de las que se encuentran en el centro de España, para el abastecimiento de Madrid y su zona industrializada, que tienen una densidad de población del orden de los 13 habitantes por km^2. Algunas se encuentran en zonas con una densidad de población comprendida entre los 300 y 510 habitantes por km^2 y en determinados casos, como el correspondiente a la central de Lemoniz, con dos unidades y una potencia eléctrica conjunta de 1 800 MW(e), está situada a una distancia de unos 20 km de la zona industrial de Bilbao con una población del orden de 600 000 habitantes.

En general, la densidad demográfica en las proximidades de las centrales nucleares (área de 30 km de radio) es del orden de 20 a 30 habitantes por km^2, si bien en las situadas en el litoral no puede eliminarse la posibilidad de que, como ocurre con la central nuclear de Vandellós, se alcance, en ciertas épocas del año, una población turista del orden de 150 000 habitantes en un área anular de radios 10 y 30 km. La figura 3 muestra la distribución de la población alrededor de algunas centrales nucleares.

Los problemas demográficos, planteados ya en la actualidad, no tenderán a mejorar en el futuro ya que los ejes de expansión demográfica, que se relaciona con la industrial y turística, coinciden precisamente con los emplazamientos de las futuras centrales nucleares. Por ello, los estudios sobre la incidencia de los efluentes radiactivos evacuados desde las centrales al medio ambiente, han de preveer que, en el futuro, ha de ser afectada una población mayor que la actual en gran parte de los emplazamientos.

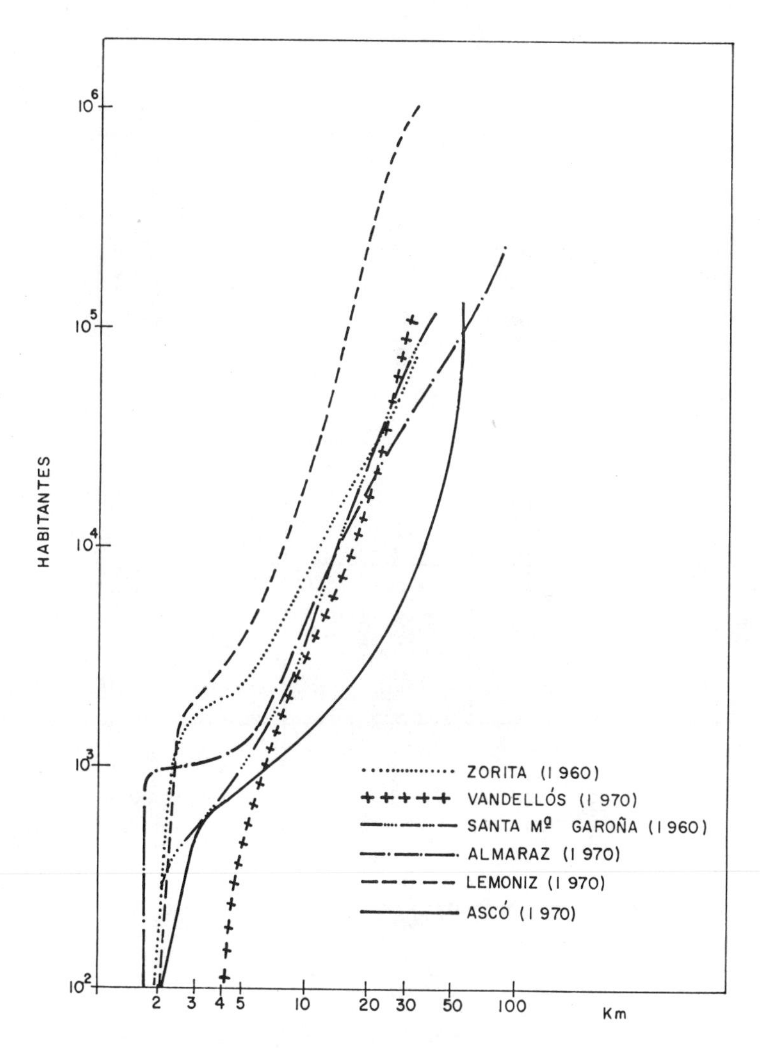

FIG.3. Población en función de la distancia a los emplazamientos de varias centrales nucleares.

2.2. Hidrología

Como se deduce de la tabla II y de la figura 1 el 50% de los emplazamientos actualmente previstos se encuentran situados en el litoral, y el otro 50% repartidos entre las cuencas de los ríos Ebro, Tajo y Júcar. El agua de estas cuencas no puede dedicarse únicamente a la refrigeración, sino que tiene otros usos más prioritarios como son el abastecimiento de poblaciones, zonas industriales y regadíos. Dados los valores expuestos en la tabla III, que corresponde a la programación prevista para el año 2000, el volumen demandado para estos fines supondrá, como mínimo, el 70% de los recursos de la cuenca del Duero, el 60% de la del Tajo, el 80% de la del Ebro y el 100%

TABLA III. APROVECHAMIENTO DE LOS RECURSOS HIDRAULICOS DE LAS CUENCAS UTILIZADAS

Cuenca	Aportación natural ($m^3 \times 10^6$/año)	Recursos disponibles ($m^3 \times 10^6$/año)	Demanda ($m^3 \times 10^6$/año)			Retornos ($m^3 \times 10^6$/año)	Balance ($m^3 \times 10^6$/año)
			Abastecimientos	Regadíos	Total		
Duero	15 860	9 764	270	6 510	6 780	1 520	4 500
Tajo	10 180	7 870	1 610	3 020	4 630	1 860	5 100
Ebro	18 842	12 720	700	9 070	9 770	2 370	5 320
Júcar	3 784	2 400	480	3 370	3 850	-	-1 450

de la del Júcar. Esto demuestra que, en nuestro caso, se acentúan las importantes implicaciones que para los aprovechamientos de las cuencas de dichos ríos tendrá el vertido de residuos radiactivos líquidos de las centrales nucleares en ellas establecidos, por sus incidencias en la dosis a la población a través del agua de suministro a las ciudades y de la cadena alimenticia que tenga su punto de partida, principalmente, en los productos vegetales cultivados mediante el empleo de dichas aguas para regadío.

En lo que respecta a las centrales situadas en el litoral, la explotación muy intensa en nuestro país de los recursos marítimos de la zona vendrán afectados por los vertidos de los residuos radiactivos y en consecuencia, el estudio detallado de los mismos será necesario para la deducción de su implicación cuantitativa en la dosis de diversos grupos de población.

2.3. Meteorología

Las diversas condiciones climatológicas, orográficas y ecológicas de nuestro país, hacen que los emplazamientos de las centrales nucleares sean dispares. Ello hace necesario que se realice un programa de estudios meteorológicos, en cada uno de dichos emplazamientos, al objeto de obtener una indicación de las zonas que más pueden verse afectadas por la contaminación atmosférica de los residuos gaseosos emitidos, y en consecuencia poder estimar las dosis recibidas por la población, mediante esta vía, y los lugares más probables de depósito de las partículas radiactivas.

El programa de estudios no se reducirá simplemente a la obtención de aquellos parámetros que conducen de una forma más o menos teórica a la determinación del valor del coeficiente de dilución en función de la distancia y del sector correspondiente, sino que se llevarán a cabo en cada emplazamiento experiencias que estimen de forma aceptable el comportamiento de los residuos gaseosos al ser sometidos al transporte atmosférico.

Dentro también de este programa, ha de evaluarse el efecto con que el medio (topografía, vegetación, etc...) incide en los valores dados por modelos clásicos de difusión de los coeficientes de dilución.

Para el estudio de los riesgos que trae consigo la explotación de las centrales nucleares y teniendo en cuenta, como se ha indicado en el apartado 2.1., que algunas tendrán en su zona de influencia poblaciones de tamaño tal que se hace necesario la aplicación del concepto de dosis genética a las mismas y que se trata de que la incidencia al respecto sea la menor posible, la medida de dichas dosis es en los momentos actuales prácticamente imposible y, en consecuencia, su cálculo requiere los parámetros locales meteorológicos perfectamente determinados.

De esta forma podremos obtener una idea aproximada, antes de comenzar la explotación de la central, de la manera con que las distintas zonas circundantes a la misma van a quedar afectadas por la contaminación gaseosa, lo que nos permitirá establecer previamente un plan de vigilancia de la radiactividad ambiental, en el que se incluyan los lugares más idóneos para la colocación de los equipos destinados a la medida de los niveles de radiación y contaminación.

3. CRITERIOS ADOPTADOS

Se encuentra en estudio un reglamento sobre vigilancia radioló gica de la población, que habrá de ser aplicado a las instalaciones nu cleares y radiactivas de nuestro país. Los criterios fundamentales en los que este reglamento está basado, en el caso de las centrales nucleares, se pueden considerar divididos en dos grupos: el que incluye aquellos que han de seguirse antes de comenzar la explotación y el que considera los que han de tenerse en cuenta después de comenzada la misma.

Aún cuando no se encuentra publicado todavía dicho Reglamento, para las centrales en funcionamiento y en construcción se exige ya el cumplimiento de los aspectos más fundamentales del mismo.

3.1. Criterios pre-operacionales

Desde la autorización previa hasta que se inician las pruebas pre-nucleares suelen transcurrir en nuestro pais de 4 a 5 años. Durante este periodo de tiempo, en la zona de influencia de la central de berán llevarse a cabo, basándose en las características meteorológicas, topográficas, hidrológicas y demográficas, estudios que conduzcan a la:

- Determinación de los radionúclidos, caminos y grupos de pobla ción que puedan ser considerados críticos como consecuencia de la explotación de la central nuclear.
- Planificación de un sistema de vigilancia de la radiactividad ambiental.
- Determinación de los niveles de fondo de la radiactividad ambiental en los lugares fijados en el plan.
- Estimación de las dosis externas o internas que los grupos crí ticos de población van a soportar.

3.2. Criterios post-operacionales

A lo largo de la vida de la central y fundándose en el plan de vigilancia radiactiva adoptado, los criterios a seguir serán:

- Determinación de los radionúclidos de los residuos radiactivos en sus puntos de emisión.
- Medida de los niveles de exposición a la radiación externa en los lugares más representativos determinados en el plan de vigilan cia de la radiactividad.
- Determinación de los niveles de contaminación de las muestras representativas correspondientes a los eslabones más característi cos de las cadenas o caminos críticos establecidos.
- Conocimiento del estado de contaminación de los individuos de la población crítica.

— Determinación de los niveles de exposición y de las dosis a la población.
— Corrección del plan de vigilancia radiactiva a la luz de la experiencia y resultados que se van obteniendo.

4. NORMATIVA

Con el fin de que pueda cumplimentarse adecuadamente el fin perseguido con los criterios expuestos anteriormente, se establecen normas que tienen por objeto fijar las bases que permitan hacer comparables los resultados obtenidos para los distintos emplazamientos y hacer, en lo posible, rutinario el proceso estimativo de las dosis, que es el objetivo esencial y último del sistema de vigilancia.

4.1. Normas pre-operacionales

Durante el periodo previo a la iniciación de las obras del emplazamiento y en la fase de construcción de la central y para cumplimentar los criterios dados se establece que:

a) Para la determinación de los posibles radionúclidos, caminos y grupos de población críticos, se sigan las normas existentes en las publicaciones al respecto del OIEA y CIPR y las recomendadas por la Administración de algunos países para los que este problema no es nuevo.

b) Respecto a la planificación de un sistema de vigilancia se determinen:
- tipos de detectores de radiación que van a ser empleados de acuerdo con la actividad y clase de radiación que se espera medir con ellos,
- lugares en donde tales detectores van a ser colocados de acuerdo con las características demográficas, hidrológicas, meteorológicas y topográficas de los emplazamientos,
- frecuencia de las lecturas de los detectores,
- proceso numérico al que se someterán las medidas para obtener valores que conduzcan a estimar el grado de irradiación o contaminación.

c) Para la determinación de la radiactividad ambiental se hagan análisis de la actividad estacional en la zona de influencia de la central con respecto a:
- los cultivos que en ella se produzcan,
- los materiales que constituyen la corteza terrestre,
- los animales que viven en la zona o que se alimentan con productos que proceden de la misma,
- el ambiente atmosférico,
- las aguas tanto superficiales como freáticas que puedan de alguna forma contaminarse por los residuos de la central y que sean utilizadas para regadío, bebida, etc.
- la flora, fauna y limos de tales aguas,

— los radionúclidos presentes en el organismo humano, siempre que su determinación sea posible.

d) Para la estimación de las dosis que estos grupos críticos de población van a soportar, las distintas compañías explotadoras pueden emplear los modelos de cálculo que estimen oportunos, pero habrán de utilizar en los mismos los parámetros propios del emplazamiento, siempre que sea posible; en caso contrario, se justificará convenientemente la aplicación de los correspondientes a otros emplazamientos.

4.2. Normas post-operacionales

Las normas que han de tenerse en cuenta durante la explotación de la central para cumplimentar los criterios establecidos en esta fase están relacionadas con:

a) Los análisis de las muestras representativas de los residuos emitidos, en cuanto se refiere a:
— continuidad o periodicidad, según las emisiones,
— determinación de la actividad total y su composición cualitativa o especificación de los radionúclidos cuya composición cuantitativa es necesario determinar; estado físico y químico en que se encuentran los componentes más importantes.

b) Instalación en las zonas más características, según el estudio previo del plan de vigilancia, de los detectores necesarios para estimar los niveles de exposición a la radiación externa correspondientes a cada zona.

c) Análisis periódicos que permitan poner en evidencia, en cada uno de los eslabones más característicos de los caminos o cadenas críticas de exposición, el efecto de los efluentes emitidos por la central en relación con los radionúclidos que, según las circunstancias, se definan como críticos fundamentales, para estimar los niveles de exposición que de ellos se derivan.

d) Vigilancia médica y determinaciones de los niveles de contaminación interna de individuos pertenecientes a los grupos críticos establecidos, con el fin de deducir la incidencia sobre el organismo de los radionúclidos absorbidos.

e) Procesamiento de los datos obtenidos, por el sistema de vigilancia de los niveles de exposición externa y los análisis de las muestras citadas, para comparar los resultados obtenidos con los teóricos calculados y poder deducir las dosis totales recibidas por la población crítica, los riesgos reales derivados del funcionamiento de la Central y la necesidad de introducir modificaciones en los límites de evacuación de efluentes y en los sistemas de vigilancia de la radiactividad.

Aunque los criterios pre-operacionales no lo indican, dado el interés por la influencia sobre el medio ambiente, se incluye la conveniencia de tomar también muestras de elementos que no están incluidas en los caminos críticos, con el fin de proceder a su estudio sistemático y deducir, a lo largo del tiempo, su grado de contaminación y los efectos de la irradiación derivada del funcionamiento de la Central.

4.3. Normas administrativas

A fin de que las autoridades competentes de la Administración, en materia de protección radiológica, estén informadas de los efectos que la explotación de una central tiene sobre el medio ambiente, cada semestre se remiten a la Junta de Energía Nuclear:

- los resultados obtenidos del análisis de las muestras de efluentes tomadas en los puntos de emisión,
- los datos meteorológicos,
- los datos hidrográficos,
- las medidas obtenidas en los distintos aparatos de la red de vigilancia de la radiactividad atmosférica,
- las medidas obtenidas en el análisis de las muestras de flora, fauna, agua y limo,
- las estimaciones de las dosis a la población.

5. PROCESO NUMERICO

Si bien el reglamento en estudio exigirá a los explotadores de las centrales nucleares una estimación de las dosis recibidas por la población, la Administración, a través de la Junta de Energía Nuclear, evaluará de nuevo estas dosis con los datos suministrados por el explotador, completados con otras medidas que estime oportuno hacer.

Actualmente, en la JEN, se dispone fundamentalmente de tres códigos de cálculo en materia de protección radiológica, cuya finalidad es la estimación de las dosis a la población. A continuación hacemos una breve descripción de los mismos.

ESDORA.- Es un código basado en modelos matemáticos clásicos y que calcula, en función de la potencia de funcionamiento del reactor, la cantidad de productos radiactivos que se forman a lo largo del tiempo.

Con los parámetros sacados de las especificaciones técnicas y de diseño de la central se determinan las cantidades, de estos productos radiactivos, que quedan confinados en los edificios de la misma o que se liberan a la atmósfera a través de los distintos sistemas de ventilación.

Sometidos estos productos a la dispersión atmosférica, según los datos meteorológicos dados por la torre meteorológica instalada (o bien empleando los valores de la dispersión horizontal y vertical $\sigma y, \sigma z$ según las categorías de estabilidad de Pasquill), se determinan las concentraciones ambientales en el aire y en el terreno (producida por la precipitación de los radionúclidos sobre el mismo) por sectores y en función de la distancia.

Definido un tipo de individuo en cuanto a su metabolismo y constitución orgánica, el código determina las dosis internas por inhalación de aire contaminado y las dosis externas por inmersión en la nube radiactiva y por efecto de la contaminación superficial del suelo.

Este mismo código nos permite deducir los lugares donde mayor riesgo de irradiación y contaminación puede existir y, en consecuencia, determinar las zonas donde deben colocarse los equipos correspondientes al plan de vigilancia de la radiactividad.

EVARISTO.- Es un código preparado para el cálculo de dosis a través de caminos críticos. Dada la fuente contaminante, sigue a los radionúclidos a través de los distintos eslabones que constituyen el camino crítico, hasta llegar al individuo. En este código, casi todos los datos de las funciones de transferencia del medio a los distintos eslabones del camino crítico, el comportamiento de los mismos en ellos, etc., son bastantes dudosos y esperamos que, con la información de otros autores y nuestro propio esfuerzo, podamos determinarlos de manera aceptable.

Una vez que los radionúclidos llegan, a través del camino crítico, al individuo y se ha definido éste orgánicamente, se determinan las dosis internas a que dan lugar.

CONCORR.- Basado en modelos de difusión en corrientes líquidas, determina, en función de la distancia, caudal, velocidad y coeficientes de dispersión, las concentraciones de los radionúclidos vertidos en la corriente de forma continua o en periodos de corta duración.

Con el valor de la concentración se determinan las dosis internas a la población por ingestión de agua, y las dosis externas que pueden ser recibidas por los que utilizan el agua como medio recreativo (baño, pesca, etc.).

Actualmente se tiene el código en estado de ampliación, a fin de determinar la contaminación producida en la vegetación cultivada que se riega con las aguas de los ríos portadores de efluentes radiactivos y en la flora y fauna de los mismos.

6. CONCLUSION.

Dado el programa nuclear planteado en nuestro país, 18 emplazamientos con 24 unidades y una potencia de aproximadamente 22 000 MW(e) en el periodo de los 10 años próximos, la Administración, a través de la JEN, tomó conciencia de la necesidad de imponer un sistema adecuado de vigilancia alrededor de ellas, para tener una estimación aceptable de las dosis que las poblaciones afectadas pueden recibir y asegurarse de que se mantengan no sólo en valores inferiores a los establecidos como no peligrosos, desde el punto de vista de la protección radiológica, según las recomendaciones de la CIPR y otros organismos internacionales, si no también en valores que contribuyan en un porcentaje pequeño a la elevación de la dosis normalmente recibida a causa del fondo radiactivo natural, de acuerdo con las normas establecidas por otros países más desarrollados y con mayor investigación básica y tecnológica.

DISCUSSION

P. SLIZEWICZ: Could you give a few more details regarding the pre-operational measurements you referred to?

F. DÍAZ DE LA CRUZ: In the oral presentation I could only give the basic considerations underlying the pre-operational measurements for the surveillance of environmental radioactivity. Section 4.1 of the printed paper indicates the instructions which are given to the nuclear power station operator. These instructions relate mainly to natural radiation background measurements and the preparation of a system of radioactivity surveillance, taking into account the demographic, hydrological and meteorological conditions of the site.

IAEA-SM-184/26

PREVISION DES CONSEQUENCES RADIOLOGIQUES DES REJETS NORMAUX D'INSTALLATIONS NUCLEAIRES A L'ECHELLE REGIONALE

Arlette GARNIER, G. LACOURLY
Association Euratom-CEA «Niveaux de pollution du milieu ambiant»,
Centre d'études nucléaires de Fontenay-aux-Roses,
Département de protection,
Fontenay-aux-Roses, France

Abstract–Résumé

PREDICTION OF THE RADIOLOGICAL CONSEQUENCES – ON A REGIONAL SCALE – OF NORMAL WASTE DISPOSAL FROM NUCLEAR FACILITIES.

With the expansion of nuclear power generation and the increasing number of possible sites for new facilities, especially near existing sites or national frontiers, the method for evaluating the radiological consequences of the normal operation of nuclear facilities is bound to need further refinement. The concept of 'dose received by the critical group' is being supplemented by the concept of 'collective dose', which combines the exposures of the different population groups involved and serves as a basis for damage evaluation. The authors examine several criteria, in particular that relating to the extent of the affected zones and the number of persons exposed. Individual and collective doses as a function of distance from the source are compared for several typical situations. The result brings out both the advantages and the limitations of the collective dose concept. The authors then consider the diversity of the characteristics of the environment and the populations liable to be affected directly or indirectly by contamination of the environment or of foodstuffs following the disposal of waste from one or more facilities. By expressing the exposure of a population in terms of collective dose per unit amount of discharged activity, it is possible to delimit zones of relative hazardousness with respect to possible sites. This approach necessitates general studies (such as the compilation of basic data), which would also be useful in evaluating the consequences of non-radioactive discharges and which should be carried out on the basis of regional or international collaboration.

PREVISION DES CONSEQUENCES RADIOLOGIQUES DES REJETS NORMAUX D'INSTALLATIONS NUCLEAIRES A L'ECHELLE REGIONALE.

La méthode d'évaluation des conséquences radiologiques du fonctionnement normal des installations nucléaires est appelée à évoluer, en raison du développement de l'énergie nucléaire et de la multiplication des sites d'implantation possibles des nouvelles installations, notamment à proximité de celles déjà existantes ou de frontières nationales. Au concept de la dose délivrée au groupe critique vient se superposer la notion de dose collective, qui intègre les expositions des différents groupes de population concernés et sert de base à l'évaluation du détriment. Plusieurs critères sont examinés, en premier lieu celui de l'étendue des zones concernées et du nombre de personnes exposées. L'évolution des doses individuelles et collectives en fonction de la distance à la source, dans plusieurs situations types, est comparée. Le résultat montre à la fois l'intérêt et les limites de validité du concept de dose collective. L'examen de ce premier critère est complété par celui de la diversité des caractéristiques de l'environnement et des populations susceptibles d'être exposées directement ou indirectement à la contamination du milieu ambiant ou des denrées alimentaires consécutive aux rejets d'une ou de plusieurs sources. L'expression de l'exposition des populations en termes de dose collective par quantité unitaire d'activité rejetée permet la délimitation de zones de risques relatifs par rapport aux points d'implantation possibles. Elle nécessite des études générales (telles que l'inventaire de données de base), utiles aussi à l'évaluation des conséquences de rejets non radioactifs, et qui devraient être réalisées dans le cadre de collaborations régionales ou internationales.

INTRODUCTION

La méthode d'évaluation des conséquences radiologiques du fonctionnement normal des installations nucléaires est appelée à évoluer, en raison du développement de l'énergie nucléaire et de la multiplication des sites d'implantation possibles des nouvelles installations, notamment à proximité de celles déjà existantes ou de fontières nationales.

L'application de la recommandation «as low as readily achievable» fait qu'au concept de la dose délivrée au groupe critique vient se superposer la notion de dose collective, qui intègre les expositions des différents groupes de population concernés et sert de base à l'évaluation du détriment.

L'examen de critères complémentaires, sur une base régionale plus élargie, apparaît indispensable et, dans certains cas, pourrait impliquer la nécessité d'une coopération internationale au stade des projets.

Un premier critère est l'étendue des zones intéressées et le nombre de personnes exposées. Un autre souvent évoqué est la superposition des rejets provenant de différentes sources plus ou moins proches les unes des autres; on peut y ajouter l'augmentation des quantités émises de radioéléments résultant de la tendance à grouper les installations nucléaires sur un même site; il y a ainsi double superposition : plusieurs sources, plusieurs radionucléides [1]. Un autre encore est la variabilité régionale des conditions de transfert ou d'accumulation des radionucléides dans le milieu et chez l'homme, susceptible de modifier les paramètres d'évaluation des doses à partir des concentrations dans le milieu ambiant. Pour toutes ces raisons il est nécessaire de procéder à l'analyse des conséquences radiologiques du fonctionnement des installations sur une base régionale, selon différents critères.

1. ETENDUE DES ZONES CONCERNEES ET NOMBRE DE PERSONNES EXPOSEES

En cas de rejet d'un produit de fission gazeux par une cheminée de hauteur hypothétique h (m), le calcul permet d'évaluer dans différents cas les concentrations atmosphériques moyennes en divers secteurs de la région environnante, à des distances x du point de rejet et en tenant compte de la dispersion transversale à l'axe du panache. Ces concentrations peuvent être représentées graphiquement, et exprimées en fraction de la concentration maximale, ce qui donne une bonne image des risques individuels relatifs, proportionnels aux concentrations atmosphériques par le jeu d'un certain nombre de paramètres de conversion. Notamment, les doses aux individus susceptibles d'être les plus exposés peuvent ainsi être évaluées. Mais le risque individuel n'est pas le seul à considérer : avec l'éloignement du site les concentrations diminuent mais l'étendue des zones contaminées augmente et, à densité de population égale, le nombre de personnes concernées va croissant [2]. Ce critère d'acceptabilité du risque lié à la dose collective devrait être considéré au départ, tant pour la prévision des rejets concertés résultant du fonctionnement normal des installations que pour celles des rejets continus.

Dans ces différents cas, on peut, à titre d'exemple, tenter une approche très grossière, pour une voie d'atteinte donnée, l'inhalation, concernant des populations de même type et de même densité telles que des populations urbaines, de densité moyenne.

En cas de rejets brefs accidentels ou en cas de rejets concertés, les zones à considérer sont situées dans la même direction, à des distances variables du point de rejet. Soient : C_i la concentration maximale observable dans le lit du vent à la distance x, et $\overline{C}_i$ la concentration moyenne à la même distance, compte tenu de la diffusion latérale.

A une distance x_i, la distance transverse à l'axe du panache le long de laquelle la concentration reste supérieure au 1/10 de la concentration maximale est $2\,\ell$ («largeur 1/10 du panache»). L'aire comprise entre deux courbes successives d'isoconcentration $\overline{C}_i$ et $\overline{C}_{i+1}$ est

$$S_i = \int_{x_i}^{x_{i+1}} 2\,\ell(x)\,dx \tag{1}$$

La concentration moyenne dans cette zone, $\overline{C}$, étant comprise entre $\overline{C}_{i+1}$ et $\overline{C}_i$ ($\overline{C}_{i+1} < \overline{C} < \overline{C}_i$), la dose collective, exprimée en hommes·rems, proportionnelle à la concentration atmosphérique et au nombre de personnes, est

$$R_i = K\,\overline{C}\,S = K \int_{x_i}^{x_{i+1}} \overline{C} \times 2\,\ell\,(x)\,dx \tag{2}$$

En admettant pour simplifier que $\overline{C} = \overline{C}_i$ (ce qui majore $\overline{C}$), il vient

$$R_i = K\,\overline{C}_i\,S_i$$

ou

$$R_i = K\,\overline{C}_i \int_{x_i}^{x_{i+1}} 2\,\ell\,(x)\,dx \tag{3}$$

et le rapport entre les doses collectives dans une zone quelconque et dans la zone de concentration maximale, C_0, toutes choses égales par ailleurs, est approximativement

$$\frac{R_i}{R_0} = \frac{\overline{C}_i}{\overline{C}_0} \cdot \frac{S_i}{S_0} \tag{4}$$

R_0 étant la dose collective au groupe critique.

Il s'agit donc de déterminer, en fonction de leur distance à la source, les aires des zones exposées à des niveaux de contamination donnés, exprimés en fractions de la contamination moyenne dans la zone la plus exposée.

Du point de rejet, la largeur ℓ_i est vue (fig. 1) sous un angle θ_i, tel que

$$\operatorname{tg} \frac{\theta_i}{2} = \frac{\ell_i}{x_i}$$

qui dépend de la durée du rejet, du temps de transport et de la persistance du vent dans la direction considérée, compte tenu du relief et des

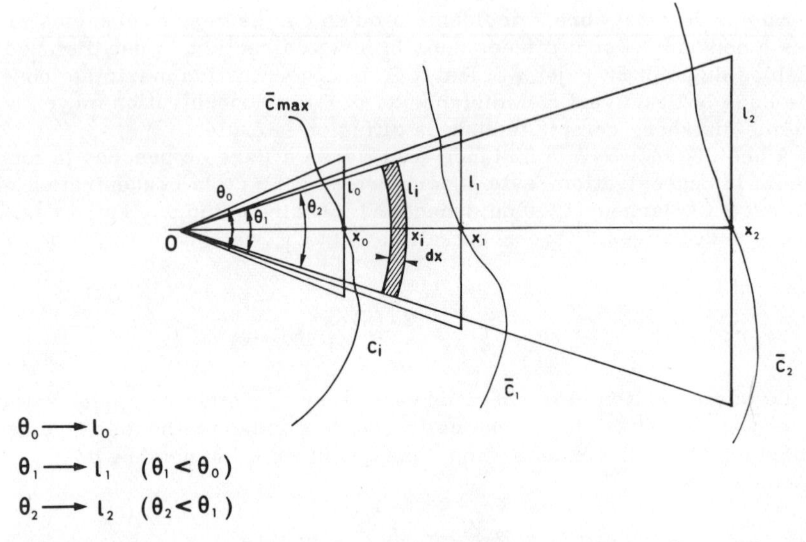

FIG.1. Largeurs des zones exposées à différents niveaux de contamination selon leur distance à la source.

particularités géographiques de la région [3]. On sait que dans des conditions données, cet angle diminue avec la distance. Pour la prévision de rejets prolongés, Beattie [4] a choisi arbitrairement $\theta_p = 30°$ à 100 m et $\theta_p = 15°$ à 10^5 m. Pour des rejets brefs, on peut représenter la variance σ_y^2 par une fonction de puissance de la distance [3]. On peut aussi déduire des courbes expérimentales de Le Quinio [5], établies pour des émissions continues d'une durée de une heure, une relation de la forme $\ell = b\,x^a$ (avec $b = 1{,}05$ et $a = 0{,}84$ pour des valeurs de x comprises entre 400 et 16 000 m). De $a < 1$, il résulte que

$$\mathrm{tg}\,\frac{\theta}{2} = \frac{b}{x^{1-a}}$$

est une fonction décroissante de x.

On peut donc déduire la valeur de ℓ_i de résultats expérimentaux indiquant ℓ ou θ, suivant les cas, en fonction de la distance, et, en donnant à $\ell(x)$ une expression algébrique, résoudre dans l'intervalle où la relation est définie les équations (3) et (4). Avec la relation $\ell = b\,x^a$, la solution est

$$R_i = \frac{2Kb}{a+1} \cdot \overline{C}_i \left[x_{i+1}^{a+1} - x_i^{a+1} \right]$$

d'où

$$\frac{R_i}{R_0} = \frac{\overline{C}_i}{\overline{C}_0} \cdot \frac{x_{i+1}^{a+1} - x_i^{a+1}}{x_B^{a+1} - x_A^{a+1}}$$

la zone de référence entourant le point de concentration maximale (C_0) étant comprise entre les abscisses x_A et x_B (choisies de telle sorte que $\overline{C}_i/\overline{C}_0 \sim C_i/C_{max}$).

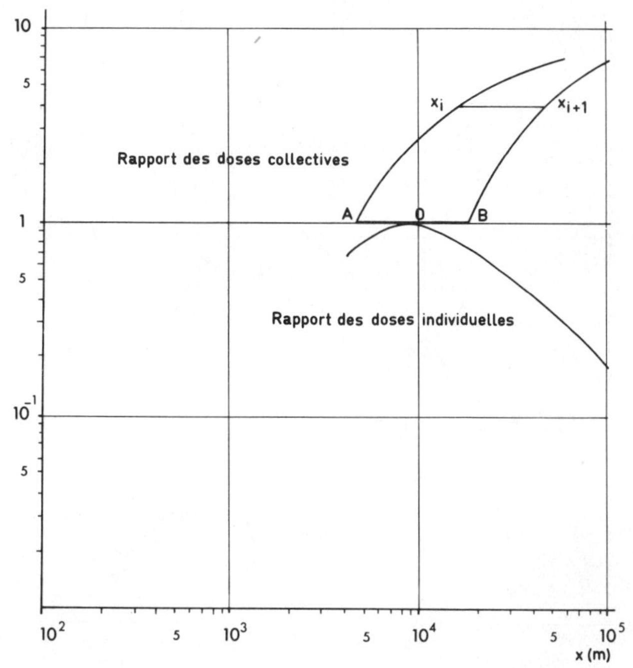

FIG. 2. Variation des doses selon la distance, par rapport à la zone de référence (toutes choses égales par ailleurs) — Rejet bref; condition F; h = 80 m.

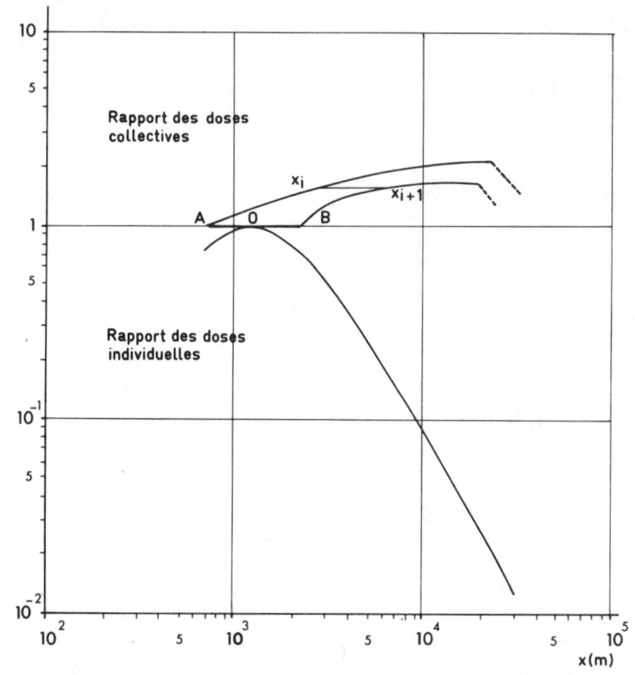

FIG. 3. Variation des doses selon la distance, par rapport à la zone de référence (toutes choses égales par ailleurs) — Rejet prolongé; conditions C-D; h = 80 m.

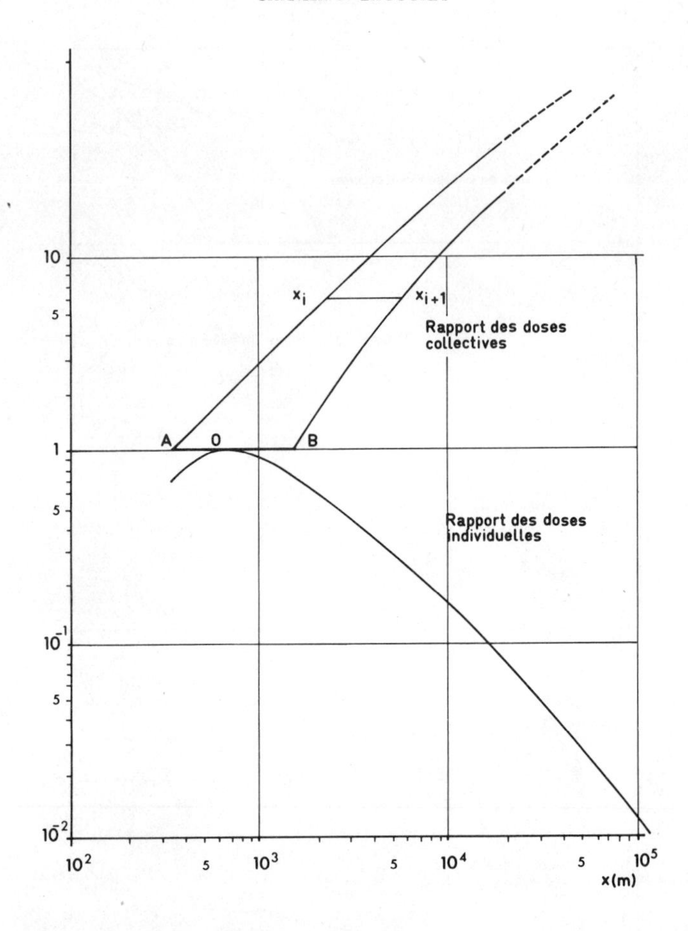

FIG.4. Variation des doses selon la distance, par rapport à la zone de référence (toutes choses égales par ailleurs) — Rejet continu; conditions pondérées; h = 80 m.

On trouve effectivement, comme le montrent la figure 2 pour un rejet bref en condition F (selon la classification de Pasquill), et la figure 3 pour un rejet prolongé en conditions C-D, le résultat suivant : bien que le rapport C_i/C_0 des concentrations atmosphériques au sol dans l'axe du panache décroisse avec la distance, le rapport des doses collectives R_i/R_0 dans le secteur concerné croît à densité de population égale.

En cas de rejets continus, on obtiendrait un résultat analogue avec des écarts encore plus prononcés, comme le montrent les figures 4 et 5, obtenues en se référant à un modèle de diffusion en conditions atmosphériques pondérées [6] et en supposant toujours une densité de population uniforme. Les distances indiquées sont alors mesurées radialement à partir de la source, dans une direction quelconque.

Ces résultats, obtenus dans le cas simple de populations homogènes exposées à une seule voie d'atteinte, l'inhalation, suggèrent une méthode

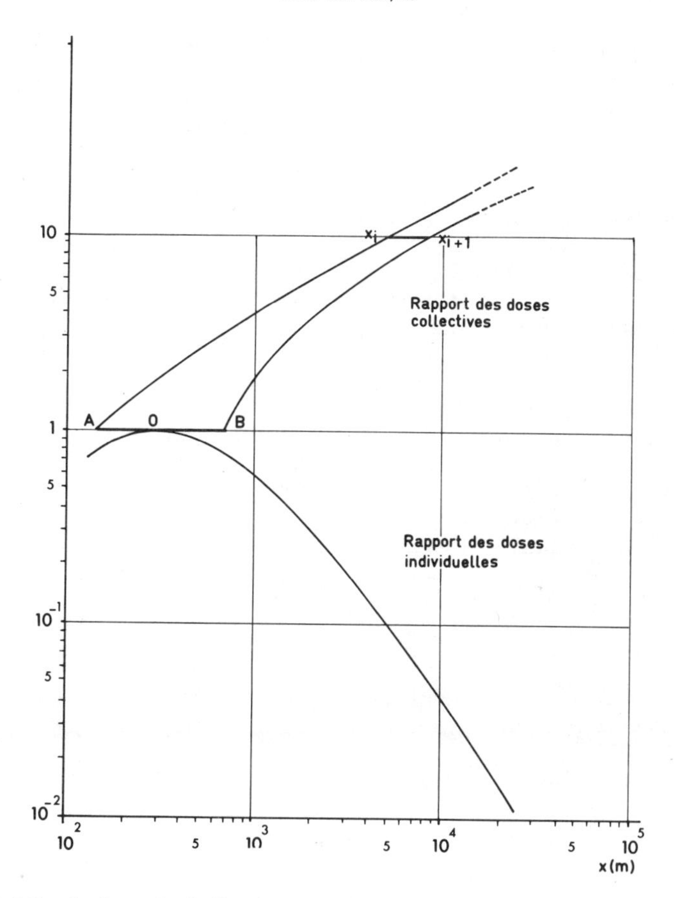

FIG. 5. Variation des doses selon la distance, par rapport à la zone de référence (toutes choses égales par ailleurs) — Rejet continu; conditions pondérées; h = 30 m.

simple permettant d'évaluer la dose collective totale à partir de la dose collective limitée au groupe de population critique en vue d'évaluer le détriment total correspondant. Elle nécessiterait l'évaluation des niveaux de contamination ambiante suivie de l'analyse comparée des diverses voies de rejet, des modes d'exposition et des caractéristiques physiques ou démographiques. Mais avant d'aborder cette analyse il faut souligner que l'évaluation des niveaux d'exposition des populations est subordonnée à celle du niveau de contamination du milieu récepteur : en cas de rejets gazeux, l'évaluation est pratiquement limitée au domaine de validité des modèles de diffusion, dont l'extrapolation au-delà des distances pour lesquelles des données expérimentales ont été obtenues présenterait peu de garanties et aboutirait à des résultats dénués de signification, tant pour les doses individuelles que pour les doses collectives. Cependant un modèle de diffusion à longue distance peut être utilisé pour des rejets de durée limitée [7].

Un schéma analogue serait valable pour les rejets liquides dilués dans des eaux susceptibles d'être utilisées en agriculture.

2. DIVERSITE DES CARACTERISTIQUES PHYSIQUES OU DEMOGRAPHIQUES

Qu'il s'agisse de dose individuelle ou collective, l'évaluation des conséquences radiologiques du fonctionnement des installations nucléaires doit tenir compte des différentes voies de rejet (air, eau, sol) et des différents modes d'exposition (irradiation externe, inhalation, ingestion d'eau et de produits alimentaires).

Si les zones considérées, les unes au voisinage du site, les autres à une certaine distance, présentaient des caractéristiques différentes, le coefficient de proportionnalité K de la relation (2) ne serait plus uniforme. Considérant le risque d'inhalation pour des catégories de population comparables, le rapport K_i/K_0 serait égal au rapport des densités de population. Dans d'autres cas, contamination de la chaîne alimentaire par exemple, il devrait tenir compte des coefficients globaux de transfert de la contamination entre les sources de pollution et l'homme.

Les doses collectives susceptibles d'être reçues en divers secteurs (x, θ) d'une zone circulaire, au centre de laquelle se trouve une source de rejets gazeux continus S, s'expriment par

$$D_N = \sum f_\theta Q K_y \, d_x \left(\frac{D}{\chi}\right) N_{\theta,x} \quad \text{hommes} \cdot \text{rems}$$

avec :

f_θ (sans dimensions) : fréquence du vent soufflant dans la direction θ

Q(Ci/s) : débit de la source

d_x(s/m^3) : coefficient de diffusion dans le lit du vent à la distance x dans les conditions météorologiques considérées

K_y : correction due à la dispersion transversale de part et d'autre de l'axe du panache

$\frac{D}{\chi}$(rem/(Ci/m^3)) : coefficient moyen de conversion des concentrations atmosphériques en doses d'inhalation ou d'ingestion à l'organe de référence; ce coefficient inclut des facteurs tels que la vitesse apparente de dépôt au sol, la fraction de radionucléide retenue par les récoltes, la production agricole, les consommations alimentaires, les paramètre anatomiques et physiologiques de l'homme

$N_{\theta,x}$(hommes) : nombre de personnes exposées par suite de la contaminati du secteur considéré.

<u>En cas de rejets liquides</u>, le calcul de la dose collective peut être effectué selon les mêmes principes, mais le schéma est un peu différent, et plus ou moins complexe selon l'utilisation ultérieure des eaux et le régime des rivières. Par exemple, pour calculer le transfert dans des eau destinées à l'irrigation, il faut se référer au débit moyen du cours d'eau pendant la saison de production : il faut donc connaître le calendrier des

cultures. D'autre part, le coefficient moyen de conversion des concentrations dans l'eau en équivalents de dose à l'homme inclut les facteurs de transfert en milieu aquatique, dont l'amplitude de variation est très importante. Ceci autorise à adopter des hypothèses simplificatrices, par exemple celle d'une diffusion homogène dans la portion du cours d'eau utilisée.

En ce qui concerne les personnes exposées, on peut être amené, selon la nature du risque, à distinguer plusieurs classes d'âges et à appliquer à chacune d'entre elles le coefficient de conversion des activités en doses D/χ approprié (comme dans le cas bien connu de la dose à la thyroïde des nourrissons).

Cette précision mise à part, le nombre de personnes exposées est égal en cas d'inhalation à la population du secteur considéré. En cas d'ingestion des produits alimentaires récoltés dans un secteur soumis à l'influence de rejets gazeux ou liquides, il en est de même si l'on admet l'hypothèse de l'autoconsommation habituellement retenue pour le calcul de la dose à l'individu le plus exposé du groupe critique. Mais on peut aussi, écartant cette hypothèse, calculer le nombre de consommateurs en divisant la production par la consommation individuelle moyenne de chaque produit, et en déduire la contribution de celui-ci à l'irradiation interne d'un groupe de population plus large en raison de la commercialisation et de la distribution des denrées : il se peut que les individus exposés ne soient pas toujours les mêmes, mais que la population susceptible d'être concernée soit celle de la région, du pays, ou d'autres régions de la Communauté européenne, par exemple dans le cas des produits laitiers (contamination atmosphérique) ou des produits irrigués (contamination du milieu aquatique). Quoi qu'il en soit, toute contamination du secteur considéré a pour résultat une certaine augmentation de l'irradiation des populations : l'évaluation du détriment correspondant compléterait l'analyse des risques liés au fonctionnement des installations projetées et pourrait entrer en ligne de compte dans la fixation des limites de rejet. Cette évaluation du détriment complétant celles des doses individuelles au groupe de population critique contribuerait à l'application de la recommandation «as low as readily achievable».

3. MULTIPLICITE DES SOURCES

Un autre cas à envisager est celui où la région considérée se trouve sous l'influence de plusieurs sources.

Rejets atmosphériques

Soient S_1, S_2,... S_i ces sources, de coordonnées polaires θ_i, x_i par rapport au centre de la zone étudiée. La dose collective est alors

$$D_N = N_0\left(\overline{\frac{D}{\chi}}\right)\sum \chi_i$$

N_0 : population de la zone étudiée.

La concentration atmosphérique résultante, pour chaque type de situation météorologique, est

$$\sum \chi_i = \sum f(\theta_i)\, Q_i\, K(y_i)\, d(x_i)$$

$f(\theta_i)$: fréquence du vent venant de la direction i

$d(x_i)$: coefficient de diffusion dans le lit du vent à la distance x_i de la source S_i

Q_i : débit de la source S_i

$K(y_i)$: correction due à la dispersion transversale de part et d'autre de l'axe du panache.

La résolution de ce problème complexe, qui nécessite évidemment une bonne connaissance du climat de la région, présente un intérêt particulier dans les zones frontalières.

Rejets liquides

Dans l'hypothèse d'une dilution homogène dans une portion d'un cours d'eau, l'ensemble de plusieurs sources rapprochées est équivalent à une seule source dont l'activité totale serait la somme des activités respectives, pour chacun des radionucléides identifiés.

CONCLUSION

Les éléments nécessaires à l'analyse prévisionnelle des conséquences radiologiques des rejets gazeux d'installations nucléaires sont, tout d'abord, les données météorologiques, à partir desquelles on peut, moyennant certaines hypothèses sur les implantations possibles, préparer des cartes d'isoconcentrations dans une zone déterminée autour des points où s'effectueraient des rejets temporaires ou prolongés de quantités unitaires. De même, les données hydrologiques de toute nature sont indispensables au calcul de la contamination des différentes parties du milieu aquatique. De telles études, habituellement effectuées par l'exploitant au stade de l'avant-projet, devraient faire partie d'études générales réalisées à l'échelle de régions à délimiter en liaison avec les autres facteurs du milieu : facteurs écologiques influençant les transferts, données démographiques et économiques, etc.

L'expression des risques en termes de doses individuelles ou collectives par quantité unitaire rejetée (rems ou hommes · rems par Ci/an de différents radionucléides) permettrait la délimitation de zones de risques relatifs, par rapport aux points d'implantation possibles. Ces études, réalisées dans le cadre d'une collaboration régionale ou internationale, prépareraient l'examen ultérieur de projets précis intéressant un ou plusieurs Etats. Elles pourraient aussi constituer un point de départ utile à l'évaluation des conséquences de rejets non radioactifs.

REFERENCES

[1] LACOURLY, G., La protection radiologique des individus du public, discussion des problèmes posés — Solutions actuelles — Perspectives futures, Health Phys. 22 (1972) 279-85.

[2] BEATTIE, J.R., Risks to the population and the individual from iodine releases, Nucl.Saf. 8 6 (1967).

[3] WATSON, E.C., SIMPSON, C.L., « Effect of wind variability on environmental consequences of prolonged releases of radioactive contaminants », Containment and Siting of Nuclear Power Plants (C.R. Coll. Vienne, 1967), AIEA, Vienne (1967) 715.

[4] BEATTIE, J.R., An Assessment of Environmental Hazards from Fission Product Releases, AHSB (S) R-64 (1963).

[5] LE QUINIO, R., Evaluation de la diffusion d'effluents gazeux en atmosphère libre à partir d'une source ponctuelle continue — Abaques et commentaires, CEA-R-3945 (1970).

[6] BRYANT, P.M., Methods of Estimation of the Dispersion of Windborne Material and Data to Assist in their Application, AHSB (RP) R-42 (1964).

[7] CAGNETTI, P., PAGLIARI, M., Long Distance Transport and Diffusion of Gaseous Effluents — Considerations and Calculations for a Health Physics Analysis, CNEN-RT/PROT (72) 32 (1972).

DISCUSSION

K.-J. VOGT: With reference to Figs 2 to 5 of your paper, I should like to know whether the cross-wind distribution of the plume — which depends on the diffusion category — has been taken into account in calculating the collective dose.

Arlette GARNIER: Yes. It was taken into account in the algebraic expression giving the "1/10 width of the plume" as a function of distance.

N.T. MITCHELL (Chairman): Can you say anything about mathematical models used for assessing liquid waste releases?

Arlette GARNIER: I have done no evaluations of that kind in the present paper. The theoretical evaluation for gaseous releases was presented only by way of example.

IAEA-SM-184/29

ETUDES COUT- AVANTAGE DANS LE DOMAINE DE LA RADIOPROTECTION
Aspects méthodologiques

G. BRESSON, F. FAGNANI, G. MORLAT
Département de protection,
CEA, Centre d'études nucléaires
de Fontenay-aux-Roses,
France

Abstract–Résumé

COST-BENEFIT STUDIES IN THE FIELD OF RADIATION PROTECTION (METHODOLOGICAL ASPECTS).

Cost-benefit analysis is a method of facilitating the decision-making process, which has been employed with varying degrees of success in connection with a number of economic and social questions. Regardless of where it is employed, it rests on certain general principles and simplifying assumptions. While these principles are relatively clear in the case of a commercial enterprise, it being possible to analyse any decision from the point of view of its financial consequences (for example, its impact on profits), this is not so when one is dealing with a broad question that has implications extending beyond this context. To some extent, radiation protection is such a question. Some radiation protection problems can be treated purely in terms of the unit of production (for example, a nuclear power station). However, if one does not take into account all the interactions between this and other units of production, which are governed by technical, economic and social relationships, one may well end by treating decision-making problems in too narrow a manner. At the general level, when formulating the problem it is scarcely possible to ignore the existence of groups and institutions, which not only are not controlled by the authority responsible for the unit of production but may even be opposed to it; people living near the site, workers employed at the site, etc. Regardless of what is included when formulating the problem, consideration must also be given to how deeply it is intended to go in the analysis of the consequences of possible decisions; one can be more ambitious or less ambitious in describing these consequences and tracing the relations of cause and effect. Different levels of ambition will entail different methods of quantification, which may range from complete evaluation in monetary terms to simple quantification of the consequences in physical terms. The relationships between these methods, which are susceptible of analysis, the difficulties involved and the results that may be expected are discussed in general terms and illustrated by an example: the radiation protection of workers engaged in uranium ore extraction.

ETUDES COUT-AVANTAGE DANS LE DOMAINE DE LA RADIOPROTECTION (ASPECTS METHODOLOGIQUES).

L'analyse coût-avantage constitue une méthode d'aide à la décision qui a été utilisée dans différents domaines économiques et sociaux avec des succès divers. Indépendamment de son lieu d'application, elle s'appuie sur un certain nombre de principes généraux et d'hypothèses simplificatrices. Si ces principes apparaissent relativement clairs lorsqu'on se situe au niveau de l'entreprise, puisque alors toute décision peut être analysée du point de vue de ses conséquences en termes financiers, ou de profit, il n'en n'est plus de même au niveau d'un problème plus large qui déborderait ce cadre institutionnel. La radioprotection apparaît dans une certaine mesure dans cette dernière catégorie. Certains problèmes de radioprotection peuvent être conçus au niveau de l'unité de production elle-même (par exemple une centrale) mais, si on ne tient pas compte de l'ensemble des interactions que cette unité entretient avec d'autres, du fait des dépendances techniques, économiques et sociales, on peut être amené à concevoir de façon trop étroite les problèmes décisionnels. Sur un plan général, il paraît difficile de ne pas inclure dans la formulation du problème l'existence de groupes et d'institutions extérieures qui non seulement ne sont pas contrôlés par l'autorité qui dirige la production mais qui peuvent se placer en opposition à celle-ci : riverains des sites utilisés, travailleurs des unités de production. Indépendamment de l'extension physique de l'analyse au sens où on l'a décrit plus haut, une autre dimension devra être également précisée selon l'approfondissement que l'on veut donner à l'analyse des conséquences des décisions possibles. On peut être plus ou moins ambitieux dans la description de ces conséquences et remonter plus ou moins loin dans leurs réseaux de causalité. A ces différentes ambitions correspondront des méthodes de quantification différentes, pouvant aller de la valorisation complète en termes monétaires à la simple quantification des conséquences en termes physiques. Les relations qui peuvent être analysées entre ces méthodes, leurs difficultés et les résultats qu'on peut en attendre, sont discutés de façon générale et illustrés à propos d'un exemple, celui de la radioprotection des travailleurs dans l'extraction des minerais d'uranium.

1. INTRODUCTION

1.1. La publication n° 9 de la CIPR adoptée le 17 septembre 1965, dans le paragraphe 52, pose le principe d'une amélioration des conditions de la radioprotection dans un sens plus large que celui de la définition d'une norme. «La Commission recommande d'éviter toute exposition inutile et de maintenir toutes les doses aux valeurs les plus faibles auxquelles l'on peut parvenir sans difficulté, compte tenu des aspects sociaux et économiques.» Le recours aux normes constituait jusqu'à présent un moyen universellement admis dans le domaine de la protection. Les normes présentent, en effet, l'avantage d'être d'un usage simple, même si leur définition pratique pose un nombre important de problèmes de toute nature. Or, avec l'interprétation nouvelle des recommandations, l'ambiguïté intrinsèque à la notion de normes et en particulier le caractère conventionnel qu'elles présentent semble réaffirmé. On considère, désormais, que la norme ne représente qu'une limite supérieure à ne pas dépasser, mais qu'il convient de faire «mieux» aussi souvent qu'il est possible. Dans ce nouveau contexte, la tâche des services chargés de faire appliquer ces principes se trouve singulièrement modifiée. En effet, les tâches qu'ils remplissent actuellement : mesures, surveillance, action sur les activités nucléaires en vue de faire respecter les normes précédentes, sont désormais incluses dans un programme plus large. Ce programme va nécessairement déboucher sur un certain nombre d'interrogations concernant les rapports qui vont s'établir entre les contraintes de protection, les nécessités d'ordre économique et les réalités sociales dans lesquelles les décisions en question s'insèrent.

Ces problèmes ne sont pas nouveaux. Ils se posaient précédemment de façon générale au niveau de l'établissement des normes elles-mêmes. Mais on pouvait alors espérer leur trouver une solution satisfaisante dans le cadre d'institutions internationales et de réunions d'experts dont les recommandations présentaient un caractère assez grand de généralité et restaient valables sur d'assez longues périodes de temps. Dans la nouvelle perspective qui est tracée, cette situation est largement transformée. Si en effet, les recommandations précédentes sont appliquées, elles impliquent une modification presque permanente des règles de protection au niveau local où celles-ci seront en vigueur. En relation avec les progrès de la technologie et des connaissances scientifiques d'une part, et d'autre part, selon la spécificité propre à chaque institution et à l'environnement dans lequel elle se situe (spécificité aussi bien sur le plan technique qu'économique et social) on devra redéfinir, cas par cas, les règles de protection dans l'esprit des recommandations précédentes. Il est évident que pour opérer cette adaptation permanente, il devient non seulement nécessaire de pouvoir s'appuyer sur un réseau d'informations très détaillées, mais encore il faut pouvoir disposer de principes d'analyse adéquats en vue d'aborder les multiples questions pratiques qui seront soulevées selon toute vraisemblance.

Le rôle des études et des informations qui permettront d'opérer les choix revêt donc une importance qui ne cessera de s'accroître. Or la nécessité d'une concertation entre les divers spécialistes de ces questions au plan national ou international est apparue depuis fort longtemps. Elle a abouti justement à la définition de politiques sur lesquelles un consensus important a pu s'établir. Pour que ce dialogue puisse exister désormais dans le cadre des nouvelles recommandations qu'on a évoquées, il semble nécessaire qu'une discussion s'engage et que de nouvelles formes de consensus se dessinent au niveau de la conception même avec laquelle peuvent être menées de telles études et de la forme des informations qui peuvent les nourrir. En effet, il serait hautement souhaitable que les résultats puissent en être rendus comparables et que les critères qu'elles utilisent puissent être dans une certaine mesure unifiés. C'est dans cette perspective que l'on peut situer les recommandations de la CIPR qui ont été publiées dans la publication 22 (Implications of Commission Recommendations that Doses be kept as Low as Readily Achievable — Avril 1973) et qui tendent explicitement à préconiser l'utilisation des études de type «coût-avantage» comme modèle général d'analyse.

Les principes sur lesquels sont basées ces études peuvent en effet apparaître satisfaisants pour aborder les problèmes de sécurité nucléaire tels qu'ils se présentent désormais.

Toutefois l'expérience que l'on a de ces études appliquées au domaine de la santé et de la sécurité dans le cadre général de ce qui est appelé en France la rationalisation des choix budgétaires (RCB), inspirée du PPBS (Programming, Projecting, Budgeting System) américain, amène à formuler quelques réserves à propos de cette approche. Les considérations qui vont suivre tenteront d'exposer ces difficultés propres à l'application de ces méthodes à la radioprotection et aboutiront en conclusion à quelques suggestions pratiques. Beaucoup de ces développements apparaîtront à coup sûr pour certains comme de simples querelles doctrinales. Il est important toutefois de se rendre compte que selon la conception générale du problème que l'on adopte, et selon la manière de le formuler, les solutions et les choix peuvent être radicalement différents. Il semble qu'en schématisant fortement la situation, on puisse dire que l'organisation globale de la protection a d'abord été conçue par des spécialistes du domaine nucléaire et médical. Puis devant l'ampleur croissante de la question, avec le développement des différentes formes d'utilisation de l'énergie nucléaire, les économistes ont été mobilisés et se sont penchés à leur tour sur le problème. L'accent qui semble avoir été mis à présent sur les méthodes coût-avantage reflète simplement la prépondérance qui est accordée au point de vue et au mode de pensée de cette nouvelle discipline. L'effort de rationalisation et de clarification qui sous-tend cette démarche est certes éminemment louable et son intérêt ne saurait être mis en doute. Encore faut-il que les représentations et les concepts utilisés soient pertinents par rapport aux réalités sociales et techniques qu'ils décrivent. Or chaque spécialiste a tendance à suivre une pente naturelle aboutissant à ramener à un problème qu'il sait traiter toutes les questions qui lui sont posées, c'est-à-dire à opérer une réduction, une traduction de ces problèmes dans les termes de sa propre discipline.

Ainsi les techniciens de la protection pensent pouvoir résoudre les problèmes, quels qu'ils soient, en réduisant l'exposition des invididus aux risques. Les économistes, quant à eux, prennent en compte le nécessaire bilan des coûts et des bénéfices qui résulte de toute décision et, extrapolant du cadre de l'entreprise où celui-ci a un sens, tentent de prolonger cette logique à la société tout entière. Si un psychologue se penchait sur le problème, il aurait probablement lui-même une autre façon de voir les choses et ainsi de suite. En face de toutes ces tendances, il y a néanmoins des décisions à prendre et qui posent des problèmes. C'est donc à une analyse de la spécificité des problèmes de protection sous l'angle décisionnel que l'on s'attachera, en essayant d'en tirer quelques conclusions quant à la pertinence des études de type coût-avantage.

1.2. Il n'est pas question ici de traiter, en général, de la définition et de la conception des normes dans le domaine de la radioprotection. On peut simplement rappeler quelques caractéristiques particulières de celles-ci qu'il paraît important de retenir dans la perspective du sujet abordé.

On peut envisager d'abord un premier axe de réflexion : par exemple, à propos du processus historique d'émergence et de définition des valeurs limites des risques considérés comme acceptables dans une société donnée et relativement à une activité économique également donnée. Pour aborder une telle analyse, il faut prendre en considération d'une part l'état de développement des connaissances scientifiques concernant les risques et leurs conséquences de toute nature, d'autre part, l'état des différentes forces sociales en présence concernées par ce risque et la façon dont ces forces sociales intériorisent et manifestent, dans leur propre stratégie, les connaissances précédentes. La résultante globale de ce processus historique peut s'analyser comme étant précisément une pratique sociale en matière de sécurité qui, en général, est sanctionnée par une définition quantitative précise appelée norme. Cette norme matérialise en quelque sorte le compromis temporaire qui s'établit entre les différentes forces sociales en présence, compte tenu de la perception que celles-ci possèdent du risque, et de la façon dont elles introduisent cette perception dans leurs objectifs respectifs. Dans son développement historique, il s'agit d'une sorte de processus d'accumulation où la prise de conscience progressive des dangers que présente une activité s'affine et s'enrichit chez ceux qui y sont exposés tandis que, dans le même temps, le développement économique et les surplus qu'il dégage permettent d'affecter une part croissante des ressources aux activités de protection ; c'est ainsi qu'on peut grossièrement expliquer le processus permanent de diminution des normes. Lorsqu'on envisage le problème au niveau d'une unité de production bien précise, à un moment donné, un ensemble complexe de déterminations particulières amène à nuancer largement ce schéma général. Toutefois, ce que l'on suggère d'en retenir, c'est essentiellement le rappel qu'une norme ne saurait être considérée comme le simple résultat d'un calcul technique que l'entreprise pourrait maîtriser, si par exemple elle possédait toutes les informations nécessaires. Une norme s'impose à l'entreprise comme une contrainte qui lui vient de l'extérieur, c'est-à-dire que la sécurité correspond, en général, à une finalité contradictoire avec celle qui régit l'entreprise dans son ensemble.

D'autres traits caractérisent la définition d'une norme à la lumière du schéma précédent : si on retient l'idée qu'il s'agit d'un indice exprimant l'équilibre momentané d'un jeu social et non le résultat d'un calcul rationnel, on peut en déduire quelques conséquences importantes. En effet, dans ce que l'on appelle ce jeu social, de multiples phénomènes de nature non technique entrent en ligne de compte et interfèrent plus ou moins librement avec les données scientifiques elles-mêmes. De plus, celles-ci sont nécessairement inachevées, elles présentent des lacunes et des incertitudes. Entre ce type d'informations et les représentations sociales qui sont attachées aux faits qu'elles décrivent, il peut y avoir des écarts considérables. En définitive, ce ne sont pas forcément les pratiques sociales les plus nocives qui sont perçues forcément comme telles et c'est ce qui explique dans une certaine mesure les différences importantes qui peuvent être notées dans les niveaux de sécurité considérés comme acceptables entre les diverses branches industrielles, ou entre diverses activités sociales. Tout un système symbolique fortement marqué par l'appartenance culturelle des groupes concernés entre en action dont il ne faut négliger ni l'importance ni la profondeur. Le domaine de l'énergie nucléaire constitue de ce point de vue un secteur privilégié pour illustrer ce type de considérations.

Enfin, un dernier élément sur lequel il est nécessaire d'insister, c'est l'hétérogénéité des domaines qu'on doit prendre en considération si on veut analyser le problème des normes. C'est faire une observation banale que de remarquer que le domaine de la sécurité se trouve à l'intersection du biologique, de l'économique et du social. Il semble toutefois qu'une telle constatation comporte des conséquences qui dépassent le simple fait de reconnaître l'existence d'une complexité particulière à ce domaine. Cela signifie en effet que pour analyser une norme, interviennent des connaissances issues d'un ensemble de sciences totalement hétérogènes. Presque tous les problèmes de décision présentent cette caractéristique à un degré plus ou moins élevé. Ce qui arrive souvent cependant, c'est que la valeur explicative fournie par l'une de ces sciences apparaisse dominante en vue de comprendre un phénomène donné. C'est ainsi que par abus de langage on parle de faits économiques ou de faits sociaux, exprimant par là la prépondérance des concepts issus de telles ou telles sciences en vue de décrire certaines réalités phénoménales. Cela n'est pas le cas de la sécurité ; il paraît nécessaire de mobiliser plusieurs disciplines des sciences sociales, économiques et biologiques en vue d'analyser les problèmes qui sont posés, aussi bien pour la connaissance que pour la décision. En ce sens, une analyse qui n'utiliserait qu'une problématique purement économique ou purement biologique ou purement sociologique posséderait une valeur explicative relativement faible.

2. RAPPEL DE QUELQUES DEFINITIONS

2.1. Etudes décisionnelles

On rassemble sous l'appellation générale d'études décisionnelles ou d'aide à la décision les méthodes d'analyse et de représentation des phénomènes naturels ou sociaux, qui sont utilisées dans le but d'éclairer les choix techniques et politiques au niveau de l'entreprise ou de l'administration. Ce type d'études se différencie donc de celles qui sont réalisées dans le cadre de la recherche fondamentale car elles ne visent pas à accroître les connaissances de l'homme ; elles partent au contraire du niveau actuel des connaissances et tentent de l'utiliser au mieux afin de faciliter l'élaboration des décisions. L'ensemble des méthodes qui ont été mises au point dans ce but ont été longtemps réunies sous le terme de «recherche opérationnelle» et ont été appliquées surtout à des problèmes techniques de nature industrielle, économique ou militaire. Ces méthodes ont connu depuis quelques années un développement important et leur usage s'est étendu à l'étude des problèmes socio-politiques.

Toute étude, toute connaissance quelle qu'en soit la nature et la forme, dans la mesure où elle constitue un discours social, peut être considérée comme rentrant de près ou de loin dans un processus de décision. Il existe donc un continuum entre les études décisionnelles proprement dites et les études et informations en général. Toutefois, les études décisionnelles ont pour spécificité d'être, en principe, directement utilisables pour la décision, c'est-à-dire de permettre une réduction de la marge laissée au jugement ou à l'appréciation subjective des décideurs. Il est évident que cette marge d'incertitude qui laisse place au jugement humain ne saurait en aucun cas totalement disparaître, surtout lorsque la décision n'est pas de nature strictement technique, et le maintien d'une telle marge est éminemment souhaitable.

2.2. Etudes coût-avantage et coût-efficacité

Dans leur généralité, les problèmes de décision peuvent tous se réduire formellement à un problème de comparaison selon certains critères entre un ensemble plus ou moins vaste d'alternatives. Représenter une telle situation implique généralement le recours à des moyens mathématiques qui peuvent, selon le cas, rendre compte des diverses caractéristiques du problème envisagé. Toutefois, dans beaucoup de situations concrètes, on ne possède pas une connaissance suffisamment approfondie des différents éléments touchés par la décision pour pouvoir parvenir à une formulation mathématique complète. On en est réduit alors à se contenter d'une analyse plus grossière où la part de l'appréciation subjective et de l'analyse qualitative devient plus importante.

C'est le cas en particulier de l'analyse dite «coût-avantage» (cost-benefit) et de ses dérivées. Elle consiste dans un premier temps à élaborer une liste de programmes, de préférence indépendants les uns des autres, constituant l'ensemble des alternatives possibles par rapport au choix étudié. Elle se propose ensuite de comparer les programmes précédents selon un ensemble de ratios rapportant leur coût à leur efficacité par rapport à un objectif fixé.

Ce qui différencie l'analyse coût-avantage de ses dérivées, c'est essentiellement le degré d'ambition avec lequel on essaie de mesurer les conséquences ou — ce qui revient au même — la nature de l'objectif que l'on fixe aux programmes.

Un exemple simple permettra d'illustrer cette différence.

Soit le cas d'un service de protection attaché à une unité de production industrielle. Supposons que l'on s'intéresse au contrôle d'une source de contamination atmosphérique liée à cette production. Supposons enfin que ce service de protection se pose comme problème d'utiliser le plus efficacement possible l'ensemble des moyens en matériel et en personnel qui sont à sa disposition.

La question immédiate qui se pose alors devant ce problème, c'est celle de savoir ce que signifie l'expression : «utilisation la plus efficace possible de ces moyens». Cette question renvoie à celle de la définition de l'objectif de ce service.

Selon les circonstances, on peut dire par exemple que le but poursuivi par le service, par rapport au problème envisagé, est :

- d'opérer un ensemble de mesures successives de contamination atmosphérique pour connaître les concentrations dégagées (mesure)
- d'avertir les services de production aussitôt qu'une évaluation de la concentration sera telle qu'une norme de sécurité donnée sera atteinte (surveillance)
- de suggérer au service de production des méthodes de travail qui permettent de réduire le débit de sources de pollution (prévention)
- de diminuer les cas de maladies professionnelles liées à l'inhalation de l'air contaminé par les travailleurs (santé publique)
- d'augmenter l'espérance de vie de la population (démographie)
- de maintenir l'intégrité de l'éco-système naturel (écologie)
- etc.

Cette énumération, dont on se rend compte qu'elle pourrait encore se poursuivre, a pour but de suggérer que tout objectif que l'on peut assigner à une activité peut toujours être inclus dans un objectif plus vaste pour lequel il devient alors un moyen.[1] Selon que l'on conviendra de considérer les objectifs et donc les conséquences induites, de façon plus ou moins large, on se trouvera dans le cadre de l'analyse coût-avantage ou de l'analyse coût-efficacité.

Ainsi, en reprenant l'exemple précédent, on sera dans le cadre de l'analyse coût-efficacité, si on se demande comment rendre minimum le coût unitaire d'une mesure de contamination (par exemple). C'est-à-dire que l'on ne s'interrogera absolument pas sur l'utilité ultérieure de cette mesure. L'objectif retenu sera celui de réaliser des mesures et «l'efficacité» sera suivie en termes physiques. Ce sera le nombre de mesures qu'on peut faire pour un coût donné.

Par contre, si on se pose le problème de savoir quelle sorte de combinaison d'activités de protection est susceptible de contribuer au bien-être social de la population, on se trouve dans le cadre de l'analyse coût-avantage; cette contribution étant mesurée en termes monétaires on peut comprendre facilement que si la définition en termes d'avantage présente une certaine élasticité, il en est de même de la notion de coût qui peut, de la même façon, recevoir des contenus plus ou moins ambitieux. C'est ainsi que la littérature portant sur ces questions parle de coût financier, de coût économique, de coût social; on peut dire cependant que ces notions et en particulier la dernière (le coût social) sont loin de recevoir des contenus parfaitement clairs.

On peut remarquer à ce propos que la langue anglaise possède une richesse supérieure au français pour qualifier ces différentes sortes d'efficacité (efficacy, effectiveness, efficiency).

Ainsi, le mot «Effectiveness», qui peut être rendu par le qualificatif «effectif», et le terme «Efficiency», qui peut être rendu par «Efficient», expriment les nuances qu'on a suggérées plus haut du point de vue du caractère plus ou moins large ou final que l'on donne aux objectifs.

Dans la suite de cette communication, on désignera par le terme «coût-avantage» ce que les anglo-saxons appellent «cost-benefit» et par «coût-efficacité» l'ensemble des méthodes «cost-effectiveness» et «cost-efficiency».

---

[1] La représentation graphique sous forme de graphe de cette propriété simple aboutit à ce qui est appelé dans le jargon de la RCB : les graphes objectif-programme.

2.3. Optimisation

On désigne sous ce terme les méthodes mathématiques qui permettent de trouver le maximum ou le minimum d'une fonction dont les coefficients et les variables sont soumis à un ensemble de contraintes.

Ces méthodes formalisent le processus de recherche de la décision qui rend maximum (ou minimum) un certain critère (expression d'un seul ou d'un ensemble d'objectifs), compte tenu de certaines contraintes.

2.4. Prise de décision

Le résumé qui vient d'être fait de différentes méthodes et concepts utilisés dans le cadre de la planification et de la prise de décisions va permettre de revenir sur le problème des normes en radioprotection et de l'utilisation possible des études de type coût-avantage. On a évoqué quelques traits spécifiques à la nature des normes de sécurité selon trois plans : une norme constitue le résultat d'un processus social et historique où s'affrontent des rationalités en conflit. Une norme intègre en ce sens aussi bien des considérations d'ordre scientifique que des représentations sociales dont la rationalité inclut des éléments de nature extra-scientifique.[2] Enfin, même si on se limite aux éléments d'ordre scientifique, ceux-ci sont totalement hétérogènes dans le sens qu'ils appartiennent à des ordres différents de connaissances : épidémiologie et médecine pour certains aspects du détriment, économie politique et gestion, sociologie industrielle, psychologie sociale, physique nucléaire, etc.

Ces difficultés sont propres à l'étude des problèmes de sûreté et de protection : toutefois il ne faudrait pas en exagérer le caractère spécifique car elles se rencontrent dans tout problème de décision. La question que l'on examine à présent est celle de savoir dans quels cas l'analyse de type coût-avantage constitue un moyen adéquat pour, non pas résoudre cette complexité, mais au moins la réduire à un degré acceptable, et selon la réponse à cette première question on envisagera ensuite d'autres démarches possibles.

Ce qu'il est utile d'évoquer avant d'entamer cette discussion ce sont les limites auxquelles s'expose toute démarche d'aide à la décision. Au départ, les problèmes sont complexes et embrouillés ; ils le sont essentiellement parce que l'on n'a pas à sa disposition un ensemble de concepts adaptés à l'objet étudié, c'est-à-dire, constitués en tant que totalité organisée. Cependant, on est contraint d'agir et de trancher malgré les incertitudes.

Il ne faudrait pas croire cependant qu'il existe une panacée, une méthode miracle qui va réduire cette complexité par ses vertus propres et qui va dicter des choix parfaits (rationnels!) à la place des hommes qui sont chargés de le faire.

Si la complexité inhérente au problème est réduite, c'est en partie au prix d'un appauvrissement qui en constitue la contrepartie. Mais à trop appauvrir le jeu des déterminations, on peut aboutir à des représentations méconnaissables, dont la simplicité n'est alors qu'illusoire. L'idéal réside

[2] Ceci ne veut pas dire qu'on ne puisse tenir un discours scientifique à leur propos ; ce discours, simplement, sera celui des sciences sociales et non celui de la biologie et de la physique.

donc en une sorte d'équilibre entre la simplification et le réalisme, et il revient à l'analyste et ensuite au décideur de juger de la pertinence de l'information qu'il utilise ou qu'il produit.

La discussion présentée dans la suite de ce document aura pour objet de démontrer les différentes hypothèses que suppose l'analyse de type coût-avantage utilisée dans un but d'aide à la décision et cela conduit à suggérer un certain nombre de principes généraux qui pourraient servir de base à une conception plus large des études, modulée selon la spécificité des problèmes qu'elles envisagent.

Certains aspects des problèmes de protection qui se posent dans le cas de l'extraction du minerai d'uranium seront utilisés à titre d'illustration.

3. L'APPROCHE DECISIONNELLE : IMPORTANCE D'UNE ETUDE APPROFONDIE

Tels qu'ils sont conçus en général, les problèmes qui apparaissent dans le champ des préoccupations des institutions sont formulés au départ de façon très vague. Dans leur réalité initiale, il n'est souvent pas immédiatement évident de prévoir quelle institution est en mesure de trouver une réponse adéquate à la crise qui se manifeste sous une forme ou une autre. Très souvent, le choix de l'organisation sociale (l'entreprise, un service particulier de celle-ci, une organisation nouvelle créée ad-hoc...) à qui est confié le soin de résoudre le problème en question, sous-entend déjà le choix d'un certain type de solutions techniques. L'analyste qui conçoit les études d'aide à la décision ne peut en aucun cas, comme on le verra à présent, se désintéresser de cette situation initiale. S'il pense par exemple que ce choix de base est déjà pratiquement déterminé et qu'il a peu de chances de pouvoir le remettre en cause, ou qu'il estime que celui-ci s'est fait judicieusement, il sera amené à s'engager dans les procédures habituelles d'analyses du type que l'on évoque et le choix de sa méthode résultera alors de la prise en considération de la rationalité de l'institution précise qui va être en situation de prendre la décision. Si, au contraire, ce choix initial ne lui semble pas encore acquis, il devra plutôt essayer de comprendre la véritable nature du problème qui lui est posé et en déduire le niveau probable auquel, selon lui, il serait préférable de l'aborder. C'est selon les résultats de ces diverses investigations préliminaires qu'une méthode d'analyse peut émerger.

3.1. Les hypothèses implicites sont-elles vérifiées?

3.1.1. Analyse

Les méthodes coût-avantage et coût-efficacité constituent, dans leur généralité, des moyens possibles pour analyser a priori ou a posteriori les conséquences des décisions politiques importantes. Elles peuvent donc constituer la base d'un langage commun pour les différentes institutions dont les décisions impliquent à l'évidence des conséquences sociales qui débordent largement le cadre initial desdites institutions : équipement collectif, localisation industrielle, urbanisme, aménagement, etc.

Ce langage commun se place dans le cadre d'une théorie de la décision fondée sur le principe de l'optimisation.

Toutefois, un tel cadre théorique est nécessairement simplificateur et suppose des hypothèses préalables et des limitations qu'il convient de mettre en évidence. Il faut bien voir que ces limitations peuvent avoir des conséquences importantes aussi bien sur la compréhension des problèmes que sur l'efficacité des décisions qui en résulteraient. Ce cadre théorique n'étant pas le seul possible, on peut être amené à en définir d'autres (faisant appel à la théorie des jeux par exemple).

3.1.2. Choix

Un premier choix s'avère fondamental: celui de l'unité qui va constituer le support de la décision qu'on prend en considération et des cellules sociales touchées par cette décision. Il faut bien voir que ce choix initial ne va pas «de soi», malgré les apparences. Il implique le recours implicite ou explicite à une analyse du système social où se jouent les évènements mis en jeu par le problème étudié. Ce système social peut se décomposer en un certain nombre d'«unités» interdépendantes, hiérarchisées, dotées d'un ensemble de «fonctions». Selon la nature du problème, son importance, son étendue, sa spécificité, des «unités» sociales de ce type doivent être définies et reconnues comme pertinentes. Ainsi pour une décision relative à l'entreprise, on devra se demander si celle-ci relève ou non de la compétence exclusive d'un service particulier, si au contraire elle implique une concertation entre des services de même niveau, si elle relève exclusivement de la direction générale, si elle demande que des contacts avec des institutions extérieures à l'entreprise soient pris, etc. Selon le nombre des «unités» concernées, la nature des liens qui les unissent et la compatibilité de leurs objectifs respectifs, un cadre d'analyse général du problème de décision peut être tracé.

Le cas le plus fréquemment décrit dans la littérature traitant de recherche opérationnelle et d'aide à la décision est celui où on considère qu'un acteur social unique, qu'il s'agisse d'un individu isolé, d'un groupe de personnes, ou d'une institution prise globalement, est en situation de choix devant un certain nombre de solutions alternatives en vue d'atteindre un objectif donné (qui n'est pas forcément unique). Cette situation se prête, en effet, assez facilement à un ensemble de représentations des diverses caractéristiques de cette réalité. C'est ce domaine que l'on appelle, de façon globale, l'optimisation. Les modèles en question ne constituent cependant qu'une approximation, qu'une simplification des phénomènes sociaux qu'ils sont censés représenter et cette réduction ne peut être considérée comme pertinente qu'à condition que le problème de décision en question satisfasse un ensemble d'hypothèses qu'on va énumérer.

3.1.3. Hypothèse fondamentale

L'hypothèse fondamentale est que l'unité en question, c'est-à-dire l'acteur social considéré, soit vraiment à même, par rapport à la décision étudiée, d'imposer ou de mettre en œuvre ses propres critères de rationalité dans le cadre des choix qui sont posés. Or, un ensemble social quel qu'il soit constitue toujours une réalité organique complexe et contradictoire où on peut considérer, a priori, que le réseau des interdépendances reliant les divers éléments qui le constituent forme un tout indissociable. L'hypothèse est donc qu'il est néanmoins légitime, dans certains cas, d'isoler dans l'analyse certains acteurs de ce contexte et que ce ne soit pas là une opération trop artificielle. Dans une telle opération, il ne faut pas croire que l'acteur dont on étudie le comportement soit considéré comme totalement isolé du contexte social car celui-ci s'introduit par le biais de ce qui sera perçu par cet acteur comme des contraintes à la fois techniques et sociales de son environnement. Mais ces contraintes seront considérées dans ce premier modèle comme données et passives. C'est-à-dire qu'on suppose qu'il est légitime de négliger (peut-être, en première approximation) l'effet en retour de la décision qui va être prise sur l'état de ces contraintes elles-mêmes. Cela signifie en dernière analyse que l'acteur en question maîtrise non seulement les divers éléments du choix initial auquel il est confronté, mais encore les diverses conséquences ultérieures de la décision qu'il a prise de façon qu'elles ne remettent pas en question le caractère «optimal» du choix primitif.

Il est bien évident que, dans la pratique, un tel modèle sera d'autant plus pertinent que la décision étudiée sera de portée limitée, déclenchera les réactions du nombre le plus petit possible d'acteurs sociaux; soit qu'il y ait unanimité générale sur l'objectif poursuivi et alors l'ensemble social se réduit à une entité unique pour ce problème, soit que la décision soit tellement infime et restreinte qu'elle n'entraîne l'intérêt de personne, hormis de celui même qui la prend.

3.1.4. Cas des études coût-avantage

Comment situer les études coût-avantage dans cette discussion? Historiquement, on peut dire de façon très schématique que les études coût-avantage sont nées de la constatation suivante : on a reconnu que les choix économiques dont la rationalité pouvait a priori être conçue comme «interne» à l'entreprise qui avait la charge de les mettre en œuvre (par exemple la localisation des unités de production industrielle, l'implantation spatiale et la nature d'un réseau de transport, le choix de normes de sécurité...) avaient des conséquences économiques et sociales «externes» dont la prise en considération était difficilement négligeable. Ainsi, créer un barrage hydro-électrique pouvait entraîner une utilisation du lac de retenue à des fins agricoles ou touristiques dont la prise en compte pouvait conduire à un choix de site tout à fait différent de celui qui aurait résulté d'un simple calcul économique en termes de production d'énergie.

On peut voir rapidement ce qui caractérise une telle démarche, ce qui fait sa force et en même temps sa faiblesse :

- d'abord elle est très générale et s'applique en fait à toutes décisions d'une certaine importance;
- si, au départ, elle constitue en quelque sorte un «enrichissement» de la décision prise au sein de l'entreprise, elle devient par contre apparemment indispensable lorsqu'il s'agit d'une décision prise dans les institutions publiques et au niveau de l'Etat; et c'est ce prolongement naturel qui lui a donné l'essor ultérieur qu'ont connu ces méthodes aux Etats-Unis dans le cadre du système PPBS.

Alors que les constatations qui précèdent n'impliquent pas nécessairement le recours à un cadre théorique unique d'analyse, les études coût-avantage ont été présentées d'emblée, et sauf exception, comme une forme particulière de calcul économique relevant de la théorie de l'optimisation. C'est-à-dire qu'elles supposent que, compte tenu du niveau auquel se situent ces analyses, les hypothèses simplificatrices qu'on a esquissées plus haut soient vérifiées. C'est là, bien sûr, l'origine des faiblesses qu'on a annoncées, faiblesses aboutissant naturellement à des analyses dont la pertinence s'avère précaire et qui servent dans la pratique souvent à donner un «habillage» scientifique à des décisions prises par ailleurs, plutôt qu'à orienter les véritables choix.

Cette situation a semble-t-il abouti à une régression assez importante dans la production de ce type d'études au profit d'analyses et d'évaluations mettant en jeu des expérimentations contrôlées en vraie grandeur.

On conçoit très bien par ailleurs les raisons de cet échec, qui résultent d'une conception illusoire de la rationalité du système social et des unités qui le composent.

Si on reprend l'exemple de l'entreprise évoquée plus haut, le problème qui se pose est celui de savoir si les critères effectifs qui président à son fonctionnement sont bien ceux d'un hypothétique optimum social qui demanderait à être précisé, ou s'ils sont d'une autre nature. En matière de sécurité par exemple, on peut se demander si la réduction des risques à un minimum constitue, en soi, l'un des objectifs de l'entreprise, ou si au contraire, on doit considérer que le niveau des normes utilisées apparaît comme une contrainte, résultant de, ou plutôt imposée par un ensemble de forces sociales extérieures à l'entreprise elle-même : spécificité particulière de la branche sur le plan technique ou psycho-sociologique, recommandations des institutions nationales, niveau général des normes dans les autres secteurs industriels, etc.

Bien entendu, on peut toujours concevoir que l'entreprise prenne l'initiative de cette réduction des normes et n'attende pas que ces forces se manifestent pour prendre les décisions correspondantes, mais cela ne changerait pas l'analyse précédente.

Il faut saisir ici les conséquences qu'impliquent les diverses considérations précédentes lorsqu'on envisage de faire une étude décisionnelle.

Vouloir dresser un bilan aussi général que possible des conséquences de toute nature qui résultent d'une décision économique ou politique peut constituer un objectif de recherche intéressant. Mais la situation est extrêmemen différente si on se place a priori (c'est-à-dire avant que les choix aient été fixés) ou a posteriori. Dans ce second cas, on se trouve dans un cadre analytique, alors que dans le premier on se situe dans un cadre prévisionnel en essayant de produire une information utile pour le décideur lui-même. Or les décideurs, et les institutions qu'ils représentent, ne sont pas concernés par les conséquences de leurs choix dans toute leur généralité et à tous les niveaux. Ce qui les intéresse, c'est surtout de savoir si les conséquences de cette décision vont être compatibles ou favorables à l'accomplissement de leurs propres objectifs, au sein du système social. Or ces objectifs sont nécessairement limités et restreints du fait de la nécessaire spécialisation des tâches et de la rationalité du système social dans lequel ces institutions fonctionnent.

Ainsi, une entreprise qui a pour objectif de produire de l'énergie selon certaines méthodes n'a pas, a priori, à se préoccuper des bienfaits sociaux qui résulteront ultérieurement de sa production de kW·h. Ce qui l'intéresse c'est de connaître les éléments d'information qui, du point de vue de la rationalité qui l'anime, seront pertinents.

Ces quelques remarques tendent simplement à rappeler que si, comme on l'a vu plus haut, au niveau d'une analyse formelle tout objectif peut être considéré comme un moyen au service d'une fin supérieure, les systèmes sociaux concrets sont finalisés, quant à eux, de manière non arbitraire et ils poursuivent des objectifs qui ne sont pas quelconques.

Le propre d'une analyse décisionnelle pertinente sera précisémment de déterminer d'abord les objectifs implicitement poursuivis par l'institution au service de laquelle elle est placée et de déduire de cette première phase le niveau de généralité des conséquences et des moyens qui devront être envisagés.

C'est là que l'on peut situer le clivage qui existe entre les analyses coût-avantage et coût-efficacité et c'est une façon de comprendre le caractère irréaliste que peut prendre la première en tant que procédure d'aide à la décision si elle est employée sans discernement.

3.2. Le vrai problème est-il traité?

Il semble important également dans ce contexte de définir avec précision la nature du problème réel qui est soulevé dans le cadre de la décision. Selon une remarque classique des praticiens de la recherche opérationnelle, il vaut mieux traiter imparfaitement un vrai problème que de résoudre parfaitement un faux problème. En matière de protection radiologique, la question est précisément de savoir si le vrai problème est celui de la rationalisation des dépenses de protection (pour un budget donné, comment définir l'allocation des ressources qui rendra minimum un indicateur global d'exposition aux risques ?) ou celui du rapport qui existe entre la valeur des normes qui sont utilisées dans la production d'énergie nucléaire et les réactions des forces sociales qui sont touchées par ces mesures de façon directe ou indirecte (travailleurs, riverains des sites, etc.). Il faut noter que la nature de la liaison qui relie les deux problèmes en question est loin d'être simple et qu'elle ne va pas nécessairement dans le sens :

réduction des normes ⟶ réduction des demandes sociales en matière de sécurité

ce qui veut dire en pratique que la résolution du premier problème ne va pas nécessairement dans le sens de la résolution du second et même qu'il s'agit de deux problèmes distincts qui présentent en fait des liens très faibles.

Alors que la solution du problème de la réduction des normes peut être basée sur des considérations techniques et économiques internes à l'entreprise et donc faire appel à des techniques classiques de recherche opérationnelle, le second présente une complexité infiniment supérieure. Avant de l'envisager sous un angle décisionnel obligatoirement simplificateur, il faudrait que la connaissance que l'on en possède soit quelque peu avancée. Or cette connaissance n'est semble-t-il pour l'instant que très partielle, et ceci pour un ensemble de raisons indépendantes : la controverse nucléaire s'est essentiellement limitée à quelques pays où elle a pris une ampleur importante ; elle ne s'est manifestée qu'à une date encore récente (1969 aux Etats-Unis par exemple); enfin l'étude de ce type de phénomènes mobilise plutôt des spécialistes des sciences sociales ; or, dans ces disciplines il y a encore assez peu d'expérience accumulée concernant les problèmes de la perception sociale des phénomènes d'environnement et de nuisance. Par conséquent, la compréhension que l'on a actuellement de ces phénomènes sociaux semble encore très limitée.

Ce qu'il paraît important de souligner à propos de ces différentes questions, c'est d'une part la constatation de leur généralité (ces aspects peuvent être analysés à partir de tous les phénomènes de santé et de sécurité quelle qu'en soit la nature) et d'autre part, l'intérêt qu'il y a à les considérer en tant que phénomènes sociaux à part entière et donc susceptibles de faire l'objet d'une explication rationnelle. «L'irrationalité» si souvent dénoncée des groupes et des individus qui tendent à proclamer de façon excessive les dangers de l'énergie nucléaire ne doit donc pas être considérée comme un phénomène particulier non susceptible de faire l'objet d'études et d'interprétations cohérentes. Une telle irrationalité ne reflète que l'ignorance relative des sciences sociales face à un problème qu'elles commencent seulement à aborder. Même si les arguments utilisés apparaissent comme peu sérieux ou peu fondés pour la majorité des spécialistes, il n'en n'est pas moins vrai que ces conduites sociales ne sauraient être considérées comme sans intérêt. Il ne faut pas oublier que c'est l'étude de la maladie mentale qui a permis le développement de la connaissance rationnelle de l'inconscient. Aussi peut-on espérer que l'analyse en profondeur des perceptions sociales relatives à l'énergie nucléaire permettra de jeter les bases d'une théorie plus générale des perceptions collectives en matière d'environnement et de nuisances. Encore faut-il que la prise de conscience de l'existence de ces dimensions du problème apparaisse dans les milieux responsables de l'orientation de la recherche et qu'il en résulte une allocation de ressources convenables qui donne une place aux sciences techniques, biologiques, économiques et sociales.

3.3. Où s'arrêter dans l'analyse des conséquences?

L'idée que l'on veut suggérer à présent est celle de l'existence d'une relation entre le niveau des objectifs auxquels on a choisi de se situer (qui, comme on l'a vu plus haut est déterminé lui-même par la nature de l'institution qui est censée agir pour résoudre le problème qui se pose) et la méthode d'élaboration correspondante des coûts et des avantages.

3.3.1. Exploration des conséquences

Le premier point qu'on peut remarquer est qu'il y a précisément une certaine latitude dans le contenu de ce qu'on qualifie de coûts et d'avantages. On avait déjà observé ce phénomène à propos des objectifs et des moyens et on aborde ici une question de même nature. En effet, dans cette exploration des conséquences (fastes et néfastes) qui résultent de la décision jusqu'à quel point peut-on aller? Même si on suppose que cette décomposition en catégories (fastes et néfastes) ait un sens (on verra cette question plus bas), on peut en effet pénétrer plus ou moins loin dans la complexité des conséquences d'une décision.

Un exemple en suggérera le principe : le cas d'une maladie professionnelle qui entraîne la disparition précoce ou l'invalidité d'un certain nombre de travailleurs au cours de leur vie active. On considère de façon évidente qu'il s'agit d'un effet néfaste dont il faut tenter d'évaluer le coût. Il y a d'abord un ensemble de coûts dont le contenu semble apparemment clair, comme ceux qui sont liés aux divers soins médicaux entraînés par l'apparition de cette maladie. On peut remarquer au passage que la définition de cette première catégorie de coût n'est déjà pas aussi évidente qu'ell paraît; on pourrait se demander si tout ce qu'on inclut généralement dans ce premier chiffre doit effectivement y entrer, car même si cette maladie était éliminée à la suite de mesures adéquates, il existerait des services de santé, des hôpitaux et des personnels médicaux. Le coût des soins prodigués à l'occasion de cette maladie doit-il alors être conçu comme la somme des facteurs dont la combinaison aboutit à produire le service en question? Ou doit-il être conçu comme la somme des facteurs précédents effectivement détruits par la production de ce service, sachant que l'autre partie aurait été de toute façon immobilisée?

Ensuite, on suppose que la disparition précoce d'un travailleur entraîne une perte de production, ou un coût supplémentaire pour l'entreprise. Là encore, la question est loin d'être claire. Cela dépend d'une multitude d'éléments supplémentaires qu'il faudrait d'abord analyser: la qualification de ces travailleurs, la difficulté qu'il y a à les remplacer, l'existence de chômage, les conditions du marché, etc.

Il n'est pas fait état bien sûr des conséquences individuelles, familiales, sociales de cette maladie et de la mortalité qui est censée en résulter...

Au total, tout événement, toute décision peuvent être conçus comme la source d'un ensemble de perturbations qui affectent virtuellement le système social dans son ensemble et à tous les niveaux. Mais cette perturbation ne se propage pas de façon linéaire. De multiples régulations, rétroactions, interactions sont mobilisées. Ce qui est néfaste à un niveau peut devenir favorable à un autre niveau.

L'hypothèse soutenue ici, c'est que cette analyse, qui pourrait se prolonger à l'infini, doit être limitée et rapportée à un système d'acteurs sociaux précis, concernés par la décision envisagée.

Ceci implique deux sortes de conséquences:

- au terme de cette analyse, on aboutit à des résultats différents, selon l'acteur social que l'on envisage[3] (l'entreprise considérée de façon isolée, la branche industrielle dans son ensemble, le personnel d'une entreprise donnée, les riverains de tels ou tels sites, etc.);
- pour chacun de ces acteurs, il y a une certaine façon d'envisager la mesure des coûts et des avantages en relation avec la rationalité propre à cet acteur et la façon dont il l'exprime en général.

3.3.2. Conception du coût

Le concept de coût se rattache en général en économie à celui de coût de production. Pour fabriquer une certaine marchandise ou produire un certain «service», une quantité donnée de travail, de matières premières,

[3] Dans la littérature traitant de ces problèmes, il existe une entité qui est souvent citée, et qui semble être considérée par certains comme pouvant servir de référence à une analyse: la société dans son ensemble.

Ainsi dans l'appendice II de la publication de la CIPR, on trouve dans un chapitre sur les analyses coût-avantage: «... que les avantages susceptibles d'être pris en compte dans le type d'analyses coût-avantage applicables dans le cadre du § 5a) (de la publication CIPR n° 9) sont définis grossièrement comme incluant tous les avantages pour la société».. Or, le problème vient du fait que la société considérée comme une totalité ne prend pas de décision et que, lorsqu'on passe aux institutions qui sont censées la représenter, celles-ci ne sont pas mues par une rationalité qui inclut les intérêts de la société tout entière mais par des motifs bien plus immédiats. Aussi la considération globale des conséquences d'une décision au niveau de la société tout entière est impropre à contribuer à la compréhension de la dynamique sociale elle-même et de sa rationalité, et ceci indépendamment de la qualité de l'information correspondante. Toute la critique qui est développée ici de l'analyse coût-avantage en tant que processus d'aide à la décision est fondée sur cette constatation.

de moyens de production est mise en œuvre. Selon la composition et la valeur de ces différents inputs on aboutit à un certain coût de production qui est consécutif à l'immobilisation des «facteurs de production» qui ont été employés. Si ces facteurs n'avaient pas été mis en œuvre dans cette production, ils auraient pu servir ailleurs, d'où la notion de «coût d'opportunité» Cette définition s'étend naturellement aux activités de l'entreprise qui ont pour objectif d'obtenir un niveau de sécurité donné (service de protection). Partant d'un niveau initial de risques qui résulte du mode de production utilisé, ce type d'activité consiste à ramener le risque en question à un niveau acceptable dans le système social considéré. Bien entendu, il reste un risque non nul qui, lorsqu'il se réalise (de façon discontinue : accident, ou continue : irradiation, par exemple) déclenche un ensemble de mécanismes de sauvetage, de secours, de soins médicaux, etc. qui peuvent être analysés de la même façon en termes de coût de production et enfin des pertes humaines et matérielles qui peuvent, quant à elles, être assimilées à une destruction de «facteurs de production». Les coûts en question doivent, en principe, se déduire des données comptables de l'entreprise. Si on envisageait le problème à une échelle plus vaste, comme par exemple le secteur de production d'énergie dans son ensemble, on pourrait se placer dans le cadre d'analyse de la comptabilité nationale.

On conçoit tout de suite que si ce premier cadre théorique présente le mérite de la clarté, il ne donne pas néanmoins toutes les réponses que l'on souhaiterait recevoir dans le cadre d'un problème décisionnel. En effet, l'imputation obtenue ainsi ne permet pas de connaître, d'une part l'affectation qui serait faite des ressources dégagées si le service en question n'avait pas été produit ou si les facteurs de production détruits avaient été sauvegardés, d'autre part les conséquences de toute nature, et en particulier externe à l'entreprise, qui résultent de cette production. On se rend compte que la notion de coût qui est utilisée dans le terme «étude coût-avantage» et qui signifie implicitement «conséquences néfastes», par opposition à avantage, est beaucoup plus riche, ou plus vague si on veut, que celle de coût de production, et demande donc à recevoir un contenu théorique précis. Or ce contenu est, semble-t-il, rarement explicité de façon satisfaisante dans la pratique des études coût-avantage. Une des ambiguïtés qui paraît la plus profonde provient de l'apparente nécessité de qualifier de l'étiquette «faste ou néfaste» toutes les conséquences prévisibles ou observables de la décision de l'opération étudiée comme si cette qualification existait «en soi». Quelle attitude prendre devant une conséquence qui apparaîtrait comme néfaste pour certains acteurs sociaux alors qu'elle serait un avantage pour d'autres? Or, loin de n'apparaître que de façon exceptionnelle, cette situation semble, au contraire, constituer le cas général. Ainsi, toute décision économique d'une certaine ampleur n'a pas seulement pour conséquence une modification de l'allocation des ressources mais également une transformation de la distribution des revenus qui, dans le cas d'un problème d'implantation industrielle, devra être considérée de plus sous un angle spatial. Comment aborder alors un tel problème dans le schéma réducteur qui est proposé en termes de coût et d'avantage?

3.3.3. Le problème de la valorisation

Les études coût-avantage sont présentées, en général, comme des procédures empiriques permettant une prise en compte très large des conséquences sociales, des décisions politiques ou économiques d'une certaine ampleur. Vues sous cet angle, il est évident que ces méthodes sortent du cadre théorique élaboré à partir des concepts économiques traditionnels. Il paraît illusoire et même dangereux (le seul danger n'étant que de réaliser des études inutiles) de vouloir à tout prix y faire rentrer des notions qui lui sont étrangères. Ainsi vouloir par exemple traduire en termes monéraires la souffrance ou le «préjudice psychologique» consécutifs à une maladie ou un accident du travail semble une entreprise contestable : ou bien le prix correspondant sera considéré uniquement comme une convention de calcul arbitraire et dans ce cas on ne voit pas pourquoi on n'arrêterait pas l'analyse à un stade antérieur, considérant l'ensemble des conséquences envisagées en termes physiques : nombre de cas de maladies intervenant chez des sujets de tel âge, ayant telle ou telle conséquence sur leur survie en nombre d'années, conduisant à telle ou telle incapacité physique, etc., ou bien on utilise le coût correspondant comme un élément particulier d'un coût «global» et on va agréger en une même entité des coûts totalement hétérogènes avec le risque que cet indice final perde toute signification aussi bien pratique que théorique.

Les réflexions qui se sont développées en France à propos de ce qu'on appelle «le prix de la vie humaine» peuvent être reprises en vue d'illustrer les considérations précédentes. S'il paraît en effet tout à fait judicieux d'opérer, dans un certain nombre de secteurs touchant à la santé et à la sécurité, le calcul a posteriori du prix implicite attaché à une vie sauvée à travers les investissements qui sont faits, l'utilisation a priori d'un tel résultat dans un calcul décisionnel paraît très contestable. D'abord, les valeurs trouvées dans la première opération sont extrêmement variables d'un secteur à un autre et il faudrait être bien naïf pour s'en étonner, car on ne voit pas par quel mécanisme social il y aurait un ajustement spontané selon les utilités marginales de ces investissements dans les différents secteurs concernés. Ces valeurs différentes expriment seulement à leur façon que la rationalité des choix qui sont faits dans le système social pour la santé et pour la sécurité ne reposent pas sur un objectif d'efficacité technique (en termes par exemple de mortalité et de maladies évitées, etc.). A vouloir à tout prix affecter une valeur monétaire à des conséquences de cette nature, on s'expose à un double danger :

- si on pense pouvoir déduire de l'observation des comportements passés un prix implicite de la vie humaine, de la souffrance... qui soit cohérent, on risque, en l'employant à nouveau dans un calcul décisionnel, de reproduire les conditions d'une situation précédente qui ne constituait en rien le modèle d'un optimum social (mauvaise information, inégalités sociales, etc.);
- si on juge ne pas pouvoir déduire ces valeurs des comportements passés, on est ramené à un choix arbitraire que seul un certain pragmatisme peut justifier; mais, c'est au décideur lui-même et non à l'analyste de prendre les responsabilités qui lui incombent dans un tel choix qui peut être déterminant sur l'issue de la décision finale.

4. EXEMPLE DE L'EXTRACTION DU MINERAI D'URANIUM

Cet exemple n'est pris ici qu'à titre d'illustration partielle d'une certain nombre de considérations qui ont été développées plus haut. Du point de vue de la sécurité nucléaire, les mines présentent une certaine spécificité dont il est utile de rappeler certains traits et qui tendent à rendre ce problème peut-être plus limité et donc mieux circonscrit que d'autres (centrales, déchets, etc.). Pourtant, malgré ces conditions favorables, les problèmes de conception et de méthodologie des études se posent de façon identique. En voici quelques cas :

Le problème de la protection dans les mines est caractérisé par le fait qu'en première approximation la population exposée est très clairement définie : il s'agit de la population des agents qui travaillent au fond. Cette population est peu nombreuse (de l'ordre de 700 personnes pour la France), elle est très stable et bien connue sur le plan à la fois médical et social. Les mesures d'exposition sont, dans le cas des mines françaises, individuelles, et sont calculées de façon systématique selon des méthodes relativement éprouvées. Cette connaissance permet donc dans une certaine mesure d'éviter de rester plongé au niveau des problèmes purement techniques liés à la production et au transfert des radionucléides à travers l'environnement jusqu'à l'homme, et d'envisager les problèmes socio-économiques sur une base technique solide.

Dans le cadre d'une mine existante, soumise à une rationalité technique et économique donnée, on peut se poser le problème décisionnel suivant : pour un niveau donné des ressources de protection, comment réduire au maximum l'exposition des mineurs? On ne vise donc pas le problème de la conception générale d'une mine nouvelle qui n'existerait pas encore.

Il est évident que l'exposition des mineurs ne résulte pas uniquement de l'efficacité relative des mesures de protection ; en fait, les méthodes d'exploitation elles-mêmes dans leur conception générale définissent un niveau de risque général. Selon que l'on travaille en chantiers mécanisés ou non, selon les méthodes de boisage et de soutainement utilisées, le niveau de risque sera, a priori, différent. Faut-il dès lors inclure parmi les actions possibles, des modifications des méthodes d'exploitation elles-mêmes?

L'étude technique générale du problème précédent conduit à élaborer une liste d'actions possibles susceptibles de réduire l'exposition des mineurs on a deux grandes catégories d'interventions possibles :

1) des actions «techniques» concernant :
   - la conception de la ventilation de la mine (pression, débit)
   - le revêtement des parois et l'obturation de zones abandonnées
   - le drainage des eaux résiduelles (à forte concentration de radon)
   - le choix des intervalles de temps pour la reprise du travail après les tirs
   - les moyens de protection individuelle du poste de travail, etc.

2) des actions concernant l'organisation du travail :
   - la rotation du personnel
   - la diminution du temps de travail.

Quant aux méthodes d'exploitation, elles se présentent à la fois comme un moyen technique et comme une forme d'organisation du travail.

Pourquoi distinguer de la sorte entre actions «techniques» et actions sur l'organisation du travail? La question peut se poser d'une autre façon: en quoi une modification du débit d'aérage serait-elle plus «technique» qu'une modification de l'organisation du travail?

Pour reprendre les analyses précédentes, il semble que la distinction à opérer à ce propos ne renvoie pas à la nature de l'action étudiée plutôt qu'à l'échelon décisionnel concerné par celle-ci. Ainsi, on peut considérer que la décision de modifier le débit d'aérage est du ressort du service de protection de la mine, qui possède une certaine autonomie pour régler ce problème. Par contre, modifier l'organisation du travail constitue une opération beaucoup plus complexe qui va mobiliser la direction générale de la mine et peut-être nécessiter de remonter encore plus haut dans la hiérarchie de l'entreprise et tenir compte de contraintes externes particulières : réactions des syndicats, réactions des autres personnels travaillant en surface, effets sur les rendements, etc. Si on envisageait une modification éventuelle des méthodes d'exploitation, le problème serait bien sûr encore plus complexe et nécessiterait de tenir compte d'un ensemble encore plus vaste d'acteurs sociaux. On voit qu'une décision est d'autant plus technique qu'elle rentre dans la compétence exclusive d'un acteur social, qui en maîtrise à un moment donné les divers éléments.

Ainsi, selon que l'on se place dans le cadre d'une étude décisionnelle pour un acteur social ou pour un autre, les termes du problème vont apparaître différemment. Ce qui est perçu à un niveau comme une contrainte rentre à un niveau supérieur comme une variable d'action possible. Pour un service de protection, le mode d'exploitation peut être considéré comme une contrainte (susceptible néanmoins d'évoluer à long terme), alors qu'au niveau de la direction générale on pourra envisager des modifications marginales de celui-ci et donc le faire rentrer dans les variables d'action.

Selon le niveau auquel on se situera, on aura également des conceptions très différentes des coûts et des avantages. Prenons par exemple le problème des coûts :

a) Evaluer le coût d'un ventilateur supplémentaire n'entraîne aucune difficulté ; mais que signifie exactement le coût d'une rotation accrue du personnel dans les postes de travail?

b) Supposons même que l'on se restreigne à l'acteur social entreprise. Changer l'organisation du travail constitue une perturbation au sein de l'entreprise, momentanée dans les meilleurs cas, durable si la nouvelle organisation ne satisfait pas à un ensemble de caractéristiques difficile à apprécier a priori. Cette organisation nouvelle perturbe en effet un certain nombre d'éléments très divers. Elle a des conséquences aussi bien économiques (baisse de la productivité par exemple) que sociales (menace pour l'équilibre des rapports sociaux dans l'entreprise, etc.). Ces éléments peuvent difficilement apparaître à partir des données comptables et il est encore plus difficile de les estimer sur le plan monétaire.

Ce problème paraît en effet se poser dans des termes tels que les instruments de l'analyse économique traditionnelle sont impropres à donner une réponse satisfaisante.

Pour le décideur, il semble qu'une analyse sociologique des rapports sociaux internes et externes à l'entreprise, de leurs déterminants et de leur élasticité par rapport à certaines formes de changement est beaucoup plus intéressante pour rendre compte d'un tel problème qu'une quelconque analyse de type coût-avantage.

Ainsi, dans l'exemple envisagé ici, on peut essayer de résumer la situation de la façon suivante : à un premier niveau d'analyse, qui s'adresse implicitement à la direction du service chargé de la protection, une première étude de type coût-efficacité est envisageable, pouvant conduire à une modélisation assez poussée de la mine et permettant de comparer l'effet des combinaisons des divers moyens techniques possibles sur l'exposition individuelle des mineurs. Les résultats d'une telle étude sont alors directement utilisables à court terme par le service de protection.

A un deuxième niveau, on suggère que des résultats peut-être équivalents et en tout cas non négligeables pourraient être obtenus par des modifications dans l'organisation du travail. Mais on s'empresse de préciser que ces modifications ne neuvent être opérées facilement et qu'elles engagent un processus de décision qui dépasse la compétence exclusive du service de protection. La décision doit remonter à un niveau plus élevé. Elle est nécessairement à plus long terme et engage des études complémentaires de type psycho-sociologique sur sa factibilité. Ces coûts sont alors plus difficiles à évaluer et, à la limite, il ne s'agit plus là que d'un emploi métaphonique de la notion de coût dans la mesure où on sort du domaine strict de l'analyse économique.

A un niveau encore plus élevé, on peut ensuite suggérer le rôle du point de vue de l'exposition au risque et des méthodes d'exploitation en usage dans les mines (mécanisation, méthode de boisage, techniques de reconnaissance et d'exploitation, etc.). A ce niveau, les aspects technologiques et économiques reprennent une forte importance. Des problèmes nouveaux de nuisance se posent : pollution provenant des engins mécaniques, risque d'accidents dus à l'extension de la taille des galeries, etc. Ici, il s'agit de choix à plus long terme et le niveau de décision est encore plus élevé que précédemment. Les différents types d'incertitude sont également plus importants. Des données comparatives entre divers systèmes peuvent devenir intéressantes.

Ainsi, l'exemple précédent constitue une bonne illustration des idées évoquées plus haut. Selon la nature des problèmes qui se posent, des approches faisant appel à des disciplines adéquates sont plus satisfaisantes que des méthodes passe-partout. La pluralité des points de vue qui résulte de la conjonction des diverses disciplines mobilisées pour analyser un problème est plus riche que la réduction artificielle de tous ceux-ci à un cadre unique dominé par une discipline, quelle qu'elle soit.

5. CONCLUSION

En définitive, il semble souhaitable de reconnaître que l'extraordinaire complexité du réseau de conséquences qui résultent d'une décision économique ne saurait s'analyser selon le schéma d'un exemple additif d'effets bénéfiques et néfastes dont il suffirait de faire le bilan pour comparer des choix alternatifs à l'échelle de la Société tout entière. L'étude d'un tel réseau de conséquences peut cependant être abordée utilement, en conjuguant la valeur explicative d'un ensemble de connaissances pratiques et théoriques issues des diverses sciences sociales, physiques et biologiques qui paraissent pertinentes par rapport aux problèmes qui se posent. A cette pluralité d'approches sur le plan des méthodes d'analyse des conséquences des décisions correspondraît par ailleurs la reconnaissance de la nature contradictoire de la réalité sociale, dans laquelle s'inscrit toute décision économique. A partir des éléments d'information et d'analyses élaborées par l'étude, il revient donc à chaque acteur social de définir à partir de ses propres critères de choix, de son propre système de valeurs, ses propres objectifs et de se situer par rapport aux différentes solutions possibles. Cette apparente régression dans le processus d'élaboration des décisions ne signifie pas cependant un retour complet à un pragmatisme sans règle. Au contraire, en vue de permettre un certain dialogue social au niveau d'un secteur de production ou sur le plan international, un certain nombre de conventions peuvent être décidées afin de rendre comparables des informations de provenances diverses. Ce type de convention, dont la nécessité est apparue par exemple pour unifier l'usage des unités de mesures, constituerait, en matière de radioprotection, une extension de la notion de «norme» et de réglementation. Afin de permettre des comparaisons dans l'évaluation des services de protection, au plan national ou international, une méthode précisant dans le détail les différents types de variables à mesurer, les unités de mesures et les méthodes d'élaboration quantitative des informations à collecter pourrait être envisagée, en tenant compte de la diversité des informations de base actuellement détenues. Il est suggéré ici que ce cadre général d'analyse ne se restreigne pas a priori à la méthodologie coût-avantage mais élabore en fonction de la spécificité du domaine de la protection radiologique ses propres règles et ses propres concepts.

BIBLIOGRAPHIE SOMMAIRE

CIPR, Protection contre les rayonnements, Recommandations de la Commission internationale de protection radiologique, Publication CIPR n° 9.

ICRP, Implications of Commission Recommendations that Doses be kept as low as Readily Achievable (adopted April 1973), Publication ICRP No.22, Pergamon Press, Oxford (1973).

BRESSON, G., COULON, R., «Analyse des principes directeurs des programmes de surveillance de l'environnement», Environmental Surveillance around Nuclear Installations 1 (C.R. Coll. Varsovie, 1973), AIEA, Vienne (1974) 35.

JAMMET, H., MECHALI, D., LACOURLY, G., «Analyse des coûts et des avantages et rapport entre le coût et l'efficacité des systèmes de contrôle» AIEA, Panel on Methods for Establishing the Capacity of the Environment to Accept Radioactive Materials, Vienne, 30 avril-4 mai 1973.

FRANÇOIS, Y., PRADEL, J., ZETTWOOG, P., «Incidence des normes de protection sur le marché de l'uranium», AIEA, Panel on Radon in Uranium Mining, the Effect of Protective Controls on Uranium Reserves and Cost, Ventilation Problems and Basic Research, Washington, D.C., 4-7 septembre 1973.

NIZARD, L., TOURNON, J., Des rapports entre perceptions sociales, demandes sociales et politiques régulatrices en matière d'environnement, Analyse socio-économique de l'environnement, problème de méthode, Mouton, Paris (1973) 153-68.

Les problèmes de décision en matière de santé, Economie et Santé n° 3, Suppl. au Bull. Statistiques de Santé et de Sécurité sociale, Ministère de la Santé publique et de la Sécurité sociale, Paris (1973).

Evaluation de l'environnement, recueil de textes, Collection «Environnement» 10, Documentation française, Paris (1973).

GUEDENEY, C., MENDEL, G., L'angoisse atomique et les centrales nucléaires, Payot, Paris (1973).

Atomic Industrial Forum and Swiss Association for Atomic Energy, The nuclear controversy in the USA, International Workshop, Lucerne, 30 April - 3 May 1972, Swiss Association for Atomic Energy, Berne.

DISCUSSION

P. RECHT: I think that for lack of time you have been obliged to pass very quickly over certain aspects of your excellent analysis which I consider very important. What is the "price of human life" to which you refer in section 3.3.3 of your paper, and to what extent can you attach a monetary value, say, to the detriment which is found in all cost-benefit studies?

F. FAGNANI: Expressing "the price of human life" in monetary terms constitutes an example of a general problem, the problem of economic quantification of elements that are outside the normal range of application of economic concepts but are nevertheless important enough to be taken into account in making decisions. In the field of health and safety, therefore, such elements as pain, after-effects, restriction of activity, etc., can be considered on the same footing as the mine itself. Any decision to allocate funds for health or protection implies that some value is attached to these elements. The a posteriori study of these implicit values presents many difficult empirical problems and is of little help to the decision makers.

D. BENINSON: In this connection perhaps I can add that various assessments of the cost of a man·rem have been made without assumptions as to the cost of life. They are based on surveys of the cost that people were willing to bear to avoid a trivial exposure. Hedgran and Lindell have made a study of this type.

F. FAGNANI: The implicit value accorded by a population to certain safety measures can of course be investigated by direct questioning of the population, but in my opinion the information obtained in this way is of little use. For one thing, the risk to which various population groups are exposed differs considerably, while the capital costs have to be borne by the total population; those who are exposed to only slight risk could consequently say that they attach no value to safety precautions. However, this is a theoretica explanation. Actually, the main problem is that very little information is available on these matters: the answers obtained in this kind of inquiry differ widely and are therefore not very informative.

H.P. JAMMET: The two types of analysis you have dealt with, cost-benefit and cost-effectiveness, differ not only in their objectives but also in their methodology. The cost-benefit analysis compares the advantages and disadvantages of an activity in an absolute manner and calls for the use of a common unit, the most practical one being the monetary unit. The cost-effectiveness analysis, however, compares the inversely reciprocal changes in the cost of the protection and the cost of the detriment and can be treated in a relative manner.

F. FAGNANI: It is true that there is a big difference between the two methods. Most of the aspects we have considered relate to cost-benefit studies. It is our view that cost-effectiveness studies are very close to the conventional techniques of operational research and are therefore ideally suited for tackling the main technical problems which arise in radiological protection. When we are concerned with less technical aspects and more wit psycho-sociological and social-economic aspects, new methods of analysis become necessary, and a fundamental choice then has to be made between the cost-benefit type of analysis or the utilization of whatever social sciences and techniques seem most appropriate. In our paper we wished to stress the many difficulties and hypotheses involved in cost-benefit analysis in order to ensure that this technique is not regarded by non-specialists as a kind of gimmick with which they can solve all problems.

MEASUREMENTS OF RADON UNDER VARIOUS WORKING CONDITIONS IN THE EXPLORATIVE MINING OF URANIUM

J. KRISTAN, I. KOBAL
Jožef Stefan Institute,
Ljubljana

F. LEGAT
Geološki zavod,
Ljubljana,
Yugoslavia

Abstract

MEASUREMENTS OF RADON UNDER VARIOUS WORKING CONDITIONS IN THE EXPLORATIVE MINING OF URANIUM.

In the northwestern part of Yugoslavia lies a uranium ore deposit. To determine the dimensions of the strata and explore the geological structure of the ore deposit, mining and drilling operation were initiated. About 30 miners are working in this operation. The paper gives the results of the radiological survey of the quality of the air in the mine and the cumulative annual inhalation doses of workers. The influence of the mine and mill waste waters on the environment and on the radiological status of the affected waters is given.

INTRODUCTION

Geological prospecting for uranium indicated some uranium ore in the northwestern part of Yugoslavia near the town of Škofja Loka. Geological studies have shown that the ore deposit of Žirovski Vrh lies in Middle Permian sediments, in one of the sheets of the imbricated structure of an overthrust, overlying Triassic autochthonic rocks. To determine the quantities and dimensions of the ore deposits, intensive mining and drilling operations started as early as 1960. At present the total length of the tunnels is about 15 km. Up to 2000 tons of U_3O_8 ore deposits have been determined and up to 8000 tons have been estimated geologically. The quality of the ore is over 0.1% of uranium.

From the very beginning the geologists have been very concerned about the working conditions in the mine and initiated measurement of the ^{222}Rn concentration in the mine atmosphere. For these determinations the direct alpha-scintillation method using scintillation cells was employed [1, 2]. The inhalation dose of the miners was estimated by measuring the concentration of radon decay products in the air by the Kusnetz field method [3] and the Tsivoglou method [4]. We considered the radon concentration measurements to be more precise and reliable and so they became the basis of radiological hazard estimations in the mine. Measurements of radon daughters are performed periodically.

QUALITY OF MINE AIR

The tunnels of the mine run in 4 main levels, with elevations at 430, 480, 530 and 580. The levels are connected by the main shaft. The mine

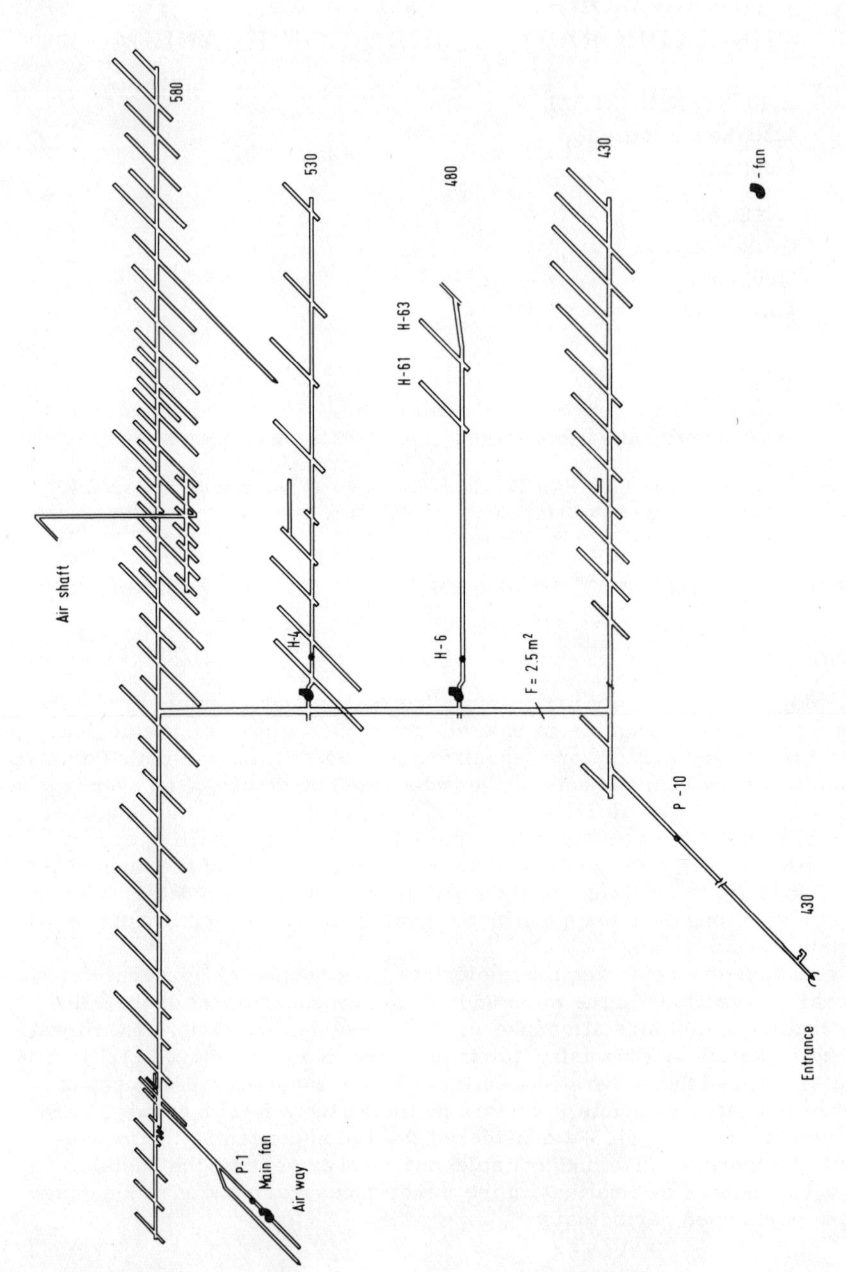

FIG. 1. Plan of the mine.

entrance is at 430, which is also where the air flows into the mine. The air is exhausted at elevation 580 at a rate of 660 m^3/min. Separate fans and ducts exhaust air from actual working places in the crosscuts and from stopes to the main stream of the air in the shaft. The rate of this ventilation is about 120 m^3/min (Fig. 1).

During normal ventilation, the average concentration of radon in the exhausted air at point P1 was 206 pCi/litre for the year 1973. The discharged air has 0.75 WL (working level) daughters on average. The total quantity of discharged radon is 0.2 Ci/d. Since this discharge point is very remote and in the forest, we have not considered it dangerous. In the intake air at point P10 we measured a radon concentration of 10 pCi/litre. In the tunnels at points H-6 and H-4 at the air inlets to the horizons 480 and 530 the average concentration of radon was 40 and 60 pCi/litre, respectively. At the actual working places in horizons H-4 and H-6 the concentration of radon increased from values of 90 pCi/litres in the tunnels up to 800 pCi/litres at the stopes with bad ventilation. Values of 0.2 up to 3.2 WL for radon daughters were measured. The highest values were observed on the crosscuts on level 530.

In the mine two types of drilling machines were used, the Yugoslavian type RGVN for pneumatic depth drilling up to 50 m, and the Longyear electrical drilling machine for boring up to 150 m in length. A comparison of radon concentrations in air for both these drilling operations and for the mining operations in the tunnels and on the stopes is shown in Table I. One can see from this table that the highest concentration of radon appears during 'Longyear' drilling, because the ore-bearing beds are small in size and in the shape of lenses, distributed at random. So the long drilling cuts more uranium ore bodies and liberates appreciable amounts of radon.

The difference in concentration of radon and radon decay products in the air under different working operations is also reflected in the cumulative inhalation dose of mine workers. For the year 1973 there were, among other workers, 11 miners, 7 drilling workers and 5 supervisors monitored and their inhalation doses were calculated as shown in Table II.

Constants efforts have been made to improve the climatic and radiological conditions of mining, such as making new air locks for abandoned

TABLE I. CONCENTRATION OF ^{222}Rn IN AIR AT THE WORK FACES

	Work performed		
	Drilling RGVN	Drilling Longyear	Mining
	(pCi/litre)	(pCi/litre)	(pCi/litre)
Minimum	30	60	30
Maximum	780	1050	450
Average	264	366	66
Number of measurements	48	34	54

TABLE II. ANNUAL INHALATION DOSES OF WORKERS IN WLM

	Miners	Dose (WLM[a])	Drilling workers	Dose (WLM)	Supervisors	Dose (WLM)
1.	B. A.	1.5	K. M.	4.7	Z. A.	1.1
2.	M. A.	2.7	C. M.	4.1	S. A.	0.8
3.	D. S.	1.9	G. B.	5.0	K. F.	1.5
4.	M. J.	1.7	P. D.	5.1	L. L.	2.1
5.	B. V.	1.5	P. V.	5.8	P. J.	1.9
6.	T. A.	1.4	M. I.	2.7		
7.	R. P.	2.9	K. C.	5.8		
8.	P. V.	4.4				
9.	B. A.	3.4				
10.	F. M.	1.8				
11.	S. S.	1.2				
Average		2.2		4.7		1.5

[a] WLM = Working level month.

TABLE III. ACCUMULATION OF ^{222}Rn IN A NON-VENTILATED MINE

Sampling points	Time after ventilation ceased (h)					
	2	8	16	24	32	40
	Concentration of radon in air (pCi/litre)					
P-10	30	30	60	60	30	30
H-6	30	60	60	60	210	210
P-1	90	150	150	150	210	210
H-61	30	150	390	480	690	720
H-63	60	330	510	690	1050	1100

drifts and crosscuts, and installing new lengthened air pipes for the ventilation of remote stopes. To obtain data on the radon emanation rate from the walls, we measured the rate of radon concentration increase at some check points in the mine during a 40-h break in mine ventilation. Owing to the natural draught in the main shaft, the concentration of radon did not increase there over a value of 200 pCi/litre, but the concentration of radon in the crosscuts H-61 and H-63 at level 480 increased in 40 hours from 30 to 720 pCi/litre and from 60 to 1100 pCi/litre, respectively, see Table III.

QUALITY OF MINE AND RIVER WATER

The only radiological impact of the mine on the environment is discharged waters. The mine is very wet, the water flowing out of numerous geological fissures and out of boreholes. The radon content of these water sources was measured and can reach up to 4 nCi/litre. The water is collected in an open trench along the drives and leaves the mine at the main entrance at the rate of about 1.3 m^3/min. This water bears up to 500 pCi/litre of radon, up to 236 μg uranium per litre and up to 2.5 pCi/litre of ^{226}Ra. After a short superficial course this water flows into the clear, fast Brebovnica brook, which has an average flow of 0.74 m^3/s. About 3 km from the mine the Brebovnica flows into the River Sora, which has an average flow of 15 m^3/s. All these surface waters were regularly sampled and the concentrations of radon, uranium and radium determined. The clear water of the Brebovnica has an average of 0.1 pCi ^{226}Ra/litre and 1 μg U/litre above the inflow. After the waters from the mine have become evenly distributed in the stream, i.e. about 500 m below the inflow, the Brebovnica water has the following average concentrations: 20 μg/litre of uranium, 0.3 pCi/litre of radium and 12 pCi/litre of radon. It is interesting to note the steady increase of uranium concentration in the water. In 1968 it was only 0.4 μg/litre, but maximum values of 24 μg/litre were observed in 1973. The River Sora has on average less than 0.1 pCi/litre of radium and less than 0.5 μg/litre of uranium. No influence of the Brebovnica waters on the quality of the Sora has been measured.

A pilot plant for uranium extraction was built near the mine entrance and has been operated periodically. Last year (1973) several tons of uranium ore were worked. The waste waters from this plant are collected, neutralized, filtered and periodically pumped into the Brebovnica at a rate not greater than 5% of the actual flow of the brook. These waste waters were analytically controlled and contained up to 5 mg/litre of uranium and up to 740 pCi/litre of radium. After sedimentation and filtration through a barytes bed, these waters contained about 200 pCi/litre radium and 1 mg/litre of uranium.

These results on the quality of water from the pilot plant made us aware of the great care needed in planning the treatment and release of the waste waters from the proposed uranium mill into this environment. After studying the radiological impact of the mill on the environment [5, 6], we considered radium to be the critical nuclide for water pollution. Because the waters of the Brebovnica and Sora are not used regularly for cattle watering and irrigation, the only possible hazard to the population living in this area is the consumption of fish taken from these waters. The concentration factor for radium in fish can be about 250 for some kinds of fish [7] and it is estimated that no person could consume annually more than 10 kg of fish caught in these waters. Taking this fact into account and assuming a maximum permissible annual intake of radium by ingestion of 9.6×10^{-3} μCi for the members of the public, the MPC for ^{226}Ra in rivers with fish could be estimated to be 4 pCi/litre, which is also the limit under the Yugoslavian regulations. We think and hope that this view is rather conservative, but nevertheless we should not tolerate a greater concentration of radium in our waters. To evaluate this possible hazard more precisely, we have started the analysis of fish, algae and river biota for raduim, uranium and gross alpha and beta contamination.

ACKNOWLEDGEMENT

This paper is partly based on work financed through the Boris Kidrič Foundation.

REFERENCES

[1] RUSHING, D.R., et al., Radiological Health and Safety in Mining and Milling of Nuclear Materials (Proc. Symp. Vienna, 1963) 2, IAEA, Vienna (1964) 187.

[2] KOBAL, I., KRISTAN, J., A modified scintillation cell for the determination of radon in uranium mine atomsphere, Health Phys. 24 (1973) 103.

[3] KUSNETZ, H.L., Radon daughters in mine atmospheres - a field method for determining concentrations, Amer. Ind. Hyg. Assoc. Quarterly 17 (1956) 85.

[4] THOMAS, J.W., Modification of Tsivoglou method for radon daughters in air, Health Phys. 19 (1970) 691.

[5] TSIVOGLOU, E.C., Radiological Health and Safety in Mining and Milling of Nuclear Materials (Proc. Symp. Vienna, 1963) 2, IAEA, Vienna (1964) 231.

[6] HAVLIK, B., Radioactive pollution of rivers in Czechoslovakia, Health Phys. 19 (1970) 617.

[7] DE BORTOLI, M., GAGLIONE, P., Radium-226 in environmental materials and in foods, Health Phys. 22 (1972) 43.

DISCUSSION

P. SLIZEWICZ: Were any γ-irradiation measurements carried out and, if so, what doses were received by the mine personnel?

J. KRISTAN: All the miners wear film badges. The average annual dose is 0.5 rem, i.e. 10% of MPD. The local gamma dose rate at working points is up to 0.1 mrem/h. In contact with ore bodies the gamma dose rate can be 5 mrem/h.

J. MARTINEZ PACHECO: Could you give a few more details on the methods used to measure radon and radium concentrations in the mine air and water?

J. KRISTAN: For radon determination we used scintillation cells, as shown in the figure. The cells were as designed by Lucas, but we used a modified version, described in Health Phys. 24 (1973). The cells are internally coated with ZnS (activated with Ag) and their volume is about 250 ml. The cells are calibrated with known amounts of ^{222}Rn in air. On average we count 0.8 cpm per pCi of radon/litre air.

For radium we use the emanation method. The water sample is tightly sealed in the emanation flask, and after a certain time the accumulated radon is expelled by bubbling pure N_2 gas through the sample. The radon is captured in a cold trap and transferred to the scintillation cell.

For low background concentrations of radium in water we use 20-litre samples. The radium is extracted by a modified ion-exchange column and, after the waiting time, the radon evolved is expelled into the scintillation cell and counted.

G. BOERI: I did not quite understand the figures given in your paper for water concentrations in discharges, e.g. 200 pCi R/litre. Could you please explain these values?

J. KRISTAN: Only a few cubic metres of waste water from the pilot plant contained radium concentrations of 200 pCi/litre. The average concentration in this waste was 60 pCi/litre. The waste was released only when the level of the river was high.

N. T. MITCHELL (Chairman): I am inclined to agree with your view that the concentration factor assumed for ^{226}Ra in fish is a rather conservative value. Have you obtained any further information from the direct monitoring to which you referred in your presentation?

J. KRISTAN: We have no results of our own yet for radium concentration in fish. In Ref. [7] of my paper De Bortoli et al. describe their measurements of radium in fish from the waters near Ispra, and they report a maximum concentration factor of 220. Radium is preferentially concentrated in fish bones, so the total radium content of the fish does not all go into man's diet.

PRACTICAL EXPERIENCE AND MONITORING

Population doses resulting from radionuclides of worldwide distribution

(Session IV cont.)

Chairman: N. T. MITCHELL (United Kingdom)

IAEA-SM-184/102

Invited Review Paper

POPULATION DOSES RESULTING FROM RADIONUCLIDES OF WORLDWIDE DISTRIBUTION

D. BENINSON
Comisión Nacional de Energía Atómica,
Buenos Aires,
Argentina

Abstract

POPULATION DOSES RESULTING FROM RADIONUCLIDES OF WORLDWIDE DISTRIBUTION.

A simple methodology for calculating collective dose commitments is presented, together with estimates of the collective dose resulting from the worldwide distribution of radionuclides produced in the fuel cycle of nuclear power generation. It appears that the contribution of this global distribution to the collective dose commitment is in the order of a few tenths of a man • rem per MW • a. An example is presented on the use of the collective dose commitment concept to assess the cumulative effect in the future of the growing nuclear power generation.

INTRODUCTION

In the fuel cycle of the nuclear power industry, and to a minor extent in the use of sub-products of nuclear activities, radioactive wastes are produced. Almost all the activity of these wastes is present in stored spent fuel elements and in well-contained fractions separated in the first cycle of reprocessing installations. Minor amounts of radioactive materials are released to the environment, complying with the relevant requirements in radiation safety.

Assessment of doses resulting from these environmental releases are likely to require substantial consideration of local situations, because they may be limiting and also because they can represent a major fraction of the collective dose to the population. Most of the nuclides released to the environment from the fuel cycle are of local concern because their half-lives are short compared with the time involved in their dispersion. A few nuclides, on the other hand, combine a substantial half-life with characteristics leading to a relatively short dispersion time and they can therefore become distributed worldwide when released to the environment.

In this category of nuclides of potential global distribution, krypton-85 and tritium are of particular interest because virtually all the activity produced during power generation is at present released to the environment when the fuel is reprocessed. Iodine-129, due to its very long half-life, is also of interest, although the fraction of the amount generated that will be released depends strongly on the waste management of the reprocessing installations.

Population doses resulting from the worldwide distribution of the three nuclides mentioned are considered in this paper, although ^{129}I is only treated marginally. The concept of dose commitment per unit practice (i.e. per MW(e) · a) is used in the paper because it is independent of predictions relating to the size of expanding nuclear power industry and also because it is deemed to be particularly relevant for radiation protection considerations.

THE COLLECTIVE DOSE COMMITMENT CONCEPT

The doses resulting from the radionuclides under consideration are presented as 'collective doses', defined as

$$S = \sum_i D_i N_i$$

where D_i is the dose received by individuals in the group i and N_i is the number of individuals in the group. The collective dose can be assessed for a group, for a sub-population or for the population as a whole. The latter assessment is required for the purpose of this paper. The unit of 'collective doses' is the product of a dose unit (rad or rem) and a unit of individuals in a group (man); the resulting units are man · rad and man · rem.

Extending the use of dose commitment as defined by UNSCEAR to collective doses, the 'collective dose commitment' of any specified practice or operation is the infinite time integral of the average dose rate caused by that practice or operation in a given population multiplied by the size of that population:

$$S_c = \int_0^\infty \overline{\dot{H}(t)} P(t) dt \qquad (1)$$

where S_c is the collective dose commitment,

$\overline{\dot{H}(t)}$ is the average dose ra.e at time t after the practice or operation

P(t) is the size of the population at time t

For krypton-85 and tritium the time scale involved allows reasonable predictions of the population size evolution P(t). It is assumed in this paper that P(t) in a short time range will be exponential, with a growth of about 2×10^{-2} per year.

Because of the very long half-life of iodine-129, most of the dose commitment would be delivered over a period of many tens of millions of years, unless the dispersed radionuclide become unavailable to exposure pathways by some unspecified process. The dose commitment concept, therefore, does not appear too meaningful for this nuclide, and only annual doses after global dispersion will be considered.

KRYPTON-85

Krypton-85 is a beta emitter with a maximum energy of 670 keV; 0.4% of the disintegrations emit a gamma photon of 514 keV. By external irradiation the gamma radiation causes whole-body exposures, while the beta radiation is responsible for skin doses only [1, 2]. Internal irradiation due to inhalation is comparatively negligible. The whole-body dose rate from a homogeneous air concentration can be readily assessed by the immersion model (1.7×10^4 rad/a per Ci/m^3, equivalent to 2×10^7 rad/a per Ci/g).

Assessments of doses to the world population from a given discharge of ^{85}Kr have been published. In one assessment [3] a two-stage dispersion model is assumed with an initial distribution in the injection latitude band (35-60°N) up to a height of 3 km lasting for about 100 days, followed by an even distribution over the troposphere (up to 10 km) of the whole northern hemisphere. Reduction in the concentration there is assumed to be caused only by decay because of slow transport to the southern hemisphere.

In the most recent UNSCEAR report [4] the distribution of ^{85}Kr is taken to be almost homogeneous over the surface of the globe and throughout the troposphere. This assumption is substantiated by results of measurements of ^{85}Kr in ground air at different latitudes [5].

Equation (1) can be used to estimate the collective dose commitment per Ci of ^{85}Kr discharged to the atmosphere. It will be assumed, as in the UNSCEAR report, that the distribution is throughout the troposphere (i.e. 4×10^{21} g of air). After dispersion, 1 Ci of ^{85}Kr will therefore cause a dose rate of 5×10^{-15} rem/a. The collective dose commitment (S_c), calculated from Eq.(1) is:

$$S_c = \frac{R_0 P_0}{\lambda - a} \tag{2}$$

where R_0 is 5×10^{-15} rem/a

P_0 is the world population at time of release (more exactly at time of uniform mixing)

λ is the decay constant of ^{85}Kr

a is the annual fractional growth of the population.

With the assumptions mentioned above, the collective dose commitment is about 4×10^{-4} man·rem/Ci. This value together with production figures for ^{85}Kr can be used to assess the collective dose commitment per MW(e)·a. Assuming a thermal efficiency of 30%, the production rates are 510 and 270 Ci/MW(e)·a for thermal fission of ^{235}U and fast fission of ^{239}Pu, respectively. The corresponding collective dose commitments, assuming that all ^{85}Kr is released to the environment when the fuel is reprocessed, are 0.2 man·rem/MW(e)·a for thermal reactors and about 0.1 man·rem/MW(e)·a for fast reactors.

As most of the world population is in the northern hemisphere, where the discharges occur, any delay in the transport of ^{85}Kr from the northern to the southern troposphere will tend to increase the collective dose commitments, by a factor not exceeding two. The thermal efficiencies are likely to improve, changing slightly the production rates. The proportion of participation of fast reactors introduces a further uncertainty when the values are used to assess the situation in the future. However, taking all factors into account one may be confident that the collective dose commitment will not exceed a few tenths of a man · rem/MW(e) · a.

TRITIUM

Tritium, in the form of tritiated water, is widely dispersed in the circulating waters, which can also include the water in body fluids. It can be shown that a concentration of 1 Ci/g in body fluids would deliver a dose rate to the whole body of about 10^8 rad/a [4]. Using this constant and some assumptions on the dispersion of released tritiated water, the collective dose commitment per Ci released can be estimated.

The model published in Ref. [3] assumes that all significant discharges in the near future will occur in the northern hemisphere and that the tritiated water will be dispersed in the circulating waters of that hemisphere (about 10^{22} g). It also postulates that the time of exchange of surface waters with the southern hemisphere and with waters below the thermocline is substantially longer than the half-life of tritium.

Using these assumptions, the collective dose commitment per Ci released can be readily calculated by an expression similar to Eq.(2), resulting in a value of 10^{-3} man · rem/Ci. This value will tend to be an overestimate if the reduction on the tritium levels is not due only to radioactive decay.

UNSCEAR estimates dose commitments for ^3H discharges by means of the relation of the water levels of natural tritium and the rate of its production [4]. Levels of tritium in water before nuclear explosions were reported to be in the range of $6\text{-}24 \times 10^{-15}$ Ci/g [6, 7], while the rate of production per hemisphere used by UNSCEAR is 0.8 MCi/a.

Using these values, the dosimetric parameter and an expression similar to Eq.(2), the collective dose commitment is in the range of $0.4 - 1.7 \times 10^{-2}$ man · rad/Ci. Values in this range are higher than the value derived above using the dispersion model of Ref. [3]. This may be due to uncertainties in the assessment of average concentration from environmental measurements and in the estimation of natural production rates, which together may amount to an order of magnitude [3]. Therefore, the value of 10^{-3} man · rem/Ci is used in this paper.

Tritium is produced in nuclear reactors by ternary fission and by activation reactions on deuterium (mainly in heavy water reactors) and on additives used for reactivity control. Production rates due to fission are 21 and 40 Ci/MW(e) · a for thermal reactors and fast reactors, respectively, again assuming a thermal efficiency of 30%. Activation reactions contribute to production an extremely variable amount, depending on, among other factors, reactor type and additives used. The contribution may range from less than 1% to about the same amount as the fission production.

Considering only the ternary fission production, the collective dose commitments result in 2×10^{-2} and 4×10^{-2} man · rem/MW(e) · a for thermal reactors and fast reactors, respectively. The additional contribution of activation reactions and variations in the projections of the participation of fast reactors in future power generation cause considerable uncertainty in the values that should properly be used. However, all being taken into account, the collective dose commitment will probably not exceed a few hundredths of a man · rem/MW(e) · a, and therefore will not add appreciably to the commitment due to ^{85}Kr.

IODINE-129

The specific activity method has been used to assess thyroid doses arising from discharges of ^{129}I [3]. A specific activity of 1 pCi/g in thyroid tissue delivers a dose rate of 2×10^{-7} rad/a. At the levels under consideration the mass of ^{129}I can be neglected in relation to the stable iodine. The thyroid exposure of the population was estimated assuming that ^{129}I discharged in liquid effluents is dispersed in the circulating waters of the northern hemisphere (about 10^{22} g) and considering that seawater has an average iodine concentration of 6×10^{-8} weight by weight. An upper limit of the dose rate can be calculated assuming that all iodine sources become contaminated at the same level as seawater. With this assumption, the dose rate in thyroid arising from the complete dispersion of 1 Ci of ^{129}I is about 3.3×10^{-10} rem/a and the real value will certainly be smaller. The upper limit of the collective dose rate in thyroid will be in the order of 1 man · rem/a per Ci discharged.

Production rates of iodine-129 are 10^{-3} Ci/MW(e) · a (thermal) and 3.4×10^{-3} Ci/MW(e) · a (fission). Assuming a complete release in liquid effluents, 1 MW(e) · a will give rise to annual collective doses to thyroid of the order of 10^{-3} man · rem. A power programme in which fast reactors participate will tend to reduce the cooling times and therefore important problems will arise in controlling the risk of ^{131}I during reprocessing. Substantial decontamination factors will have to be instrumented and these will automatically reduce the dose rate resulting from 1 MW(e) · a, probably by several orders of magnitude [3].

LIMITATION OF COLLECTIVE DOSE COMMITMENTS

Collective doses are used in the application of the radiation protection requirement that exposures should be kept as low as reasonably achievable, taking into account social and economical considerations [8]. Collective dose commitments per unit practice can also be used as a method of controlling the future average population dose. This is the mathematical consequence of the fact that the annual dose D_a in a future equilibrium situation, provided that the population size remains relatively constant, will be:

$$D_a = \frac{1}{p} \sum_j S_{cj}$$

where S_{cj} is the collective dose commitment per year of practice involving the existence of source j

P is the population size.

In the case of nuclear power generation, the sum $\sum_j \cdot S_{cj}$ can be represented by $S_c P \cdot W$, where W is the installed capacity per person in the world. The above equation, therefore, reduces to $D_a = S_c W$ and S_c has to be limited according to the value deemed acceptable for D_a. As an exercise let us postulate that the nominal risk rate of 10^{-4} rem^{-1} applies and that a risk of the order of 10^{-6} per year (lower than the risk of public transport) is acceptable for members of the population. These assumptions imply that a value of 10^{-2} rem/a would be acceptable for D_a.

The optimistic forecasts for world nuclear programmes, together with population size projections, show that the world average per capita installed nuclear power may be in the order of 1 kW/man in the first decade of the next century. For the time being, therefore, the collective dose commitment per unit practice in the proposed exercise should be limited to a value of 10 man · rem/MW(e) · a.

This value covers all collective dose contributions. However, even taking into account estimates of the other contributions [3], the results presented above for ^{85}Kr and ^{3}H are minute compared to the limit. It seems there reasonable to assume that the other requirements, namely to comply with individual dose limits and with the concept of "as low as reasonably achievable" will be the main radiation protection considerations in the development of nuclear power.

REFERENCES

[1] DUNSTER, H.J., WARNER, B.F., The disposal of Noble Gas Fission Products from the Reprocessing of the Nuclear Fuel, Rep. AHSB (RP) R 101 (1970).

[2] HENDRICKSON, M., "The dose from Kr-85 released to the earth's atmosphere", Nuclear Power Stations (Proc. Symp. New York, 1970), IAEA, Vienna (1971) 237.

[3] BRYANT, P.M., JONES, J.A., The Future Implications of Some Long-Lived Fission Product Nuclides Discharged to the Environment in Fuel Reprocessing Wastes, Rep. NRPB-R 8 (1972).

[4] UNSCEAR, Ionizing Radiation: Levels and Effects, A report of the United Nations Scientific Committee on the Effects of the Atomic Radiation to the General Assembly, UN, New York (1972).

[5] PANNETIER, R., Distribution, transfert atmosphérique et bilan du Krypton-85, Rep. CEA-R-3591 (1968).

[6] KAUFMAN, S., LIBBY, W., The natural distribution of tritium, Phys. Rev. 93 (1954) 1337.

[7] VON BUTLER, H., LIBBY, W., Natural distribution of cosmic ray produced tritium, J. Inorg. Nucl. Chem. 1 (1955) 75.

[8] INTERNATIONAL COMMISSION ON RADIOLOGICAL PROTECTION, Publication 22, Pergamon Press, London (1973).

DISCUSSION

P. SLIZEWICZ: Do you not think it necessary to purify the gaseous effluents and eliminate ^{85}Kr?

D. BENINSON: From the information available it appears that ^{85}Kr would not have to be removed for global collective dose reasons, at least in the near future. Local considerations may make it necessary in some cases

P. SLIZEWICZ: Would you agree that although one can tolerate the global dispersion of ^{85}Kr, the situation may be rather different for ^{129}I?

D. BENINSON: Yes, I would. The assumption of dispersion was made to get a very conservative assessment of the collective dose from a world-wide distribution. Reduced dispersion — or localized retention — would cause higher local doses, but this was not within the scope of the paper.

F.D. SOWBY: Your paper indicates that nuclear power may contribute a collective dose of a few tenths of a man·rem per MW·a from radioactive releases. To this must be added a few man·rems per MW·a for occupational exposure at reactor sites and fuel reprocessing plants. The total collective dose from nuclear power should thus be only a few man·rems per MW·a in the foreseeable future, and it is therefore perhaps inappropriate to consider a limit as high as 10 man·rems per MW·a in this case.

D. BENINSON: You are quite right. If an operational, authorized limit is considered, it should be of the order of a few man·rems per MW·a. The calculation in the paper aims at showing that in the near future the addition of sources is not likely to result in average doses corresponding to risk levels. The value obtained is an upper limit; 'optimization' will lower this 'authorized' limit.

P. RECHT: To what extent can the concept of population dose influence national decisions relating to particular sites and result in reduced concern for the prevention of releases of ^{85}Kr, ^{3}H and ^{129}I? I think a precise explanation is required of how the concept can be applied and what it can be applied to.

D. BENINSON: Population dose considerations do not obviate the need for considering individual and critical group limits. The local situation will be limiting in many cases and will be most important in influencing decisions. On the other hand, collective doses will be used for applying the justification and optimization concepts. In the near future, however, the addition of collective dose commitments will not be limiting for any reasonable acceptable level of risk.

POPULATION DOSE CONSIDERATIONS FOR THE RELEASE OF TRITIUM, NOBLE GASES AND ^{131}I IN A SPECIAL REGION

A. BAYER, T.N. KRISHNAMURTHI*, M. SCHÜCKLER
Institut für Neutronenphysik und Reaktortechnik,
Kernforschungszentrum Karlsruhe,
Federal Republic of Germany

Abstract

POPULATION DOSE CONSIDERATIONS FOR THE RELEASE OF TRITIUM, NOBLE GASES AND ^{131}I IN A SPECIAL REGION.

Radiation exposure and risk to the population of the Upper Rhine river basin resulting from the operation of nuclear facilities are analysed in this paper. The study-year is chosen as 1985, by which time an aggregate nuclear generating capacity of 40 000 MW(e) is planned to be installed in this region. Tritium, noble gases and iodine-131, being the main contributors to the radiological burden, are considered as the critical isotopes. Both atmospheric and hydrospheric transport paths of the nuclides are investigated in the study. Isopleths of whole body, lung, gonad, skin and thyroid doses are shown. Population dose and the resulting risk for certain radiation induced diseases are estimated and compared with natural incidence of the diseases.

1. INTRODUCTION

The Upper Rhine river basin constitutes a special region in the nuclear map of the middle Europe owing to its selection as a site for locating several nuclear power stations. An aggregate nuclear generating capacity of about 40,000 MWe is planned to be installed in this region by the year 1985 and this power output will be derived mainly through light water reactor systems. The power stations are to be located along the Rhine river and its tributaries. In addition, a reprocessing plant for treating 40 tonnes of spent fuel per annum is already in operation in the middle of this region. For the time span up to 1985, the sites of the nuclear power plants are known definitely. Beyond this date, a further increase in nuclear installations is visualised in view of the forecasted energy demand and the assumption that a major part of this demand will be met by nuclear plants.

The recommendations of the "Advisory Committee on the Biological Effects of Ionising Radiation" [1], in regard to the effects on population, fit in aptly to this situation of multiple sources within a limited region. The Committee recommends that every effort should be made to estimate and predict the radiation dosage from existing and planned sources; further, studies should be improved and strengthened to answer among other things the following questions : how much, where and what type of radioactivity is released ? How are these materials transported through the environment and what is their effect when they contact man ? Lindell [2] also emphasises the need for such an assessment, the main objective of which is to check that the detriment from the given sources is not unjustifiable and the future situation will continue to be acceptable even if the number of sources increases.

---

* On Fellowship from Bhabha Atomic Research Centre, Bombay, India.

Table I : Sites of Nuclear Plants

No.	Site / Plant	River	Country	Type	Year of Commiss.
1	Biblis 1	Oberrhein	D	PWR 1200 MWe	1974
	Biblis 2			PWR 1300 MWe	1976
	Biblis 3			PWR 1300 MWe	
2	Ludwigshafen 1	Oberrhein	D	PWR 600 MWe	1980
	BASF 2			PWR 600 MWe	
3	Philippsburg 1	Oberrhein	D	BWR 900 MWe	1975
	Philippsburg 2			BWR 900 MWe	1977
	Philippsburg 3			BWR 900 MWe	
4	Karlsruhe	Oberrhein	D	RR-FR2-40 MWt	1961
	Nucl. Research			PHWR 60 MWe	1966
	Centre			Repr. Plant-WAK 40 t/a	1971
				LMR 20 MWe	1973
5	Neupotz 1	Oberrhein	D	1000 MWe	
	Neupotz 2			1000 MWe	
6	Gambsheim 1	Oberrhein	F	1200 MWe	
7	Whyl 1	Oberrhein	D	PWR 1300 MWe	1979
	Whyl 2			PWR 1300 MWe	
8	Fessenheim 1	Oberrhein	F	PWR 900 MWe	1975
	Fessenheim 2			PWR 900 MWe	1976
	Fessenheim 3			PWR 1200 MWe	1981
9	Kahl 1	Main	D	BWR 15 MWe	1961
10	Hörstein 1	Main	D	BWR 1200 MWe	1979
11	Grafenrheinfeld 1	Main	D	PWR 1200 MWe	1979
	Grafenrheinfeld 2			PWR 1200 MWe	
12	Bamberg +)	Main	D	1000 MWe	
13	Obrigheim 1	Neckar	D	PWR 300 MWe	1969
14	Neckarwestheim 1	Neckar	D	PWR 800 MWe	1976
	Neckarwestheim 2			PWR 1300 MWe	
15	Kaiseraugst	Hochrhein	CH	BWR 900 MWe	1979
16	Schwörstadt 1		D	1200 MWe	
	Schwörstadt 2			1200 MWe	
	Schwörstadt 3			1200 MWe	
17	Leibstadt	Hochrhein	CH	BWR 950 MWe	1979
18	Beznau 1	Aare	CH	PWR 350 MWe	1969
	Beznau 2			PWR 350 MWe	1971
19	Gösgen-Däniken 1	Aare	CH	PWR 920 MWe	1978
20	Graben 1	Aare	CH	BWR 1100 MWe	1980
	Graben 2			BWR 1100 MWe	1983
21	Mühleberg 1	Aare / Saane	CH	BWR 310 MWe	1972
22	Inwil 1	Reuss	CH	1000 MWe	
	Inwil 2			1000 MWe	
23	Rüthi	Vorder-rhein	CH	PWR 900 MWe	1979
24	Gundremmingen 1	Donau	D	BWR 250 MWe	1967
	Gundremmingen 2			BWR 1200 MWe	1979
	Gundremmingen 3			BWR 1200 MWe	1980

+) = not shown in the map - is situated about 60 km east of site No. 11

Remarks : Oberrhein = Rhein between Basel and Mainz
Hochrhein = Rhein between Konstanz and Basel
Vorderrhein = Rhein upstream from Konstanz

D = Germany
F = France
CH = Switzerland

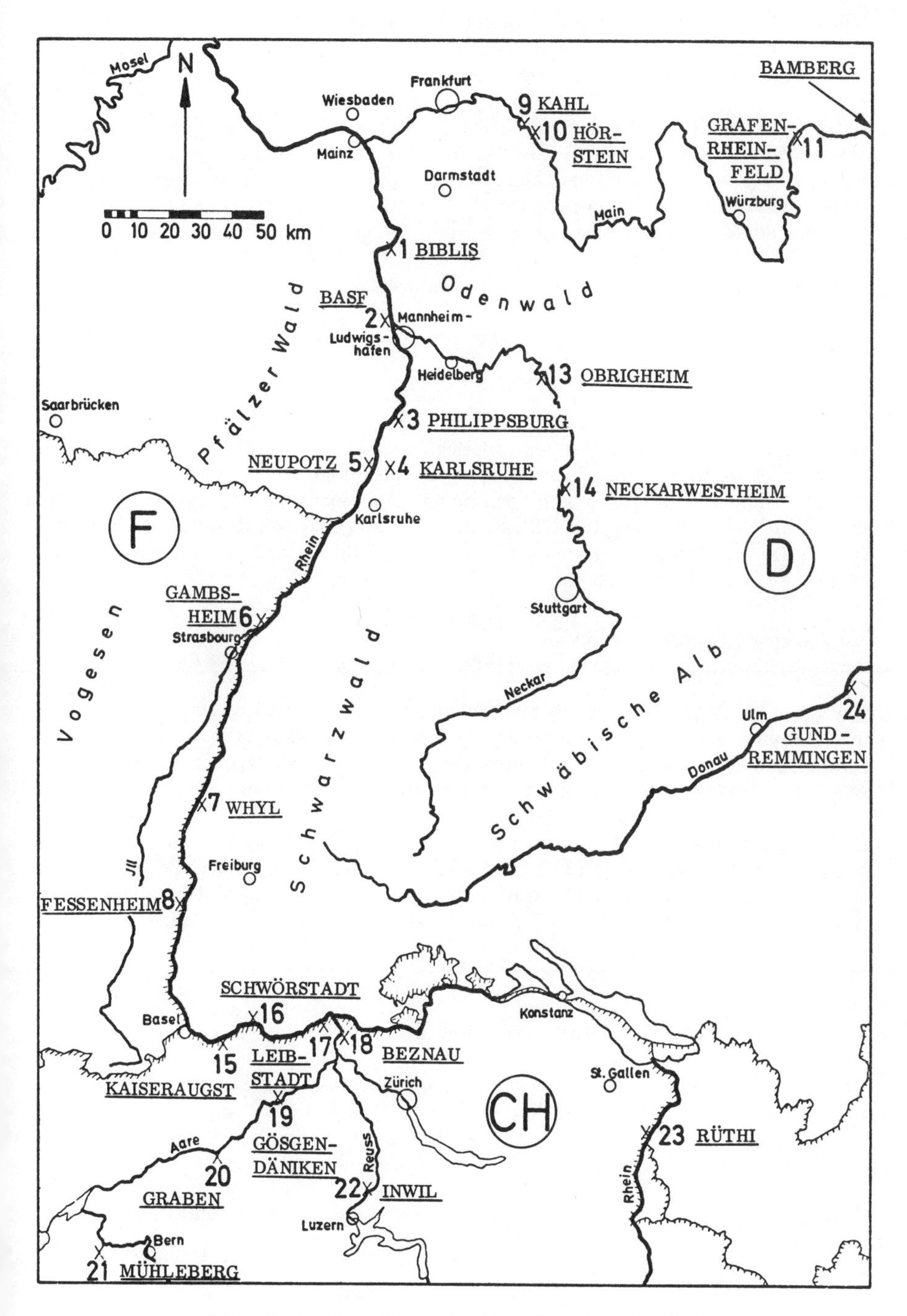

FIG.1. Study region: Upper Rhine Basin with nuclear plant sites.

Table I [3] lists the nuclear plants already in operation together with those under construction and planning. The sites of the various facilities are indicated in Fig. 1.

Using the projected power plant sites and the strength of radionuclide releases associated with normal operation of the facilities, the present study investigates, through the year 1985, the potential radiation exposures to the population of the Upper Rhine basin and the consequent risk. Tritium, noble gases and iodine-131 are considered as the critical isotopes, being the main contributors to the radiological hazard. Amongst the noble gas nuclides, ^{41}Ar, ^{85}Kr, ^{88}Kr, ^{131m}Xe, ^{133}Xe and ^{135}Xe are considered for this evaluation.

2. EXPOSURE LIMITS

The radiation protection guidelines, applicable to the Federal Republic of Germany, stipulate a genetically significant dose of not more than 2 rem in 30 years from the nuclear industry [4]. For long term planning purposes this dosage has been allocated equally between the radionuclides released with gaseous effluents and those with liquid effluents. Thus, the recommended exposure limits for licensing the operation of nuclear power plants are :

Individual dose rate	via gaseous discharges	- 30 mrem/yr
	via liquid discharges	- 30 mrem/yr

The limit for liquid effluents has been further subdivided equally between the three major exposure modes, namely, drinking of water, ingestion of food and external irradiation. For individual members of the public, this corresponds to an average exposure of 10 mrem/yr for each of these exposure modes.

In the case of radioiodines, a dose rate of 90 mrem/yr for the thyroid of children has been laid down as a guiding limit for the grass-cow-milk exposure pathway. For tritium, a maximum concentration of not more than 3000 pCi/l in the receiving water body, based on medium water flow has been adopted as the guiding value.

3. CONCENTRATION, DOSE AND RISK RELATIONS

For a source of atmospheric activity release at location $\vec{r}_\nu$, the concentration in air at position $\vec{r}$ is

$$C(\vec{r}, \vec{r}_\nu) = J(\vec{r}, \vec{r}_\nu) \cdot A_\nu \cdot f(\vec{r}, \vec{r}_\nu, v_d, T_{1/2}) \quad (1)$$

where			
A_ν	(Ci/sec)	=	activity release rate
$J(\vec{r}, \vec{r}_\nu)$	(sec/m^3)	=	atmospheric dispersion factor
$C(\vec{r}, \vec{r}_\nu)$	(Ci/m^3)	=	concentration
$f(\vec{r}, \vec{r}_\nu, v_d, T_{1/2})$		=	correction factor for deposition and radiological decay

In the case of hydrospheric activity release at position $\vec{r}_\nu$ of the river, the concentration in the river, downstream at $\vec{r}$ is

$$C(\vec{r}, \vec{r}_\nu) = \frac{A_\nu}{F(\vec{r})} \cdot d(\vec{r}, \vec{r}_\nu) \cdot f(\vec{r}, \vec{r}_\nu, v_d, T_{1/2}) \qquad (2)$$

where $F(\vec{r})(m^3/sec)$ = flow rate

$$d(\vec{r}, \vec{r}_\nu) = \begin{cases} 0 & \text{if } \vec{r} < \vec{r}_\nu \\ 1 & \text{if } \vec{r} \geq \vec{r}_\nu \end{cases}$$

When there are several sources of release, the resulting concentration due to the superposition of the different sources is

$$C(\vec{r}) = \sum_\nu C(\vec{r}, \vec{r}_\nu) \qquad (3)$$

For an individual at position $\vec{r}$, the resulting dose rate is

$$D(\vec{r}) = g \cdot C(\vec{r}) \qquad (4)$$

where $g\left(\frac{rem \cdot m^3}{Ci \cdot sec}\right)$ = dose factor

$D(\vec{r})\left(\frac{rem}{sec}\right)$ = dose rate

The dose factor g is characteristic of the organ of reference and depends on the path the isotope takes to reach the organ, e.g. inhalation and / or ingestion etc. In the case of ingestion Eq. (4) is valid only for individuals who are self-suppliers; for those supplied by a central supply system, the concentration needs to be averaged over the area of produce Ap.

$$\overline{C(\vec{r})} = \frac{1}{Ap} \int_{Ap} C(\vec{r})\, dA \qquad (5)$$

Average gamma dose to a ground level receptor from the radioactive cloud, arising from continuous release, is estimated using Eq. (6). The integration is carried out over the cloud volume V, using the approximation cited in reference [5].

$$D_\gamma(\vec{r}) = \int_V I_\gamma(\varrho, E_\gamma) \cdot C(\vec{r})\, dV \qquad (6)$$

where $I_\gamma(\varrho, E_\gamma) = k \cdot \frac{\exp(-\mu\varrho)}{4\pi\varrho^2} B(\mu\varrho)$

E_γ (MeV) = energy of gamma photon

ϱ (m) = distance from elemental volume dV

μ (m^{-1}) = linear attenuation coefficient

k $\left(\frac{rem \cdot m^2}{sec \cdot Ci}\right)$ = conversion factor

$B(\mu\varrho)$ = dose build-up factor

The whole body, skin, lung and gonad dose rates due to beta and gamma radiation are calculated for immersion in noble gases dispersed in the atmosphere. External dose from the cloud and internal dose from the gas dissolved in body tissues and the gas in lung are considered . The amount of noble gas activity in the human body due to absorption of the inhaled gas in the blood and its subsequent diffusion into the tissues is estimated using the experimentally measured partition coefficients for krypton and xenon [6]. For argon the partition coefficient is derived, assuming it to behave like krypton.

The gamma doses from the gas in tissues and in lung are evaluated using the method developed by the Medical Internal Radiation Dose (MIRD) Committee [7, 8]. The lung is assumed to have the same concentration of noble gas, as in the atmosphere. A lung mass of 1000 gms and an effective volume of 3.5 litres are used. External cloud beta dose estimate is based on an infinite cloud model because of the short range of betas in air. Beta depth dose is evaluated using the empirical equation of Loewinger et al. [9].

The risk rate, defined as the probability per year that an individual receiving a certain dose will suffer a radiation induced disease, is obtained by multiplying the total dose rate by the risk factor r, by assuming a linear dose-risk relationship.

$$R(\vec{r}) = r \cdot D(\vec{r}) \tag{7}$$

where $D(\vec{r})\ (\frac{rem}{yr})$ = dose rate

$r\ (\frac{cases}{man \cdot rem})$ = risk factor

$R(\vec{r})\ (\frac{cases}{man \cdot yr})$ = risk rate

Sometimes a risk factor $r(t)$ is defined as the yearly probability per rem during the latent period L (e.g., $L \approx 20$ years in the case of thyroid carcinoma). Then the risk rate is

$$R(\vec{r}) = r(t) \cdot L \cdot D(\vec{r}) \tag{8}$$

where $r(t)\ (\frac{cases}{man \cdot rem \cdot yr})$ = risk factor

$L\ (yr)$ = latent period

In both cases, for children with age τ less than the latent period L, a correction factor K has to be applied to the risk rate $R(\vec{r})$

$$K = \frac{\tau}{L} \quad (K \leqslant 1) \tag{9}$$

For a given population distribution $P(\vec{r})$, integration over the individual dose rates gives the population dose rate

$$D_P = \int_A P(\vec{r}) \cdot D(\vec{r})\ dA \tag{10}$$

where $P(\vec{r})$ $\left(\frac{man}{m^2}\right)$ = population density

D_P $\left(\frac{man \cdot rem}{yr}\right)$ = population dose rate

Similarly the population risk rate is obtained by integration over the individual risk rates.

$$R_P = \int_A P(\vec{r}) \cdot R(\vec{r})\, dA \qquad (11)$$

where R_P $\left(\frac{cases}{yr}\right)$ = population risk rate

Calculational models for this study are explained in detail in reference [10].

4. SITE AND RADIOLOGICAL PARAMETERS

4.1 Meteorological Data

For most of the reactor sites, the mean wind velocity, percentage frequency of its distribution and direction, as determined from measurements, are available for 12 compass sectors. As for the frequency of occurence of the different Pasquill weathers categories A to F, the values reported by Nester [11] for the Karlsruhe site are considered as representative for the entire study area. These frequencies are listed in Table II. Mixing heights for the weather categories are also indicated.

Table II : Frequencies and Mixing Heights of Weather Categories

Category (Pasquill s)	Frequency (%)	Mixing Height (m)
A	2.00	1500
B	7.27	1500
C	13.93	1000
D	41.36	500
E	21.17	200
F	14.27	200

Table III : Activity Release Rates from Nuclear Power Plants

Nuclear Plants / Type	Gaseous Release (Ci/yr)								Liquid Release (Ci/yr)	
	^{3}H	^{41}Ar	^{85}Kr	^{88}Kr	^{131m}Xe	^{133}Xe	^{135}Xe	^{131}I	^{3}H	^{131}I
Karlsruhe - RR FR-2 a)	800	150,000								
Karlsruhe - PHWR a)	1200	600							1000	
Karlsruhe - Repr. Plant WAK b)	1000		250,000					0.5	3000	
Kahl - 1 c) - BWR				150		2700		0.003		0.025
Obrigheim - 1 c) - PWR		45	5			4000	400	0.035	300	0.3
Gundremmingen - 1 c) - BWR				450		9000		0.2	20	0.2
PWR d) - 1000 MWe	10		1300	75	75	3200	250	0.06	950	0.2
BWR d) - 1000 MWe	10		1200	4	180	2400	120	0.3	130	0.5
XWR e) - 1000 MWe	10		1250	40	130	2800	190	0.2	550	0.35

a) Derived from reference [12]

b) Estimated by the authors at design throughput of the Reprocessing Plant; Tritium wastes will be stored mostly in dried-up oil wells

c) Derived from references [13, 14, 15, 16]

d) Derived from references [14, 15, 16]

4.2 Release Rates

Release rates of the nuclides, considered in this study, for the various plants are shown in Table III. In the case of plants under operation, the values are based on measured release rates ; for those under construction or being planned, the emission rates are derived on the basis of plant design features.

4.3 Dose Factors

The inhalation and ingestion dose factors for the whole body due to tritium-intake and for the thyroid due to iodine-131-intake [10, 17] are shown in Table IV.

Table IV : Dose Factors for ^{3}H and ^{131}I

Exposure mode	Tritium (whole body)	Iodine-131 (thyroid)
	rem·m³ / Ci·sec	
Inhalation	5 (-2) a)	1200 a)
Ingestion	4 (-6) b) (drinking water)	90, 000 a) (air-grass-cow-milk pathway)

e.g., 5 (-2) = $5.0 \cdot 10^{-2}$

a) related to concentration in air
b) related to concentration in water

The ingestion dose factors for tritium and iodine are with regard to water and milk consumption respectively. Iodine figures in Table IV apply to infants who are considered as the critical group of the population in respect of iodine intake.

4.4 Partition Coefficients for Noble Gases

Tissue / air partition coefficients for the average tissue for krypton, xenon and argon are listed below for normal body types:

Krypton - 0.2 [8]
Xenon - 0.6 [8]
Argon - 0.07 (approximate)

4.5 Risk Factors

Risk studies, in general, assume conservatively that the consequences of radiation exposure are linear with dose and that threshold or rate effects and repair after radiation damage are absent. In fact, these effects do exist and lower the actual risk. The risk factors summarized in Table V therefore provide upper limits for the various risks considered. Risk factors are defined in two ways, as can be seen from Table V.

Table V : Risk Factors

Risk from	r (t) [1] cases / man• rem•yr		r [18] cases / man•rem	
Leukemia	children	2 - 3 (-6)		20 (-6)
	adults	1 - 2 (-6)		
Thyroid cancer	children	2 - 10 (-6)	children	30 (-6)
			adults	10 (-6)
Skin cancer	lower than for thyroid - no sufficient data			
Lung cancer		1 (-6)		10 (-6)
Genetic risk [19] (Deformation at birth)	r = 1 - 2 (-4) cases/birth / 1 rem/30 yrs			

5. RESULTS AND DISCUSSION

5.1 Exposure Rates from Atmospheric Release

In the calculation of immersion dose from a noble gas cloud, ICRP [20] makes no distinction relative to external gamma and beta ($E_{max} \geqslant 0.1$ MeV) dose to the whole body ; the beta dose that is actually delivered to the surface of the body has been treated as dose received by the whole body as a conservative measure, e.g. at a depth of 2 mm in tissue, beta dose from ^{85}Kr falls by about 6 orders of magnitude from the body surface dose [21]. Our calculations consider the external beta dose contribution for skin and gonad doses only and exclude it for whole body dose. Irradiation by the noble gas dissolved in the body tissues and by the gas in the lung has been accounted for in the dose estimates for all organs.

The various contributing factors to the total dose to different organs are indicated below :

Whole body dose = A + D + E + H

Skin dose = A + B + D + E + H

Gonad dose = A + C + D + E + H

Lung dose = A + D + F + H

Thyroid dose = G

where

A = external gamma dose from the noble gas cloud

B = external beta dose from the noble gas cloud to the surface of the body

C = external beta dose from the noble gas cloud to tissue at 2 mm depth

D = beta and gamma dose to the whole body from dissolved gas in the body tissues

E = gamma dose to the whole body from the noble gas in the lung

F = beta and gamma dose to the lung from the noble gas in the lung

G = inhalation and ingestion dose to the thyroid from ^{131}I

H = inhalation and ingestion dose to the whole body from ^{3}H

Thyroid dose rate (G) from ^{131}I intake is indicated separately as it applies to children below 5 years of age; further, the German Radiation Protection Ordinance (R.P.O.) recommends a separate dose limit for the thyroid from iodine intake. Total dose rate to the thyroid of children will be slightly less than the sum of G and whole body dose rate.

The male gonads are taken to be at a depth of 2 mm, though the genetically sensitive region is normally at a depth of 10 - 20 mm. The female gonads are at sufficient depth and the beta exposure is negligible compared to exposure from penetrating gamma radiation.

Isopleths of total dose rates to the whole body, skin and thyroid are shown in figures 2, 3, and 4 respectively for continuous release to atmosphere. It has been observed from the present study that the dose from inhaled noble gases is insignificant compared to the dose arising from external radiation due to immersion. Consequently, the total dose rate to the gonad and lung are almost the same as that received by the whole body. J.T. Whitton [22] also showed the contribution from dissolved ^{85}Kr in the body tissues as 1 % of the dose from external penetrating radiation.

The regional pattern of the dose distribution shows that the atmospheric release from the nuclear facilities of the Karlsruhe Research Centre will contribute most to the dose even after the commissioning of other nuclear installations. The areas adjoining the Karlsruhe site record the maximum dose rates which lie above the average values by a factor of 10 to 100.

Release of argon-41 from the research reactor FR-2 (150,000 Ci/yr) continues to be the dominant factor in spite of the large krypton-85 release from the neighbouring reprocessing plant (250,000 Ci/yr) WAK. This can be established from the ratio of skin to whole body dose; near the Karlsruhe site, this ratio is 1.44, which lies very near the ratio, due only to ^{41}Ar, whereas for ^{85}Kr skin to whole body dose ratio works out to be about 200. +)

+) The argon-41 releases of the Nuclear Research Centre Würenlingen /Switzerland (near the Beznau-site) could not be considered due to the lack of published data.

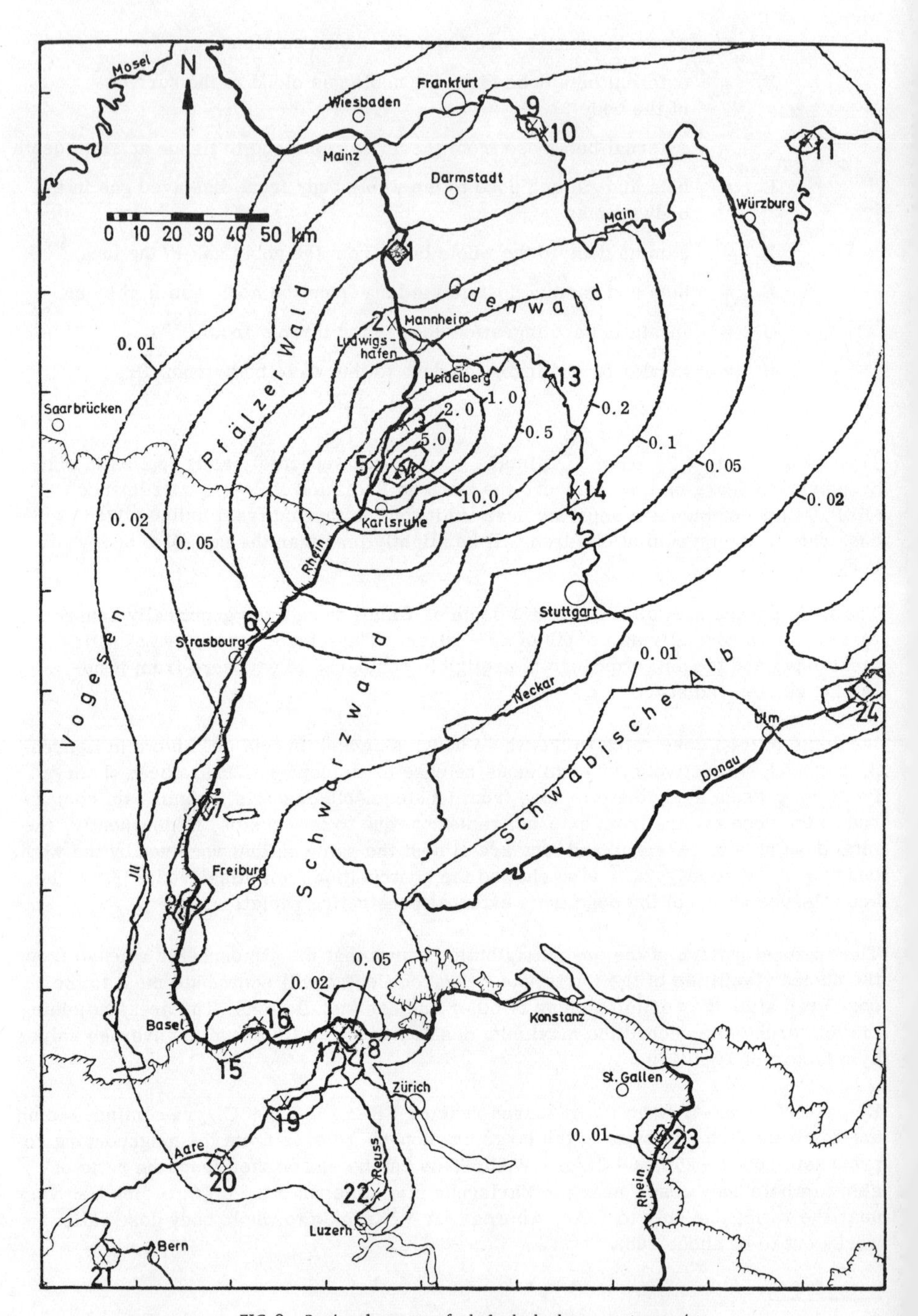

FIG.2. Regional pattern of whole-body dose rate (mrem/a).

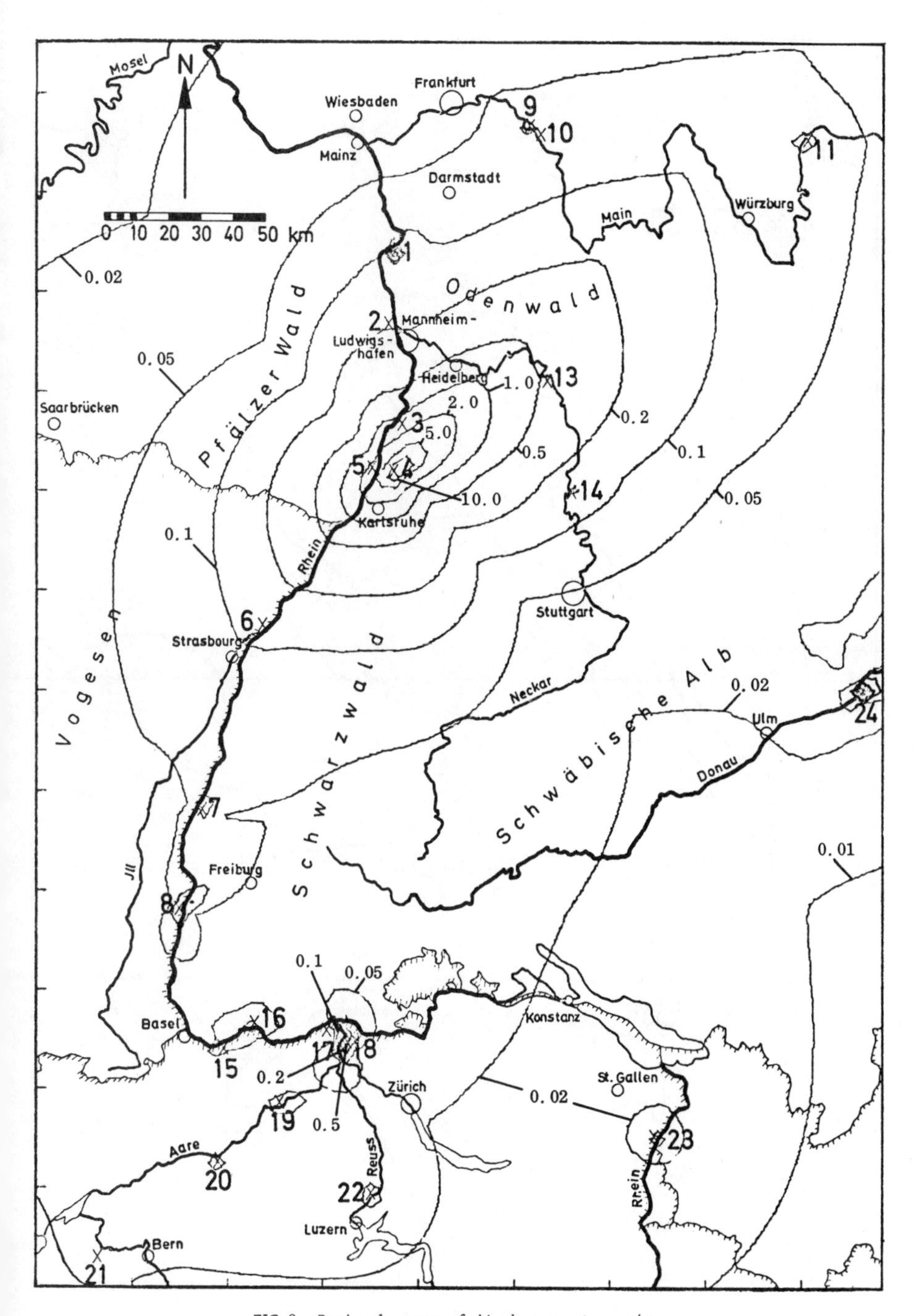

FIG. 3. Regional pattern of skin dose rate (mrem/a).

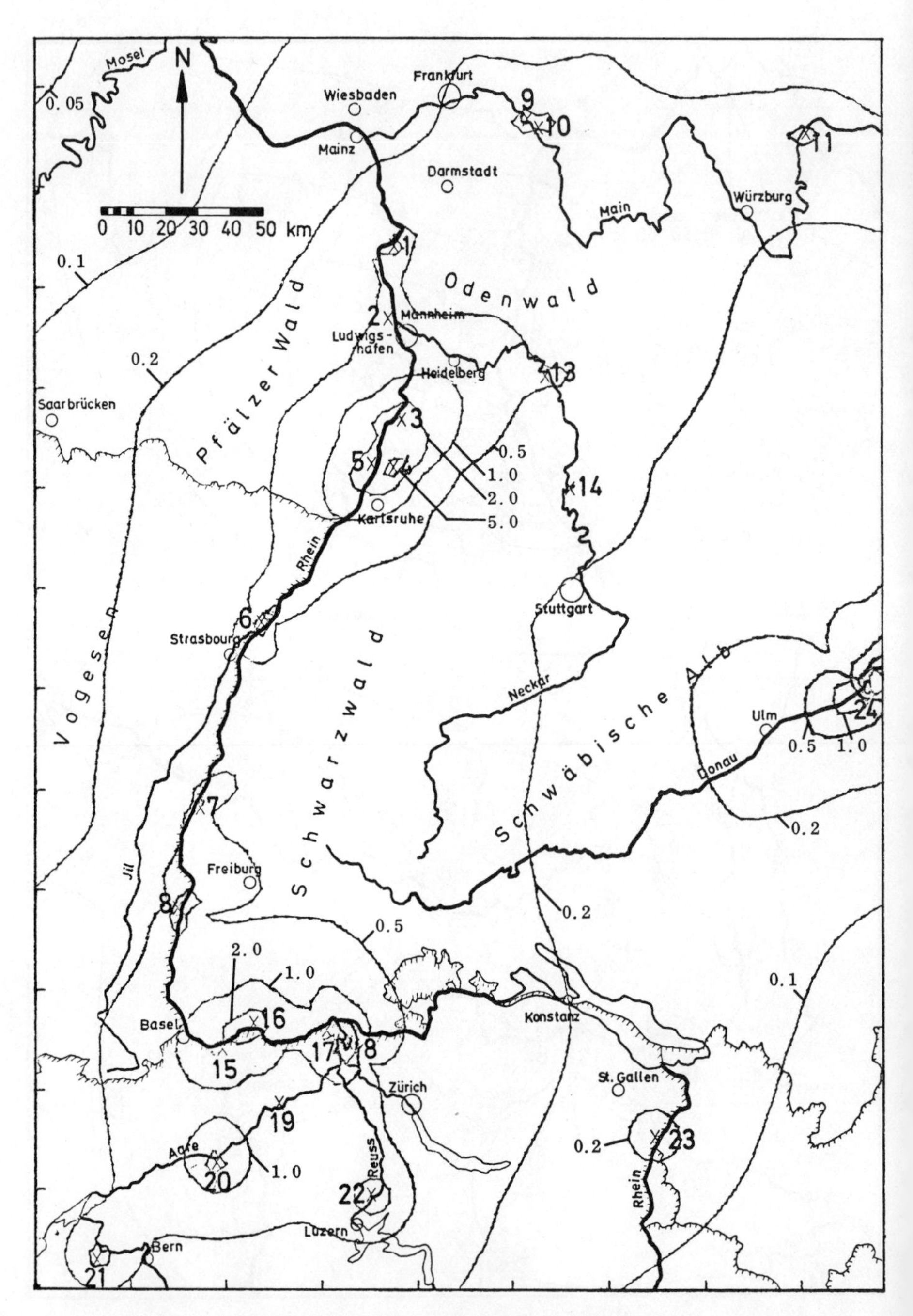

FIG.4. Regional pattern of thyroid dose rate (mrem/a).

The computational methods used here assume the receptor to be in open plains and the shielding offered by structures and buildings have been ignored: The exposure rates are thus conservative to a certain extent. Dose reduction due to shielding effects can be as high as 50% in the case of buildings with about 20 cm thick brick walls and roof and will not be significant for wooden structures.

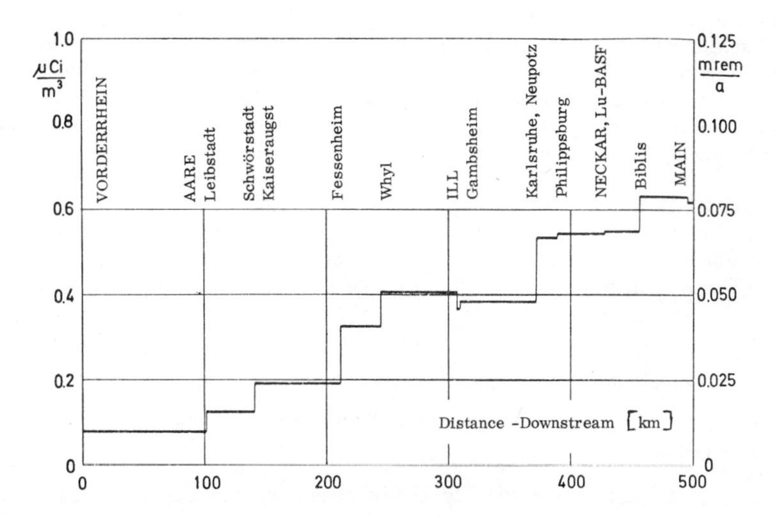

FIG.5. Tritium concentrations in the River Rhine in 1985 due to regional releases from nuclear facilities.

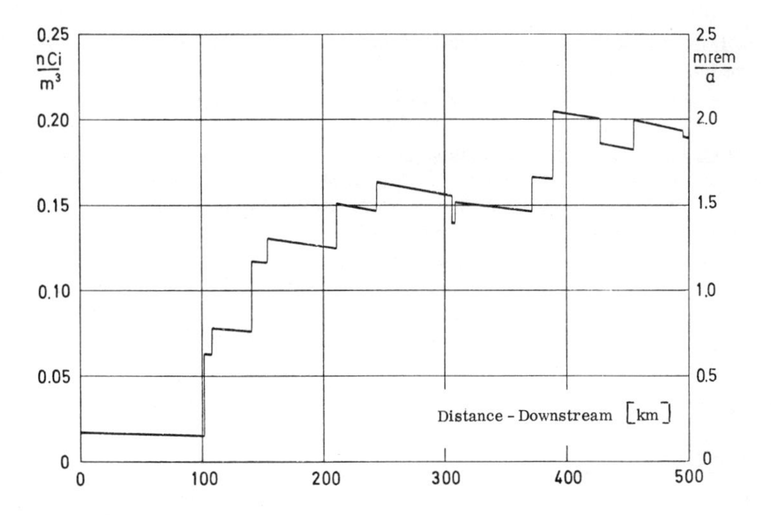

FIG.6. Iodine-131 concentrations in the River Rhine in 1985 due to regional releases from nuclear facilities.

5.2 Radiation Exposure from Liquid Discharges

Discharge of radioactive effluents into the Rhine and its tributaries results in additional radiological burden in humans through drinking water and food products. It is foreseen that in the coming years about 10 to 20 percent of the drinking water requirements in Germany will be met through purification of river water [23].

Figures 5 and 6 show the computed concentrations of tritium and iodine-131 in the Rhine for all downstream locations within the study region from Konstanz. The dose rates indicated alongside in these figures are based on a gross assumption that drinking water needs of the study area is met solely by the Rhine river. With this assumption, tritium contamination results in a maximum whole body dose of 0.08 mrem/yr; for iodine, the maximum thyroid dose to children is slightly higher than 2 mrem/yr. These values should be scaled down appropriately, depending on the extent of the use of Rhine water.

Estimation of the concentrations in food products and the consequent hazards are under investigation. Accuracy of these studies will however be limited owing to lack of information on the behaviour of radionuclides in soils, plants and animals; further agricultural practices and dietary habits cannot be exactly modelled.

5.3 Effects of Global Sources

Besides the regional sources, there exist other extraneous sources of global nature. These sources are :

(i) natural production of radionuclides

(ii) production by detonation of nuclear weapons

and (iii) production and release of activity by the world-wide distributed nuclear facilities

Among the radionuclides considered in this paper, only tritium and krypton-85 assume importance. The concentrations that would be attained in the coming decades have been computed as far as the development of the nuclear industry can be foreseen [14, 16, 24]. In the case of tritium, the concentrations in water shown in Fig. 7, are mainly governed by the nuclear weapon tests. All tests up to the year 1962 after which such studies have become much less intense, have been accounted for in this evaluation. The tritium concentration levels indicated in Fig. 7 are valid for surface water. Measurements on drinking water pumped from deep layers in the study region lie partially below a concentration of 0.1 $\mu Ci/m^3$ [12]. In the case of ^{85}Kr, release from the world-wide nuclear industry has already overwhelmed all other sources. Fig. 8 shows ^{85}Kr levels in air in the next few decades with and without isolation of the noble gas from the off-gas stream from the nuclear plants. For 1985, the year of consideration for this study, the concentrations of ^{3}H and ^{85}Kr and the resulting doses are shown in Table VI.

5.4 Population Dose

A region comprising the areas within 50 km east and west of the Rhine river is considered for population dose studies. The estimated total population in this study area is 12.7 million.

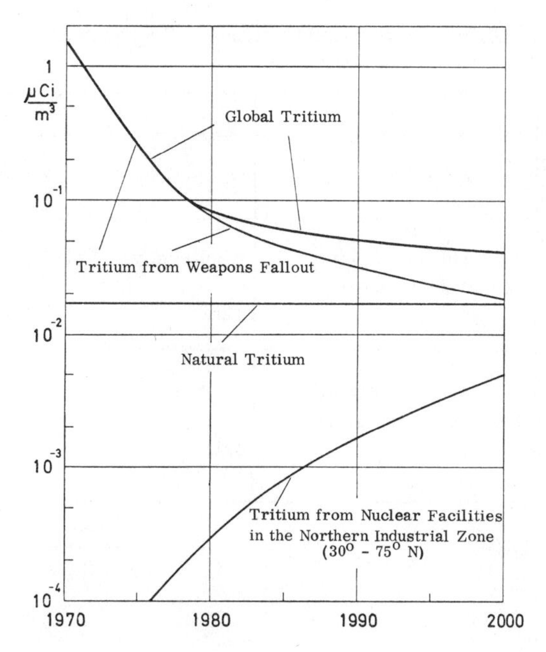

FIG.7. Tritium concentration in water due to nuclear weapon tests, natural production and releases from nuclear facilities.

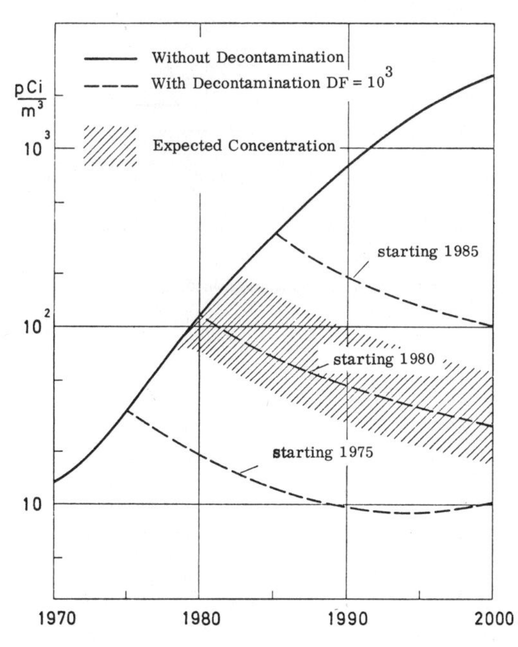

FIG.8. Krypton-85 concentrations in air due to releases from nuclear facilities (natural and weapons krypton are negligible).

Table VI : Global Tritium and Krypton-85 Concentrations and Dose Rates in the Year 1985

Nuclide	Concentration	Dose Rate	
^{3}H	6 (- 2) μCi/m^{3}	whole body	8 (-3) $\frac{mrem}{yr}$
^{85}Kr	100 pCi/m^{3} (with decontamination)	whole body	1.7 (-3) $\frac{mrem}{yr}$
		skin	1.7 (-1) $\frac{mrem}{yr}$
	400 pCi/m^{3} (without decontamination)	whole body	6.8 (-3) $\frac{mrem}{yr}$
		skin	7.0 (-1) $\frac{mrem}{yr}$

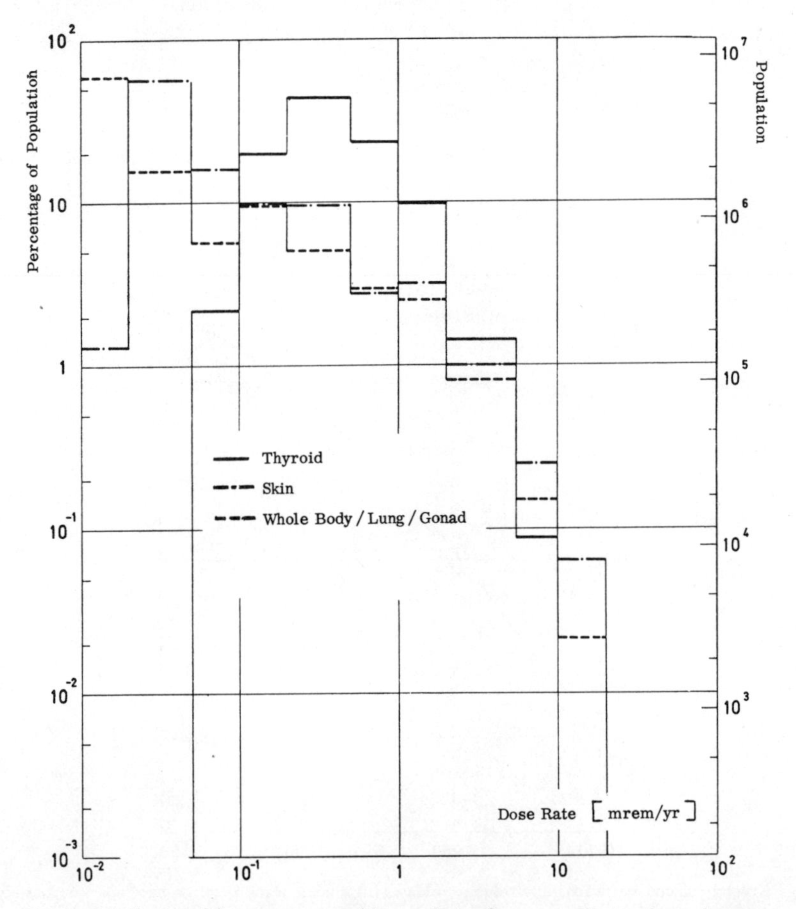

FIG. 9. Distribution of dose rates received by population from atmospheric release, by organ.

Table VII : Individual and Population Dose Rates

Organ		Regional Sources					
		Atmospheric Releases		Hydrospheric Releases		Total	
		AIDR mrem/yr	PDR man·rem/yr	AIDR mrem/yr	PDR man·rem/yr	AIDR mrem/yr	PDR man·rem/yr
Whole Body		0.13	1713	0.005	64	0.14	1777
Gonads		0.13	1715	0.005	64	0.14	1779
Lung		0.13	1716	0.005	64	0.14	1780
Skin		0.19	2462	0.005	64	0.20	2526
Thyroid from ^{131}I	0 - 5 yr	0.53	673	0.10	127	0.63	800
	5 - 15 yr	0.15	354	0.03	67	0.18	421
	Adults	0.02	209	0.004	40	0.03	249
			1236		234		1470

Organ	Global Sources (without decontamination)	
	AIDR mrem/yr	PDR man·rem/yr
Whole Body	0.015	190
Gonads	0.015	190
Lung	0.030	380
Skin	0.71	9000

AIDR = Average Individual Dose Rate

PDR = Population Dose Rate

Table VIII : Individual and Population Risk Rates

Disease		Regional Sources AIRR cases/man·yr	Regional Sources PRR cases/yr	Global Sources AIRR cases/man·yr	Global Sources PRR cases/yr	Total AIRR cases/man·yr	Total PRR cases/yr	Individual Natural Incidence [26] cases/man·yr	
Leukemia		3.0 (-9)	36 (-3)	0.3 (-9)	3.6 (-3)	3.3 (-9)	40 (-3)	1 (-5)	ICD-Pos. 200-209
Lung cancer		1.5 (-9)	18(-3)	0.3 (-9)	3.6 (-3)	1.8 (-9)	22 (-3)	6 (-4)	ICD-Pos. 160-163
Thyroid cancer	0 - 5 yr	25 (-9)	30 (-3)	0.45 (-9)	0.54 (-3)	25 (-9)	31 (-3)	0 - 30 yr ca. 2 (-6)	ICD-Pos. 193
	5 - 15 yr	10 (-9)	25 (-3)	0.45 (-9)	1.1 (-3)	10 (-9)	26 (-3)		
	adult	2 (-9)	16 (-3)	0.15 (-9)	1.2 (-3)	2 (-9)	17 (-3)	adult ca. 2 (-5)	
			71 (-3)		3 (-3)		74 (-3)		
Genetic risk (Deformation at birth) case / birth		0.6 (-6)		0.06 (-6)		0.7 (-6)		1 (-2) [19]	

AIRR - Average Individual Risk Rate

PRR - Population Risk Rate

Figure 9 shows the percentage of population as well as the number of people receiving a certain dose rate from atmospheric release to specific organs - whole body (or lung or gonads), skin and thyroid. The curve for thyroid in Fig. 9 applies to children of age up to 5 years, who constitute about 10% of the population. The number of children receiving a certain dose rate to the thyroid is 1/10th of the values obtained from Fig. 9 (scale on the right).

The population dose rate and the average individual dose rate for specific organs are given in Table VII. Age dependency with regard to thyroid dose is also indicated.

A realistic analysis of the population exposure from liquid discharges is not feasible due to the existance of different sources of drinking water in the study area. Considering that 10% of the drinking water needs is met by river water, a rough estimate of the average annual dose to the whole body and the thyroid yields 0.005 mrem/yr and 0.1 mrem/yr respectively.

5.5 Risk Estimation

Risk evaluations in general are fraught with uncertainties and this is particularly true in the region of very low doses. Harmful effects, if any, due to exposures in the neighbourhood of natural background levels have come to be accepted as tolerable and the risk presented is relatively small and is offset mostly by the benefits.

The absence of a threshold and the linearity of dose and effect have to be regarded as working hypotheses and not as proven facts [25]. Risk analysis is carried out in the present study with the presumption that the results will not be regarded as precise but represent only the magnitudes involved.

In Table VIII population and average individual risk rates are presented for certain radiation induced disorders. For cancer incidence in thyroid, total dose from all sources (iodine, tritium and noble gases) has been considered. Risk rates from natural incidence are also indicated for purposes of comparison.

6. CONCLUSION

An investigation of the radiological impact due to the growing nuclear industry in the Upper Rhine basin by the year 1985, is carried out. Exposure to the individual members of the public is well within the limits recommended by the Radiation Protection Ordinance and the accompanying risk is also far below that arising from natural incidence. The study reveals that with existing methods of effluent treatment and control, exposure levels can be maintained at a small fraction of the natural background radiation level and there is scope for reducing it further with improved effluent handling technology.

ACKNOWLEDGEMENT

The authors are indebted to Mrs. R. Kalckbrenner for evaluating the risk due to natural incidence and for preparing many of the figures in this paper.

REFERENCES

[1] ADVISORY COMMITTEE ON THE BIOLOGICAL EFFECTS OF IONIZING RADIATION , The Effects on Population of Exposure to Low Levels of Ionizing Radiation, BEIR-Report, Washington (1972)

[2] LINDELL, B., Assessment of Population Exposures, Symp. Environmental Behaviour of Radionuclides Released in the Nuclear Industry (Aix-en-Provence 1973), Proc. IAEA Vienna (1973)

[3] Neue Kernkraftwerke in Europa, Atomwirtschaft 18 (1973), 127, 179 (Main reference)

[4] SCHWIBACH, J., Radiation Protection Guides and Technical Limits for the Discharge of Radionuclides in Effluents as Currently Applied in Licensing of Nuclear Power Plants in the Federal Republic of Germany, II. International Summer School on Radiation Protection (Herceg Novi - 1973)

[5] SLADE, D.H., (Ed.). Meteorology and Atomic Energy, Report TID 24190, (1968), p. 350

[6] LASSEN, N.A., Assessment of Tissue Radiation Dose in Clinical Use of Radioactive Inert Gases, with Examples of Absorbed Doses from $^{3}H_2$, ^{85}Kr and ^{133}Xe, Minerva Nucleare 8 (1964), 211

[7] BROWNELL, G.L., ELLETT, W.H., REDDY, A.R., Absorbed Fractions for Photon Dosimetry, Journal of Nuclear Medicine, MIRD, Supplement No. Pamphlet No. 3 (1968)

[8] SNYDER, W.S., FORD, M.R., WARNER, G.G., FISHER, H.L., Estimate of Absorbed Fractions for Monoenergetic Photon Sources Uniformely Distributed in Various Organs of a Heterogeneous Phantom, Journal of Nuclear Medicine, MIRD, Supplement No. 3, Pamphlet No. 5 (1969)

[9] HINE, G.J., BROWNELL. G.L. (Eds.), Radiation Dosimetry, Academic Press, (1956), p. 694

[10] BAYER, A., KALCKBRENNER, R., KRISHNAMURTHI, T.N., SCHÜCKLER M., Radiologische Belastung durch kerntechnische Anlagen bei Normalbetrieb, KFK-Report, to be published

[11] NESTER, K. Statistische Auswertungen der Windmessungen im Kernforschungszentrum Karlsruhe aus den Jahren 1968/69, Report KFK-1606, (1972)

[12] KIEFER, H., KOELZER, W., (Eds.),
ASS-Jahresbericht 1970, Report KFK 1365 (1971)
ASS-Jahresbericht 1971, Report KFK 1565 (1972)
ASS-Jahresbericht 1972, Report KFK 1818 (1973)

[13] ABE-Ausschuß des Deutschen Atomforums
Jahresbericht 1970, Atom und Strom 17 (1971), 97
Jahresbericht 1971, Atom und Strom 18 (1972), 73
Jahresbericht 1972, Atom und Strom 19 (1973), 45

[14] GRATHWOHL, G., Erzeugung und Freisetzung von Tritium durch Reaktoren und Wiederaufarbeitungsanlagen und die voraussichtliche radiologische Belastung bis zum Jahr 2000, Report KFK-Ext 4/ 73 -36, (1973)

[15] PORZ, F. Erzeugung und Freisetzung von Jodisotopen durch Reaktoren und Wiederaufarbeitungsanlagen und die voraussichtliche radiologische Belastung bis zum Jahr 2000, Report KFK-1912, (1974)

[16] HILBERT, F., Erzeugung und Freisetzung von Krypton- und Xenonisotopen durch Reaktoren und Wiederaufarbeitungsanlagen und die voraussichtliche radiologische Belastung bis zum Jahr 2000, KFK-Report, to be published

[17] BAYER, A., Dose and Risk Considerations for the Release of ^{131}I at Special Sites, Symp. Principles and Standards of Reactor Safety (Jülich 1973), Proc. IAEA Vienna (1973)

[18] DOLPHIN, G.W., MARLEY, W.G., Risk Evaluation to the Protection of the Public in the Event of Accidents at Nuclear Installations, Report AHSB (RP) R 93, (1969)

[19] ALDER, F., Sicherheit und Risiko von Kernkraftwerken, Neue Technik 12, (1970), 201

[20] ICRP, Report of Committee II on Permissable Dose for Internal Radiation, ICRP Publication 2, (1960)

[21] HENDRICKSON, M.M., The Dose from ^{85}Kr Released to the Earth's Atmosphere, Report BNWL-SA 3233A, (1970)

[22] WHITTON, J.T., Calculations of Whole Body Dose from Absorption of an Inhaled Noble Gas, Health Physics 23, (1972), 573

[23] CLODIUS, S., Wasserversorgung im Jahre 2000, Kommunalwirtschaft 40, (1970), 216

[24] BAYER, A., KALCKBRENNER, R., KRISHNAMURTHI, T.N., Die zukünftige Tritium-Belastung des Oberrheins, Reaktortagung Berlin 1974, ZAED Karlsruhe, (1974)

[25] JAMMET, H. Aims and Needs for Environmental Programs Related to the Nuclear Industry-Panel Discussion, Symp. Environmental Behaviour of Radionuclides Released in the Nuclear Industry (Aix-en-Provence, 1973) Proc. IAEA Vienna, (1973)

[26] Statistisches Landesamt Hamburg, Hamburger Krebsdokumentation 1956 - 1971, Hamburg, (1973)

DISCUSSION

R. LE QUINIO: You say that the burden due to gaseous effluents is much greater than that due to liquids. Why, then, do you specify the same value of 30 mrem/a for both liquid and gaseous effluents?

A. BAYER: 30 mrem/a via gaseous releases and 30 mrem/a via liquid discharges are guide limits for the normal operation of nuclear reactors in Germany. The maximum dose in the neighbourhood of a modern light-water power reactor via gaseous releases lies in the region of 1 mrem/a (see Ref. [16] of the paper). The radiological impact via liquid discharge is actually smaller, but both doses are very small compared with the guide limits. Perhaps these guide limits look rather arbitrary but at the present time they are very useful.

O. ILARI: In this same connection I should like to say that I am not convinced that gaseous releases are more important than liquid discharges in terms of radiological risk; in fact, with present waste-treatment technology, and with the introduction of zero release systems for radioactive gases, I think that population doses receive a higher contribution from liquid than from gaseous effluents.

R. LE QUINIO: What is your personal opinion regarding the number of mrem/a that could be attributed to gaseous releases resulting from accidents?

A. BAYER: I am not certain that I understand your question but perhaps I can answer that the case of accidental releases of radionuclides is dealt with in publications by Schwibach and Pohl.

O. ILARI: In my opinion dose limits are intended only for normal conditions and should not be apportioned between normal situations and accidental situations. The accident situation raises different evaluation problems and cannot be treated in the same terms as the normal situation.

Turning to Table VII of the paper, I think the collective doses for hydrospheric releases presented therein are misleading, because they take into account only drinking water and not the other, probably more important, exposure pathways connected with liquid releases.

A. BAYER: To prevent any misunderstanding I would repeat that the guiding limits in our paper refer only to normal operation.

The consumption of Rhine fish is not very great, and a rough estimate has shown that this contribution is not significant.

S.O.W. BERGSTRÖM: Your population doses are evidently only integrated over a population of 13 million people. Have you made any estimate how these collective doses compare with world population totals?

A. BAYER: We calculated the population dose rates only for the people living in the Upper Rhine region. For the world population a dose rate estimate can easily be performed with the data of Table VI. Taking the 1985 data for ^{85}Kr without decontamination and assuming a world population of 5000 million at that time, the corresponding population dose rates would be about 34 000 man·rem/a for the whole body and about 3 500 000 man·rem/a for the skin. Owing to its more complicated geophysical paths, a corresponding estimate for tritium would require more extensive investigations.

K.-J. VOGT: Could you give us some information on the regional and global models you have used in calculating the ^{85}Kr concentrations?

A. BAYER: In the regional case, at distances close to a facility in the z-direction (height) we used the Gaussian formula. For greater distances we used a homogeneous mixture within the mixing layer given in Table II. More details can be obtained from Ref. [10].

In the case of global sources a homogeneous mixture within the troposphere (height ~ 10 km) was taken as being the most realistic model. The different models are discussed in Ref. [16].

CANADIAN EXPERIENCE IN ASSESSING POPULATION EXPOSURES FROM CANDU REACTORS

A.K. DAS GUPTA, H. TANIGUCHI, Mary P. MEASURES
Radiation Protection Bureau,
Health Protection Branch,
Department of National Health and Welfare,
Ottawa, Ontario,
Canada

Abstract

CANADIAN EXPERIENCE IN ASSESSING POPULATION EXPOSURES FROM CANDU REACTORS.

The Department of National Health and Welfare has carried out a programme of environmental radioactivity surveillance around nuclear reactors since 1961 to assess the effect of low levels of radioactivity releases on public health. This has been in co-operation with the department of health of the province in which the reactors are located and includes the research reactors on the Ottawa River and the Winnipeg River, the Nuclear Power Demonstration station on the Ottawa River, the prototype generating stations at Douglas Point, Ontario and Gentilly, Quebec and the 4-unit, 2000 MW(e) station at Pickering, Ontario. These reactors are of the Canada Deuterium Uranium (CANDU) type, using natural uranium dioxide as fuel, heavy water as moderator and pressure tubes in the heat transport system. The features of this type of reactor which minimize potential releases of radioactivity to the environment are briefly discussed. The surveillance programme includes measurement of external radiation from gaseous effluents, tritiated atmospheric water vapour, radioactivity on airborne particulate matter, and in milk, water and biota from reactor environs. The current levels of these are presented. Estimates are made of the radiation dose being received by a hypothetical individual at the station boundary and is compared with the contribution from radioactive fall-out and natural background. It is shown that the CANDU reactors are capable of being operated reliably such that the resulting radiation dose to the population living in the vicinity is a few per cent of the natural radiation background.

INTRODUCTION

The place of nuclear power stations in meeting Canada's energy requirements has been firmly established in the relatively short period since the 20 MW(e) Nuclear Power Demonstration reactor at Rolphton on the Ottawa River first went critical in April, 1962. Anticipating the successful demonstration of the Canada Deuterium Uranium (CANDU) concept, the design and construction of a medium size (208 MW(e)) prototype plant on the shores of Lake Huron, at Douglas Point, was started in 1959. The experience and the knowledge generated by these encouraged the Hydro-Electric Power Commission of Ontario (Ontario Hydro) to have its first large scale nuclear generating station built at Pickering, only 32 kilometers from the centre of Toronto. The station consists of 4 units each capable of producing 508 MW(e) net. This makes it the largest known nuclear electric station presently operating in the world.

A further complex for producing 3000 MW(e) (4 units of 750 MW(e) each) is under construction at the Bruce Generating Station near Douglas Point. A second provincial utility, Hydro-Quebec, is operating a 250 MW(e) unit using boiling light water for heat transport at Gentilly, Quebec. Plans for the

immediate future are for doubling the generating capacity at both the Pickerin
and Bruce sites and building an additional 600 MW(e) unit of the Pickering typ
at Gentilly.

THE CANDU SYSTEM

The Canadian power reactor design is based on the use of natural uranium fuel. Uranium dioxide in the form of compacted and sintered clylinders is sheathed and sealed in zirconium alloy tubes. Several of these are put together between end plates to form fuel bundles which are placed end to end in fuel channels running inside horizontal tubes of a calandria. The calandri which contains heavy water as the moderator and reflector, is housed in a concrete reactor vault. High pressure and temperature generated as a result o fission is confined to the fuel channel coolant tubes and the calandria is not subjected to these stresses. The heat is carried by a flow of pressurized heavy water and eventually transferred to natural water in heat exchangers for the production of steam to run the turbines.

Typically, boosters and control rod absorbers are used to regulate reactivity but other control mechanisms are also provided. Apart from such process systems designed for the normal operation of the reactor and associate turbines, a functionally independent protective system is included to limit th adverse consequences of any failure of the process system. The protective system, has, among other features, capabilities for emergency reactor shut dow emergency core cooling, and containment. The newer reactors have two emergenc shut down systems which are engineered to be independent of each other and of the process regulatory system. The design philosophy for containment is to provide pressure relief and adequate space to contain safely the expanded air and steam that may arise from the worst accident situation postulated. For multiple unit stations, this is achieved by building reinforced concrete housi around individual units and connecting them to a single vacuum reservoir equip with water dousing.

From the environmental safety point of view the CANDU system provides several barriers to the movement of fission products from the fuel elements to the outside. These are:

(a) the low rate of diffusion from the ceramic pellets of fuel
(b) the sheathing of the fuel
(c) the confinement of heat and pressure to the fuel channels, and,
(d) the structural containment.

REGULATORY CONTROL OF NUCLEAR ENERGY

In Canada, all dealings in atomic energy are subject to the federal Atomic Energy Control Act which is administered by the Atomic Energy Control Board. Proposals for constructing and operating nuclear reactors received by the Board are referred to its Reactor Safety Advisory Committee. This Commit is composed of Board staff and experts drawn from various agencies having interests in safety and other aspects. The federal department of health is represented along with a representative from the health department of the province in which the generating station is to be located.

Following exploratory discussions of potential sites, the applicant provides sufficient details of the selected site, the reactor and its associa equipment to support a favourable independent safety analysis by the Board st and this Committee. On the recommendations of this Committee, the Board issu a Construction Licence. An Operating Licence is granted after the reactor an auxiliary systems have been satisfactorily tested and commissioned, including aspects of the operating procedure. These licencing requirements have been adequately described elsewhere[1,2] and will not be discussed further here.

Great emphasis is placed on component quality and the reliability of performance of each system separately and independently. This principle of "defence-in-depth" reduces the probability of any accidents of serious consequence to the public. From a health point of view, therefore, the main interest is in the long-term effects of normal effluents.

The regulatory levels for exposure of the population at risk are based on the current recommendations of the International Commission on Radiological Protection. For an individual member of the public at the exclusion boundary these are:

0.5 rem/year for the whole body and
3 rem/year to the thyroid.

In addition, a maximum population dose for each reactor site has been set at 10^4 man.rem/year.

ENVIRONMENTAL SURVEILLANCE

As technology improves and the need for energy increases, power reactors tend to become larger in size. Further, because it is economically attractive, the arguments for building them close to population centres become stronger. The need for systematic environmental surveillance, therefore, assumes great importance. Such functions in Canada are undertaken by a number of groups according to its interests, including the operator of an installation and public authorities such as health and environment departments.

For developing a monitoring program the ICRP recommends[3] that the critical group concept be followed. A critical group is a homogenous group of people whose location, age, eating habits, life style and personal habits may cause them to receive a relatively higher dose than the average received by others. In a built up site like Pickering where the dietary habits are typical of most urban areas in Canada, the influence of local produce is not very substantial. Nevertheless, as the radiation dose that can be received by the population is mainly from air- and water-borne activities and their associated pathways, it is appropriate to consider a hypothetical group which lives close to the exclusion area and which uses locally produced vegetables, fruits, milk, meat and fish.

The Department of National Health and Welfare conducts a nation-wide program of radiological surveillance to determine the extent and nature of radioactive pollution of the environment. The Department also carries out radioactivity surveillance in the environs of nuclear reactors in cooperation with the Ontario Ministries of Health and Environment. Similar arrangements exist with other provinces. The data obtained permit estimation of radiation exposure to the population due to reactor operations. The radiation exposure to a population is from inhalation, ingestion and external exposure. These factors are monitored as follows:

Air

Air-borne particulate matter is collected on filters at both the cross-country sampling sites (Fig. 1) and the reactor sampling sites. Filters are changed every seven days and analyzed for gross β and gamma activity.

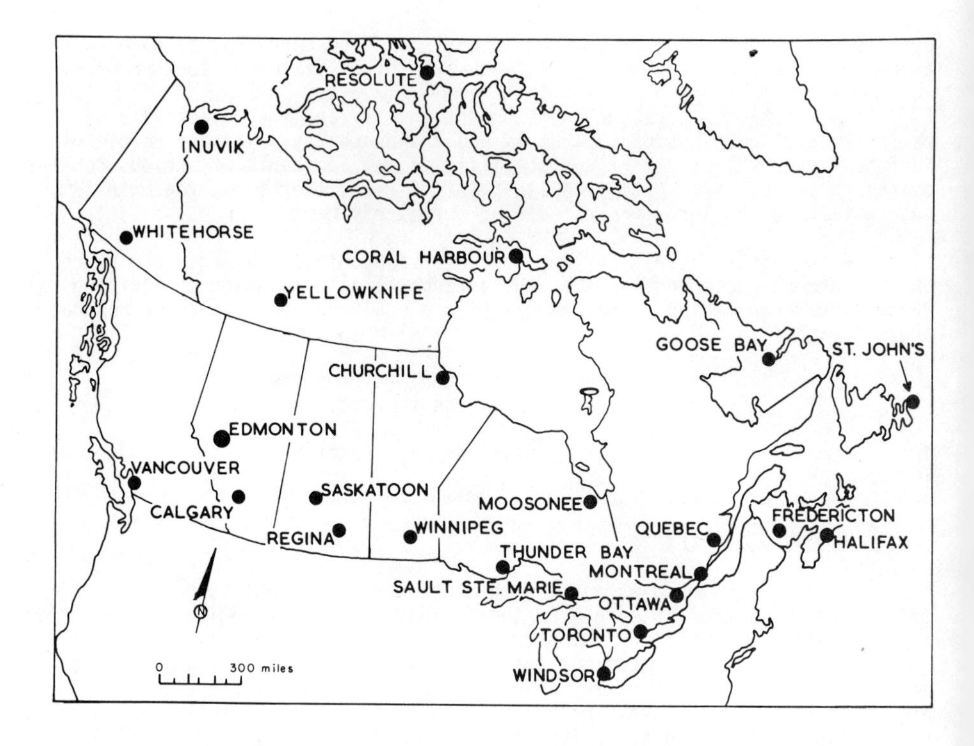

FIG. 1. Canadian air and precipitation sampling stations.

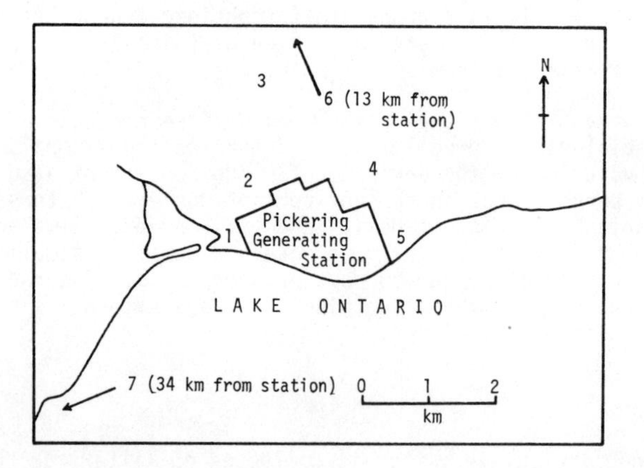

FIG. 2. Air water vapour sampling locations in the vicinity of the Pickering generation station.

The tritium concentration of air water vapour in the vicinity of the generating stations is also measured since tritium is a major component of releases from Canadian reactors. Continuous monthly samples are collected by means of molecular sieve traps [4] set up in populated areas near the reactor sites. At Pickering, for example, five sampling sites are located outside the boundary of the generating station (Fig. 2); one is located 13 km northwest of the station and a control site is located 34 km southwest from the station.

External Exposure

External radiation dose from air-borne activity and ground deposits is measured by manganese-activated calcium fluoride thermoluminescent dosimeters with a minimum detectable level of 0.5 mR. Dosimeters in replicate are placed at most of the cross-country sampling sites and at the tritium sampling sites in reactor areas. They are changed monthly.

Milk

Samples of fresh milk are obtained weekly from sixteen Canadian cities and from dairy farms located in the vicinity of generating stations. In reactor areas, the atmosphere→pasture→cow→milk→man pathway is an important food chain leading to population exposure. Samples from reactor areas are analyzed weekly for ^{3}H, ^{131}I and ^{137}Cs. Samples from the fallout network are analyzed monthly for ^{137}Cs and quarterly for ^{90}Sr.

Water and Precipitation

Rain and snow samples are collected at twenty-four sampling sites. Precipitation is accumulated in special collecting pots for one month and then analyzed for gross-β radioactivity.

Samples of unfiltered water are collected daily from municipal filtration plants near all reactor sites. Monthly composites are analyzed for gross α, gross β, ^{3}H, ^{60}Co, ^{89}Sr, ^{90}Sr and ^{137}Cs levels. In addition, grab samples from surrounding waters are taken periodically for gross α, gross β and ^{3}H analysis. Since raw water is used in all analyses for greater sensitivity, the activity measured represents concentrations in the lakes and rivers rather than lower levels reaching the population after water treatment. Where applicable, well waters are also analyzed.

Biota

Biota taken from the environs of nuclear reactors may be consumed by members of the public. Dominant fish species are, therefore, collected each spring and autumn from the vicinity of the power stations and analyzed for ^{90}Sr levels in the bone and ^{137}Cs in the flesh. Rabbits are collected and analyzed to indicate the levels found in a land animal which may be used as food. Levels of ^{90}Sr and ^{137}Cs in a filamentous alga are determined as a measure of their concentrations in water. This grows attached to submerged rocks in the vicinity of the Pickering Generating Station.

A summary of the Federal-Provincial monitoring program in Ontario is given in Table I.

The operator of a nuclear generating station has the responsibility of demonstrating compliance with the release limits specified in the Operating Licence through on-site measurements of the gaseous and liquid effluents.

TABLE I. FEDERAL-PROVINCIAL MONITORING PROGRAM IN ONTARIO

Sample	Location	Frequency	Analyses	By
Air				
Water vapour	Outside station boundary and reference location	Continuous monthly	^{3}H	F
TLD	As for water vapour	Dosimeters changed monthly	Integrated monthly gamma dose	F
Particulates	Outside station boundary	Filters changed every 7 days	gross β, γ	P
Milk	Farms near station	Weekly	^{131}I, ^{137}Cs, γ, ^{3}H	F-P
Water				
Drinking water	Municipal pumphouses	Monthly composite	^{89}Sr, ^{90}Sr, ^{137}Cs	F
		Weekly composite	gross α, gross β, ^{3}H, γ	P
Surface water	Lake and river	Lake - annual grab	gross α, gross β, ^{3}H	P
		River - monthly grab	gross α, gross β, ^{3}H	P
Well water	Vicinity of station	Annual grab	gross α, gross β, ^{3}H	P

TABLE I. (cont.)

Sample	Location	Frequency	Analyses	By
Biota				
Fish	Ottawa River - above and below reactor	twice/yr	^{90}Sr, ^{137}Cs	F
	Lakes - near reactor areas	twice/yr	^{90}Sr, ^{137}Cs, γ, ^{3}H	F-P
Rabbit	Reactor environs - Chalk River	twice/yr	^{90}Sr, ^{137}Cs	F
Algae	Lake - near reactor area	twice/yr	^{90}Sr, ^{137}Cs	F
Misc. (fish, crayfish, hare, clams, seagulls, snails, etc.)	Reactor environs	twice/yr	gross α, gross β	P
Urine	Local Hospital	bi-weekly	^{3}H, γ	F

F - Federal

P - Provincial

TABLE II. ONTARIO HYDRO MONITORING PROGRAM [5]

Sample	Location	Frequency	Analyses
Air			
Water vapour	At station boundary and reference location	Continuous monthly	^{3}H
TLD	At air water vapour sites	Dosimeters changed quarterly	Integrated quarterly gamma dose
Integrating dose rate meter	At one suitable TLD site per station	Quarterly	Integrated quarterly gamma dose
Precipitation	As for air water vapour	Quarterly	^{3}H
		Quarterly composite of site buckets	Gross β
Milk	Composite from farms within 10 km of station	Monthly in Summer (April - October)	^{3}H, ^{131}I
Water			
Surface water	Cooling water effluent	Weekly composite	Gross β
		Quarterly composite	^{3}H, specific nuclides
Drinking water	Municipal pumphouse	Quarterly composite	^{3}H, Gross β (specific radionuclides if $>10^{-7}$ μCi/ml gross β)
Fish	Near station outfall	Semi-annually	Gamma spectrometric analysis

In addition, it is in his interests to verify independently the absence of any unsuspected releases of radioactivity and to verify dispersion parameters through an environmental monitoring program. The studies which are being carried out by Ontario Hydro at its largest nuclear station at Pickering are summarized in Table II [5].

CURRENT RADIOACTIVITY LEVELS AND POPULATION DOSE

Any attempt to measure the contribution of nuclear power reactors to environmental radioactivity and any estimate of the resulting radiation dose to the general population must take into account the presence of residual fission products in the environment from nuclear weapons tests. The national fallout measurement program of the Radiation Protection Bureau provides convenient reference data for the evaluation of radioactivity from nuclear power reactors.

The levels of radioactivity in the various media sampled near the Pickering Generating Station during 1973 are given in Table III and compared with values observed for control stations and for fallout measurements. (Since the values are subject to seasonal fluctuations, the standard deviations are high.) Of all the specific radionuclides measured, tritium, in the absence of thermonuclear weapons tests, can be clearly associated with the release of activated heavy water. The average concentration during 1973 is estimated to have been 150 pCi/m^3 of air at ground level at stations about 1 km from the station.

The levels of radionuclides found in milk sampled from farms in the vicinity of nuclear power stations were uniformly low. ^{131}I arising from defective fuel elements was not detected and the tritium concentration was

TABLE III. ENVIRONMENTAL RADIOACTIVITY LEVELS, PICKERING GENERATING STATION, 1973

Sample	Radionuclide	Average Concentration	
		Control Station	Reactor Environs
Air (pCi/m^3)	^3H	< 1	150 ± 259
Milk (pCi/ℓ)	^3H	<1000	<1000
	^{131}I	< 10	< 10
	^{137}Cs	6.7 ± 3.5	7.1 ± 3.3
Water (pCi/ℓ)	^3H	<1700	<1700
	^{60}Co	< 10	< 10
	^{89}Sr	0.04 ± 0.04	0.05 ± 0.06
	^{90}Sr	0.95 ± 0.09	0.87 ± 0.07
Fish (pCi/g fresh wt)	^{137}Cs	0.048 ± 0.044	0.044 ± 0.011

The high standard deviations are mainly due to seasonal fluctuations generally less than the routine detection limit of 1000 pCi/ℓ. Low Levels of ^{137}Cs continue to be found in milk. It is, however, clear from the measurements of milk taken from several large milksheds distant from the reactor area that the levels being observed can be attributable to previous nuclear weapons testing and the continued testing in the atmosphere on a reduced scale. The average value for Southern Ontario milk was 6.2 pCi/ℓ during 1973.

Water supplies for communities drawing from the same body of water providing station cooling water were analyzed for ^{3}H, ^{89}Sr, ^{90}Sr and ^{137}Cs and were scanned for gamma radioactivity. Comparisons with values observed for control stations show no differences due to operational releases from generating stations.

The use of fish caught in these waters by sports fisherman or for commercial purposes contribute to the intake of radioactivity. Because of the potential for concentration of radionuclides along the food-chain, it is important to estimate the contribution from this source. For purposes of calculation of radiation dose to the hypothetical individual, a consumption rate of 15 grams/day/person is assumed.

ESTIMATES OF POPULATION DOSE

One of the primary objectives of the various measurements which have been described is to assess the radiation dose which the hypothetical individual may receive from the operation of a nuclear generating station and to estimate the integrated dose to the critical group. Health authorities may attempt to do this through representative sampling techniques at the station/population interface and sensitive analytical methods.

TABLE IV. DOSE TO STANDARD MAN AT THE BOUNDARY OF PICKERING GENERATING STATION, 1973

Source	Radionuclide	Dose (mrem/yr)	
		Control Station	Reactor Environs
Air	^{3}H	0	0.33
Milk*	^{137}Cs	0.03	0.03
Water	^{90}Sr	0.25	0.23
	^{137}Cs	0.002	0.003
Fish**	^{137}Cs	0.016	0.015

* adult consumption rate of 0.2 ℓ/day

** adult consumption rate of 15 g/day

The estimates of radiation dose are summarized in Table IV. The measurement of tritiated water vapour concentrations in the atmosphere collected at off-station sampling points has enabled an estimate of the radiation dose due to inhaled tritium to be made. It is apparent that this is the major source of population dose. For Standard Man at the station boundary, this dose is calculated to be 0.33 mrem/yr.

The operators of the station may estimate radiation doses to the population from their knowledge of the specific radionuclides released, their quantities and their pathway to man. For such calculations, several factors have to be considered and various assumptions made. Experimental verification of such assumptions is not always possible when release levels are very low. In the case of tritium, however, the levels are measurable. Using values of the amount of tritium released from the stack each month[6] and the monthly measurement of tritium in air concentrations made as described, it is possible to calculate the dilution factor. The annual average value of this factor, K_a, ranged from 0.6 x 10^{-7} to 5 x 10^{-7} sec/m^3 for sites 1 to 5 shown in Fig. 2. The average K_a value for the 4 sites approximately 1 km from the station was 2 x 10^{-7} sec/m^3 over a 2 year period.

This direct measurement of the dilution factor may be compared with the value of 1 x 10^{-7} sec/m^3 for distances between 1.6 and 9.6 km obtained by Barry[7] from a survey of experimental measurements. For meteorological conditions in the Pickering area, the weighted mean dilution factors presented by Bryant[8] are considered to be applicable. For an effective release height of 20 m, a value of 9 x 10^{-7} sec/m^3 is estimated for K_a at 1 km.

Assuming that deposition losses and wash-out do not significantly affect the value of K_a derived from tritium oxide, the same factor could be applied to the radioactive isotopes of argon, krypton and xenon. From the known releases of radioactive noble gases, the external radiation dose at the station boundary is expected to be approximately 3 mrem/yr above background. Such small increases are not detectable by the TLD monitors due to normal fluctuations in background.

Thus, the total radiation dose that may be received by the hypothetical individual is less than 4 mrem/yr. The current estimate of dose to the same individual due to fallout is approximately 2.3 mrem/yr[9].

CONCLUSION

The construction of nuclear power reactors in Canada and in the United States of America is proceeding at a rapid pace. The recent indications of shortage of sources of energy can be expected to accelerate the trend further. Conservative estimates of installed nuclear capacity in the province of Ontario alone suggest approximately 6000 MW(e) by 1980 and more than 20000 MW(e) by 1990 [10].

A special consideration is that many of the installations are likely to be along the shores of the Great Lakes. Where many reactors discharge into a common watershed and each reactor releases radioactivity at a rate permitted by the ICRP recommendations for maximum permissible doses to the individual in the population, the levels of contamination in and around the lakes would eventually approach the maximum permissible level. This would not be desirable for the large population living around the Great Lakes.

Quantitative assessments of health effects of radiation at very low levels are subject to many uncertainties. Recently, the National Academy of Sciences (U.S.A.) published a report entitled "The Effects on Populations of Exposure to Low Levels of Ionizing Radiation" [11]. This report cites evidence that lethal cancers as well as leukemia are important long-term consequences of low levels of radiation. Many of its findings support the contention that any exposure carries with it some risk of somatic or genetic damage. In view of this, it seems essential to follow ICRP's more general recommendation that radiation doses should "be kept as low as is readily achievable, economic and social considerations being taken into account" [12].

The average annual whole body dose to an individual of the population in the immediate vicinity of the Pickering reactor site shows that it is possible to operate a large nuclear generating station with only a few mrem/yr exposure to the population. It can be concluded that the average dose to over two million persons living in the adjoining cities is within the range of normal fluctuations in natural background.

It is suggested that the average population exposure from power reactors should not be more than a small fraction of natural background radiation and should preferably be within the normal range of its fluctuations. Canadian reactors intend to operate to a target not exceeding 1% of the regulatory limit. The results of the surveillance programs as outlined in this paper confirm that this intention is reasonably achievable. There seems no reason to require a poorer standard in the future.

ACKNOWLEDGEMENTS

The authors are thankful to Dr. A.H. Booth, Director, Radiation Protection Bureau for his suggestions and continued interest in limiting population exposures. They are thankful to Dr. A.H. Aitken, Ontario Ministry of Health, for providing data from his programs. The technical assistance from the staff of the Bureau is gratefully acknowledged. The varied samples were received through the cooperation and assistance of many people at the sampling sites.

REFERENCES

[1] HURST, D.G., BOYD, F.C., Reactor Licensing and Safety Requirements, Proc. Ann. Conf. Can. Nuclear Assoc., 1972, Paper 72-CNA-102 (1972).

[2] JENNEKENS, J.H., Safety Aspects of Nuclear Plant Licensing in Canada, Atomic Energy Control Board Rep. AECB-1062 (1973).

[3] International Commission on Radiological Protection, Principles of Environmental Monitoring Related to the Handling of Radioactive Materials, ICRP Publication 7, Pergamon Press, Oxford (1966).

[4] VASUDEV, P., MEASURES, M.P., FENNING, J.M., TANIGUCHI, H., Collection and Measurement of Tritium in Atmospheric Water Vapour, Dept. National Health and Welfare, Internal Rep. IR-134 (1973).

[5] WONG, K.Y., KNIGHT, G.B., Routine Environmental Monitoring Program for Ontario Hydro's Nuclear Power Stations, Ontario Hydro Rep. (1971), revised to April, 1974.

[6] WONG, K.Y., Ontario Hydro, Personal Communication (1973).

[7] BARRY, P.J., Maximum Permissible Concentrations of Radioactive Nuclides in Airborne Effluents from Nuclear Reactors, Atomic Energy of Canada Limited Rep. AECL-1624 (1962).

[8] BRYANT, P.M., Methods of Estimation of the Dispersion of Windborne Material and Data to Assist in Their Application, United Kingdom Atomic Energy Authority Rep. AHSB (RP) R.42 (1964).

[9] TANIGUCHI, H., MCGREGOR, R.G., MEASURES, M.P., VASUDEV, P., Environmental Radioactivity Surveillance for January - June, 1973, Dept. of National Health and Welfare, Internal Rep. IR-126 (1973).

[10] SMITH, H.A., Nuclear Power in Ontario, Proc. 13th Ann. Int. Conf. Can. Nuclear Assoc., 1973, Paper CNA '73-201 (1973).

[11] The Effects on Populations of Exposure to Low Levels of Ionizing Radiation, Rep. of the Advisory Committee on the Biological Effects of Ionizing Radiation, National Academy of Sciences - National Research Council (1972).

[12] International Commission on Radiological Protection, Recommendations of the International Commission on Radiological Protection, ICRP Publication 9, Pergamon Press, Oxford (1966).

DISCUSSION

R. BITTEL: In one of your tables the ^{3}H contents of milk and water are shown as less than 1000 pCi/litre. Could you specify these values more exactly? Are there any significant differences between the control zones and the exposed zones near to nuclear installations?

A.K. DAS GUPTA: For routine measurements we use the 1000 pCi/litre level; however, we do have the possibility of tritium enrichment for estimating levels more accurately. The concentrations of tritium in air at stations close to the boundary are significantly different from those at the control stations located several miles away.

P. SLIZEWICZ: Is any kind of co-operation provided for between the various operators located on Lake Ontario, especially the United States operators?

A.K. DAS GUPTA: All programmes in Canada are co-ordinated by Federal authorities in Ottawa under the Atomic Energy Control Act. As regards contamination levels in the Great Lakes, talks are currently in progress between United States and Canadian authorities for the purpose of setting standards.

D. BENINSON: I would just like to mention that the limit of 10^4 man·rem/a quoted in your paper is consistent with my tentative figure of 10 man·rem/MW·a for reactors of the order of 1000 MW(e) capacity.

Monitoring programmes for the determination of population doses

(Session V)

Chairman: W.J. BAIR (United States of America)

СОВРЕМЕННЫЕ ПРОБЛЕМЫ И НЕКОТОРЫЕ НОВЫЕ КОНЦЕПЦИИ В ОБЛАСТИ НОРМИРОВАНИЯ ОБЛУЧЕНИЯ ПРОФЕССИОНАЛЬНЫХ РАБОТНИКОВ И НАСЕЛЕНИЯ

М.М.САУРОВ, В.А.КНИЖНИКОВ, А.Д.ТУРКИН
Институт биофизики Министерства
здравоохранения СССР,
Москва,
Союз Советских Социалистических Республик

Abstract–Аннотация

CURRENT PROBLEMS AND NEW CONCEPTS IN STANDARDS FOR THE EXPOSURE TO RADIATION OF WORKERS AND WHOLE POPULATIONS.

The availability of information on the occurrence of somato-stochastic effects in persons exposed to radiation has created a need for new approaches in establishing a sound basis for the standards for permissible exposure. The authors make a critical analysis of the concept of acceptable risk and introduce for consideration the concept of the impermissibility of actual risk. This concept is, essentially, that permissible risk should be interpreted as such risk of increased mortality from malignant neoplasms as cannot actually be detected in a given group of people throughout one generation. Adoption of the number of deaths from malignant neoplasms as the criterion for permissible risk requires partial reappraisal of the present classification of the critical organs. It is proposed that the division of the critical organs into groups should be governed by the damage caused by exposure of the organ to radiation as manifested by the development of neoplasms. It follows from the data on the dose/effect relationship given in the paper that, depending on the tumour yield per unit dose, the organs can be divided into three groups: Group I: red bone marrow, stomach, lungs and uterus; Group II: mammary glands, ovaries, kidneys, bladder, liver, intestine, pancreas, central nervous system, and any other organs except those contained in Groups I and III; Group III: bone, thyroid and skin. The mean permissible yearly dose (mean dose received by each member of a group or population) and yearly dose limit are established for workers and whole populations as a function of the numbers involved. The yearly dose limit makes allowance for probable non-uniformity of exposure. The authors quote mathematical expressions for determining the permissible risk and dose, and also give numerical values for permissible doses to the whole body and critical organs, both for occupationally exposed groups of different size and for the population as a whole.

СОВРЕМЕННЫЕ ПРОБЛЕМЫ И НЕКОТОРЫЕ НОВЫЕ КОНЦЕПЦИИ В ОБЛАСТИ НОРМИРОВАНИЯ ОБЛУЧЕНИЯ ПРОФЕССИОНАЛЬНЫХ РАБОТНИКОВ И НАСЕЛЕНИЯ.

В связи с появлением информации о возникновении соматикостохастических эффектов под влиянием облучения появилась необходимость в изыскании новых подходов к обоснованию нормативов допустимого облучения людей. В докладе критически анализируется концепция приемлемого риска и выносится на обсуждение концепция недопустимости реального риска. Суть этой концепции сводится к тому, что допустимым следует считать такой риск увеличения смертности от злокачественных новообразований, который не может быть реально обнаружен на данном контингенте людей в течение всей жизни поколения. Принятие в качестве критерия допустимого риска числа смертей от злокачественных новообразований требует частичного пересмотра действующей классификации критических органов. Распределение критических органов по группам предлагается поставить в зависимость от того ущерба, который приносит облучение органов с точки зрения развития новообразований. Из данных о зависимости "доза-эффект", представленных в докладе, следует, что в зависимости от выхода опухолей на единицу дозы органы распределяются на 3 группы. I группа: красный костный мозг, желудок, легкие, матка. II группа: молочные железы, яичники, почки, мочевой пузырь, печень, кишечник, поджелудочная железа, центральная нервная система и другие органы, кроме входящих в I и III группы. III группа: кость, щитовидная железа, кожа. Для профессиональных работников и населения в зависимости от численности устанавливается средняя допустимая годовая доза (средняя доза, получаемая каждым

членом коллектива, популяции) и предел годовой дозы. Последний учитывает вероятную неравномерность облучения людей. В докладе приведены математические выражения для определения допустимого риска и допустимых доз и числовые значения допустимых доз облучения всего тела и критических органов как для различных по численности профессиональных групп, так и для населения.

Материалы последних лет, относящиеся к проблеме отдаленных соматических последствий воздействия малых доз ионизирующего излучения, указывают на то, что выход злокачественных опухолей в результате облучения носит фактически беспороговый характер, т.е. при достаточно большой популяции облучение в самых малых дозах может индуцировать дополнительные случаи рака. В частности, весьма обстоятельные аргументы в пользу беспорогового характера бластомогенных эффектов радиации были представлены от имени 50 ведущих ученых США на III Международном конгрессе по радиационной защите (Вашингтон, 9-15 сентября 1973 г.) в докладах Комара, Клоу и Аптона. Концепция беспорогового характера бластомогенных эффектов облучения поддерживается также публикациями МКРЗ № 9 [1] и № 14 [2], материалы которых позволяют дать удовлетворительную по своей точности количественную оценку зависимости "доза-эффект" для индуцирования опухолей, если и не во всем диапазоне доз, то во всяком случае в практически важной области профессиональных уровней облучения.

Таким образом, в свете современных представлений следует констатировать, что концепция беспорогового риска увеличения смертности от злокачественных заболеваний с научной точки зрения является наиболее обоснованной платформой для нормирования радиационных воздействий. Это положение представляется справедливым как для профессиональных работников, так и для населения, поскольку риск увеличения смертности от рака примерно равен или даже больше риска, обусловленного генетическими повреждениями (материалы Национальной Академии Наук США и МКРЗ).

Из указанных положений вытекает, что существующие теоретические основы нормирования нуждаются в принципиальном пересмотре, поскольку принятая ныне МКРЗ и большинством стран система базируется, как известно, на концепции порога действия радиации. Существующие пределы доз для профессиональных работников установлены ниже предполагавшегося порога повреждающего действия радиации. Тем самым принято считать, что существующие нормативы гарантируют полную безопасность и не сопряжены с риском проявления каких-либо эффектов облучения. Нормативы для отдельных лиц из населения установлены на аналогичной основе с введением дополнительного коэффициента безопасности.

Принятие концепции беспороговости неизбежно означает отказ от привычного представления о полной безвредности облучения в пределах установленных норм. Возникает новая для практики нормирования и сложная проблема — обоснование допустимой степени ущерба (риска).

При решении данной проблемы логично исходить из универсального принципа "вред-польза", нашедшего применение в практике нормирования дозовых нагрузок на пациентов при радиоизотопных процедурах [3].

Однако пока возможно оценить с той или иной степенью точности лишь отрицательные последствия облучения. Выразить в сравнимых с вредом показателях пользу от применения ядерной энергии в народном хозяйстве пока затруднительно. Использование излучений в медицинских целях представляет собой в этом отношении исключение, поскольку риск облучения и польза диагностических процедур в конце концов могут быть выражены в одинаковых показателях общественного здоровья. Например, риск учащения смертности от рака легких за счет доз, полученных при профилактической флюорографии, может быть соотнесен с пользой, выражающейся в своевременном выявлении туберкулеза, а также случаев "спонтанного" рака легких. Изложенное выше позволяет нам рекомендовать, чтобы нормативы облучения за счет медицинских процедур устанавливались особо, в соответствии с медицинскими показателями, возрастом и полом пациента на основании соотношения ожидаемой пользы и возможного риска от процедуры.

К сожалению, как было отмечено, в отношении профессиональных работников и населения применить принцип "польза-вред" из-за отсутствия соответствующих критериев пока не представляется возможным. В этих условиях появились предложения нормировать облучение на основании концепции "приемлемого риска" [4,5]. Согласно этим предложениям, риск от облучения может быть принят на уровне риска от других факторов, действующих в быту, на производстве, на транспорте, с которым общество примирилось и считает его приемлемым. Например, для профессионалов предлагается принять риск облучения, соответствующий риску умереть от несчастного случая в различных других отраслях человеческой деятельности, считающихся "благополучными". Поскольку риск этот существенно колеблется (от 1-10 случаев на 1 млн. человек в год в обувной, текстильной промышленности и до 1000-10 000 случаев в авиации, рыболовецком флоте), выбор "благополучного" уровня представляется достаточно произвольным.

Не может не вызвать возражения также и то, что по концепции приемлемого риска облучение как бы навязывается. Для профессиональных работников это означает, что уровень риска, который для них допускается, не зависит от их действий, соблюдения ими инструкций. В то время как несчастные случаи на производстве, хотя в известной степени и закономерны для данного рода деятельности, не являются чем-то заранее заданным и в значительной мере зависят от действий конкретных лиц, состояния техники безопасности и т.д.

Подобный подход к определению допустимого риска не приемлем и для населения, ибо, хотя бытовые несчастные случаи, аварии на транспорте и др. пока еще неизбежны, тем не менее общество ведет с ними постоянную борьбу, а индивидуум имеет возможность выбора. По концепции же приемлемого риска общество должно сознательно идти на людские потери, а отдельный индивидуум – на риск, причем в значительной части это касается людей, деятельность которых не имеет никакого отношения к источникам облучения, и, следовательно, риск облучения в этом случае непосредственно не компенсируется.

Эти обстоятельства делают концепцию приемлемого риска непривлекательной, что также не может не учитываться при разработке новых теоретических основ нормирования радиационных факторов. Возникает необходимость изыскания иных концепций и путей обоснования допустимого риска облучения.

Одна из них выносится для обсуждения на данном семинаре МАГАТЭ. Суть этой концепции сводится к тому, что за допустимый принимается такой риск увеличения смертности от злокачественных новообразований, который не может быть реально обнаружен на данном контингенте людей в течение всей жизни поколения. Иначе говоря, риск облучения считается допустимым, если он укладывается в пределы статистической ошибки показателя смертности от спонтанных новообразований. В этом случае риск облучения становится по существу теоретической величиной, не поддающейся достоверному обнаружению. В этой концепции сохраняет свою силу известное положение, лежащее в основе нормативов для пороговых агентов, согласно которому для населения, а в СССР – и для профессиональных работников, допустимыми считаются лишь уровни воздействия, полностью безопасные для организма человека.

Разница заключается лишь в том, что в предлагаемой нами концепции, в соответствии со стохастическим характером вызываемых облучением эффектов, возможность повреждения рассматривается на популяционном уровне, а не на уровне индивидуума.

Для определения конкретных величин допустимого риска можно использовать простейшее математическое выражение, по которому в демографической статистике принято оценивать ошибку показателя смертности (m):

$$m = \pm\sqrt{\frac{P \cdot Q}{n \cdot M}}$$

где m – статистическая ошибка показателя смертности (допустимый риск облучения);

P – величина показателя смертности от "спонтанных" новообразований (число умерших в расчете на $10^3 - 10^7$ человек в 1 год в зависимости от численности облучаемых контингентов);

n – конкретная численность облучаемых контингентов;

M – средняя продолжительность жизни. Для профессиональных работников с учетом их поступления на работу с 18-20 лет средняя продолжительность предстоящей жизни принимается равной 50 годам, для населения – 70 годам.

Предлагаемый способ оценки риска облучения учитывает численность облучаемых контингентов, что весьма важно, так как общий ущерб, обусловленный повышенной смертностью от злокачественных новообразований, определяется не только уровнем дозы облучения, но и числом облучаемых людей. Не безразлично, подвергается ли риску облучения часть людей или, например, все население. При небольших численностях риск дополнительного возникновения опухолей будет менее 1,0, т.е. становится величиной теоретически нереальной.

В действительной жизни радиационному воздействию могут подвергаться лишь определенные контингенты профессиональных работников и населения. Численность профессиональных работников в разных странах в ближайшее десятилетие по примерным подсчетам может варьировать в пределах $10^3 - 10^6$ человек, а численность населения, которое подвергнется воздействию повышенных уровней радиационного воздействия – $10^4 - 10^7$ человек. Для этих контингентов и целесообразно вводить нормативы.

По найденной величине допустимого риска, на основе зависимости "доза-эффект", определяется уровень средней допустимой годовой дозы (СДД). Устанавливается также предельно допустимое значение годовой дозы (ПДД). ПДД учитывает вероятную неравномерность облучения людей в реальных условиях. Если СДД меняется в зависимости от численности облучаемого контингента, то ПДД сохраняется во всех случаях не более: для профессиональных работников – 6 бэр/год, для населения – 0,5 бэр/год. При этом получается, что для контингентов малой численности разрыв между СДД и ПДД невелик, для контингентов большой численности – значителен (порядок и более).

При расчете СДД используется следующее выражение:

$$СДД = \frac{m}{N}$$

где N – число случаев смерти от злокачественных новообразований за 1 год при облучении данной численности профессиональных работников или населения в дозе 1 бэр/год.

Величина N может быть получена из соотношения "доза-эффект" для злокачественных новообразований. Суммирование накопленной по этому вопросу информации позволяет определить дозовую зависимость для наиболее распространенных новообразований. При современных естественных уровнях смертности облучение в дозе 1 бэр обусловливает следующее число смертей от различных новообразований (табл.I).

ТАБЛИЦА I. ЧИСЛО СМЕРТЕЙ ОТ РАЗЛИЧНЫХ НОВООБРАЗОВАНИЙ ПРИ ОБЛУЧЕНИИ В ДОЗЕ 1 БЭР

Наименование новообразования	Естественный уровень смертности, число случаев на 10^6 чел. в год	Число смертей при облучении в дозе 1 бэр на 10^6 чел. в год
Лейкозы и другие лимфо- и гемобластозы	50	0,7 - 3,0
Злокачественные опухоли желудка	500	0,5 - 2,0
Злокачественные опухоли костей	10	0,04
Злокачественные опухоли поджелудочной железы	40	0,3
Злокачественные опухоли легких	300	1,8 - 3,0
Злокачественные опухоли матки	200	1,4 - 1,6
Злокачественные опухоли грудной железы	30	0,3
Злокачественные опухоли органов мочевой системы	40	0,5
Злокачественные опухоли щитовидной железы	5	0,05
Опухоли центральной нервной системы	15	0,5
Опухоли кожи	10	0,05
Прочие новообразования	~300	1
Все новообразования	~1500	7 - 12

Наиболее значителен выход на единицу дозы новообразований кроветворной системы, опухолей желудка, легких и матки. В то же время, например, злокачественные опухоли щитовидной железы, возникающие со сравнительно высокой частотой (1%), в связи с низким естественным уровнем смертности дают всего 0,05 случая в год на 10^6 человек.

Общее число смертей от злокачественных новообразований при облучении всего тела в дозе 1 бэр с учетом современных уровней смертности от "спонтанных" новообразований, равных 1500-2000 случаев смерти на 10^6 человек в год, можно принять равным в среднем 10 случаям на 10^6 человек в год. При переходе от этой величины к значению N, определяемому для различных по численности контингентов профессиональных работников и населения, необходимо считаться с тем, что облучение профессиональных работников начинается с 18-20-летнего возраста. С учетом латентного периода (при действии малых доз в среднем 20 лет) появление опухолей можно ожидать с 40 лет. Облучение же после 50 лет уже не даст ощутимого прироста злокачественных новообразований, так как подавляющая часть людей доживает только до 70-75 лет, и канцерогенный эффект в полной мере не успеет проявиться. Согласно расчетам при этих условиях облучение всего тела в дозе 1 бэр/год может дать около 50 случаев смерти на 10^6 человек в год. Для населения же с учетом облучения с периода новорожденности и до 70 лет (средняя продолжительность жизни) это даст 150 случаев смерти на 10^6 человек в год.

ТАБЛИЦА II. ЧИСЛО СЛУЧАЕВ СМЕРТИ ОТ ЗЛОКАЧЕСТВЕННЫХ НОВООБРАЗОВАНИЙ В ГОД ПРИ ОБЛУЧЕНИИ ВСЕГО ТЕЛА В ДОЗЕ 1 БЭР/ГОД

Численность контингента	Профессиональные работники	Население
10^3	0,05	-
10^4	0,5	1,5
10^5	5	15
10^6	50	150
10^7	-	1500

ТАБЛИЦА III. ЧИСЛО СЛУЧАЕВ СМЕРТИ ОТ ЗЛОКАЧЕСТВЕННЫХ НОВООБРАЗОВАНИЙ НА 10^6 ЧЕЛОВЕК В ГОД ПРИ ОБЛУЧЕНИИ В ДОЗЕ 1 БЭР/ГОД

Группа критических органов	Профессиональные работники	Население
I	10	30
II	3	10
III	0,3	1

ТАБЛИЦА IV. РАССЧИТАННЫЕ УРОВНИ СДД ОБЛУЧЕНИЯ ВСЕГО ТЕЛА И КРИТИЧЕСКИХ ОРГАНОВ ДЛЯ ПРОФЕССИОНАЛЬНЫХ РАБОТНИКОВ И НАСЕЛЕНИЯ (БЭР/ГОД)

Численность профессиональных работников и населения	СДД облучения всего тела		СДД облучения критических органов					
			I группа		II группа		III группа	
	Профессиональные работники	Население	Профессиональные работники	Население	Профессиональные работники	Население	Профессиональные работники	Население
$\leqslant 1.10^3$	3	-	15	-	30	-	30	-
1.10^4	1	0,3	5	1,5	15	3	30	3
1.10^5	0,3	0,1	1,5	0,5	5	1,5	30	3
1.10^6	0,1	0,03	0,5	0,15	1,5	0,5	15	3
$\geqslant 1.10^7$	-	0,01	-	0,05	-	0,15	-	1,5

Примечание: Верхняя граница СДД облучения критических органов принята равной для профессиональных работников 30 бэр/год, для населения – 3 бэр/год.

Исходя из этих соотношений, величина N соответствует величинам, приведенным в табл.II.

Аналогичным образом определяется величина N для случая облучения отдельных органов. В зависимости от выхода опухолей на единицу дозы все органы можно распределить на три группы. В первую группу входит костный мозг, желудок, легкие, матка; в третью группу – щитовидная железа, кости, кожа; во вторую группу – все остальные органы. При современных уровнях смертности средний выход опухолей в расчете на 1 бэр облучения составляет для I, II и III групп критических органов соответственно 2; 0,5 и 0,05 случая на 10^6 человек в год. При перерасчете на величину N, т.е. при облучении в дозе 1 бэр/год это дает для профессиональных работников и населения следующие величины (табл.III).

Пока еще трудно оценить риск облучения гонад. Проводимые определения возможного ущерба, обусловленного развитием генетических последствий, условны. По-видимому, при определении допустимого уровня облучения гонад целесообразно сохранить те требования, которые заложены в действующих нормативах. Допустимая доза облучения гонад должна быть не выше допустимой дозы облучения всего тела.

По рассчитанным величинам допустимого риска и значениям N определяются СДД для профессиональных работников и населения (табл.IV).

Рекомендуемый способ определения допустимого риска и доз облучения создает возможности для разработки как единых международных нормативов, так и национальных норм, учитывающих состояние и перспективы национального ядерного развития, а также региональные особенности распространения злокачественных новообразований.

Исходя из указанных положений и прежде всего из численности облучаемых контингентов профессиональных работников и населения, нормы радиационной безопасности для большинства стран, рассчитанные на основе предлагаемой концепции недопустимости реального риска, очевидно, уложатся в следующие значения (бэр/год):

	СДД	ПДД
1. Профессиональные работники	0,3 - 1,0	5,0
2. Население	0,01 - 0,1	0,5

Предлагаемая концепция недопустимости реального риска построена на определенной объективной основе и психологически представляется более привлекательной, чем концепция приемлемого риска. Предлагаемый в докладе подход к нормированию, с одной стороны, гарантирует общество от любого реально ощутимого ущерба, с другой – не создает препятствий для развития ядерной энергетики и использования атомной энергии в мирных целях.

ЛИТЕРАТУРА

[1] Публикация МКРЗ № 9, 1966 г.

[2] Публикация МКРЗ № 14, 1969 г.

[3] АГРАНАТ, В.З., КНИЖНИКОВ, В.А., ЛЯСС, Ф.М., Допустимые уровни облучения организма обследуемых в радиоизотопной диагностике, Медицинская Радиология, 6 (1971) 3.

[4] МОСКАЛЕВ, Ю.И. и др., Концепция биологического риска воздействия ионизирующего излучения. М., Атомиздат, 1973.

[5] SEELENTAG, W., Working paper for development of dose guidelines for normal and emergency situations, Berlin-Dahlem, 1967.

DISCUSSION

F. D. SOWBY: For a number of years the ICRP has clearly stated that its recommendations are based on the conservative assumption of a linear dose-effect relationship, without a threshold. In the future there will be an increasing tendency to base dose limits on this concept, using a comparison with other societal risks to assess an acceptable upper level of risk.

M. M. SAUROV: In this paper we support the views of the ICRP regarding the linear, no-threshold dose-effect relationship.

F. D. SOWBY: It is difficult to see how dose limits could be selected so that the effects could not be detected in addition to the 'normal' incidence. There is in fact no 'normal' level of cancer incidence — it can vary enormously for particular types of cancer in various countries. This would imply that dose limits would have to be tailored to individual countries, taking account of the incidence existing in those countries. Even supposing that this were practicable, it would still raise the question of the propriety of permitting ill health provided it cannot be detected!

M. M. SAUROV: We thought it necessary to introduce two concepts. The average permissible annual dose (APAD) and the maximum permissible annual dose (MPAD).

The MPAD concept serves to safeguard the individual members of a particular occupational group or population: it is a limiting dose and depends on the applicable standards (5 rem/a for workers and 0.5 rem/a for the population).

The APAD is determined by calculation and for this it is necessary to know the number of persons expected to be irradiated and the level of mortality from 'spontaneous' tumours. The value of the APAD is determined mainly by the number of persons irradiated, not by the mortality from 'spontaneous' tumours. Using this approach it is possible to determine common standards for all countries, and (temporary or permanent) national norms can be specified if and when necessary.

E. KUNZ: How do you determine the size of the irradiated groups?

M. M. SAUROV: We take the numbers expected in the different countries during the next few decades.

E. KUNZ: What risk coefficients did you use to allot the organs to the various groups you mentioned?

M. M. SAUROV: We used world-wide data on the dose-effect relation for malignant tumours. These relations are known for all the main types of tumour. On this basis it was possible to select the three groups of critical organs.

E. KUNZ: Do you not think that the acceptable level of population exposure will be affected by the spontaneous frequency of malignant tumours in the non-irradiated groups? This number may differ greatly depending on the organ involved.

M. M. SAUROV: As I think I already mentioned, in this method of determining the acceptable risk the main factor is the number of irradiated people, not the level of mortality from spontaneous tumours.

IAEA-SM-184/103

Invited Review Paper

MONITORING PROGRAMMES FOR THE DETERMINATION OF POPULATION DOSES

O. ILARI, C. POLVANI
Divisione Protezione Sanitaria e Controlli,
CNEN, Rome, Italy

Abstract

MONITORING PROGRAMMES FOR THE DETERMINATION OF POPULATION DOSES.

A review is presented of the principal problems of environmental monitoring aimed at the assessment of the population exposure, jointly with a discussion of their possible practical solutions. The two fundamental types of dose assessment (source-related and people-related) are recalled, also with reference to the cases of well-defined and of ill-defined sources of irradiation of the population. The problem of the cut-off point in the integration process of the individual doses to obtain the population dose is briefly discussed. The various approaches to the monitoring for the assessment of population exposure (direct approach, indirect approach, concentration factor method and systems analysis method) are illustrated, with reference to possible applications to actual situations.

1. DEFINITION AND OBJECTIVES OF MONITORING

1.1. Definition of monitoring

A correct principle in modern society is that any course of action capable of producing a risk to the public has to be coupled with a monitoring system able to check the correct performance of the operations and to assess the relevant risks for the public. In the nuclear field this monitoring, commonly called environmental monitoring, may be defined as all the surveys, measurements, investigations and assessments carried out on man and his environment and capable of satisfying the general principle mentioned above.

Radiation protection specialists have widely discussed environmental monitoring, its meaning and its objectives in recent years. The ICRP has taken up an official position on this problem, the major radiation protection experts have expressed their views in recent congress and publications with a wide convergence of opinions, and the matter appears at last to be sufficiently settled to allow a synthesis to be attempted.

1.2. Objectives of a monitoring programme

1.2.1. Principal objectives

The starting point for a definition of the objectives of an environmental monitoring programme is represented by the following statement from Paragraph 2 of ICRP Publication 7 [1]:

"The principal objectives of an environmental monitoring program can be summarized as follows:

(a) assessment of the actual or potential exposure of man to radioactive materials or radiation present in his environment or the estimation of the probable upper limits of such exposure;

(b) scientific investigation, sometimes related to the assessment of exposures, sometimes to other objectives;

(c) improved public relations."

On these broad objectives all authors are in complete agreement, but other objectives have been suggested by several authors, and different emphases and priorities for the various objectives, reflecting the different points of view and professional interests of such authors according to whether they are nuclear operators, representatives of the regulatory or public health authorities, scientists, etc.

1.2.2. The role of nuclear operators

For example, particular emphasis is given by the nuclear operators to the role of environmental monitoring as a means of verifying the compliance with norms and regulations in the matter of radioactive discharges and of radiation protection of the public [2-5]. This implies effluent monitoring, to check the compliance with statutory discharge limits, and environmental monitoring, to verify that the levels of radioactivity in the environment induced by the operation of the given installation are in conformity with the appropriate environmental radiation protection standards (MPC, DWL, reference levels, etc.) [3]. More generally, such a monitoring programme is intended to provide information on the adequacy and effectiveness of waste treatment and control practices and to reassure the operator on his ability to comply with norms and regulations.

Several health physicists also consider environmental monitoring as a useful tool to facilitate the detection of unforeseen or accidental releases due to leaks, failures or malfunctions of the installations, or to pick out incorrect waste management practices [5-7]. Though this objective is mainly fulfilled by careful effluent monitoring, unintentional releases can be discovered by environment monitoring oriented on indicator materials capable of sensitive and timely detection of abnormal trends in the environmental situation [7].

Another reason that justifies the carrying out of an environmental monitoring programme by a nuclear operator is the need to single out and unequivocally distinguish the real contribution from his installation to the radioactive pollution of the environment from the contributions from other sources that do not come within his responsibility [4], or of maintaining a continuing record of the environmental status as a valid basis to face any legal action to recover damages claimed by members of the public [5].

1.2.3. The role of public authorities

This last problem greatly concerns also the regulatory authorities who through the very character of their vigilance, must be able to identify selectively the different sources of contamination of a given environment, in order to allocate responsibilities correctly and to intervene adequately with those requiring corrective or enforcement action.

The regulatory authorities are deeply interested in the problems of environmental monitoring for at least two other reasons. The first is that an environmental monitoring programme, correctly performed by the operator and independently assessed and controlled by the regulatory authority, represents one of the instruments that the authority can utilize to be satisfied that the public is adequately protected against the environmental risks associated with nuclear activities. The second reason is that in many cases the contamination of a given environment may be due to the sum of contributions caused by several small or ill-defined sources. In this case environmental monitoring programmes carried out by specific operators cannot be envisaged, but the public authorities have to assume directly the responsibility of carrying out, or co-ordinating, environmental monitoring programmes of a general nature not referred to specific sources of radioactive pollution, but rather applied to survey the environmental situation as such [2, 4].

1.2.4. The role of scientific research

Scientific research into the behaviour of radionuclides in the environment is quoted as one of the principal objectives of environmental monitoring. The presence of radioactivity in the environment has provided environmental scientists with a powerful tool for a better knowledge of the environment and the ecocycle, through a study of the phenomena of dispersion of radionuclides in the physical environment and of their transfer through the different links in the environmental chains [7].

In addition, operators and regulatory bodies attach importance to the research aspects of environmental monitoring, though for more practical aims, because an adequate programme of studies of the physical and ecological behaviour of radioactive substances in the environment may play an invaluable role in making possible better and more reliable estimates of the environmental capacity and in verifying the physical and mathematical models used in making preoperational estimates of the environmental impact of radioactive discharges or releases [7].

1.2.5. The public information aspect

The public information aspect of environmental monitoring programmes is largely agreed to be another significant objective of such programmes, but a considerable breadth of opinion exists on the degree of relative emphasis that this aspect should be given in the organization of monitoring programme. Some people believe that extensive and rigorous surveillance can reassure the anxiety of a sensitive and sometimes distrustful public opinion, reassuring it about the safety of the nuclear activities. Others, on the contrary, believe that the strictness of surveillance might be automatically associated in the minds of laymen with the seriousness of the risks, with the sole result of unduly alerting public opinion.

Probably the most correct choice is that the environmental monitoring programme should be primarily set up to satisfy its technical and statutory objectives, and that on this technically unquestionable basis an adequate programme of public information be carried out on the meaning of the surveillance and its results; this information should be such as to instil trust that all that is needed to guarantee the safety of the public is actually being carried out [2, 4, 7]. Nevertheless, many people agree that where a monitoring programme is not technically needed or only a very limited programme is required, it may be worth establishing an additional modest programme of monitoring to provide the public with unequivocal evidence of the safety of discharges [7].

1.2.6. Economic aspects

Some authors have also attempted to attribute an economic role to environmental surveillance. In fact, they believe that one might balance the cost of expanding an environmental monitoring programme against the savings that this expansion could produce, avoiding the installation of additional waste treatment and containment facilities that would lower the discharge of radioactive effluents to levels that make environmental surveillance unnecessary [4, 5].

This approach does not fully agree with the principle that the exposure of population, and therefore the discharge of radioactive effluents, should be as low as readily achievable. In fact, it is highly probable that a purely economic balance would make it generally more convenient to expand the environmental surveillance rather than the waste treatment systems [4].

1.3. Scope of the paper

As we have seen, environmental monitoring can be considered appropria and necessary for several reasons, and its objectives can be numerous and quite varied. Nevertheless, there is general agreement that a fundamental objective is the assessment of the dose to the population for radiation protection purposes. In fact, the main parameter for the evaluation of the biological risk from radiations is the dose, and therefore only a knowledge of the dose received by the population allows a meaningful estimate of the impact of the nuclear operations on the human environment, in terms of healt risks, to be made.

In the early days of atomic energy it was not possible to relate the monitoring data to a parameter that could be used to express the actual population risk. At that time environmental monitoring was simply able to check the radioactivity concentration levels in various parts of the environment in terms of compliance with certain DWLs, such as the Maximum Permissible Concentration. Today, the development of techniques and methodologies and a better understanding of the environmental and biological behaviour of radioactive substances should make it possible to relate radioactivity in the environment to radiation exposure of people, and thus to evaluate the possible consequences of this radioactivity better [7].

It is obvious that in principle the most correct way of assessing the dose to the population should be that of measuring the dose absorbed by the individual members of the public. Such individual monitoring is generally infeasible for several practical reasons: the large numbers of people to be individually monitored; the insufficient sensitivity of methods and of measuring instruments for doses generally insignificant; the impossibility of measuring directly the doses from internal irradiation, etc. Therefore, only exceptionally groups of people are directly monitored by measurement on the person, while there has been considerable development of methods of indirect assessment of the exposure of man based on measurement of the radiation and radioactivity levels within his environment.

The purpose of the present paper is to review some problems of environmental monitoring aimed at the assessment of the population exposur and to discuss their possible practical solutions.

2. THE RELATION BETWEEN MONITORING PROGRAMMES AND DOSE ASSESSMENTS

2.1. Definitions of people-related and source-related assessments

The assessment of the population exposure has the task of verifying whether the public is exposed to radiological risks that are acceptable. Necessary elements for this check are a knowledge of the maximum individual risk to which persons of the population are subject and of the total detriment within the population. The first concept is connected with the individual dose to the most highly exposed person in the population, while the detriment is proportional to the sum of individual doses in the population, expressed in man · rads or man · rems, defined as collective dose.

If one limits oneself to an evaluation of the individual and collective doses in a given population, without specific reference to the sources of such irradiation, i.e. if one performs what is called a people-related assessment (or population-related assessment, as quoted by Lindell [8]), one is only able to verify the following objectives:

(a) No individual in the given population is subject to a risk that is not acceptable

(b) The total detriment within the given population does not exceed acceptable levels [8].

Nevertheless, it is not sufficient to check the absolute acceptability of a given risk or detriment, but, as stated in the ICRP recommendations [9, 10], it is also necessary to verify that any unnecessary exposure is avoided, that sources are not accepted unless it has been ensured that the expected doses are justified in terms of benefits that would not otherwise have been received, and that all doses are kept as low as is readily achievable, economic and social considerations being taken into account [8, 11]. As Lindell clearly recalls [8, 11], the avoidance of unnecessary exposures requires risk-benefit evaluations, while the attainment of doses as low as readily achievable, that is the avoidance of undue exposures, involves cost-benefit, or, more correctly, cost-effectiveness evaluations of further dose reductions.

Obviously a cost-effectiveness evaluation is meaningful only if it refers to a certain source of population exposure. On the other hand, a people-related assessment is not suitable for this purpose, and at the same time it does not allow any risk or detriment estimated projected into the future, as required in many decisions of radiation protection policy. Therefore, Lindell [8] introduced the concept of source-related assessment, whose objectives were to verify that:

(a) The detriment associated to a given source, both in terms of individual risk and of total detriment within the population exposed, is not unjustifiable

(b) The future situation will continue to be acceptable, even if the source continues to be in operation or if the number of sources increases.

It is interesting to note the different meaning of collective dose in these two different types of exposure and detriment assessment. In the case of a people-related assessment the collective dose is the sum of all the individual doses in the given population from multiple sources and irrespective of them. In this case the collective dose may be defined as community dose and is proportional to the total detriment, within the given population group, due to radiation irrespective of source. In the case of a source-related assessment we are interested in knowing the global detriment associated with the source considered, which is proportional to the collective dose calculated over the world population. In this case the collective dose is more commonly referre to as population dose. The corresponding dose-rate is the population dose rate, expressed in man · rad/a or man · rem/a.

The detriment associated with a given practice or course of action, to be completely meaningful, has to be assessed on the basis not only of the present situation, but also of the future expected exposures due to that operation. In fact a given annual dose represents different detriments according to whether it is due to short half-life nuclides or to long half-life nuclides remaining in the environment for a long time [8, 11]. To take into account this exigency, the concept of dose commitment was introduced, as defined in Ref. [8].

For the assessment of global detriment the analogous concept of population dose commitment may be introduced, as the product of the dose commitment by the number of individuals in the population.

These concepts have been mentioned because of their implications for problems of the environmental monitoring. A more exhaustive discussion can be found in the papers presented by Lindell at the IAEA Symposiums at Aix-en-Provence [8] and Warsaw [11] in 1973.

2.2. Source-related monitoring and assessment

The definition of a monitoring program for the assessment of population exposure depends on the type of dose assessment chosen (people- or source-related). This choice, in turn, is influenced by the purpose of the assessme and by the practical possibilities.

Anyone controlling a given source, intended as a single installation or a single activity (as, for example, the production of electrical energy by a nuclear reactor, the medical use of radiation, or the sale of radioluminesce timepieces), is interested in the assessment of the risk and detriment associated with that source or activity, in order to carry out the risk-benefi and cost-effectiveness evaluations necessary to take appropriate decisions o a health, technical or economic nature. This is a typical source-related assessment, and the related monitoring programme must allow an estimate to be made of the dose to the most highly exposed individuals and of the global detriment associated with the given source, in terms of population dose and of dose commitment.

This type of monitoring and assessment is the only one that concerns an operator up to the point where he is able to keep under his control an installation with its effluents or any activity involving risks for the public. Pertinent examples of this case are the environmental surveillance networks around nuclear installations and the related assessments of the exposure of critical groups of population, the monitoring and assessment of the total exposure of the patients and the public from the medical practice in a given hospital, etc.

The national public authorities are greatly interested in source-related assessments as the starting point for risk-benefit and cost-effectiveness balances, on the basis of which they may take the appropriate decisions in terms of acceptance and licensing of a given source and of operational limits to impose upon the operators. Typical cases of this kind are: the assessme of the total detriment associated with the production of electrical energy in a given country, based on the results of local monitoring programmes and of mathematical models; the nation-wide assessment of the genetically significant dose from the medical use of X-rays, based on campaigns of dosimetric measurement and on surveys of the frequency of different types of radiological examinations and of the sex and age distribution of the population; the national networks for the surveillance of fall-out from nuclear explosions in the atmosphere and the related assessments of population dose commitment; dosimetric surveys and assessments of the population exposure from the natural radiation and radioactivity.

The international organisms also make use of source-related assessments for risk and detriment evaluations on a global scale and for the consequent radiation protection recommendations. A typical example of source-related assessment of international interest is the evaluation of the dose commitment due to the world-wide distribution of ^{85}Kr and ^{3}H from the nuclear industry; this assessment is based on measurement and estimates of the discharges of these nuclides to the atmosphere, on the measurement of concentrations in air in various parts of the world, and on mathematical models for the transport and diffusion of pollutants into the atmosphere [12]. A similar international concern has been addressed in the last few years, particularly during the periods of nuclear weapons tests in the atmosphere, to the assessment of the dose commitment to the world population from fall-out of nuclear debris, on the basis of the results of monitoring programmes carried out in several countries [12].

In a modern view of the problem of detriment assessments it should be deemed that the estimate of the global detriment associated with a given source is exhaustive only if it takes into account the contribution to the population dose due to the collective dose absorbed by the workers engaged in the operation of that source. Such workers, in fact, are at the same time members of the population and therefore their exposure contributes to the total genetic and somatic burden of the population. In some cases this contribution is found to be a significant, or even predominant, fraction of the population dose. A typical case is the production of electrical energy by nuclear reactors, where the annual collective dose to the workers at the nuclear power stations is sometimes of the same order of magnitude or even higher than the collective dose to the public. For example, recent estimates concerning a certain nuclear power reactor in Italy show a total-body collective dose to the workers of the order of 100 - 200 man · rem/a [13], while the collective dose to the surrounding population — assessed up to a distance of about 50 km, beyond which the individual doses are insignificant — is of the same order of magnitude. A situation still more remarkable in this respect is probably that of mine workers. Therefore, it should be necessary for any source-related environmental monitoring programme to be completed with the assessment of the collective dose to the workers involved.

2.3. People-related monitoring and assessment

As said before, the global detriment associated with any source or course of action involving radiation and radioactivity can be determined on the basis of appropriate source-oriented monitoring programmes and by means of a source-related assessment of the population dose and the dose commitment. Nevertheless, when the irradiation of a certain group of population is due to the contribution of several sources, the public authority needs to know the overall detriment within that specific group of population due to radiation from all sources, just to verify whether this detriment exceeds any level considered acceptable. In other words, the public authority has to check, by means of people-related monitoring and assessment, whether the commun dose (as previously defined) of that population group from all sources remain below a value stipulated as acceptable for that population group. On these premises, it is obvious that people-related monitoring and assessment is not of interest to individual operators of nuclear installations or of radiation sources, and is also beyond their practical possibility, but concerns almost exclusively public health authorities.

A people-related assessment is needed not only to evaluate the overall detriment within a given population due to radiation, but also to identify whether any cases of individual risks might be found in the same community that are too high, due to anomalous concentrations of contributions of irradia tion from different sources. In this case people-related monitoring has the task of detecting the existence of individuals or groups who are 'critical' fro this point of view. Such critical groups or individuals (people-related) are sometimes different from those (source-related) associated with the single sources and discussed in ICRP Publication 7 [1, 15]. A case of people-related assessment is the study published by the US Environmental Protectio Agency in 1972, concerning the total radiation exposure of the US population from 1960 to the year 2000 from all radiation sources existing in the country [14]. This assessment is based on the results of various monitoring programmes or campaigns specifically concerning the different sources, and on a series of source-related assessments for such different sources of irradiation. The conclusive result of this study, that is the collective dose to the US population from now to the year 2000, is a typical community dose, as previously defined.

2.4. The evolution of the assessment of population exposure

In the early period of the peaceful applications of atomic energy the sources of population irradiation were relatively few in number and concentrated in well-defined sites. Under these conditions it was fairly easy to correlate the environmental pollution and the irradiation of population groups with the pertinent sources, and it was quite uncommon that an overlapping of contamination due to different sources took place in any given area and that groups of people subject to irradiation from multiple sources were found. The only exceptions were represented by natural radioactivity, fall-out from nuclear explosions and by medical uses of X-rays; but this was not a problem because they were global sources relatively homogeneous over large areas, and their contribution is quite easily distinguishable from those from other sources. Therefore, the most natural approach to the evaluation of the radiological risks to the public was the source-related assessment, and the environmental monitoring programmes were also source-related. In fact, the first environmental surveillance programmes, in the 1950s and 60s, wer the local networks around the nuclear reactors and the national fall-out networks, which are typical source-related monitoring programmes.

However, in recent years, with the rapid expansion of the nuclear industry and of the use of radiation in medicine, industry and research, we are witnessing a multiplication and diversification of the sources of irradiation of the public. As a consequence, an increase in and diversification of the exposure pathways and a more and more frequent overlapping of contributions of irradiation from different sources to the same population groups are being experienced. For this reason increasing emphasis is being given to the concept of people-related assessment. In fact, only on this basis may the public authorities be able to evaluate the impact of the different activities involving the use of radiation and to take decisions like the following:

What residual acceptable detriment from radiation, within a given area or population, can be stipulated in addition to existing sources;

What is the most appropriate apportionment of different sources of radiation that can be planned to satisfy the needs of a given population;

What is the most appropriate territorial planning for the siting of large nuclear plants from the viewpoint of 'optimization' of population exposure.

2.5. Well-defined sources and people-related monitoring

In some cases the sources contributing to the irradiation of a certain group of population are few and well defined. In such a case the dose to that population group, due to each individual source, may be estimated by means of the appropriate source-related monitoring and assessment. Afterwards, the community dose to that group of population (people-related assessment) may be obtained simply by the sum of the collective doses due to the individual sources. Similarly, evaluating and conveniently combining the distributions of individual doses, due to the different sources throughout the population group under examination, it is possible to identify a possible overall critical group for such combination of sources and to calculate the average dose to its individuals.

When the exposure of this new overall critical group is estimated, a people-related monitoring programme, in addition to the source-related programmes pertaining to the single sources, may be needed to keep under control the environmental contamination and the radiation exposure of the particular population group thus identified. This additional monitoring programme could be the result of a co-ordinated effort jointly effected by the operators of the single sources, or it could be the responsibility of the public authorities.

A quite typical example of such a situation is that of the gaseous discharges of several nuclear reactors located in the same area at distances of some kilometres. For each reactor a critical group for external irradiation by the plume of radioactive gases does exist, generally placed at a short distance from the stack (from some hundreds of metres to some kilometres). But it can happen that the isodose curves for the effluents of the various reactors will combine in such a way as to give rise to an exposure of another group of population, different from such critical groups, that is higher than those of the individual critical groups. Under these conditions it is evident that the source-related environmental radiation monitoring networks around the single reactors should be supplemented by people-related monitoring of the radiation levels in the area of maximum overlapping of the isodose curves.

A more complex case could be that of a certain small town receiving its drinking water supply from a river where, several kilometres upstream, a nuclear plant A discharges its effluents, and, at the same time, consuming preferentially some foodstuffs coming from a rural area, also very far from the town, affected by the airborne discharges from another nuclear plant B. For each of the two plants a preoperational monitoring programme may have identified a local critical group; it may have also estimated the exposure of the population of our town due to the consumption of drinking water or, respectively, of agricultural products coming from the above-mentioned area; and, if the waste treatment systems of the two plants are well designed from the radiation protection viewpoint, each of those two studies will have concluded that the exposure of our population group, so remote from the two plants in question, is well within acceptable limits and in any case lower than the exposure of the respective critical groups. On the basis of these conclusions the environmental monitoring programmes routinely carried out by the two plants would have probably neglected the small town here examined being specifically oriented to the local critical groups. But if, on the contrary, the two preoperational monitoring programmes were carried out in a co-ordinated manner, they could reveal that the total exposure of the population of our town due to both the above-mentioned pathways is not negligible, but requires careful consideration and monitoring.

In this case an ad hoc monitoring programme may be needed to permit an assessment to be made of the total exposure of the population of our small town. Such a programme should consist of:

Measurement of radioactivity concentrations in drinking water at the intake of the town water supply system;

Measurement of radioactivity concentrations in foodstuffs coming from the area affected by the discharges of the nuclear plant B, together with an estimate of the dilution factor of these foodstuffs on the market, or, alternatively, measurement of radioactivity concentrations carried out directly on samples of the typical diet of the population involved.

Many other examples of this type could be envisaged. At present they appear to be very uncommon and rather theoretical, but cases of this kind are likely to happen not infrequently in the future.

2.6. Ill-defined sources and people-related monitoring

If the sources contributing to the contamination of a given area or to the irradiation of a given population are not sufficiently defined, or if their specific contributions cannot be clearly discriminated by means of appropriate source-related monitoring programmes, the sole basis for the evaluation of the exposure of that population and of the relevant community dose is a people-related monitoring programme carried out without consideration of sources. This means that the monitoring programme has to be addressed directly to the population group under examination and to the area where they live.

For example, we might need to know the total exposure of a certain group of population by external irradiation. This population is exposed to external radiation from X-rays for medical purposes, building materials, miscellaneous products of common use containing radioactivity, fall-out from nuclear explosions and natural radioactivity; at the same time this particular population group could live in the vicinity of a nuclear installation discharging radioactive effluents, they could possess colour television receivers and cook their meals by means of natural gas obtained by stimulation with nuclear explosives, and finally several members of this population could frequently travel by air! In such a situation it might be impractical, if not impossible, to carry out source-related assessments of the individual contributions to the population exposure. It is therefore necessary to carry out measures of total exposure, irrespective of source, directly on the members of this group of population or, more frequently, on their environment.

Analogously, the population of a given area might be exposed to internal irradiation both by inhalation of contaminated air and by ingestion of contaminated foods; these latter coming from the same area or from other regions. In this case it would be extremely difficult to discriminate the contributions to the intake of radioactivity from the different sources, both because the contamination of air and foods may be due to the same radionuclides and because home consumption represents an increasingly smaller fraction of the human diet; this last occurrence renders it ever more difficult to identify the place of origin of each foodstuff, due to the increasing complexity of commercial distribution. Under these conditions the exposure of such a population by internal irradiation can be estimated only on the basis of their total intake of radioactivity, irrespective of source. This intake can only be determined by means of direct measurement, on each individual, of the body burden of the various radionuclides, or, more frequently, by the measurement of radioactivity content of air, water and diet that they incorporate.

3. THE PROBLEM OF THE CUT-OFF

If we remember the definition of population dose (see section 2.1), we have to assume that in principle the integration of the dose over the number of persons should be extended up to the point where the effect of the source is still different from zero. This means that individual doses should be added up even if they are assessed to be very low, because even very small doses to a large number of individuals may still contribute to the global population dose [8]. But in practice it is necessary to assume an appropriate cut-off point in the integration process of the individual doses.

No specific recommendations have yet been made as to the level at which this cut-off should be set, and therefore we may only recall the following general principle stated in paragraph 17 of ICRP Publication 22 [10]:

> "... at levels of individual dose that are small fractions of the relevant dose limit, there will be no need to pursue the summation beyond the point where it becomes clear that the further contribution to the sum will not change the estimate of population dose by more than a factor of about 2 or 3, or, alternatively, beyond the point where it becomes clear that the remaining detriment is insignificant in comparison with the benefits expected from the source".

Some practical proposals have been raised to solve this problem. A possible solution is that assumed in the United States of America to limit the detriment associated with the discharge of radioactive effluents from light water power reactors; in that country, in fact, the distance from the plants within which the collective dose is assessed is generally limited to 50 miles, on the assumption that the individual doses beyond that distance would not add substantially to the integral defining the population dose [16]. According to other proposals, the individual doses might be excluded from the integration when they fall to a very small fraction of the dose due to the natural background. A practical proposal, not too far from the above-mentioned principles, could be to set the cut-off point at a value of the individual dose rate between 0.1 and 1 mrem/a. As is evident, assessment of the collective dose consequent to a cut-off operation gives rise not to the global population dose, but to a fraction of it, which could be called the partial (or area) population dose.

Similar concepts have to be applied to the formulation of an environmenta monitoring programme. In fact, the effort and expenditure devoted to enviro mental monitoring should be proportional to the value of the information drawn from it, taking into account the specific objectives of the monitoring programme itself. In other words, the amplitude and depth of a monitoring programme should be limited at a level beyond which the further improvemer of the information obtainable would not counterbalance the cost and effort needed to achieve it.

It is appropriate to recall another order of considerations influencing the size of a monitoring programme. In fact, the organizational and economic commitment to be devoted to a monitoring programme should be established on the basis of a well-balanced appraisal of the risks actually incurred by the public [17]. If the actual risks and detriment, i.e. the individual and collective doses, are small, the execution of detailed ecological investigation and of extensive and sophisticated measurements may be unnecessary and unduly expensive, and large resort may be made to mathematical models of the environment, supported by limited programmes of environmental measur ment and investigations. If, on the other hand, the estimated doses to the public exceed some pre-established values, the assessment of the population exposure has to be based to a much greater extent on the results of actual studies and measurement in the field. These pre-established values could be set at such a level that the relevant monitoring programme guarantees a dose assessment affected by an uncertainty that is compatible with the cut-of criteria previously mentioned.

In conclusion, the amplitude and degree of sophistication of a monitorin programme should be established on the basis of a cost-effectiveness analys jointly with the consideration of the estimated level of risk and detriment within the population.

4. VARIOUS APPROACHES TO ENVIRONMENTAL MONITORING

The quantity most directly associated with possible damage to man from the presence of radiation and radioactivity in the environment is the resulta radiation dose. For this reason the ideal approach for a monitoring programme intended to evaluate the population exposure would be the direct measurement of the individual external exposure, by means of personal dos meters worn by the people involved, and the direct measurement of their radioactive body burdens by means of whole-body counting or other techniq In this way the external dose to individuals is directly obtained and the internal dose can be easily calculated using the available dosimetric models

When the individual doses are obtained, the collective dose can be immediately evaluated by the sum of such individual doses within the population involved. From the distribution of the individual doses the average individual dose, which is useful for subsequent elaborations (genetic dose, dose commitment), can also be calculated. It is clear that this approach is equally applicable to both source-related and people-related assessments, as previously defined.

This approach, which we may call the direct approach, can be applied only rarely for several practical reasons, and in any case only for small groups of people. It is therefore necessary to have recourse to less direct methods of monitoring and assessment. With regard to indirect methods various approaches have been envisaged, which are more or less indirect, depending on the type and degree of information available, the complexity of the exposure pathways involved, and on the measurements that are feasible. These approaches can be characterized by the varying 'distance' of the environmental links subject to measurement from the final links of the biological chains, that is from the target of the dose assessment, i.e. man.

The less indirect approach is certainly that consisting of the measurement of radiation levels in the environment and of radioactivity concentrations in the last links of the biological chains, that is those having immediate contact with man (inhaled air, drinking water, foodstuffs). This experimental information, which can frequently be supplied by an accurately conceived monitoring programme, has to be processed in the light of the results of a socio-economic enquiry concerning the living and dietary habits of the population involved. This process leads to the assessment of the external doses, while, for the internal contamination, it is only able to supply data on the intake of radioactivity; the latter data, introduced into appropriate metabolic models, allow an assessment of the body burdens and, consequently, of the internal doses to be made. As may easily be understood, this approach, which may be called the indirect approach, is also applicable to both source-related and people-related assessments.

Still more indirect is the approach called the 'concentration factor method' (CFM). The physical measurements are usually limited to the first links of the biological chains, that is to the first stage of the exposure pathways. For sources releasing radioactivity into the environment these first links are almost invariably the effluent recipient media, that is the air and the water, rarely the soil. On the assumption that an equilibrium exists between the rate of discharge and the steady-state concentrations of radioactivity in the environment, the concentrations of radioactivity actually measured in these media are now combined with the concentration factors for radionuclides between the last links of the biological chains – for instance, foodstuffs – and the air or water, to assess the equilibrium concentrations in the last stages of the exposure pathways. Processes of deposition or sedimentation in the environment from the primary medium are analogously considered as relevant aspects for the assessment of external irradiation. When the concentrations of radioactivity in the last links of the biological chains or the radiation exposure rates in the environment are evaluated, the next steps of the procedure for the assessment of individual and collective doses are the same as in the indirect approach.

The CFM is typically applicable to source-related assessments, while its applicability to a people-related assessment seems to be less frequent. It is adequate for most situations involving routine releases satisfying the following conditions:

The release of activity should be intense enough to give rise to levels of radioactivity in the recipient medium that are measurable with sufficient accuracy;

The release rate should be reasonably uniform during the period in order to allow the establishment of equilibrium conditions between the discharge rate and the radioactivity concentrations through the biological chains;

The specific concentration factors for equilibrium conditions must be known;

The situation to be monitored should be sufficiently simple in terms of nuclides discharged and of possible exposure pathways.

When these conditions are not satisfied or the sources are diffused, a more comprehensive and more indirect method may be required. This is the 'systems analysis method' (SAM), which consists of an overall simulation of the dynamic behaviour of radionuclides in the environment by

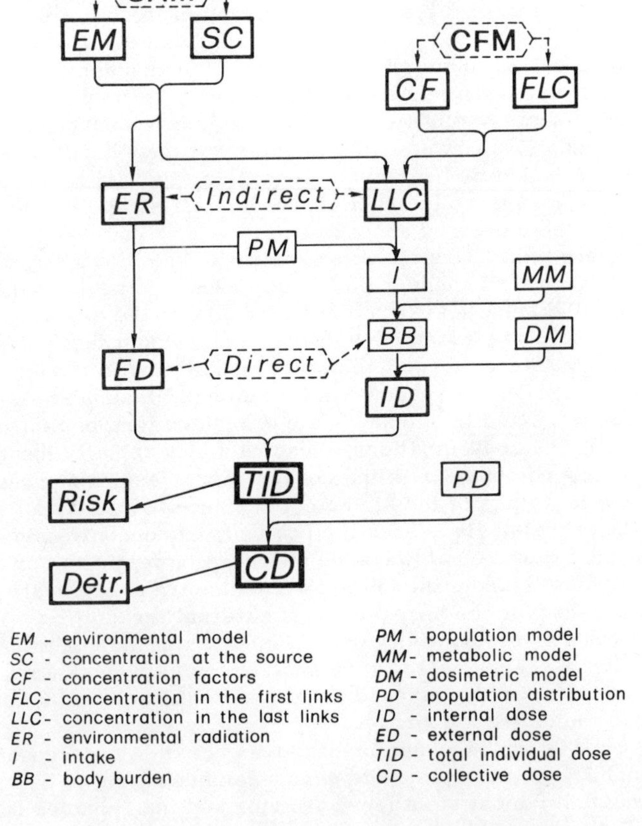

FIG. 1. Logical scheme of the four approaches to environmental monitoring.

TABLE I. MONITORING APPROACHES

Approach	Source	First links	Last links	Man	Applicable assessment
Direct	-	-	-	M	SR or PR
Indirect	-	-	M	A	SR or PR
CFM	-	M	A	A	SR
SAM	M	A	A	A	SR

M = measurement; A = assessment; SR = source-related; PR = people-related.

means of an environmental model based on a system of compartments through which the transfer of radionuclides may be evaluated with mathematical equations. In the application of this approach the measurement component of the monitoring programme is generally reduced to the measurement of the effluent discharge rate. This effluent monitoring allows an evaluation to be made of the source term to be introduced as an input function into the systems analysis model. When the source term and the environmental model have been established, the next steps of the calculation procedure up to the dose assessment follow the same logic already discussed for the indirect approach. The SAM is applicable only to source-related assessments. It is worth noting that the SAM and CFM approaches are not substantially different in principle, but they do differ essentially in complexity and flexibility of application. In fact, they actually overlap when simple situations at equilibrium are dealt with.

A general outline of the procedures followed in the four approaches already discussed is summarized in Fig. 1, while Table I gives a summary of the measurement and assessment components required in each approach, and of their field of application.

In general not all the exposure pathways are equally important, but it is frequently possible to identify one or more sources or exposure pathways that are predominant over the others in terms of population exposure, just as it is possible to identify one or more groups of people whose exposure is predominant in the total exposure of the population involved. In these cases the environmental monitoring may be greatly simplified by the application of the 'critical path approach', which was introduced by ICRP in its Publication 7 [1] and discussed in detail by several authors [6, 7, 18-20]. It is worthwhile to remark that this approach is compatible with any one of the general approaches previously described and may be applied to routine environmental monitoring as a simplification of any of those approaches.

5. CONCLUSIONS

A very important objective of an environmental monitoring programme is the assessment of population exposure. Several other reasons can justify or require the set up of a monitoring programme; these various reasons were examined in section 1. Anyone who is responsible for an environmental monitoring programme has to define clearly in advance the particular objectives of his specific monitoring programme.

In any case one must establish if and up to what degree the assessment of the dose to the public is really important and necessary. If the assessment of population exposure is deemed necessary, the definition of the features of the monitoring programme is affected by the type of dose assessment that is required. In particular, it is necessary to identify in due time the kind of dose (individual dose, collective dose, dose commitment) to be assessed and the type of dose assessment (source-related or people-related) to be made, according to the situations and the principles highlighted in Section 2.

When these points are established, and taking into account the peculiar problems raised by the particular sources and environment concerned, it is possible to select the most appropriate approach for the monitoring programm from among those discussed in section 4. On this basis the analytical method the techniques and the instrumentation needed to carry out the environmental programme may be selected.

Because environmental monitoring is easily susceptible to incorrect definition that gives rise to waste of effort or meaningless information, great care has to be given to use correct professional judgement, balancing the cost of the monitoring with its ability to fulfil the desired objectives.

REFERENCES

[1] INTERNATIONAL COMMISSION ON RADIOLOGICAL PROTECTION, Principles of Environmental Monitoring related to the Handling of Radioactive Materials, Publication 7 Pergamon Press, Oxford (1965).

[2] DUNSTER, H.J., "The objectives of environmental monitoring in the vicinity of nuclear facilities in the United Kingdom", Proc. Symp. Environmental Surveillance in the Vicinity of Nuclear Facilities, Charles C. Thomas Publisher, Springfield (1970).

[3] BILES, M.B., et al., "The objectives and requirements for environmental surveillance at US Atomic Energy Commission facilities", Environmental Surveillance around Nuclear Installations (Proc. Symp. Warsaw, 1973), IAEA, Vienna (in press).

[4] BRESSON, G., COULON, R., "Analyse des principes directeurs des programmes de surveillance de l'environment", Environmental Surveillance around Nuclear Installations (Proc. Symp. Warsaw, 1973), IAEA, Vienna (in press).

[5] HONSTEAD, J.F., "Bases for environmental survey design", Proc. Symp. Environmental Surveillance in the Vicinity of Nuclear Facilities, Charles C. Thomas Publisher, Springfield (1970).

[6] CNEN, Considerazioni per l'aggiornamento delle reti locali attorno ad importanti centri nucleari – Report of a Working Group, Doc. PROT.SAN/09/70 (1970).

[7] MITCHELL, N.T., et al., "Principles and practice of environmental monitoring in the United Kingdom", Environmental Surveillance around Nuclear Installations (Proc. Symp. Warsaw, 1973), IAEA, Vienna (in press).

[8] LINDELL, B., "Assessment of population exposures", Environmental Behaviour of Radionuclides Released from the Nuclear Industry (Proc. Symp. Aix-en-Provence, 1973), IAEA, Vienna (1973) 25.

[9] INTERNATIONAL COMMISSION ON RADIOLOGICAL PROTECTION, Recommendation of the International Commission on Radiological Protection, Publication 9, Pergamon Press, Oxford (1966).

[10] INTERNATIONAL COMMISSION ON RADIOLOGICAL PROTECTION, Implications of Commission Recommendations that Doses be kept as Low as Readily Achievable, Publication 22, Pergamon Press, London (1973).

[11] LINDELL, B., "The establishment of standards and working limits for radioactive contaminants", Environmental around Nuclear Installations (Proc. Symp. Warsaw, 1973), IAEA, Vienna (in press).

[12] UNITED NATIONS SCIENTIFIC COMMITTEE ON THE EFFECTS OF ATOMIC RADIATIONS, Ionizing Radiation: Levels and Effects, UN, New York (1972).

[13] COMETTO, M., TOCCAFONDI, G., ILARI, O., TAGLIATI, S., "Radiation protection experience in Italian Nuclear power plants", Peaceful Uses of Atomic Energy (Proc. Conf. Geneva, 1971) 11, UN, New York, and IAEA, Vienna (1972) 91.

[14] US ENVIRONMENTAL PROTECTION AGENCY, Estimates of Ionizing Radiation Doses in United States, 1960-2000, Rep. ORP/CSD 72-1 (1972).
[15] JAMMET, H., "Analyse critique du colloque", Environmental Behaviour of Radionuclides Released in the Nuclear Industry (Proc. Symp. Aix-en-Provence, 1973), IAEA, Vienna (1973) 703..
[16] UNITED STATES ATOMIC ENERGY COMMISSION, Final Environmental Statement Concerning Proposed Rule Making: Numerical Guides for Design Objectives and Limiting Conditions for Operation to Meet the Criteria "As Low as Practicable" for Radioactive Material in Light-Water-Cooled Nuclear Power Reactor Effluents, Rep. WASH-1258 (1973).
[17] BRAMATI, L., et al., "Organizational and economic aspects of environmental investigations and monitoring in the vicinity of nuclear facilities", Proc. Symp. Radioecology Applied to the Protection of Man and his Environment, Rome, 1971, Rep. EUR 4800.
[18] FOSTER, R.F., et al., "Evaluation of human radiation exposure", Radioactivity in the Marine Environment, National Academy of Sciences, Washington, DC (1971) 240-60.
[19] PRESTON, A., "The United Kingdom approach to the application of ICRP standards to the controlled disposal of radioactive waste resulting from nuclear power programs", Environmental Aspects of Nuclear Power Stations (Proc. Symp. New York, 1970), IAEA, Vienna (1971) 147.
[20] MITCHELL, N.T., "The roles of effluent and environmental monitoring in surveillance of radioactive wastes released from nuclear installations", Environmental Surveillance around Nuclear Installations (Proc. Symp. Warsaw, 1973), IAEA, Vienna (in press).

DISCUSSION

R.F. BARKER: You mentioned the two uses of the term "collective dose". It describes the sum of doses to a specific population group from all sources and also the dose to a population group from a single source. In the USA, we have used the term "cumulative dose" to describe the sum of doses to invididuals in various population groups from a single source.

O. ILARI: We use the term "collective dose" only in a general sense, as a sum of individual doses, without reference to the various situations. To avoid the confusion which is frequently found in the literature, we prefer to use the term "community dose" for the collective dose to a specific population group from multiple sources, and the term "population dose" for the collective dose to the entire world population due to a single source. We have not explicitly mentioned the case you referred to, namely the collective dose to a specific population, other than the world population, from a single source; I think that your "cumulative dose" would be perfectly compatible with the two terms we have introduced and would complete the terminology in this field.

R.F. BARKER: In the paper it is suggested that doses from single sources to individuals not exceeding 0.1 to 1 mrem/a may be neglected in assessing "collective doses". Does this not depend on the number of persons exposed to these low doses? For example, individual doses from wrist watches with luminous dials may be much less than 1 mrem/a per individual, but with a million persons exposed the "cumulative" or "collective" dose from that source is not negligible.

O. ILARI: I think you are right. In fact, in proposing cut-off values ranging from 0.1 to 1 mrem/a, we realized that this cut-off has to be used with great care, attention being paid not only to the number of persons exposed but also, and particularly, to the maximum value that the individual dose from a given source can attain. For example, if the maximum individual dose estimated for a certain source is already lower than the cut-off level that we have indicated, we think that another appropriately lower cut-off is required — to avoid the questionable result that the collective dose due to that source would be zero. This could be the case, for instance, witl the luminous-dial watches you mentioned.

A. BAYER: In our Upper Rhine study (paper IAEA-SM-184/1) we compared the radiological impact on critical groups in the neighbourhood of nuclear facilities with the radiological impact on the 'average' individual living in this district, which has an area of 150 km × 300 km. The burden to the critical group is higher by a factor of about 100 than the burden to the average individual.

Can you give any tentative figures relating to the scheme shown in Fig. : of the paper? What expense is justified at the moment?

O. ILARI: In the Direct Approach the actual measurement corresponds to ED and B; the value of ID is obtained from B by calculation.

In the Indirect Approach the physical measurement is performed on ER and LLc (measurement of radiation exposure rates in the environment and of radioactivity concentrations in the diet, in drinking water, or in inhaled air); an experimental component is represented by the social-economic investigation of population habits (PM). The remaining procedures are deal with partly by calculation and partly by estimation.

In the CFM Approach physical measurement is carried out only on FLc that is on the radioactivity concentrations in the medium directly receiving the effluents, generally air and water, after dilution; the concentration factors (CF) may be the result of experimental investigations, while sometimes they are the result of appropriate assumptions based on current literature. In this approach, of course, the population habits and diet resu from field investigations or sometimes from conservative assumptions. Tl remaining part of the process is calculation.

Finally, in the SAM Approach, the only radioactivity measurements ar made at the source (concentrations of radioactivity in the effluents before release, total activity released in a given time), while for the remaining part of the process the same considerations as above can be applied. As fa as the expense and its justification are concerned, some general considerations are reported in section 4 of the paper.

D. BENINSON: Perhaps I may comment here that "collective dose" is quantity whose dimensions are dose x persons and whose unit is the man·re or man·rad. Collective dose could apply to groups, sub-populations, or to the world population. It could be contributed by one or by many sources. All these factors should be indicated when giving a collective dose.

O. ILARI: I quite agree, but I think that an acceptable way of indicatin the factors that you require for the different situations would be to use the terms "community dose", "population dose" and "cumulative dose" with the specific meanings previously mentioned.

RESULTS OF MEASUREMENTS RELATING TO THE POPULATION DOSE

H. KIEFER, W. KOELZER, G. STÄBLEIN
Health Physics Division,
Karlsruhe Nuclear Research Center,
Karlsruhe,
Federal Republic of Germany

Abstract

RESULTS OF MEASUREMENTS RELATING TO THE POPULATION DOSE.

The Health Physics Division of the Karlsruhe Nuclear Research Center carries out a variety of radiation protection measurements for research purposes and as a routine procedure. The paper deals with the results obtained from such measurements that allow information to be generated about the exposure of members of the public.

A monitoring program for control of the environment of the Karlsruhe Nuclear Research Center with respect to radioactivity includes measurements and analyses, which by far exceed normal radiation protection requirements. The results provide a good survey of the radiation dose in the environment and, hence, of the external radiation exposure of the population in the environment. The external radiation exposure of all the persons working at the Center is monitored by the distribution of personnel dosimeters. In addition to persons occupationally exposed to radiation, all persons not exposed to radiation as a result of their work have also been equipped with dosimeters as a voluntary measure. Over the past ten years the average number of persons monitored was about 3000. For areas not involving significant occupational radiation exposure this resulted in values between 60 and 80 mrem/a, including natural radiation background.

In mid-1973 a comprehensive measuring program was started to determine the radiation dose in private homes. Results of measurements conducted in more than 2500 private homes in Southern Germany are available. An influence of building material upon radiation exposure can be detected.

1. Radiation Burden in the Environment of the Karlsruhe Nuclear Research Center

The environment of the Karlsruhe Nuclear Research Center is monitored for radioactivity under a broad program whose measurements and analyses by far exceed standard radiation protection requirements and the conditions imposed by licensing authorities, both in type and in number. In this way it is possible to generate detailed information about any radiation dose to which the public may be exposed in addition to the natural radiation background.

Firstly, the monitoring program is carried out to determine potential incorporation hazards to man by determining the radioactivity content of various media directly or indirectly incorporated by man. Secondly, radiation is measured directly by means of counter tube stations and the accumulated radiation dose is determined by solid state dosimeters distributed throughout the environment. The results especially of these measurements carried out over the past few years provide a good survey of the radiation dose in the environment of the Center and thus of the external radiation burden of the population living around the Center.

The solid state dosimeters used to assess the accumulated radiation dose in the environment of the Karlsruhe Nuclear Research Center are installed in concentric rings of 1 km, 2 km and 3 km radius around the border of the site (Fig. 1) [1].

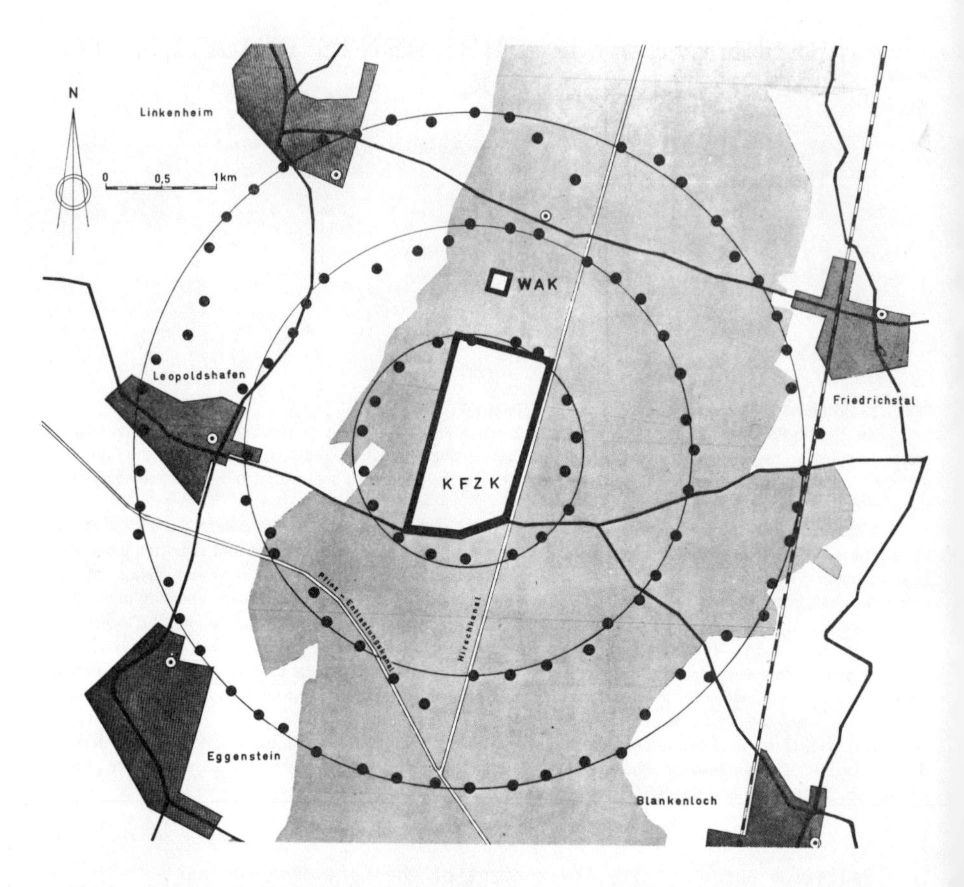

FIG. 1. Environment of the Karlsruhe Nuclear Research Center; points indicating solid-state dosimeter.

These are types of dosimeter used:

- Phosphate glass dosimeters in spherical capsules; within $\pm$ 8% energy independent measurement of the γ-exposure for energies > 40 keV.
- LiF-dosimeters, within $\pm$ 40% energy independent measurement of γ-exposure for energies > 15 keV and of β-radiation.

The dosimeters are sealed in plastic bags to protect them from influence of the weather. A lightproof packing of the LiF-dosimeters is ensured by sealing them in an additional black plastic foil. The dosimeters are installed on trees or aluminum poles some 3 m above ground level [2,3].

97 measuring points are arranged at distances of 50 m along the 5 km long borderline of the site of the Nuclear Research Center, another 6 are arranged along the fence of the WAK Reprocessing Plant. The concentric circles around the Nuclear Research Center hold 18 measuring points (1 km radius), 36 (2 km radius) and 54 (radius 3 km). Each measuring point is equipped with two phosphate glass dosimeters, the measuring points along the borderline of the site and those installed on the ring of a radius of

TABLE I. AVERAGE ANNUAL DOSE OUTSIDE AND ALONG THE FENCE OF THE KARLSRUHE NUCLEAR RESEARCH CENTER

Time of exposure *	average annual dose in mR					
	Phosphate Glass-Dosimeters				LiF-Dosimeters	
	fence of KFZK	1 km-circle	2 km-circle	3 km-circle	fence of KFZK	3 km-circle
1966/67	74	61	67	52	-	-
1967/68	102	93	93	86	-	-
1968/69	51	38	38	42	-	-
1969/70	58	34	36	47	76	-
1970/71	67	48	52	55	81	72
1971/72	74	64	56	57	69	70
1972/73	66	57	62	67	87	72
mean value	70	56	58	58	78	71

* from May to May

3 km in addition hold two LiF-dosimeters each. Moreover, there are seven measuring points installed at the counter tube stations in the villages around the Nuclear Research Center which are also equipped with two phosphate glass and LiF-dosimeters each.

This adds up to a total of 218 measuring points with 768 dosimeters available to determine the accumulated radiation dose in the environment. Since some of these measurements have been performed for many years already and especially the measuring points located on the ring with 3 km radius include inhabitated areas in the network of measurements, pertinent information is obtained about the radiation dose in the environment of the Karlsruhe Nuclear Research Center and the external radiation exposure of the population of this area. Table I shows values of the average annual dose outside of and along the fence of the Karlsruhe Nuclear Research Center.

It is seen from Table I that the annual dose found for the ring zones between 1 km and 3 km distance in the radial direction is practically the same for all zones, i.e., approximately 60 mR, which may be regarded as the external radiation background. The dose values indicated by the dosimeters arranged on one ring show no distinct peak values for certain directions which might be correlated with the two main wind directions encountered on the site of the Nuclear Research Center.

Such correlation, perhaps as a result of the discharge of radioactive effluents into the atmosphere, is not expected after all. Calculations of the maximum expected radiation dose in the environment caused by the activities discharged with effluent air, e.g. for 1973 indicate the annual doses for the villages in the environment as shown in Table II (for the locations of these villages, see Fig. 1) [1].

TABLE II. CALCULATED LOCAL DOSES OWING TO RADIOACTIVE EMISSION OF THE KARLSRUHE NUCLEAR RESEARCH CENTER IN 1973

place	dose-values in mrad	
	γ - dose	β - dose
Friedrichstal	2.7	o.5
Blankenloch	o.6	o.2
Karlsruhe, center	o.5	o.3
Eggenstein	4.4	1.3
Leopoldshafen	2.5	1.1
Linkenheim	1.4	o.4

It is evident from the data on Tables I and II that the radiation dose in the environment of the Karlsruhe Nuclear Research Center is not increased above the natural background radiation dose as a consequence of the operation of nuclear facilities and the scientific institutes. Merely along the fence of the Center there is a slight increase of some 10 mR/a.

2. External Radiation Burden of Occupationally Non-Exposed Staff of the Karlsruhe Nuclear Research Center

The external dose burden of all the staff of the Karlsruhe Nuclear Research Center is controlled by means of personnel dosimeters. The Radiation Protection Ordinance valid in Germany, which corresponds to the Basic Euratom Standards, includes a legal requirement for the control of all persons occupationally exposed to radiation. In the area of the Nuclear Research Center, these would be persons working in restricted areas because of their handling radiation sources or unsealed radioactive materials. The dosimeters used for this group of personnel are provided and evaluated by an official measuring agency which also determines the type of dosimeters used. In many cases film badges are still employed for the time being.

In addition, the whole staff of the Center are equipped with phosphate glass dosimeters. A valuable contribution to the measuring scene was made by the decision to equip also those persons with dosimeters voluntarily whose place of work is more on the periphery or who will be in restricted areas only occasionally without actually handling radioactive substances. These groups include persons doing theoretical work, maintenance crews, guards and administrative staff, and also visitors of the Center.

The choice of phosphate glasses for monitoring purposes offers a number of advantages:

- The dosimetry system functions with a generally recognized degree of reliability.
- The dosimeter elements are sufficiently sturdy.
- The broad range of measurement allows doses to be detected in the range of natural background radiation as well as accident doses.
- In their function as long time dosimeters these glasses can be assigned to a person over a monitoring period of several years with a possibility of obtaining as many interim readings as desired.

Dosimeters were distributed voluntarily to all persons working at the Center for the following reasons:

- Experience was to be collected in routine monitoring and evaluation of a large number of persons monitored.
- Comparable readings were to be obtained from various groups of personnel, especially those not occupationally exposed to radiation.
- Unknown potential radiation exposures which might exist in some work places were to be detected.
- A broadly distributed number of significant local dose readings were to be collected for incident situations.

Under the present method of measurement monthly evaluations will be made only for a limited group of persons. For the others a monitoring period of six months is required so that the limit of detection ascertained to be 40 ± 10 mrem is exceeded by a clearly measurable level also as a result of natural background radiation. Statistically better founded information is obtained over longer periods of observation. The glass dosimeter assigned to a person should not be erased before a period of approximately five years and will actually furnish an integral reading for this period not encumbered with the errors produced by the addition of single results. Moreover, the annual values obtained may be regarded as mean values for a specific group and can be investigated for any significant deviations.

It is known from practical experience that personnel dosimeters outside the restricted area are often locked away at the place of work and not worn by a person all the time; this does not impede the purpose of the control procedure if the reading is explained as the local dose at the work place. Wearing dosimeters all the time, i.e. taking them home, is not feasible or even meaningful with respect to information about the local radiation burden.

Table III is a survey of the evaluations made within this procedure including special measurements and the number of persons monitored.

The dose readings obtained as the sum total of natural radiation background and the dose due to handling ionizing radiation were investigated statistically. In order to pinpoint statistical peculiarities, only those persons were admitted for evaluation who had not changed their type of work, division and work place within the period under consideration. With personnel fluctuation taken into account this leaves some 2 000 persons per annum.

The frequency distribution of all the doses measured in the years under review 1963 to 1973, for example, see the distribution for 1973 in Fig. 2, shows a clear peak at about 60 - 80 mrem/a with a distribution around a

TABLE III PHOSPHATE-GLASS-DOSIMETRY, 1963 - 1973
KARLSRUHE NUCLEAR RESEARCH CENTER

year	number of measurements	number of surveyed persons
1963	1,56o	91
1964	2,244	137
1965	5,5oo	415
1966	9,94o	726
1967	12,741	3,114
1968	15,o93	3,36o
1969	13,1oo	3,244
197o	14,1oo	3,o54
1971	16,75o	3,572
1972	15,9o4	3,1oo
1973	17,57o	3,2oo
1963 - 1973	124,5o2	-

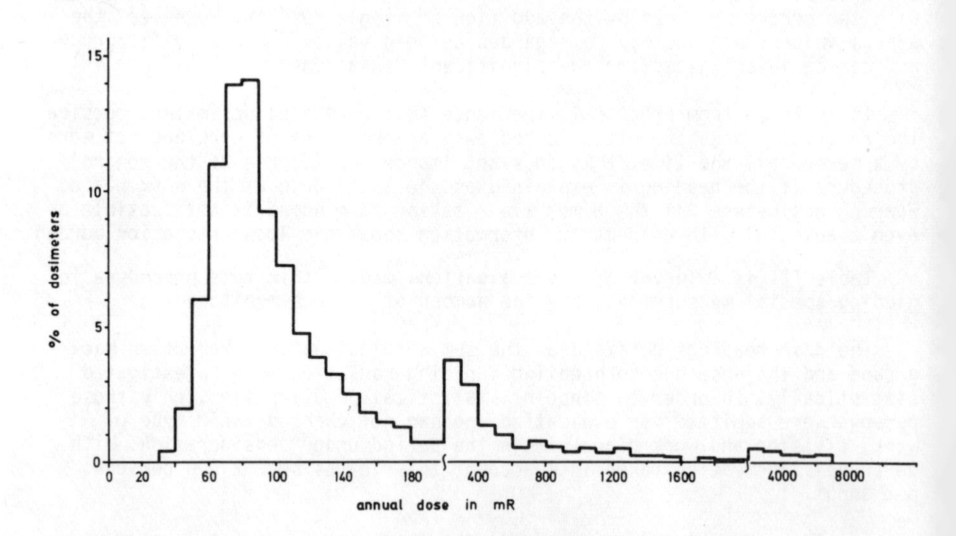

FIG. 2. Dose-distribution, Karlsruhe Nuclear Research Center, 1973.

TABLE IV MEAN ANNUAL DOSE IN DIFFERENT INSTITUTES OF THE KARLSRUHE NUCLEAR RESEARCH CENTER

	% of all surveyed persons	mean annual dose per person in mrem 1968	69	7o	71	72	mean value mrem
Decontamination	3.5	559	693	465	753	1.144	723
Radiation Protection	2.o	4o8	217	328	31o	5o1	353
Accelerator	1.8	266	37o	37o	424	273	341
Reactors	8.2	365	3o2	367	271	376	336
Chemical Institutes	7.o	145	2o7	25o	244	256	22o
Supply Facilities	9.o					19o	19o
Physical Institutes	14.2	1o5	85	87	81	95	93
Biolog./med.Institutes	2.o	75	74	71	78	81	76
Others	52.3	82	7o	79	83	89	81

mean value due to uncertainties in the measurements and statistical scatter, which slightly flattens towards the top as a result of some single values caused by occupational exposure. The values of 60 - 80 mrem/a largely correspond to the natural external background radiation dose (see Section 1). Hence, compared with this natural background dose, the doses of most of the staff are increased either not at all or only slightly so.

Also a breakdown by personnel groups into various areas of activity, as shown in Table IV, clearly shows the significantly lower radiation burden, almost corresponding to the natural radiation dose, of those persons who did not work in restricted areas.

If, in addition, the groups of personnel not occupationally exposed to radiation are broken down by the type of building in which they have their places of work, a clear tendency becomes apparent which indicates a slight shift in the annual dose depending upon whether the work is performed in a wooden house (40 mrem/a), the library (48 mrem/a), or a concrete building (64 mrem/a) (data measured in 1968/69). The higher values in the concrete building are due to the natural radiation of the material, which is partly shielded by bookshelves in the library.

For reasons outlined above the monitoring program is continued and is expected to provide statistically increasingly better founded data as the years go on.

3. Measuring Dose Rates in Private Homes

Since mid-1973 comprehensive statistical measurements have been carried out under the environmental protection program in the Federal Republic of Germany. This project serves the purpose of producing representative information about the radiation burden caused by natural background radiation and radioactive substances contained in building materials in private homes of the German people. Within the framework of these measurements extending throughout Germany, the Health Physics Division of the Karlsruhe Nuclear

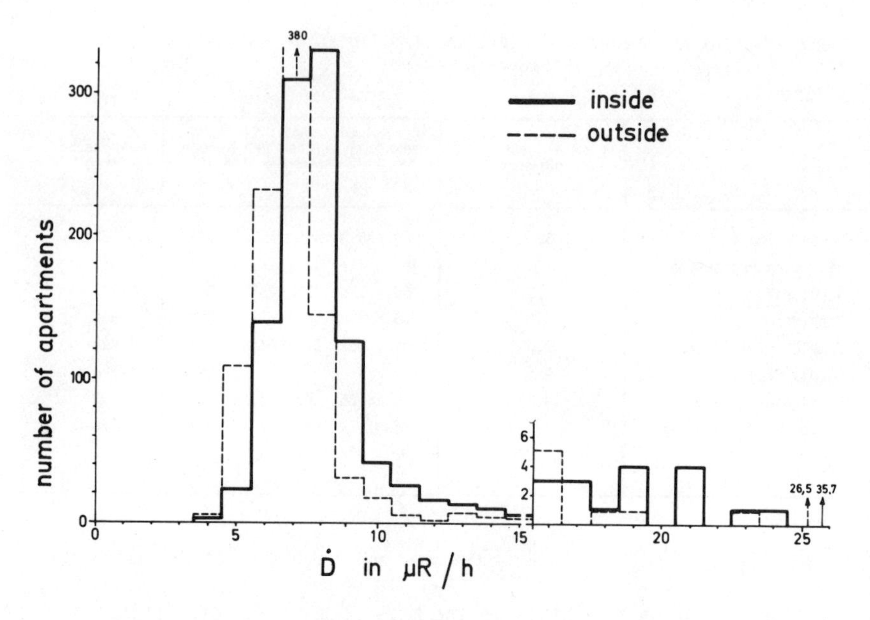

FIG.3. Dose rate inside and outside buildings, Schwaben, Gov. District, FRG.

Research Center so far has conducted dose rate measurements in Southern Germany, especially in the Regierungsbezirke Schwaben und Mittelfranken (districts of Suebia and Middle Franconia).

In the months of July and August 1973 dose rate measurements were carried out in a total of 1 054 private homes and in the open air in the 14 cities and rural districts of the Regierungsbezirk Schwaben. The results are shown in Fig. 3. 352 out of the 1 054 measurements conducted in private homes were performed in homes built before the year 1900 (old buildings), 327 measurements were conducted in houses built between 1901 and 1948 (medium old buildings), and 375 measurements were conducted in houses built after 1948 (new buildings).

Table V shows the differences in dose rates within and outside of buildings.

In the months of October, November and December 1973 the corresponding measurements were performed in 1 487 private homes in 12 cities and rural districts of the Regierungsbezirk Mittelfranken. The results are shown in Fig. 4. 364 of these measurements were performed in old buildings, 522 in medium old buildings, and 601 measurements in new buildings.

Table VI shows the relation of the mean values of dose rate measured in the houses and those measured in the open air.

Fig. 5 and 6 show the respective frequency distributions of dose rate values for solid buildings of different ages and of framework houses, wooden houses and prefabricated houses.

The mean dose rate in private homes is some 20% higher than in the open air, according to these measurements. In some instances, much higher values

TABLE V. MEAN VALUE OF DOSE RATE; GOVERNMENTAL DISTRICT SCHWABEN, FED. REP. OF GERMANY

county	mean value µR/h inside of buildings $\bar{\dot{D}}_H$	mean value µR/h outside of buildings $\bar{\dot{D}}_F$	$\bar{\dot{D}}_H / \bar{\dot{D}}_F$	number of measurements
Kaufbeuren Stadt	7.4	7.3	1.o1	75
Lindau Land	7.3	7.1	1.o3	68
Kempten Stadt	7.3	6.9	1.o6	79
Günzburg Land	7.8	7.2	1.o8	77
Memmingen Stadt	7.9	7.2	1.o9	75
Dillingen Land	8.1	7.3	1.11	76
Donauwörth Land	8.2	7.2	1.14	75
Oberallgäu Land	8.o	6.8	1.18	75
Ostallgäu Land	7.9	6.5	1.21	77
Friedberg-Aichach	8.3	6.9	1.2o	77
Neu-Ulm Land	8.4	6.9	1.23	75
Mindelheim Land	8.4	6.4	1.31	75
Augsburg Stadt	9.1	6.9	1.32	86
Augsburg Land	9.o	6.2	1.45	64
governmental district Schwaben	8.1	6.9	1.17	1o54

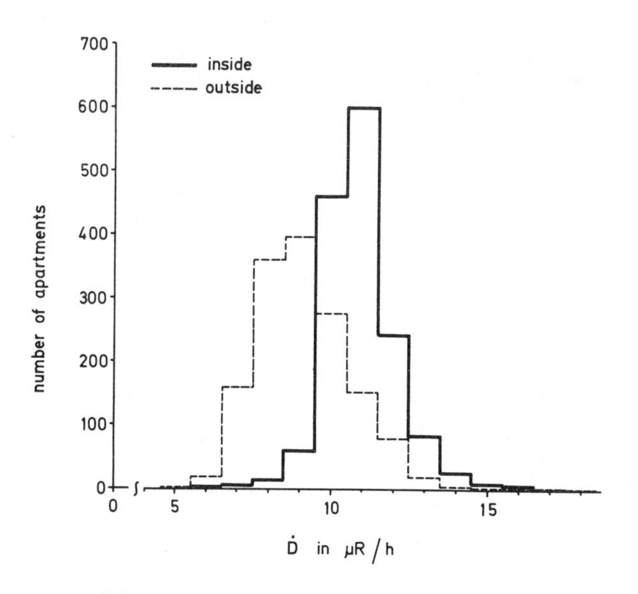

FIG.4. Dose rate inside and outside buildings, Mittelfranken, Gov. District, FRG.

TABLE VI. MEAN VALUE OF DOSE RATE; GOVERNMENTAL DISTRICT MITTELFRANKEN, FED. REP. OF GERMANY

	mean value µR/h inside of buildings $\bar{\dot{D}}_H$	mean value µR/h outside of buildings $\bar{\dot{D}}_F$	$\bar{\dot{D}}_H / \bar{\dot{D}}_F$	number of measurements
Landkr.Weißenburg	1o.1	8.4	1.2o	11o
Landkr. Roth	1o.6	8.6	1.23	11o
Stadt Schwabach	1o.9	8.9	1.22	75
Stadt Nürnberg	11.o	8.6	1.27	186
Stadt Fürth	11.1	9.o	1.23	152
Stadt Erlangen	1o.9	8.3	1.31	154
Stadt Ansbach	1o.4	9.1	1.14	12o
Landkr.Erlangen	1o.9	9.o	1.21	99
Landkr.Fürth	11.6	9.6	1.2o	75
Landkr.Ansbach	1o.6	9.3	1.13	18o
Landkr.Neustadt an der Aisch	11.6	9.9	1.17	13o
Landkr.Lauf	1o.7	9.1	1.17	96
governmental district Mittelfranken	1o.9	9.o	1.2o	1487

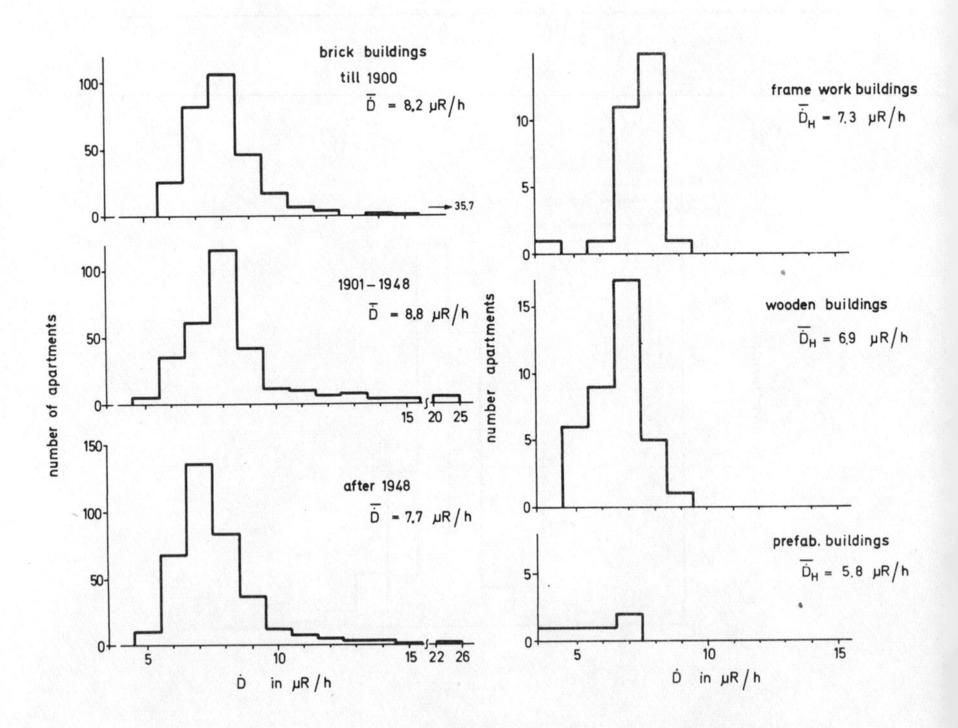

FIG. 5. Dose rate inside different types of buildings, Schwaben, Gov. District, FRG.

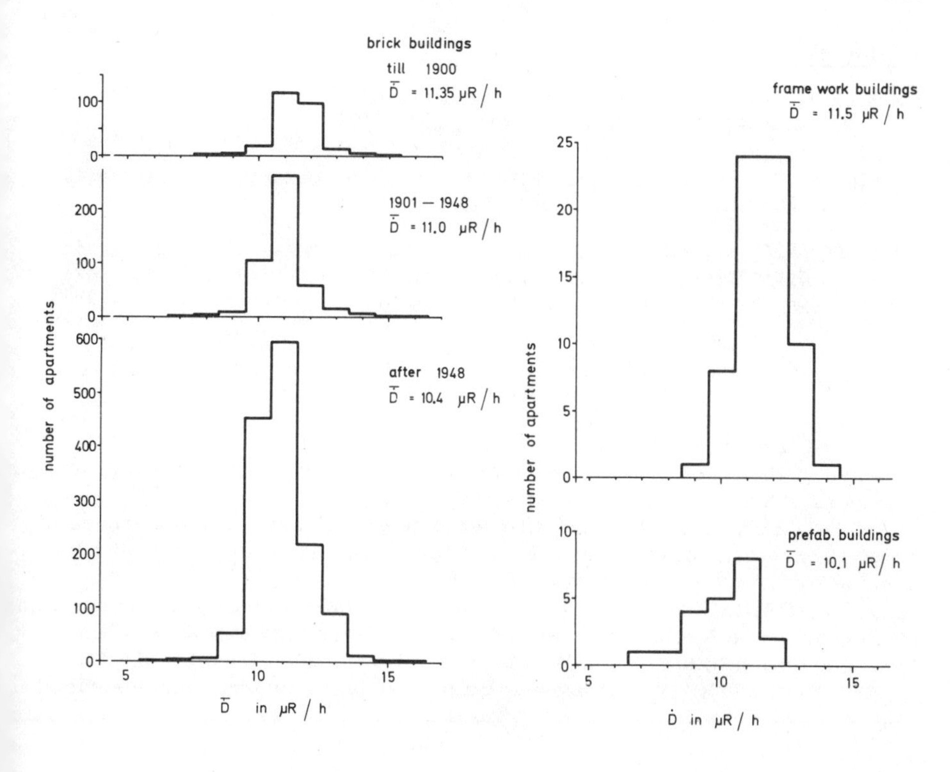

FIG. 6. Dose rate inside different types of buildings, Mittelfranken, Gov. District, FRG.

were found, especially in brick buildings, the maximum being 35.7 µR/h. The lowest dose rate value was measured to be 4.2 µR/h in a prefabricated house. The data measured in various rooms of a house differ only slightly, except for tiled rooms.

In the light of all the measured data the average external radiation burden for the population in the areas investigated is 10 µR/h in private homes, which corresponds to 86 mR/a. This value is 20% higher than the dose rate determined in the open air, which was an average of 8.4 µR/h, corresponding to 73 mR/a.

The dose rates were measured with a scintillation dose rate counter. This unit indicates not only terrestrial radiation but also some fraction of cosmic radiation. This first statistical evaluation of the results measured was made without any correction for the fraction of cosmic radiation.

A comprehensive evaluation of all the data measured in private homes by various agencies throughout the Federal Republic of Germany will be made by the Federal Health Office after termination of the measurements.

The authors are indebted to Mr. Piesch performing the measurements.

References

[1] KIEFER, H., KOELZER, W., Eds., Jahresbericht 1973 der Abteilung Strahlenschutz und Sicherheit, KFK-Report 1973 (1974).

[2] WINTER, M., Proc. Int. Symp. Rapid Methods for Measuring Radioactivity in the Environment, IAEA, Vienna (1971) 525.

[3] BURGKHARDT, B., PIESCH, E., WINTER, M., Long-term use of various solid-state dosimeters for environmental monitoring of nuclear plants, - experience and results, 3rd Int. Congr. IRPA, Washington D.C., Sept. 1973.

DISCUSSION

P. SLIZEWICZ: From Table I it appears that the measurements obtained at the fence of a nuclear installation are higher than those at greater distances. Are these differences significant?

G. STÄBLEIN: Yes, the difference is significant. The difference of about 10 mrem/a between the values at the fence and at a distance of 1 km are due to the nuclear installations near the fence.

W.J. BAIR (Chairman): Do you have a similar comprehensive programme at the Karlsruhe Nuclear Research Centre to assess the exposure of the local population to radionuclides that might be released from the Centre?

W. KOELZER: No, at present we have only the programme described.

ASSESSMENT OF ENVIRONMENTAL RADIATION DOSE FROM AIRBORNE EFFLUENTS USING REAL TIME METEOROLOGICAL DATA

V. SITARAMAN, P.L.K. SASTRY, P.V. PATEL, V.V. SHIRVAIKAR
Health Physics Division,
Bhabha Atomic Research Centre,
Trombay, Bombay, India

Presented by S.J. Supe

Abstract

ASSESSMENT OF ENVIRONMENTAL RADIATION DOSE FROM AIRBORNE EFFLUENTS USING REAL TIME METEOROLOGICAL DATA.

Airborne radioactive effluents from nuclear plants should be limited such that the total environmental dose to individual members of the public does not exceed the limit set by the ICRP or a fraction of it apportioned to dose via air. The dose can either be evaluated from real time meteorological data or be measured directly by continuous recorders. In India we find the first approach has the advantages that: (i) the entire environmental dose field can be computed in contrast to only few, localized measurements that could be made using radiation recorders; (ii) the low level of external dose rates (μrem/h or less) makes continuous radiation monitoring difficult; (iii) with the recent trend of reducing the permissible environmental dose levels increasing reliance has to be kept on indirect estimates; and (iv) in the case of multiple installation sites monitoring stations will not be able to give contributions from individual plants. This information might sometimes be useful in the assessment and limiting of release rates of any individual installation.

In the paper the measured environmental doses from release of ^{41}Ar from the CIRUS reactor are satisfactory compared with the estimates obtained from meteorological data and the site characteristics.

1. INTRODUCTION

Gaseous radioactive effluents from nuclear plants and fuel processing plants are often a major source of environmental radiation exposure. The ICRP has laid a limit of 500 mrem/a for the dose an individual member of the public can receive from all the nuclear operations at a site. In India the above limit is used for evolving the environmental safety policy. The radiation exposure to a member of the public would arise from the gaseous and liquid effluent releases. In certain cases a contributary dose may also arise from land operations, e.g. waste burial land transport of irradiated fuel or radiation sources. To keep the total dose around a power station site within the prescribed limit, it is necessary [1]:

(a) To apportion dose limits to each of the routes, viz. air, water and land. The total should not exceed 500 mrem/a;
(b) To apportion further, fractions of these apportioned limits, to each plant taking into account the nature of the effluents from each;
(c) To derive effluent discharge limits for each plant based upon the limits in (b).

If we confine ourselves to gaseous effluents (though the approach in the case of liquid effluents is similar), it is necessary to evaluate the

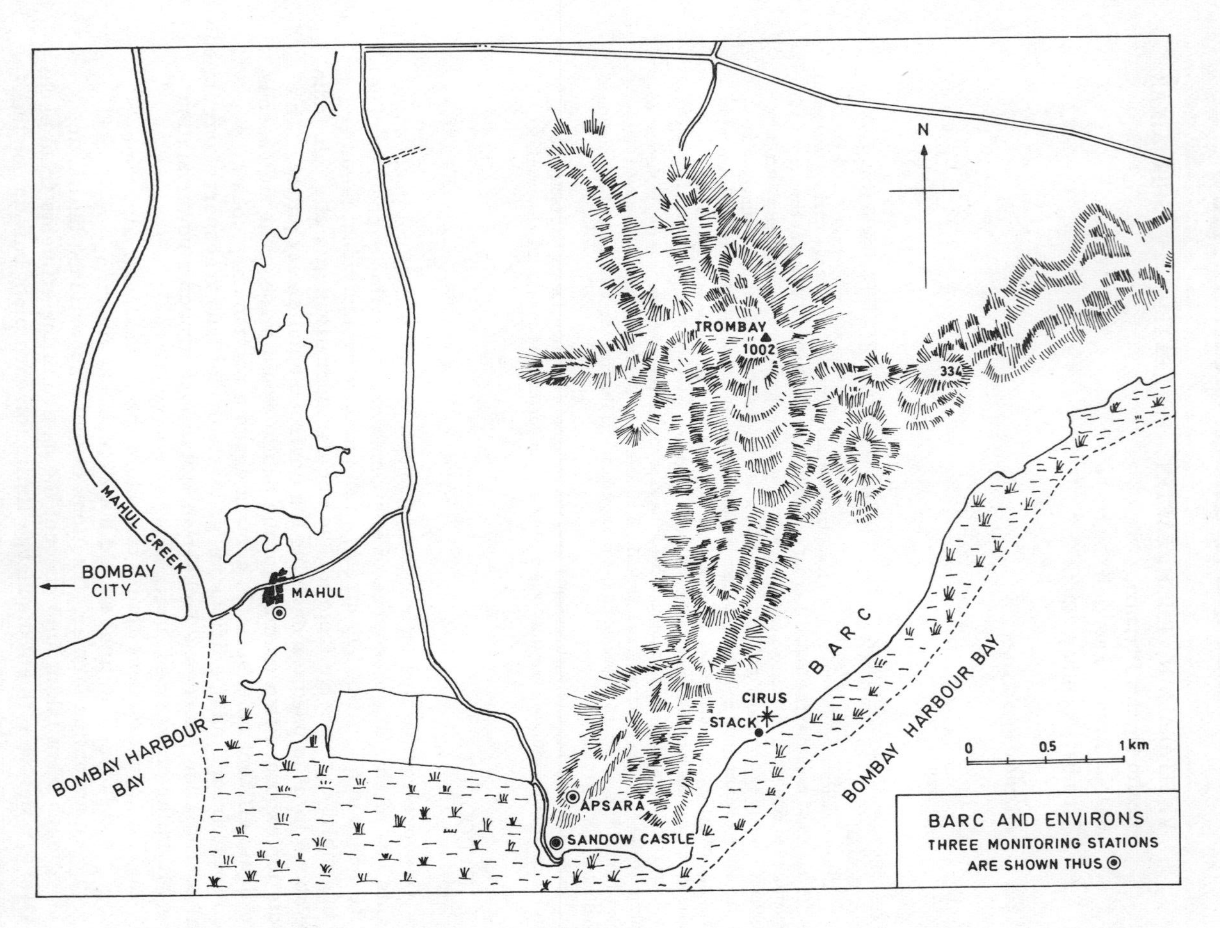

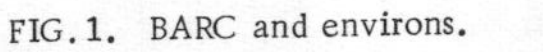
FIG. 1. BARC and environs.

possible annual off-site doses at the design stage itself from the design estimate of the effluent release rates and the local meteorological data. During the operational phase it is further necessary to ascertain that the plants actually comply with the off-site dose limit. This would involve operation of continuous radiation monitors at suitable locations near the site boundary where the dose rates would generally be of the order of a few μrem/h. While it is feasible to measure such low levels of activity with the currently available radiation recorders, the maintenance of such equipment over prolonged periods poses operational difficulties. In addition, at such low levels of activity it will be difficult to distinguish them from the background level. The latter difficulty is one of the principal problems in the use of thermoluminescent dosimeters for estimating the off-site doses from plant operations. Where there are multiple nuclear installations in a site, contributions from individual plants cannot be assessed by integrating-type radiation monitors.

As an alternative, in this paper we consider the use of the real time meteorological data available at a site to obtain an indirect estimate of the environmental radiation field. An obvious advantage with this method is that the monitoring of the activity discharge rate at stack level can be done with better accuracy than off-site radiation measurements. Though the estimates are indirect and involve atmospheric diffusion models that are being progressively improved upon, the currently accepted models of Pasquill [2] as modified by Gifford [3] and others are known to give reliable estimates of the concentration of effluents in field conditions. In the early stages of our environmental safety work at Trombay we made a comparison of the off-site doses due to ^{41}Ar released from the CIRUS reactor at Trombay (estimated at 640 Ci/d [4]) as computed from the site meteorological data with that measured at a fixed monitoring station using continuous recorders. The paper describes the above work and its extrapolation to similar computations at power reactor sites.

2. SITE CONDITIONS AND LOCATIONS OF THE MONITORING STATIONS

The theoretical models available for atmospheric diffusion [2, 5] are valid for level uniform terrain and generally require modification to take into account local site characteristics. Some idea of the problem can be had from the site map in Fig. 1, which shows the site characteristics of the Trombay environment. The terrain is uneven, with the Trombay ridge stretching from north of the reactor stack through west to the southwest of it. Its maximum elevation is about 300 m to the northwest of the stack. At the southwest end the ridge splits into two branches.

Three monitoring stations were operated for periods of from 2 to 4 years. Two of these were very near the western site boundary about one km south-west of the reactor, on the south-western leg of the ridge. These are marked Apsara and Sandow Castle in Fig. 1. The Apsara station also had a wind recorder, data from which was used in dose computation. The third station was located 4.5 km west of the reactor midway between the Trombay Establishment and Bombay city. The terrain at this station is level.

3. DOSE COMPUTATIONS

Wind speed and direction recorded at the monitoring station for a period of three years are shown in the form of the wind rose in Fig. 2. Table I presents the micrometeorological and diffusion parameters used in the computations. These are based on Sutton's atmospheric diffusion model [6]. Values of Sutton's vertical virtual diffusion parameter C_z were obtained after a number of smoke photography studies carried out at various times of the day and night. Values of C_y, the cross-wind diffusion parameter, were obtained from wind direction fluctuation measurements. The stability classes were defined for certain typical seasons and time of day. Values of stability index 'n' for the appropriate classes of diffusion were taken from published data.

Gamma doses were calculated using Holland's nomograms [7]. The gamma dose in any sector over a year is the sum of the dose during the period when the plume is in that sector plus the contributions to the dose when the plume is in other sectors. As suggested by Holland, the latter doses were computed using the effective stack height equal to $(h^2 + y^2)^{\frac{1}{2}}$, where h is the effective height of the stack for the plume with respect to ground and y is the cross-wind distance between the plume sector and the sector in which the dose is being computed. The stack height is 120 m. The plume rise is primarily due to efflux velocity and is calculated to be only about 10 m. From Fig. 1 it may be noted that on the western half circle of the stack the ridge is higher than the stack height and therefore when crossing the ridge the plume will travel nearly at the surface. The effective stack height in these sectors was therefore assumed to be zero for the distances of interest. This was also supported by the observation of activity at Apsara station where the observed concentrations can be explained only if the effective stack height is a few metres. For other sectors, however, this assumption was not made. The radioactive decay correction factor in the above computations was found to be small. Even for the Mahul station the correction factor is 0. 8 when the wind speed is low as 2 m/s. Annual isodose curves were plotted based on these calculations, as shown in Figs 3 and 4, on two different scales.

4. MEASURED DOSES

Continuous measurement of γ and β, γ dose rates were made at the three monitoring stations. Apsara and Mahul monitoring stations were operated for four years, 1964-67, and the Sandow Castle station operated for two years, 1964-65. Each station had on its roof a pair of thin-walled G-M counters, one of which was shielded by a close fitting shield of 2.3 mm thick brass to cut off β-radiation [8] and register only γ-radiation. The counters were calibrated with respect to a ^{60}Co source, which has a γ-energy very close to that of ^{41}Ar. A typical counter gave 300 counts/min for a dose rate of 0.1 mrem/h. The count rate was recorded on a continuo strip chart recorder. Figure 5 shows a trace of γ and β, γ radiation. Pea in the count rate occurred in the early morning hours during the regime of easterly land breeze and were definitely attributable to ^{41}Ar from CIRUS.

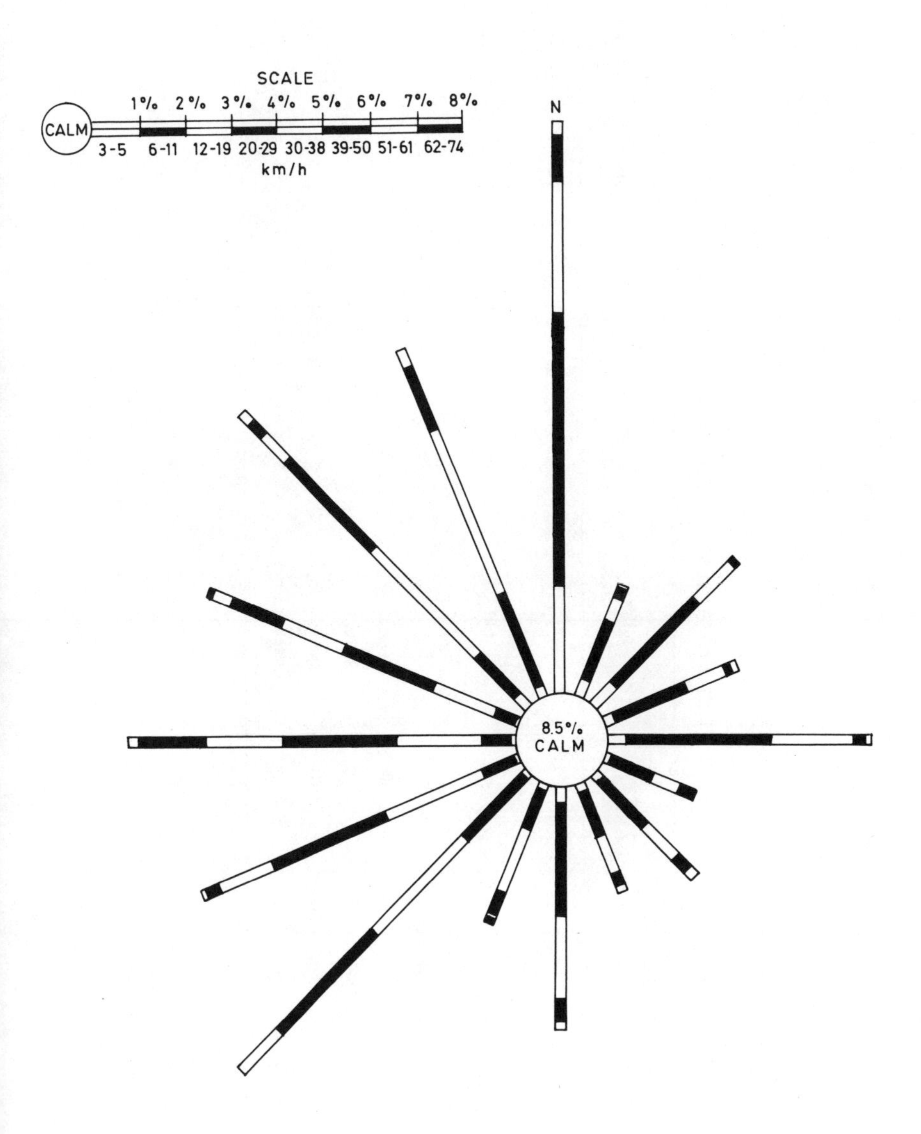

FIG.2. Annual wind rose, Apsara, Trombay.

TABLE I. MICROMETEOROLOGICAL AND DIFFUSION PARAMETERS OF THE TROMBAY AREA

	Period	Time of day (h)	Type	Stability	Wind speed U (m/s)	n	C_y $(m)^{n/2}$	C_z $(m)^{m/2}$
1.	March to May	1000 to 1800	A1	Unstable	5	0.20	0.35	0.25
2.	Sep. to Feb.	1000 to 1800	A2	Unstable	4	0.20	0.35	0.25
3.	March to May	0000 to 0700 and 0800 to 2400	B1	Neutral	2.5	0.25	0.25	0.25
4.	Sep. to Feb.	"	B2	Neutral	2.5	0.25	0.25	0.25
5.	June to Aug.	0000 to 2400	B3	Neutral	5	0.25	0.25	0.25
6.	March to May	0000 to 0700	C1	Stable	2	0.50	0.25	0.10
7.	Sep. to Feb.	"	C2	Stable	2	0.50	0.25	0.10

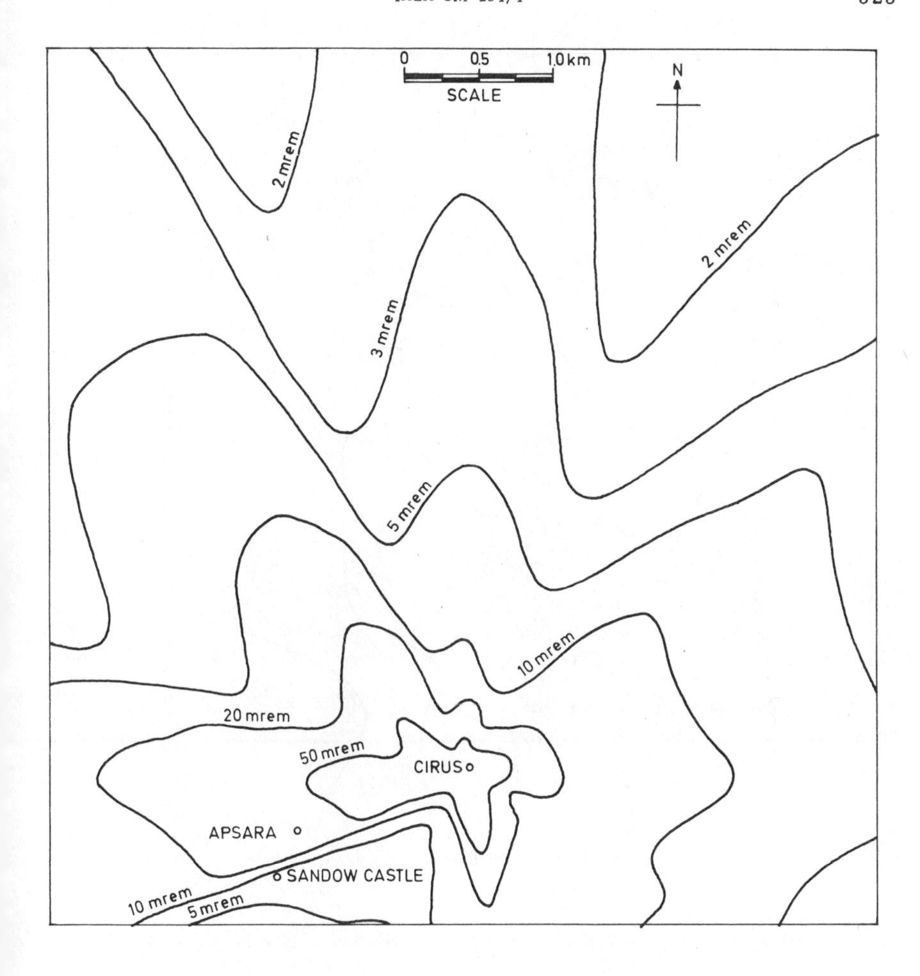

FIG. 3. Annual isodose curves from ^{41}Ar.

From the trace it could be seen that the background γ-level was of the order of 10-20 μrem/h. The mean hourly dose rates were computed from the charts and the average monthly values are tabulated in Table II. The computed annual dose is also given in the last row of Table II.

5. COMPARISON AND DISCUSSION

The agreement between the estimated and measured values is satisfactory, despite the assumptions made in the computations to take into account the uneven nature of the terrain. This has encouraged us to use the meteorological data for routine assessment of off-site doses at power reactor sites. It must be noted that the work described earlier was carried

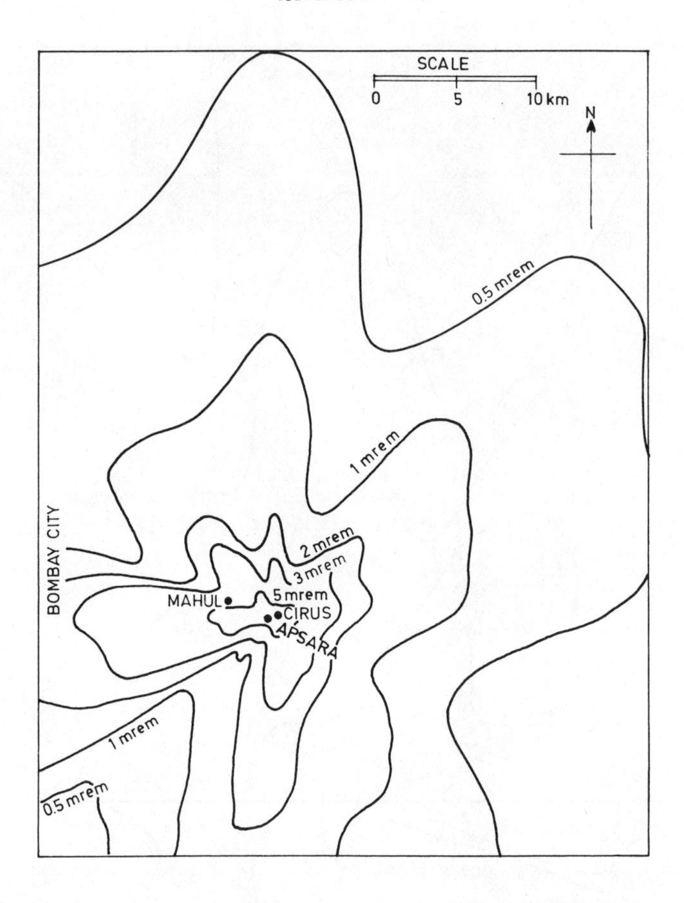

FIG. 4. Annual isodose curves from ^{41}Ar for a larger area than that covered in Fig. 3.

out primarily by hand computation using Holland's nomograms at a time when the computer was not available. At present we have used the meteorological data for estimating off-site doses at Tarapur. These estimates have been based on Pasquill's model and the integration of the doses from the plume has been done using a CDC-3600 computer. The procedure followed at present is to obtain stability classification through statistics of σ_θ (root mean square deviation of wind direction fluctuation) as recommended by Slade [9]. It was found, however, at Tarapur that the numerical values of σ_θ ranges suggested by Slade for various stability classes could not be applied. Experiments using the smoke photography method were made to reclassify values with respect to stability at Tarapur. Doses thus computed are now being incorporated in evolving the environmental safety policy at Tarapur, including establishing discharge limits for gaseous radioactive effluents for other plants at the site. Unfortunately, systematic measure-

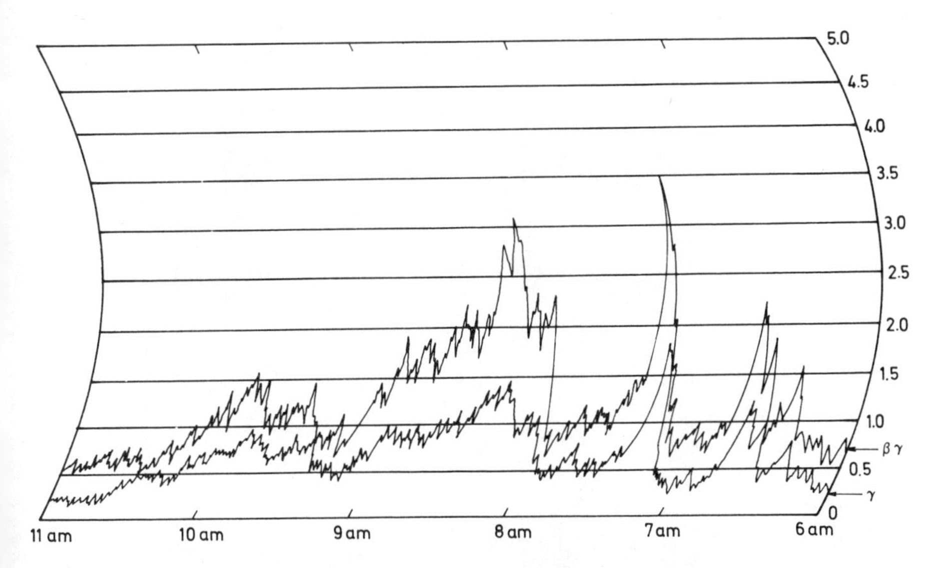

FIG. 5. A typical trace of β, γ and γ activity from ^{41}Ar from the Apsara monitoring station. Full-scale deflection corresponds to 600 counts/min (0.2 mrem/h). The two traces are mapped on a single chart for comparison.

ments of the environmental dose are not available at Tarapur and hence a direct comparison with computed data could not be made. However, the values calculated by us agree well with those estimated with the method given by May and Stuart [10], which was checked against data from Brookhaven National Laboratory.

6. CONCLUSIONS

From our experience we now feel that in future continued reliance will have to be placed in the estimates of external radiation dose made using real time meteorological data, which have the following advantages:

(1) From measurements at a single station combined with the monitoring of activity at the stack itself the entire environmental dose field can be computed, in contrast to only a few localized measurements that could be made using radiation recorders.
(2) The external dose rates in the environment around a nuclear plant are usually low, i.e. of the order of a few μrem/h or less. This makes continuous radiation monitoring difficult and the downtime of such equipment is high compared to that of meteorological equipment.
(3) With the recent trend of reducing the permissible environmental dose levels to values of the order of a few mrem/a, increasing reliance has to be kept on such indirect estimates rather than on direct estimates.
(4) In the case of multiple installation sites monitoring stations will not be able to assign contributions to individual plants. This information might sometimes be useful in the assessment and limiting of release rates from an individual installation.

TABLE II. COMPARISON OF MEASURED VALUES OF GAMMA DOSAGES (mrem) WITH THE COMPUTED DOSAGES AT THE THREE MONITORING STATIONS

Station	Apsara				Sandow Castle		Mahul			
Year	1964	1965	1966	1967	1964	1965	1964	1965	1966	1967
Jan.	1.84	3.85	6.73	1.53	M	4.10	0.11	0.15	1.28	0.08
Feb.	1.21	3.06	3.07	5.25	M	2.40	0.29	0.08	0.57	1.03
March	0.82	2.53	1.97	6.14	0.74	1.28	0.50	0.44	0.04	0.79
Apr.	0.46	1.45	0.58	1.28	0.48	1.23	0.17	0.26	0.05	0.37
May	0.41	0.04	0.39	0.71	0.27	0.24	M	0.06	M	M
June	0.36	0.48	M	0.32	M	0.12	M	M	0.12	M
July	0.40	0.54	0.45	0.32	M	M	M	M	M	M
Aug.	M	3.48	0.68	0.95	0.13	M	M	M	0.17	M
Sep.	3.84	2.07	1.62	0.77	M	M	M	0.03	1.03	0.17
Oct.	2.44	2.65	4.46	3.46	M	M	M	0.81	1.63	0.58
Nov.	5.67	7.19	5.13	6.21	1.19	3.46	M	0.78	0.46	0.98
Dec.	10.15	9.30	7.78	4.58	6.18	0.05	M	0.13	0.82	0.71
Total	27.6	36.64	32.86	31.42	8.99	12.88	1.07	2.74	6.17	4.71
Computed value		41.2			8.5			4.5		

Note: 'M' indicates missing data. During the period May to September the plume direction is usually away from the three monitoring stations.

In addition to these advantages we must also record certain limitations of the system. As already mentioned, the current knowledge on atmospheric diffusion in non-uniform terrain is limited and the modifications of the basic diffusion expressions valid in uniform terrain conditions can be subjective. Another aspect, as noted earlier, is that the use of real time meteorological data requires classification of the data into stability categories based on some governing parameters (e.g. wind direction fluctuation, temperature profile). It has been our experience, which is further corroborated by others (e.g. Skibin [11]), that the classification, such as that given by Slade, in terms of σ_θ are not universally applicable at all sites. Nevertheless, there is no doubt that the approach suggested by Slade is a very pragmatic one.

ACKNOWLEDGEMENTS

Thanks are due to Shri S.D. Soman for encouragement and useful discussions in the preparation of the paper.

REFERENCES

[1] SHIRVAIKAR, V.V., GANGULY, A.K., "Safety criteria for operating power reactors", presented at Symp. Nuclear Science and Engineering, Bhabha Atomic Research Centre, Bombay, 1973.
[2] PASQUILL, F., Meteorol. Mag. 90 (1961) 33.
[3] GIFFORD, F.A., Nucl. Safety 2 (1961) 47.
[4] SOMAN, S.D., ABRAHAM, P., Health Phys. 11 (1965) 497.
[5] SUTTON, O.G., Micrometeorology, McGraw Hill, New York (1953).
[6] SUTTON, O.G., Quart. J. Roy. Meteorol. Soc. 73 (1947) 426.
[7] HOLLAND, J.Z., Peaceful Uses atom. Energy (Proc. Conf. Geneva, 1956) 13, UN, New York (1956) 110.
[8] WEISS, M.M., Area Survey Manual, Rep. BNL-344 (1955).
[9] SLADE, D.H., "Meteorology and atomic energy", USAEC Rep. TID-24190 (1968) Chap.4.
[10] MAY, M.J., STUART, I.F., "Comparison of calculated and measured long term gamma doses from stack effluent of radioactive gases", presented at Health Physics Mid-year Symp., Augusta, Georgia, 1968.
[11] SKIBIN, D., J. Meteorol. Soc. Japan 50 (1972) 501.

THE DERIVATION OF WORKING LIMITS FOR THE CONTROLLED DISCHARGE OF RADIOACTIVE WASTES FROM NUCLEAR INSTALLATIONS

T. SUBBARATNAM, S.D. SOMAN
Health Physics Division,
Bhabha Atomic Research Centre,
Trombay, Bombay, India

Presented by S.J. Supe

Abstract

THE DERIVATION OF WORKING LIMITS FOR THE CONTROLLED DISCHARGE OF RADIOACTIVE WASTES FROM NUCLEAR INSTALLATIONS.

The paper outlines the policy and methodology adopted in India for the routine control of liquid and gaseous radioactive wastes and describes the factors and considerations used in applying ICRP recommended dose limits to the establishment of derived working limits for the controlled releases of radioactive effluents to the environment. Specific examples of the methods used in obtaining derived working limits are given. The impact of such controlled releases on the radiation dose to the population is discussed.

1. INTRODUCTION

In India the task of laying down limits for the discharge of radioactive liquid and gaseous effluents to the environment is the responsibility of the Health Physics Division of the Bhabha Atomic Research Centre. This paper outlines the policy and methodology for control of routine discharges to the environment and the considerations under which limits for such discharges are arrived at. These limits are subject to revision, pending work connected with the identification of critical pathways and the critical group of the population. Since such environmental survey programmes involve considerable effort in view of the widely varying living and food habits of the people of India, an attempt has been made to calculate these working limits from experience elsewhere and the expected conditions obtainable in India based on the data collected so far.

2. POLICY REGARDING CONTROLLED DISCHARGES OF RADIOACTIVE EFFLUENTS

The criteria adopted in India governing the exposure to ionizing radiations of both occupational workers and members of the public are based on the standards of radiation protection recommended by the International Commission on Radiological Protection (ICRP) [1]. The routine discharges of radioactive effluents to the environment cause exposure to members of the public through the air, land and water routes and through intake of food likely to be contaminated. Pending the establishment of the critical pathways for each major installation, the dose limits, which include both internal as well as external exposure, are arbitrarily apportioned to either air and water or to air, water and land, equally, depending on the site and its features. For instance, at sites where all three routes are expected to contribute to population exposure the apportionment is 1/3 of the corresponding dose limits, with a small margin for future expansion if necessary.

It is the practice in India to group several large installations in a single site for economic grounds. For purposes of the task of setting derived working limits (DWL) such multiple installation sites are treated as a single unit and the land, water or air-route dose-limit apportionments are further apportioned to individual units, based on the normal quantities of radionuclides expected to appear at the discharge points.

3. THE DERIVATION OF WORKING LIMITS

3.1. General considerations

The DWL values presented here are the result of an attempt to translate the ICRP dose limits and their apportionment into permissible total releases from a nuclear installation under normal conditions of operation and for short-term high releases, both for safety design criteria and preliminary operation of the installation.

The method of calculation is identical to that used in the report of the ICRP Committee II on permissible dose for internal radiation [2], except that the values of the parameters used are those appropriate to the assumed critical section of the population. Venkataraman et al. [3] have shown that considerable variation exists in the standard man data for Indian conditions as compared with ICRP Standard Man data. Thus, Indian Standard Man data have been used throughout.

3.2. Discharges to the atmosphere

3.2.1. Atmospheric diffusion characteristics

For purposes of computing the annual average integrated ground level concentration of a radionuclide from a continuous point source, the site is divided equally into sixteen $22\frac{1}{2}°$ sectors. The time-integrated concentration at any given sector is given by [4]:

$$\chi_{TIC} = \left(\frac{2}{\pi}\right)^{\frac{1}{2}} \frac{Qf_i}{\sigma_z \bar{u} \left(\frac{\pi}{8}\right) x} \exp\left[-\frac{h^2}{2\sigma_z^2}\right]$$

where

χ_{TIC} is the time-integrated concentration at ground level (Ci/m^3)

Q is the rate of discharge of material from a continuous point source (Ci/s)

σ_z is the standard deviation width of the distribution of the material in a plume in the z direction m

$\bar{u}$ is the mean-wind velocity in the X direction (m/s)

h is the physical stack height (m)
f_i is the frequency of wind in the sector
x is the distance downwind from the stack (m).

The value of x is held to be the exclusion distance from the installation, taken to be 1000 m. No credit is given for the plume rise. In cases where downwash effects are expected to be present appropriate reductions in the value of h to be used are made. The values of f_i, $\bar{u}$ and σ_z for different stability conditions are computed from a 3-year average of the meteorological observations made at the site, based on which the worst sector and its values of f_i are established.

3.2.2. Derived air concentrations

Derived air concentration values (DAC) have been calculated for both the inhalation and milk routes for ^{131}I, ^{137}Cs, ^{90}Sr and ^{89}Sr. The organ masses, transfer coefficients and other relevant parameters used are summarized in Table I. Only the assumptions used and the considerations involved in obtaining the DAC values for the milk route are presented here. DAC (inhalation route) for the radionuclides listed above have been obtained in a manner identical to that of the ICRP [2], using parameters applicable to Indian conditions. Values of DAC for the radionuclides ^{131}I, ^{137}Cs, ^{90}Sr and ^{89}Sr are listed in Table II.

Iodine-131

According to Kamath et al. [5], the average milk consumption by an Indian child is 0.2 litre/d. Using the data presented in Table I, the DWL (milk) for ^{131}I calculates to 788 pCi/litre, which has been rounded off to 750 pCi/litre.

TABLE I. PARAMETERS USED IN CALCULATION OF DERIVED AIR CONCENTRATIONS FOR ^{131}I, ^{137}Cs, ^{90}Sr AND ^{89}Sr

Parameter	Iodine-131	Caesium-137	Strontium-90	Strontium-89
Critical group	6-month-old child	Adult	6-month-old child	6-month-old child
Organ mass (g)	Thyroid 1.09 [3]	Whole body 4.6×10^4 g	Bone 370 g	Bone 370 g
Effective energy (MeV)	0.18 [7]	0.59 [7]		
f_w	0.35 [7]	1.0		
t_{eff} (d)	6.0 [7]	140 [7]		50.4
DWL (milk) (pCi/litre)	750	5×10^4	1500	4.6×10^4
DWL (grass) (pCi/m^2)	4500	1.0×10^5	-	-
Transfer coefficient pasture to food	0.17 pCi/litre per pCi/m^2	0.5 pCi/litre per pCi/m^2 [8]	50 pCi/d per mCi/km^2 per month [8]	8.3 pCi/d per mCi/km^2 per month [8]

TABLE II. DERIVED AIR CONCENTRATION VALUES

Pathway	^{131}I (Ci/m^3)	^{137}Cs (Ci/m^3)	^{90}Sr (Ci/m^3)	^{89}Sr (Ci/m^3)	^{3}H (Ci/m^3)	^{85}Kr (Ci/m^3)
Milk route	3.01×10^{-13}	3.82×10^{-12}	1.42×10^{-12}	2.12×10^{-9}	-	-
Inhalation	3.06×10^{-11}	2.23×10^{-9}	2.63×10^{-11}	3.97×10^{-9}	4.75×10^{-8}	-
Immersion						3.61×10^{-6}

Using the data provided by Garner for UK conditions [6], the transfer coefficient from pasture to milk for Indian conditions is calculated to be 0.17 pCi/litre per pCi/m^2. Assuming a deposition velocity of 2×10^{-2} m/s and an effective mean life in grass of 6 days [7], the DAC (milk route) is obtained as 3.01×10^{-13} Ci/m^3.

Strontium-90

For ^{90}Sr the maximum permissible continuous rate of intake is so set that the maximum permissible body burden (MPBB) would be reached only after 10 years for a member of the public exposed from birth. The dose rate to bone during the first few years of such exposure is very much less than the dose limit and the smaller skeletal weight of the child does not make it necessary to lower the rate of intake during the early years [8]. Thus the MPBB and DAC are assumed to be independent of age and are taken to be 1/10 the ICRP values modified to Indian conditions. Since the ICRP MPC_w for members of the public is 4×10^{-7} μCi/cm^3, the maximum permissible daily intake for Indian conditions calculates to

$$\text{MPDI (Indian)} = 4 \times 10^{-7} \times 2.2 \times 10^3 \times \frac{46}{70} = 5.49 \times 10^{-4}\ \mu\text{Ci/d}$$

which has been rounded off to 550 pCi/d. Assuming that the daily intake is obtained through milk only, the DWL (milk) calculates to 1500 pCi/litre. Using the transfer coefficient of 50 pCi/d per mCi/km^2 per month [8] and V_g as 3×10^{-3} m/s, the DAC (milk route) calculates to 1.42×10^{-12} Ci/m^3.

Strontium-89

In the case of ^{89}Sr the body reaches equilibrium during the first year of exposure if the intake is at the rate of 1/10 ICRP MPC_w for continuous occupational exposure. Because of this, the ICRP MPBB is scaled down to Indian conditions as

$$\text{q (Indian child)} = \frac{4 \times 370}{7000} = 0.216\ \mu\text{Ci}$$

where 370 represents the skeletal weight of an Indian child and 7000 that of an ICRP adult. Using the same parameters as applicable to ^{90}Sr yields the DWL (milk) of 4.6×10^4 pCi/litre and DAC (milk route) of 2.12×10^{-9} Ci/m

Caesium-137

For ^{137}Cs the adult is found to be the limiting case. Using a body weight of 4.6×10^4 g [3] and an effective energy of 0.59 MeV [7], the MPBB for an Indian adult is 2.18 μCi. For an effective half-life of 140 days [7] the DWL (milk) calculates to 5.4×10^4 pCi/litre, which has been rounded off to 50 000 pCi/litre per pCi/m^2 [8], the DWL (grass) is obtained as 1×10^5 pCi/m^2. V_g is taken to be 3×10^{-3} m/s and for a half-life in grass of 70 days the DAC (milk route) is calculated as 3.82×10^{-12} Ci/m^3.

DAC for ^3H

Because of the lack of data, it is assumed that the body-fluid weight of an Indian child is in the same proportion as that of the body-fluid to body weight of an Indian adult. This value is obtained as 3.53 kg. Using this and the effective energy of tritium as 0.0059 MeV, the MPBB of an Indian child calculates to 16.75 μCi, giving a uniform concentration in the body fluid of 4.75×10^{-3} μCi/g. Assuming tritium concentration in the atmosphere, the DAC can be obtained, influenced as it were by the humidity of the atmosphere. If the average humidity is taken to be 10 g/m^3 of air, the DAC calculates to 4.75×10^{-8} Ci/m^3.

DAC for noble gases

Under normal operating conditions of the reactors in India the stack effluents contain ^{41}Ar besides ^3H in small quantities. Where fuel leaks are present fission product noble gases also appear in the stack. However, for purposes of design and control it is conservatively assumed that the noble gases from reactor stacks consist only of ^{41}Ar.

Effluents from fuel reprocessing facilities contain predominantly ^{85}Kr.

In the case of ^{41}Ar the DWL (stack) are based on two considerations: (1) the immersion dose due to the plume reaching the ground level; and (2) the gamma dose due to the passage of the plume. The immersion dose has been evaluated from the time-integrated ground level concentration at the sector under study. The gamma dose due to plume passage includes in addition to the plume dose in the relevant sector, the dose contributions due to the plume being in the adjacent two sectors on either side. The total dose is thus the sum of the immersion dose and the gamma dose of plume in the relevant sector and those adjacent to it.

In the case of ^{85}Kr it has been shown that [10] 1 pCi/cm^3 of ^{85}Kr in air at 15°C and 760 mm gives

2.08 rad/a to skin
1.73×10^{-2} rad/a to shallow tissues and gonads
1.41×10^{-2} rad/a to the whole body.

It is readily seen that skin is the limiting case. Thus the DAC (immersion) for skin calculates to 3.61×10^{-6} μCi/cm^3. Table II presents the DAC values.

DWL (stack) values

Since the DAC values of Table II are equated to the full quota of the corresponding dose limit, it would be necessary to obtain the stack discharge limits, i.e. DWL (stack) for specific installations based on the apportioned dose limits. This is illustrated by a typical example.

For the MAPP reactors the common stack of 100 m gives a TIC value of 6.5×10^{-8} s/m^3 at the worst sector. The DAC (milk route) for ^{131}I is 3.01×10^{-13} Ci/m^3. Out of the total dose limit of 1500 mrem per year to a child one-half, or 750 mrem/a, is assigned to the twin-reactor MAPP station. Thus the DWL (stack) for MAPP is

$$\frac{3.01 \times 10^{-13} \times 8.64 \times 10^4}{6.5 \times 10^{-8}} \times \frac{400}{1500} = 0.106 \text{ Ci/d}$$

Similar considerations have been applied to other radionuclides and other installations.

3.2.3. Applicability of DWL (stack) values

The derived working limits so calculated represent the average release rates on a continuous basis - the actual releases may fluctuate over a limited range. To provide flexibility in operation and maintenance, higher discharge rates may be permitted for shorter periods, provided the average release rates do not exceed the DWL (stack) values. It is, however, necessary to set an upper limit for such short-term high releases. The philosophy adopted is to permit 10% of the DWL (stack) values to peak discharges, provided such high releases do not occur for more than $2\frac{1}{2}$ hours in any calendar fortnight. The normal release rates are correspondingly reduced by a proportionate factor.

Where several radionuclides are discharged at the same time, the rule of mixtures as suggested by the ICRP [2] is applied. Thus if C_1, C_2, C_3 ... are the concentrations of radionuclides 1, 2, 3 ..... then

$$\frac{C_1}{DWL_1} + \frac{C_2}{DWL_2} + \frac{C_3}{DWL_3} + \ldots\ldots \leq 1$$

3.3. Discharges to aquatic environment

Ganguly and Pillai [11] have calculated the MPC_w values for several radionuclides applicable to a freshwater environment on the basis of concentration factors measured in local samples of food items. The same analogy has been extended to obtain the MPC_w of several radionuclides for Indian standards. For example, the ICRP MPC_w for unidentified radionuclides for continuous occupational exposure is 10^{-6} μCi/cm^3, provided that ^{226}Ra or ^{228}Ra is absent [2]. One-tenth of this value is applied to members of the public. Raghunath and Soman [12] have established that an Indian Standard Man consumed 4.1 litres of water per day as against the ICRP value of 2.2 litres/d. Use of this data results in the scaled-down value of $MPCU_w$ of 4×10^{-8} μCi/cm^3 for Indian conditions. If one-half the dose limit is assigned to exposure through the water route, the $MPCU_w$ for a freshwater environment is 2×10^{-8} μCi/cm^3.

Discharges to the marine environment are based on the consideration that exposure of the public results through consumption of harvest from the sea or the contamination of near-shore waters as the case may be. No credit is taken for dilution in the environment after the discharge. Moreover, the discharged wastes should contain no particulate material.

Since the separation of tritium from radioactive liquid wastes from nuclear processes poses a difficult problem, it has been assigned a special limit, particularly for the freshwater environment. The ICRP MPC_w for this isotope for members of the public is 5×10^{-3} $\mu Ci/cm^3$, assuming an effective energy of 0.0059 MeV. One third of this value is the genetic dose limit. It has been proposed that 1% of the genetic dose limit may be assigned to tritium in the environment due to discharges by the nuclear industry. On the basis that body fluid is more limiting than whole body for tritium and since an Indian Standard Man has 28.4 kg of body fluid [9], the permissible tritium concentration calculates to 6000 pCi/litre. If 50% of the dose limit is assigned to the water route, then the permissible tritium concentration for a freshwater environment is 3000 pCi/litre.

4. IMPACT OF RADIOACTIVE RELEASES ON THE POPULATION DOSE

It will be seen that the DWL for stack discharges and releases into the aquatic environment have been so set that it is extremely unlikely that any individual off-site will incur an exposure exceeding the dose limits prescribed by the ICRP either through the air route or the water route or a combination of both. Assuming that the nuclear installations are operating very close to the set DWL values, it has been estimated that a person living at 3 km from any installation will not receive more than 30 to 40 mrem per year through gamma-emitting isotopes. In actual practice, the releases from the plants are between 1/12 to 1/50 lower than the values permitted. A rough estimate of the theoretical annual average dose to the population living in the neighbourhood of nuclear installations, based on the population distribution shows that the average population dose is not likely to exceed a few microrems per year.

REFERENCES

[1] INTERNATIONAL COMMISSION ON RADIOLOGICAL PROTECTION, Recommendations, ICRP Publication 9, Pergamon Press, Oxford (1966).
[2] INTERNATIONAL COMMISSION ON RADIOLOGICAL PROTECTION, Report of ICRP Committee II on Permissible Dose for Internal Radiation (1959), Health Phys. 3 (1960).
[3] VENKATARAMAN, K., SOMASUNDARAM, S., SOMAN, S.D., Health Phys. 9 (1963) 647.
[4] SLADE, D.H., Ed., Meteorology and Atomic Energy 1968, Rep. TID 24190.
[5] KAMATH, P.R., et al., "Preoperational search for baseline radioactivity, critical food and population group at Tarapur Atomic Power Station", Proc. First IRPA Congr. 1966, Pergamon Press, Oxford.
[6] GARNER, R.J., Health Phys. 9 (1962) 597-605.
[7] MORLEY, F., BRYANT, Pamela M., "Basic and derived radiological protection standards for the evaluation of environmental contamination", Environmental Contamination by Radioactive Materials (Proc. Symp. Vienna, 1969), IAEA, Vienna (1969) 225.
[8] BARRY, P.J., Maximum Permissible Concentrations of Radionuclides in Airborne Effluents from Nuclear Reactors, Rep. AECL 1624 (1963).
[9] RAGHUNATH, V.M., personal communication (1971).

[10] DUNSTER, H.J., WARNER, B.F., The Disposal of Noble Gas Fission Products from the Reprocessing of Nuclear Fuel, UKAEA Rep. AHSB (RP)R 101 (1970).
[11] GANGULY, A.K., PILLAI, K.C., "Secondary standards derived from ICRP basic safety standards, Radiation Protection Monitoring (Proc. Seminar Bombay, 1968), IAEA, Vienna (1969) 27.
[12] RAGHUNATH, V.M., SOMAN, S.D., Environ. Health 11 (1969) 1-7.

POPULATION EXPOSURE EVALUATION BY ENVIRONMENTAL MEASUREMENT AND WHOLE-BODY COUNTING IN THE ENVIRONMENT OF NUCLEAR INSTALLATIONS

I.S. BHAT, A.A. KHAN, A.G. HEGDE, S. SOMASUNDARAM
Health Physics Division,
Bhabha Atomic Research Centre,
Trombay, Bombay,
India

Presented by S.J. Supe

Abstract

POPULATION EXPOSURE EVALUATION BY ENVIRONMENTAL MEASUREMENT AND WHOLE-BODY COUNTING IN THE ENVIRONMENT OF NUCLEAR INSTALLATIONS.

Measurements of the extent of environmental radioactive pollution from the operation of nuclear installations in India are carried out at the installation sites. The paper describes in detail environmental measurements and evaluation of annual dose to individual members of the public, mainly in the environment of Tarapur Atomic Power Station due to the radioactive waste releases from the plant.

Samples of all kinds of food materials are collected at regular frequency and analysed for radioactivity. Demographic survey of the microenvironment of the nuclear site has given the spectrum of local diet and per capita consumption of food items. From these environmental data the per capita daily intake of critical radionuclides and the resultant internal radiation exposure to individual members of the general population are evaluated. The external population exposure due to stack discharges from nuclear installations is measured by thermoluminescent dosimeters. The increase in dose observed during operation over the natural background and its variation as seen during the pre-operational period is evaluated as the direct exposure. Whole-body counting of the general population has been carried out to evaluate internal contamination of individuals. Representative members of each village are counted about 3 times a year and the average annual body burden is evaluated. The annual dose due to the body burden is estimated for the important radionuclides found in the villagers. The population dose commitment due to nuclear operation in the country is assessed on the basis of present observation for the environment of the nuclear installations.

INTRODUCTION

Low-level radioactive waste from the major nuclear installations in the country is dispersed to the environment under controlled conditions after the required dilution. Certain food products grown in the environment concentrate the released radionuclides and so contribute to the radiation exposure of general population. The radioactive gaseous wastes released through high stacks cause direct radiation exposure to the surrounding population with an intensity that varies with the weather conditions. The location of major nuclear installations like nuclear power reactors and power reactor fuel production and reprocessing plants in India are spread out over the country. The evaluation of population exposure from the operation of nuclear plants is the major task of the Environmental Survey Laboratories (ESL) of the Health Physics Division of the Atomic Energy Department. The paper describes the population exposure evaluation

TABLE I. ANNUAL AVERAGE ACTIVITY IN SEAFOOD FROM TAPS ENVIRONMENT

Year	Edible soft tissue (pCi/kg)		
	^{131}I	$^{134+137}Cs$	^{60}Co
1970	84.98 ± 9.5	39.0 ± 7.1	Below detection
1971	508.6 ± 242.2	512.1 ± 129.5	4.86 ± 2.9
1972	58.1 ± 6.2	533.7 ± 85.5	71.86 ± 12.4
1973	48.2 ± 4.4	454.2 ± 25.6	89.5 ± 15.6

TABLE II. DAILY INTAKE OF SEAFOOD BY THE TARAPUR POPULATION

Population groups	Per capita daily intake of seafood (g)
1. Fishermen	180.0
2. Farmers	43.5
3. Others	67.0

TABLE III. DAILY INTAKE OF RADIONUCLIDES BY THE TARAPUR POPULATION DURING 1970-1973

Population group	Year	Per capita daily intake (pCi)		
		^{131}I	$^{134+137}Cs$	^{60}Co
1. Fishermen	1970	11.47 ± 1.3	5.3 ± 1.0	Below detection
	1971	68.67 ± 3.6	69.13 ± 17.4	0.66 ± 0.38
	1972	6.88 ± 0.83	72.05 ± 11.54	9.70 ± 1.67
	1973	6.50 ± 0.59	61.32 ± 3.45	12.68 ± 2.1
2. Farmers	1970	2.71 ± 0.32	1.27 ± 0.23	Below detection
	1971	16.59 ± 7.90	16.71 ± 4.23	0.16 ± 0.09
	1972	1.89 ± 0.2	17.41 ± 2.79	2.34 ± 0.40
	1973	1.57 ± 0.14	14.82 ± 0.83	2.92 ± 0.51
3. Others	1970	4.27 ± 0.47	1.96 ± 0.35	Below detection
	1971	25.56 ± 12.1	25.73 ± 6.5	0.24 ± 0.14
	1972	2.92 ± 0.31	26.81 ± 4.29	3.61 ± 0.62
	1973	2.42 ± 0.22	22.82 ± 1.29	4.50 ± 0.78

experience at the Tarapur Atomic Power Station (TAPS), which has been under commercial operation from 1970 with an annual release of about 400 curies of liquid radwaste to the sea and about 50 mCi/sec gaseous waste to the atmosphere through its 110-metre high stack.

RADIOACTIVITY IN FOOD SAMPLES AT TARAPUR

Seafood, cereals and pulses, vegetables, milk, meat and condiments produced in the TAPS environment have been tested for radioactivity with a frequency of once a month. The radionuclides of signficance in TAPS liquid releases are ^{131}I, ^{134}Cs, ^{137}Cs and ^{60}Co. Comparing the levels of radionuclides in these food products during the pre-operational [1] and operational periods [2-4] it has been observed that only seafood is contributing to the intake of radionuclides released from the power station. No other food material shows any detectable amount of reactor-released activity. The annual average radionuclide content of seafood from 1970 to 1973 is given in Table I.

DEMOGRAPHIC STUDY

To evaluate the critical population group for radiation exposure from power plant release a demographic survey was conducted in the environment [1, 4]. This provided information on the professional and dietary habits of the population and data on the per capita consumption of food components. These are essential to an evaluation of the radiation exposure and are not available for the small local region of the power reactor environment from national census data.

The population in the TAPS environment has been classified into three occupational groups: fishermen, farmers, and others (consisting of people from skilled professions). Since seafood is the significant diet component for exposure evaluation, the daily intake of seafood by the individual in the three population groups is given in Table II.

DAILY INTAKE OF RADIONUCLIDES AND RESULTANT ANNUAL RADIATION EXPOSURE OF INDIVIDUALS

From the annual average seafood radioactivity data and per capita daily intake of seafood the daily intake of radionuclides of power station origin can be evaluated. It has been observed that on average only 75% by weight of the fish is used when preparing food and 25% is discarded as non-edible. The activity in the edible portion only should be considered for daily intake evaluation. The data on daily intake of radionuclides for individual members of three population groups are given in Table III for 1970 to 1973.

The radiation dose from the estimated daily dietary intake can be calculated from dose factors evaluated from ICRP data as in the method used at Hanford [5, 6]. The dose factors used for the nuclides considered are given in Table IV. These are evaluated from the ICRP's value for MPC_w and the standard man's water intake data and maximum permissible

TABLE IV. RADIATION DOSE FACTORS

Nuclide	Critical organ	Dose factor mrem/μCi intake
^{131}I	Thyroid	1860
^{134}Cs	Whole body	69.0
^{137}Cs	Whole body	31.2
^{60}Co	Whole body	6.23
	G-I tract	37.4

TABLE V. ANNUAL DOSE FROM SEAFOOD INTAKE TO INDIVIDUALS OF TARAPUR POPULATION CALCULATED ON ICRP's MAXIMUM PERMISSIBLE INTAKE VALUES

Population group	Year	Dose to the critical organ from (mrem/a)				
		^{131}I Thyroid	^{134}Cs Whole body	^{137}Cs Whole body	^{60}Co Whole body	^{60}Co G-I tract
Fishermen	1970	7.80	0.052	0.037	-	-
	1971	46.7	0.683	0.778	0.0015	0.009
	1972	4.67	0.716	0.498	0.024	0.132
	1973	4.41	0.607	0.424	0.0274	0.765
Farmers	1970	1.84	0.0125	0.0087	-	-
	1971	4.48	0.1664	10.115	0.0004	0.002
	1972	1.28	0.1739	0.119	0.0053	0.032
	1973	1.06	0.1465	0.1031	0.0066	0.039
Others	1970	2.90	0.19	0.0135	-	-
	1971	17.35	0.257	0.176	0.0005	0.0033
	1972	1.98	0.264	0.186	0.008	0.0493
	1973	1.64	0.226	0.157	0.0102	0.0614

intake (MPI) values. The ICRP's permissible dose values taken are 0.5 rem for the whole body, 1.5 rem/a for the G-I tract and 3 rem/a for the thyroid. Table V gives the evaluated doses.

DOSE EVALUATION BY WHOLE-BODY COUNTING

Whole-body counting of the general public gives an exact measurement of the body burden of all gamma-emitting radionuclides. Since exposure of the general public through dietary intake is gradual and continuous, body counting once in 3 months would be sufficient to give the annual average

TABLE VI. ANNUAL RADIATION DOSE TO INDIVIDUAL VILLAGERS EVALUATED FROM BODY-BURDEN MEASUREMENTS

Village and distance from reactor site and direction (Population)	Average individual body burden [a] (nCi)				Annual radiation dose from these body burdens [a] (mrem)			
	Radiocaesium		Cobalt-60		Radiocaesium		Cobalt-60	
	1972	1973	1972	1973	1972	1973	1972	1973
1. Ghivali (1.6 km N.) (2100)	11.88	13.83	18.44	32.4	3.54	4.12	10.56	18.5
2. Tarapur (5 km N.) (4800)	12.94	21.78	21.01	22.5	3.85	6.45	12.03	12.9
3. Chinchani (6 km N.) (2850)	8.76	14.88	10.49	16.2	2.61	4.4	6.0	9.25
4. Varor (10 km N.) (2850)	4.3	6.5	4.6	7.5	1.28	1.93	2.63	4.3
5. Kudan (2.5 km N.E.) (1050)	14.25	15.7	13.6	24.5	4.25	4.68	7.77	14.05
6. Uchali (4.8 km S.) (1200)	5.43	6.50	8.37	9.00	1.62	1.94	4.79	5.14
7. Navapur (5.5 km S.) (2150)	7.90	14.80	9.48	26.8	2.35	4.4	5.43	15.4
8. Betagaon (15 km S.E.) (2552)	8.14	12.76	7.07	12.3	2.42	3.78	4.05	7.02
9. Umroli (12 km S.E.) (2605)	8.5	9.2	8.2	10.1	2.53	2.74	4.68	5.8
10. Nandgaon (12 km S.) (1525)	8.1	9.8	6.78	15.1	2.59	3.12	3.88	8.68
Average for the area	8.8	13.9	11.5	17.4	2.72	4.12	6.56	9.95

[a] The body burden and dose values have 25% standard error.

body burden. At Tarapur about 0.1% of the village population (aged 12 years and above) are counted in the fixed shadow-shield whole-body counter at the ESL. This shows the contamination from ^{134}Cs + ^{137}Cs, ^{60}Co and ^{58}Co. The level of ^{131}I in the thyroid of these people was below the detection limit of the counting system. Slight differences in the body burdens have been noted with age groups and also on the basis of profession. Fishermen had the highest body activities.

From the annual average body burden, the average body weight and the effective energy $\Sigma E(RBE)n$ the radiation dose to the individual of the general public can be calculated. The values for 1972 and 1973 for some of the villages in the Tarapur environment are given in Table VI for radiocaesium and radiocobalt. Radioiodine has not been detected in thyroid counting of the villagers.

TABLE VII. DIRECT RADIATION DOSE MEASURED WITH TLDs IN THE TAPS ENVIRONMENT

Village and distance from TAPS site	Dose (mrem/a)					
	1968	1969	1970	1971	1972	1973
1. Ghivali (1.8 km north)	58.50	57.8	59.5	64.75	65.0	68.5
2. Akkarpatti (1.8 km south)	62.20	59.6	61.2	68.0	70.0	82.0
3. Panchamarg (2.5 km east)	58.00	61.0	60.00	59.0	58.0	60.5
4. Kudan (2.5 km N.E.)	57.00	56.0	57.0	58.00	60.0	60.0
5. Pamtembhi (5 km S.E.)	56.00	58.0	57.0	58.00	59.0	57.0

TABLE VIII. TOTAL OCCUPATIONAL EXPOSURE (ANNUAL) AT TAPS

Year	Total No. of persons exposed	Total exposure dose (rem)
1969	399	43.312
1970	550	153.411
1971	622	444.684
1972	1503	2455.037
1973	1883	2732.477

The body burdens averaged for the area weighted to the population of the villages and the average individual annual dose for the affected environment are given at the end of Table VI.

DIRECT RADIATION EXPOSURE

The direct exposure of the general public around the power station from stack gas releases is measured by using thermoluminescent dosimeters (TLDs) in the village population centres. These dosimeters, kept in duplicate, are read once every three months. The observed annual dose as measured with TLDs for the Tarapur environment is given for the nearby villages in Table VII.

Thus only the two nearby villages within 2 km of the site had a dose about 5 to 10 mrem/a above the natural background radiation levels observed up to 1969. There was no detectable direct exposure dose for people living beyond 2 km from the site.

TABLE IX. DOSE TO WHOLE POPULATION (166 500) GROUP WITHIN 32 km FROM TAPS SITE

Year	Total occupational exposure (rem)	Total dose to affected public from dietary contamination (rem)	Total dose to external exposure above natural background (rem)	Total dose (rem)	Average dose for population within 32 km (mrem)
1972	2455.0	825.0	30.5	3310.5	19.9
1973	2732.5	1255.0	61.8	4049.3	24.3

OCCUPATIONAL EXPOSURE CONTRIBUTION OF POPULATION EXPOSURE

The exposure of occupational workers in the power plant has been measured regularly right from the commissioning stage. The total annual occupational exposure recorded [7] from 1969 to 1973 is given in Table VIII, and this has to be considered in the evaluation of the average population dose.

EXPOSURE OF THE GENERAL POPULATION

The internal contamination of people from dietary intake has been observed in the TAPS environment up to a distance of 16 km till the end of 1973. The total population in this region is about 89 500. The average annual whole-body exposure to the individual from $^{134+137}Cs$ and ^{60}Co was observed to be 9.2 mrem in 1972 and 14 mrem in 1973. The total non-occupational dose and occupational dose are added together to obtain the total population exposure. The direct exposure, wherever significant, is also added.

Genetic mixing in the country is very poor even at the state level. It would be sufficient to consider population within 32 km of the site to evaluate the population dose due to the operation of the nuclear installation rather than to consider the state population or country's population as whole. The doses to the population in the TAPS environment during 1972 and 1973 are shown in Table IX.

DISCUSSION

There is a significant difference between the dose estimated from the dietary intake of radionuclides and that estimated from whole-body counting (cf. Tables V and VI). The difference cannot be attributed to intakes through pathways other than seafood since neither other dietary components nor air filter samples have shown detectable amounts of radionuclides released from the power station.

The difference in dose may be mainly due to the uncertainty in the human metabolic uptake of nuclides incorporated in different dietary components. The dose evaluation from quarterly whole-body counting can be considered to be more accurate than that from daily dietary intake.

The radiation exposure level observed at the TAPS site is unfortunately due to abnormal operation and maintenance conditions and may not persist continuously. The expanding nuclear power programme of the country is based on heavy water natural-uranium (CANDU) reactors and not on the BWR type at Tarapur. Future evaluation of exposure at Kota, Rajasthan, where India's second nuclear power station has started operating, will help in assessing dose commitments from the country's nuclear power programme.

ACKNOWLEDGEMENTS

Authors are indebted to Dr. A.K. Ganguly, Director, Chemical Group, for encouragement and permission to publish the article. Thanks are due to Shri S.D. Soman and Shri P.R. Kamath for helpful suggestions and discussions.

REFERENCES

[1] KAMATH, P.R., BHAT, I.S., KHAN, A.A., GANGULY, A.K., "Preoperational search for baseline radioactivity critical food and population group at TAPS site", IRPA Congress, Rome, Pergamon, Oxford (1966).

[2] BHAT, I.S., "Planning and management of environmental pollution control at Tarapur Atomic Power Station site", Proc. Seminar Pollution and Human Environment, Bombay, BARC (1970).

[3] BHAT, I.S., KHAN, A.A., CHANDRAMOULI, S., "Radioactivity measurement at Tarapur Nuclear Power Station environment", Proc. Symp. Radiation Physics, BARC, Bombay, 1970, 493-505.

[4] BHAT, I.S., KAMATH, P.R., GANGULY, A.K., Dispersal and Uptake of Radioactive Elements in the Tarapur Environment, Rep. BARC-644 (1972).

[5] ESSIG, T.H., Environmental Surveillance in the Vicinity of Nuclear Facilities (RENIG, W.E., Ed.), Charles Thomas Publisher (1970).

[6] HONSTEAD, J.F., ESSIG, T.H., "Evaluation of environmental factors affecting population exposure", Proc. Conf. Environmental Radiation Protection from Nuclear Power Plants, USEPA BRP/SID/72-4 (1971) 29-57.

[7] PATNAIK, D., Communications from the annual reports of operational health physics at Tarapur Atomic Power Station, 1969-73, private communication.

DISCUSSION

ON PAPERS IAEA-SM-184/4, 5 AND 6

F.D. SOWBY: The results given in Table IX of paper SM-184/6 show once again that occupational exposure makes the major contribution to the collective dose. It would be interesting to know the electrical output of the station in the relevant years, so that we could estimate the man · rems per megawatt-year.

I.S. BHAT[1]: I think that expressing exposure as man · rems per megawatt-year is misleading since exposure is often more significant during outages than when the reactors are operating. However, the electrical outputs were as follows:

1972	870 411 MW · h (total)
1973	20 068 810 MW · h (total)

S.O. BERGSTRÖM: I should like to add in this connection that I do not think it meaningful to calculate occupational man · rad dose per MW(e) for stations that have not been running for a reasonably long time. In the early years man · rad dose per MW(e) might often be inversely proportional to power production.

I also have a question. What is the basis for assuming that the main part of the total population dose to the public is delivered within a distance of 32 km?

I.S. BHAT: The exposure pathway is via the ingestion of fish from coastal waters. Our study of local demography has shown that only people within 32 km are of interest in the dose evaluation. Some oysters and lobsters from this region are exported to Bombay but only in small quantities. The 32 km boundary is of practical significance. The values obtained are indicative of the kind of exposure expected and are not necessarily exact.

P. SLIZEWICZ: Taking account of the date the releases started, can you assume that the caesium-137 content of the fish flesh for the year 1970 is due solely to fall-out caesium?

I.S. BHAT: No. The level of fall-out caesium-137 in fish flesh during 1970 was about 10 to 20 pCi/kg. The releases from the Tarapur plant had started in 1969 during the commissioning period.

W.J. BAIR: Do you think that the differences between the doses estimated from the dietary intake of radionuclides and those estimated by whole-body counting could be resolved using another metabolic model, perhaps one which is more appropriate to the population studied?

I.S. BHAT: We have not yet attempted this. There are difficulties due to paucity of observational data. We cannot work with known models because the daily consumption is not regular. I think it should be possible to resolve this difference when more metabolic data become available, and we will keep this in view for further studies.

G. BOERI: Have you calculated concentration factors for ^{131}I in the water to seafood pathway, and could you correlate iodine discharges with

[1] Answer submitted by Mr. Bhat, who was not present at the Seminar.

measured concentrations? Can you give any iodine discharge data for the years 1970-71-72?

I.S. BHAT: We reported the concentration factor for ^{131}I from seawater to fish in several earlier papers. Only the fresh catch from the coastal region will have iodine radioactivity; the bulk of the marketed fish comes from the off-shore region and has no ^{131}I.

The total activity discharged during 1972-73 was about 41.55 curies, normalized as per 10 CFR 20, and this is well below the authorized limit. About 14% of the total activity was radioiodine. The values could be less for other years. Considering its short life compared with ^{137}Cs and ^{60}Co and the fact that it is released only under defective fuel conditions, ^{131}I is not a continuous hazard.

IAEA-SM-184/28

PROGRAMME DE SURVEILLANCE DE L'ENVIRONNEMENT MARIN DU CENTRE DE LA HAGUE

J. SCHEIDHAUER, R. AUSSET
Direction des productions,
CEA, Centre de La Hague,
Cherbourg

J. PLANET, R. COULON
Département de protection,
CEA, Centre d'études nucléaires
de Fontenay-aux-Roses,
Fontenay-aux-Roses,
France

Abstract–Résumé

PROGRAMME FOR MONITORING THE MARINE ENVIRONMENT AROUND THE LA HAGUE CENTRE.

The La Hague Centre, the main function of which is the reprocessing of irradiated fuel, is located by the sea at a site which is particularly favourable as regards the safety of the local population. Before the reprocessing plant went into operation, extensive studies were carried out in order to determine how the effluent from it would disperse and by what major pathways it could reach the local population. The results of these studies have led to the formulation of a monitoring programme covering 12 zones and various elements of the marine environment: seawater, sediments, algae, fauna which is not involved in the food chain, fauna of importance in the food chain. Only after 1968 did the findings reveal certain effects attributable to the centre, the extent of the effects depending on the type of sample and the zone. Algae proved to be the best indicators: the measured activities correlated fairly well with the amounts of effluent discharged and the concentration factors were in agreement with the basic hypotheses. As regards the fauna, the activity levels are generally too low for a correlation with the effluent discharges to be established.

PROGRAMME DE SURVEILLANCE DE L'ENVIRONNEMENT MARIN DU CENTRE DE LA HAGUE.

Le Centre de La Hague, dont la vocation essentielle est le retraitement des combustibles irradiés, est implanté sur un site marin particulièrement favorable du point de vue de la sécurité des populations. La mise en œuvre de l'usine a été précédée d'un important programme d'études visant à déterminer les conditions de dispersion des effluents et les principales voies d'atteinte de la population. Ces études ont conduit à la définition d'un programme de surveillance portant sur douze zones différentes et divers éléments du milieu marin: eau de mer, sédiments, algues, faune fixée, faune intéressant la chaîne alimentaire. Les résultats obtenus n'ont mis en évidence une certaine influence du Centre qu'après 1968, influence plus ou moins marquée selon les types d'échantillons et selon les zones. Les algues se sont révélées être les meilleurs indicateurs: les activités mesurées sont en assez bon accord avec les rejets effectués et les facteurs de concentration obtenus sont conformes aux hypothèses de base. Pour ce qui est de la faune, les niveaux sont généralement trop faibles pour qu'il soit possible d'établir une corrélation avec les rejets.

INTRODUCTION

La vocation essentielle du Centre de La Hague est le retraitement des combustibles irradiés. Son activité l'inclut donc dans le cycle du combustible des générateurs nucléaires d'électricité qui connaissent le développement que l'on sait.

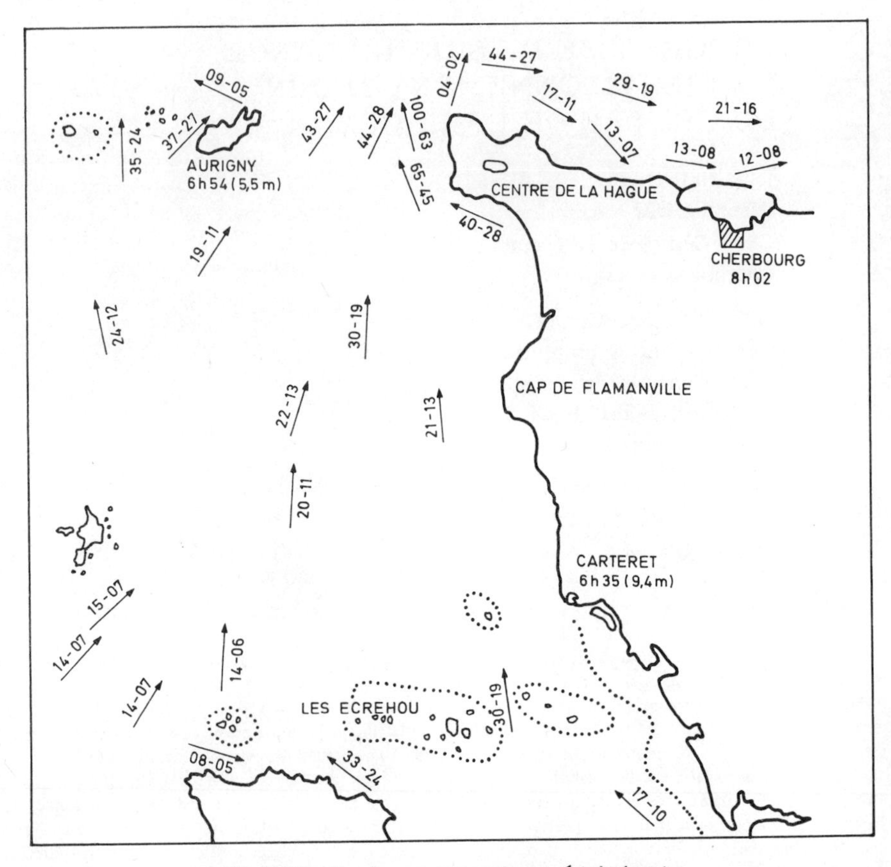

FIG. 1. Situation des courants marins en période de rejet.

Etudié et réalisé dans le schéma de la filière graphite-gaz, une récente adaptation technique aux réacteurs de la filière à eau ordinaire en fait un outil important de l'effort nucléaire européen.

Situé en France, à l'extrémité nord-ouest du Cotentin, le Cap de La Hague, le Centre est implanté sur un site marin exceptionnel au point de vue de la sûreté. Les résidus industriels gazeux ou liquides du procédé sont assurés d'un stockage ou pour ceux de très faible activité d'une élimination en toute sécurité dans les milieux atmosphérique et marin par dispersion.

La séparation et la purification de l'uranium et du plutonium à partir des combustibles irradiés actuellement dans les réacteurs à uranium naturel de l'Electricité de France s'effectuent par voie liquide essentiellement. Les opérations chimiques, outre la séparation et le stockage des produits de fission, produisent des effluents de faible et moyenne activité. Ces solutions sont acheminées vers la Station de traitement des effluents. Des techniques très étudiées, spécifiques et en amélioration permanente, faisant appel à des traitements par coprécipitation, chimisorption, échange d'ions minéraux en solution, échange d'ions sur résine organique et filtration permettent une épuration radioactive très satisfaisante. Les eaux résiduaires obtenues à la suite de ces opérations peuvent alors être éliminées en mer.

Ces rejets sont effectués par l'intermédiaire d'une conduite terrestre et marine dont l'extrémité place le point d'émission à 2000 m de la côte, au sud-ouest du Nez de Jobourg, au sein d'une zone de turbulence exceptionnelle due aux forts courants de marée (environ 10 nœuds) dans le Raz Blanchard (fig. 1).

Les rejets font l'objet d'un contrôle très strict. Tout d'abord de la part des Autorités gouvernementales qui donnent au Centre de La Hague, chaque année, une autorisation préfectorale, après examen du dossier permanent par les Services spécialisés du Ministère de la Santé publique. Ensuite, avant chaque rejet en mer, une autorisation est délivrée après échantillonnage et analyse des eaux résiduaires, isolées dans une cuve, par le Service de contrôle radioactif du Centre. Ce Service s'assure en permanence, au nom du Directeur du Centre, que les activités éliminées et les procédures de rejet (heure, débit) sont bien conformes à la réglementation. Il assure également la comptabilité détaillée des rejets qui est ensuite transmise aux Services centraux du Commissariat à l'énergie atomique et aux organismes gouvernementaux de contrôle. Néanmoins, le respect des conditions de rejet n'exempte pas le Centre de La Hague, comme tout industriel, de s'assurer du devenir dans l'environnement marin des éléments radioactifs rejetés dans le milieu.

Cette surveillance a débuté sur le Centre il y a plusieurs années, avant les premiers rejets actifs, pour établir un «point zéro» et définir l'organisation d'une surveillance de l'environnement efficace.

La surveillance des rejets et de l'environnement marin a permis de suivre avec souplesse l'influence de l'émission des eaux résiduaires sur la radioactivité de la flore et de la faune voisines du Centre.

ETUDES PRELIMINAIRES - POINT ZERO

L'étude de sûreté du site marin de La Hague a mis en jeu des moyens extrêmement importants. L'ensemble des études théoriques et appliquées qui ont fait appel à des organismes très divers ont abouti en 1964, après la mise en place de l'émissaire, à une simulation en vraie grandeur des rejets à l'aide de Rhodamine. Une série de trente rejets (deux par vingt-quatre heures) a été effectuée à partir du 30 octobre 1964. La dispersion de la Rhodamine au large et le long des côtes a pu être étudiée grâce au concours de la Marine nationale. Les résultats de ces études, publiées en leur temps, ont permis de déterminer les cheminements des eaux rejetées, les points de retour éventuels à la côte, les points de stagnation, etc.

Parallèlement, le Département de protection du Commissariat a effectué une enquête écologique et sanitaire qui a permis de déterminer la capacité théorique de réception du milieu en fonction de l'incidence des activités rejetées sur l'homme à travers la chaîne alimentaire et les activités de pêche ou de tourisme.

Enfin, les mesures d'activité de la flore et de la faune marines débutèrent sur le Centre en 1964. Elles prenaient la suite, en ce qui concernait le Commissariat à l'énergie atomique, des déterminations effectuées dès 1961 dans cette région par le Groupe d'études atomiques de la Marine nationale.

L'ensemble des résultats obtenus et leur évolution jusqu'en 1967 constitua le point zéro du site et de l'environnement. Ces mesures s'étendirent à de nombreux constituants de la faune et de la flore marines, aux sables et sédiments, et évidemment à l'eau de mer à la suite de prélèvements effectués en de nombreux points de la côte.

L'ensemble de ces études et de ces déterminations permit de définir l'organisation d'une surveillance satisfaisant à plusieurs exigences.

TABLEAU I. RADIOACTIVITE DES ELEMENTS SURVEILLES (pCi/kg – Quatrième trimestre 1973)

Echantillon	Ru-Rh-106	Ce-Pr-144	Cs-Ba-137	Zr-Nb-95	Autres radioéléments
Eau de mer					
points proches (2, 5, 6)	2,3	NS	2,8	NS	Sb-125: 0,9
points éloignés (11)	0,9	NS	0,8	NS	Cs-134: 0,12
Sables					
points proches (3, 5)	390	NS	NS	NS	
Sédiments					
points proches (2, 5, 6)	500	NS	NS	NS	
Chondrus crispus					
points proches (2, 5, 6)	6400	NS	120	NS	
points éloignés (11)	430	NS	85	NS	
Corallina officinalis					
points proches (2, 5, 6)	3500	NS	170	NS	
points éloignés (11)	690	NS	NS	NS	
Fucus serratus					
points proches (2, 5, 6)	2150	NS	90	100	
points éloignés (11)	240	NS	NS	NS	
Laminaria flexicaulis					
points proches (2, 5, 6)	1330	NS	90	NS	
Laminaria digitata					
points éloignés (11)	180	NS	40	NS	
Porphyra linearis					
points proches (5)	8020	NS	100	NS	
Balanus balanoides	NS	NS	NS	NS	
Patella vulgata					
points proches (2, 5, 6)	760	150	NS	NS	Ag-110: 105
points éloignés (8)	200	NS	20		
Cancer pagurus					
points proches (6)	1130	NS	100	NS	Ag-110: 75
Gryphea angulata	NS	NS	NS	NS	
Mytilus edulis	NS (380+1)	NS	NS	NS	
Poisson	NS	NS	100-700	NS	

ORGANISATION DE LA SURVEILLANCE

La surveillance du site marin de La Hague présente plusieurs finalités:

- suivre en permanence l'évolution de la radioactivité d'indicateurs vivants ou inertes bien choisis pour leurs facteurs de concentration élevés et leur faible temps de réponse à une pollution; ce suivi permettra d'apporter toute correction aux conditions d'épuration ou de rejet;
- fournir aux Services spécialisés du Commissariat les éléments nécessaires à la vérification des conclusions de l'étude de sûreté préliminaire et l'évaluation des expositions possibles des hommes vivant dans le milieu environnant;
- fournir aux organismes de contrôle ou même au public des informations quantitatives sur le niveau radioactif des éléments vecteurs de la radioactivité vers l'homme.

Ces buts étant définis, il était aussi nécessaire de tenir compte des réalités. En l'occurrence, par exemple pour assurer un constat continu, dont une des qualités majeures doit être l'intercomparaison dans l'espace et dans le temps, il est nécessaire de s'adresser à un élément de la flore ou de la faune présentant une abondance et une permanence satisfaisantes en chaque point ou zone de prélèvement choisi. D'autre part, cet élément doit assurer à l'analyste un facteur de concentration suffisant pour obtenir des mesures significatives. Un compromis permanent a donc dû être consenti entre ces diverses qualités, toutes nécessaires. Il a abouti à un plan et un programme de surveillance qui ont été adoptés durant les huit années d'exploitation du Centre.

a) Eléments surveillés

Après l'étude sur le terrain d'une quinzaine d'algues, de cinq éléments de la faune fixée, de plusieurs éléments de la faune intéressant la chaîne alimentaire, des sédiments et des sables et enfin évidemment de l'eau de mer, les échantillons indiqués sur le tableau I ont été sélectionnés.

b) Zones de surveillance et fréquence

Les zones de surveillance du littoral, les plus importantes pour des prélèvements reproductibles, sont au nombre de douze. Les zones les plus proches du Centre, donc de l'émissaire, sont resserrées et de relativement petites dimensions. Leur étendue et leur espacement croissent approximativement avec leur éloignement. Elles sont portées sur la carte sous le titre «zones de prélèvement» (fig. 2). Ce sont celles de:

Barneville-Carteret	n° 1	Anse St-Martin	n° 7
Siouville	n° 2	Urville	n° 8
Anse de Vauville	n° 3	Cherbourg	n° 9
Herquemoulin	n° 4	Cap Lévy	n° 10
Ecalgrain	n° 5	St-Vaast-la-Hougue	n° 11
Goury	n° 6	Granville	n° 12

Outre ces zones côtières, des zones «pleine mer» ont été retenues pour des contrôles particuliers (sédiments, pêche).

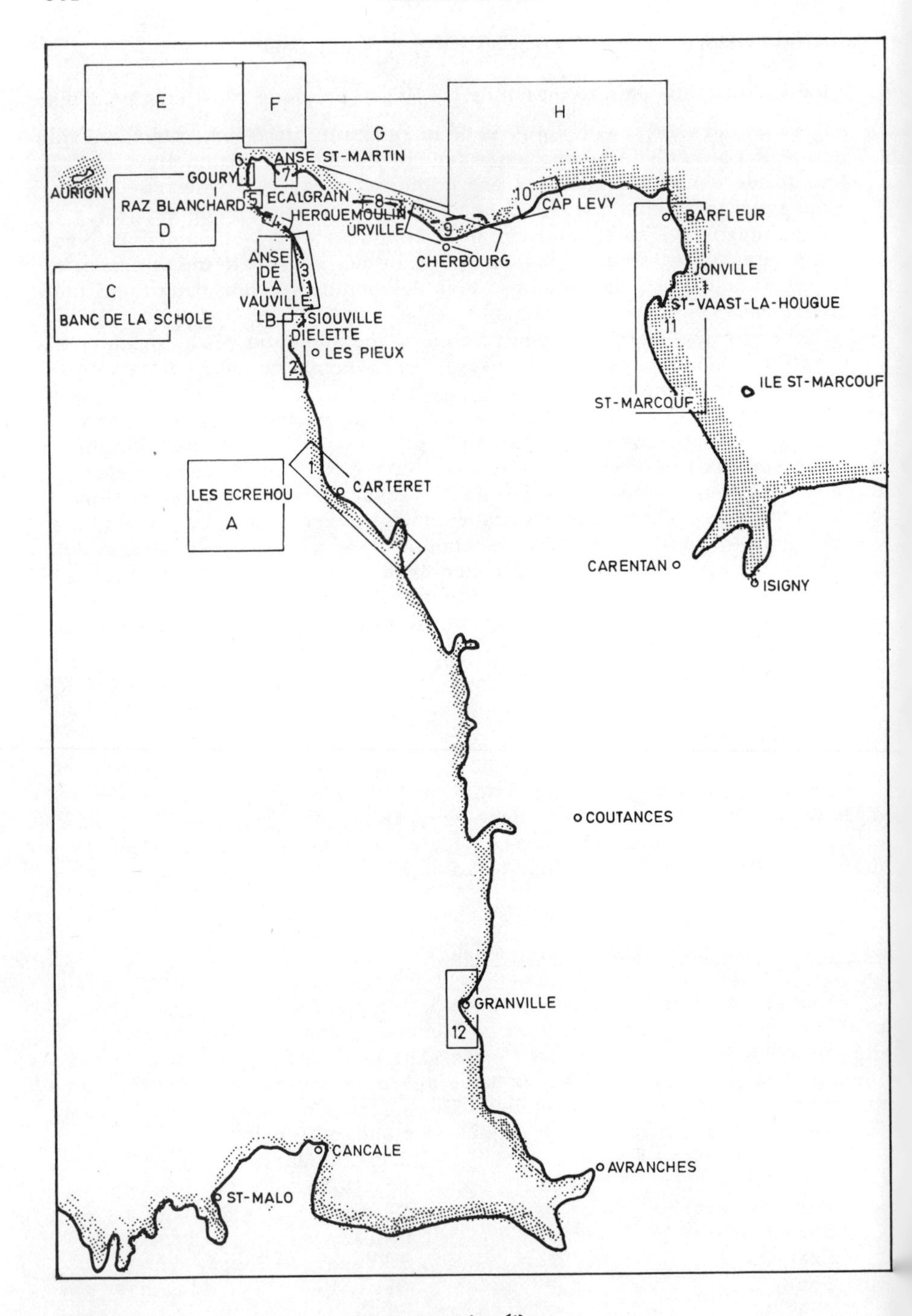

FIG.2. Zones de prélèvement.

Les fréquences des prélèvements systématiques sont trimestrielles ou semestrielles. Mais de nombreux prélèvements de confirmation, d'évolution, de recherche, d'origine, d'évaluation, de comparaison sont effectués hors programme.

c) Techniques

La connaissance permanente de la composition des rejets permet d'orienter la recherche et la mesure des radioéléments dans les échantillons marins. Les éléments principaux rejetés sont les:

ruthénium-rhodium-103 et 106
césium-134 et 137
zirconium-95
niobium-95
cérium-praséodyme-144

en moindre quantité l'antimoine-125 et les strontium-89 et 90,

et à très faible teneur, on trouve les:

cobalt-60
mercure-203
iode-131
antimoine-124
argent-110-110m
zinc-65.

La collecte des algues et animaux fixés s'effectue évidemment manuellement et avec beaucoup de soins. Ces échantillons sont lavés à l'eau de mer au point de prélèvement; au Laboratoire ils sont égouttés, décoquillés éventuellement, et séchés à l'étuve après élimination de tous corps étrangers. Après séchage, un broyage réduit les échantillons en poudre fine conditionnée en boîte de plastique pour la mesure par spectrométrie gamma.

Les poissons sont préparés de la même façon après séparation de la chair et des viscères.

L'activité des sables et sédiments est mesurée après séchage.

Enfin l'eau de mer, sous forme d'échantillon de 400 litres, est traitée par passage sur colonne de ferrocyanure de cobalt et potassium, chimisorption sur le dioxyde de manganèse préparé dans le milieu et éventuellement coprécipitation avec du sulfure de cobalt.

La mesure d'activité s'effectue systématiquement par spectrométrie gamma.

RESULTATS

Les résultats de la surveillance n'ont mis en évidence l'influence du Centre qu'à partir de 1968. L'évolution de l'activité des éléments surveillés a été en relation directe et en bon accord avec l'importance des rejets d'eaux résiduaires. L'examen des résultats montre également une nette influence de la proximité du Centre, malgré une dilution immédiate des effluents au voisinage de l'émissaire atteignant 10^6, identique à celle trouvée lors des expériences de rejet de colorants. Toutes les valeurs trouvées sont faibles mais permettent cependant, compte tenu des techniques employées, l'acquisition de résultats significatifs.

a) Evolution de l'activité dans le temps

Nous avons retenu pour mettre en évidence cette évolution un point d'échantillonnage proche du Centre. Il s'agit de la zone n° 5, l'anse d'Ecalgrain, située au sud du point de rejet. Les échantillons étudiés et les radioéléments mesurés sont respectivement

Eau de mer	césium-137 – ruthénium-rhodium-106
Sédiments immergés	ruthénium-rhodium-106
Sable de plage	ruthénium-rhodium-106
Flore:	
Chondrus crispus Corallina officinalis Fucus serratus Laminaria flexicaulis	ruthénium-rhodium-106 césium-137 – cérium-praséodyme-144 zirconium-niobium-95
Porphyra linearis	ruthénium-rhodium-106 césium-137 – zirconium-niobium-95
Faune fixée:	
Balanus balanoides	ruthénium-rhodium-106 césium-137 – zirconium-niobium-95
Patella vulgata	ruthénium-rhodium-106 césium-137 – cérium-praséodyme-144 argent-110-110m.

Sur les graphiques[1] sont portés les quantités de chaque radioélément rejeté par trimestre ainsi que le niveau de radioactivité dû à ce radioélémen dans chaque échantillon. L'examen de ces graphiques montre des évolutions de la radioactivité des indicateurs retenus en assez bon accord avec l'importance des rejets. Pour l'eau de mer, la concordance est bonne pour le césium-137 malgré une représentativité discutable d'un échantillonnage instantané et ponctuel.

Les sédiments immergés n'ont pas permis de mettre en évidence et évidemment d'observer une évolution significative de radioactivité artificiell avant 1972. La relation avec l'importance des rejets ne peut être retenue. Seuls un phénomène de très lente accumulation et aussi, il faut le dire, le progrès des techniques ont permis cette détection.

En ce qui concerne la flore, les résultats sont divers mais en général montrent une corrélation satisfaisante avec les rejets. L'évolution parallèl des rejets de ruthénium, césium, cérium et zirconium et celle de la radioactivité de Chondrus crispus due à ces radioéléments est remarquable, moi bonne pour Corallina officinalis. Pour Fucus serratus et Laminaria flexicaulis, les résultats sont décevants en ce qui concerne le cérium.

[1] Voir annexe.

Pour la faune représentée par Balanus balanoides et Patella vulgata, la comparaison est satisfaisante, mais avec une évolution de l'activité cérium et zirconium-niobium peu explicable en 1972 et 1973.

Pour Patella vulgata, la détection d'argent-110-110m n'est intervenue qu'en 1972. Cette détection a donné lieu à des séparations dans les eaux résiduaires rejetées qui ont fait apparaître que l'activité rejetée due à ces radioéléments est de l'ordre de moins de 10^{-3} de l'activité totale. Un facteur de concentration de l'ordre de 10^4 est responsable d'une teneur mesurable, bien que très faible, dans Patella vulgata.

b) Situation actuelle

Les derniers résultats disponibles sont rassemblés dans le tableau I.

Les activités spécifiques les plus élevées se trouvent dans les algues et concernent, comme il fallait s'y attendre, le ruthénium-rhodium-106.

La chaîne alimentaire se caractérise par une valeur, très relativement, élevée, trouvée pour Cancer pagurus avec 1, 3 pCi par gramme en ruthénium-106, ordre de grandeur de l'activité naturelle (1,7 pCi par gramme en potassium-40).

Les patelles, peu consommées, atteignent une valeur de 0, 76 pCi par gramme et présentent comme Cancer pagurus une teneur caractéristique en argent-110-110m de l'ordre de 0, 1 pCi par gramme.

Par contre, les huîtres et moules, collectées à St-Vaast-la-Hougue (n° 11) pour les premières, et à St-Vaast et Granville pour les secondes, points d'élevage les plus proches, ne présentent pas d'activité artificielle mesurable à part un taux de ruthénium-rhodium-106 de 0, 4 pCi par gramme pour des moules de Granville.

CONCLUSION

Le bilan établi après plusieurs années de fonctionnement du programme de surveillance du site marin du Centre de La Hague se révèle positif.

Le choix des zones de prélèvement, comme celui des types d'échantillons surveillés, a été fondé sur les études préalables relatives à la dispersion des effluents et aux modalités d'atteinte de la population: ce choix permet de répondre aux objectifs assignés à ce programme.

Si, d'une façon générale, l'influence des rejets a pu être mise en évidence pour les principaux vecteurs, les niveaux observés sont demeurés faibles, souvent à la limite de la sensibilité des appareils de mesure.

Il a été vérifié que les algues de différentes espèces constituent les meilleurs indicateurs de pollution. Les facteurs de concentration observés sont de l'ordre de 10^3 pour le ruthénium et de 10^2 pour le césium: ces valeurs sont en assez bon accord avec les résultats de laboratoire et les hypothèses de calcul utilisées.

Leur présence généralisée sur le site concerné et la facilité de leur récolte ont conduit à les inclure largement dans le programme de surveillance.

En conséquence, les mesures effectuées sur les algues peuvent être considérées comme un bon moyen d'estimation des risques d'exposition du public, qu'il s'agisse des pêcheurs, des consommateurs de produits marins ou des touristes.

Or, l'étude des résultats obtenus, jointe aux informations portant sur les différents vecteurs intervenant dans les voies de transfert pour l'homme, permet d'assurer que les hypothèses initiales utilisées pour l'estimation des conséquences sanitaires des rejets du Centre de La Hague sont convenables et qu'aucun risque ne peut être encouru du fait de ces rejets.

ANNEXE

QUANTITES DE RADIOELEMENTS REJETEES PAR TRIMESTRE ET NIVEAU DE RADIOACTIVITE DANS LES ECHANTILLONS

BAIE D'ECALGRAIN
Eaux de mer

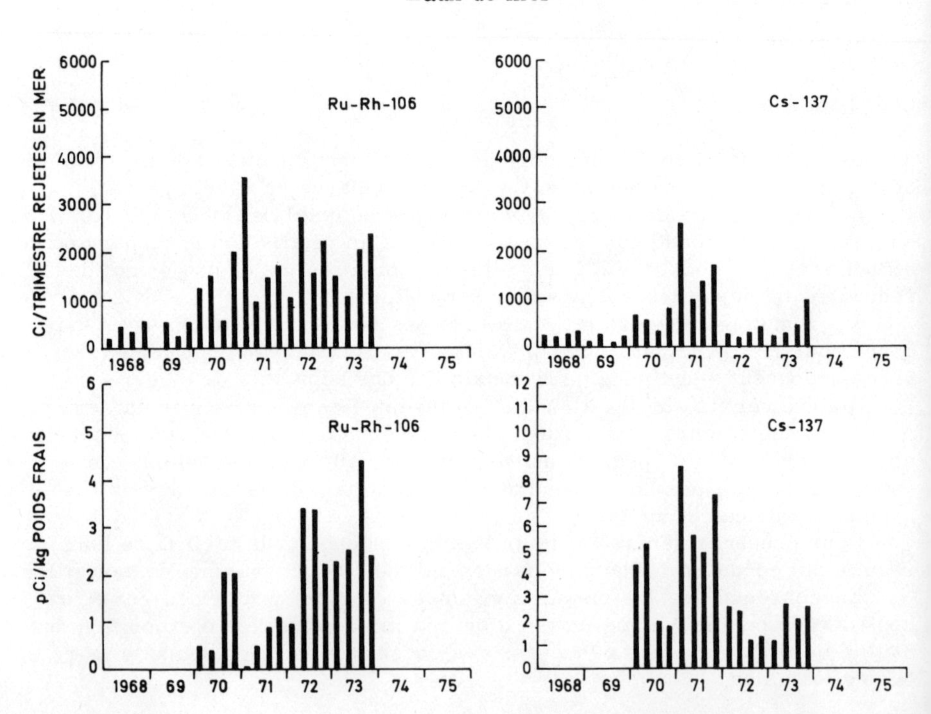

BAIE D'ECALGRAIN

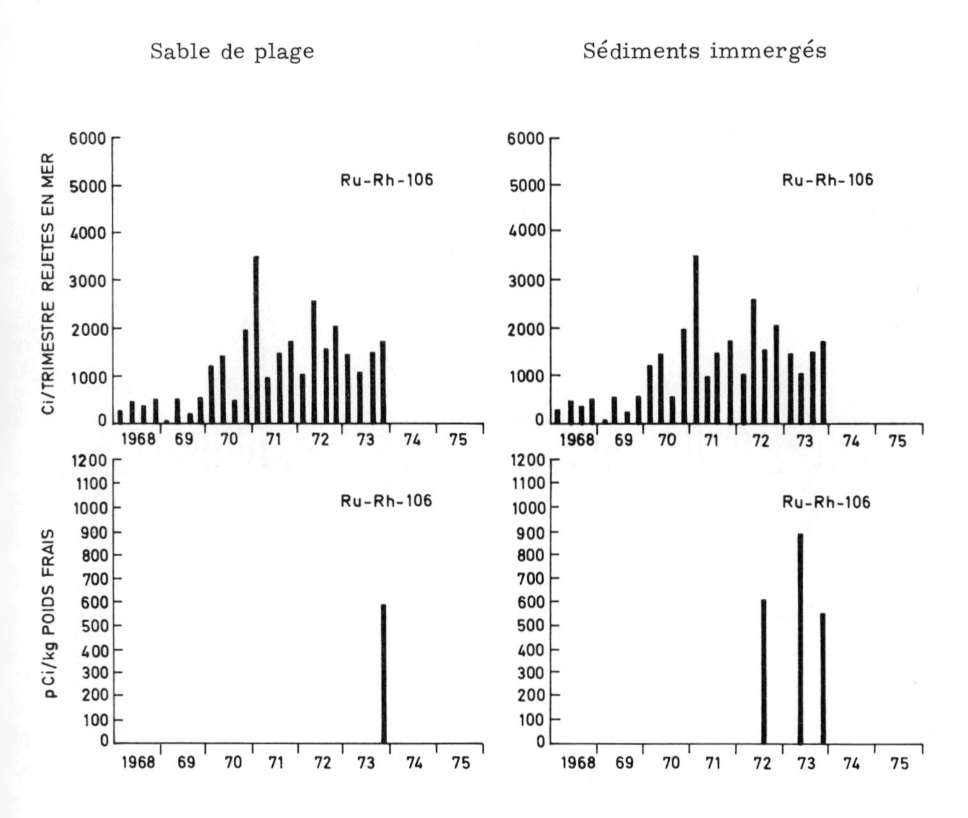

BAIE D'ECALGRAIN
Chondrus crispus

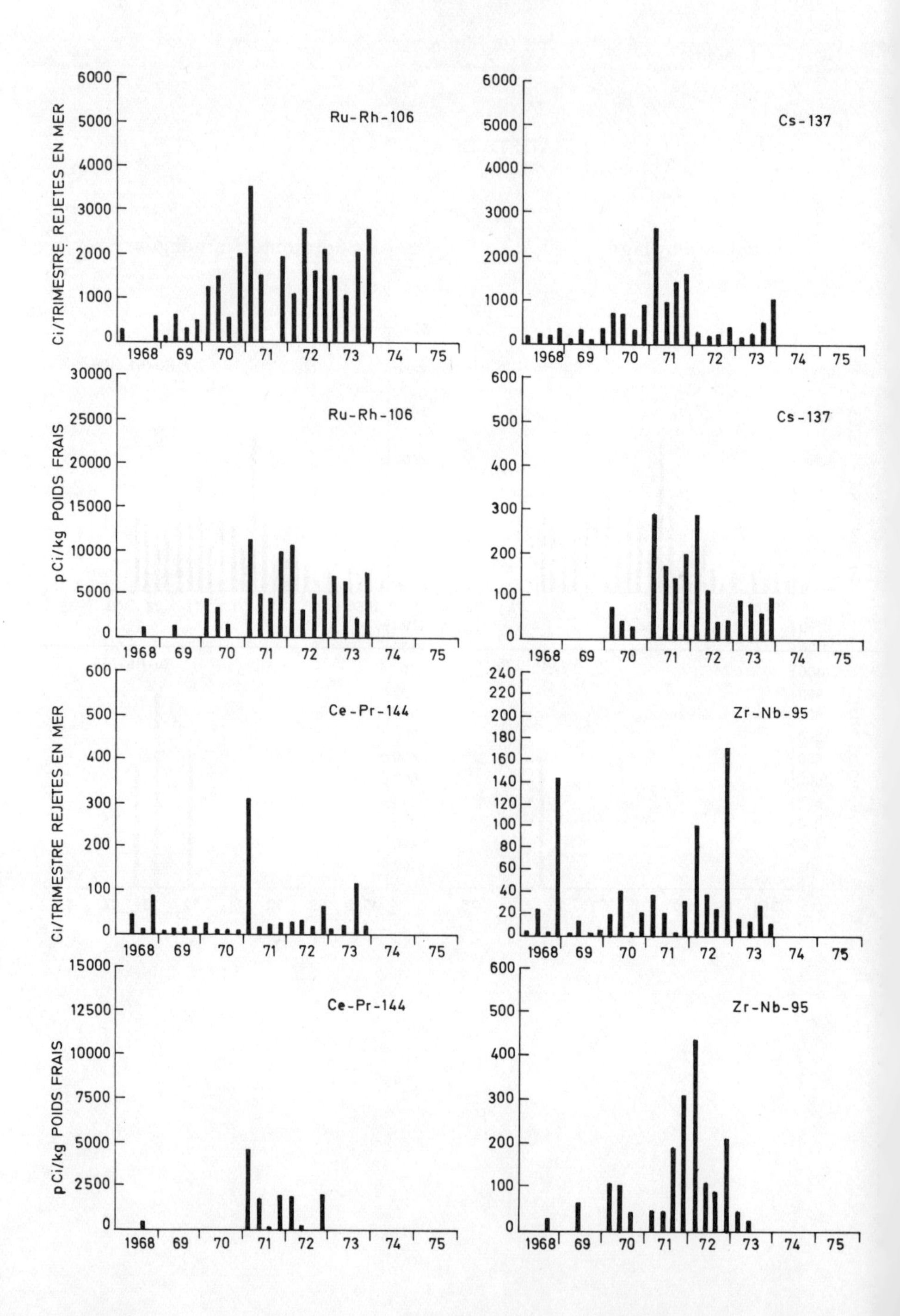

BAIE D'ECALGRAIN
Corallina officinalis

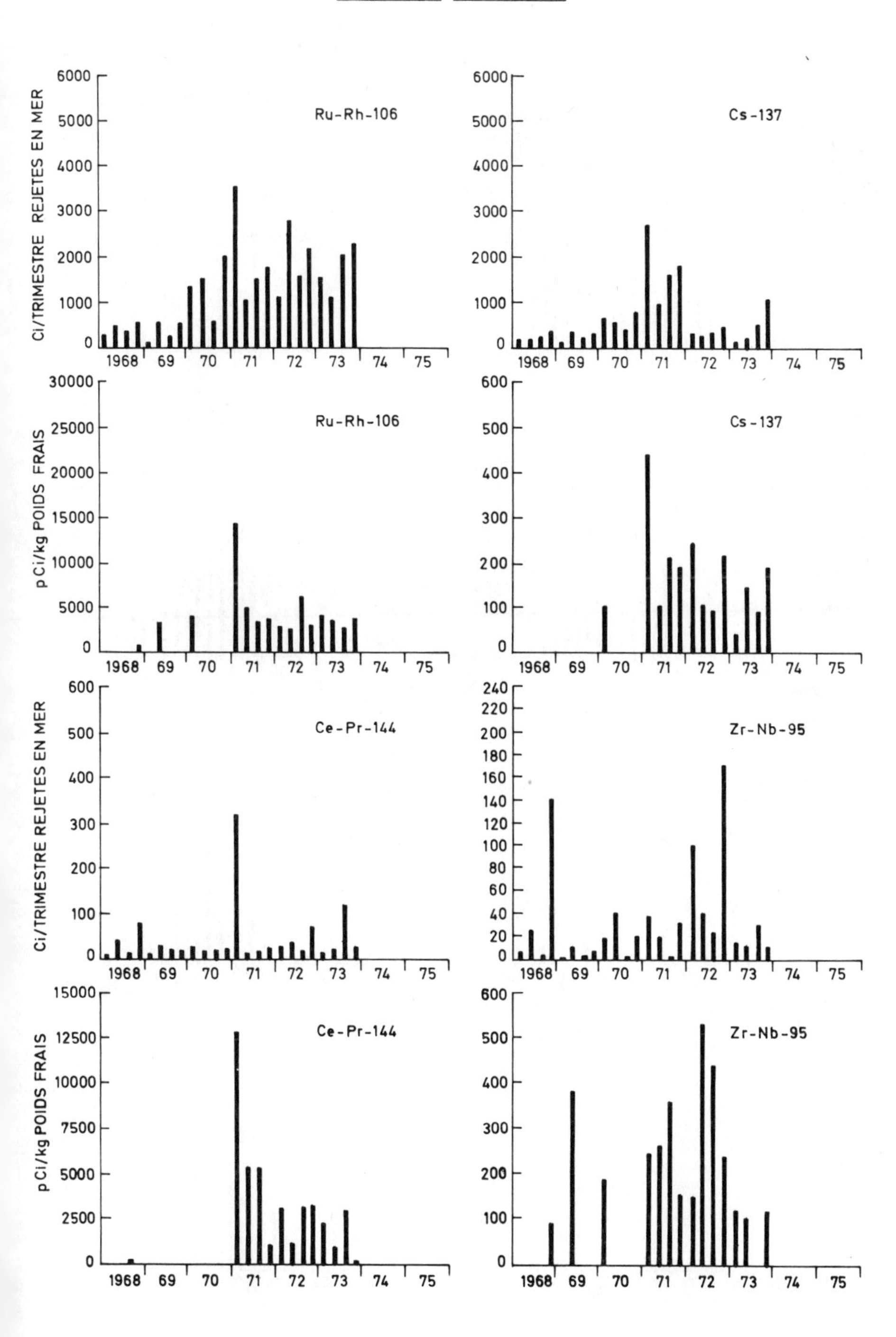

BAIE D'ECALGRAIN

Fucus serratus

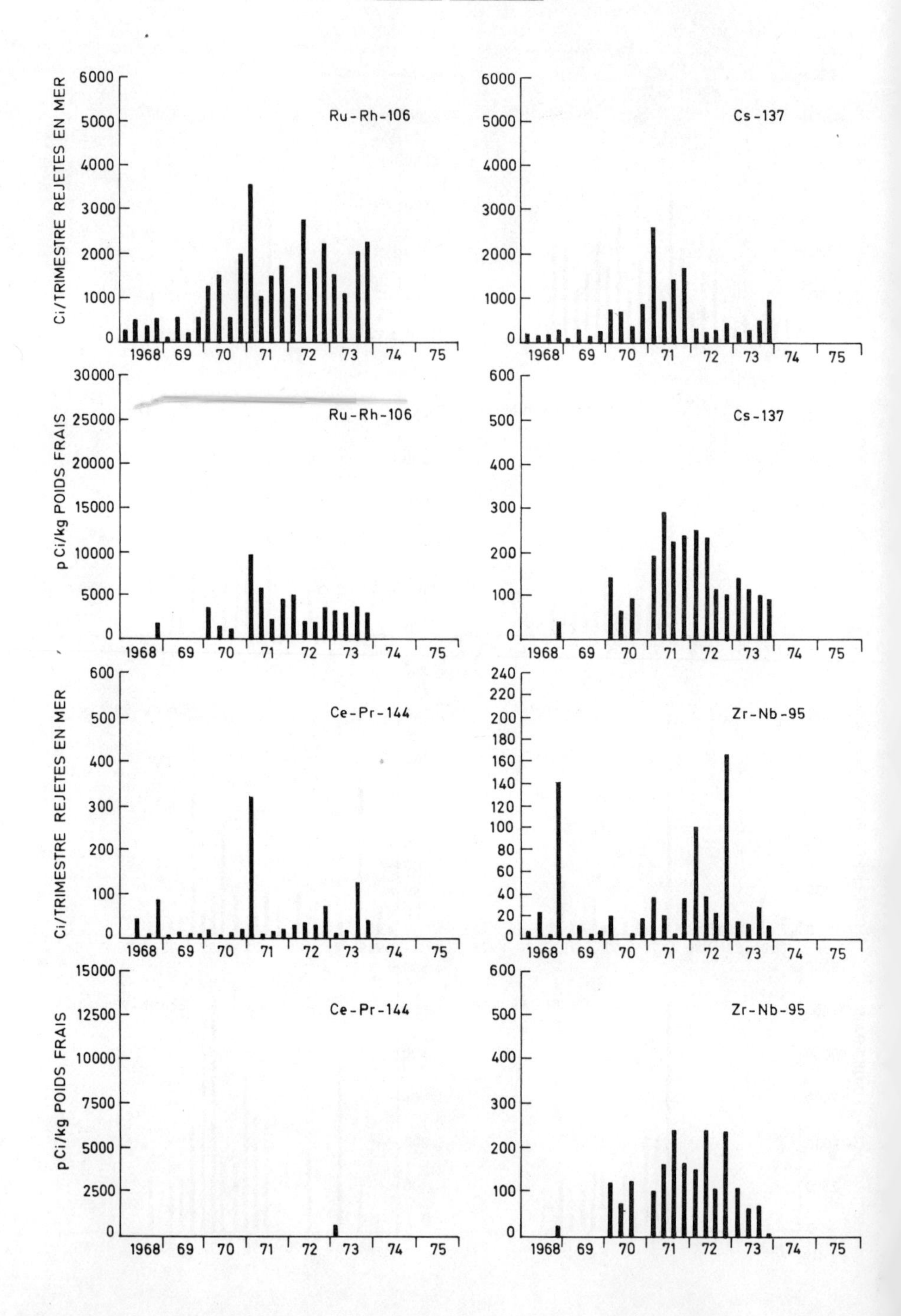

BAIE D'ECALGRAIN
Laminaria flexicaulis

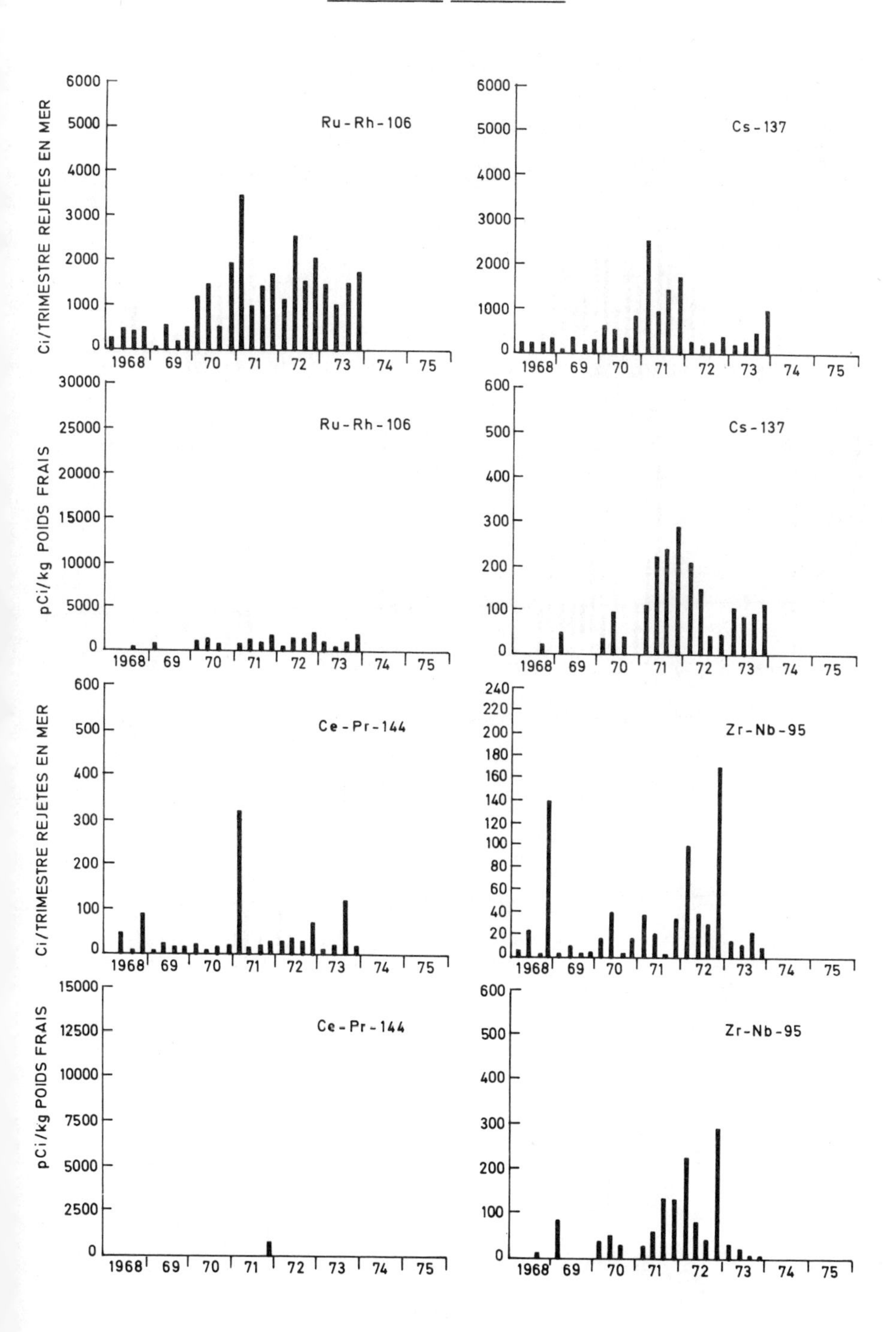

BAIE D'ECALGRAIN
Porphyra linearis

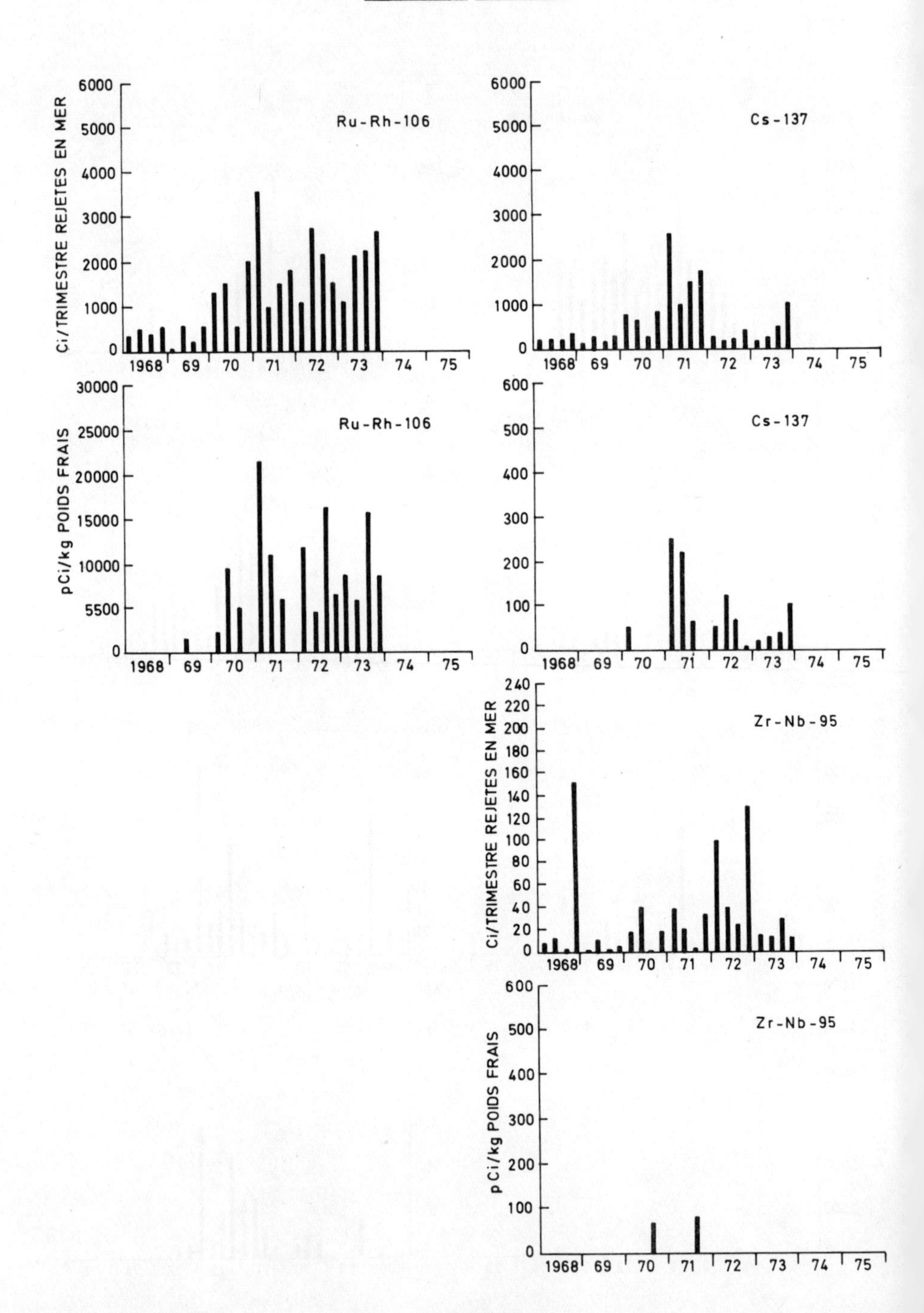

BAIE D'ECALGRAIN

Balanus balanoides

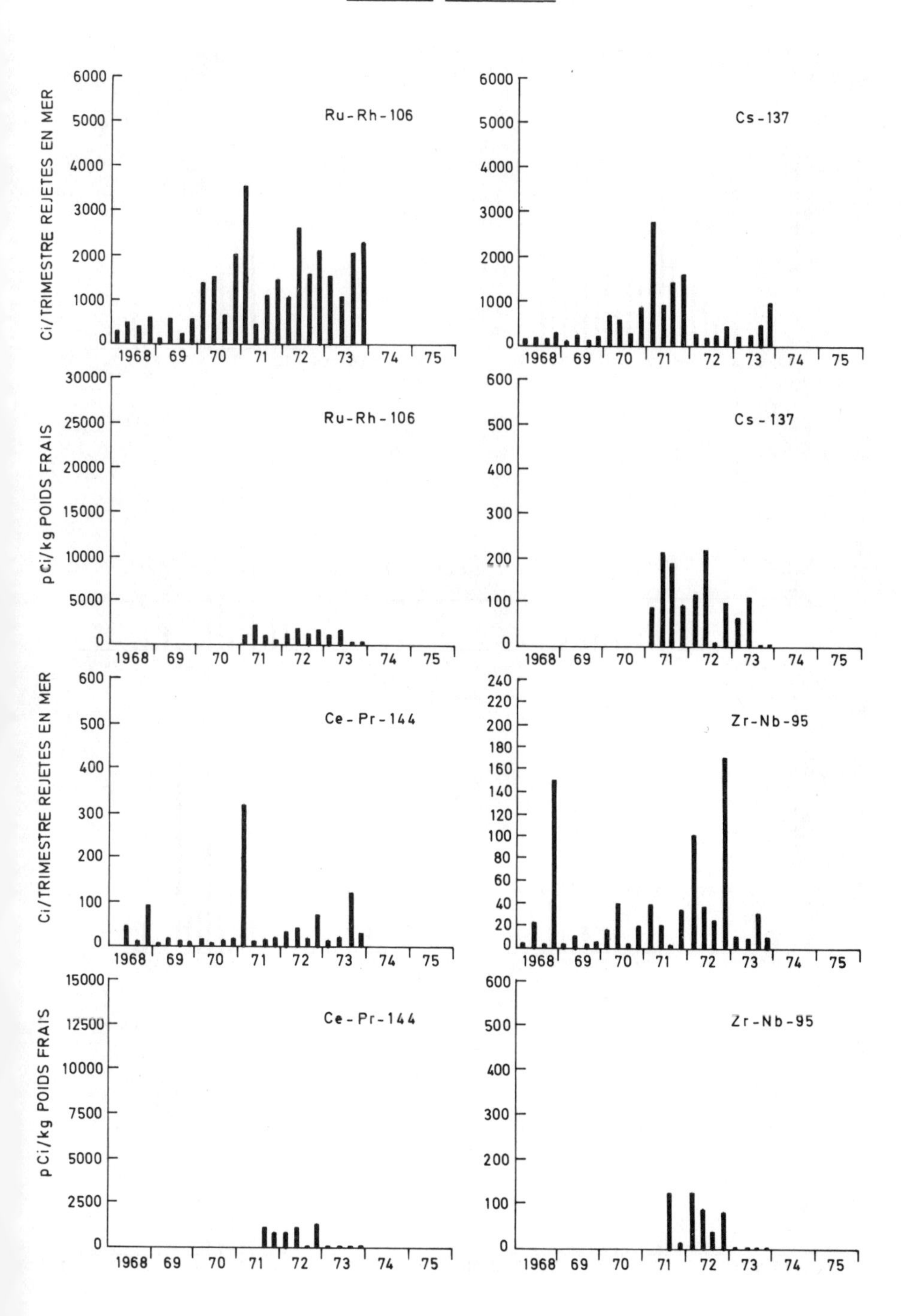

BAIE D'ECALGRAIN
Patella vulgata (chair)

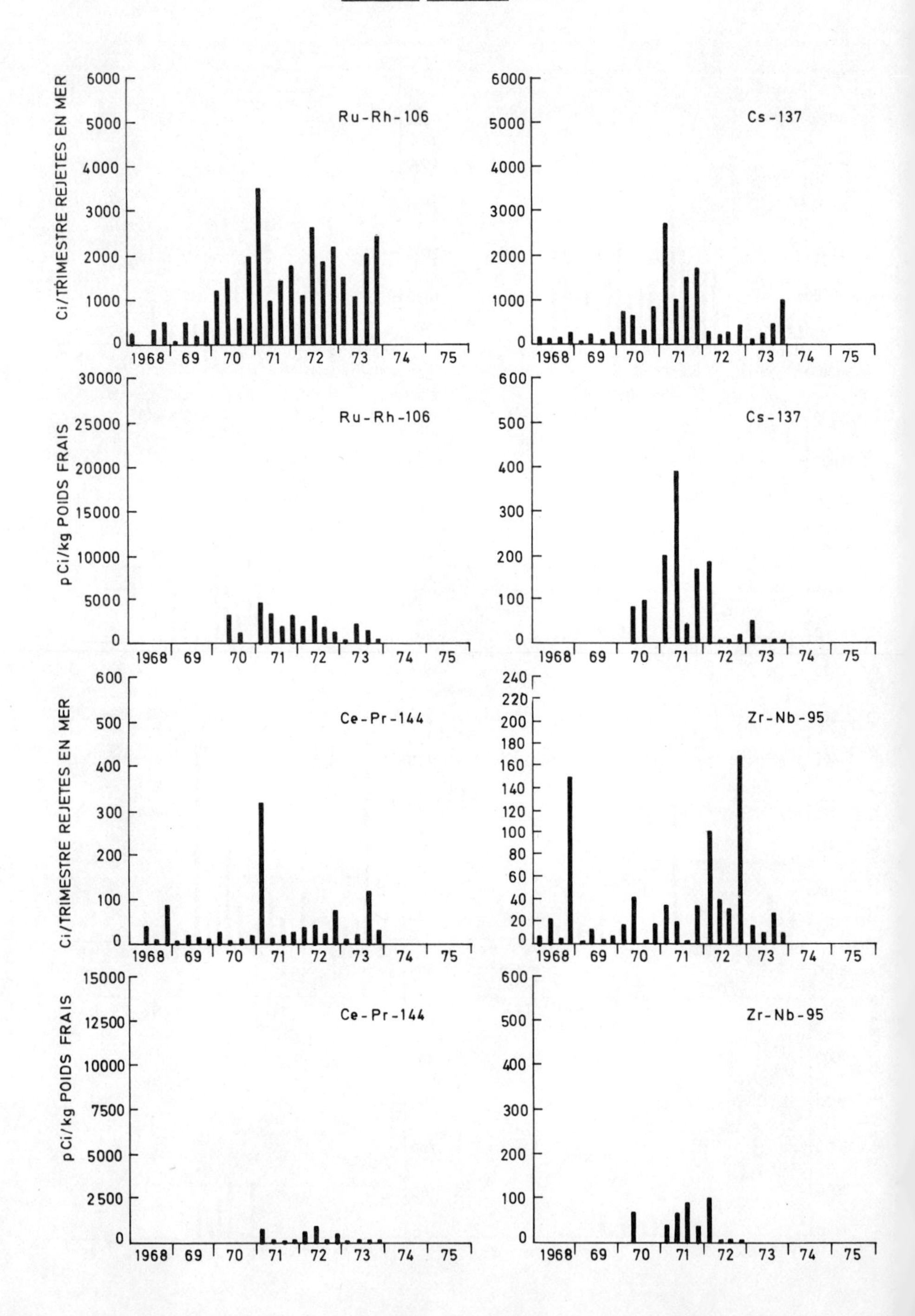

BAIE D'ECALGRAIN

Patella vulgata (chair)

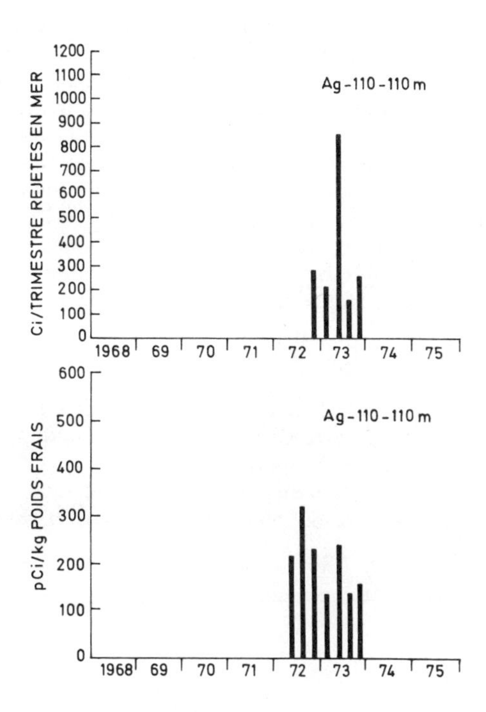

DISCUSSION

D. MECHALI: I should like to add some additional information regarding the estimation of population exposure resulting from marine disposals at the la Hague Centre. The critical group is a small group of fishermen using small boats along the coast in the area affected by discharges. These fishermen naturally consume the seafood which they catch and, in addition, they are subjected to external irradiation by contamination of their tackle. Since these coastal fishing catches are distributed throughout the region, a study has also been made of two other population groups: the rural population in the hinterland and the population of the town of Cherbourg. In the latter case, investigations have shown that coastal fishing makes only a small contribution to the total supply of seafood; the main contribution comes from the trawlers which fish further away from the Cotentin coast. The consumption of various seafoods by these three population groups was determined in a survey covering a period of a year.

A marine radioecology laboratory has been established at la Hague and has determined the concentration factors for various species in respect of the most important radioisotopes, so that it has been possible to check the values found in the literature.

It is difficult to compare the results of the monitoring programme with the model used to predict population exposure, since the discharges and sampling are not continuous and it is therefore necessary to take account of the kinetics of the phenomena. A study using all the available data will be performed in this connection. Until the results of this study are available, we can only say that measurements appear to confirm the model adopted.

The exposure of the critical group, which is of very small size, can at present be estimated at less then 1% of the dose limit.

M. J. A. DELPLA: I have a question for Mr. Ausset. What is the mean daily amount of fish consumed by the Norman fishermen? It would be interesting to compare this figure with that for the Indian fishermen referred to in the preceding paper.

R. AUSSET: Perhaps Mr. Planet, one of my co-authors, would care to answer this question.

J. PLANET: The daily quantities of seafood consumed by sea fishermen in the neighbourhood of Cap de la Hague are of the order of 75 g of fish, 12 g of molluscs and 5 g of crustaceans.

The consumption of crustaceans is very low because the fishermen do not consume their own catches of lobsters and crayfish, but sell them to neighbouring restaurants.

Population doses

(Session VI)

Chairman: M.M. SAUROV (Union of Soviet Socialist Republics)

Invited Review Paper

POPULATION DOSES FROM MEDICAL EXPOSURES

L.-E. LARSSON
National Institute of Radiation Protection,
Stockholm,
Sweden

Abstract

POPULATION DOSES FROM MEDICAL EXPOSURES.

During the past twenty years great effort has been devoted to the determination of genetically significant doses caused by medical radiological procedures while few corresponding surveys have been made to determine the appropriate population doses to such tissues and organs as the active bone marrow, thyroid, breast, lungs etc., that might be of interest in view of the present information on radiation-induced neoplasms. The available data on GSDs and somatic doses are discussed and the need for more information on somatic doses is demonstrated.

INTRODUCTION

During the last 30 years great effort has been expended on determining population doses from medical exposure. Extensive epidemiological studies have resulted in estimates of the magnitude of risk due to irradiation for long-term somatic malignancies, especially cancer, and for genetic damage. These estimates of genetic risks are not based on human data but are derived from information obtained from animals. The information on radiation-induced cancer originates mainly from the survivors of the atomic bombing of Hiroshima and Nagasaki but these results are complemented by studies on groups of patients who have undergone radiotherapy or X-ray diagnosis.

For many years the interest in population doses arising from medical radiological procedures was focussed on the determination of genetic doses. In 1962, when the second comprehensive report of the UNSCEAR was published [1], the national and regional studies on genetic radiation doses covered about 6-7% of the population of the whole world. In 1972, when the latest report of the UNSCEAR was published, the surveys on genetic doses covered almost 20% of the world population.

Among the various types of radiation-induced malignancies leukaemia is the the best know. However, only three major surveys of radiation doses to active bone marrow had been published by 1972.

GENETICALLY SIGNIFICANT DOSE

The genetically significant dose (GSD) is defined as "the dose which, if received by every member of the population, would be expected to produce the same total genetic injury to the population as do the actual doses received by the various individuals".

The GSD is expressed in the following way

$$D = \frac{\sum_{jk} (N_{jk}^{(F)} \times W_{jk}^{(F)} \times d_{jk}^{(F)} + N_{jk}^{(M)} \times W_{jk}^{(M)} \times d_{jk}^{(M)})}{\sum_{k} (N_{k}^{(F)} \times W_{k}^{(F)} + N_{k}^{(M)} \times W_{k}^{(M)})}$$

where D is the genetically significant dose
N_{jk} is the number of individuals of age group k, subject to class j exposure
N_k is the total number of individuals of age group k
W_{jk} is the future number of children expected by an individual of age group k subsequent to an exposure of class j
W_k is the future number of children expected by an average individual of age group k
d_{jk} is the average gonad dose per class j exposure of individuals of age group k
(F) and (M) denote female and male, respectively.

The GSD is not a per capita dose but can be regarded as the average dose to those gametes that will be effective for child production. The denominator is equal to the total number of effective gametes in the population.

Most national and regional surveys of the GSD from X-ray diagnosis have been collected in the UNSCEAR Reports. The 26 surveys mentioned report SSDs ranging between 5 and 58 mrad for the year of investigation. The annual numbers of X-ray examinations per 1000 of total population vary considerably between the countries and region, from 39 to more than 1000.

The highest contribution to the GSD comes from six to seven types of examination (lumbosacral spine, pelvis, hip and upper femur, urography, colon (barium enema) and obstetric examinations). In general, they are responsible for 80 - 90% of the GSD, but constitute together only 10 - 15% of the total number of examinations. Examinations of the chest, including mass surveys, contribute less than about two per cent of the total GSD.

Because of differences in the survey technique and differences between countries in the distribution of the various types of examination and performance procedure, it is difficult to draw conclusions from comparisons of the results. There have been attempts to interpret the results from later investigations as a tendency towards lower GSD in spite of the increasing annual numbers of diagnostic examinations.

TABLE I. DISTRIBUTION OF GSDs IN SURVEYS MADE IN THE FIFTIES AND SIXTIES

Period of survey	Distribution of surveys on range of dose (mrad)		
	0-20	21-40	41-60
1950-59	6	5	2
1960-69	6	5	2

Half the number of the 26 surveys mentioned were performed during the 1950s, while the other 13 surveys originate from the 1960s. In Table I the distribution of the GSDs obtained in the earlier surveys are compared with the ones from the later surveys. In general there is no support for an optimistic hope that the GSDs have decreased.

Japan may be an exception. The first survey was made in 1958 - 1960 and the second one in 1969. The frequency of examinations had increased from 730 to 1429 per 1000 individuals, but the GSD had decreased from 39 to 27 mrad. An analysis of the increased frequency of examinations seems to indicate that the increment depends on a much higher frequency of chest examinations, which does not considerably increase the GSD. In fact, the frequency of chest examinations has increased by 724 per 1000 individuals, which really means a decrease by 25 in the frequency of other types of examination.

The results of the national surveys are in general presented in such a way that the frequency of each type of examination is given per 1000 individuals. For several countries the frequency figures for obstetric examinations are 1 - 3. These low figures given per 1000 individuals unfortunately lead us to a false conclusion that the obstetric examinations are very few. However, if the comparison is made with the total number of pregnancies or live births it can be demonstrated that up to about 20% of pregnant women undergo an obstetric X-ray examination.

In Sweden we have recently measured individual gonad doses on patients undergoing X-ray examinations of the types that contribute most to the GSD. Unfortunately, the results from the majority of these examinations do not show significantly lower doses to the gonads than the results of the Swedish survey in 1955 - 1957. Since that period almost 20 years ago the annual number of X-ray examinations has doubled. This indicates the need for a new assessment of the GSD in Sweden.

Surveys of the diagnostic use of radionuclides have given GSDs that are far below one millirad. The frequencies of examinations with radionuclides are between 2 and 10 per 1000 individuals. However, if examinations with radionuclides were as frequent as X-ray examinations, the GSDs for the two types of examination would be of the same magnitude. It is hoped that the increase in the use of radionuclides for morphological and functional studies will be counterbalanced by a corresponding decrease in the number of X-ray examinations.

Radiation treatment with external sources is reported to cause GSDs of a few up to about ten millirads. The treatments of non-neoplastic diseases contribute about two thirds of the total GSD from radiotherapy. The therapeutic use of radio-pharmaceuticals gives a GSD of a few millirad.

SOMATIC DOSES

The active bone marrow is regarded as the tissue of interest for radiation-induced leukaemia. About twenty years ago the concept of mean marrow dose was introduced. It is expressed as the radiation dose to the irradiated part of the active marrow multiplied by that fraction of the total active marrow that was irradiated. The mean marrow dose can then be averaged over the population of interest and is then called the per capita mean marrow dose (CMD).

As mentioned earlier, there have been three comprehensive surveys on the CMD caused by X-ray examination. They were made in Japan, the district of Leiden in the Netherlands and in the United Kindgom. In addition, some information is available from Czechoslovakia.

Most of the data are based on phantom measurements. Thus, factors have been obtained for the transformation of skin doses to bone marrow doses for various types of examination. The data on the distribution of the active bone marrow originate (except for the Japanese data) from the figures published by Mechanik in 1926 and based on 13 cadavers. For various reasons the Mechanik data are uncertain. More reliable data are needed.

TABLE II. PER CAPITA MEAN MARROW DOSE FROM X-RAY DIAGNOSIS

	(mrad)
Japan	189
Leiden (Netherlands)	30
United Kingdom	32
Czechoslovakia	68 - 184 [a]

[a] Preliminary estimate.

The results of the surveys on the per capita mean marrow dose are summarized in Table II. The highest contributions come from examinations of the chest and the trunk. In the United Kingdom mass miniature examinations of the chest contribute 25% and in Japan the stomach surveys 55%. It should be observed that these types of examination contribute only a minor part to the genetically significant dose and therefore do not constitute a problem from that point of view. However, these and similar types of examination cause the major contribution to the per capita mean marrow dose. Information is also available on the mean marrow doses caused by radiation treatment. The doses range from below 100 rads up to about 400 rads, with the highest dose figures for the alimentary tract and the respiratory system.

Recent information on the incidence of radiation-induced cancer is collected in the UNSCEAR 1972 Report [2]. These incidence data are presented in Table III. As can be seen, the risks for lung and thyroid cancer are about the same as for leukaemia, while the risk for breast cancer gives half that figure. It is remarkable that there are almost no comprehensive studies on per capita mean doses to the breast, lungs and thyroids arising from medical radiological procedures.

In the Japanese survey on per capita mean marrow dose attempts were made to assess a leukaemia-significant dose in order to account for the shap of the time-incidence curve of radiation-induced leukaemia and survival tim after irradiation. This weighting procedure seems to be of importance wher the periods of observation in the epidemiological studies on various cancers amounts to 25 years.

TABLE III. EXPECTED NUMBERS OF RADIATION-INDUCED CANCER AFTER IRRADIATION OF 10^6 INDIVIDUALS WITH ONE RAD

Leukaemia	15-40
Lung	10-40
Thyroid	40
Breast	6-20
Others	40

Period of observation: 25 years.

At present risk-benefit judgements on various types of diagnostic and therapeutic radiological procedures are based on radiation-induced leukaemia and on the magnitude of the genetic significant dose. To establish 'a full picture' of the hasards for neoplasm, it is essential that the relevant tissue doses are determined. At present our knowledge about the relevant doses is not sufficient.

REFERENCES

[1] UNSCEAR, 1962 Report to the UN General Assembly, General Records: 17th Session, Suppl. No.16 (A/5216).

[2] UNSCEAR, Ionizing Radiation: Levels and Effects, II: Levels, UN, New York (1972).

The original papers on which this review is based (more than 50) are found in these two reports.

DISCUSSION

E. KUNZ: I should like to add some information on the estimate of per capita mean bone-marrow dose in Czechoslovakia, which was mentioned in your paper. During a gonad dose survey, the exposure of the skin in the centre of the irradiated field was measured in the course of more than 5000 examinations. Information obtained in Great Britain (reports of Lord Adrian's Committee) was used in two different ways to calculate mean marrow dose: we took either the same mean marrow dose per film as in the British data, or the same ratio of skin dose to mean marrow dose. These alternative approaches gave us a probable range of values of per capita mean marrow dose (68-184 mrad in 1966). A number of assumptions had of course to be made, the most important of them concerning the influence of geometrical factors, e.g. the assumption that there are no significant differences — as regards the current practices in various countries — in the frequency distribution of deviations from 'good standard practice'. This assumption is supported in the literature. About a third of all our examinations were related to mass chest radiography, which accounted for more than 40% of the per capita bone marrow dose.

Since a number of countries intend to perform population studies on the doses to various tissues resulting from X-ray diagnosis, information on appropriate methods for performing such studies would be of great value. Perhaps ICRP and WHO could assist in this work.

F.D. SOWBY: Emphasis on indices such as genetically significant dose or on per capita mean marrow dose alone may hide the true impact of medica radiographic procedures. The GSD from chest examinations, for example, is very low, but such an examination involves irradiation of some of the most radiosensitive tissues (lung, breast, marrow and possibly thyroid), so that the total impact of a chest examination may be much greater than that indicate by the value of the GSD. For this reason it would be useful to consider makir assessments of the total impact of various examinations, rather than continuing to perform GSD and CMD surveys.

J.C. VILLFORTH: I also believe that we should look at other doses as well as the gonad dose and the resultant GSD. We must get on with the task of reducing the unnecessary component of medical irradiation. Too much attention has been paid to GSD as the end product of our studies. In Israel recent studies on irradiations to treat ringworm of the scalp performed durin the 1940s and 1950s suggest that there is a five-fold increase in thyroid abnormalities as a result of a dose of about 6.5 rads to the thyroid from such irradiations. Thyroid scanning with ^{131}I can contribute doses to the thyroid of up to 100 rads; heart catheterization procedures in children can also produce high doses to the thyroid as well as excessive bone-marrow doses. These special procedures must be examined and action programmes develope to reduce their unnecessary components. Other types of examinations should be looked at in a similar way.

K.J. KOREN: If you were able to reduce medical doses substantially, would you be willing to allow higher doses to the population from nuclear energy?

L.-E. LARSSON: Your question gives me an opportunity to say that I strongly oppose the tendency of radiation producers to try to excuse themselves by pointing the finger at others who cause higher contributions to the population dose. There are cases, for example, of nuclear power people attacking medical radiology. In my opinion each radiation producer should clean up his own area. As for the possibility of donating dose-savings in medical work to nuclear energy, I think that this is a political matter.

O. ILARI: I agree that a lot can still be done in the medical use of radiations in terms of dose reduction. However, I would like to emphasize what you briefly mentioned, namely that these reductions can be obtained not only by technical means but also, and in particular, by properly training the medical people in the correct use of radiation sources.

I should also like to comment on Table III of your paper, and to point out the varying significance, from an overall radiation protection point of view, of the risk factors for cancer induction given therein. The mortality risk associated with various forms of cancer are not identical. For example the risk of death from leukaemia is 100%, while the risk associated with thyroid cancer is generally assumed to be 10%. Forty cases of leukaemia and 40 cases of thyroid cancer for 10^6 man· rems would thus correspond to 40 and to 4 deaths respectively. There is a big difference here as far as concerns radiation protection and public health evaluations extending over entire populations.

L.-E. LARSSON: I agree with what you say about Table III. Cancer is not necessarily fatal and the possibility of a cure for thyroid and breast cancer is much greater than in the case of lung cancer and leukaemia.

E. SHALMON: I fully support your emphasis on the importance of reducing doses from medical exposures and ensuring that there exposures are properly warranted. A first step in assessing the benefits of medical radiation practices should be to follow up the results, evaluate their usefulness and compare them with non-radiation alternatives. Could you say something more about what has been accomplished in various countries in this area?

L.-E. LARSSON: Radiological procedures should of course be beneficial to the patient. I would refer you to the ICRP Publication 16, on the protection of patients in X-ray diagnosis, where the problem is dealt with. It is certain that some radiological procedures are carried out without adequate justification. One should realize that patients are referred to radiological departments by non-radiologists and I think that the referring physicians should be more adequately informed on radiation matters than is at present the case.

M.J.A. DELPLA: Perhaps I can once more draw attention to a persistent confusion. Training doctors to think in terms of a certain risk value for 1 rad (for a given organ) is liable to frighten them and make them dispense with radiation diagnosis or treatment, even though this would be in the patient's interest. In other words, I think it is dangerous to talk about a theoretical risk (for 1 rad), and particularly so in connection with medical radiology as opposed to industrial nuclear energy.

L.-E. LARSSON: I should like to point out that our present information on radiation-induced malignancies is based partly on epidemiological studies performed on groups of patients who have undergone radiological procedures for diagnostic or therapeutic purposes. Since cases of radiation-induced malignancies have been found in these studies, it is obvious that they will be found also as a consequence of day-to-day medical practice. In the medical field the per capita dose of 1 rad does not mean that every individual has received the same dose. It might mean that 99 individuals receive no dose and one receives 100 rads.

W.D. ROWE: In the United States X-ray examinations are often performed for the benefit not of the patient but of his physician or some other organization or institution. Examples of this are X-ray photographs taken to offset a potential malpractice suit, or the pre-appointment scans of job applicants. It would seem prudent to minimize or eliminate these doses irrespective of the exact nature of the dose-effect relationship.

L.-E. LARSSON: Examinations involving radiation which is not beneficial to the irradiated individual but is performed purely for administrative or legal purposes should of course be avoided regardless of the magnitude of the radiation dose.

GENETICALLY SIGNIFICANT DOSE FROM THE USE OF RADIOPHARMACEUTICALS

H.D. ROEDLER, A. KAUL
Klinikum Steglitz der Freien Universität Berlin

G. HINZ, W. PIETZSCH, F.E. STIEVE
Bundesgesundheitsamt Berlin,
Federal Republic of Germany

Abstract

GENETICALLY SIGNIFICANT DOSE FROM THE USE OF RADIOPHARMACEUTICALS.

To estimate the genetically significant dose (GSD) from the use of radiopharmaceuticals, a survey of the applications of radionuclides in nuclear medicine in Berlin (West) was performed for the years 1953 - 1970. In addition, the values of gonad doses were recalculated on the basis of the unitarily demonstrated biokinetic data and the recently extended dose concept of absorbed fractions. The recalculations proved to be necessary since an extensive compilation of published absorbed doses had yielded variations of up to a factor of 100. This is due to the incompleteness and uncertainty of biokinetic data and the choice of different biokinetic models and mathematical methods for absorbed dose calculation.

The GSD from the use of radiopharmaceuticals proved to be 0.2 mrad for 1970. It does not change significantly if conventional radionuclides are replaced by short-lived radionuclides or pure gamma emitters aimed at critical organ dose reduction. Even if the present 20% yearly increase in the number of applications of radiopharmaceuticals is supposed to remain constant in the future, the contribution of diagnostic nuclear medicine to the total GSD will be small compared to that from the diagnostic use of X-rays.

INTRODUCTION

The introduction of radioactive substances in medical examination and treatment methods has given medical science the opportunity both to diagnose and localize organic defects more accurately and to gain quantitative information about metabolism and organ functions. Since it is to be expected that these methods will be used to an even greater extent within clinical routine and that further development of new methods will progress steadily, a further expansion in the number of examinations per year will certainly be forthcoming.

However, the progress resulting from the employment of radioactive substances in medicine has to be carefully considered against the genetic and somatic radiation risk to both the individual patient and to the entire population. Exact knowledge of radiation burden caused by various nuclear medical methods is, therefore, a prerequisite.

1. DISCUSSION OF PUBLISHED DOSE VALUES AND BIOKINETIC DATA

1.1. Dose values

To be able at least to evaluate the dimension of the nuclear-medically caused radiation burden, a number of authors have compiled published

TABLE I. FLUCTUATION OF PUBLISHED GONAD DOSES

Radionuclide	Radiopharmaceutical	Gonad dose (mrad/μCi)	
		Lowest value	Highest value
^{131}I	Iodide	0.056	8.5
^{131}I	o-iodohippurate	0.016	0.32
^{198}Au	Colloid	0.25	1.4
^{99m}Tc	Pertechnetate	0.006	0.3
^{99m}Tc	Sulphur colloid	0.01	0.2
^{57}Co	Vitamin B_{12}	0.06	140
^{58}Co	Vitamin B_{12}	2.6	150
^{131}I	HSA	1.7	9
^{131}I	MAA	0.074	2
^{197}Hg	BMHP	0.05	0.4
^{75}Se	L-selenomethionine	1	15
^{133}Xe		0.0002	0.04
^{51}Cr	Na-chromate	0.03	3
^{51}Cr	EDTA	0.01	0.013
^{85}Sr	nitrate, chloride	2.9	40
^{59}Fe	chloride, citrate	6	350
^{203}Hg	Chlormerodrine	0.02	1.5

dose values [1 - 3]. In continuation of these papers we have sifted the up-to-date available literature on the medical application of radioactive substances for an extensive study with the goal of not only tabulating the dose values for single organs but also reporting examination methods, including special examination conditions and the activities applied [4]. This compilation recognizes a substantial fluctuation in published dose values. To demonstrate this fluctuation we extracted and tabulated the highest and lowest gonad doses of all radiopharmaceuticals applied in Berlin (West) during 1970 (Table I). As the table shows, the dose values reported by individual authors may differ by up to a factor of 100. The reason for this lies in the large biologically and experimentally caused inaccuracy of those biokinetic data that are available for the calculation of radiation burden (distribution and length of persistence of radioactive substances within the organism) and also in the application of differentiating mathematical patterns for dose calculations.

1.2. Biokinetic data

The insecurity of biokinetic data has been shown by another study we conducted and which was aimed towards the compilation of all available data on the kinetics of radiopharmaceuticals for the purpose of absorbed dose recalculations [5]. For the most important radiopharmaceuticals used in nuclear medicine those biokinetic data were compiled that had been gained from human and/or animal tests and that seemed most applicable for dose calculations.

2. DOSE RECALCULATION

2.1. The expanded dose concept

The dose calculations were conducted according to our expanded concept of absorbed fractions [6 - 8]. This is based on the calculation scheme of the Medical Internal Radiation Dose Committee [9, 10] and permits, as well as the calculation of energy doses for given source-target pairs, the consideration of residual body activity influence. The output resulting from the expanded-dose concept for the formally exact target dose reads:

$$D_\tau = \frac{A_0}{m_\tau \ln 2} \sum_{\sigma,\nu} F_\sigma Q_{\sigma\nu} T_{\sigma,\nu} \sum_i \Delta_i \phi_{(\tau\leftarrow\sigma),i} + \frac{A_0}{m_\tau \ln 2}$$

$$\times \left(\sum_\nu Q_{GK,\nu} T_{GK,\nu} - \sum_{\sigma,\nu} F_\sigma Q_{\sigma,\nu} T_{\sigma,\nu} \right)$$

$$\times \sum_i \Delta_i \left(\frac{m_{GK}}{m_{RK}} \phi_{(\tau\leftarrow GK),i} - \sum_\sigma \frac{m_\sigma}{m_{RK}} \phi_{(\tau\leftarrow\sigma),i} \right)$$

where A_0 is the administered activity (μCi)

m_σ, m_τ, m_{GK}, m_{RK} are the masses of source (σ), target organ (τ), total body (GK) and residual body (RK); m is in grams

F_σ is the distribution factor of a source organ σ

$Q_{\sigma,\nu}$, $Q_{GK,\nu}$ are the quotas of the νth retention function component extrapolated to application time of a source organ σ or total body GK

$T_{\sigma,\nu}$, $T_{GK,\nu}$ are the νth components of effective half-life for a source organ σ and the total body GK; T is in hours

Δ_i is the mean energy per disintegration of i-tieth radiation component; Δ_i is in g · rad/μCi · h

$\phi_{(\tau\leftarrow\sigma),i}$, $\phi_{(\tau\leftarrow GK),i}$, $\phi_{(\tau\leftarrow RK),i}$ are the fractions to mean energy of radiation type i emitted from source organ σ, total body GK and residual body RK, absorbed by a target organ τ.

The calculation of formally exact target doses according to the expanded-dose concept is especially demanding. Therefore, a computer

```
1.   APPLIED ACTIVITY    1000.000 MICRO-CURIES OF    F-18     ION                    BIOL DATA FROM ICRP PUBLICATION 2 (1959)

2.   TARGET ORGAN    OVARIES                  MASS      8.8 GM

3.   TOTAL BODY (HOMOGENOUS DISTRIBUTION)
               DF/MX/C     FR/M/PR    ERM HT      ALLTYP     GAMMA    NONPEN    ELMAGN    ELECTR     BETA     INCON     AUGER     X-RAY
     COMP 1    1.00000      1.0000      1.87        58.7      38.8      19.8      38.8      19.8      19.8      0.0       0.0       0.0
     %           1.000        1.87                 100.0      66.1      33.8      66.1      33.8      33.8      0.0       0.0       0.0

4.1  TOTAL SKELETON             19
               DF/MX/C     FR/M/PR    ERM HT      ALLTYP     GAMMA    NONPEN
     COMP 1    0.53000      1.0000      1.87         9.8       9.8       0.0
     %           3.678        0.99                 100.0     100.0       0.0

5.   SUM OF SOURCE ORGANS
               DISFACT        MASS    EFF HT      ALLTYP     GAMMA    NONPEN    ELMAGN    ELECTR     BETA     INCON     AUGER     X-RAY
     SUM       0.53000     10091.0      1.87         9.8       9.8       0.0       9.8       0.0       0.0      0.0       0.0       0.0
     %                                             100.0     100.0       0.0     100.0       0.0       0.0      0.0       0.0       0.0

6.1  REST  OF BODY
               DISFACT        MASS    EFF HT      ALLTYP     GAMMA    NONPEN    ELMAGN    ELECTR     BETA     INCON     AUGER     X-RAY
     SUM       0.47000     59945.0      1.87        29.1      18.2      10.9      18.2      10.9      10.9      0.0       0.0       0.0
     %                                             100.0      62.5      37.4      62.5      37.4      37.4      0.0       0.0       0.0

7.1  TOTAL DOSE
               SOURCES        REST                ALLTYP     GAMMA    NONPEN    ELMAGN    ELECTR     BETA     INCON     AUGER     X-RAY
     SUM           9.8        29.1                  39.0      28.1      10.9      28.1      10.9      10.9      0.0       0.0       0.0
     %            25.2        74.7                 100.0      72.0      27.9      72.0      27.9      27.9      0.0       0.0       0.0

6.2  REST  OF BODY
               DISFACT        MASS    EFF HT      ALLTYP     GAMMA    NONPEN    ELMAGN    ELECTR     BETA     INCON     AUGER     X-RAY
     SUM       0.47000     59945.0      1.87        29.1      18.2      10.9      18.2      10.9      10.9      0.0       0.0       0.0
     %                                             100.0      62.5      37.4      62.5      37.4      37.4      0.0       0.0       0.0

7.2  TOTAL DOSE
               SOURCES        REST                ALLTYP     GAMMA    NONPEN    ELMAGN    ELECTR     BETA     INCON     AUGER     X-RAY
     SUM           9.8        29.1                  39.0      28.1      10.9      28.1      10.9      10.9      0.0       0.0       0.0
     %            25.2        74.7                 100.0      72.0      27.9      72.0      27.9      27.9      0.0       0.0       0.0

6.3  REST  OF BODY
               DISFACT        MASS    EFF HT      ALLTYP     GAMMA    NONPEN    ELMAGN    ELECTR     BETA     INCON     AUGER     X-RAY
     SUM       0.47000     59945.0      1.87        32.6      19.8      12.7      19.8      12.7      12.7      0.0       0.0       0.0
     %                                             100.0      60.8      39.1      60.8      39.1      39.1      0.0       0.0       0.0

7.3  TOTAL DOSE
               SOURCES        REST                ALLTYP     GAMMA    NONPEN    ELMAGN    ELECTR     BETA     INCON     AUGER     X-RAY
     SUM           9.8        32.6                  42.5      29.7      12.7      29.7      12.7      12.7      0.0       0.0       0.0
     %            23.2        76.7                 100.0      69.9      30.0      69.9      30.0      30.0      0.0       0.0       0.0
```

FIG. 1. Absorbed dose calculation, computer output.

TABLE II. COMPARISON OF GONAD DOSES

Radionuclide	Radiopharmaceutical	Gonad dose (mrad/μCi)		
		Wolf [14]	ICRP [3]	Recalculation
^{131}I	Iodide	0.2	2.0	0.2
^{131}I	o-iodohippurate	0.16	0.023	0.007
^{198}Au	Colloid	0.6	0.18	0.25
^{99m}Tc	Pertechnetate	0.016	0.015	0.02
^{99m}Tc	Sulphur colloid	0.02	-	0.007
^{57}Co	Vitamin B_{12}	150	1.4	0.11
^{58}Co	Vitamin B_{12}	150	5.5	0.55
^{131}I	HSA	2	1.7	2.0
^{131}I	MAA	0.15	0.29	0.3
^{197}Hg	BMHP	0.13	0.019	0.3
^{75}Se	L-selenomethionine	10	10	10
^{51}Cr	Na-chromate	0.5	0.22	0.3
^{51}Cr	EDTA	0.013	(0.22)	0.06
^{85}Sr	nitrate, chloride	20	2.9	3
^{87m}Sr	nitrate, chloride	0.0067	0.02	0.02
^{113m}In	Colloid	0.016	(0.006)	0.003
^{113m}In	EDTA	(0.55)	(0.006)	0.015
^{59}Fe	chloride, citrate	20	22	25
^{203}Hg	BMHP	(2.7)	(0.01)	6
^{203}Hg	Chlormerodrine	1.35	0.01	2.4

Values in parentheses were not given by Wolf [14] or the ICRP [3], but were assumed by Hinz and Weil [21] for calculating the GSD.

program was established considering all radiation types (gamma, röntgen, fluorescent and beta radiation, Auger and conversion electrons) and including all source-target pairs for which the absorbed fractions were published by the Medical Internal Radiation Dose Committee [10].
Necessary input data:

The decay data of a radionuclide (e.g. Refs [11 - 13])
The administered activity
The biokinetic data of the radiopharmaceutical.

The computer output (Fig.1) contains basically the following information:

Target organ dose under the assumption of homogeneous distribution of the radiopharmaceutical in the entire body
Target organ dose from the activity in the single source organ as well as in the residual body
Total dose in target organ.

TABLE III. GONAD DOSES FROM ADMINISTRATION OF ^{131}I IODIDE CALCULATED ACCORDING TO DIFFERENT BIOKINETIC DATA TAKEN FROM HUMANS

Biokinetic data from	Gonad dose in (mrad/μCi)		
	Normals	Hypothyroidism	Hyperthyroidism
Colard et al. [15]	0.095	-	0.20
Pope [16]	0.24	0.23	0.31
Oberhausen and Muth [17]	0.18	-	0.18

All dose values are separately listed according to radiation type. In addition, they are tabulated as a percentage of the respective total dose.

2.2. Results of dose recalculations

As a necessary requirement to specify the radiation exposure of the individual patient as well as of the entire population from diagnostic and therapeutic application of readily available radioactive substances in nuclear medicine, recalculations of the gonad dose on the basis of the expanded dose concept were conducted for those radiopharmaceuticals utilized in Berlin (West) during the year 1970. The biokinetic data required for this purpose were gained from the above-mentioned compilation [5]. The results of dose calculations are summarized in Table II and compared with earlier results by Wolf [14] and the ICRP [3]. The comparison of these values indicates that the gonad doses reported by the individual authors may vary by a factor of more than 100.

Since at the present time ^{131}I is the radiopharmaceutical most used, it was deemed especially important, considering the factor of 10 between the gonad doses by Wolf and the ICRP, to conduct dose recalculations on the basis of biokinetic data reported by various authors. The results are given in Table III. The comparison shows only minor differences in single values - even for various functional conditions of the thyroid - and demonstrates that the average gonad dose is about 0.2 mrad/μCi. This leads to the assumption that the gonad dose value as reported by the ICRP [3] and Myant [18], without indication of biokinetic data, had been set too high.

Except for ^{131}I iodide it was only possible to conduct dose calculations, based upon several sets of apparently dependable biokinetic data, for colloid ^{198}Au, ^{99m}Tc pertechnetate and ^{75}Se selenomethionine. The deviations of the single values obtained from earlier reported gonad doses are insignificant when compared with such discrepancies as occur for ^{131}I o-iodohippurate, ^{57}Co and ^{58}Co vitamin B_{12}, ^{197}Hg and ^{203}Hg BMHP, ^{51}Cr EDTA and ^{203}Hg chlormerodrine.

3. GENETICALLY SIGNIFICANT DOSE (GSD)

3.1. Calculation foundation

The radiation burden for the entire population from medical application of ionizing radiation to individuals is described by the 'genetically significant dose' (GSD). The calculation of GSD goes back to Osborn and Smith [19] and takes into consideration the birth probabilities within the various population age groups. Those gonad doses received by each age group are multiplied by the factors of child expectancy of the respective age groups and divided by the total number of children that may be born to currently living men and women. This definition of the concept of a genetically significant dose is justified by the fact that for the genetic population burden only those radiation doses are taken into account that are received at a regenerative age considering the still existing child expectancy of each age group.

The formal expression for the GSD is:

$$D = \frac{\sum_k \sum_j \left(N_{jk}^{(m)} \cdot w_k^{(m)} \cdot d_j^{(m)} + N_{jk}^{(f)} \cdot w_k^{(f)} \cdot d_j^{(f)} \right)}{\sum_k \left(N_k^{(m)} \cdot w_k^{(m)} + N_k^{(f)} \cdot w_k^{(f)} \right)}$$

where D is the yearly genetically significant dose (mrad)

N_{jk} is the yearly number of individuals in age group k exposed to radionuclide application of type j

w_k is the child expectancy of an age-group k individual

d_j is the gonad dose received by an individual exposed to radionuclide application j

N_k is the number of individuals in age group k

m and f are male and female.

3.2. Statistics for the development of nuclear medical application of radioactive substances in Berlin (West)

To gain a quantitative and qualitative survey of radionuclide application for the entire medical field during the time period 1953-1968, necessary to the population GSD calculation, the Federal Department of Health conducted surveys in Berlin (West) in 1969 [20]. The county of Berlin, with 2.18 million inhabitants and a 15-year history of adequate nuclear medical care, offers an advantageous possibility, also influenced by its geographical situation, for the purpose of a statistical evaluation of a defined population group. Because of the formation of several key hospitals, examination and treatment methods with radioactive substances can be applied to a larger extent than in other parts of the Federal Republic. It may be assumed, therefore, that the population radiation burden from this is higher in the county of Berlin than in large parts of the Federal Republic of Germany where practising physicians generally have less opportunity to conduct nuclear medical examinations of their patients.

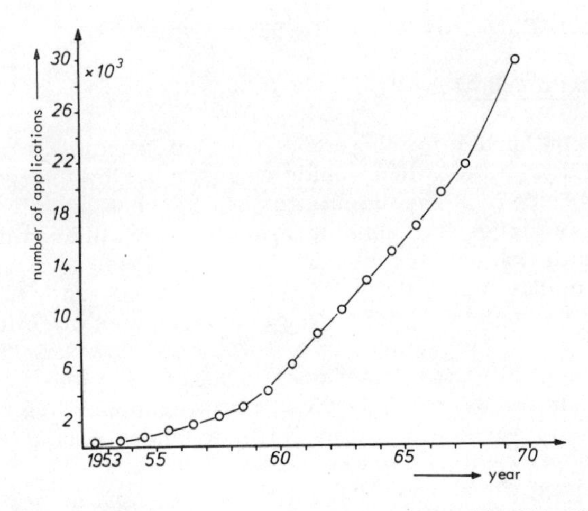

FIG. 2. Number of medical applications of readily available radionuclides 1953-1970 (Berlin/West).

The statistics include all users of readily available radioactive substances in the medical field, which means seven hospitals and four general practices for the year 1969. Because of the large amount of applications within the total period from 1953 to 1968 it was impossible to include all applications. Therefore, using patients' records, those with first letter names A - D were evaluated as a representative fraction, constituting 14.7% of the total patient bulk.

The control sheets utilized collected such information as:

Year of birth
Sex
Year of treatment
Diagnostic or therapeutic application
Repeat examinations
Outpatient or hospitalized treatment
Function or localization tests
Organ examined
Radionuclide
Chemical form
Administered activity.

A total number of 98 000 patients had been collected for the period 1953 - 1968. With the inclusion of repeat examinations the total number of applications is 131 000. The first letters A - D (14.7%) correspond to 14 415 main and first examinations and to 4850 repeat examinations. Of these 19 262 applications 18 821 were transferred to key punch cards and evaluated by an IBM 1130 data-processing system. The difference of 441 applications could not be evaluated due to incomplete information. 398 applications of rare short-lived radionuclides were conducted in connection with the scientific questionnaires and were not considered

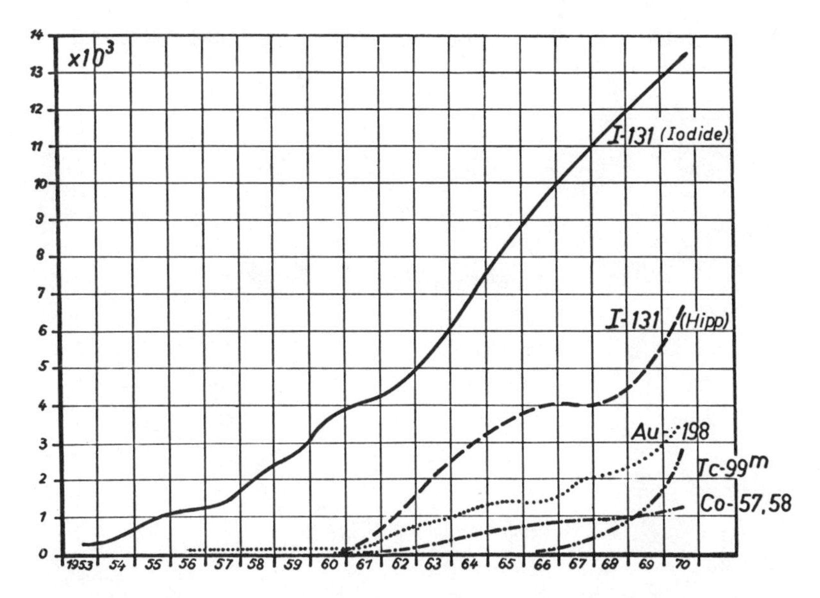

FIG. 3. Number of yearly applications of several radionuclides 1953-1970 (Berlin/West).

for the calculation of GSD. Therefore, a factual evaluation of 18 429 applications took place. The statistical Yearbook of 1968 was used to ascertain the population structure. The evaluation results showed that an average of each 100th citizen of Berlin (West) received a nuclear medical application.

The remarkable increase in the application of readily available radioactive substances in medicine was the cause for a comparative second statistical control as to the number of examinations and genetically significant dose. The control was conducted in 1971 and covered the first quarter of the year 1970. For this particular quarter all applications were registered and evaluated in the 13 hospitals and 6 general practices applicable. There appeared to be an increase of about 40% for the period of 1968-1970, which amounts to a yearly rate of increase of about 20%. The total application number in 1970 was 30 312, i.e. an average of each 70th citizen of Berlin (West) received a nuclear medical application in 1970. The data evaluation was processed in the same manner as in 1969. With regard to population structure, the information sources of the 1970 State Statistic Records Office Berlin were utilized for the second statistical control.

Figure 2 demonstrates the rapid increase in application in those ten years, while Fig.3 shows the increase in the number of applications of several essential radionuclides. The application frequency of ^{131}I iodide and ^{131}I o-iodohippurate has increased steadily and the same applies, to a lesser degree, to ^{198}Au and ^{57}Co/^{58}Co. As a short-lived generator product, ^{99m}Tc shows the largest increase rate.

Table IV demonstrates the application frequency of individual radionuclides for the year 1968 as compared to 1970. It is evident that, on

TABLE IV. NUMBER OF DIAGNOSTIC APPLICATIONS OF OPEN RADIOACTIVE SUBSTANCES IN BERLIN (WEST) FOR THE YEARS 1968 AND 1970

Radionuclide	Radiopharmaceutical	Number of applications	
		1968	1970
^{131}I	Iodide	11300	12980
^{131}I	o-iodohippurate	4200	6636
^{198}Au	Colloid	2100	3396
^{99m}Tc	Pertechnetate	430	2720
^{57}Co, ^{58}Co	Vitamin B_{12}	770	1284
^{131}I	HSA, MAA	550	884
^{197}Hg	Chlormerodrine, BMHP	210	492
^{87m}Sr	nitrate, chloride	-	340
^{75}Se	L-selenomethionine	50	292
^{133}Xe		15	176
^{113}In	Colloid	-	168
^{51}Cr	Na-chromate	95	112
^{85}Sr	nitrate, chloride	10	84
^{59}Fe	chloride, citrate	100	76
^{203}Hg	Chlormerodrine, BMHP	80	32
^{132}I	Iodide	760	-
		20670	29672

the one hand, short-lived substances, such as ^{99m}Tc, ^{87m}Sr and ^{113m}In, are of greater importance but, on the other, it is equally evident that the application of two radionuclides with a relatively high radiation burden to the patient, such as ^{85}Sr and ^{75}Se, has increased further. ^{203}Hg has been progressively superceded by ^{197}Hg compounds.

3.3. Calculations of genetically significant dose

On the basis of these statistical enquiries into the nuclear medical application of radioactive substances in Berlin (West), as well as those single values of gonad doses reported by Wolf [14] and ICRP [3], together with our recalculated single values of gonad dose, the entire GSD for the years 1968 and 1970 was determined. These, as well as the percentage GSD portion of the most important radiopharmaceuticals for the respective values of total GSD, have been compiled in Table V. While all diagnostic applications were included in the GSD calculations of all therapeutic measures, only the ^{131}I therapy was considered since for other therapeutic processes employing radionuclides a genetic burden could not be applied because of indication or age of patients.

TABLE V. $(GSD)_j$ FROM INDIVIDUAL NUCLEAR-MEDICALLY APPLIED RADIOPHARMACEUTICALS AS A PERCENTAGE OF THE RESPECTIVE TOTAL GSD
Calculated according to the survey for the years 1968 and 1970 in Berlin (West), based on different gonad dose values

Radionuclide	Radiopharmaceutical	$(GSD)_j/GSD(\%)$					
		for 1968			for 1970		
		based upon gonad doses from					
		Wolf	ICRP	recalcul.	Wolf	ICRP	recalcul.
^{131}I	Iodide						
	diagnostic appl.	29.8	68.5	30.7	11.2	47.3	13.8
	therapeut. appl.	8.63	19.9	8.9	6.9	29.3	8.52
^{131}I	o-iodohippurate	2.52	0.08	0.11	7.2	0.44	0.4
^{198}Au	Colloid	14.7	0.10	6.29	13.3	1.70	6.1
^{99m}Tc	Pertechnetate	3.44	0.74	4.44	13.8	5.5	21.25
^{99m}Tc	Sulphur colloid				0.63	0.27	0.06
^{57}Co	Vitamin B_{12}	0.63	<0.01	<0.01	0.97	<0.01	<0.01
^{58}Co	Vitamin B_{12}	5.99	0.05	<0.01	7.7	0.12	0.4
^{131}I	MAA	8.98	3.99	18.6	0.35	1.7	0.87
^{131}I	HSA	0.63	0.15	0.65	0.03	0.01	0.04
^{197}Hg	BMHP	2.83	0.1	7	2.1	0.13	6.15
^{197}Hg	Chlormerodrine	0.17	<0.01	0.21			
^{87m}Sr	nitrate, chloride				0.22	0.3	0.81
^{75}Se	Selenomethionine	19.6	4.51	20.2	29.8	12.6	36.7
^{113m}In	Colloid				0.31	0.05	0.07
^{51}Cr	Na-chromate				0.19	0.04	0.14
^{51}Cr	EDTA	0.07	0.26	0.33	0.25	0.04	0.03
^{85}Sr	nitrate, chloride	0.47	<0.01	<0.01	4.4	0.27	0.81
^{59}Fe	chloride, citrate	2.00	0.49	2.51	0.6	0.03	0.94
^{203}Hg	BMHP				<0.01	<0.01	<0.01
^{203}Hg	Chlormerodrine	0.02	<0.01	0.04	<0.01	<0.01	<0.01
Total GSD (mrad)		0.11	0.47	0.11	0.25	0.60	0.20

If a comparison is made between the values of total GSD for the year 1968, it becomes evident that the value calculated by Wolf does not differ from the one recalculated by the authors. This agreement is caused by the fact that for radiopharmaceuticals with highest gonad doses, or those most often employed, the single gonad dose values show only slight variation. The total GSD value calculated according to the ICRP data as a factor of 4 higher resulted essentially from the gonad dose of ^{131}I iodide value of 2 mrad/μCi, assumed from literature, as compared with 0.2 mrad/μCi according to Wolf and our own recalculations. The above

TABLE VI. DOSE REDUCTION FOR EXAMINED OR CRITICAL ORGAN AND GONADS BY SELECTION OF SUITABLE RADIOPHARMACEUTICALS

Radiopharmaceuticals and mean administered activities (μCi) Conventional	Replaced by	Diagnostic method	Dose reduction coefficient for: Examined or critical organ		Gonads
^{131}I iodide (43)	^{99m}Tc pertechn. (1000)	Thyroid scanning	Thyroid	0.013	2.6
	^{132}I iodide (25)	Function test (30%)	Thyroid		
^{198}Au colloid (240)	^{99m}Tc S-coll. (1500)	Liver scanning	Liver	0.063	0.13
^{58}Co vitamin B_{12} (0.9)	^{57}Co vitamin B_{12} (0.5)	Schilling test	Liver	0.12	0.11
^{131}I MAA (220)	^{99m}Tc MAA (3000)	Lung scanning (66%)	Lungs	1.1	0.071
	^{133}Xe (15 000)	Lung scanning (34%)	Lungs		
^{131}I HSA (10^+)	^{99m}Tc HSA (100)	Blood volume	Total body	0.088	0.1
^{131}I HSA (10^+)	^{99m}Tc HSA (500)	Placental localization	Total body	0.44	0.5
^{131}I HSA (100^+)	^{99m}Tc HSA (1500)	Myelography	Total body	0.13	0.15
^{197}Hg BMHP (360)	^{99m}Tc S-coll. (1500)	Spleen scanning	Spleen	0.15	0.069
^{85}Sr -nitrate (330)	^{99m}Tc polyphosphate (10 000)	Bone scanning	Skeleton	0.11	0.2
^{203}Hg BMHP (400)	^{99m}Tc DTPA (3000)	Kidney scanning	Kidneys	0.0009	0.0096
^{203}Hg Chlormerodrine (180)	^{99m}Tc DTPA (3000)	Kidney scanning	Kidneys	0.017	0.053

reflections are equally valid for the interpretation of the total GSD for 1970 as for the various parts of the GSD for each single radiopharmaceutical of the appropriate total GSD.

Comparing the respective values of the total GSD for the years 1968 and 1970, it is evident that the GSD for the year 1970 based on recalculation has doubled itself. The slight gain in the total GSD calculated according to ICRP resulted essentially from the gonad dose, which was fixed too high for ^{131}I iodide. The slightly stronger gain of Wolf's calculated total GSD in comparison with the recalculation is caused by the comparatively high gonad dose of ^{131}I o-iodohippurate, ^{58}Co vitamin B_{12} and ^{85}Sr nitrate, as well as the partially above average gain of the examination frequency of these pharmaceuticals.

4. DOSE REDUCTION

4.1. Discussion of the possibilities of dose reduction

As can be deduced from the equation for the calculation of formally exact target doses, the energy absorbed by a target depends on the following physical and biological quantities:

Administered activity
Physical half-life of a radionuclide
Mean emitted energy per disintegration
The fraction of emitted energy per disintegration absorbed in a target
Distribution factors or activity concentration of a radiopharmaceutical in the organs and in the residual body
Biological half-lives of a radiopharmaceutical in total body and in the organs.

By reduction of administered activity (e.g. by application of the whole-body counter for the purpose of functional examinations, such as a radioiodine test for children or examinations of kidney function with ^{51}Cr EDTA and ^{169}Yb EDTA) and/or by selection of radionuclides with short physical half-life, it is basically possible to reduce the patient's radiation burden. The use of radionuclides with minor mean emitted energy per disintegration ($g \cdot rad/\mu Ci \cdot h$) (e.g. 2.171 for ^{58}Co, 0.3071 for ^{57}Co; 1.2176 for ^{131}I, 0.1324 for ^{125}I) in conjunction with the smallest possible portion of non-penetrating radiation components also facilitates a reduction in radiation burden. The criterion of a very diminished portion of the non-penetrating radiation is granted by the isomers (e.g. ^{99m}Tc, ^{113m}In, ^{87m}Sr and ^{109m}Ag). Selection possibilities, however, are narrowed by the energy range of gamma radiation appropriate for scintigraphy of between about 50 and maximum 500 keV, as well as by technical (e.g. availability of a cyclotron or reactor), radiochemical (length of production of a radiopharmaceutical in relation to physical half-life) and financial considerations.

Further possibilities for the reduction of radiation burden exist in the alteration of the distribution pattern and kinetics of the radiopharmaceutical. Examples are the blocking of critical organs (e.g. of thyroid with iodine in order to avoid intake of ^{131}I iodide as a splinter portion of a ^{131}I-labelled pharmaceutical, or of kidney with inactive chlormerodrine in the application of radioactive chlormerodrine in brain scintigraphy) and excretion intensification after termination of examination.

4.2. Calculation of dose reduction

According to reports by Pfannenstiel [22], conventionally applied radiopharmaceuticals were substituted with those that correlate to the above-discussed criteria for the selection of radionuclides and radiopharmaceuticals for dose reduction. Dose calculations were conducted for the respective critical or examined organs as well as for the gonads, whereby the mean administered activities of the conventional radiopharmaceuticals were extracted from the Berlin findings of 1970. Results of the calculations have been compiled in Table VI. Accordingly, the

TABLE VII. COMPARISON OF GSD (BERLIN/WEST) FOR 1970 FOR CONVENTIONALLY APPLIED AND ASSUMEDLY SUBSTITUTED RADIONUCLIDES AND RADIOPHARMACEUTICALS

Radionuclide	Radiopharmaceutical	Diagnostic method	Mean admin. activity (μCi)	D_j (mrad) for convent. radionucl.	D_j (mrad) for substit. /radiopharm.
^{131}I	Iodide	Thyr. funct. /scan	43	2.84 E-2	
^{99m}Tc	Pertechnetate	Thyr. scan.	1 000		6.65 E-2
^{132}I	Iodide	Thyr. funct. (30%)	25		
^{198}Au	Colloid	Liver scan	240	1.41 E-2	
^{99m}Tc	S-colloid		1 500		1.59 E-3
^{58}Co	Vitamin B_{12}	Schilling test	0.9	3.00 E-6	
^{57}Co	Vitamin B_{12}		0.5		3.70 E-4
^{131}I	MAA	Lung scan	220	1.80 E-3	
^{99m}Tc	MAA	Lung scan (66%)	3 000		
^{133}Xe		L. funct. /scan (34%)	15 000		2.60 E-4
^{131}I	HSA	Blood volume	10 [a]	7.16 E-5 (Blood volume, Placental local., Myelography combined)	5.00 E-6 (Blood volume, Placental local., Myelography combined)
^{99m}Tc	HSA		100		
^{131}I	HSA	Placental local.	10 [a]		
^{99m}Tc	HSA		500		
^{131}I	HSA	Myelography	100 [a]		
^{99m}Tc	HSA		1 500		
^{197}Hg	BMHP	Spleen scan	360	1.27 E-2	
^{99m}Tc	S-colloid		1 500		8.10 E-4
^{85}Sr	nitrate	Bone scan	330	1.68 E-3	
^{99m}Tc	Polyphosphate		10 000		1.06 E-2
^{203}Hg	BMHP	Kidney scan	400	1.75 E-5	
^{99m}Tc	DTPA		3 000		1.60 E-7
^{203}Hg	Chlormerodrine	Kidney scan	180	7.48 E-6	
^{99m}Tc	DTPA		3 000		4.00 E-7
			Total:	0.059	0.080

TABLE VII. (cont.)

Radionuclide	Radiopharmaceutical	Diagnostic method	Mean admin. activity (μCi)	D_j (mrad) for convent. radionucl.	substit. /radiopharm.
^{131}I	o-iodohippurate	Kidney function	73	7.99 E-4	
^{99m}Tc	Pertechnetate	Thyroid scan Brain scan Salivary gland scan Stomach scan	4500	4.39 E-2	
^{99m}Tc	S-colloid	Liver scan Spleen scan	980	1.13 E-4	
^{87m}Sr	chloride, nitrate	Bone scan	1600	1.68 E-3	
^{75}Se	Selenomethionine	Pancreas scan	310	7.58 E-2	
^{113m}In	Colloid	Liver scan	1100	1.49 E-4	
^{51}Cr	EDTA	Kidney function	80	6.59 E-5	
^{59}Fe	chloride	Iron metabolism	23	1.95 E-3	
			Total:	0.124	
			Total GSD:	0.183	0.204

(%) portion of the total number of respective examinations.

[a] Mean values from published data [4].

substitution of radionuclides and radiopharmaceuticals leads to a reduction in energy dose in the critical or examined organ of up to a factor of 1000. Thereby the dose reduction is expressed by a dose reduction coefficient d as a proportion of dose values for the substituted and the conventional radionuclides and radiopharmaceuticals. The maximum dose reduction coefficient for gonad doses resulted in 0.01; by substitution of ^{131}I iodide with ^{99m}Tc pertechnetate and ^{132}I iodide; however, dose reduction in the thyroid of about 1% is connected with a gonad dose increase of more than twice the amount. For this reason, and because of the relatively large number of thyroid examinations in the total number of nuclear medical examinations, the GSD remains practically unchanged (increase about 10%; see Table VII).

SUMMARY AND DISCUSSION

According to the findings in Berlin (West) on the application frequency of readily available radioactive substances in nuclear medicine, a yearly increase rate of about 20% in examinations can be expected. Thereby the GSD has just about doubled within the period 1968 to 1970.

Even with the assumption that by 1970 conventional radionuclides and radiopharmaceuticals could have been substituted from the viewpoint of dose reduction, the value of GSD for 1970 would have remained practically unchanged. Acknowledging a constant examination increase rate, it is to be expected that the portion of nuclear-medically dependent GSD, now approximately 0.2 mrad/a, resulting from the medical application of ionizing radiation of the present total GSD of about 30 to 50 mrad/a will remain insignificant in the future. Nevertheless, it must remain the goal of endeavours to reduce substantially the radiation burden of the individual patient. Calculations of energy dose for critical or examined organs have demonstrated the size of possible dose reductions by the suitable selection of radionuclides or radiopharmaceuticals. This dose reduction can definitely assume a value of 1000 and should be considered more than ever by selection of radiopharmaceuticals.

REFERENCES

[1] GARBY, L., Straldoser fran Radioaktiva Ämnen i Medicinskt Bruk, Information till sjukhusens isotopkommittéer, Statens Stralskyddsinstitut Stockholm.

[2] HINE, G.J., JOHNSTON, R.E., Absorbed dose from radionuclides, J. Nucl. Med. 11 (1970) 468-69.

[3] INTERNATIONAL COMMISSION ON RADIOLOGICAL PROTECTION, Protection of the Patient in Radionuclide Investigations, Radiation Protection, Publication 17, Pergamon Press, Oxford (1971).

[4] KAUL, A., OEFF, K., ROEDLER, H.D., VOGELSANG, T., Die Strahlenbelastung von Patienten bei der nuklearmedizinischen Anwendung offener radioaktiver Stoffe, Informationsdienst für Nuklearmedizin, Klinikum Steglitz der Freien Universität Berlin (1973) 700 pp.

[5] KAUL, A., OEFF, K., ROEDLER, H.D., VOGELSANG, T., Radiopharmaceuticals - Biokinetic Data and Results of Radiation Dose Calculations, Informationsdienst für Nuklearmedizin, Klinikum Steglitz der Freien Universität Berlin (1973) 654 pp.

[6] ROEDLER, H.D., KAUL, A., BERNER, W., KOEPPE, P., GLAUBITT, D., "Development of an extended formalism for internal dose calculation and practical application to several biologically distributed radioelements", Assessment of Radioactive Contamination in Man (Proc. Symp. Stockholm, 1971), IAEA, Vienna (1972) 515-41.

[7] ROEDLER, H.D., KAUL, A., Neuere Berechnungen der Strahlendosis durch inkorporierte radioaktive Stoffe nach dem erweiterten Konzept der absorbierten Bruchteile: Formal exakte und Näherungslösung, Atomkernenergie 21 4 (1973) 249-53.

[8] ROEDLER, H.D., Strahlenbelastung durch Radiopharmaka - Entwicklung eines mathematischen Dosiskonzepts und Ergebnisse von Neuberechnungen der Energiedosis, Dissertation, Freie Universität Berlin (1974).

[9] LOEVINGER, R., BERMAN, M., A Schema for Absorbed-Dose Calculations for Biologically-Distributed Radionuclides, MIRD, Pamphlet 1, J. Nucl. Med. 9, Suppl. 1 (1968).

[10] SNYDER, W.S., FISHER, H.L., FORD, M.R., WARNER, G.G., Estimates of Absorbed Fractions for Monoenergetic Photon Sources Uniformly Distributed in Various Organs of a Heterogeneous Phantom, MIRD, Pamphlet 5, J. Nucl. Med. 10, Suppl. 3 (1969).

[11] DILLMAN, L.T., Radionuclide Decay Schemes and Nuclear Parameters for Use in Radiation-Dose Estimation, MIRD, Pamphlet 4, J. Nucl. Med. 10, Suppl. 2 (1969).

[12] DILLMAN, L.T., Radionuclide Decay Schemes and Nuclear Parameters for Use in Radiation-Dose Estimation, part 2, MIRD, Pamphlet 6, J. Nucl. Med. 11, Suppl. 4 (1970).

[13] DILLMAN, L.T., personal communications (1972).

[14] WOLF, R., personal communications (1969).

[15] COLARD, J.F., VERLY, W.G., HENRY, J.A., BOULANGER, R.R., Fate of the iodine radioisotopes in the human and estimation of the radiation exposure, Health Phys. 11 (1965) 23-35.

[16] POPE, R.A., Whole body counter in routine tests of hyperthyroidism using iodine-131, Acta Radiol. 4 (1966) 187-208.

[17] OBERHAUSEN, E., MUTH, H., Ganzkörperbelastung bei Verabreichung von J-131, Technical Report, Institut für Biophysik, Universität des Saarlandes Homburg (Saar) (1968).

[18] MYANT, N.B., The radiation dose to the body during treatment of thyrotoxicosis by I-131, Minerva Nucl. 8 (1964) 207-10.

[19] OSBORN, S.B., SMITH, E.E., The genetically significant radiation dose from the use of X-rays in England and Wales, A preliminary survey, Lancet 16 (1956) 949.

[20] AURAND, K., HINZ, G., Erhebungen über die Entwicklung der Anwendung offener Radionuklide in Diagnostik und Therapie, Bundesgesundheitsblatt 13/3, (1970) 30-33.

[21] HINZ, G., WEIL, H., "Entwicklung der Anwendung offener Radionuklide in Berlin (West)", 53. Tagung der Deutschen Röntgengesellschaft, Gesellschaft für medizinische Radiologie, Strahlenbiologie und Nuklearmedizin e.V., Stuttgart, 1972, Deutscher Röntgenkongress 1972, Suppl. of Fortschr. Röntgenstr., Stuttgart: Thieme (1973).

[22] PFANNENSTIEL, P., personal communications (1973).

IAEA-SM-184/33

GENETICALLY SIGNIFICANT DOSE TO THE POPULATION IN INDIA FROM X-RAY DIAGNOSTIC PROCEDURES

S.J. SUPE, S.M. RAO, S.G. SAWANT,
G. JANAKIRAMAN, A. GANESH
Division of Radiological Protection,
Bhabha Atomic Research Centre,
Trombay, Bombay,
India

Abstract

GENETICALLY SIGNIFICANT DOSE TO THE POPULATION IN INDIA FROM X-RAY DIAGNOSTIC PROCEDURES.

The genetically significant dose from X-ray diagnostic procedures has been measured in many countries. In the national programme for evaluation of the genetically significant dose to the population of India it was decided to estimate the same for the population of Eastern and Northern Regions of India and of Maharashtra and Tamil Nadu States, representing Western and Southern Regions. From these regions at least one major city, two urban cities and four to five district hospitals were selected for data collection so that the various types of medical facilities and radiographic techniques are represented. Age and sex distribution of patients undergoing various diagnostic examinations was collected from five sample districts of Tamil Nadu State, nine sample districts of Maharashtra, eight sample districts of Northern Region and eleven sample districts of Eastern Region. So far the age and sex distribution for more than 120 000 patients has been collected. Gonad dose measurements were carried out in major radiodiagnostic centres in these selected places. Gonad dose measurements have been carried out on as many as 15 000 patients. The child-expectancy factors were estimated from the data of Indian Census results on fertility rates and mortality rates. Evaluation of the annual genetically significant dose from the data yielded a much lower value compared to other countries. The annual genetically significant doses and gonad doses in a few examinations for India are compared with those from other countries.

INTRODUCTION

The genetically significant dose arising from diagnostic radiology procedures has been measured in many countries [1 - 24]. No attempt has hitherto been made to measure this dose in India. A study of the diagnostic techniques currently in vogue in India, coupled with the non-availability of sophisticated radiodiagnostic equipment such as image intensifiers, suggest that the genetically significant dose could be much higher in this country. On the other hand, the fact that a much smaller percentage of the population undergoes diagnostic X-ray examinations would tend to lower the annual genetically significant dose. For evaluating the genetically significant dose to the population of India it was decided to assess the same for the population in a few randomly selected states. Two states, viz. Maharashtra and Tamil Nadu, and the Northern and Eastern Regions were chosen for these studies. The results for these states and regions are presented below.

The annual genetically significant dose to the population is defined as the average of the individual gonad doses, each weighted for the expected number of children conceived subsequent to the exposure. This is mathematically expressed as

$$G = \frac{\sum_i \sum_j \left[D_{ij} N_{ij} P_i^{(M)} + D_{ij} N_{ij} P_i^{(F)} \right]}{\sum_i \left[N_i P_j^{(M)} + N_i P_j^{(F)} \right]}$$

where, D_{ij} is the average gonad dose received by the patients in the ith age group as a result of undergoing the jth type of examination, N_i is the total number of persons in the ith age group, P_i is the average child expectancy factor for the persons in the ith age group and N_{ij} is the number of persons in the ith age group undergoing the jth type of examination per year. (M) and (F) denote male and female, respectively. As for foetuses, half are assumed to be males and other half females and the annual genetically significant dose is evaluated separately taking the frequency of examination of pregnant mothers into account.

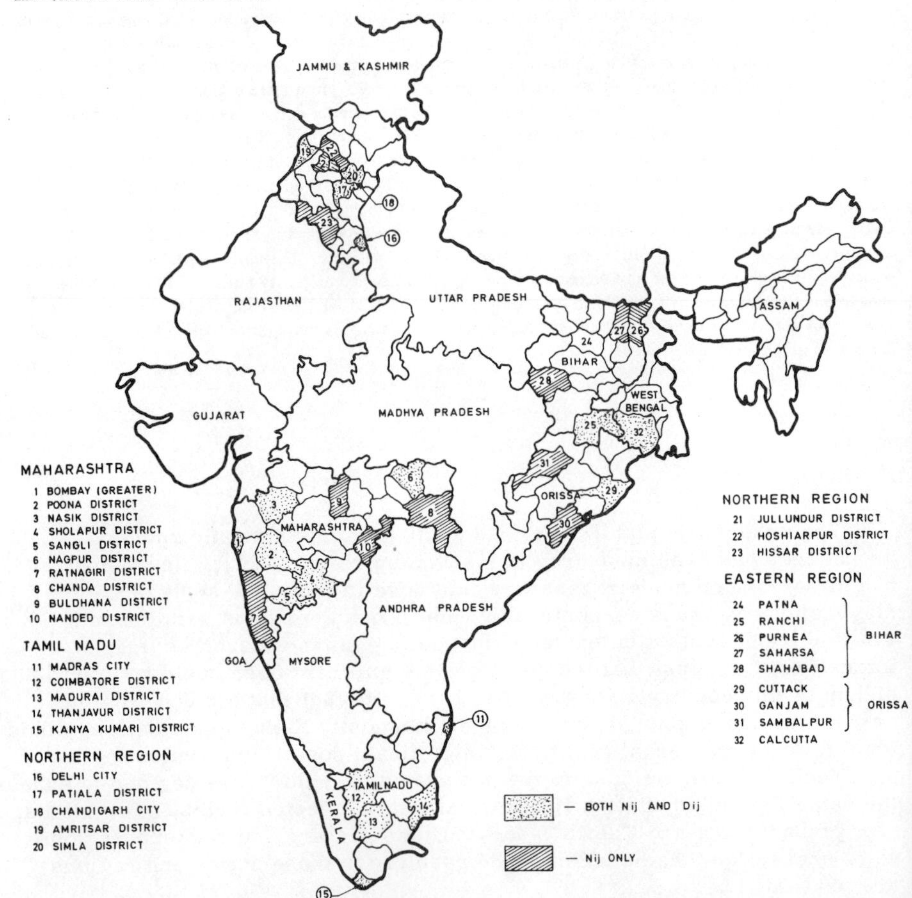

FIG. 1. Map of India showing various districts selected for collection of data on dose to patients and number of patients undergoing various types of examinations.

GONAD DOSE MEASUREMENTS

The measurements of gonad doses were carried out in the following manner. For the State of Maharashtra investigations were confined to three hospitals from the capital city of Bombay, two hospitals each from Poona and Nagpur districts and one major hospital from each of the remaining three sample districts (Fig. 1). In the State of Tamil Nadu work was done in four hospitals from the capital city of Madras and in one hospital each from four sample districts. In the Northern Region four hospitals from Delhi, one hospital each from four sample districts, and from the Eastern Region three hospitals in Calcutta and one hospital each from three sample districts were selected for these measurements. All of these were general hospitals except for one in Bombay, which was a children's hospital. It was assumed that the radiographic techniques followed in each of these hospitals were more or less representative of the techniques followed in other hospitals in the same district. For radiographic examinations the measurements were made, as far as possible, with patients during actual radiography. In line with procedures adopted by earlier workers for field measurements [8, 12, 25], ionization-chamber type X-ray dosimeters developed in the Division of Radiological Protection were used for gonad dose measurements. Whenever the gonads were in the direct beam, personnel monitoring films were used in place of the ionization chamber in order to avoid the shadow of the chamber on the radiograph. In the case of males the chamber was placed directly in contact with the scrotum. In the case of females, however, the chamber was placed directly on the skin above the ovaries and the skin dose measured. The position of the ovaries was assumed to be 5 centimetres to the side of the navel and 5 centimetres below it at a depth of three fifths of the thickness of the body in that section. The ovary doses were obtained by multiplying the values of the skin doses by appropriate ovary-to-surface dose conversion factors. Four homogeneous tissue-equivalent Mix-D phantoms representing four age groups (viz. 0-4, 5-9, 10-14 years and adult) were constructed and used for obtaining the conversion factors under various operating conditions and for different examinations. During the field measurements an attempt was made to collect data for at least 10 examinations of each type so that a fairly representative average value could be obtained for the value of the exposure involved in each type of examination. The results of the measurements and the number of cases for which these measurements were carried out are presented for a few types of examinations in Table I. Whenever gonad dose data could not be obtained directly on patients because a particular type of examination was not being done during the period of measurements, they were determined by measurements on phantoms. Ionization chambers could not be used directly on patients undergoing fluoroscopic examination because continuous movement of the patient caused serious fluctuations in the dosimeter readings. Hence the measurements were made on phantoms under identical conditions. As in earlier work [3, 26, 27], the foetal dose was assumed to be the same as the ovary dose of the mother.

VITAL STATISTICS

The age and sex distribution of patients undergoing various types of examination was determined by collecting data on the entire radiological

TABLE I. GONAD DOSE (mrem) RECEIVED BY ADULTS IN VARIOUS TYPES OF EXAMINATIONS

Type of examination	Sex	Bombay City only		Madras City only		Delhi City only		Calcutta City only	
		Average dose	No. of patients	Average dose	No. of patients	Average dose	No. of patients	Average dose	No. of patients
Lumbar spine AP	M	399.0	76	24.3	9	109.7	10	30	1
	F	292.0	16	119.0	9	67.1	9	92.8	5
Lumbar spine Lat.	M	170.0	41	847.0	25	10.5	11	30.5	4
	F	389.0	23	378.0	24	158.1	10	359.8	5
Pelvis	M	1210.0	14	539.0	34	432.3	16	953.9	8
	F	191.0	12	55.8	18	26.2	6	95.8	7
Abdomen	M	179.0	90	69.4	35	30.5	39	70.6	33
	F	249.0	59	92.4	15	38.4	39	98.1	27
Chest PA	M	0.28	125	0.21	83	0.36	42	0.39	244
	F	3.13	46	4.52	63	8.1	33	5.9	134

TABLE I. (cont.)

Type of examination	Sex	Maharashtra (excluding Bombay)		Tamil Nadu (excluding Madras)		Northern Region (excluding Delhi)		Eastern Region (excluding Calcutta)	
		Average dose	No. of patients	Average dose	No. of patients	Average dose	No. of patients	Average dose	No. of patients
Lumbar spine AP	M	37.5	21	75.6	17	78.1	9	51.3	11
	F	89.4	15	64.6	4	80.7	5	158.8	6
Lumbar spine Lat.	M	267.0	31	23.8	16	62.9	9	77.0	12
	F	195.0	18	248.0	4	333.5	8	654.9	7
Pelvis	M	732.0	21	335.0	8	581.9	26	1654.2	26
	F	73.8	15	47.0	4	96.2	16	155.3	22
Abdomen	M	79.6	85	56.8	36	43.6	176	59.7	70
	F	1020.0	33	244.0	39	96.2	151	204.2	36
Chest PA	M	0.49	171	0.39	279	1.41	389	0.20	148
	F	1.62	97	8.96	162	4.0	281	5.67	75

TABLE II. GONAD DOSE MEASUREMENTS CARRIED OUT IN VARIOUS DISTRICTS OF MAHARASHTRA AND TAMIL NADU AND IN THE NORTHERN AND EASTERN REGIONS

Sl. No.	District	Total No. of X-ray institutions contacted for N_{ij} data	Total No. of X-ray institutions supplying N_{ij} data	Total No. of patients for whom N_{ij} data was obtained		Total	Total No. of patients on whom gonad dose measurements were carried out
				Supplied by institutions	Obtained from oral enquiries		
				MAHARASHTRA			
1.	Bombay	215	83	9990	8173	18 163	2207
2.	Poona	53	33	4502	1056	5558	701
3.	Nagpur	22	19	4970	207	5177	913
4.	Nasik	22	15	1435	358	1793	225
5.	Sangli	21	11	660	325	985	286
6.	Sholapur	14	10	535	534	1069	141
7.	Nanded	5	2	231	232	463	-
8.	Buldhana	5	2	121	73	194	-
9.	Ratnagiri	9	7	515	72	587	-
	TOTAL	366	182	22 959	11 030	33 989	4473
				TAMIL NADU			
1.	Madras	56	39	2892	2171	5063	1035
2.	Coimbatore	45	34	3119	385	3504	592
3.	Madurai	56	43	4378	541	4919	1330
4.	Thanjavur	24	20	3365	196	3561	517
5.	Kanyakumari	15	15	1445	-	1445	227
	TOTAL	196	151	15 199	3293	18 492	3701
				NORTHERN REGION			
1.	Delhi	76	45	14 799	3888	18 687	1016
2.	Amritsar	17	12	2492	99	2591	659
3.	Patiala	18	11	3171	596	3767	800
4.	Chandigarh	5	4	6493	51	6544	1091
5.	Himachal Pradesh	10	8	1301	52	1353	450
6.	Hissar	8	8	786	-	786	-
7.	Hoshiarpur	4	3	270	-	270	-
8.	Jullundar	13	8	2387	96	2483	-
	TOTAL	151	99	31 699	4782	36 481	4016
				EASTERN REGION			
1.	Cuttack	9	9	2108	-	2108	603
2.	Patna	19	13	4328	497	4825	894
3.	Ranchi	16	15	2937	75	3012	382
4.	Calcutta	139	99	16 382	2364	18 746	1406
5.	Saharsha	5	2	123	-	123	-
6.	Shahabad	8	3	791	-	791	-
7.	Ganjam	6	2	374	-	374	-
8.	Sambalpur	7	4	742	-	742	-
	TOTAL	209	147	27 785	2936	30 721	3285

work of the sample districts for three randomly chosen weeks. A questionnaire designed to give information on the age, sex, type of examination and number of radiographs taken for each patient was sent to all the hospitals in the sample districts (Fig. 1). The number of hospitals from various districts to which the forms were sent and that returned them after completion is presented in Table II. These hospitals gave information on 22 595 cases in Maharashtra State, 15 199 cases in Tamil Nadu, 31 699 cases from the Northern Region and 27 785 cases from the Eastern Region. An auxiliary survey was conducted subsequently to collect the information from those hospitals that did not return the forms. The earlier data were supplemented by the data subsequently collected during personal interviews. It was found that these non-response data are about 18% of the total N_{ij} data. A differential distribution of age, sex and type of examination was calculated from the completed data, which in turn yielded the distributions for the entire year. The final distribution for two of the important examinations are shown in Figs 2-5. These figures indicate that the age and sex distributions for these examinations are nearly identical in the four regions.

Not all hospitals supplied data on the pregnancy of the female patients. To evaluate the number of foetuses corresponding to each type of examination, the records for one year from a large hospital in Maharashtra that

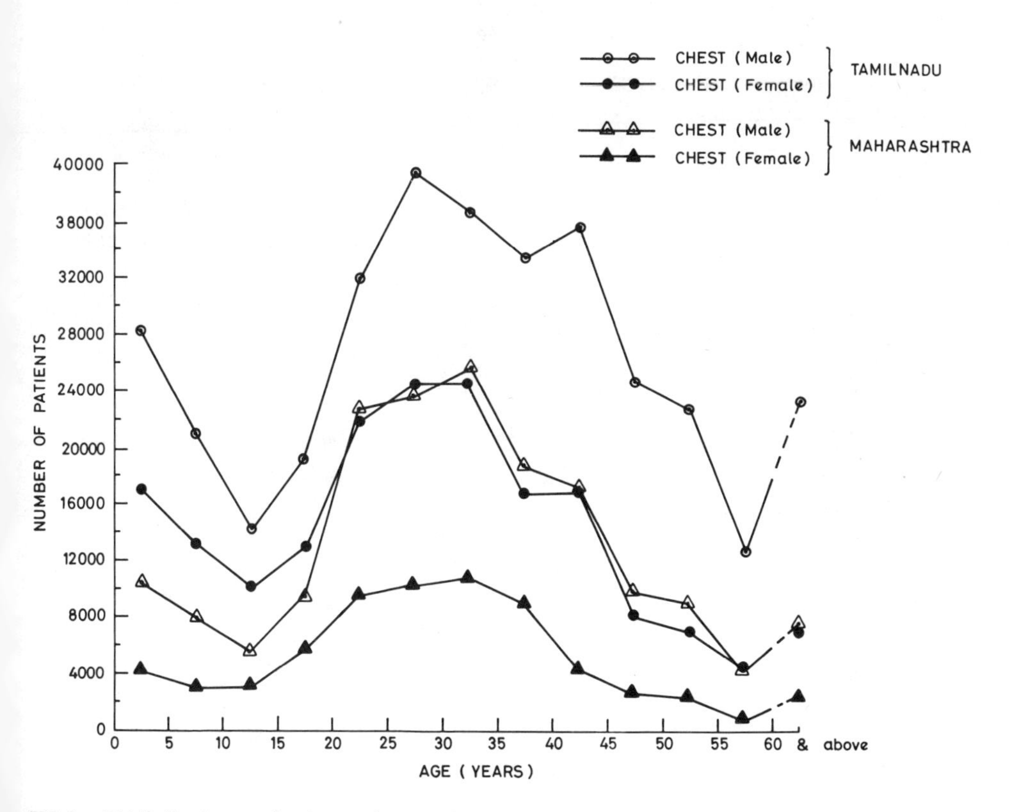

FIG. 2. Distribution by age of patients undergoing chest X-ray examination in Maharashtra and Tamil Nadu.

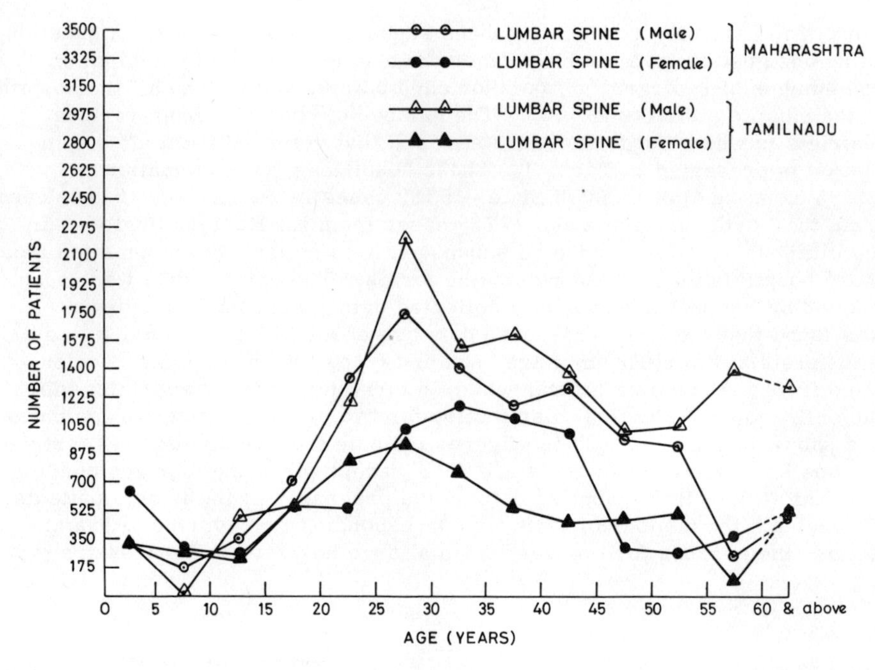

FIG. 3. Distribution by age of patients undergoing lumbar spine X-ray examination in Maharashtra and Tamil Nadu.

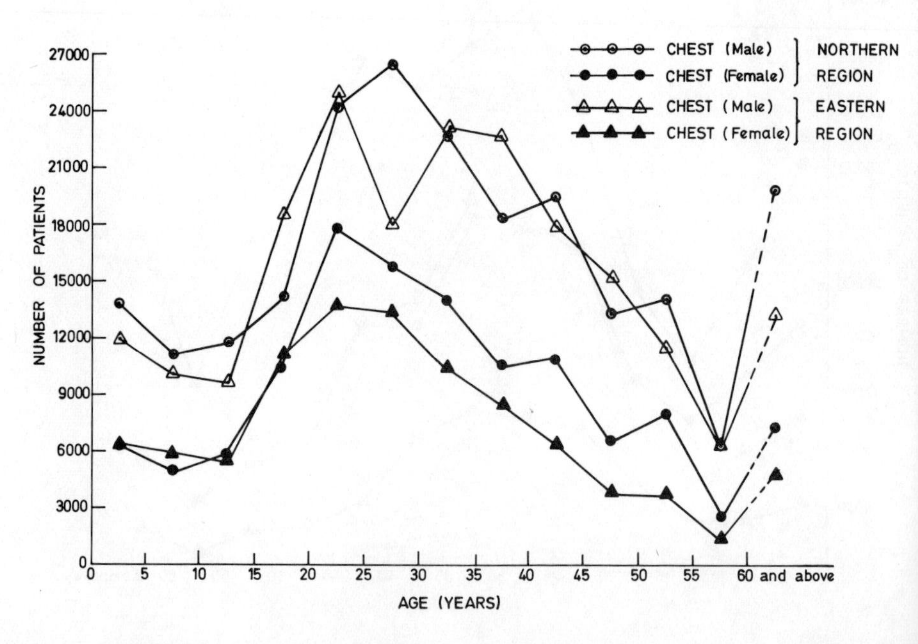

FIG. 4. Distribution by age of patients undergoing chest X-ray examination in the Northern and Eastern Regions

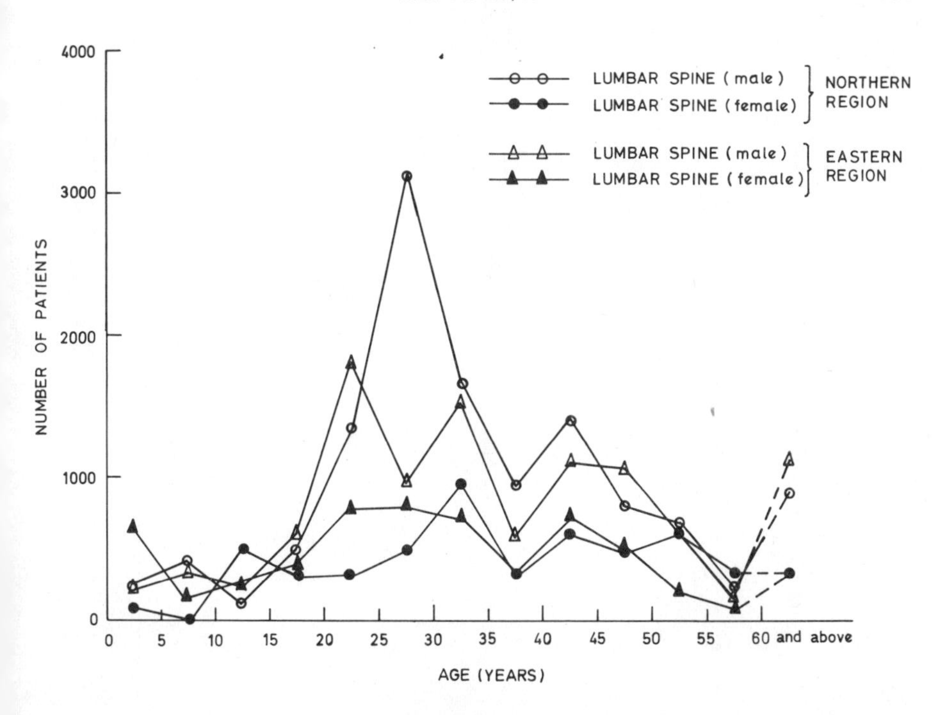

FIG. 5. Distribution by age of patients undergoing lumbar spine X-ray examination in the Northern and Eastern Regions.

maintained this information were scanned and the ratio of the total number of pregnant patients to the total number of female patients was found for each type of X-ray examination. These ratios were assumed to be the same for the hospitals in Maharashtra and were used to evaluate the number of pregnancies present in the females undergoing various types of examinations. Apparently this information was not kept in any of the hospitals from Tamil Nadu and two other regions. It was not considered correct to assume that this ratio would be the same for Maharashtra State and the other regions as the child expectancy factors for females in these regions differ appreciably. The ratios for the other regions were therefore obtained in the following fashion. The fertility factor as given in the census results gave the fraction of pregnant females in a particular age group per year. This factor was multiplied by 0.77 (ratio of period of pregnancy to period of one year) to give the ratio of pregnant women at any time to the total number of women in that age group. A summation of these ratios was carried out over all age groups, weighting the factor by the number of female patients in various age groups, divided by the total female patients of the region. This evaluation gives a value of 8.0 pregnant patients out of every 100 patients for Tamil Nadu, 9 for the Eastern Region and 10 for the Northern Region, which compares well with the value of 9.1 (220 out of 2414) for Maharashtra.

TABLE III. NUMBER OF MALES AND FEMALES IN VARIOUS AGE GROUPS (N_i) AND THEIR CHILD EXPECTANCY FACTORS (P_i)

Age group	Maharashtra N_i (thousands)		Tamil Nadu N_i (thousands)		Maharashtra P_i		Tamil Nadu P_i	
	Male	Female	Male	Female	Male	Female	Male	Female
0 - 4	2994	2943	2623	2510	4.577	4.116	3.2785	3.3937
5 - 9	2890	2845	2622	2624	4.836	4.368	4.0837	4.1844
10 - 14	2328	2084	2212	2126	4.958	4.486	4.2967	4.3130
15 - 19	1679	1533	1589	1538	4.926	4.214	4.4539	4.0507
20 - 24	1718	1781	1589	1710	4.745	3.396	4.1797	3.1349
25 - 29	1774	1690	1576	1652	3.815	2.651	3.2548	1.9872
30 - 34	1511	1332	1329	1313	2.625	1.660	2.0487	1.0325
35 - 39	1298	1097	1260	1222	1.613	0.9016	1.351	0.3805
40 - 44	1067	931	1055	1009	0.874	0.0874	0.3744	0.0639
45 - 49	897	764	888	802	0.342	0.0594	0.0627	0.0029
50 - 54	744	649	740	674	0.0579	0.0082	0.0028	0.0011
55 - 59	505	415	411	411	0.0075	0.00	0.0011	0.0000
60 and over	1023	1061	918	878	0.00	0.000	0.000	0.000

TABLE III. (cont.)

Age group	Northern Region N_i		Eastern Region N_i		Northern Region P_i		Eastern Region P_i	
	Male	Female	Male	Female	Male	Female	Male	Female
	(thousands)		(thousands)					
0 - 4	3506	3161	9311	9166	4.96	4.825	4.2465	4.1026
5 - 9	3729	3280	10582	10218	5.316	5.362	4.9274	4.8868
10 - 14	2949	2561	7460	6392	5.376	5.417	5.0118	4.9935
15 - 19	2071	1694	4934	4644	5.417	5.390	5.0653	4.8535
20 - 24	1930	1686	5016	5239	5.384	4.795	4.9127	4.1156
25 - 29	1840	1555	5482	4969	4.777	3.50	4.1391	2.9236
30 - 34	1522	1236	4688	4094	3.482	2.134	2.9155	1.7844
35 - 39	1217	1009	3928	3341	2.116	1.056	1.7630	0.8990
40 - 44	1130	926	3220	2973	1.048	0.368	0.8801	0.3593
45 - 49	866	655	2608	2224	0.365	0.062	0.3517	0.0913
50 - 54	873	591	2225	1917	0.062	0.0	0.0903	0.0
55 - 59	436	286	1323	1133	0.0	0.0	0.0	0.0
60 and over	1377	914	2702	2940	0.0	0.0	0.0	0.0

The distributions giving the total number of males and females in various age groups of the general population were obtained from the data supplied by the Commissioner, National Census Survey, Maharashtra State [28]. For other Regions these data were obtained from census results of India [29 - 31] These are presented in Table III.

The values of subsequent child expectancy factors for males and females for various age groups were calculated from the data available in a National Sample Survey for Maharashtra State [32]. For evaluating these values for other Regions use was made of fertility factors published by the National Census. These values are also tabulated in Table III.

DISCUSSION

It is observed that there are large variations in the doses resulting from any one particular type of examination. These variations in gonad dose from case to case in any particular type of examination could result from variations in techniques such as kV, filtration used, beam angulation, shielding provided and other operating conditions.

From the data presented in Table I it is seen that for most of the examinations the average values for Bombay are significantly higher than those for other cities. The measurements in Bombay City [33] were made a few years prior to those in other cities [34] and the significant reduction in the average gonad dose values in other cities can be attributed to a country-wide radiation safety programme during the intervening period, which emphasized the importance of reducing the dose both to the patient and to the doctors and technicians. The effectiveness of the radiation safety programme has manifested itself in a sizeable reduction in the average gonad dose resulting from each type of examination.

Table IV shows the contribution to the annual genetically significant dose from males and females for radiographic and fluoroscopic examinations in various districts from the two states and two regions. In general, the dose contribution from examinations is lower for the female population than for the male population in the case of radiography. This is partly due to the fact that the male gonad doses for most of the radiographic examinations are larger than the corresponding ovary doses. In addition, under Indian conditions female patients do not undergo X-ray examination as often as males do. The contribution from fluoroscopic examinations, however, shows a reverse picture. This is because the female gonads are always in the direct beam during fluoroscopy in special procedures like barium meal and barium enema and these special procedures contribute more than 90% of the dose due to fluoroscopy. It can also be observed from this table that districts like Madras, Bombay, Nagpur, Sholapur, Delhi, Simla, Chandigarh and Calcutta contribute larger annual genetically significant doses. This is because of the greater frequency of radiographic examinations resulting from the larger number of radiological facilities available in these urban districts. In fact, a significant proportion of radiological examinations conducted in these urban centres consist of the examination of a sizeable number of patients from the adjoining smaller districts. The annual genetically significant dose for various districts was weighted according to the population and from these data the annual genetically significant dose for each of the two states and the two regions was estimated

TABLE IV. CONTRIBUTION TO THE ANNUAL GENETICALLY SIGNIFICANT DOSE IN MILLIREMS DUE TO DIAGNOSTIC RADIOLOGY

District	Male		Female			Annual genetically significant dose	District contribution to states' annual genetically significant dose
	Radio-graphy	Fluoro-scopy	Radio-graphy	Fluoro-scopy	Foetal dose		
				TAMIL NADU			
Madras	4.1469	0.0048	1.4391	0.1646	0.2605	6.0159	0.3098
Madurai	0.3064	0.0003	0.4502	0.0418	0.0865	0.8852	0.0845
Coimbatore	0.6209	0.0005	0.2171	0.0714	0.0475	0.9574	0.1010
Thanjavur	0.1403	0.0004	0.0534	0.0022	0.0074	0.2037	0.0196
Kanyakumari	0.1137	0.0023	0.1215	0.2185	0.0780	0.5340	0.0158
Tamil Nadu State						0.7664	
				MAHARASHTRA			
Bombay	6.059	0.447	1.308	2.592	0.240	10.646	1.118
Poona	0.725	0.043	0.158	0.743	0.201	1.870	0.117
Nagpur	1.875	0.001	1.259	0.482	0.435	4.052	0.155
Nasik	0.822	0.006	0.083	0.009	0.037	0.957	0.152
Sangli	0.281	0.004	0.157	0.041	0.007	0.490	0.015
Nanded	0.212	0.014	0.032	0.081	0.023	0.362	0.043
Ratnagiri	0.023	0.013	0.018	0.081	0.006	0.141	0.016
Buldhana	0.014	0.003	0.001	0.001	0.002	0.021	0.003
Sholapur	0.110	0.147	0.036	2.619	0.014	2.926	0.138
Maharashtra State						1.894	
				NORTHERN REGION			
Delhi	1.9751	0.0153	1.0166	0.1785	0.2590	3.4445	0.3257
Amritsar	0.7500	0.0031	0.2000	0.0037	0.0417	0.9985	0.0426
Patiala	1.0474	0.0156	0.5999	1.0706	0.3271	3.0606	0.0868
Chandigarh	7.1332	0.7060	3.0894	8.5605	3.5231	23.0122	0.1377
Himachal Pradesh	0.6457	0.0017	0.2240	0.0114	0.0359	0.9187	0.0739
Hissar	0.0533	0.0008	0.0426	0.0004	0.0077	0.1048	0.0052
Hoshiarpur	0.0064	0.0009	0.0003	0.0018	0.0003	0.0097	0.0002
Jullundar	0.4307	0.0224	0.1337	0.0017	0.0273	0.6158	0.0208
Northern Region						0.8772	
				EASTERN REGION			
Calcutta	5.4412	0.0089	3.4016	0.6293	0.1908	9.6718	0.2478
Cuttack	1.2770	0.0080	0.5430	0.1184	0.0098	1.9562	0.0611
Patna	0.7365	0.0161	0.4203	0.0298	0.0257	1.2284	0.0356
Ranchi	0.2963	0.0038	0.0785	0.0018	0.0065	0.3869	0.0082
Shahabad	0.0254	0.0078	0.0101	0.0454	neg	0.0887	0.0028
Ganjam	0.7832	0.0008	0.1531	neg	neg	0.9371	0.0175
Sambalpur	1.2204	0.0004	0.4168	0.0012	neg	1.6388	0.0246
Saharsha	0.0006	neg	0.0103	neg	neg	0.0109	0.0002
Eastern Region						0.9337	

TABLE V. ANNUAL GENETICALLY SIGNIFICANT DOSE FROM DIAGNOSTIC RADIOLOGY IN VARIOUS COUNTRIES

Country	Year	Annual number of X-ray examinations per person of the population (excluding dental and fluoroscopy)	Total annual genetically significant dose (mrem)	Annual genetically significant dose for one examination per person
Argentina (Buenos Aires)	1950-59	0.270	37.0	137.0
Denmark	1956-58	0.260	27.5	105.8
Egypt				
Alexandria	1956-60	0.036	7.0	194.4
Cairo	1955-61	0.040	7.0	175.0
Fed. Rep. Germany (Hamburg)	1957-58	0.560	17.7	31.6
France	1957-58	0.150	58.2	388.0
Italy (Rome)	1957	0.500	43.4	86.8
Japan	1958-60	0.410	39.0	95.1
Japan	1969	0.610	25.7	42.1
Netherlands (Leiden)	1959-60	0.350	6.8	19.4
Netherlands	1967	0.810	20.0	24.7
New Zealand	1969	0.400	13.7	34.3
Norway	1958	0.390	10.0	25.6
Sweden	1955-57	0.290	37.8	130.3
Thailand	1970	0.039	5.2-1.3	133.3-33.3
United Kingdom	1957-58	0.280	14.1	50.4
United Kingdom (Sheffield)	1964	0.310	8.6	27.7
USA (National)	1964	0.475	55.0	115.8
USSR	1964	0.171	27.0	157.9
Yugoslavia (Slovenia)	1960-63	0.594	9.1	15.3
India	1967-72	0.035	1.11	31.7
Maharashtra	1967-68	0.014	1.89	135.0
Tamil Nadu	1969-70	0.025	0.77	30.8
Northern Region	1970-71	0.024	0.88	36.7
Eastern Region	1971-72	0.051	0.93	18.2

These values are 1.89, 0.77, 0.88 and 0.93 mrem/a per person for the States of Maharashtra [35], Tamil Nadu [34] and the Northern and Eastern Regions, respectively. The higher value for Maharashtra State is due to (1) the higher values of gonad doses for this state, and (2) the higher percentage of fluoroscopic examinations for Maharashtra due to a severe shortage of radiographic films during the period of investigation. The annua genetically significant dose value for the state as a whole is close to the values obtained for the smaller districts, which show the predominance of overall conditions approximating those typical of a rural population.

Table V compares the values of annual genetically significant doses in various countries. The values for the two states and two regions of India are quite small compared with those of other countries. This is probably due to the relatively lower frequency of radiological examinations undertaken in India.

ACKNOWLEGEMENTS

We owe a great deal to Shri G. Subrahmanian for providing many facilities for the extensive field work undertaken and S/s. R. V. Dhond, Shiv Dutta, Masood Ahmad, P.S. Viswanathan, S.K. Gupta, R.N. Kulkarni, V. Sundararaman, J.C. Gupta, N.J. Sahajwala, R.V. Patil, M.P. Ghonge, N.S. Iyer and M.L. Bhutani for their help in the arduous tasks of collecting and processing the statistical data. We also wish to express our indebtedness to the authorities, the radiologists and staff of the hospitals and clinics of Maharashtra, Tamil Nadu, and the Northern and Eastern Regions whose co-operation has made it possible for us to complete this project.

REFERENCES

[1] MARTIN, J.H., Radiation doses to gonads in diagnostic radiology and their relation to long term radiation hazard, Med. J. Aust. 2 (1955) 806-10.

[2] STANFORD, R.W., VANCE, J., The quantity of radiation received by the reproductive organs of patients during routine diagnostic X-ray examinations, Br. J. Radiol. 28 (1955) 266-73.

[3] OSBORN, S.B., SMITH, E.E., The genetically significant radiation dose from the diagnostic use of X-rays in England and Wales, Lancet (1956) 949-53.

[4] BILLINGS, M.S., NORMAN, A., GREENFIELD, M.A., Gonad dose during routine roentgenography, Radiology 69 (1957) 37-41.

[5] LAUGHLIN, J.S., PULLMAN, I., Gonadal Dose Produced by the Medical Use of X-rays, Rep. prepared for the Genetics Committee of the US Acad. Sci. Study of the Biological Effects of Atomic Radiation; UN doc. A/AC.82/G/R.74.

[6] MARTIN, J.H., The contribution to the gene material of the population from the medical use of ionizing radiations, Med. J. Aust. 45 (1958) 79-84.

[7] LINCOLN, T.A., GUPTON, E.D., Radiation dose to gonads from diagnostic X-ray exposure, J. Amer. Med. Assoc. 166 (1958) 233-39.

[8] LARSSON, L.E., Radiation doses to the gonads of patients in Swedish roentgen diagnostics: Studies on magnitude and variation of the gonad doses together with dose reducing measures, Acta Radiol., Suppl. 157 (1958) 7-127; UN doc. A/Ac.82/G/R.182.

[9] SEELENTAG, W., et al., Zur Frage der genetischen Belastung der Bevölkerung durch die Anwendung ionisierender Strahlen in der Medizin. IV. Teil: Die Strahlenbelastung durch die Röntgendiagnostik in Kinderkliniken, Strahlentherapie 107 (1958) 537-55.

[10] REBOUL, J., et al., Doses gonades resultant de l'utilisation des radiations ionisantes en France. I. Radiodiagnostic, Ann. Radiol. 2 (1959) 179-96; 571-84; UN doc. A/AC.82/G/L.341.

[11] NORWOOD, W.D., et al., The gonadal radiation dose received by the people of a small American City due to the diagnostic use of roentgen rays, Amer. J. Roentgenol. 82 (1959) 1081-97.

[12] MINISTRY OF HEALTH, Department of Health for Scotland, Radiological Hazards to Patients, 2nd Rep. of the Committee, H.M.S.O. (1960); UN doc. A/AC.82/G/L.557.

[13] BIAGINI, G., BARILLA, M., MONTANARA, A., Zur genetischen Strahlenbelastung der Bevölkerung Roms durch die Röntgendiagnostik, Strahlentherapie 113 (1960) 100-9.

[14] REBOUL, J.G., et al., Doses gonades resultant de l'utilisation des radiations ionisantes en France. II. Radioscopie. Ann. Radiol. 3 (1960) 89-99; UN doc. A/AC.82/G/L.341.

[15] HAMMER-JACOBSEN, E., Genetically significant radiation doses from diagnostic radiology in Denmark, J. Belge Radiol. 44 (1961) 253-76.

[16] HOLTHUSEN, H., LEETZ, H.H.K., LEPPIN, W., Die genetische Belastung der Bevölkerung einer Grosstadt (Hamburg) durch medizinische Strahlenanwendung, Schriftenreihe des Bundesministers für Atomkernenergie und Wasserwirtschaft, Strahlenschutz 21 (1961).

[17] PLACER, A.E., Dosis geneticamente significativa debida al radiodiagnostico medico, Comision Nacional de Energia Atomica, Inf. No.49, Buenos Aires (1961); UN doc. A/Ac.82/G/L.485.

[18] RESEARCH GROUP on the Genetically Significant Dose by the Medical Use of X-rays in Japan, The genetically significant dose by the X-ray diagnostic examination in Japan, as quoted in UNSCEAR 1962 Report to the UN General Assembly, General Records: 17th Session, Suppl. No.16 (A/5216).

[19] ZUPPINGER, A.W., et al., Die Strahlenbelastung der schweizerischen Bevölkerung durch röntgendiagnostische Massnahmen, Radiol. Clin. 30 (1961) 1-27.

[20] MAHMOUD, K.A., et al., Gonadal and bone marrow dose in medical diagnostic radiology, UAR Scientific Committee on the Effects of Atomic Radiation on Man, Vol. 3-1 (1961); UN doc. A/AC.82/G/L.661.

[21] VIKTURINA, V.P., TROITSKII, Ye., PASYNKOVA, I.Ye., Doses to which patients are exposed during roentgenological examinations, Vestn. Rentgenol. Radiol. 36 (1961) 44-49; U.S.J.P.R.S. trans. 9120.

[22] MAHMOUD, K.A., et al., Report on genetically significant dose from diagnostic radiology in Cairo and Alexandria, UAR Scientific Committee on the Effects of Atomic Radiation on Man, Vol. 4-2 (1962).

[23] FLATBY, J., "Genetically significant dose in X-ray diagnosis in Norway", as reported in UNSCEAR Suppl. 16 (A/5216) (1962).

[24] BEEKMAN, Z.M., WEBER, J., "The genetically significant dose from diagnostic radiology to a defined population in the Netherlands", as reported in UNSCEAR Suppl. 16 (A/5216) (1962).

[25] OSBORN, S.B., BURROWS, R.G., An ionisation chamber for diagnostic X-radiation, Phys. Med. Biol. 3 (1958) 37.

[26] CLAYTON, C.G., FARMER, F.T., WARRICK, C.K., Radiation dose to the foetal and maternal gonads in obstetric radiography during late pregnancy, Br. J. Radiol. 30 (1957) 291-94.

[27] BAKER, P.M., The Genetically Significant Radiation Dose from Diagnostic X-rays in Canadian Public Hospitals, Department of National Health & Welfare, Ottawa, Rep. R.P.D.-31 (1963).

[28] CENSUS OF INDIA, 1961, Vol. X. Maharashtra, Part II - C - (1) Social and Cultural Tables (1965).

[29] CENSUS OF INDIA, 1961, Vol. IX, Madras, Part - C (1) Cultural Tables.

[30] CENSUS OF INDIA, 1961, Vol. XIII, Punjab, Part II - C (1) Social and Cultural Tables.

[31] CENSUS OF INDIA, 1961, Vol. IV, Vol. XII, Vol. XVI, Part II-C(1) Social and Cultural Tables.

[32] NATIONAL SAMPLE SURVEY, Indian Govt. Publ. 89 (1961).

[33] SUPE, S.J., DHOND, R.V., MASOOD AHMAD, VISWANATHAN, P.S., RAO, S.M., "Genetically significant dose to the population of Bombay resulting from diagnostic procedures", presented Indian Congress of Radiology, Vellore, 1966.

[34] SUPE, S.J., RAO, S.M., SHIV DATTA, VISWANATHAN, P.S., SAWANT, S.G., KULKARNI, R.N., "Genetically significant dose to the population of Tamil Nadu resulting from diagnostic X-ray procedures", presented 24th Indian Congress of Radiology, Calcutta, (1971).

[35] SUPE, S.J., MASOOD AHMAD, VISWANATHAN, P.S., RAO, S.M., SHIV DATTA, GUPTA, J.C., GUPTA, S.K., SAWANT, S.G., "Genetically significant dose to the population of Maharashtra resulting from diagnostic X-ray procedures", presented Indian Congress of Radiology, Panaji, 1969.

DISCUSSION

L.-E. LARSSON: In the final column of your Table V you have illustrated just what I warned against in my own paper. When you extrapolate the results from national or regional surveys to an examination frequency of one examination per individual, you anticipate an unchanged distribution of type of examination. That means that a country with a low frequency of examinations, and these mainly chest examinations, will get a better figure for their extrapolated GSD than, say, another country which also has a low total frequency but a high contribution from urography. Such comparisons have no meaning. Unfortunately the latest report of UNSCEAR presented the same type of comparison.

S. J. SUPE: Actually, the table has been taken largely from UNSCEAR reports and the values for India have been added. I would like to point out, however, that for countries where the frequency of radiological examinations is low because of scarce facilities, the increase in frequency with more extensive facilities is expected to have the same proportionate distribution over the various examinations. I think this distribution may be different when the frequency reaches the value of one examination per person per year.

F. D. SOWBY: Can you explain the very high contribution made by fluoroscopic examinations of females in Chandigarh?

S. J. SUPE: The GSD is high for Chandigarh as a whole compared with other Indian cities. This could be due to many factors. The institution in which the work in question was done is very well known and attracts patients from all over the region; this would make the frequency of examinations much larger than one would expect for the population of Chandigarh alone. In the case of fluoroscopic examinations we also found that the female gonad doses are quite large compared with those for other cities; this may be due to longer fluoroscopy times during special procedures such as barium meals and barium enemas. The operating conditions of the X-ray tubes may also have caused this higher dose.

A PRACTICAL METHOD OF ASSESSING RADIATION DOSES TO THE POPULATION

The use of reference levels and some results of their application to Italian sites

G. BOERI, Carla BROFFERIO
Divisione di Protezione Sanitaria e Controlli,
CNEN, Rome,
Italy

Abstract

A PRACTICAL METHOD OF ASSESSING RADIATION DOSES TO THE POPULATION: THE USE OF REFERENCE LEVELS AND SOME RESULTS OF THEIR APPLICATION TO ITALIAN SITES.

The authors present some results on the use of derived reference levels, as adapted to evaluate environmental surveillance data collected around Italian nuclear installations. A comparison is made between theoretically derived reference levels and actually measured environmental concentrations of radioactivity in order to assess radiation doses to the population and to investigate the environmental behaviour of radioactive effluents.

Among the purposes of environmental surveillance ICRP Publication No.7 includes the collection of data to evaluate radiation doses to the individuals of the population living in the vicinity of nuclear facilities.

In a recent paper [1] we discussed the use of reference levels for contaminated foodstuffs and environmental samples as a feasible means of controlling the environmental contamination around nuclear installations and evaluating radiation doses to the population. Here we shall present some observations resulting from the application of these reference levels to an actual case in Italy.

Reference levels can be defined as the concentration of radioactivity in environmental samples delivering a certain dose to the population. They can be derived for each plant on the basis of a computation model describing the behaviour of radioactivity in the environment and making use, whenever possible, of actual environmental parameters and socio-economic data.

In our opinion, reference levels can be looked at from two different points of view: one more strictly connected to environmental processes and to the amount of discharged activity and the other to the utilization of the environment and to dose assessments. This is shown in Fig.1.

To derive reference levels for a given nuclide at a given site, information is needed on its behaviour in the environment (processes of dilution, dispersion, accumulation, etc.) and on the uses man makes of that environment (working and living habits). All these data are fitted to a model that gives a concentration in an environmental sample corresponding to a certain dose to members of the population and to the release of a certain amount of radioactivity.

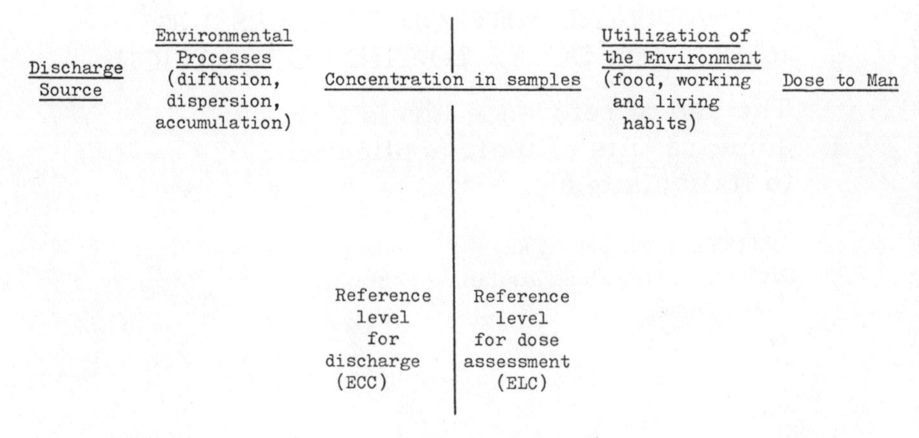

FIG. 1. Evaluation of surveillance data.

When, in an actual instance, radioactivity is released from an installation, the concentration data obtained from the radiological surveillance network may or may not, depending on its adequacy, agree with the values predicted in the computing model. Anyhow, if the radioactive concentrations are sufficiently high to be detected, by comparing these concentrations with the corresponding reference levels it is possible to obtain a quick, though rough, evaluation of the dose to the population. The reference levels themselves should be carefully checked on the basis of the experimental data, so that they correspond as closely as possible to the actual situation.

Until all the main environmental processes are known to a good extent, it may be useful to have two different kinds of reference levels in order to follow the effluents behaviour (Environmental Compliance Concentration (ECC)) and to assess doses to the public (Environment Limit Concentration (ELC)) [1].

The ECC corresponds to the concentration of a given nuclide expected in an environmental segment if the authorized discharge is released; this serves the purpose of directly verifying the plant's compliance with the authorized limits. The second level, ELC, represents the concentration of a given nuclide expected in a segment corresponding to the limiting radiological capacity; this allows a prompt evaluation of the exposure of the public to be made. The reference level for discharges (ECC) is particularly suitable for checking the assumptions about environmental processes, because it does not depend upon the data on the socio-economic utilization of the environment, and if there are more exposure pathways, its numerical value does not depend upon their mutual interrelations.

The reference level for dose assessments (ELC) allows first approximation doses to the population to be evaluated. Since each ELC is referred to one nuclide, to have a more elaborate and precise assessment of the consequences of a radioactive discharge, it is essential to have detailed data on the composition of the discharge itself.

We shall now deal with some applications of reference levels to the interpretation of surveillance data and present some observations on their use in the case of a PWR power plant discharging its effluents

(I, Cs, Mn, Co and Sr) into a river. For a short period the reactor had to discharge activity in the liquid effluents at a rate above the normal range. At the same time the environment was controlled with an 'ad hoc' radiological surveillance campaign that was superimposed on the usual surveillance network. The licensee had to take samples of water, fish and sediments in the river, besides keeping a strict record of its flow rate during this period.

The site had already been studied in radioecological and socio-economic surveys and potentially critical pathways and population groups had been indicated in people living in the area irrigated with the river water and consuming local fish.

On the basis of the results of these studies and making use of a computation model [2] we calculated the reference levels (ELCs and ECCs) for the main radionuclides discharged by the reactor (see Table I). These measurements were carried out over a period of six months until the environmental radioactivity had dropped to undetectable levels and plant discharges had long resumed their normal course.

Comparing the measurement results with the ECCs, we have tried to follow the environmental behaviour of I, Cs, Co and Mn and the corresponding processes of concentration in the food chain.

The values that can be drawn from experimental results cannot be taken as proper concentration factors because they do not correspond to the equilibrium conditions that are generally assumed as a basis for the determination of concentration factors; they are nonetheless very useful because they give a clear picture of the actual environmental behaviour of discharged activity that is actually released by the reactor in a discontinuous way and they are more closely related to the operational life of the installation. Therefore we call them 'operative transfer and/or concentration factors'.

The dose resulting from the measured environmental contamination can be evaluated making use of the reference levels (ELC) in the following way:

$$D_1 = \sum_n \frac{C_n}{ELC} (DL)_{weekly} \qquad (1)$$

where C_n is the mean weekly concentration (pCi/kg)

$(DL)_{weekly}$ is equal to 1/52 of the annual dose limit for the population

$\sum_n$ indicates a sum made over the whole period taken into consideration.

A more precise assessment of dose was also carried out directly from the exposure and intake levels derived from the experimental concentration values:

$$D_2 = \sum_n C_n K_f x \qquad (2)$$

where K_f is the fish consumption factor (kg/week)

x is a conversion factor (mrem/pCi)

C_n and $\sum_n$ have the same meaning as above.

TABLE I. REFERENCE LEVELS FOR THE PO RIVER

Nuclide	Environmental segment	ELC (pCi/litre or pCi/kg)	ECC (pCi/litre or pCi/kg)	Discharge rate (Ci/a)	Measurement sensitivities (pCi/litre or pCi/kg)
^{131}I	Water	1.4×10^2	1.2	5×10	1.6
	Fish [a]	1.4×10^3	1.2×10^2		35
	Milk	1.5×10^3	1.2×10^2		1.0
	Vegetables	1.5×10^3	1.2×10^2		3.0
^{137}Cs	Water	2.1×10^2	1.2×10^2	5×10	0.30
	Milk	1×10^3	5.5×10^2		1.0
	Fish	3.1×10^5	1.8×10^4		10
	Sediment	2.3×10^6	1.3×10^5		0.50
^{60}Co	Water	1.7×10^3	2.4×10^2	1×10^2	0.10
	Fish	1.7×10^5	2.4×10^4		100
	Sediment	2.3×10^6	3×10^5		0.50
^{54}Mn	Water	1.2×10^3	2.4×10	1×10^2	0.07
	Fish	50×10^5	1.0×10^4		50
	Sediment	1.3×10^7	3.0×10^5		0.50

[a] This value was included only during the surveillance campaign, because fish samples had not previously been taken in the surveillance network. As a result of the measurements all the ELCs values had to be modified.

This estimate can provide a comparison for the assumptions taken for the reference levels.

We should like to mention that the numerical values derived from this survey can be taken only as an indication of the order of magnitude for the parameters involved in a realistic assessment of the dose because the data were not always available in sufficient numbers and some measurements were below the detection limits.

The levels of ^{131}I discharges have allowed operative water-fish concentration factors to be determined. This was not possible for the water-vegetables and water-milk pathways because most of the activity was discharged into the river during a season when there was no irrigation; on the contrary, during the period of irrigation ^{131}I discharges were very low and consequently the concentration in vegetables and milk was below the detectable range. The curves of Figs 2-12 summarize the results of the surveillance campaign.

Figures 2, 3 and 4 show the estimated monthly mean values for the ^{131}I concentration in water and weekly measured values, together with the ^{131}I concentration in fish. The curves show the interrelation between discharge and concentration in fish. Experimental data indicated that the actual concentration in fish was five times (mean value) greater than expected. We therefore assumed an 'operative' concentration factor of 50, discarding the value of 10 used in the preliminary estimates.

The value of 10, assumed before, compared with the concentrations factors for vegetables and milk, rendered the fish exposure pathway a minor one; for this reason fish sampling had not been previously included in the routine surveillance network.

Applying Eq.(1) to our experimental results, we obtained a dose (D_1) to the thyroid of individual members of the population:

$$D_1 = 40 \text{ mrem}$$

If we evaluate the dose to the thyroid due only to the intake through consumption of fish, on the basis of the measured values we have:

$$D_2 = 2.8 \text{ mrem}$$

It should be mentioned that the value of 40 mrem includes the contribution from ingestion of milk and vegetables in correspondence with concentrations in these samples higher than those actually measured. In fact when we derived the reference levels we assumed that a given concentration in fish would imply a concentration in vegetables and milk of the same order (see Table I, column 4). On the other hand, these concentrations would deliver a much higher dose (about 35 out of 40 mrem) than fish in relation to the different ratios of the consumption rates.

Modifying the reference levels by taking into consideration experimental evidence, we obtain an ELC for fish about 4 times larger; the corresponding dose (D_1) is reduced to 10 mrem. This figure is now comparable with D_2, considering that the possible vegetable and milk contribution, if there was irrigation, would be about 7 mrem.

In a similar way, we have followed ^{137}Cs concentrations in water, sediments and fish (Figs 5 - 8). Comparison of the estimated values with the measurements shows fairly good agreement, remembering that the estimated concentration is a monthly mean value.

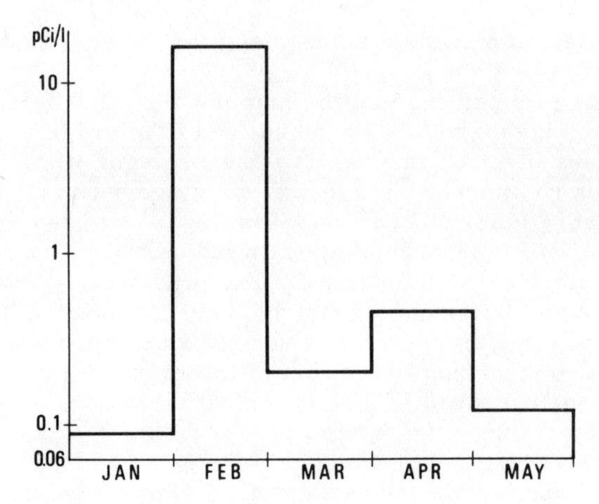

FIG. 2. ^{131}I water concentration, calculated monthly mean.

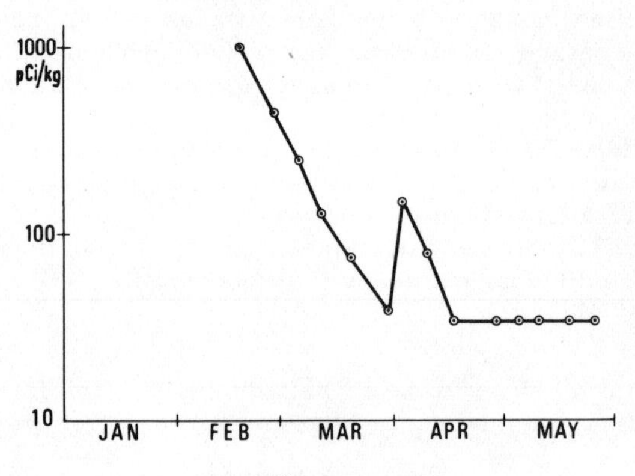

FIG. 3. ^{131}I fish concentration.

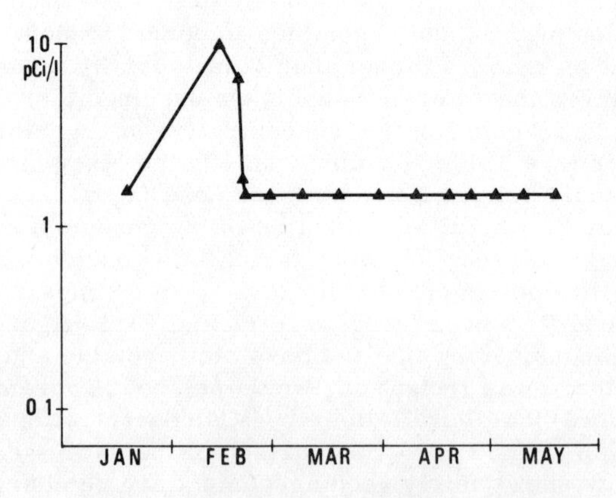

FIG. 4. ^{131}I water concentration, measured.

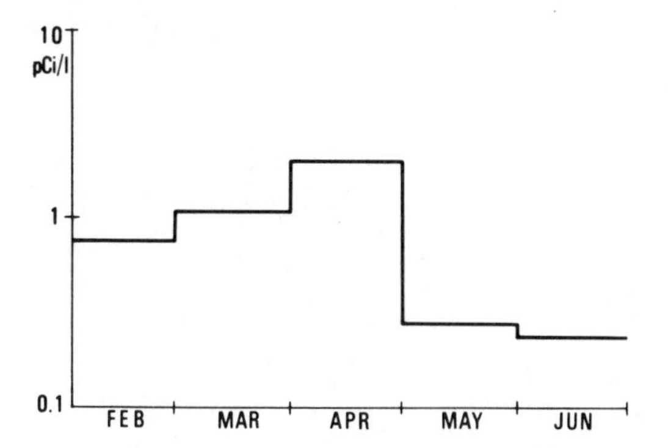

FIG. 5. ^{137}Cs water concentration, calculated monthly mean.

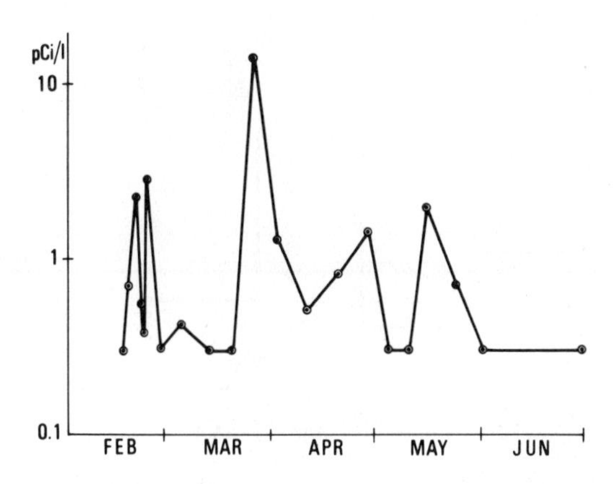

FIG. 6. ^{137}Cs water concentration, measured.

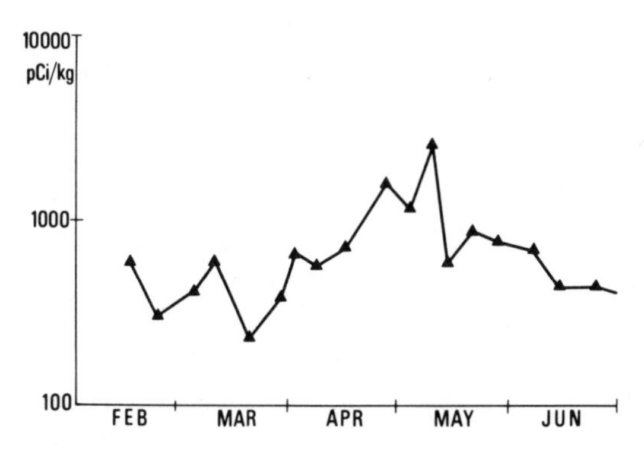

FIG. 7. ^{137}Cs fish concentration.

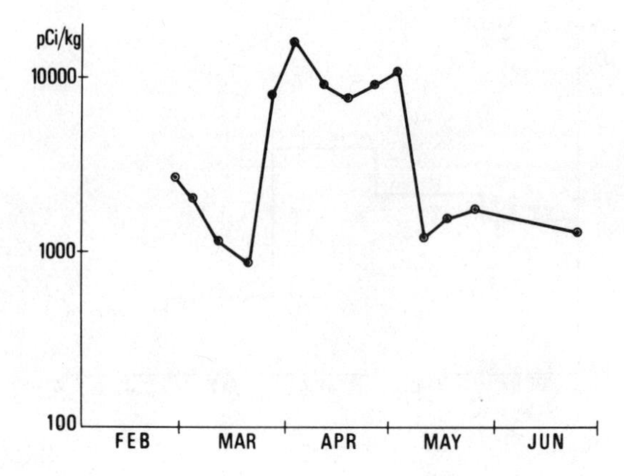

FIG. 8. ^{137}Cs sediment concentration.

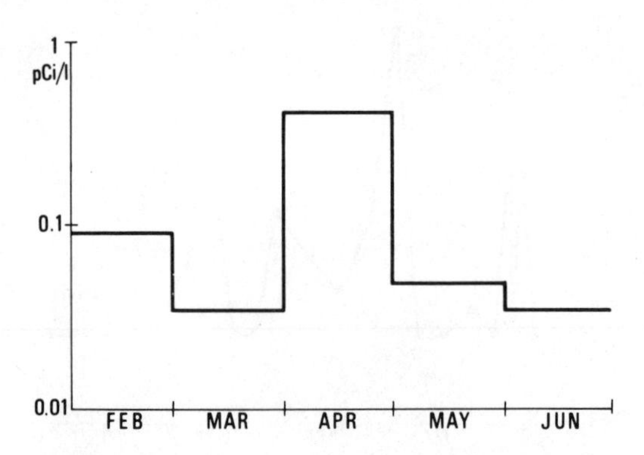

FIG. 9. ^{54}Mn water concentration, calculated monthly mean.

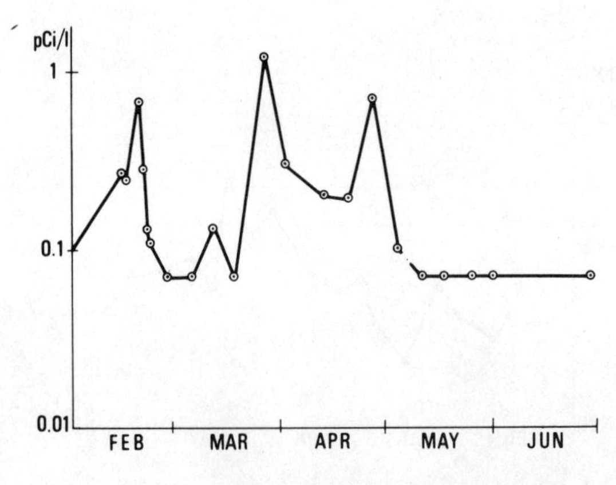

FIG. 10. ^{54}Mn water concentration, measured.

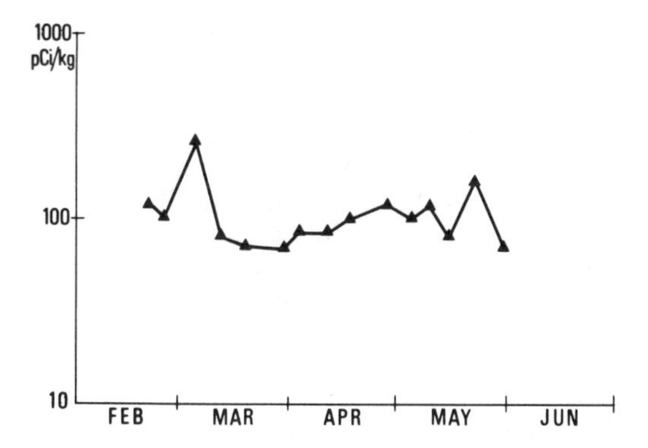

FIG. 11. ^{54}Mn fish concentration.

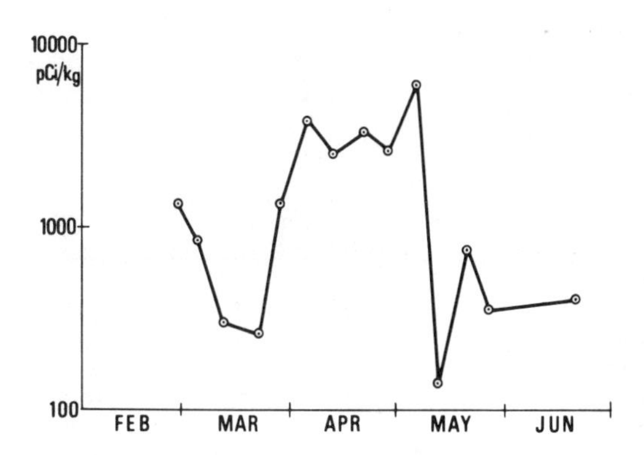

FIG. 12. ^{54}Mn sediment concentration.

We have noticed that the ^{137}Cs peak in fish seems to appear about 30-45 days after the discharge. The operative water-fish concentration factor can be taken to be about 1000, which, incidentally, corresponds to the figure assumed in the preliminary evaluations. ^{137}Cs accumulation in sediments can be easily followed on its relative curve. The corresponding concentration factor is near 10^4. Since in this case there is good agreement between the expected and the measured concentrations, the assessment of dose through the reference level method and direct calculation produces very similar results.

For ^{54}Mn (Figs 9 - 12) we also found good agreement between the doses assessed with the use of reference levels and the values calculated directly from environmental data. In this case the operative concentration factor for fish can be assumed to be 10^3 and 10^4 for sediments.

COMMENTS

The data collected in the survey have been very helpful for testing the feasibility of reference levels and their application to the prediction of doses.

The discussion has shown the importance of checking reference levels, derived for a given installation, on the basis of experimental results. In fact, incorrect assumptions about accumulation or transfer factors might lead to one underestimate the role of one exposure pathway, paying undue relative importance to others, which may ultimately produce a wrong assessment of the doses.

We recall here the case of ^{131}I: too low a value for the concentration factor in fish resulted in this important and direct exposure pathway being omitted from the routine surveillance network. Therefore if the radioactivity levels in vegetables and milk are below detectable limits and consequently the evaluation of doses with the use of the ELC related to these pathways would give no significant value, doses to the population coming from fish ingestion could not be accounted for.

It can be concluded that reference levels can be usefully employed for interpreting environmental surveillance measurements and also for making dose estimates, provided that the assumptions upon which they have been derived are verified with the help of field results.

REFERENCES

[1] BOERI, G., BROFFERIO, Carla, "Practical reference levels for radioactive contamination in environmental surveillance", Environmental Surveillance around Nuclear Installations (Proc. Symp. Warsaw, 1973), IAEA, Vienna (1974).

[2] BRAMATI, L., MARZULLO, T., ROSA, I., ZARA, G., "Vadosla: a simple code for the evaluation of population exposure due to radioactive discharges", presented 3rd IRPA Congress, Washington, D.C., 1

[3] I.N.N., Informazioni ambientali per la Centrale di Trino Vercellese, Rome (1973).

IAEA-SM-184/39

THE PROBLEM OF THE ASSESSMENT OF THE RADIOLOGICAL IMPACT ON POPULATIONS FROM RADIOACTIVE DISCHARGES
Some considerations on the concept of global risk

G. BOERI, F. BREUER, Carla BROFFERIO
Divisione di Protezione Sanitaria e Controlli,
CNEN, Rome,
Italy

Abstract

THE PROBLEM OF THE ASSESSMENT OF THE RADIOLOGICAL IMPACT ON POPULATIONS FROM RADIOACTIVE DISCHARGES: SOME CONSIDERATIONS ON THE CONCEPT OF GLOBAL RISK.

The authors discuss some particular aspects of the problem of evaluating the radiological impact on the population from radioactive effluent discharges. Generally, when this evaluation is made it is customary to consider dose commitments to the critical organs, neglecting doses to other organs. Since radioactive effluents from nuclear installations generally consist of many radionuclides with different critical organs and since even for a single radionuclide it is sometimes difficult to define a single predominant critical organ, the authors suggest considering in some cases a global risk factor that takes account of all the contributions to doses to the affected organs.

Current practice in restricting to acceptable levels the risks connected with ionizing radiation is derived from international guides and recommendations, which are based mainly on the compliance with the dose limits proposed by ICRP. These values, which have been obtained after long study of the effects of radiations, have been modified and up-dated many times to ensure that radiation risks be kept to such a low level that they do not result in a statistically significant increase in harmful consequences to the exposed populations [1-3].

Scientific interest in the effects on man of radiation exposure at low doses is always alive [4, 5]; recently the attention of research into the various problems of health protection has centred on evaluating the probability of harmful effects rather than only on dose calculations. In the following we shall treat essentially the problem of the evaluation of the risk to the population from radioactive discharges from nuclear plants; the approach that we shall present can be also applied to the other problems of public health and radiation risk assessment.

Discharge limits for radioactive effluents customarily take into account all the processes of diffusion and concentration in the environment and in the food chain so that it would be highly improbable that even children in the critical group of population would receive doses greater than the ICRP dose limits. From an operative point of view, discharge limits may in some instances be expressed by an equation establishing that the sum of the ratios between the discharged activity of each radionuclide and its relative discharge limit does not exceed unity or a stipulated fraction of it (discharge formula).

Discharge limits are normally based on the concept of the limiting radiological capacity of the environment, which is referred for each nuclide to the population dose limits for the critical organ. Considering all the radionuclides composing the discharge mixture of a nuclear plant, in many cases the critical organ is not a single one and, because of the various exposure pathways, it is not always easily identifiable.

We should like to make some criticism of the approach generally followed. First, we observe that the sum of the terms in the discharge formula is equivalent to a sum of doses to different critical organs; this has a radiologically weak meaning and it does not even ensure that dose limits for each organ are not exceeded.

Secondly, in most cases no account is made of the different physico-chemical forms of radionuclides in the effluents and along the food chain.

Furthermore, following this method, we cannot verify, in accordance with the statement contained in ICRP Publication 6, that:

"When the radioactive isotopes in a mixture are taken up by several organs and the resultant tissue doses in such organs are of comparable magnitude, the combined exposure is considered essentially whole-body exposure. Accordingly the permissible levels of exposure will be those applicable to the gonads and the blood-forming organs".

A more comparable method of calculation that allows for all ICRP recommendations would be rather complicated and would result in a set of discharge limits each related to a different critical organ. In any case, discharge formulas, so evaluated, could not enable us to express directly a real estimation of the risk to the population deriving from a given discharge, although the ICRP, when establishing dose limits, had undertaken a great deal of effort in accounting for several factors that contribute to the real risk deriving from an irradiation and a contamination.

Compliance with the ICRP dose limits may be considered a valid way of keeping the risk within acceptable limits, but public authorities sometimes need to have more direct data for assessing the risk to population. Accordingly it seems more useful to derive discharge formulas on the basis of the evaluation of the risk that, as outlined by other authors [7], could be expressed by the sum of the probabilities of harmful effects deriving from a given exposure or contamination ('global risk' concept).

This global risk approach could be developed along the following main lines:

(1) Assess the external exposure and the intake of the various radionuclides in the discharge mixture
(2) Evaluate the external and internal dose to each organ and the corresponding probabilities of harmful effects
(3) Make the sum of probabilities of harmful effects
(4) Choose a given level of global risk as 'reference risk'
(5) Verify that the global risk from the allowable radioactive discharge does not exceed the 'reference risk'.

Operatively the following simplified method could be employed. For a given nuclide the reference intake corresponding to the reference risk can be given by the equation:

$$I_R = \frac{1}{\frac{1}{I_{1R}} + \frac{1}{I_{2R}} + \ldots \frac{1}{I_{iR}}} \tag{1}$$

where I_{iR} is the intake that delivers a probability equal to the reference risk to the ith organ.

In the same way, when we consider the external exposure from that nuclide it is possible to evaluate the cloud dosage or the time integral of surface activity giving the reference risk. On this basis the limiting radiological capacity (A_{Ri}) for a radionuclide can be calculated.

Finally discharge limits for a radioactive mixture may be expressed in the following formula:

$$\frac{A_1}{A_{R1}} + \frac{A_2}{A_{R2}} + \ldots + \frac{A_n}{A_{Rn}} \leq 1$$

where A_n and A_{Rn} represent respectively the activity of the nth nuclide and the reference (or limiting) activity of that nuclide, which involves the attainment of the reference risk for exposed individuals of the critical group of the population.

Furthermore, if we evaluate the relative risk for all the population groups exposed to radiation in consequence of a discharge, it is possible to obtain an assessment of the global risk for the population as a whole. In this way it is also possible to verify whether the proposed discharge limits might be considered acceptable in relation to a chosen level of the global risk to the population.

We must note that unfortunately at the moment some difficulty stands in the way of the immediate application of such method. In the first place, there are still severe problems in the correlation between risk and dose, both regarding the organs under consideration and the public health implications.

In fact most statistics are based on observations on animals and made at rather high doses. We should bear in mind here the difficulty met in extrapolating dose-effect curves to zero. Their behaviour at low doses and the existence of a threshold value, below which irradiations would eventually result in no effects, are still uncertain.

According to Vol. II of the UNSCEAR Report [4], only in the case of leukaemia and breast cancer does the lack of any threshold and a relative linearity of the effect curve at low doses seem possible.

Other uncertainties and difficulties derive from the response of dose-effect curves with the irradiation rate and from a correct interpretation of the difference in effects observed between external irradiation and internal contamination at equal absorbed doses.

Similar uncertainties arise in the choice of an appropriate RBE to apply to each internal emitter, in relation to its effects, taking into account differences of localization, metabolism, transport of the radionuclide and the application of the results to man. We have little knowledge of many aspects of the metabolism of radionuclides, essential for a correct evaluation of doses to all the organs of the body, even without considering the degree of individual variability in the exposed persons. The evaluation of risks due to external exposure entails some unsolved problems.

No small point is the limited knowledge we have of the genetic effects of radiation since observations have been carried out exclusively on animals as well as the difficulty in translating these results into terms of hazard to human beings.

We should like to remark that the proposed operative method for the assessment of the global risk can be applied only if linearity of the dose-effects curve and lack of threshold are proved. Only then would there be a linear, and therefore summable, response for the effects deriving from each nuclide in each organ.

If we had instead to accept the existence of a threshold value for a given effect on a given organ (for instance, 100 rads), we could not presume that a delivered dose of 40 rads from the one nuclide and 80 rads from a second would give a zero effect, even if either dose, taken alone, would have no effect

It may be observed, moreover, that the same criticism can be considered in the evaluation of the probability of harmful effects to populations as can be made on the basis of the collective dose approach that is most commonly used in the assessment of risks to the population.

CONCLUSIONS

The method of assessing the global risk connected with a radioactive discharge to the environment can supply public authorities with more significant data than can simple discharge limits based on the values of dose limits as suggested by ICRP, even though they have comparable meanings. In this paper we have suggested a general criterion for making such assessment, pointing out the importance, in many instances, of calculating doses and their relative risk to different organs of the body in addition to the critical organ.

A method for evaluating in practice discharge limits has been proposed that can be used under the assumption of the linearity of the dose-effect curve. The problems met in the application of this method are the same as are met in any other method for the evaluation of the radiological risk.

REFERENCES

[1] INTERNATIONAL COMMISSION ON RADIOLOGICAL PROTECTION, Recommendations, Publication 2, Pergamon Press, London (1959).

[2] INTERNATIONAL COMMISSION ON RADIOLOGICAL PROTECTION, Recommendations, Br. J. Radiol. Suppl. 6 (1955).

[3] INTERNATIONAL COMMISSION ON RADIOLOGICAL PROTECTION, Recommendations, Publication 9, Pergamon Press, Oxford (1966).

[4] Ionizing Radiation: Levels and Effects, A Report of the United Nations Scientific Committee on the Effects of Atomic Radiation, UN, New York (1972).

[5] The Effects on Populations of Exposure to Low Level of Ionizing Radiation, Report of the Advisory Committee on the Biological Effects of Ionizing Radiation, National Academy of Sciences – National Research Council, Washington, D.C. (1972).

[6] MORGAN, K.Z., "Proper use of information on organ and body burdens of radioactive material", Assessment of Radioactive Contamination in Man (Proc. Symp. Stockholm, 1971), IAEA, Vienna (1972) 3.

[7] MARLEY, W.G., "Radiological hazards associated with internal contamination of the body by radionuclides", Proc. I.R.P.A. 2nd Europ. Cong. Radiation Protection, Budapest, 1972.

DISCUSSION

C.B. MEINHOLD: I have just a minor comment. In the paper you use the term RBE when discussing differences in localization, metabolism and transport for various nuclides. I believe this term has a rather restricted meaning, which does not cover this variety of differences between radionuclides.

G. BOERI: We are using the term RBE in a more general sense.

ENVIRONMENTAL CHARACTERISTICS OF THE DANUBE RIVER SYSTEM AND THE PROBLEMS OF RADIOLOGICAL SAFETY STANDARDS

T. TASOVAC, R. RADOSAVLJEVIĆ, M. ZARIĆ
Boris Kidrič Institute of Nuclear Sciences,
Belgrade,
Yugoslavia

Abstract

ENVIRONMENTAL CHARACTERISTICS OF THE DANUBE RIVER SYSTEM AND THE PROBLEMS OF RADIOLOGICAL SAFETY STANDARDS.

As many environmental parameters affect the behaviour of different radionuclides introduced into the environment, it is necessary in evaluating the population dose commitment to have at one's disposal, among others, the results of rather complex environmental studies. Such complex environmental studies are especially important in assessing the doses from and defining the operating conditions of nuclear installations located beside large international rivers. By analysing the results of radioecological studies carried out on the River Danube in Yugoslavia and other riparian states, the problems of setting the maximum permissible concentrations of radionuclides in this aquatic system are discussed.

INTRODUCTION

The basic principles of radiation protection require (1) that all unnecessary exposure be avoided; (2) that any exposure be kept as low as is readily achievable; and (3) that under no circumstances should the exposure exceed the dose limits set up by the International Commission on Radiological Protection and other national and international bodies.

When applying these basic radiation protection principles to the larger population groups, one faces in practice many problems. Because many factors affect the evaluation of the population dose commitment and different radionuclides from different practices contribute to the population dose commitment, it is in some cases rather difficult to set practical radiation protection guides and technical limits to meet the above requirements. As, for example, many environmental parameters affect the behaviour of different radionuclides introduced into the environment, it is necessary in setting practical radiation protection guides, i. e. in decision making for normal operating conditions of nuclear installations as well as emergency situations, and in assessing the population dose commitment, to have at one's disposal the results of rather complex environmental studies. Such complex environmental studies are especially important in decision making for the operating conditions of nuclear installations located beside large international rivers. In such systems very often the hydrological, biogeochemical and other characteristics of the river flow and the surrounding environment differ very much in various parts of the river course, resulting in different biogeochemical and ecological processes. Consequently, to establish the exact capacity of the environment to receive radionuclides, i. e. to determine the safe limits for the discharge of radioactive wastes, to define the normal operating conditions of nuclear installations and to assess population doses, it is necessary to have at one's disposal the results of studies along the whole water course. The results of radioecological studies carried out on the River Danube in Yugoslavia and other riparian states and the existing differences in some parameters point out the necessity of promoting and co-ordinating further more complex investigations of the river aquatic system and its tributaries, to evaluate the environmental capacity of the Danube and to define the conditions for the safe operation of nuclear facilities that are or shall be located in the River Danube catchment area.

RESULTS AND DISCUSSIONS

The radioecological characteristics of the Danube river system have been studied and the results published by Ruf [1-4], Wachs [5], Frantz [6-8], Mayer [9], Stelczer [10], Meszner [11-13], Keleman [14], Tasovac [15-19] Radosavljević [20-25], Radovanović [26], Drašković [27-29], Vukmirović [30, 31], Petrović [32, 33], Filip [34], Stankova [35] and Gavrishova [36].

According to the published results, the types and intensities of radioecological investigations carried out in the Danube riparian states are very different, ranging from the measurement of the total beta radioactivity to the determination of particular radionuclides in specific components of aquatic media and the determinations of their concentration factors in suspended material, sediments and hydrobiological material. The majority of the published results refer to investigations carried out in West Germany on the river course from km 2547 to km 2203 and in Yugoslavia on the river course from km 1433 to km 945, while the results of radioecological studies referring to other parts of the water course are less numerous. However, even the results of the simple total β-activity measurements carried out in the Danube riparian states in the period 1958 - 1973 show large differences. The highest total β-activities in this period, especially in the period of intensive nuclear weapons testing, i. e. from 1958 to 1965, were found in the water samples collected in Yugoslavia. While the minimum activities were relatively similar, ranging from 1 to 7 pCi/litre in W. Germany, 3 to 5 pCi/litre in Austria, 2 to 6 pCi/litre in Czechoslovakia, 0 to 4 pCi/litre in Hungary and 2 to 7 pCi/litre in Yugoslavia, the maximum values varied widely, reaching 35 pCi/litre in W. Germany, 50 pCi/litre in Austria, 10 pCi/litre in Czechoslovakia, 37 pCi/litre in Hungary and as high as 818 pCi/litre in Yugoslavia. For the same period the average values of the total β-activity of the Danube in Yugoslavia were 2 to 7 times higher than in the other riparian states. This could be partly explained by different hydrological conditions caused by a great increase in the water flow through the influx of three big tributaries, the Drava, Tissa and Sava, which cover a catchment area of 293 956 km^2 and in view of the high correlation coefficien between fall-out activity and the total β-activity of the Danube (0,56-0,96) in the period of intensive experiments with nuclear weapons, and would contribute their considerable activity to that of the Danube.

The picture becomes more and more complicated if we analyse the content and behaviour of particular radionuclides along the course of the Danube and its tributaries. According to the results published by Ruf for the West German part of the Danube, the content of ^{60}Co in water, bed sediment, seston, water plants and fishes ranges from 0,03 to 1,3 pCi/litre in 1969 and 0,02 to 0,29 pCi/litre in 1970. The averages for the ^{60}Co content of sediments for 1967, 1968, 1969, 1970, 1971 and 1972 were 0,05, 0,10, 0,11, 0,07, 0,12, and 0,32 pCi/g, respectively, with concentration factors of 3300, 380, 550, 1400 and 2000. The average ^{60}Co concentrations in seston in the same area for these years were 0,10, 0,05, 0,08, 0,07 and 0,28 pCi/g. The concentration factor for seston reached a value of 20 000, for fish flesh a value of 80, for fish bone 310 and for water plants 90. In the same period the ^{60}Co content of the Danube in Yugoslavia was 4,2 to 14 pCi/litre and the average value was about forty times higher than in West Germany. The ^{60}Co concentrations in plankton in the Tissa at the junction of the Tissa and the Danube at high water April 1970, according to our measurements, reached the enormous value of 1500 pCi of ^{60}Co per gram of plankton. According to these results, the calculated value of the concentration factor for plankton was 357 000.

The difference in the behaviour of ^{60}Co in the upper part of the Danube and in its middle was demonstrated by investigating also the capture of ^{60}Co by suspended particles in the Danube. Our investigations of the ^{60}Co capture by suspended particles were carried out on samples of water with suspended material collected in isokinetic conditions in an experimental zone of 25 km between the 1144th and 1120th kilometres of the water course, i.e. in the upper part of the new Djerdap Lake, which had not been formed at the time these experiments were carried out. According to these results the ^{60}Co concentration factors for suspended material ranged from 2000 to 400 000, the most frequent values being about 100 000.

Very high values for concentration factors were also found for stable cobalt by determining the content of this element by non-destructive neutron activation analysis of samples of water, suspended material, bed sediments, plankton and other biomaterial. According to these results the concentration factors for suspended material reached a value of 177 000, for bed sediments 44 500 and for plankton 38 100. In the case of the Sava, one of the big tributaries of the Danube, the concentration factor for plankton was 360 000. A comparison of the published results on the ^{137}Cs content in water of different parts of the Danube (West Germany, Hungary, Yugoslavia) showed less significant differences.

The picture, however, has become more complex since the completion of the Djerdap Dam. By comparing the results before and after the closing of the Djerdap Dam we found that the average increase in ^{137}Cs in bed material was over one order of magnitude, indicating that new hydrological conditions completely change the behaviour of particular radionuclides in an aquatic system. The calculated concentration factors for ^{137}Cs in bed sediments range up to 12 000 and the relative increase in activity in bed material under the new hydrological conditions was 27.

The changes in the ^{137}Cs concentrations in suspended and bed material depend very much on the distribution of ^{40}K in water and bed material. The same tendency, i.e. the increase in activity in bed material, was also found for other radionuclides, ^{106}Ru, ^{95}Zr - ^{95}Nb, ^{144}Ce etc.

A comparison of the calculated preliminary river standards for particular radionuclides for the upper Danube, published in IAEA Safety Series No. 36, with the values calculated on the basis of the experimental data referring to the middle part of the Danube shows significant differences.

CONCLUSIONS

It is obvious that all environmental characteristics, which differ along the water course, influence the behaviour of radionuclides in an aquatic system considerably. The existing data referring to hydrological, hydro-ecological, geochemical, biogeochemical and other characteristics of the river, together with the data on the present and future use of the water, are not sufficient at present to establish safe limits for the discharge of radionuclides into the Danube. The necessity of establishing radiation safety norms and technical standards on the basis of more complex and more numerous data referring to the whole water course of the river becomes more and more urgent, especially since all riparian states are building or intend to build nuclear power stations beside the Danube or its tributaries. In addition, it is necessary to take into account the synergetic effects of other non-radioactive pollutants.

For these reasons we think that it would be necessary to co-ordinate all efforts and actions to establish interdisciplinary and multidisciplinary approaches to the problem, methods of investigations and the unification of criteria and the interpretation of results. We would therefore suggest that the IAEA and other specialized UN agencies and existing research groups working on the problems of the Danube should continue and intensify their work on establishing environmental criteria for the Danube.

REFERENCES

[1] RUF, M., Die Kontamination der Fliessgewässer durch radioaktive Substanzen mit besonderer Berücksichtigung der Flussstaufen, Wasserwirtschaft 1 (1968) 16.

[2] RUF, M., Arbeitsbericht über Untersuchungen an der bayerischen Donau in der Zeit vom 1 Juli 1969 bis Juni 1970, Bayerische Biologische Versuchsanstalt, Munich (1970).

[3] RUF, M., "Radioökologische Analyse der oberen Donau", Proc. Symp. Radioecology Applied to the Protection of Man and his Environment, Rome, 1971 (1972) 477.

[4] RUF, M., "Radioactive Waste Water Release from Nuclear Power Plants into the upper Danube", European Study Group Meeting on Radiological and Environmental Protection, Budapest, 1973.

[5] WACHS, B., Arbeitsbericht über Untersuchungen an der bayerischen Donau in der Zeit vom Juli 1971 bis Juni 1972, Bayerische Biologische Versuchsanstalt, Munich (1972).

[6] FRANTZ, A., Die Radioaktivität in der österreichischen Donau, Arch Hydrobiol. Suppl. 30 (Donauforschung II) (1967) 4, 340.

[7] FRANTZ, A., Limnologie der Donau, Pts III, IV, Schweizbartsche Verlagsbuchhandlung, Stuttgart (1967).

[8] FRANTZ, A., TASOVAC, T., DRAŠKOVIĆ, R., RADOSAVLJEVIĆ, R., PETROVIĆ, G., Ein Vergleich des Gehaltes an Spurenelementen in Donauwasser bei Wien und Beograd für 1961-1970, Arch. Hydrobiol. Suppl. 44, Donauforschung 5 (1973) 2, 258-62.

[9] MAYER, J., Unveröffentlichte Berichte des Hygienischen Institutes in Bratislava über die Bestimmung der Radioaktivität in Wasserproben aus dem tschechoslowakischen Abschnitt der Donau, 1964/1965.

[10] STELCZER, K., DARAB, K., Unveröffentlichte Berichte der Forschungsanstalt für Wasserwirtschaft in Budapest über die Messung der radioaktiven Verunreinigung der Oberflächengewässer an der ungarischen Donaustrecke von 1957-1965.

[11] MESZNER, J., "Radioactive pollution in the Hungarian reach of Danube from 1969 to 1971", Panel on Capacity of the Environment to accept Radioactive Materials, Vienna, 1973.

[12] MESZNER, J., "Source and release of radioactive effluents in the Hungarian nuclear programme", Panel on Capacity of the Environment to accept Radioactive Materials, Vienna, 1973.

[13] MESZNER, J., "Transport and dispersion of radionuclides, their interactions with sediments and biological accumulation in rivers", Panel on Capacity of the Environment to accept Radioactive Materials, Vienna, 1973.

[14] KELEMEN, et al., "Proposal for the international control of radioactive pollution of the Danube in connection with the nuclear programmes of the countries on the river", European Study Group Meeting on Radiological and Environmental Protection, Budapest, 1973.

[15] TASOVAC, T., RADOSAVLJEVIĆ, R., Die Verteilung der gesamt beta Radioaktivität auf die flüssige Phase und die suspendierten Teilchen im Donauwasser, Arch. Hydrobiol. Suppl. 36 Donauforschung IV (1969) 92-97.

[16] TASOVAC, T., RADOSAVLJEVIĆ, R., Ergebnisse systematischer Messungen der alpha und beta Radioaktivität in der Donau im Gebiet von Vinča, Arch, Hydrobiol. Suppl. 36 Donauforschung IV (1969) 85-91.

[17] TASOVAC, T., RADOSAVLJEVIĆ, R., "Investigation of the radioactivity of the Danube", Environmental Contamination by Radioactive Materials (Proc. Seminar Vienna, 1969), IAEA, Vienna (1969) 427-33.

[18] TASOVAC, T., RADOSAVLJEVIĆ, R., DRAŠKOVIĆ, R., FILIP, A., VUKMIROVIĆ, V., RADOJČIĆ, M., VUKOTIĆ, R., "Nuclear techniques in studies of dispersion and some other properties of the Danube", Isotope Hydrology 1970 (Proc. Symp. Vienna, 1970), IAEA, Vienna (1970) 497-507.

[19] TASOVAC, T., RADOSAVLJEVIĆ, R., ZARIĆ, M., "Beitrag zur Kenntnis der K^{40} Radioaktivität im Donau und Sava Wasser bei Beograd", 16 Arbeitstag. Int. Arbeitsgemeinschaft Donauforschung Bratislava, 1973.

[20] RADOSAVLJEVIĆ, R., TASOVAC, T., "Capture of cobalt-60 by suspended particles in the Danube", Proc. Int. Symp. Radioecology, Cadarache 1 (1969) 83-92.

[21] RADOSAVLJEVIĆ, R., TASOVAC, T., Die Untersuchung der Radioaktivität der Donau, Arch. Hydrobiol. Suppl. 36, Donauforschung IV (1971) 402-4.

[22] RADOSAVLJEVIĆ, R., TASOVAC, T., DRAŠKOVIĆ, R., ZARIĆ, M., MARKOVIĆ, V., Complex behaviour of cobalt in the Danube River, Arch. Hydrobiol. Suppl. 44, Donauforschung 5 (1973) 241-48.

[23] RADOSAVLJEVIĆ, R., et al., Radioekologija K^{40} u Dunavu VI jugoslovenski simpozijum o radiološkoj zaštiti, Ohrid (1972) 297.

[24] RADOSAVLJEVIĆ, R., TASOVAC, T., DRAŠKOVIĆ, R., ZARIĆ, M., "Beitrag zur Kenntnis des biogeochemischen Verhaltens von Cr in der Donau, der Save und der Theiss", 16 Arbeitstag. Int. Arbeitsgemeinschaft Donauforschung, Bratislava, 1973.

[25] RADOSAVLJEVIĆ, R., TASOVAC, T., "Application of ergodic principe to estimation of regularity of radioactive material transport in the Danube", (in Serbocroat), Proc. V Jug. Simp. Radiološkoj Zaštiti, Bled, 1970, No. 12.

[26] RADOVANOVIĆ, R., "Bilans radioaktivne kontaminacije teritorije SFRJ, 1962-1969", Proc. V Jug. Simp. Radiološkoj Zaštiti, Bled, 1970.

[27] DRAŠKOVIĆ, R., TASOVAC, T., RADOSAVLJEVIĆ, R., "Neutron activation analysis of the aquatic environment in the Danube", Nuclear Techniques in Environmental Pollution (Proc. Symp. Salzburg, 1970), IAEA, Vienna (1971) 329-34.

[28] DRAŠKOVIĆ, R., RADOSAVLJEVIĆ, R., TASOVAC, T., ZARIĆ, M., "Interaction between water, trace elements and different components in the Danube River", Proc. Symp. Radioecology Applied to the Protection of Man and His Environment, Rome, 1971, pp. 1167-74.

[29] DRAŠKOVIĆ, R., RADOSAVLJEVIĆ, R., TASOVAC, T., ZARIĆ, M., "Neutron activation analysis of aquatic environment in the Sava River", Proc. Symp. Nuclear Activation Techniques in the Life Sciences, Bled, 1972.

[30] VUKMIROVIĆ, V., RADOJČIĆ, M., TASOVAC, T., RADOSAVLJEVIĆ, R., "Research on dispersion and other characteristics of the Danube River near Belgrade", Proc. 6th Conf. Danube Countries in Hydrological Forecasting, Kiev, 1971, 287-93.

[31] VUKMIROVIĆ, V., et al., "Contribution to the kinetic theory of bed material discharge", Isotopes in Hydrology (Proc. Symp. Vienna, 1966), IAEA, Vienna (1967) 279.

[32] PETROVIĆ, G., TASOVAC, T., RADOSAVLJEVIĆ, R., Beitrag zur hydrochemischen Untersuchung der Donau bei Vinča, km 1144, unweit Beograd, Arch. Hydrobiol. Suppl. 36 Donauforschung IV (1969) 71-84.

[33] PETROVIĆ, G., TASOVAC, T., RADOSAVLJEVIĆ, R., Der Schwebestoffgehalt der Donau bei Beograd, Arch. Hydrobiol. Suppl. 36 Donauforschung IV (1969) 98-108.

[34] FILIP, A., Investigation of dispersion and dilution of suspended species in river flow by radiotracer techniques, Int. J. Appl. Radiat. Isotopes 23 (1971).

[35] STANKOVA, E., KOVATSCHEV, D., PENTSCHEV, A., GEORGIEVA, M., Beobachtungen über die Radioaktivität der Donau längs des bulgarischen Ufers, Limnologische Berichte der X Jubileumstagung Donauforschung, Bulgarische Akademie der Wissenschaften, Sofia (1968).

[36] GAVRISHOVA, E., Limnologische Berichte der X Jubileumstagung Donauforschung, Bulgarische Akademie der Wissenschaften, Sofia 1968.

DISCUSSION

S.O.W. BERGSTRÖM: How have you defined 'water' when determining water concentrations for concentration factor calculations? Usually cobalt is not in an ionic form but is only attached to suspended material of some kind. Water treatment such as filtering might therefore greatly affect the concentration factor.

T. TASOVAC: The methods used for the investigation of cobalt are described in our paper "Capture of cobalt-60 by suspended particles in the Danube", Proc. International Symposium on Radioecology, Cadarache, September 1969, Vol. I, pp. 83-92, as well as in the paper "Capture of cobalt-60 on the bed material of the Danube river", Report IBIC-1012.

S.O.W. BERGSTRÖM: Was the value 1500 pCi/g measured in more than one sample?

T. TASOVAC: Yes, we found this high value in several samples.

M.M. SAUROV (Chairman): Have you determined the quantities of radionuclides in the sediments at the bottom of the Danube?

T. TASOVAC: We measured the activity of cobalt-60 in suspended material, a lot of which will have become bed material as a result of settling caused by the construction of a dam.

M.M. SAUROV: Did you measure the rate at which radionuclides flow past the point where you made your measurements?

T. TASOVAC: No.

POTENTIAL HARM AND ECOLOGICAL ASPECTS AND CONTROL

Potential harm to populations from exposures

(Session VII)

Chairman: J. C. VILLFORTH (United States of America)

CONSIDERATIONS IN ASSESSING THE POTENTIAL HARM TO POPULATIONS EXPOSED TO LOW LEVELS OF PLUTONIUM IN AIR*

W.J. BAIR
Biology Department,
Battelle Pacific Northwest Laboratories,
Richland, Wash.,
United States of America

Abstract

CONSIDERATIONS IN ASSESSING THE POTENTIAL HARM TO POPULATIONS EXPOSED TO LOW LEVELS OF PLUTONIUM IN AIR.

Results from experimental animal studies have provided extensive knowledge of the disposition and biological effects of plutonium. While these laboratory studies involved levels of plutonium much higher than the predicted dose commitments from nuclear power plant operation, the results of these experiments along with the limited data from accidental human exposures can provide a basis for preliminary assessment of the harm to populations from plutonium that might be released to the environment. Experimental animal studies have identified lung and bone cancer as the most sensitive effects of plutonium exposure. Drawing upon the results from experiments with several animal species and several plutonium compounds the possibility of deriving risk estimates for plutonium-induced lung cancer on a toxicological basis was explored, thus avoiding the need to consider the spatial distribution of absorbed radiation dose in lung, latent period for cancer induction, etc. Values for the risk of lung cancer resulting from inhaled plutonium average about 2×10^{-4} per nCi/g lung·days for ^{239}Pu. A value of 2×10^{-3} was calculated for ^{238}Pu from the results of one experiment with rats.

1. INTRODUCTION

Plutonium is a component of global fallout originating from nuclear weapons testing and burnup reentry of a satellite that contained a plutonium power source. According to the 1972 report of the United Nations Scientific Committee on the Effects of Atomic Radiation, the total integrated level of plutonium in surface air from the beginning of the nuclear tests through 1970 is of the order of 4 fCi years per M^3 [1]. Inhalation of air containing this concentration of plutonium as particles of 0.5 μm AMAD over this period would give integrated doses over 50 years of about 2, 400, 0.8, and 0.2 millirads to the pulmonary region of the lung, thoracic lymph nodes, liver, and bone, respectively [1]. Recently, estimates have been made of the dose commitment from plutonium entering the environment as a result of current and predicted worldwide proliferation of nuclear power plants including the plutonium-fueled breeder reactors [2]. What is the significance of these estimated dose commitments in terms of harm to the population? To answer this question it is necessary to draw upon the results of numerous laboratory studies in which experimental animals were exposed to air contaminated with plutonium particles or administered plutonium compounds by injection. While these laboratory studies involved levels of plutonium much higher than the dose commitment levels predicted, the results of these experiments along with the limited data from accidental human exposures can provide a basis for a preliminary assessment of

* Based on work performed under US Atomic Energy Commission Contract AT (45-1)-1830.

the harm to human populations that might result from plutonium released to the environment. Experimental animal studies have identified pulmonary and bone cancer as the most sensitive effects of plutonium contamination. However, no specific physical injury attributed to plutonium has been identified in persons occupationally exposed to plutonium contamination, some exposed more than 25 years ago.

This paper will deal with the problem of assessing the harm to a populatio exposed to levels of plutonium in air of the order predicted for nuclear energy electrical power generation. First, it is important to understand the behavior of inhaled plutonium in the body. This has been the subject of recent reviews and will only be summarized here [3,4,5]. Consideration also will be given to the biological effects observed in experimental animal studies and the problem of determining risk estimates for man.

2. BEHAVIOR OF INHALED PLUTONIUM

Airborne plutonium particles are similar to most other particles when they are inhaled in that deposition in the respiratory tract is primarily dependent upon the physical properties of the particles and the respiratory characteristics of the individual inhaling the particles. After deposition in the respiratory tract, plutonium may be retained in the lung for a long time, be translocated to other tissues in the body, accumulated in the lymph nodes associated with the respiratory tract or excreted in urine and feces. The actual disposition of inhaled plutonium depends largely upon the physical and chemical characteristics of the inhaled material.

2.1. Clearance of plutonium from lung

Within the first week after exposure, a fraction of the deposited plutonium is cleared from the respiratory tract and excreted. The magnitude o the fraction cleared within the first week depends upon the fraction of readil soluble material present and also upon the distribution of the deposited plutonium within the respiratory tract. Plutonium particles deposited on the ciliated epithelium of the upper respiratory tract are trapped in mucus, propelled to the esophagus, and swallowed. Plutonium deposited in the lower regions of the lung, in the alveoli, is not readily available for clearance an may be incorporated into the cellular structures of the lung and retained for long time.

The kinetics of the clearance of plutonium from lung are complicated and difficult to quantitate. Since the clearance of plutonium from the lower lung is generally assumed to be exponential with time over a reasonably long period after exposure, retention half-times are estimated. Animal experiments and limited human data provide a range of values for the retention half-times of several plutonium compounds [6], Figure 1. The retention half-times for organ complexes of plutonium, plutonium nitrate and plutonium fluoride range from le than 100 days to about 300 days in rats and dogs. The retention half-times fo PuO_2 are substantially longer, ranging from 200 to 500 days in rats, 300 to 1C days in dogs and 250 to 300 days in human beings. The wide range of values fo dogs is largely due to extensive experimentation with a variety of plutonium oxides with different particle size characteristics. For example, PuO_2 calcin at high temperatures is cleared more slowly than air oxidized plutonium; PuO_2 comprised of large particles ($\sim$ 3 μm AMAD) tend to be cleared more slowly than aerosols of small PuO_2 particles ($\sim$ 0.1 μm AMAD)[7]; and $^{238}PuO_2$ has a much shorter lung retention time than $^{239}PuO_2$. The relatively low values for human beings, compared with dogs, suggest either that man clears plutonium particles from his lungs more rapidly than do dogs or that the materials inhaled in the human accident cases were more soluble than plutonium dioxide.

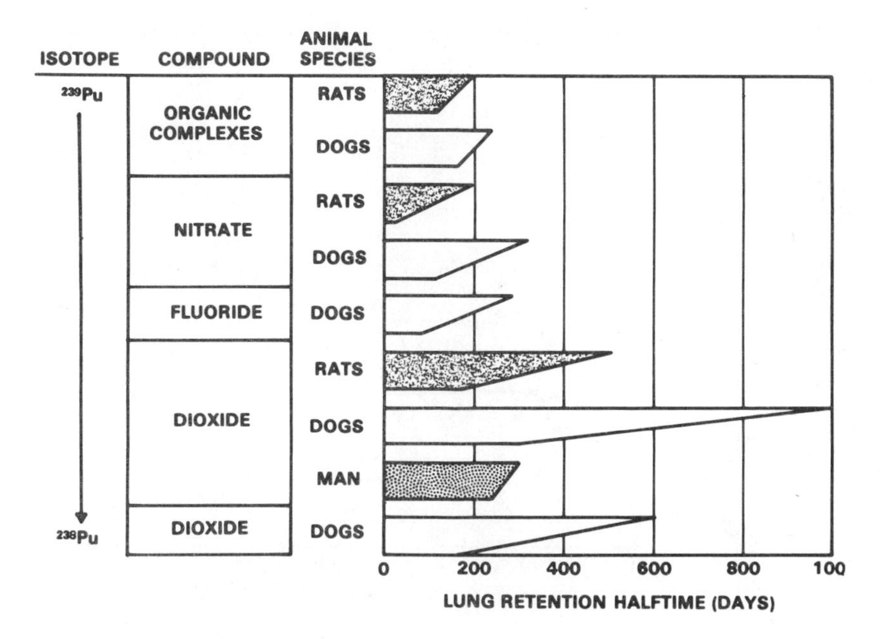

FIG. 1. Retention of plutonium in pulmonary region of lung. Ranges of published values for retention half-times are indicated for each animal species and plutonium compound.

2.2. Translocation

The relative distribution among body tissues of plutonium translocated from lung is essentially the same for all plutonium compounds, but may differ quantitatively depending upon the chemical and physical state of the plutonium inhaled.

In beagle dogs within several months after inhalation of plutonium nitrate, the fraction remaining in lung decreased to 40% or less of the amount deposited in the lower respiratory tract, Figure 2 [8]. Translocation of plutonium from lung resulted in bone accumulating about 30% and liver, about 10%, of the alveolar-deposited plutonium. A small percentage was found in spleen, lymph nodes, and other soft tissues and the remainder was excreted.

When plutonium dioxide is inhaled, the lymphatic system accounts for a large fraction of plutonium cleared from lung. Data from a 11-year study with beagle dogs shows that after 11 years the lung burden had decreased to about 9% and the thoracic lymph nodes had accumulated 40% of the alveolar-deposited plutonium. Translocation of plutonium from lung resulted in levels of about 10% in liver, 5% in bone and 7% in abdominal lymph nodes, Figure 3 [9].

The relative mean radiation doses to these tissues bear the same relationship as the plutonium concentrations. The mean concentration of plutonium was highest in the thoracic lymph nodes and next highest in the abdominal lymph nodes, Table I.

In experiments with dogs translocation of ^{238}Pu from lung to other tissues in the body was greater than for ^{239}Pu after inhalation of oxides with similar

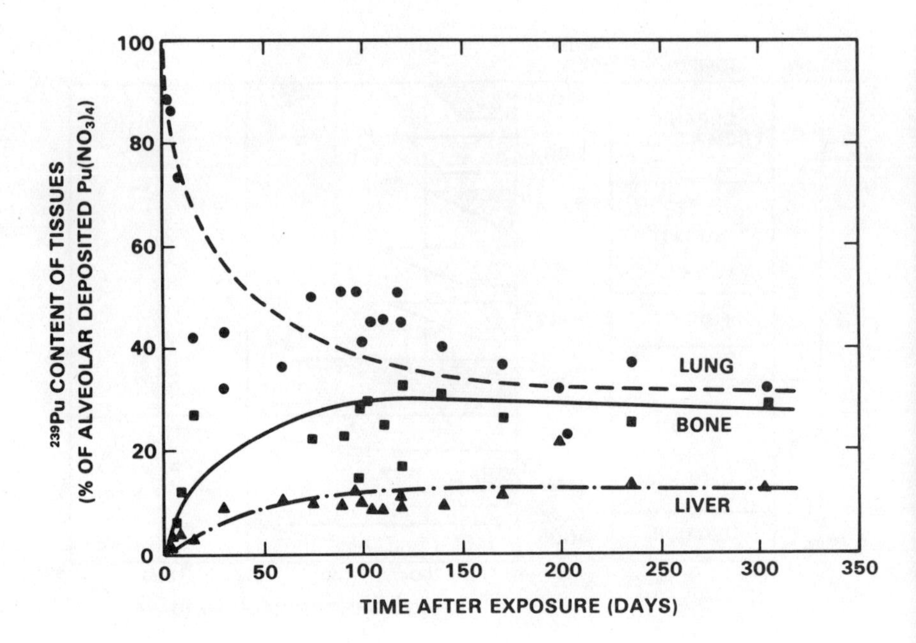

FIG. 2. Distribution of plutonium in beagle dogs after inhalation of $^{239}Pu(NO_3)_4$ [8].

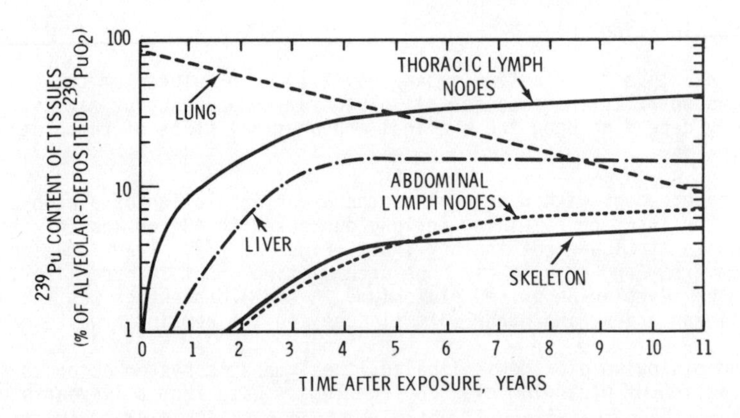

FIG. 3. Distribution of plutonium in beagle dogs after inhalation of $^{239}PuO_2$ [9].

physical properties [10]. Five years after exposure only 10% of the body bur of the ^{238}Pu was in lung compared with 46% for ^{239}Pu. Accumulation in thorac lymph nodes was three times greater for ^{239}Pu (29%) than for ^{238}Pu (10%); however, the bone burden of ^{238}Pu (48%) was 12 times that of ^{239}Pu (4%). PuO particles composed of ^{238}Pu with a high specific activity of 17.4 Ci/g compar with 6.14×10^{-2} Ci/g for ^{239}Pu appear to be unstable in aqueous media, possibly due to radiolysis [10].

TABLE I. RELATIVE CONCENTRATIONS OF PLUTONIUM IN TISSUES OF DOGS 7-9 YEARS AFTER INHALATION OF $^{239}PuO_2$

Tissue	Relative Concentration of Plutonium[a]
Lung	1
Thoracic Lymph Nodes	1400
Abdominal Lymph Nodes	100
Liver	0.5
Spleen	0.2
Bone	0.06

(a) Normalized to the value for lung.

Mean of five dogs calculated from Park et al., 1972 [9].

TABLE II. RELATIVE CONCENTRATIONS OF PLUTONIUM IN HUMAN TISSUES

Tissue	Type of Exposure	
	Fallout[a]	Occupational[b]
	Relative Concentration of Pu[c]	
Lung	1	1
Thoracic Lymph Nodes	8	60
Hepatic Lymph Nodes	-	0.4
Liver	3	0.1
Kidney	2	-
Bone	1	0.2

(a) Calculated from Campbell et al, 1973 [11].
(b) Calculated from Nelson et al., 1972 [12].
(c) Normalized to the values for lung.

Human contamination has occurred as a result of occupational exposures in the nuclear industry and as a result of exposure to fallout plutonium from nuclear weapons testing. Tissues obtained from a number of human autopsy cases have been analyzed for plutonium, Table II. The concentrations of plutonium in thoracic lymph nodes, liver and kidney are higher than in lung and bone. This suggests that fallout plutonium contained a relatively soluble fraction, e.g., the particle size may have been quite small. The data from an occupational case suggest the individual was exposed to a relatively insoluble plutonium compound because the relative tissue concentrations are similar to those in dogs exposed to PuO_2.

2.3. Spatial distribution of plutonium within tissues

Neither soluble nor insoluble plutonium compounds are uniformly distributed throughout the tissues in which they are deposited. Plutonium particles do not remain at or near the site of deposition for the duration of their

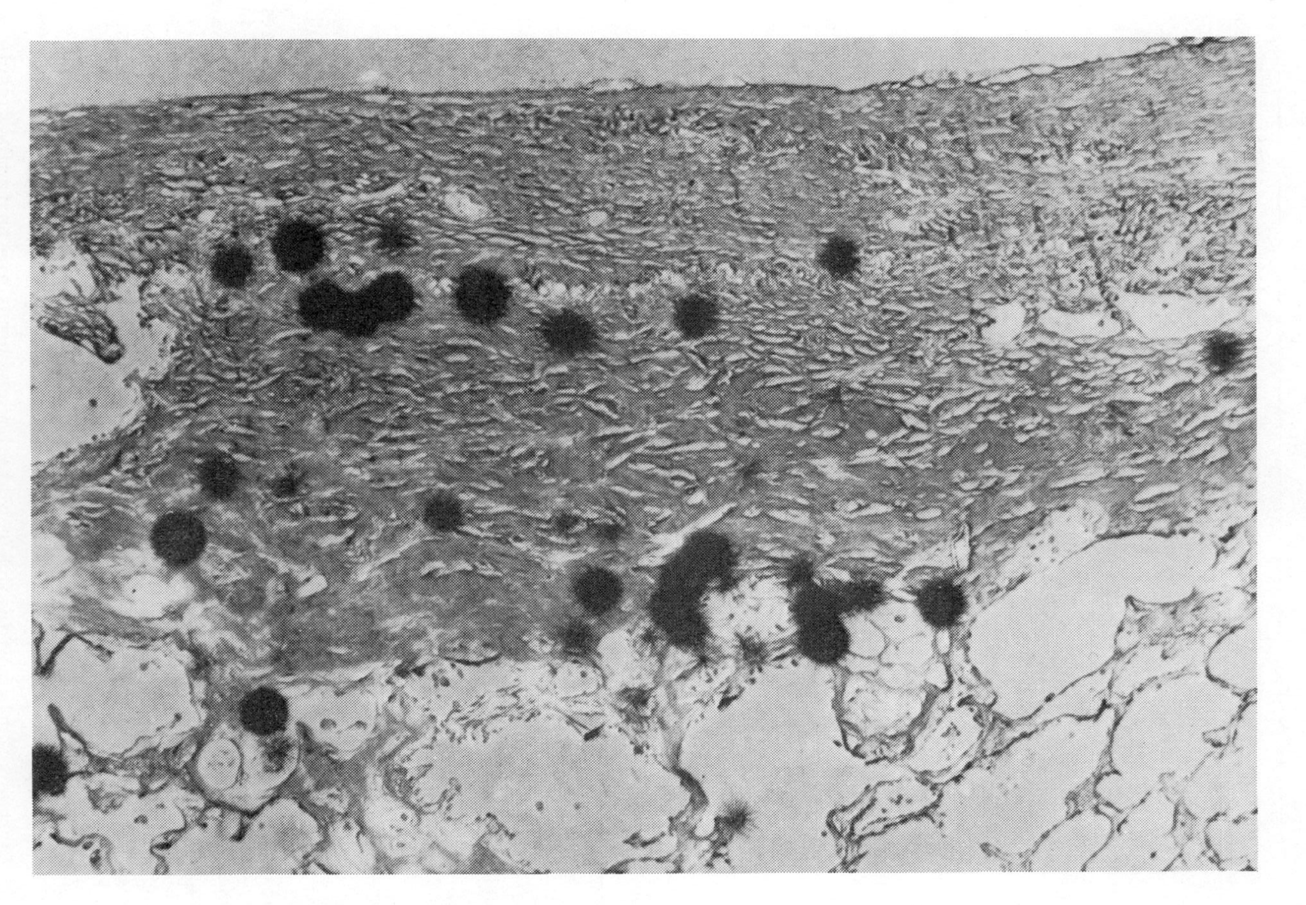

FIG. 4. Autoradiograph of lung section from dog several months after inhalation of $^{239}PuO_2$, showing subpleural concentration of plutonium particles. 320X. (Provided by G.E. Dagle, Battelle Pacific Northwest

residence time in lung. Both soluble and insoluble plutonium compounds tend to become further aggregated in lung tissue as a result of phagocytosis by macrophages and Type I alveolar cells and are mobilized to peribronchiolar and subpleural areas (Figure 4). The most obvious changing distributions of plutonium within lung are those associated with the accumulation of plutonium in lymphatics and transport to thoracic lymph nodes and with the continued slow clearance of plutonium from lung by the mucociliary pathway.

Plutonium in thoracic or other lymph nodes is concentrated in lymphatic sinuses of subcapsular and medullary area and may be eventually sequestered in scar tissue. Plutonium translocated to liver is at first uniformly distributed among the hepatic cells. However, with time the plutonium becomes localized in the reticuloendothelial cells lining the sinusoids and in connective tissue [13]. Occasionally, large alpha stars are seen in liver suggesting the possibility of intact particles being translocated from lung to liver. Plutonium translocated to bone is deposited nonuniformly on bone surfaces where it may be buried by apposition of new bone or removed by resorption. In the latter case plutonium aggregates are redistributed to osteoclasts on bone surfaces or to marrow spaces by macrophage activity.

The dynamic nature of plutonium in the principal tissues in which it is deposited, lung, liver, lymph nodes and bone, results in a complex distribution of absorbed alpha radiation energy from the radioactive decay of the plutonium. This causes considerable uncertainty in identifying the tissue components and radiation doses that are associated with resulting long term biological effects such as cancer. When the amount of plutonium deposited in tissue such as lung is relatively large, the radiation dose will be distributed throughout a large part of the lung, although nonuniformly. When only a few particles are deposited in lung, only a small fraction of the lung will be irradiated. However, it is unlikely the radiation dose will be completely localized unless the plutonium is immobilized in scar tissue. Because of all the uncertainties regarding the distribution of plutonium in lung, it has been customary to assume uniform distribution of the absorbed energy throughout the total lung tissue and a mean dose calculated. This has appeared to be adequate for quantitative descriptions of biological effects in experimental animal studies. However, there is some concern that averaging the dose over the total lung is not adequate for low level lung burdens of a few hundred or thousand small particles.

3. BIOLOGICAL EFFECTS

It has been shown that the distribution of plutonium and the radiation dose among the tissues in the body will vary depending upon the physical and chemical characteristics of the plutonium inhaled. The biological effects that occur will depend upon the radiation exposure and the relative radiation sensitivity of each tissue into which plutonium is transported and deposited. These are primarily blood, bone, liver, lung, and lymphatic system. The biological effects of greatest interest are those that might occur at low doses, such as cancer. Since no detrimental biological effects have been observed in persons occupationally exposed to plutonium that can be unequivocally related to plutonium, all our knowledge of the biological effects of plutonium have come from animal experiments.

3.1. Blood

Plutonium is cleared from the circulating blood within a few days after absorption. Therefore, the effects seen in blood cells are probably due to irradiation of hematopoietic tissue in which plutonium is deposited or to irradiation of blood circulating through plutonium-containing tissues. Although plutonium deposits in bone and lymphatic tissue, leukemia has not been a common finding in animal experiments. The most sensitive effect seen in blood is a reduction of the number of circulating lymphocytes. In current experiments with dogs this appears to be a very sensitive indicator of a biological effect occurring after inhalation of $^{239}PuO_2$. The possible health consequences of this reduced level of circulating lymphocytes are not yet known.

3.2. Bone

Radiation-induced osteogenic sarcoma is the most sensitive effect of plutonium deposited in bone. Because plutonium deposits on bone surfaces, a large fraction of the alpha energy is absorbed in sensitive cells. Although osteogenic sarcoma has been reported in rats that accumulated mean doses of less than 10 rads, Figure 5, statistically significant increases in tumor incidence have not occurred at doses less than 30 to 50 rads. The available data suggest that the dose-effect curve for dogs is different from that for rodents. Mays and Lloyd [14] conclude that the incidence of plutonium-induced osteogenic sarcoma is 0.38%/rad for beagle dogs, 0.10%/rad for mice, and 0.06%/rad for rats.

Osteogenic sarcoma has been observed in mice, rats, rabbits, and dogs after intravenous injection, inhalation, or intratracheal injection of several plutonium-239 compounds including organic complexes and plutonium nitrates. Osteogenic sarcoma has also been observed in rats after inhalation of $^{238}PuO_2$, but has not been reported in any animal species after inhalation of $^{239}PuO_2$.

3.3. Liver

Liver accumulates levels of plutonium similar to bone. However, liver tumors have not been a common finding in experimental animal studies. Bile duct tumors have been observed to occur in beagle dogs at doses as low as 60 rads [13]. However, not only was the incidence very low, but bile duct tumors also occurred in control dogs.

3.4. Lung

Inhalation of relatively soluble plutonium compounds such as organic complexes, plutonium nitrate, and $^{238}PuO_2$ has resulted in primary pulmonary neoplasia in rodents, rabbits, and dogs in addition to osteogenic sarcomas already mentioned, Figure 6. Pulmonary neoplasia has also been observed in beagle dogs, baboons [22] and rodents after inhalation of $^{239}PuO_2$. Tumors were observed in animals that had lung doses of less than 10 rads. Statistica significant increases in pulmonary neoplasia occurred at doses of about 30 rad and above. In a 11-year study of $^{239}PuO_2$ in beagle dogs, nearly all animals which deposited between about 3 nCi and 50 nCi per gram of lung had lung tumor The mean dose to the lungs of the dogs that developed pulmonary neoplasia rang from 1200 to 4000 rads. As in the case of osteogenic sarcoma, the dose-effect curve for pulmonary neoplasia in dogs appears to differ from the rodent dose-effect curve.

3.5. Lymph nodes

It was shown that plutonium accumulates in lymph nodes following depositi in the respiratory tract and may attain concentrations many times the average concentrations in lung. Although dogs have been studied for 11 years after inhaling $^{239}PuO_2$ and rodents have been studied in life-span experiments after inhalation of a variety of plutonium compounds, primary neoplasia of lymphatic tissue has not been a common observation. Metastases of tumors from lung to

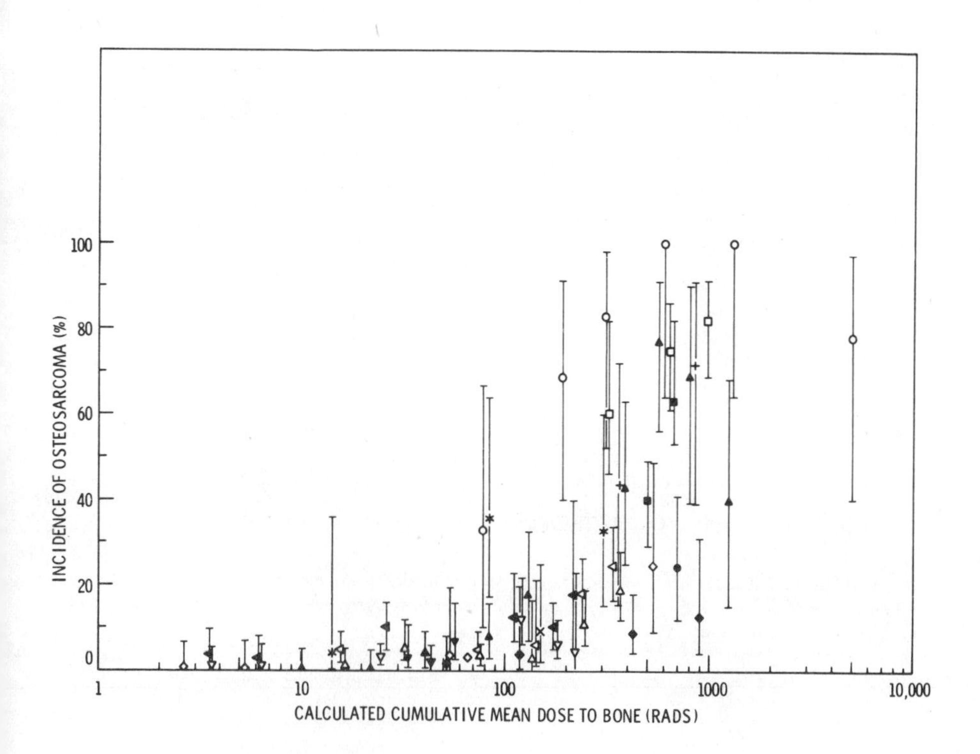

FIG. 5. Plutonium-induced osteosarcoma in experimental animals. Mean incidence and radiation dose values are those reported in the literature. Binomial confidence limits were calculated from data included in the referenced literature.

○ ^{239}Pu citrate, monomeric - IV - dogs [15]
△ ^{239}Pu citrate - inhaled - rats [16]
▽ ^{239}Pu plutonylpentacarbonate - inhaled - rats [16]
◇ ^{239}Pu nitrate - sub- and intracutaneous - rats [17]
▼ ^{239}Pu citrate - oral (daily) - rats [3]
● ^{239}Pu plutonyltriacetate - I. T. - rats [18]
▲ ^{239}Pu citrate - IV - mice [19]
□ ^{239}Pu citrate, monomeric - IV - mice [20]
■ ^{239}Pu citrate, polymeric - IV - mice [20]
× $^{238}PuO_2$ - inhaled - rats [29]
◆ ^{239}Pu nitrate - I. T. - rats [18]
\+ ^{239}Pu nitrate - I. T. - rabbits [21]
✶ ^{239}Pu (pentacarbonate) - inhaled - rabbits [21]
◁ ^{239}Pu citrate - inhaled - rats [21]
◀ ^{239}Pu pentacarbonate - inhaled - rats [21]

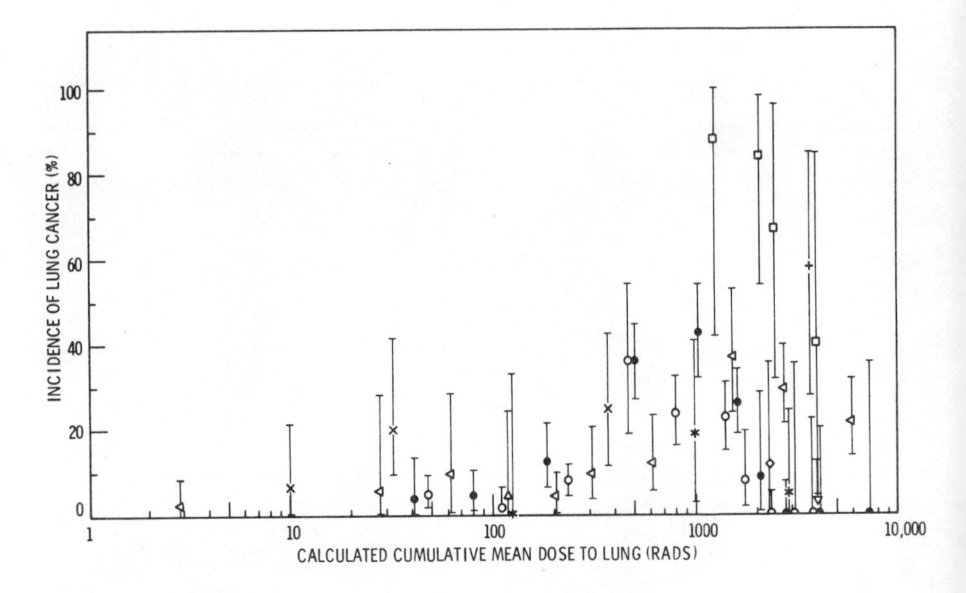

FIG. 6. Plutonium-induced lung cancer in experimental animals. Mean incidence and radiation dose values are those reported in the literature. Binomial confidence limits were calculated from data included in the referenced literature.

□ $^{239}PuO_2$ - dogs [23]
▽ $^{239}PuO_2$ - mice [24]
△ $^{239}PuO_2$ - mice [24]
◇ $^{239}PuO_2$ - mice [25]
○ ^{239}Pu citrate - rats [16]
● ^{239}Pu plutonylpentacarbonate - rats [16]
× ^{238}Pu - rats [29]
◁ ^{239}Pu - rats - $Pu(NO_3)_4$ [18]
+ ^{239}Pu - rabbits - $Pu(NO_3)_4$ [21]
✱ ^{239}Pu - rabbits - NH_4 Pu pentacarbonate [21]

lymph nodes have occurred, but none have been the primary cause of death. Numerous soft tissue tumors have been reported including epithelial tumors of hypophysis and adrenal after intratracheal injection of $^{239}Pu(NO_3)_4$ [3] and mesenchymal tumors of the visceral peritoneum after intraperitoneal injection of $^{239}PuO_2$ [26].

Studies of the biological effects of inhaled plutonium indicate that bone and lung are the most susceptible tissues to the carcinogenic action of low levels of plutonium. However, the experiments have been with higher level of plutonium than would be expected to occur as a result of most accident situations involving human exposures and many orders of magnitude higher than predicted for population exposures. All tissues which accumulate plutonium mu remain suspect as possible sites of neoplasia or other effects detrimental to the individual or his progeny.

4. APPLICATION OF EXPERIMENTAL ANIMAL DATA TO RISK ASSESSMENT

The United Nations Scientific Committee on the Effects of Atomic Radiati concluded that existing experimental animal data on radiation carcinogenesis a quantitatively inadequate for estimating cancer risks to the human population [1]. Therefore, most estimates of risk are based on results from studies of patients given radiation therapy and from studies of accidental whole-body exposures at high dose rates. These data can be applied to many kinds of radiation exposure, but their application to the continuous irradiation of lung tissue with high LET alpha particles can be questioned.

Because there are no human data on the biological effects of plutonium, it is not known exactly how to extrapolate experimental animal data to man. However, since lung and bone have been shown to be susceptible to the carcinogenic action of plutonium in several animal species, the assumption that these tissues in man would respond in the same way to plutonium is reasonable.

From results of experiments with intravenously injected plutonium citrate in dogs, Mays et al. [27] calculated that average skeletal and liver dose rates of 8.8 and 8.0 rad/year were associated with a 33% incidence of osteosarcoma and a 17% incidence of liver tumors, respectively. Applying these data to a 50-year average dose rate of 0.014 rad/nCi to the skeleton of man and 0.057 rad/nCi to liver, and using a non-threshold linear extrapolation, estimated tumor risks of 5×10^{-4} per nCi for bone and 1.2×10^{-3} per nCi for liver were calculated for man. Thompson et al. [28] used data from a life-span study of inhaled $^{239}PuO_2$ in dogs to calculate a risk estimate of 2×10^{-4} per nCi for pulmonary neoplasia in man. This estimate was based on the finding that dogs which deposited more than about 3 nCi $^{239}PuO_2$ per gram lung (blood free) would have a high probability of developing lung cancer. This method of

TABLE III. RISK OF LUNG CANCER FROM INHALED PLUTONIUM ESTIMATED FROM EXPERIMENTAL ANIMAL STUDIES

Animal Species	Plutonium Compound	Plutonium Deposited (nCi/g lung)	Lung Cancer Incidence (fraction)	Lung Cancer Risk(a) (incidence/ nCi/g·days) x 10^{-4}	Reference from which Experimental Data Were Obtained
Mouse	$^{239}PuO_2$	6.	.12	0.6	[25]
Rat	^{238}Pu (hydroxide)	7.2	.233	20.	[29]
Rat	^{239}Pu Citrate	2.6	.05	4.	[16]
Rat	^{239}Pu Penta-carbonate	15.	.364	2.	[16]
Rat	$^{239}Pu(NO_3)_4$	1.4	.059	5.	[18]
Rat	^{239}Pu-Plutonyl triacetate	333.	.376	0.2(b)	[18]
Rabbit	^{239}Pu Penta-carbonate	11.	.167	0.5(b)	[21]
Rabbit	$^{239}Pu(NO_3)_4$	43.	.583	0.5(b)	[21]
Dog	$^{239}PuO_2$	3.5	.88	2.	[4]

(a) Calculated from experimental group showing statistically significant increased cancer incidence and giving maximum lung cancer risk.

(b) Life-span of experimental group was significantly reduced because of high plutonium dose. The risk would probably have been higher at plutonium doses that allowed animals to survive longer.

extrapolating the results from animal experiments to man avoids calculating radiation doses and making assumptions about energy absorption and critical tissues. It deals directly with the toxicological effects of plutonium and is essentially the method used to evaluate the toxicity of non-radioactive materials. A modification of this approach can also be applied to the experiments from which the lung cancer incidence data in Figure 6 were derived. By assuming exponential rates of loss of plutonium deposited in the alveolar regions of the lung and integrating over the mean life-span of the animals in each group, doses expressed as nCi/g lung · days can be calculated.[1] Using a non-threshold linear extrapolation of the observed cancer incidence for each group of animals in the plutonium experiments, risk estimates for lung cancer were calculated, Table III. Low risk estimates of the order of 10^{-6} per nCi/g · days were obtained using data from the experimental groups given sufficiently large plutonium doses to cause an appreciable shortening of life-span, probably because the animals did not live long enough for the full cancer potential to be expressed. Risk estimates calculated from experimental groups which showed little or no life shortening were of the order of 10^{-4} per nCi/g · days. Values ranged from only about 0.6 to 20 x 10^{-4} per nCi/g · days for several plutonium compounds in the mouse, rat, and dog. If this risk estimate is applied to man, the pulmonary deposition of the permissible occupational lung burden of 16 nCi would result in a risk of lung cancer of about 1 in 1000 for ^{239}Pu and 1 in 100 for ^{238}Pu. These values may very likely be overestimates of the risks because of the conservative approach taken in extrapolating from the results of animal experiments.

While the data available for this kind of exercise are limited, the agreement among four animal species and several plutonium compounds is promising. Additional experimental data are required to develop confidence in estimating the risk of lung cancer from inhaled plutonium by direct application of toxicological data.

REFERENCES

[1] United Nations, Scientific Committee on the Effects of Atomic Radiation, "Ionizing Radiation: Levels and Effects". A report of the Committee to the General Assembly, with annexes. UN, New York (1972).

[2] KLEMENT, A.W., Jr., et al., Estimates of Ionizing Radiation Doses in the United States 1960-2000, USEPA Rep. ORP/CSD 72-1 (1972).

[3] BULDAKOV, L.A., et al., Problemy Toksikologii Plutoniya, Atom Publications Moscow (1969) (translated as "Problems of Plutonium Toxicology", U. S. Atomic Energy Commission LF-TR-41, pp. 149-178, 1970).

[4] BAIR, W.J., et al., "Plutonium in soft tissues with emphasis on the respiratory tract", Ch. 11, Handbook of Experimental Pharmacology; Uranium, Plutonium, Transplutonic Elements, Vol. 36 (HODGE, H.C. STANNARD, J.N., HURSH, J.B. Eds), Springer-Verlag, Berlin (1973).

---

[1] I elected to express the plutonium dose in nCi/g · days rather than calculate radiation doses, to avoid complicating this discussion with concern about the importance of the spatial distribution of the absorbed radiation dose from plutonium in lung tissue.

[5] BAIR, W.J., "Toxicology of Plutonium", pp. 255-315, Advances in Radiation Biology, Vol. 4 (LETT, J.T., ADLER, H., ZELLE, M., Eds), Academic Press, Inc. (1974).

[6] BAIR, W.J., THOMPSON, R.C., Science 183 (1974) 715-722.

[7] BAIR, W.J., "Plutonium Inhalation Studies", BNWL-1221 (1970), Battelle-Northwest, Richland, Washington.

[8] BALLOU, J.E., PARK, J.F., MORROW, W.G., Health Phys. 22 (1972) 857-862.

[9] PARK, J.F., BAIR, W.J., BUSCH, R.H., Health Phys. 22 (1972) 803-810.

[10] PARK, J.F., et al., "Proceedings of the Third International Congress of the International Radiation Protection Association", Washington, D.C., September 9-14, 1973 (in press).

[11] CAMPBELL, E.E., et al., "Plutonium in autopsy tissue", USAEC Doc. LA-4875, 47 p. (1973).

[12] NELSON, I.C., et al., Health Phys. 22 (1972) 925-930.

[13] TAYLOR, G.N., et al., "Hepatic changes induced by ^{239}Pu", pp. 105-127, Radiobiology of Plutonium (STOVER, B.J., JEE, W.S.S., Eds), J. W. Press, University of Utah, Salt Lake City (1972).

[14] MAYS, C.W., LLOYD, R.D., "Bone sarcoma incidence vs. alpha particle dose", pp. 409-430, Radiobiology of Plutonium (STOVER, B.J., JEE, W.S.S., Eds), J. W. Press, University of Utah, Salt Lake City (1972).

[15] JEE, W.S.S., Health Phys. 22 (1972) 583-595.

[16] BULDAKOV, L.A., LYUBCHANSKY, E.R., "Experimental basis for maximum allowable load (MAL) of ^{239}Pu in the human organism, and maximum allowable concentration (MAC) of ^{239}Pu in air at work locations", Report ANL-trans-864, Argonne National Lab., Argonne, Illinois. (Translated from a Russian Report, 32 pp.) (1970).

[17] BULDAKOV, L.A., et al., "Biological effect of plutonium-239 with cutaneous and intracutaneous injection", pp. 350-355, Remote Aftereffects of Radiation Damage (MOSKALEV, Y.I., Ed), Atomizdat, Moscow (1971)(Translated in AEC-tr-7387, 381-387).

[18] EROKHIN, R.A., et al., "Some remote aftereffects of intratracheal administration of chemically soluble plutonium-239 compounds", pp. 315-333, Remote Aftereffects of Radiation Damage (MOSKALEV, Y.I., Ed) Atomizdat, Moscow (1971) (Translated in AEC-tr-7387, 344-363).

[19] FINKEL, M.P., BISKIS, B.O., Health Phys. 8 (1962) 565-579.

[20] ROSENTHAL, M.W., LINDENBAUM, A., Radiat. Res. 31 (1967) 506-521.

[21] KOSHNURNIKOVA, N.A., LEMBERG, V.K., LYUBCHANSKY, E.R., "Remote after-effects of inhalation of soluble plutonium-239 compounds", pp. 305-314, Remote Aftereffects of Radiation Damage (MOSKALEV, Y.I., Ed) Atomizdat, Moscow (1971) (Translated in AEC-tr-7387, 334-343).

[22] METIVIER, H., et al., C. R. Acad. Sci. Ser. D 275 (1972) 3069-3071.

[23] PARK, J.F., BAIR, W.J., Personal communication (1972).

[24] TEMPLE, L.A., et al., Nature (London) 183 (1959) 408-409.

[25] WAGER, R.W., et al., "Toxicity of radioactive particles IA. Intratracheal injection of radioactive suspensions", pp. 61-72, Biology Research Annual Report for 1955, HW-41500 (1956), Hanford Atomic Products Operation, Richland, Washington.

[26] SANDERS, C.L., JACKSON, T.A., Health Phys. 22 (1972) 755-759.

[27] MAYS, C.W., et al., Health Phys. 19 (1970) 601-610.

[28] THOMPSON, R.C., PARK, J.F., BAIR, W.J., "Some speculative extensions to man of animal risk data on plutonium", pp. 221-230, Radiobiology of Plutonium (STOVER, B.J., JEE, W.S.S., Eds), J. W. Press, University of Utah, Salt Lake City (1972).

[29] SANDERS, C.L., Radiat. Res. 56 (1973) 540-553.

DISCUSSION

A. LOPEZ BARRIOS: What monitoring systems exist for Pu?

W.J. BAIR: Your question is very broad and could be the subject of a separate meeting! I assume you are asking about monitoring for plutonium in the human body. To identify occupational exposure, nasal smears are collected and analysed for Pu. A bioassay programme based on routine collections of urine and faecal samples is used to quantify suspected plutonium exposure. In addition, whole-body and chest counting techniques are employed. These usually involve the use of thin (1 mm) sodium iodide crystals in photomultiplier tubes to detect the 15-keV X-rays emitted during the decay of Pu and the 60-keV photons emitted from the decay of ^{241}Am, which is present as a contaminant of most plutonium. Other monitoring systems under development include an intragastric probe to measure the amount of plutonium in the thoracic lymph nodes and proportional counters to increase the sensitivity of chest counters. The minimum amount of plutonium which can be detected in the human lung by external monitoring systems is about 5 nCi (5×10^{-9} Ci).

A. LOPEZ BARRIOS: Are there any special regulations for the users of pacemakers containing Pu?

W.J. BAIR: The use of plutonium-238 in pacemakers was approved in the United States of America on condition that the ^{238}Pu was adequately contained and would not be released in an accident situation, e.g. automobile collision, etc. The manufacture and use of pacemakers containing ^{238}Pu is regulated to ensure that they meet the containment requirements.

J. SUŠNIK: One of your tables showed the relative concentrations of Pu in different human tissues, due to fall-out and due to occupational exposure. Could you give us some values for the actual Pu concentration in these tissues resulting from fall-out?

W.J. BAIR: Analyses of autopsy samples indicate that the body burden from fall-out is about 2 pCi (2×10^{-12} Ci). The lung burden is about 0.4 pCi.

H. P. JAMMET: The data you have given show that in spite of very high Pu concentrations in lymphatic ganglia it has not been possible to discern any pathological changes. Does this mean that the lymphatic ganglia are not a critical organ for Pu?

W. J. BAIR: Results from many animal experiments in which lymph nodes were exposed to a wide range of plutonium doses do not lead to the conclusion that lymph nodes are a critical tissue for plutonium contamination. Although histological changes have been observed in lymph nodes and tumours have formed in the lymphatic vessels of animals after inhalation of plutonium, these effects were secondary to pulmonary neoplasm, which was the cause of death. However, I believe continued attention should be given to the possibility that lymph nodes might show carcinogenic responses at very low plutonium doses.

ASSESSMENT OF POTENTIAL HEALTH CONSEQUENCES OF TRANSURANIUM ELEMENTS*

N.F. BARR
Division of Biomedical and Environmental Research,
United States Atomic Energy Commission,
Washington, D.C.,
United States of America

Presented by M.B. Biles

Abstract

ASSESSMENT OF POTENTIAL HEALTH CONSEQUENCES OF TRANSURANIUM ELEMENTS.

The results are presented of a study to provide quantitative estimates, per unit of electrical power generated, of potential consequences to human health from inhaled or ingested transuranium elements released during the generation of electricity by the liquid metal fast breeder reactor (LMFBR) fuel cycle. Soon after release direct inhalation of downward descending airborne particles is the dominant route of entry of transuranic elements into man. The fraction of material entering man in this way is estimated to be 4×10^{-6}. A comparable fraction enters man through inhalation of material resuspended into the atmosphere after deposition.

Estimates for the fraction of released material that enters man through ingestion processes from the time of release to complete decay of the activity are made for the most significant radioisotopes of plutonium, americium and curium. ^{241}Am is the largest contributor to dose by this pathway. Health effects estimates based on current metabolic data, ICRP dose models and the report of the committee on Biological Effects of Ionizing Radiation of the US National Academy of Sciences, when combined with these pathway analyses, indicate that there are unlikely to be more than about 10^{-3} potential health effects for each 1000 megawatt electric year of power generated.

One of the major concerns regarding nuclear operations is the release of transuranic elements to the environment. The concern is based largely upon two considerations:

1. the well documented toxic effects observed in experimental animals exposed to sufficiently high levels of alpha-emitting nuclides; and

2. the exposure of many hundreds of human generations to long-lived transuranic radionuclides which persist and accumulate in the environment.

One perspective on the magnitude of this problem is provided by an analysis of total population exposures and potential health consequences which might occur to all future generations as the result of releases of transuranic elements.

---

* Much of this paper is based upon an analysis undertaken in connection with the preparation of the Draft Environmental Impact Statement for the US Liquid Metal Fast Breeder Program. The author wishes to acknowledge the contributions of numerous USAEC staff and laboratory scientists to this analysis.

The results of such an analysis for the operation of the U.S. Liquid Metal Fast Breeder Reactor (LMFBR) are summarized in this paper. The analysis does not include consideration of the extremely important matter of individual exposures and the distribution of exposures amongst population groups since these are largely dominated by conditions specific to actual locations of release.

The analysis estimates were made in three categories:

1. The amount of transuranic elements expected to be released during operation of LMFBRs and associated fuel cycle facilities,

2. The fraction of that amount released to the environment which may find its way to man. And,

3. The possible health consequences occurring from the incorporation of this material into man.

Each of these estimates involves many complex items. All of the information necessary to make accurate predictions is not available. In the absence of complete information estimates were made using the best current knowledge. Where there were deficiencies of knowledge conservative assumptions were made. The assumptions make it likely that the estimates of population exposures and potential health consequence are well above the most likely values.

Source Term

The dominant source of release of transuranic elements from the LMFBR fuel cycle is expected to be airborne releases from fuel reprocessing plants. It is expected that releases from these plants will be approximately 1 millicurie of alpha-emitting transuranic activity for each 1000 megawatt electrical year (MWe-year) equivalent of fuel reprocessed, and that airborne releases from other components of the LMFBR fuel cycle (fuel fabrication, reactor, waste storage) and from transportation will be smaller by a factor of ten or more. The probabilities for accidental releases have generally not been estimated. However, calculations of releases for a variety of hypothetical accidents indicate that accidents will make a negligible addition to this source term. For these reasons the source term used in this analysis is the release of 1 millicurie of alpha-emitting radioisotopes per 1000 MWe-year of energy generation. The activity is assumed to have the specific radiological composition of high burnup fuel shown in Table 1; it is further assumed to be released as small airborne particles with a size distribution having an activity median aerodynamic diameter (AMAD) of 0.3 micrometer (μm).*

Pathways to Man

Figure 1 is a summary of estimates for the fraction of released activity which enters man. There are two paths to man -- inhalation and ingestion. The cumulative fraction inhaled by all individuals before the airborne material is deposited on the ground was estimated to be 4×10^{-6} of the

* In such a distribution one half of the radioactivity is associated with particles whose settling speeds in air are less than the settling speed of 0.3 μm, unit density spheres and half the radioactivity is associated with particles of larger settling speeds.

TABLE 1. SOURCE TERM

Isotope	Millicuries released per 1000 MW(e)·a	
	beta	alpha
^{238}Pu		0.18
^{239}Pu		0.04
^{240}Pu		0.05
^{241}Pu	5.4	
^{241}Am		0.01
^{242}Cm		0.70
^{244}Cm		0.01
	5.4	0.99

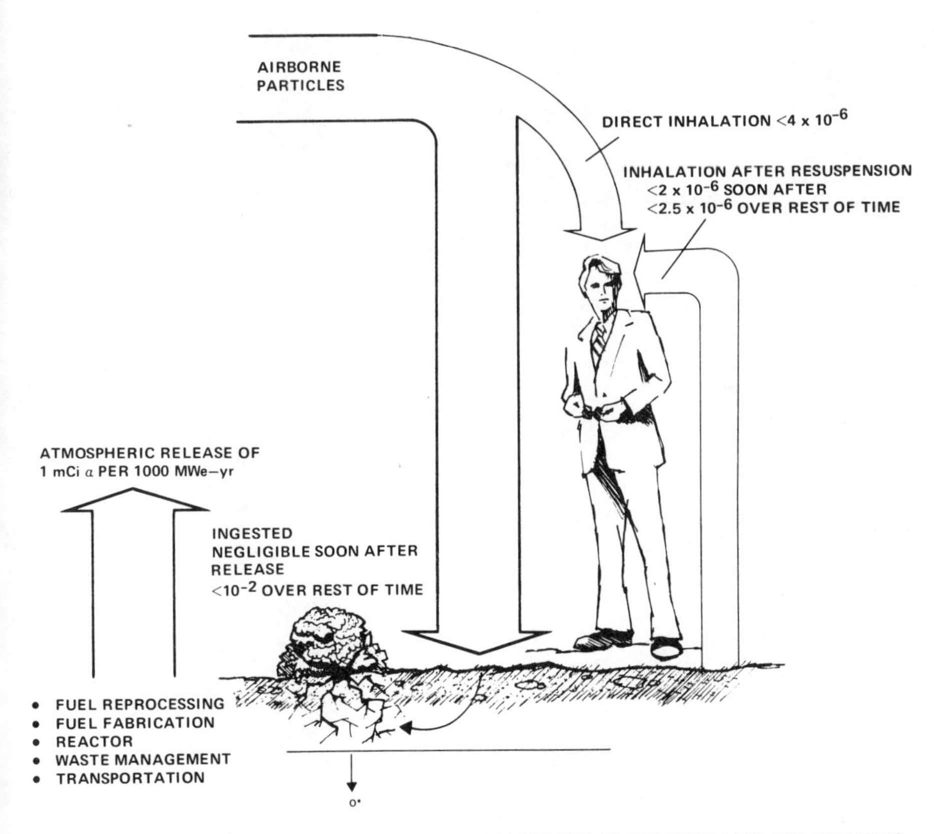

FIG. 1. Summary of major routes to man.

TABLE 2. INHALATION OF TRANSURANIC ISOTOPES (nanocuries from 1 mCi release)

Element	Isotope	Resuspension			Direct	Total
		Early	Late	Total		
Plutonium	238	0.44	0.028	0.47	0.72	1.2
	239	0.10	1.76	1.86	0.16	2.0
	240	0.12	0.6	0.72	0.20	0.92
	241 [a]	(19.6)	(0.14)	(19.7)	(21.6)	(35)
Subtotal (alpha only)		0.66	2.39	3.05	1.08	4.12
Americium	241	0.04	0.12	0.17	0.04	0.21
Curium	242	1.2	0.0	1.2	2.81	4.0
	244	0.012	0.0	0.012	0.04	0.05
Total alpha		1.91	2.51	4.43	3.97	8.4

[a] β-emitter, not included in subtotal and total for α-emitters.

amount released for a source in the North Central United States. The location of the source relative to population centers and the number of people exposed to the airborne material influence the value of this fraction.

Material reaching the ground may be resuspended in the air and inhaled. The fraction of the released alpha activity which, for all future years, enters man in this way is estimated to be approximately equal to the amount inhaled before deposition. The estimate assumes deposited material has an initial resuspension factor** of 10^{-5} per meter which decays with a half-life of 50 days and finally reaches a minimum value of 10^{-9} per meter. The amounts of various isotopes entering man through inhalation are shown in Table 2.

Of the total fraction of the alpha activity released we estimate that man inhales approximately 1×10^{-5}, the sum of 4×10^{-6} (fraction inhaled directly) and 4×10^{-6} (fraction inhaled after resuspension). Most of this (70 percent according to the model used) enters man within two years after release. The remainder would enter during many years afterward. Observations on global fallout from nuclear weapons tests indicate that the model and parameters used do not underestimate the resuspended transuranics which might enter man soon after release.

Ingestion is the other pathway to man. The various ingestion pathways are shown in Figure 2. Our analysis indicates that the dominant ingestion pathway commences with plant root uptake. The fraction entering man through this pathway is estimated by assuming a uniform distribution of released material across the surface of the United States. This assumption underestimates the fraction entering man for points of release upwind of land used for intensive agriculture and overestimates the fraction for points of release upwind of lands used predominantly for other purposes. The estimate also assumes that plant material achieves a concentration of transuranics equal to one-tenth of the concentration in the soil in which the plants grow, that there is no downward movement in the soil beyond the root zone and that there is no runoff. The amounts of various isotopes entering man under these assumptions are presented in Table 3. Because of their long half-lives, Pu-239 and Pu-240 are the principal radioisotopes ingested. On the other hand, because of larger uptake from the gastrointestinal tract, Am-241 is the largest contributor to dose.

Observations on the behavior of plutonium falling out from nuclear weapons tests indicate that these latter assumptions probably lead to large overestimation of the fraction entering man through ingestion.

Calculations of doses from inhaled transuranics were based on the lung model developed by the International Commission on Radiological Protection (ICRP) applied to both class W (retained in lungs for moderate times) and class Y (retained in lungs for relatively long times) particles; the results are summarized in Table 4. Following ICRP guidance, the fraction of ingested radioactivity absorbed from the GI tract was taken as 3×10^{-5} for plutonium. A conservative value of 10^{-3} was assumed for americium and curium. For inhalation, the lung and bone dose equivalents delivered by most of the transuranics are similar when integrated over seventy years. For a particle distribution with an AMAD of 0.3 μm these amount to approximately one millirem per picocurie inhaled.

**The resuspension factor is the ratio of the concentration in air (Ci/m^3) to the amount deposited per unit area (Ci/m^2).

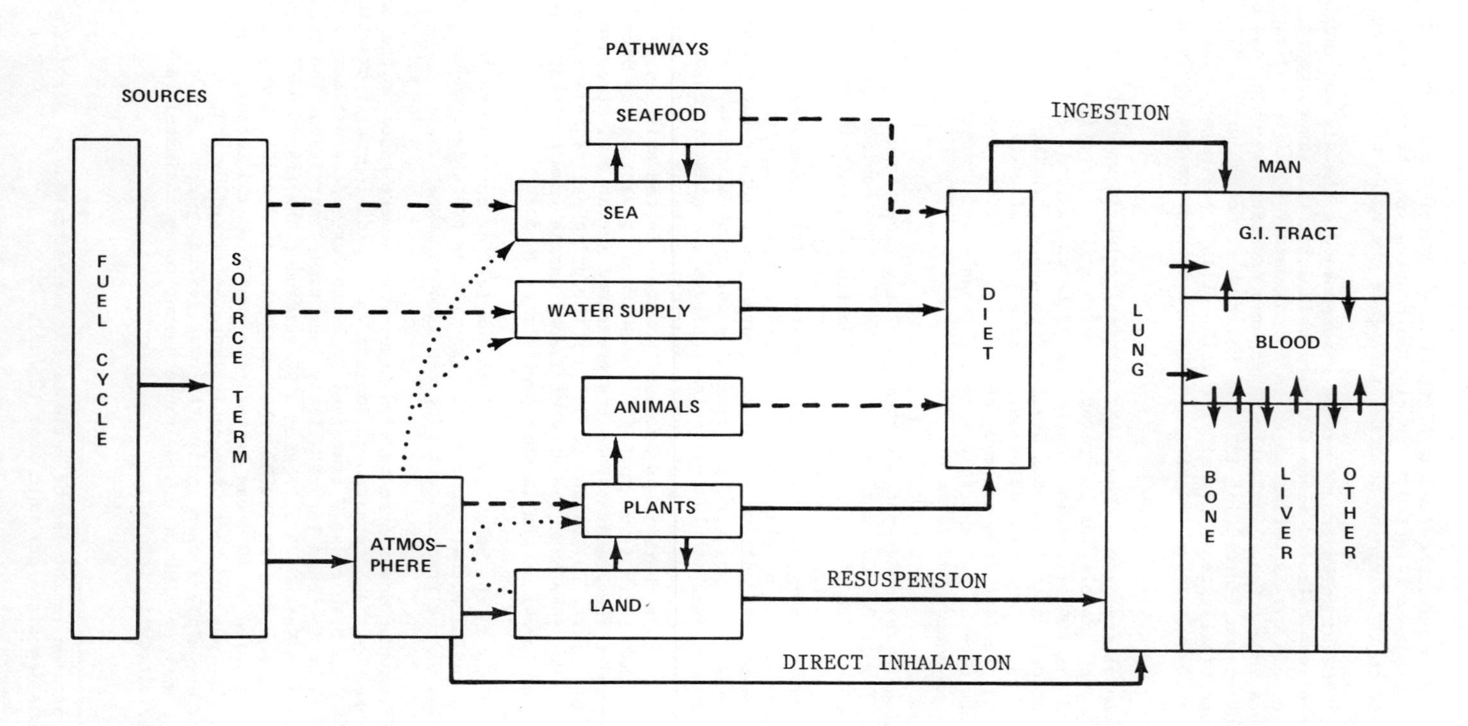

FIG. 2. Pathways to man.

TABLE 3. RELATION OF INGESTED CURIES FROM SOIL VIA ROOTS (C) TO CURIES RELEASED (S)

Element	Isotope	$S\left(\frac{Ci}{1000\ MW(e) \cdot a}\right)$	$t_{\frac{1}{2}}$ (a)	$C\left(\frac{Ci}{1000\ MW(e) \cdot a}\right)$	$\frac{C}{S}$
Plutonium	238	1.8×10^{-4}	87	5×10^{-8}	3×10^{-4}
	239	0.4×10^{-4}	24400	3×10^{-6}	8×10^{-2}
	240	0.5×10^{-4}	6600	1×10^{-6}	2×10^{-2}
Subtotal				4.05×10^{-6}	
Americium	241	2.0×10^{-4} [a]	433	3×10^{-7}	2×10^{-3}
Curium	242	7.0×10^{-4}	0.4	9×10^{-10}	1×10^{-6}
	244	0.1×10^{-4}	18	6×10^{-10}	6×10^{-5}
Total alpha-emitters				4.35×10^{-6}	

[a] 0.19 millicuries of this amount arise as a daughter nuclide from complete decay of ^{241}Pu, as a beta emitter.

TABLE 4. DOSE EQUIVALENT (rem) PER MICROCURIE INHALATION AND INGESTION

Isotope	Class W compounds								Class Y compounds	
	Inhaled				Ingested					
	Bone	Liver	Kidneys	Gonads	Bone	Liver	Kidneys	Gonads	Lung	Lymph nodes
^{238}Pu	2730	1140	148	44	0.6	0.3	0.03	0.01	913	38 600
^{239}Pu	3250	1320	170	46	0.7	0.3	0.04	0.01	863	43 700
^{240}Pu	3270	1330	172	46	0.7	0.3	0.04	0.01	871	44 000
^{241}Pu	67	25	3	0.7	0.02	0.01	-	-	2	369
^{241}Am	3330	1360	176	48	25	10	1	0.4	924	45 100
^{242}Cm	50	26	3	1	0.4	0.2	0.03	0.01	257	819
^{244}Cm	1400	645	83	30	11	5	0.6	0.2	907	23 500

Particle distribution of 0.3 nm AMAD; deposition probabilities of 0.08 (N-P), 0.08 (T-B) and 0.42 (P). Doses include contributions from both parent and daughter isotopes for ^{241}Pu, ^{242}Cm and ^{244}Cm.

TABLE 5. ALPHA EMITTING RADIOISOTOPES IN HUMANS

	From natural radioactivity	From 1000 MW(e)·a - LMFBR
Total curies in US population	>0.1 [a]	$<10^{-8}$
Total man·rem/a		
To bone	$>1 \times 10^7$ [a]	<0.2 [b]
To lung	$>1 \times 10^7$ [a]	<0.04 [b]

[a] Derived from: Ionizing Radiation: Levels and Effects, 1: Levels, UN, New York (1972).
[b] Approximately one half of the total man·rem doses to these organs are delivered over 70 years. The remainder is delivered at much lower rates over very much larger periods. The rates presented here are for the first 70 years.

When applied to the amounts of transuranics entering man these values yield total population exposures of approximately 5 lung rems, 25 bone rems, 10 liver rems and 0.5 gonadal rems per 1000 MWe-year.

The quantities of alpha-emitting radioisotopes which might enter man over their entire lifetime in the environment subsequent to release from the LMFBR fuel cycle and the population doses they deliver are small compared to the quantities of naturally occurring radionuclides which normally occur in man. These estimates are summarized in Table 5 along with quantities and dose equivalents for naturally occurring alpha radioactivity found in man.

These estimates of the released material which enters man provide a substantial basis for believing that potential health consequences associated with the release, persistence and accumulation of transuranic elements which are expected to be released by the LMFBR fuel cycle will be small in comparison to many other accepted and unavoidable risks. Nevertheless, to provide additional perspective, numerical estimates were made of potential health consequences possibly attributable to the entry into man of these exceedingly small quantities of alpha-emitters (over the many generations they will exist in the environment). Such risk estimates, it must be emphasized, have no experimentally established basis at the very low levels of exposure expected to result from transuranic elements released from the LMFBR fuel cycle.

Risk estimators for cancer in lung, bone and liver and genetic risks were derived from the BEIR report and are summarized in Table 6 which also shows total estimated potential risks per 1000 MWe-year of electrical energy generated by LMFBRs. These estimates of potential risk are subject to great uncertainty for a number of reasons. There are no observations in man or in animals which are directly relevant to estimating potential health effects at the exceedingly low doses and dose rates being considered here. The estimates assume that health consequences are directly proportional to the quantity and residence time of alpha-emitters in the organs of man, and most probably lead to great overstatement of the maximum potential health consequences. The estimates are presented solely to provide some perspective on the long term health consequences to large populations which might arise as the result of the persistence and accumulation of transuranic elements released by the LMFBR fuel cycle. The method, results and conclusions are not generally applicable or appropriate to other specific questions

TABLE 6. SUMMARY OF RISK ESTIMATORS AND TOTAL POTENTIAL HEALTH CONSEQUENCES

Organ system	Risk estimators	
	Cancer death probability per man·rem	
	Minimum	Maximum (range)
Pulmonary	0	$40\ (15-110) \times 10^{-6}$
Skeletal	0	$6\ (2-17) \times 10^{-6}$
Liver	0	$2\ (1-5) \times 10^{-6}$
	Genetic defects per man·rem (at genetic equilibrium)	
Reproductive		
Specific genetic defects:	30×10^{-6}	300×10^{-6}
Defects of complex origin:	6×10^{-6}	600×10^{-6}
	Total risks	
	Cancer death probability per 1000 MW(e)·a	
Lung cancer	0	$200\ (75-550) \times 10^{-6}$
Bone cancer	0	$160\ (52-180) \times 10^{-6}$
Liver cancer	0	$20\ (11-55) \times 10^{-6}$
Total	0	$380\ (138-785) \times 10^{-6}$
Genetic risks		
Specific defects:	12×10^{-6}	120×10^{-6}
Complex conditions:	2×10^{-6}	240×10^{-6}
Total	14×10^{-6}	360×10^{-6}

TABLE 7. GONADAL POPULATION EXPOSURES

	Curies released / 1000 MW(e)·a	Gonadal man·rems / curie	Gonadal man·rems / 1000 MW(e)·a
Transuranics	10^{-3}	4×10^{2}	0.4
Krypton	5×10^{5}	2×10^{-4}	100
Tritium	2×10^{4}	2×10^{-4}	4
Radium	10^{-1}	10^{3}	100

involving exposure to alpha-emitting transuranic radioisotopes. These estimates indicate that there may be from near zero to less than 10^{-3} potential cancer and genetic defects of all types produced as a result of environmental releases of transuranics for each 1000 MWe-year of electrical power generated by the LMFBR fuel cycle. These estimates were made for the entire period of decay of the released transuranics.

Looking ahead to the turn of the century, when perhaps as much as 400,000 MWe of LMFBR capacity may be installed in the U.S. (about equal to total installed electric capacity in the U.S. at present), the foregoing estimate translates to less than 0.4 cancer deaths over all future years attributable to transuranium effluents from the operation of all of these plants and associated facilities for one year. Similarly, we would expect less than 2.2 cancer deaths over all future years due to one year's operation of 2,200,000 MWe of LMFBR capacity in the year 2020. Using the same kind of conservative estimates, up to several thousand potential cancer deaths per year in the United States would be attributable to natural background radiation.

For comparison, gonadal population dose equivalents for several radionuclides released in generating electrical power using nuclear fuel and coal are presented in Table 7. It is assumed that all krypton-85 and tritium produced in nuclear reactors and 10% of the fly ash produced in coal-fired stations is released. With the release term and models used in this study a population dose equivalent of approximately 0.5 gonadal man rem per 1000 MWe-year of electrical power generated is obtained. The bulk of this population dose equivalent arises from radioactivity incorporated into man promptly after release. Most of the exposure is to the population within a few hundred miles of the point of release.

For krypton-85, a model which assumes complete mixing in the atmosphere prior to decay, leads to a population exposure of 100 gonadal man rems per 1000 MWe-year. Using a dispersion model similar to that used for the airborne transuranics, but without a deposition term, it can be shown for specific locations that only a few percent of the population exposure are delivered locally. The largest portion is delivered more or less uniformly to the entire population in the hemisphere in which the release occurs, and at a rate which decreases with a half-life of ten years.

Tritium is similar except that under certain conditions of release local exposures can add significantly to the total population exposure. For radium, only the exposure due to external radiation is calculated assuming that once the radium in fly ash is deposited it does not penetrate into the soil.

DISCUSSION

P. SLIZEWICZ: What is the experimental basis of your estimated release of 0.1 mCi per 1000 MW(e) per year from plants reprocessing sodium-cooled reactor fuel?

M.B. BILES: On the basis of experience it was assumed that 10^{-9} of the ^{239}Pu throughput would be released, which corresponds to 80 mCi/a for a 5 tonne per day reprocessing plant. This is probably higher than will actually be the case, because at present we are releasing only 3 mCi/a from our Savannah River plant, which is not based on current technology.

P. SLIZEWICZ: How do you measure the 3 mCi/a release at Savannah River, and how do you distinguish it from the 2 mCi/a bomb fall-out?

M. B. BILES: I am not sure how these measurements are performed.

O. ILARI: In connection with the question just asked by Mr. Slizewicz, I believe the value of 3 mCi/a quoted by Mr. Biles for the Savannah River release is estimated at the source and not in the environment. The problem of identifying it therefore does not exist.

R. BITTEL: Living plants become contaminated via both the root and the aerial parts. Can one estimate the relative importance of these two pathways for plutonium and the transplutonians?

M. B. BILES: The foliage of plants will indeed become contaminated by both primary fall-out and resuspended material. The analysts have assumed that the estimate of root uptake is sufficiently liberal to include foliage contamination.

R. BITTEL: Plutonium incorporated in animal and human food is in the form of organic and mineral complexes. Can one assess the intestinal absorption of these complexed forms?

M. B. BILES: Perhaps Mr. Bair could answer this question.

W. J. BAIR: Experiments are now in progress aimed at providing answers to questions such as this. Early data on the absorption of plutonium from the gastro-intestinal tract were obtained mainly in experiments where animals were given rather large quantities of relatively pure plutonium compounds. It is not known whether these data apply to plutonium incorporated in plant materials, since there is a possibility that the Pu might be present as soluble complexes. Moreover, earlier studies of plutonium uptake by plants did not take into consideration the possible long-term action of bacteria and other biological entities on plutonium deposited in soil, which might make plutonium more available for incorporation in plants. This is especially important in relation to the possible long-term residence time of long half-life Pu isotopes in soils. Further, few experiments have looked at the foliar deposition and absorption of plutonium, primarily because of technical problems associated with exposing growing plants to plutonium aerosols. It is now apparent that the uptake of plutonium in plants by different pathways and the gastro-intestinal absorption of plutonium incorporated in foodstuffs require further in-depth study to provide information relevant to the health hazards of environmentally dispersed plutonium.

O. ILARI: I should like to point out that the dose assessment procedure used in this paper, starting from a source-term (as the only quantity capable of measurement) and introducing it in a system analysis mathematical model, is a typical application of the SAM approach I presented yesterday in my paper IAEA-SM-184/103.

H. P. JAMMET: The estimates given in this paper show that the collective dose due to alpha-emitter releases represents 10^{-7} to 10^{-8} of the collective dose due to natural alpha-emitters. On this basis, is the suppression of all releases in new installations justified by cost-benefit analysis?

M. B. BILES: We have not performed cost-benefit analyses for our new plants. However, judging from experience Pu releases can be held to very low values without high cost.

HEALTH EFFECTS OF ALTERNATIVE MEANS OF ELECTRICAL GENERATION*

K.A. HUB, R.A. SCHLENKER
Argonne National Laboratory,
Argonne, Ill.,
United States of America

Abstract

HEALTH EFFECTS OF ALTERNATIVE MEANS OF ELECTRICAL GENERATION.

The objective of the paper is to provide a discussion and quantitative display of health effects resulting from electrical energy generation. Three broad categories of health effects are considered: disease mortality, gene mutation and genetic death, and accidental injury and death. These effects for a nominal 1980 period were defined for the nuclear, coal, and oil energy systems that included all activities from mining of fuel through disposal of waste from the generating plants. The health impacts were expressed in terms of man-days lost for an annual production of 6.6×10^9 kW·h of electrical energy from the system. For the items quantified the oil-fired system had the fewest number of man-days lost and the coal-fired system the highest value. The predominant contribution to health and accident impacts for nuclear systems comes from the effects on employees and their progeny and not from the effects on the public at large.

I. INTRODUCTION

The objective of this paper is to provide a discussion and quantitative display of health effects resulting from electrical energy generation using various fuel systems. The information contained herein is mainly obtained from a "social costs" study[1,2] carried out at the Argonne National Laboratory; social cost is used in the context of the total cost to society, i.e., the health, environmental, and dollar costs. These costs were defined for energy systems that are based on a nominal 1000-MWe generating plant.

Effects from each system are assigned based on a one-year period of operation to produce 6.6×10^9 kWh of electrical energy using a 1000-MWe plant. Estimates are made of not only the health effects occurring during the year of operation, but also for all those future effects that can be assigned to the year of operation. It should be noted here, that although many health impacts were quantified there are a significant number that were not quantified in the study.

Three broad categories of health effects are considered: (1) disease mortality, (2) gene mutations and genetic death, and (3) accidental injury and death. Disease mortality may result from exposure to radiation or to system-emitted pollutants. Gene mutation and genetic death may also result from exposure to radiation or pollutants. Accidental injury and death effects are caused by a direct physical interaction between man and the energy system. Health effects are measured in terms of man-days lost.

The paper includes a description of the energy systems, the methods of analysis, quantification of health impacts, the results, and observations for the study.

* Work sponsored by the US Atomic Energy Commission.

II. ENERGY SYSTEMS

The nuclear, oil, and coal energy systems are undergoing change. In order to describe system conditions for a decade starting in 1980, a scenario describing the future was necessary. Projections were made based on discernable trends and developments such as the additions and changes resulting from implementation of environmental health and safety laws. Performance of energy systems was selected to represent U. S. industry conditions. Generating plants were base-load types.

A. Coal

For the coal-fired plant, 50% of the 2.3 million annual coal tonnage was from underground mines, 40% from area-type strip mines, and 10% from contour strip mines. Based on the coal tonnages from each of three types of mines, health effects are estimated. Working conditions for the miners are those associated with the Federal Coal Mine Health and Safety Act of 1969. Coal was transported by train from the mine to the generating plant.

The average sulfur content of the coal is 2.5%; 80% of the sulfur is assumed to be removed by flue-gas treatment. No particular type of sulfur removal process is assumed other than that the removed sulfur can be stored in elemental form. Thus any health effects associated with some processes which involve large volumes of stored wastes are not accounted for. Ash content of the coal is 11%; only 0.8% of the ash is assumed to be emitted from the plant stack; the rest is placed in waste storage or used in the industry.

There are many trace elements present in coal besides the major impurities (aluminum, silicon, iron and sulfur). About 30 trace elements occur widely in coals; concentrations vary from about 2 up to nearly 100 ppm averaged over several coal provinces[3]. Health effects of six toxicologically important elements (mercury, beryllium, arsenic, cadmium, lead, and nickel) were considered. Estimates of the releases of these elements were made based on physical and chemical characteristics of the elements. Their impact on health at environmental concentrations were not considered due to lack of dose-response data.

B. Oil

The oil-fired plant burns about nine million barrels of residual fuel oil per year. Approximately 90% is estimated to come from foreign refineries, the rest from the U.S.A. The sulfur and ash content are 0.7 and 0.1 weight-percent, respectively. Particulate matter is removed from the flue gas with a 90% efficiency electrostatic precipitator. Hydrocarbon emissions due to evaporation, oil spills, and waste discharges were estimated. It was assumed that the injury frequency and severity in foreign oil operations were the same as experience in the domestic oil industry.

C. Nuclear

For the nuclear energy system, the characteristics of a PWR plant were assumed; it was found that the health effects of the PWR system were almost the same as those for the BWR system. For the purposes of this paper, the system will be called a light water reactor (LWR) energy system.

About 130 metric tons of U_3O_8 need to be mined for one year of operation and supplied to the fuel cycle. Fifty percent of the ore is assumed to come from open pits, which have a very low accident rate compared with underground mining[4].

It was assumed that the low-hazard-potential solid wastes were shipped to a commercial burial ground. High-level wastes were shipped to a hypothetical federal repository where the wastes were kept separated from the biosphere. Health effects were not quantified for these wastes; however, the results of a recently published study on health impacts from activides are briefly stated.

Health effects due to transportation accidents and accidents at the fabrication and reprocessing plants were estimated. The effects were tabulated in terms of annual expectancies.

III. STUDY APPROACH AND METHODS OF ANALYSIS

The investigation of social costs of electricity generation initially started as a risk cost-benefit study; however, the benefits are those which result from the ready availability of electricity to the consumer and were assumed to be the same for each energy system. The risk for a health effect is expressed in terms of an annual expectancy and is placed on the cost side of the ledger. From society's point of view, for any type of risk, the method of amelioration or abatement means can be balanced, where possible, by equating the marginal cost of the risk against the marginal cost for the abatement means. This balancing is, of course, carried out on a present-worth basis.

The approach in the dollar quantification of the costs throughout the energy system was to use a present-worth type calculation. This approach was not used in the health effects because other than in the accident category of health effects, most of the impacts occur a long time after the year's operation of the energy system. The impact on society for these health effects is not clear. In particular, the genetic impacts to society will occur for generations and a present-worth calculation would not be meaningful because the impact of the added genetic burden in future years is not clear.

For deaths, genetic deaths, and accidents, the impacts caused by the annual operation of the energy system were expressed in man-days lost. The assignment of the magnitude of man-days lost for each activity in energy system is a large task that was simplified by making some gross approximations. For example, for all diseases and accidents resulting in death, the American National Standards Institute value of 6000 disability days for a loss of life was used regardless of the age or other life-factor condition of the victim. Thus, the impact of a system-caused death of a 12-year old was assumed equal to that of a 60-year old person.

As is the case with virtually all productive activities, there are certain costs associated with electrical energy generation for which the impacted individuals are not fully compensated. Health effects suffered as a result of air-pollution emissions are an example of such costs. Because a market system does not exist for these activities, the utility does not compensate the affected parties for these damages. Such costs are referred to as external costs or diseconomies. The internal costs are those costs which are borne by the utility and are therefore reflected in the bill for electricity. In this study the man-days lost were assigned to internal and external categories. Because of the complexity of the actual situation it was necessary to simplify the assignment of costs. Briefly, the health costs were placed either entirely into the external cost category or were placed 50% into the external and 50% into the internal category.

Several methods of analysis were applied in the study. Some brief comments will be made about some of these in order to relate the overall flavor of the analyses. Computer codes were used to aid analysis in only a few instances, such as in estimation of the transport and diffusion of certain gaseous releases. Other gaseous-related potential impacts were estimated in a simpler fashion.

An approximate estimate of airborne trace element concentrations from fossil-fueled power plant operation is an example of this. Representative trace element concentrations in coal were used in combination with the expected fraction released from the stack to provide the exit airborne concentration from the plant. A dilution factor of ten thousand was used to estimate the concentrations of the element in the ambient air. On this basis of all the elements considered only beryllium appeared important; it was only a factor of five below a suggested limiting community ambient concentration level. This generalized story, however, may not be appropriate for some coals in the western U.S. where trace element concentrations may be 100 to 1000 times greater than the approximate natural averages used in the estimates.

The radiation doses received by humans as a result of one year's operation of the reactor plant with its support fuel system were broken into six categories: the general population dose calculated using a global model, the general population dose using a local model, general population dose due to transportation of radioactive nuclides, uranium miners' dose, transportation workers' dose, and occupational dose in all other steps in the nuclear fuel cycle.

The airborne source for the local population dose is based on computerized calculations and is determined by the releases from the fuel reprocessing plant and the generating plant; local dose accounts for an exposure up to a 50-mile radius. The waterborne source of local population dose is only due to liquid releases at the generating plant and is derived from information given in Environmental Impact Statements for specific nuclear plants. The global population dose is based only on ^{85}Kr and ^{3}H isotopes and represents the whole-body commitment resulting from complete decay of these isotopes.

The genetic effects due to radiation were estimated on the basis of mutations per rad of gonadal radiation to certain classes of individuals. It was assumed that each mutation introduced into the active gene pool resulted in a genetic death.

IV. HUMAN HEALTH EFFECTS

A. Categories of Health Effect Studied

Three broad categories of effect on the public and on persons employed in the energy system have been considered: (1) disease mortality, (2) production of gene mutation and genetic death, and (3) accidental injury and death.

Disease mortality may result from exposure to radiation or to air pollutants. Specific consideration was given to the radiation exposure of the public from nuclear power plant and fuel reprocessing plant releases, both routine and accidental, the exposure of the public to air pollutants from fossil fuel plants, the radiation exposure of employees in the nuclear energy systems, to lung cancer in uranium miners and to black lung disease in coal miners.

Gene mutation and genetic death from radiation exposure of the public and of employees has been considered along with cancer mortality. The genetic impact of air pollutants has not been considered because of the paucity of pertinent data.

Accidental injury and death affect the public as well as the employee. The public suffers mainly from accidents in the transportation phase of the energy system and from nuclear power plant accidents, while the employee suffers accidents throughout the system.

B. Radiation Exposure

1. Dose-Response Functions

Linear dose-response functions were used to estimate cancer mortality following whole-body irradiation and lung cancer mortality following exposure to radon daughter products in mine atmospheres. Whether or not linear extrapolations from high doses and high dose rates are applicable to the low doses encountered in exposure of the public or the low dose rates encountered in occupational exposures is a moot point. In accordance with current radiation protection philosophy, it is assumed that, if incorrect, the linear functions overestimate the actual cancer incidence.

Dose-response functions for specific types of malignancies and for all malignancies taken together following whole body irradiation were based on reviews of human external exposure data[6-13]. Both post-natal and in utero irradiation were considered. Table I lists the dose-response functions adopted for whole body exposure. Breast and thyroid cancers are excluded because of the high initial survival[13]. Only the function for total malignancies due to postnatal irradiation was used to determine the impact, which was expressed in man-days lost. (The man-days lost for leukemia and lung cancer can be found by taking the ratios of the function for these diseases to the function for total malignancies and multiplying by the total loss given in Table V.) In utero irradiation was ignored in the computation because only a small percentage of the population is in utero at any one time and the excess risk is only a factor of 3 higher than for postnatal exposure.

A dose-response function for uranium miner exposure to radon daughter products was based on the U. S. Public Health Service dose-response model for the Colorado Plateau miners[14]. Because of the strong dependence of lung cancer risk on smoking, it was assumed that the smoking habits of future miners will be the same as the smoking habits of the miners in the study. Implicit in this is the assumption that on-the-job and off-the-job smoking affect lung cancer risk in the same way. In the application of radon-daughter and whole-body dose-response relationships to mortality prediction, the errors inherent in the linear extrapolation to low exposures probably far exceed the errors which result when the relationships are applied to populations with a different age structure from the study populations on which they are based.

Gene mutations were used as the measure of genetic damage and, under the assumption that all mutations are unconditionally harmful, were assumed to lead to genetic death[6,15]. Population size was assumed to be constant for as many generations as required to remove all mutant genes. Each mutation was thus assumed to lead to a reproductive failure and a factor was applied to convert each such genetic death into an equivalent number of man-days lost.

Table I. Radiation Dose-Response Functions

A. Death from malignancy	
1. Whole-body irradiation	Excess Deaths/per rad
Postnatal:	
Leukemia	0.00002
Lung cancer	.00004
All cancers	.00020
In utero	.00060
2. Radon daughter product exposure	Excess Deaths/WLM[a]
in uranium miners	0.00010
B. Gene mutation	
Whole-body irradiation	Mutations/offspring/rad
of father	0.00200

[a]Working level month

A dose-response function was determined from the size of the human genome and from the mutation frequencies at specific loci in mice exposed to chronic gamma irradiation[6,7,16]. Mutations in offspring were considered to be contributed solely by the father because of the small mutation rate in the oöcyte when exposed to gamma radiation at low dose rates. The dose-response function for mutations in offspring per rad dose delivered to the father in a population of constant size are given in Table I. The estimate was based on data given in the 1966 UNSCEAR report[16] and exceeds estimates given in the 1972 report[7]. Inasmuch as the estimate takes no account of polygenic mutations, which, on the basis of data collected on Drosophila, may be very abundant[17], a downward revision toward the more current values was considered unnecessary.

2. Population Doses

a. Global

Tritium and ^{85}Kr distribute widely throughout the environment[18,19], tritium in water and ^{85}Kr in the atmosphere. For these isotopes models of global distribution were developed.

Over 90% of the current environmental inventory of ^{3}H has come from atmospheric testing of hydrogen bombs. The residence time of ^{3}H in the troposphere is 20-40 days, and about 50% of the ^{3}H released by bomb tests has been deposited between 30° and 50° north latitude. Movement out of this region over periods of 5-10 years does not appear to be large. On the basis of these data, gaseous ^{3}H releases were assumed to distribute immediately throughout surface waters and to be retained--50% between 30° and 50° north latitude and 50% in the rest of the world. Liquid releases may be held locally for long periods if they are emitted into a reservoir which turns over slowly, such as Lake Michigan. However, they are a small fraction of the total ^{3}H release and were assumed to distribute in the same manner as the gaseous release. The world population was assumed to be that projected for 1990[20] and it was divided between 30° and 50° north latitude, and the rest of the world, according to the current population distribution. On the assumption that the population size is constant until the ^{3}H has decayed to a small fraction of its initial level, the whole body dose commitment for ^{3}H releases from the PWR reactor and reprocessing plant is 22 man-rad.

The concentration (pCi/g air) of environmental ^{85}Kr is nearly uniform over the surface of the earth and throughout the troposphere[19,21,22]. For the purpose of dose calculation ^{85}Kr was assumed to distribute uniformly throughout the troposphere, immediately upon release. The same population assumptions were used as for ^{3}H, to obtain a whole body dose-commitment of 130 man-rad from X-rays and bremsstrahlung. This dosage excludes the contribution from β rays and, as a consequence, is a factor of approximately 70 less than the value which would have been obtained had the β radiation been included, as it is when dose factors are based on ICRP recommended air concentration limits[23] or those promulgated by the USAEC[24]. Dose information is summarized in Table II.

Although we did not model or calculate the health impacts due to actives, the U. S. Environmental Protection Agency has calculated the radiation dose commitment from actinides for the U. S. nuclear power industry[25]. Doses and health effects were estimated for 100 years after

Table II. Radiation Dose for Nuclear Energy System

General population	
Global model (man-rad)	150
Local model (man-rad)	
Water	4
Air	4
Transportation (man-rad)	3
Occupational	
Miners' exposure (man-WLM)	110
Transportation workers (man-rad)	4
Reactor (man-rad)	400
All other fuel cycle steps (man-rad)	50

release, which, of course, is a short time period in view of the long half lives of some isotopes. The results depended upon assumptions and ranged from a trivial effect compared to that of ^{85}Kr and ^{3}H, to a health impact of about three times that of ^{85}Kr and ^{3}H.

b. Local

By local dose for airborne nuclides, we mean the population dose within a 50-mile radius of the LWR reactor and reprocessing plant. A computerized atmospheric transport model was used to make the dose estimate[26]. For simplicity the reactor and reprocessing plant were assumed to be located at the same site and make their gaseous releases through the same stack. Only doses from ^{3}H and noble-gas releases were computed since doses from other releases have been estimated to make a small contribution to total dose.* The noble-gas doses included contributions from short-lived daughter products. The dose was computed for three different population distribution representing low (214,000 in 50-mile radius), medium (2,720,000) and high (18,400,000) densities. The population doses for the three densities were 0.61, 4.2 and 42 man-rad respectively. The medium density value was assumed to be typical and was used for cancer mortality projections in conjunction with dose response relationships.

The local dose from liquid releases from the reactor is estimated to be about 4 man-rads in USAEC Environmental Impact Statements[28]. Dose information is summarized in Table II.

c. Population Exposure from Transportation of Fuel

The general public is exposed to radiation during the transport of spent fuel. This group receives about 3 man-rad[28].

d. Reactor Accidents

Noncatastrophic reactor accidents (Classes 1-8) which release radioactive material to the environment were estimated to give 2 man-rad dose to the population in a 50-mile radius. The large nuclear accident (Class 9) was not included in the original "Social Costs" study[1,2] because the "Reactor Safety Study" conducted under N. Rasmussen was intended as the companion study. This study has not been published, but information given in the press indicates a very small probability associated with a major catastrophe (once in 10^9 to 10^{10} reactor years)[29]. With a frequency of once in a billion years, even with an equivalent loss of life of one hundred thousand people only 0.0001 deaths would be assignable to one year of operation. This compares with 0.03 deaths world-wide for a year of operation from normal releases. On a man-days lost basis, it corresponds to less than one man-day as compared with 3200 given in Table V. Hence reactor accidents appear to contribute little to the total health impact.

---

*Preliminary results of the year 2000 study[27] were scaled to our release assumptions and indicated that ^{3}H and ^{85}Kr would contribute 85% of the total-body dose to adults in the Upper Mississippi River Basin from nuclear power generated there.

3. Occupational Doses

a. Reactor

Dose estimates for reactor personnel were based on the operating experience of U. S., British, and Canadian power reactors[30,31]. For U. S. power reactors during 1969 and 1970 the personnel doses per calendar year range from 12 man-rad to 1470 man-rad[32]. Four U. S. reactors in the 75-292 MWe range showed an average of about 170 man-rad per reactor-year averaged over 33 reactor-years of operation[31]. The exposure in British and Canadian reactors in the 220 to 660 MWe range was 237 man-rad per reactor year averaged over 35 reactor-years. The averages for the individual stations ranged from 72 to 462 man-rads per reactor year. A comparison of the U. S. exposures with the British and Canadian exposures suggests that with increasing reactor size the fractional increase in exposure is less than the fractional increase in power level. On the assumption that man-hours for refueling and maintenance will be twice as high in a 1980 model 1000-MWe reactor as in 1960's power reactors, a value of 300 man-rad was estimated for the annual staff dose in a 1000-MWe reactor. This compares with an estimate of 400-500 man-rad currently used in USAEC environmental impact statements[33]. A 400 man-rad value was used in the computations in this paper.

b. Reprocessing Plant

Dose estimates for reprocessing plant personnel were based on the operating experience of a U. S. commercial reprocessing plant and on projections of plant employment and throughput for 1980. During the period 1968-1970 average annual whole-body exposure to reprocessing plant personnel increased from 2.7 rad to 6.7 rad[34] with annual throughput maintained approximately constant at 100 metric tons uranium per year. On the basis of these exposure data the conservative assumption was made that each reprocessing plant employee will receive 5 rad annually in 1980. Reprocessing plants in 1980 are projected to have 825 employees and an annual throughput of 3400 metric tons uranium, giving 1.2 man-rad per metric ton uranium. For the 1000-MWe LWR in 1980 we project 25 metric tons uranium being sent to the reprocessing plant for a total of 30 man-rad from reprocessing.

c. Other Fuel Cycle Steps

For the other LWR fuel cycle steps, except mining, exposure estimates were based on data which were not specific to the employment groups involved. For AEC licensee employees the average annual exposure is about 200 mrad[35], and about 80 man-years are required to carry out the activities of the remaining fuel cycle steps, exclusive of mining. Based on these figures a value of 20 man-rad was adopted for the occupational dose in these other steps. The annual dose to transportation employees is about 4 man-rad[36].

The radon daughter product concentrations in underground mines have dropped steadily in recent years due to improvements in ventilation. From 1965 to 1968, the average concentrations dropped from 4.1 WL to 1.1 WL[37]. Current concentration and exposure limits are 1 WL and 4 WLM/year.* Inasmuch as further decreases in concentration may be difficult to attain, it is assumed that underground miners will receive the maximum exposure of 4 WLM/year in the years through 1980. In open pit mines, the concentrations are at background levels[38]. The annual exposure in the uranium

*Working level months.

mining was estimated to be 110 man WLM, on the assumption that 50% of the uranium comes from underground mines and 50% from open pit mines. Dose information is summarized in Table II.

C. Accidental Injury and Death

1. Occupational Accidents

The voluminous statistics published by the Bureau of Mines, Bureau of Labor Statistics and the Department of Transportation and others [39-48] on accidental injury and death were used to estimate the nonfatal injury and death associated with each major component of the fuel cycles and with transportation of materials through the cycles. Combined statistics for the oil and natural gas industries were separated by allocating to each industry a fraction of the total dollar value of production which was assignable to each.

Table III presents the expected numbers of injuries and deaths for the coal, oil and LWR cycles. For presentation here the components of the cycles have been grouped into three categories, each of which has the same function within its cycle: fuel production, power plant operation and maintenance and transportation. In both the coal and LWR cycles, mining contributes more than 90% of the expected injury and death. Because strip mining and open pit mining are much safer than underground mining, the contribution from mining injury and death depends strongly on the amounts of strip and underground mining done. Here it was assumed that 50% of the annual coal or uranium ore tonnage was from underground mining and 50% from strip or open pit mining.

Table III. Expected Numbers of Annual Injuries and Deaths due to Occupational Accidents per 1000 MWe Plant for 1980 Fuel Cycles

Cycle	Ruel Production[a]	Power Plant Op. & Maint.	Transportation	Total
A. Injury (Nonfatal)				
Coal	41	1.5	5.2	48
Oil	10	1.5	1.1	13
PWR	5.2	1.3	0.042	6.5
B. Death				
Coal	0.98	0.037	0.055	1.1
Oil	0.11	0.037	0.030	0.18
PWR	0.10	0.011	0.0017	0.11

[a]For coal, this category includes mining and preparation. For oil, it includes production and refining. For PWR, it includes all steps from mining through fuel element fabrication and reprocessing.

2. Accidents Involving the Public

The public is involved in accidents with vehicles which are transporting coal and nuclear materials. Department of Transportation statistics on rail and truck accidents were used to estimate accidental injury and death for the public[47,48]. The expected numbers of public injuries and deaths due to transportation accidents for 1980 energy systems are:

Cycle	Injury	Death
Coal	1.2	0.55
LWR	0.081	0.0087

Note than in the coal fuel cycle, where the transportation is all by rail, the public suffers a much greater proportion of fatal accidents than do occupational personnel. This is a reflection of the fact that most rail accidents involving the public are at grade crossings.

D. Air Pollution Exposure

1. Dose-Response Functions and Population Exposure

In contrast to radiation exposure, dose-response functions for air pollution are obtained from studies of human population response to environmental levels. Because of the lack of long-term data on pollution levels extending over many years, dose-response studies have focused on the effects of incremental exposures. For example, a number of studies have sought to determine the excess deaths coincident with one day of additional exposure to urban air pollution[49-52]. Since these studies account only for acute response, the long-term response to air pollutants is not established by such effect; therefore, they probably underestimate the total effects of air pollution.

A number of dose-response estimates have been made using various pollutants as indices of air pollution level[49-64]. The relationships do not imply that the specific substances used as indices cause ill health. For example, by studying the mortality in 89 cities in relation to average pollution levels in 1969, Lave and Freeburg find that the annual additional deaths for each additional $\mu g/m^3$ of SO_2 is 3.9 per million persons exposed and for each $\mu g/m^3$ of particulates is $8.5/10^6$ people. Glasser and Greenburg[49] quote an additional 10-20 deaths/day in New York City on days when the SO_2 concentration is above 0.4 ppm compared with days when it is below 0.2 ppm. In the range between 0.2 and 0.4 ppm the excess death rate is linear with SO_2 concentration. Buechley et al.[51], find an additional 14 deaths in the New York--New Jersey metropolis on days with SO_2 concentration greater than 500 $\mu g/m^3$ as compared with days with SO_2 concentrations less than 30 $\mu g/m^3$. Over the range of 30 $\mu g/m^3$ to 1300 $\mu g/m^3$ the relationship is approximately linear in the logarithm of SO_2 concentration.

The air pollution levels on which dose-response relationships are based are usually measured at only one or a few locations in large urban areas and thus do not represent average population exposure. To make an accurate estimate of the health effects from power plant operation the effect of the emissions on the concentrations of the index pollutants at the monitoring stations should be determined. This requires a model which accounts for plant siting, pollutant transport and chemical reactions between the emissions and pollutants from other sources which are present in the air.

Studies which compare fossil and nuclear plants have not employed such modeling to date. Instead, rough comparisons have been made of effects from the two types of plants. For example, Starr et al[62], estimate that 60 times as many deaths would be caused by an oil-fired plant operating at the regulated exposure limits as by a PWR plant operating at the limits. Lave and Freeburg[53] estimate that an oil-fired plant would produce between 2300 and 39,000 times more deaths than a PWR plant operating at the same site, dependent on the sulfur content of the oil and the amount of sulfur removal from the stack gas.

As a simple model which gives insight into the relative effect of fossil and LWR cycles, we assumed that a coal-fired plant and a combined LWR reactor and reprocessing plant were operating on the same site (Indian Point) near New York City. The same atmospheric conditions were assumed for both nuclear and coal-fired plant releases and air concentrations were computed as a function of distance from the site. The SO_2 concentrations in mid-Manhattan, the approximate location of the monitoring station from which SO_2 data were taken for the Buechley et al. and Glasser and Greenburg studies, were used to estimate mortality. The dose-response function based on the Buechley et al. study predicts 0.2 deaths per million exposed people due to one year of the plant operation, and the function from the Glasser and Greenburg study predicts a five-fold greater value. Values in this range can be compared with a prediction of 0.0004 per million inhabitants within 50 miles of a combined nuclear power and fuel processing plant for one year of operation (airborne emissions only). Thus, the coal plant produces 500 to 2500 times as many deaths as the LWR combined plant. Since SO_2 releases are nearly the same for the oil-fired plant as for the coal plant, the relative deaths would be approximately the same. This neglects deaths which occur outside of the New York City area.

These three estimates all predict that the mortality from a fossil plant is much greater than the mortality from a nuclear plant, under normal operating conditions. None of the models is well founded enough, however, to confidently estimate fossil-plant mortality. Nevertheless, it appears that fossil-plant health effects may be large.

2. Coal Dust Exposure

Coal miners suffer from coal workers pneumoconiosis, or black lung disease, as a result of inhalation of coal and other mine dusts in the respirable particle-size range. The disease has several stages, the most severe of which, progressive massive fibrosis, leads to disability and death. In past years about 0.2% of the miner work force, or 200 to 300 miners, has become totally disabled each year from progressive massive fibrosis[65]. As a result of the Federal Coal Mine Health and Safety Act of 1969[66], air concentrations are being reduced in coal mines, and by the end of 1975 a respirable dust limit of 2 mg/m^3 will be in effect throughout the country. It is believed that thirty-five years of exposure at this limit will produce at worst category 1 of simple pneumoconiosis, a mild nonprogressive form of the disease, and will rarely cause progression of existing disease[67]. Thus, for the 1980 coal fuel cycle it appears that there will be little or no excess mortality from black lung disease in miners.

Table IV. Man-days Lost per Death, Injury and Genetic Death

A. Death: 6000

B. Injury

System	Fuel Production	Power Plant Op. & Maint.	Transportation[a]	
			Occupational	Public
Coal	60	83	50	200
Oil	50	83	38	-
PWR	50	32	50	120

C. Genetic death: 2000

[a]For coal, transportation is all by rail. For oil, transportation is by pipeline and tanker. For LWR 90% of transportation mileage is by truck, 10% by rail.

E. Man-days Lost per Injury, Death and Genetic Death

The American National Standard Institute recommendation of 6000 man-days lost per death was used in our work[68]. Average man-days lost per injury were computed from statistics on total man-days lost for both fatal and nonfatal injuries, by subtracting out the contribution from fatalities and averaging the remainder over all injuries. This was done for each component in the fuel cycle and for transportation. The numbers of man-days lost for death, and for injury in the fuel production, power plant operation and maintenance, and transportation categories for coal, oil and PWR fuel cycles are given in Table IV.

The number of man-days lost per genetic death was determined by first estimating the dollar cost per genetic death and then transforming this into time lost through the use of the dollar cost per man-day lost. The number of genetic deaths per million recognized pregnancies[69] was estimated to be 150,000 and thus we expect 600,000 genetic deaths for the approximately 4,000,000 recognized pregnancies in the U. S. per year. Using Lederberg's[70] estimate that 20% of the annual U. S. health bill of $80,000,000,000 is of genetic origin we figure about $27,000 per genetic death in the late 1960's and early 1970's. Escalating this to account for the rapidly rising medical costs and assuming the value of lost work time to equal the medical cost, we calculated a value of $100,000 per genetic death for 1980. We estimated the value of a man-day lost in 1980 at $50. The division of the dollar value of genetic death by this figure yielded the value of 2000 man-days lost per genetic death which we used.

Table V. Man-days Lost in the 1000-MWe Energy Systems for 1980

	LWR	Coal	Oil
Internal			
Public injuries in transportation	30	1750	-
Occupational			
Accidents	490	4620	860
Chronic diseases	280	S	S
Total internal man-days lost	800	6370	860
External			
Public			
Transportation accidents	30	1750	-
Disease mortality	200	L	L
Genetic effects	320	U	U
Occupational			
Accidents	490	4620	860
Disease mortality	280	S	S
Genetic effects	1080	U	U
Total external man-days lost	2400	6370	860
Total man-days lost (rounded)	3200	13000	1700

L=possibly very large
S=small
U=unevaluated
-=approximately zero
Basis: 6.6×10^9 kWh/yr output from system

Above information summarized according to class of people

Public impact	580	3500	-
Occupational impact	2620	9240	1760

V. Results and Observations

The results of the study and the observations made during the study are mingled because of a natural linking of each; however, the reader should consider the limitations of the study and the importance of perspective in interpreting the information presented herein. Table V summarizes and classifies the health impacts for the three energy systems. The following discussion will pertain mainly to coal and nuclear energy systems because oil system is not a viable alternative in the U. S.

1. On the basis of the accidental injury and death category alone, a greater impact is associated with the coal energy system than with the total of all categories for either of the other two energy systems given in Table V.

2. For the nuclear system, the predominant contribution to health and accident impacts comes from the effects on employees and their progeny

and not from the effects on the public at large. Perhaps society is not sufficiently emphasizing this aspect of the health problem in our expenditure of effort.

3. The health effects are but one part in the balancing of costs and benefits in man's social activities for the generating and use of electrical energy. The thrust of government and industry has been towards more abatement equipment and improved operational procedures in the generation of electrical energy; the cost-benefit balancing indicates that some of this thrust has apparently been justified to reduce impact to the individual. The guidelines for establishing allowable costs are not clear.

The age-old question reappears . . . How important to society is a shortened life or one with increased morbidity?

4. Health effects due to generating plant emissions for fossil-fueled energy systems are difficult to quantify either in terms of effects for specific pollutants or collective effects. A comparison of the coal and nuclear plant alternatives for the U. S. does not suffer from this lack of quantification because nuclear plants appear to have lower social costs than the coal-fired plants, without including this fossil plant penalty (the adverse impacts from these emissions); however, as better models for health impacts from storage of radioactive wastes and for probability estimates of the large nuclear accident become available, nuclear plant potential impacts might increase and a knowledge of the magnitude of this penalty might be required. From the information presented herein the penalty is significant. In an absolute sense, the gathering of information for health impacts from fossil-type energy system is more essential because of the large number of such plants in operation and for the use of such plants in non-base-load electrical energy generation.

Acknowledgements

The authors thank J. G. Asbury, W. A. Buehring, P. F. Gast, J. T. Weills, K. P. DuBois and J. L. Gardner, whose contributions to "A Study of Social Costs for Alternate Means of Electrical Power Generation for 1980 and 1990"[1] were drawn upon for the preparation of this report. We also thank D. Grahn and K. Eckerman, who, as interested members of the Argonne National Laboratory Staff, advised us on important details of the study.

REFERENCES

[1] HUB, K.A., ASBURY, J.G., BUEHRING, W.A., GAST, P.F., SCHLENKER, R.A., WEILLS, J.T., Summary Report (Draft), "A Study of Social Costs for Alternate Means of Electrical Power Generation for 1980 and 1990," Special Reactor Study Group, Argonne National Laboratory (February, 1973).

[2] HUB, K.A., ASBURY, J.G., BUEHRING, W.A., GAST, P.F., SCHLENKER, R.A., WEILLS, J.T., "Social Costs for Alternate Means of Electrical Power Generation for 1980 and 1990," Special Reactor Study Group, Argonne National Laboratory (March 1973) a 4 vol. ref. report (Draft).

[3] ABERNETHY, R.F., et al., "Spectrochemical Analyses of Coal Ash for Trace Elements," U. S. Bureau of Mines (July 1969).

[4] "Statistical Data of the Uranium Industry," GJO-100 (January 1972) USAEC Grand Junction Office.

[5] MOLE, R.H., "Radiation Effects in Man: Current Views and Prospects," Health Physics 20 (1971) 485-490.

[6] "The Evaluation of Risks from Radiation," A report prepared for Committee 1 of the International Commission on Radiological Protection. ICRP Publication 8, Pergamon Press, Oxford (1966).

[7] "Ionizing Radiation: Levels and Effects," A report of the United Nations Scientific Committee on the Effects of Atomic Radiation to the General Assembly, with annexes. United Nations, New York (1972).

[8] "The Effects on Populations of Exposure to Low Levels of Ionizing Radiation," A report of the Advisory Committee on the Biological Effects of Ionizing Radiation. Division of Medical Sciences, National Academy of Sciences National Research Council, Washington, D.C. (1972).

[9] DOLPHIN, G.W., MARLEY, W.G., "Risk Evaluation in Relation to the Protection of the Public in the Event of Accidents at Nuclear Installations," in Environmental Contamination Proceedings of a Seminar, Vienna 24-48 March 1969, pp. 241-254. Proceedings Series, International Atomic Energy Agency Vienna (1969).

[10] TAMPLIN, A.R., GOFMAN, J.W., "Population Control Through Nuclear Pollution," Nelson-Hall Co., Chicago (1970).

[11] STEWART, ALICE, KNEAL, G.W., "Radiation Dose Effects in Relation to Obstetric X-rays and Childhood Cancers," Lancet 1 (June 6, 1970) 1185-1188.

[12] MAC MAHON, BRIAN, "Prenatal X-ray Exposure and Childhood Cancer," J. Natl Cancer Institute 28 (1962) 1173.

[13] Ref. 7, p. 413, ¶89 Thyroid Cancer; Ref. 7, p. 415, ¶106 Breast Cancer.

[14] LUNDIN, F.E., Jr., WAGONER, J.K., ARCHER, V.E., "Radon Daughter Exposure and Respiratory Cancer, Quantitative and Temporal Aspects," U. S. Dept. of Health, Education and Welfare, Public Health Service, National Institute for Occupational Safety and Health, National Institute of Environmental Health Sciences Joint Monograph No. 1 (June 1970).

[15] MULLER, H.J., "Radiation and Heredity" Am. J. Public Health 54 (1964) 42-50.

[16] Report of the United Nations Scientific Committee on the Effects of Atomic Radiation, General Assembly, official records: twenty-first session, supplement no. 14, document A/6314 (1966).

[17] Ref. 7, p. 258, ¶642.

[18] JACOBS, D.G., "Sources of Tritium and Its Behavior upon Release to the Environment" TID-24635 (1968).

[19] KAROL, I.L., IVANOV, V.M., KOLOBASHKIN, V.M., LEYPUNSKIY, O.I., NEKRASOV, V.I., GUDKOV, A.N., USHAKOV, N.P., "Global Contamination of the Atmosphere by Krypton-85 from Worldwide Nuclear Power Plants and the Radiation Danger" Joint Publications Research Service, Washington, D.C., Report No. JPRS-53174 (May 1971).

[20] FISHER, J.L., POTTER, N., "World Prospects for Natural Resources--Some Projections of Demand and Indicators of Supply to the Year 2000" Resources for the Future (1964).

[21] SCHRODER, J., MÜNNICH, K.O., EBHALT, D.H., "Krypton-85 in the Troposphere" Nature 223 (1971) 614-615.

[22] PANNETIER, R., "Distribution, Transfert Atmospherique et Belán de Krypton-85" Rapport CEA-R-3591 Saclay (1968).

[23] "Recommendations of the International Commission on Radiological Protection" ICRP, Publication 2, Pergamon Press (1959).

[24] "Concentrations in Air and Water Above Natural Background" Code of Federal Regulations, Title 10, Part 20, Appendix B. Superintendent of Documents, U.S. Government Printing Office, Washington (1972).

[25] "Environmental Radiation Dose Commitment: An Application to the Nuclear Power Industry" U. S. Environmental Protection Agency, Office of Radiation Programs, EPA-520/4-73-002 (February 1974).

[26] FRIGERIO, N.A., ECKERMAN, K.F., STOWE, R.S., "The Argonne Radiological Impact Program (ARIP)" ANL/ES-26 (1973).

[27] STRAUCH, SAUL, "Year 2000--Nuclear Power and Man" in D.J. Nelson(ed.) Radionuclides in Ecosystems, Springfield, Va. NTIS (CONF-710501) (1973) 53-62.

[28] Draft Environmental Statement by the United States Atomic Energy Commission related to the proposed Byron Station Units 1 and 2. Docket Nos. STN 50-454 and STN 50-455.

[29] INFO, Atomic Industrial Forum, INFO No. 66 (January 1974).

[30] WILSON, R., "Man-rem Economics and Risk in the Nuclear Power Industry" Nuclear News 15(2) (1972) 28-30.

[31] Reference 7, p. 183 Table 12.

[32] Reference 2, p. V-75.

[33] Reference 28, p. 5-19.

[34] Reference 2, p. V-80.

[35] "Estimates of Ionizing Radiation Doses in the United States, 1960-2000" Report of the Special Studies Group, Division of Criteria and Standards Office of Radiation Programs, Environmental Protection Agency, Rockville, Maryland, ORP/CSD72-1 (August 1972).

[36] Reference 28, p. 5-18.

[37] Reference 14, p. 32, Table 18.

[38] Reference 14, p. 33.

[39] "Coal-Mine Injuries and Worktime," December and Annual Summary 1970. Mineral Industry Surveys. U.S. Dept. of Interior, Bureau of Mines (1971).

[40] "Bituminous Coal Facts 1970," National Coal Association, Washington, D.C.

[41] MOYER, F.T., "Injury Experience and Worktime in the Mineral Industries," in Minerals Yearbook 1968, Vol. I-II, U. S. Department of Interior, Bureau of Mines, U. S. Government Printing Office, Washington, D.C. (1969) 125-138.

[42] "Coal Mine Fatalities in 1970," Mineral Industry Surveys. U.S. Dept. of Interior, Bureau of Mines (1971).

[43] "Fatality Rates for Surface Freight Transportation 1963 to 1968," U. S. Dept. of Transportation, National Transportation Safety Board, Special study report STS-71-4 (1972).

[44] "Injury Rates by Industry, 1969," U. S. Dept. of Labor, Bureau of Labor Statistics, Report No. 389 (1971).

[45] "Disabling Work-Injury Experience of the Oil Industry (All Activities) and the Natural Gas Industry (Excluding Distribution Activities) in the United States, 1969," Mineral Industry Surveys, U. S. Dept. of the Interior, Bureau of Mines (1970).

[46] U. S. Bureau of the Census, Statistical Abstract of the United States: 1971 (92nd edition), Washington, D.C. (1971).

[47] "1969 Accidents of Large Motor Carriers of Property," U. S. Dept. of Transportation, Bureau of Motor Carrier Safety (1970).

[48] "Summary and Analysis of Accidents on Railroads in the United States, Calendar Year 1970," U. S. Dept. of Transportation, Bureau of Railroad Safety, Accident Bulletin 139 (1970).

[49] GLASSER, M., GREENBURG, L., "Air Pollution, Mortality, and Weather, New York City, 1960-1964," Arch. Environ. Health 22 (1971) 334-343.

[50] HODGSON, T.A., Jr., "Short-term Effects of Air Pollution on Mortality in New York City" Environ. Sci. and Tech. 4 (1970) 589-597.

[51] BUECHLEY, R.W., REGGAN, W.B., HASSELBLAD, V., VAN BRUGGEN, J.B., "SO_2 Levels and Perturbations in Mortality, A Study in the New York--New Jersey Metropolis" Arch. Environ. Health 27 (1973) 134-137.

[52] SCHIMMEL, H., GREENBURG, L., "A Study of the Relation of Pollution to Mortality, New York City, 1963-1968" J. Air Poll. Control Assoc. 22 (1972) 607-616.

[53] LAVE, L.B., FREEBURG, L.C., "Health Effects of Electricity Generation from Coal, Oil, and Nuclear Fuel" Nuclear Safety 14 (1973) 409-428.

[54] LAVE, L.B., SESKIN, E.P., "An Analysis of the Association between U. S. Mortality and Air Pollution" J. Amer. Statis. Ass. 68 (1973) 284-290.

[55] LAVE, L.B., SESKIN, E.P., "Health and Air Pollution, The Effect of Occupation Mix" Swed. J. of Economics 73 (1971) 76-95.

[56] LAVE, L.B., SESKIN, E.P., "Air Pollution, Climate, and Home Heating: Their Effects on U. S. Mortality Rates" Am. J. Public Health 62 (1972) 909-916.

[57] HICKEY, R.J., SCHOFF, E.P., CLELLAND, R.C., "Relationship between Air Pollution and Certain Chronic Disease Death Rates" Arch. Environ. Health 15 (1967) 728-738.

[58] HICKEY, R.J., BOYCE, D.E., HARNER, E.B., CLELLAND, R.C., "Ecological Statistical Studies Concerning Environmental Pollution and Chronic Disease" IEEE Trans. on Geosci. Electron. GE-8 No. 4 (1970) 186-202.

[59] HICKEY, R.J., "Air Pollution" in W.W. Murdock (ed.) Environment, Resources, Pollution and Society. Sinauer Associates, Publishers. Stamford, Conn. (1971) Chapter 9.

[60] HICKEY, R.J., BOYCE, D.E., HARNER, E.B., CLELLAND, R.C., "Exploratory Ecological Studies of Variables Related to Chronic Disease Mortality Rates" Institute for Environmental Studies, Graduate School of Fine Arts, Univ. of Pennsylvania, Philadelphia, Pa. (July 30, 1971).

[61] CARNOW, B. W., MEIER, P., "Air Pollution and Pulmonary Cancer" Arch. Environ. Health 27 (1973) 207-218.

[62] STARR, C., GREENFIELD, M.A., HAUSKNECHT, D.F., "A Comparison of Public Health Risks: Nuclear vs. Oil-fired Power Plants" Nuclear News 15(10) (Oct. 1972) 37-45.

[63] HAUSKNECHT, D.F., "Approximate Mortality Risk from SO_2 and Particulates" in C. Starr, M.A. Greenfield and D.F. Hausknecht. Public Health Risks of Thermal Power Plants. Appendix VI, Univ. Calif. Los Angeles report UCLA-ENG-7242 (May 1972).

[64] LARSEN, R.J., "Relating Air Pollution Effects to Concentrations and Control" J. Air Poll. Control Assoc. 20 (1970) 214-225.

[65] Reference 2, p. V-152.

[66] Federal Coal Mine Health and Safety Act of 1969. Public Law 91-173. 91st Congress S. 2917 (Dec. 30, 1969).

[67] KERR, L.E., Director, Dept. of Occupational Health, United Mine Workers of America, Washington, D.C., personal communication (April 26, 1972).

[68] American National Standard Method of Recording and Measuring Work Injury Experience. American National Standards Institute. ANSI Z16.1-1967.

[69] GRAHN, D., Argonne National Laboratory, personal communication.

[70] LEDERBERG, J., "Squaring an Infinite Circle, Radiobiology and the Value of Life," Bulletin of the Atomic Scientists 27 (7) (Sept. 1971) 43-45.

DISCUSSION

D. BENINSON: Your assessment of the genetic contribution takes into account all genetic deaths, most of which correspond to failure to implant and not to disease or death causing social cost. Counting only those effects which cause social costs would substantially reduce the entry for genetic effects.

K.A. HUB: We attempted to estimate the social costs of genetic effects by basing the cost on values used by Lederberg in Ref. [70].

D. BENINSON: In view of the long latent period, the value of 6000 days per death is probably too pessimistic in the case of radiation effects. Would you agree with this?

K.A. HUB: Yes. The need for improved values for the various classes of mortality is pointed out in Refs [1] and [2].

W.D. ROWE: Did your figures for coal effluents take account of the emission controls which will be required by EPA around 1980?

K.A. HUB: Yes.

J.G. HÉBERT: You say that you were unable to take account of the genetic effect of atmospheric contaminants released by coal-fired and oil-fired power stations owing to lack of data, and your Table V contains a genetic effect entry (a relatively substantial one) only in respect of nuclear stations. It is worth noting that in the case of coal-fired stations any identified genetic effects of the contaminants released will further increase the gap between coal and LWR man-day losses, while in the case of oil-fired stations they would bring the oil figure up closer to LWR. Thus any refinement of the data would modify the comparison in favour of nuclear.

F.D. SOWBY: Have you any information about the health effects resulting from the use of natural gas?

K.A. HUB: The Social Costs Study did consider a natural gas energy system, but we did not quantify gas explosions. I would refer you in this connection to Refs [1] and [2] of the paper.

J.C. VILLFORTH (Chairman): Have you estimated the impact of the reduction in accidents and occupational disease in coal mines resulting from the Coal Workers Health and Safety Act? If this Act is effective, it will certainly reduce the number of lives lost in coal mines.

K.A. HUB: The influence of the Coal Workers Health and Safety Act was indeed considered in the study. I'm not sure what effect this Act will have on accidents per ton of coal, because mining productivity decreases with the new standards. At all events, accidents may be expected to decrease in number, and it is anticipated that such things as progressive massive fibrosis will disappear.

Transfer of radionuclides to man through environmental pathways

(Session VII cont. and Session VIII)

Chairmen: J. C. VILLFORTH (United States of America)
L. -E. LARSSON (Sweden)

Invited Review Paper

TRANSFER OF RADIONUCLIDES TO MAN THROUGH ENVIRONMENTAL PATHWAYS

N. T. MITCHELL
Ministry of Agriculture, Fisheries and Food,
Fisheries Radiobiological Laboratory,
Lowestoft, Suffolk,
United Kingdom

Abstract

TRANSFER OF RADIONUCLIDES TO MAN THROUGH ENVIRONMENTAL PATHWAYS.

The most widely accepted method in current use for the evaluation of environmental impact from releases of radioactivity that may cause human radiation exposure used an environmental pathway model. The more important aspects of the model involve the use of critical pathway techniques to set controls on the releases of radioactive material to the environment and to assess the human radiation exposure arising as a consequence.

The basic concepts of the environmental pathway model and its application are discussed. The model depends on achieving an understanding of radionuclide behaviour in the environment, from the moment of release up to the time when contamination results in radiation exposure to the public. Exposure may be as a result of contaminated material entering the body — as air, water or a foodstuff; alternatively, it may be due to radiation whose source is external to the body. For each of these types of exposure pathway the sequence of events can be divided up into a number of compartments, between which transfer takes place. Transfer along the pathway is a dynamic process and can be described mathematically in terms of transfer functions between interacting compartments. Alternatively, for the situation of a regular discharge resulting in a steady-state condition the relationship between compartments can be expressed as a 'concentration factor'. The derivation of both of these terms is described and some aspects of their use are discussed.

The paper then goes on to discuss the types of environmental pathway that are encountered as a result of release of radionuclides to the atmospheric and aquatic environments. Disposal to the ground is regarded as a special case of release to the aquatic environment. Some of the literature on the subject is reviewed in discussing the better known pathways showing what, to date, have been the most important mechanisms of transfer of radionuclides to man.

1. INTRODUCTION

Concern at radiation exposure caused by the presence of man-made radioactivity in the environment has led increasingly to a need for methods to evaluate the consequences of such releases. The first releases of artificial radioactivity to have environmental impact on a large scale were from the detonation of nuclear weapons, and the quantity present in the biosphere as a result still far exceeds the total amount from controlled disposal operations, though it is now fairly evenly distributed and its contribution to radiation exposure of an average individual is small compared with exposure from natural sources[1]. Experience gained from measuring fallout has made an invaluable contribution to our knowledge of environmental transfer mechanisms and has generated a considerable store of information which has been supplemented by data from the controlled disposal of radioactive materials.

The production of large quantities of waste radioactivity is an inevitable consequence of the use of nuclear energy, and although the vast majority of these wastes can be stored to decay away, effectively isolated from human contact, complete avoidance of disposal to the environment is neither a requirement nor a practicable possibility. A means of evaluating the consequences of controlled release of radioactive waste is therefore necessary and is of particular importance to the economic exploitation of nuclear energy, for without it steps might be taken, at considerable and unnecessary expense, to reduce discharges to far lower levels than cost-risk considerations would merit. A solution has been provided by the development of environmental pathway models, based on a knowledge of the behaviour of radionuclides as they move through the environment. One of the major features of such models lies in the use of critical pathway techniques[2, 3] to set controls on releases of radioactive material to the environment. Models used in this way provide a direct basis for decisions regarding the installation of plant and equipment to restrict discharges. In addition, environmental pathway modelling has an important part to play in evaluating the possible consequences of nuclear accidents[4] and thus in deciding the nature and scale of contingency planning required to cope with emergency situations.

This paper is concerned with the transfer of radionuclides through environmental pathways which culminate in radiation exposure of humans and human populations ('exposure pathways'), since in setting controls on radioactive waste disposal, it has generally been found that the radiological hazard to environmental resources is of a lower order than that to human populations[5]. Nevertheless, the concept of environmental pathway models is equally applicable to evaluation of the effects on resources.

2. THE ENVIRONMENTAL, EXPOSURE PATHWAY, MODEL

2.1. Basic concepts

The basic concept of the model is that of a sequential transfer process, beginning with the release of a radionuclide into the environment, and proceeding through a number of stages or 'compartments' through which the radionuclide moves until it reaches a material via which the human population is exposed to radiation. Amongst the several ways in which pathways may be classified, perhaps the most fundamental is according to whether the source of deposited radioactivity responsible for exposure is either internal or external to the human body. Internal exposure occurs when radionuclides are either inhaled or ingested, whilst external exposure is due to radioactivity in our surroundings.

Fig. 1 is a simplified representation of the model in block diagrammatic form, each of the boxes signifying compartments, and the arrows signifying transfer of the radionuclide between them, with the exception of the last stage in each chain which is the irradiation of man. In practice, pathways can often be broken down into many more stages than are displayed in Fig. 1, processes which may describe the true behaviour more faithfully, for instance the food chains (or food webs as they are sometimes called) which are involved in internal exposure pathways. Although it can be profitable to explore them - for instance, if knowledge of the mechanism of the transfer process is desired - a simplified approach is usually quite adequate for control purposes, particularly when the overall aim in

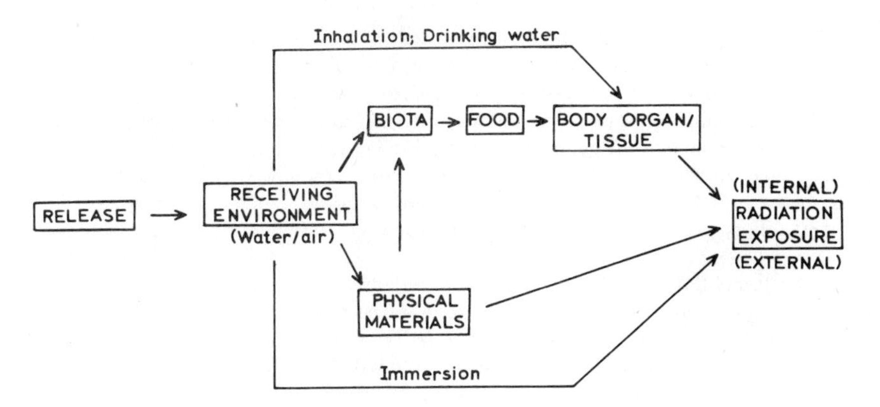

FIG. 1. Compartmental model of an environmental exposure pathway.

applying information on these transfer processes is to relate human radiation exposure to the rate of release of the responsible radionuclide(s). Indeed, some pathways can be expressed quite adequately in very simple terms, such as the release of a liquid waste to a river which is a source of drinking water, and in fish consumption problems it is usually quite unnecessary to understand the way in which activity dispersed in water enters fish, a process which may be made up of several components, such as swallowing of water, and entry through the gills and through the food eaten.

The first stage in a pathway will normally be recognized as dilution into the receiving environment, to be followed by further dispersion before transfer occurs into physical or biological materials further along the chain. In the case of a controlled release of a radionuclide, dilution can be and often is deliberately enhanced so as to minimize the concentrations which will be realized in the materials from which human radiation exposure subsequently occurs. To this end, liquid radioactive wastes at nuclear power stations are diluted into the large volume of cooling water outflow, and releases to atmosphere are diluted with air. Further steps can be taken to ensure subsequent dilution after release by judicious positioning of the point of release. Natural dispersive forces can be employed to advantage, such as air currents, by using tall stacks, or tidal forces, by using a long pipeline into deeper water.

Represented at its simplest, therefore, the exposure pathway model can be visualized as consisting of three stages:

(a) an input stage, comprising dilution/dispersion of the radionuclide into the receiving environment, from which radioactivity is distributed along perhaps many pathways;

(b) a chain of events through which the radionuclide passes, ultimately reaching the material which is the responsible vehicle for human radiation exposure;

(c) the irradiation of man from ingested or inhaled material, or from material which is external to the body.

This final stage requires the derivation of dosimetric models which, in the case of internal exposure pathways, must include the metabolism of ingested or inhaled nuclides[6]. Dosimetry models for external exposure are much simpler and, usually on the basis of uniform exposure, develop the required relationship from basic physical concepts of geometry, radiation energy and attenuation by the medium through which the radiation passes before reaching man. In neither of these cases is the process a transfer of a radionuclide; hence it is outside the remit of this paper, which concentrates on the true radionuclide transfer processes.

The immediate dilution which occurs on entry of an effluent into the environment will depend on mainly physical factors. The positioning of a pipeline itself will be important, as well as the outlet velocity of the waste. Further factors which will influence dilution at this early stage will be differences in temperature and density between the effluents and the water or air masses receiving them, together with the position of the outlet point in respect of parameters such as tidal currents, river flow, wind speed, etc.

The parallel situation for radionuclides introduced through burial of solids or flow of liquids into the ground is the transfer into an aquifer or other watercourse. In the case of liquids, in addition to physical factors such as the distance which must be travelled before the radionuclide reaches water and the adsorptive characteristics of the strata, some chemical characteristics will influence the transfer process and affect the degree of dilution obtained, and it is mainly chemical properties which affect the adsorptive processes. For transfer to occur from buried solid waste, a leaching process must be envisaged; thereafter the flow of leached activity may be regarded as though it was a liquid effluent which had been released into ground. Following the initial diluting process, further dispersion may occur. A number of models have been developed (tailored to different types of environment) to describe these processes, thereby enabling the rate of introduction to be correlated with concentrations in water or air at the stage where transfer is about to take place into biota or physical materials. These models need not be complex; simple conceptual models have usually been found to be adequate, based on, for instance, tidal excursions or gaussian-plume distributions in the atmosphere.

The next stage involves the transfer from water or the atmosphere to a physical material (river/marine sediment, soil, etc.) or biota (crops, fish, etc.) and is often described as a concentration or reconcentration process. This term has come into common use because of the application of this type of model to waste disposal control using critical pathway techniques, for an increase in concentration between compartments often occurs. This nomenclature is perhaps a little unfortunate, for there are pathways where either no change or even a reduction in concentration take place but where the ultimate human exposure is still high enough to warrant identification of the transfer processes leading to it. In consequence a term which is more widely applicable would simply be 'transfer'.

It is at this stage that many pathways become much more complex than the schematic diagram of Fig. 1 suggests. Quite apart from the complex mechanism along animal food chains leading to a resource exploited as a human foodstuff, there is often an interplay between physical materials and biota. On the one hand there may be competition - for instance, between physical materials, such as estuarine sediments[7], and animal or plant life - for activity in the water mass which would otherwise be available, whilst in other circumstances the sediment may provide a reservoir of activity to be metabolized by biota. The atmosphere/ soil/plant sequence can be cited as an example of this latter type of process[8].

The scheme described here has been evolved primarily for wastes disposed of directly to the atmospheric and aquatic environments. However, evaluation of radioactive wastes buried in the ground, whether in liquid or solid form initially, can be accommodated by the environmental pathway model as a special case of the aquatic system. The usual aim in disposing of wastes in this way is to ensure that they are isolated from human contact and remain so to a high degree; the only conceivable means of entry into an environmental pathway leading to human radiation exposure would be via contamination of an aquifer or other source of water.

2.2. Mathematical treatment of the model

Methods of evaluation of the environmental pathway model depend very much on the use to which it is being put, but also on personal attitudes and preferences. The problem is very different, and obviously more difficult to solve, if it is the impact of an event in the future which is being predicted as opposed to evaluating the consequences of a release which has already occurred. Steady-state conditions, such as may occur from a continuing controlled release from nuclear power operation, permit some important simplifications, and provide a situation which is relatively easily and accurately predicted and therefore produces a basis for planning waste disposal control. Methods of solution will also depend on the type of dose assessment required - for example, the dose received by a particular individual or by a group of people, for each of which a time limitation is specified, or the source-related dose commitment to infinite time, limited only by radioactive decay unless some time qualification is introduced.

Coupled compartment models, such as the one displayed schematically in Fig. 1 but often with more competing chains than are shown there, are amenable to analysis by study of the transfer function between interacting compartments. With complex situations, such as assessment of the consequences of excavating a new Pacific-Atlantic oceanic canal with nuclear explosives[9], and the evaluation of the impact of the growth of nuclear power in the Upper Mississippi River Basin[10], a detailed systems analysis approach has been developed.

2.2.1. Transfer functions

Expressed at its simplest, the transfer of radioactivity from one compartment to the next can be described in terms of a transfer coefficient, that is the fraction of the radioactivity which is transferred. A good example of the development and use of this approach is in the work of the United Nations Scientific Committee on the Effects of Atomic Radiation[11], which estimated exposure from world-wide fallout through mainly agricultural pathways, as a dose commitment to the world population.

In the UNSCEAR method the symbol P represents the transfer coefficient, so that P_{ij} would signify the coefficient for transfer from compartment i to compartment j. P_{ij} represents the fraction of the radionuclide in compartment i which reaches compartment j; hence it can never exceed unity, and because of radioactive decay will never quite equal it, though it may come very near to it for nuclides of very long half-life compared with the rate of transfer. Because there may be other transfer processes competing for the activity in compartment i, P_{ij} may be much less than unity, indicating that this pathway is of only minor importance. In the UNSCEAR example, P_{ij} is the ratio of the integrals to infinite time of the amounts in the two compartments, since the Committee was concerned to deduce the total dose commitment from fallout from weapons-testing:

$$P_{ij} = \left(\int_{-\infty}^{\infty} A_j(t)dt\right) \bigg/ \left(\int_{-\infty}^{\infty} A_i(t)dt\right)$$

$A_i(t)$ and $A_j(t)$ are the amounts in compartments i and j at time t. In practice, integration would only be carried out to some finite time T, and a time T_o can be defined (the beginning of the release) when A_i and A_j due to the release are each zero, so that the equation for the transfer coefficient P_{ij}, to time T, becomes:

$$P_{ij} = \left(\int_{T_o}^{T} A_j(t)dt\right) \bigg/ \left(\int_{T_o}^{T} A_i(t)dt\right)$$

Applied to evaluation of a controlled release of radioactivity and its impact on sections of the public, use of the transfer function system will indicate branching ratios along competing pathways; an example would be the fraction removed to a sediment from a marine water mass which is a significant factor in determining the relative importance of pathways into which radioactivity is introduced via water.

2.2.2. Concentration factors

Universal as the principle of the transfer function may be, it is not an easy quantity to estimate, requiring environmental data which may be difficult to acquire or, in the situation where it is necessary to predict the consequences of an unprecedented event, non-existent. An alternative but very much simplified means of evaluating the transfer process depends on the changes in concentration of radioactivity within the related compartment, the equivalent transfer function being a concentration factor (CF). It only strictly refers to the steady-state condition resulting from a continuous release but, nonetheless, has been found to be a most useful tool in evaluations using the critical pathway approach to planning controls on waste disposal. The concentration factor, as usually defined for a nuclide X, is simply the ratio of its concentration at equilibrium in the materials of which compartments i and j are made up:

$$CF_X = [X]_j / [X]_i \ .$$

Complete assessment of a multi-stage pathway would require a number of concentration factors which, when multiplied together, would enable the concentration in the material responsible for human exposure to be related to discharge rate. The first stage in the pathway is not normally expressed as a concentration factor but as the relationship between discharge rate and concentration in the receiving environment, and is based on the use of dilution/dispersion models. Breakdown of the fine detail of multi-compartment pathways is rarely needed for waste control purposes, though knowledge of the intermediate processes will help in understanding the mechanism by which radioactivity is transferred along the pathway. The use of concentration factors to describe the relationship between water and fish may be cited as an example. It is quite valid for purposes of calculating the permissible limit to the release of the radionuclide concerned, in order to calculate a concentration factor relating two materials, even though it is clear that most of the contamination in the fish does not enter it directly from the water but via a food chain which is itself made up of several stages.

3. ENVIRONMENTAL PATHWAYS IN PRACTICE

3.1. Atmospheric pathways

Much of our experience of transfer processes which follow the release of radioactivity to the atmosphere has accrued from the testing of nuclear weapons. As a result, most of the information has been on fission products and some extensive reviews are available[12,13]. A few simple pathways involve direct exposure of the human population from the atmosphere itself, though a majority are much more complex and are the product of deposition on to the ground, of uptake by crops, etc., growing on it, or deposition on to water surfaces. Apart from the processes of dilution and dispersion within the atmosphere, which are common to all these more complex pathways, aspects of deposition on surface water are covered in section 3.2; it is to be noted, however, that they have so far been of little importance, most of the contamination of water by activity of atmospheric origin reaching it via the ground as run off.

Direct exposure from the atmosphere can occur either externally, due to a person being immersed in it, as in a 'cloud', or internally as a result of inhalation, though neither of these pathways has proved to be of any great significance compared with those which involve terrestrial materials, crops and other foodstuffs. For instance, in the case of iodine-131 a factor of 700[14] has been established for the relative importance of the pathway resulting in exposure from milk contamination as compared with the inhalation pathway. In certain circumstances, however, the inhalation pathway might still be a force to be reckoned with, for example following an emergency release, when the milk pathway could be eliminated by withdrawal of contaminated supplies.

Nuclides which have been considered as sources of direct exposure from the atmosphere include tritium, krypton-85 and radioiodine, particularly iodine-131. Tritium can be absorbed through the skin as well as by direct inhalation[15], though due to its low β energy external exposure is ignored. In contrast, exposure from krypton-85 is almost entirely external, that from inhalation or ingestion being insignificant, and its uptake by plants is regarded as being too low to contribute to any relatively important extent. Discharges of both krypton-85[16] and tritium will increase as the use of nuclear power expands, though neither poses a serious threat for many years to come, interest resting mainly on their persistent behaviour. Iodine-131 has been studied intensively as an important component of fresh fallout and as potentially the most hazardous radionuclide from an airborne release after a nuclear reactor accident[17].

The importance of the atmosphere is mainly as the medium from which transfer occurs on to the earth's surface, leading to very many pathways of potential significance to man; this stage in the process can be considered as firstly dilution and dispersion, followed by deposition on to the earth's surface or to something associated with it. In the most widely adopted model for quantifying the dispersion process for a point source release, it is assumed that the concentration distribution is gaussian with respect to both vertical and horizontal transects of the plume travelling downwind from the point of release. The basic idea was first proposed by Sutton[18], though this has been refined and extended to take account of variables such as changing wind speed and direction. As in modelling of the aquatic environment, it is usually long-term average concentrations in the sector that are important, rather than short-term variations. Modifications of the model have been developed, for instance by Bryant[19], by Soldat et al.[20] and others.

Several mechanisms have been conceived for deposition from the atmosphere to the earth's surface. In addition to dry deposition, there is washout by rain falling through a cloud of activity in the atmosphere, or rainout in which the activity is incorporated into rain drops at the time of their formation. All the important pathways are internal, the result of ingestion of activity which contaminates food or water supplies; direct external exposure from soil which is contaminated is negligible. The role of soil is as an important intermediary providing a means by which radioactivity is introduced into plants via their root systems. Re-suspension of dust from the earth's surface and subsequent deposition on to plant surfaces has been considered but is less important than the deposition of particles already present in the atmosphere, the inflorescences providing effective traps, particularly for the smaller particles. A factor which will influence the importance of such deposition is the utilization of the crop: for instance, with wheat, the significantly higher level of ^{90}Sr in dark bread compared with white bread[21].

Uptake by crops provides a further diversity of pathway, activity finding its way into many different materials such as grain, forage, vegetable and root crops. Some, as direct human foodstuffs, will provide a simple pathway to man; for others, where the crop is used as an animal feedstock, the pathway will be longer, extending through further transfer processes. Both animals and poultry may yield either meat as a foodstuff in itself or a product such as milk or eggs. Transfer into meat has been extensively studied[22], especially for the longer-lived fallout nuclides. There is a marked difference between caesium and strontium, reflecting the difference in metabolism which exists in man; caesium-137 is transferred into muscle and other edible parts, sometimes with a small-degree reconcentration, whilst there is discrimination against strontium-90. Most of the strontium-90 enters the bone, whereas there is virtually uniform exposure of the body from caesium-137, and also from intake of tritium.

A well-known example of the contamination of an animal product is the sequence by which iodine-131 reaches cow's milk, an important basic human food in many countries. The thyroid is the critical organ, and infants are usually the group of the public at greatest risk, due to a combination of their high rate of milk consumption and the sensitivity of their thyroid. Other nuclides are also transferred to milk and in the absence of iodine-131 from a release (because of its impact, strong measures are taken to eliminate it from routine discharges) the other longer-lived fission products caesium-137 and strontium-90 are of principle concern. These last-named nuclides are also found in milk products; iodine-131 is usually ignored because of its short half-life. Whilst there is discrimination against strontium-90 and caesium-137 in butter compared with the milk from which it is made, the concentration ratio for cheese:milk varies and in some varieties a small concentration factor (up to 6 or 7) occurs.

3.2. Aquatic pathways

Pathways in the aquatic environment vary somewhat according to the type of water mass concerned and can be divided up into fresh, brackish/estuarine and the sea, though further subdivision can be useful, for instance between soft and hard fresh waters and also coastal sea water compared with the deep ocean. The most important source of controlled introduction is liquid wastes, solid wastes having very little impact whether buried in the ground[23], from which activity may be leached to enter ground water, or deposited on the sea bed[24]. Fallout has also been an important source of radioactivity; radionuclides may enter the ocean directly, or primarily through run-off from the land masses too. Run-off is almost wholly responsible for the activity in rivers due to fallout and until recently that in estuarine and near-coastal sea water too. However, in contrast to experience with terrestrial pathways, where almost all the information has come from fallout, a significant and useful amount of the data on the aquatic environment has accrued from the controlled introduction of radioactive wastes, especially in the sea and in estuaries, and some important reviews are available[25, 26, 27]. Critical pathway techniques have been developed and applied extensively to aquatic environments and transfer is often quoted in terms of the concentration factor for the equilibrium situation.

The role of water in aquatic pathways parallels that of air in atmospheric/terrestrial pathways quite closely, for the most important pathways are those for which water is the transport medium, in which dispersion occurs and from which activity is transferred further along the pathway. The only pathway where contamination of water is directly responsible for any significant exposure is drinking water; others which could be envisaged - such as bathing, or inhalation of spray - can be discounted in real terms.

Mathematical models[28, 29] are extensively used to describe the dilution/dispersion/transport phenomena in the receiving water mass; they are usually fairly simple and describe the long-term average situation rather than any short-term changes. The type of model varies with the environment; most can be described as conceptual, representing the processes which are at work, though often in a simplified manner. An adequate description of a lake may be a completely mixed body of water, or alternatively stratification may occur[30], a river may be likened to a pipe for purposes of describing the mixing processes, mixing processes in estuaries can be represented by the concept of a tidal 'plug', and tidal processes often provide the means for understanding processes in coastal waters and so on. In some situations - for instance in seas with little tidal action, or the open ocean - natural diffusion[31] may be the rate-controlling process, and many of the classical diffusion equations have been applied, such as those of Joseph and Sendner[32], Okubo[33] and Osmidov[34], to mention only a few; these have been reviewed by Smith[35]. Diffusion models alone may adequately cover the transport process within some water masses, but allowance must be made for other parameters, particularly the role of sediment[7] in inland and estuarine environments.

3.2.1. Fresh water

The end product of the more important pathways in freshwater rivers and lakes may be human drinking water, foodstuffs such as fish and shellfish, or some of the materials which have been discussed in section 3.1 and to which radioactivity has been transferred, as a result of the use of water for irrigation or livestock drinking purposes.

Sediments can have a profound effect on the fraction of activity remaining in the water mass and available in drinking water or to be transferred along other pathways. Caesium-137, for instance, is strongly removed by suspended matters and river muds[36], whilst radiostrontium tends to stay up with the water[37]. Much depends on the nature of the river as well as the characteristics of the element, and experience in two American river systems, the Columbia[38] and the Clinch[39], has been well documented. These accounts have included a number of fission and activation-product nuclides, amongst them strontium-90, ruthenium-106, rare-earths, caesium-137, phosphorus-32, chromium-51, cobalt-60 and zinc-65. Work in the Rhone delta[40] - the Camargues region of south-eastern France - has provided a good illustration of pathways through which strontium-90 and caesium-137 are introduced through irrigation to arable crops.

Transfer of radioactivity into fish has shown radiocaesium to be the most important of the fission-products, with data on activity from both fallout[41,42] and controlled releases[43]. These studies have shown an inverse relationship between the concentration factors for caesium-137 and potassium; a similar relationship exists for strontium-90 and calcium[42]. The concentration factor calculated for caesium-137 in a very soft water area[42] (K 0.4 ppm) was 4000, though this should not be taken as indicating the mechanism by which it enters the fish, which was not directly from water but almost entirely via the food chain. In these circumstances the potential exposure from consumption of the fish (trout) far outweighed that from drinking the lake water.

3.2.2. Saline environments

Although estuaries and the sea have different characteristics, they can be grouped together for the purpose of illustrating exposure pathways, on the common basis that the water is saline. Salinity apart, there are many ways in which saline environments contrast with freshwater environments. On the one hand, there is no significant use of the water for drinking or irrigation - desalination has yet to become an important source of potable water; on the other hand, external exposure pathways can be important due to the uptake of activity by sediment, particularly mud banks left bare by the receding tide. The sediments composed of fine mud or silt often found in estuaries take up a number of fission products very readily and zirconium-95/niobium-95, ruthenium-106 and the rare earths are removed to a high degree[44], behaviour which is also displayed by some activation products such as cobalt-60 and zinc-65. Contamination of sediment in this way generates an external radiation field for those who occupy areas where mud collects, the most important of the radioactive components being the gamma-emitting radionuclides. In another example of external pathways, where suspended matter is trapped by fishing gear, it is beta rather than gamma radiation that is of greater significance.

The important types of foodstuff in internal exposure pathways are seaweeds, fish and shellfish, each with markedly different characteristics. Whilst seaweed is a rather unusual seafood, it has provided perhaps one of the best-known pathways to public exposure - that from the marine discharges from Windscale, UK[45]. The seaweed species concerned, Porphyra, is manufactured into a food known as laverbread, the chief amongst a number of contaminants being ruthenium-106 with a concentration factor of about 2 x 10^3. The Porphyra also concentrates cerium-144 (CF 10^3), zirconium-95/niobium-95[32] (CF 4 x 10^2) and traces of other nuclides, though none of these carries the same radiological significance as ruthenium-106.

In marked contrast to its behaviour in freshwater regimes radiocaesium is much better represented in the water compartment of the marine environment, less being removed by biota, suspended matter and sediments. Concentration factors into marine and estuarine biota are thus much lower than in fresh water, with values in the region of 50 for the edible parts of commercial species of fish and shellfish. In addition to evidence from world-wide reporting of data on caesium-137 from fallout, caesium-134 from controlled waste disposal from Windscale has been followed over considerable distances, confirming the conservative behaviour of these radionuclides[46]. The behaviour of strontium-90 is also conservative and it is widely distributed in the oceans of the world, but uptake into the flesh of fish and shellfish is even lower than that of radiocaesium, the 'concentration factor' being less than unity for most species of fish; it can therefore be disregarded as a source of human radiation exposure. Among other conservative fission-product nuclides are iodine-131, tritium and technetium-99. In sharp contrast to its importance in the terrestrial environment the significance of iodine-131 in the sea is low, the only materials for which it shows any marked affinity being seaweeds[32]. This is also true of technetium-99, which displays a remarkable selectivity[47] for the Fucus seaweeds, and of the activation product manganese-54, an element which has similarities to technetium and iodine.

The non-conservative nuclides in the estuarine and marine environments tend to be metals which are both polyvalent and higher in the periodic classification than group IIA metals; examples are ruthenium-106, zirconium-95/niobium-95, cerium-144, iron-55, cobalt-60, zinc-65 and the transuranics. Whilst high concentration factors are observed into some biota, the fraction of radioactivity in this compartment is very small - the size of the compartment itself, of course, being small compared with water and sediment which between them contain nearly all the radioactivity introduced into the saline environment.

Some of the more scientifically interesting pathways in estuarine and coastal waters involve shellfish, which generally appear to have a higher potential than fish for concentrating radionuclides. Whilst concentration factors for fission products such as caesium-137 and strontium-90 in the edible parts of shellfish are a little higher than those in fish, concentration factors for the majority are much higher and there is a wide range between crustacea and molluscs and within species in these groups. As a further - though perhaps rather extreme - example of this phenomenon, some of the American and British experience with oysters may be quoted. From the Hanford, USA discharges to the Columbia River, data relating to the Willapa Bay oysters suggested a concentration factor of 1.1 x 10^4[48], and UK experience from the discharges from the Bradwell nuclear power station to the Blackwater estuary indicate a value in excess of 10^5 for the locally-grown oyster species[49]. Silver-110m has also been found in the Blackwater oysters, the concentration factor inferred being at least as high as that for zinc-65, yet its behaviour in the water mass of the estuary is unlike that of zinc-65, since it behaves conservatively[50]. Traces of iron-55 and cobalt-60 have also been detected in these oysters but at much lower significance; occasionally some phosphorus-32 is also found, but to a lower extent than in the Willapa Bay oysters, for which the concentration factor was suggested as several thousand.

4. CONCLUSIONS

This paper has presented a brief review of the environmental pathway model leading to the radiation exposure of man and has pointed to some of the more significant pathways which have been identified so far. A great deal of this information has come from fallout studies, especially in atmospheric and terrestrial environments, though this has been supplemented, particularly in recent years and especially in the aquatic environment by a growing literature on the distribution of radioactivity from the controlled introduction of waste from the exploitation of nuclear power. Data of this kind have a vital role to play in the control of waste disposal, for they provide the means by which the consequences in terms of public radiation exposure can be accurately assessed, so ensuring that adequate control measures are taken.

REFERENCES

[1] ANON., Ionising Radiation Levels and Effects, Ch. 1, Levels, Report of the United Nations Scientific Committee on the Effects of Atomic Radiation, General Assembly Official Records, 27th Session, Suppl. No. 25 (A/8725), United Nations, New York (1972).

[2] PRESTON, A., "Site evaluations and the discharge of aqueous radioactive wastes from civil nuclear power stations in England and Wales", Disposal of Radioactive Wastes into Seas, Oceans and Surface Waters (Proc. Symp. Vienna, 1966), IAEA, Vienna (1966) 725.

[3] PRESTON, A., "Critical path analysis applied to the control of radioactive waste disposal to aquatic environments", Disposal of Radioactive Wastes (Proc. NEA Information Meeting, 1972), OECD/NEA, Paris (1972) 121.

[4] BEATTIE, J. R., BRYANT, Miss P. M., Assessment of environmental hazards from reactor fission product releases, UKAEA Rep. AHSB(S) R135 (1970).

[5] WOODHEAD, D. S., "Levels of radioactivity in the marine environment and the dose commitment to marine organisms", Interaction of Radioactive Contaminants with Constituents of the Marine Environment (Proc. Symp. Seattle, 1972), IAEA, Vienna (1973) 499.

[6] ICRP, Report of Committee II on Permissible Dose for Internal Radiation (1959), Recommendations of the International Commission on Radiological Protection, ICRP Publication 2, Pergamon Press, Oxford (1959).

[7] DUURSMA, E. K., CROSS, E. G., "Marine sediments and radioactivity", Ch. 6, Radioactivity and the Marine Environment, National Academy of Sciences, Washington DC (1971).

[8] SCOTT RUSSELL, R., "Entry of radioactive materials into plants", Ch. 5, Radioactivity and Human Diet (SCOTT RUSSELL, R., Ed.), Pergamon Press, Oxford (1966).

[9] KAY, S. V., BALL, S. J., Systems analysis of a coupled compartment model for radionuclide transfer in a tropical environment (Proc. Second Internat. Symposium on Radioecology Ann Arbor, 1967), USAEC, TID-4500 (1969) 731.

[10] FLETCHER, J. F., DOTSON, W. L., PETERSON, D. E., BETSON, R. P., "Modelling the regional transport of radionuclides in a major United States river basin", Environmental Behaviour of Radionuclides Released in the Nuclear Industry (Proc. Symp. Aix-en-Provence, 1973), IAEA, Vienna (1973) 449.

[11] ANON., Report of the United Nations Scientific Committee on the Effects of Atomic Radiation, General Assembly Official Records, 24th Session, Suppl. No. 13 (A/7613), United Nations, New York (1969).

[12] SCOTT RUSSELL, R. (Ed.), Radioactivity and Human Diet, Pergamon Press, Oxford (1966).

[13] GARNER, R. J., Transfer of radioactive materials from the terrestrial environment to animals and man, Chemical Rubber Company, Critical Reviews in Environmental Control 2 (1971) 337.

[14] BURNETT, T. J., A derivation of the 'factor of 700' for iodine-131, Hlth Phys. 18 (1972) 73.

[15] ANSPAUGH, L. R., KORANDA, J. J., ROBINSON, W. L., MARTIN, J. R., The dose to man via food-chain transfer resulting from exposure to tritiated water vapour, USAEC Rep. UCRL-73195 Rev. 1 (1972).

[16] DUNSTER, H. J., WARNER, B. F., The disposal of noble gas fission products from the reprocessing of nuclear fuel, UKAEA Hlth and Saf. Branch Rep. AHSB(RP)R101, HMSO, London (1970).

[17] GARNER, R. J., MORLEY, F., Agricultural implications of a release of fission products from a criticality accident, Hlth Phys. 13 (1969) 465.

[18] PASQUILL, F., The estimation of the dispersion of wind-borne material, Met. Mag. 90 (1961) 33.

[19] BRYANT, P. M., Methods of estimation of the dispersion of wind-borne material and data to assist in their application, UKAEA Rep. AHSB(RP) R42, HMSO, London (1964).

[20] SOLDAT, J. K., BAKER, D. A., CORLEY, J. P., "Applications of a general computational model for composite environmental radiation doses", Environmental Behaviour of Radionuclides Released in the Nuclear Industry (Proc. Symp. Aix-en-Provence, 1973), IAEA, Vienna (1973) 483.

[21] HARLEY, J. A., "Radionuclide in food", Biological Implications of the Nuclear Age, USAEC Symposium Series 16 (1969) 189.

[22] COMAR, C. L., "Radioactive materials in animals - entry and metabolism", Ch. 7, Radioactivity and Human Diet (SCOTT RUSSELL, R., Ed.), Pergamon Press, Oxford (1966).

[23] MAWSON, C. A., "Consequences of radioactive disposals into the ground", Progress in Nuclear Energy, Series XII, Hlth Phys. 2 (DUHAMEL, A. M. F., Ed.), Pergamon Press, Oxford (1969) 461.

[24] WEBB, G. A. M., MORLEY, F., A Model for the Evaluation of the Deep Ocean Disposal of Radioactive Waste, Rep. NRPB-R14, HMSO, London (1973).

[25] MAUCHLINE, J., TEMPLETON, W. L., "Artificial and natural radioisotopes in the marine environment", Annual Review of Oceanography and Marine Biology (BARNES, H., Ed.), George Allen and Unwin Ltd., London 2 (1964) 229.

[26] PRESTON, A., JEFFERIES, D. F., "Aquatic aspects in chronic and acute contamination situations", Environmental Contamination by Radioactive Materials (Proc. Symp. Vienna, 1969), IAEA, Vienna (1969) 183.

[27] BOVE, A., VETTER, R. (Eds), Radioactivity in the Marine Environment, National Academy of Sciences, Washington DC (1971).

[28] INTERNATIONAL ATOMIC ENERGY AGENCY, Radioactive Waste Disposal into the Sea, IAEA Safety Series No. 5, IAEA, Vienna (1961).

[29] PRITCHARD, D. W., REID, R. O., OKUBO, A., CARTER, H. H., "Physical processes of water movement and mixing", Radioactivity in the Marine Environment, National Academy of Sciences, Washington DC (1971) 90.

[30] BRANCA, G., BREUER, F., CIGNA, A. A., AMAVIS, R., Applications of a Derived Formula for the Discharge of Radioactive Liquid Wastes, Rep. EUR 4897e, Commission of the European Communities, Luxembourg (1973).

[31] GYLLANDER, Christina, "Water exchange and diffusion processes in Tvaren, a Baltic bay", Disposal of Radioactive Wastes to Seas, Oceans and Surface Waters (Proc. Symp. Vienna, 1966), IAEA, Vienna (1966) 207.

[32] JOSEPH, B., SENDNER, H., Uber die Horizontale Diffusion in Meere, Dt. Hydrogr. Z. 11 (1958) 49.

[33] OKUBO, A., Horizontal diffusion from an instantaneous point source due to oceanic turbulence, Tech. Rep. No. 32, Chesapeake Bay Institute, The Johns Hopkins University, Baltimore, Maryland (1962).

[34] OSMIDOV, R. V., On the calculation of horizontal turbulent diffusion of the pollutant patches in the sea, Dokl. Akad. Nauk. SSSR 120 (1958) 761.

[35] SMITH, D. R., Comparison of Theoretical Ocean Diffusion Models, US Naval Radiological Defense Lab., San Francisco, Calif., Rep. USNRDL-TR-67-50 (1967).

[36] MORGAN, A., ARKELL, Mrs G. M., Radioactive Effluent Discharged from AERE Harwell into the River Thames (1) Preliminary Survey, UKAEA Rep. AERE-R3555 (1961).

[37] MITCHELL, N. T., EDEN, G. E., Radioactive strontium in the River Thames, J. Inst. Wat. Eng. 16 (1962) 175.

[38] NELSON, J. L., PERKINS, R. W., NIELSEN, J. M., HAUSHILD, W. L., "Reactions of radionuclides from the Hanford reactors with Columbia River sediments", Disposal of Radioactive Wastes to Seas, Oceans and Surface Waters (Proc. Symp. Vienna, 1966), IAEA, Vienna (1966) 139.

[39] PARKER, F. L., CHURCHILL, M. A., ANDREW, R. W., FREDERICK, B. J., CARRIGAN, P. H., Jr., CRAGWALL, J. S., Jr., JONES, S. L., STRUXNESS, E. G., MORTON, R. J., "Dilution, dispersion and mass transport of radionuclides in the Clinch and Tennessee Rivers", Disposal of Radioactive Wastes into Seas, Oceans and Surface Waters (Proc. Symp. Vienna, 1966), IAEA, Vienna (1966) 33.

[40] BOVARD, P., GRAUBY, A., FOULQUIER, G., PICART, Ph., "Radio-ecological study of the Rhône basin: strategy and balance sheet", Environmental Behaviour of Radionuclides Released in the Nuclear Industry (Proc. Symp. Aix-en-Provence, 1973), IAEA, Vienna (1973) 507.

[41] KOLEHMAINEN, S., HÄSÄNEN, E., MIETTINEN, J. K., "Caesium-137 in fish, plankton and plants in Finnish lakes during 1964-65", Radioecological Concentration Processes (ABERG, B., HUNGATE, F. P., Eds), Pergamon Press, Oxford (1967) 913.

[42] PRESTON, A., JEFFERIES, D. F., DUTTON, J. W. R., The concentrations of caesium-137 and strontium-90 in the flesh of brown trout taken from

rivers and lakes in the British Isles between 1961 and 1966: the variables determining the concentrations and their use in radiological assessments, Wat. Res. 1 (1967) 475.

[43] MITCHELL, N. T., JEFFERIES, D. F., "Control of low-level liquid radioactive waste disposal: UK experience in the derivation and use of 'environmental capacity'", Environmental Behaviour of Radionuclides Released in the Nuclear Industry (Proc. Symp. Aix-en-Provence, 1973), IAEA, Vienna (1973) 633.

[44] JEFFERIES, D. F., "Exposure to radiation from gamma-emitting fission-product radionuclides in estuarine sediments and the north-east Irish Sea", Environmental Surveillance in the Vicinity of Nuclear Facilities (REINIG, W. C., Ed.), Charles C. Thomas, Springfield, Illinois (1970) 205.

[45] PRESTON, A., JEFFERIES, D. F., The assessment of the principal public radiation exposure from, and the resulting control of, discharges of aqueous radioactive waste from the United Kingdom Atomic Energy Authority factory at Windscale, Cumberland, Hlth Phys. 13 (1967) 477.

[46] JEFFERIES, D. F., PRESTON, A., STEELE, A. K., Distribution of caesium-137 in British coastal waters, Mar. Pollut. Bull. 4 (1973) 118.

[47] MITCHELL, N. T., Radioactivity in Surface and Coastal Waters of the British Isles, Tech. Rep. Fish. radiobiol. Lab., Lowestoft, FRL 9 (1973) 34.

[48] SEYMOUR, A. H., "Accumulation and loss of zinc-65 by oysters in a natural environment", Radioecological Concentration Processes (ABERG, B., HUNGATE, F. P., Eds), Pergamon Press, Oxford (1967) 605.

[49] PRESTON, A., DUTTON, J. W. R., HARVEY, B. R., Detection, estimation and radiological significance of silver-110m in oysters in the Irish Sea and the Blackwater Estuary, Nature, Lond. 218 (1968) 689.

[50] PRESTON, A., The control of radioactive pollution in a North Sea oyster fishery, Helgoländer wiss. Meeresunters. 17 (1968) 269.

DISCUSSION

G. BOERI: I want to ask a question about some environmental values to which you referred in your paper. How do you explain the different concentrations in samples taken at the point of fall-out and at a distance 0.5 km from the fall-out. Are they just due to dilution, or are there any other factors involved?

N.T. MITCHELL: The fall in concentration is a reflection of two processes — dilution and physical removal from the water mass, mainly to sediments. It is the second process which I particularly wanted to illustrate; in short, the conservative or non-conservative behaviour of individual radionuclides and their parent elements.

R. BITTEL: Do you not think that it would be of value to consider transfers of contamination between harvested raw materials and food at the stage of consumption by man, in particular movements of food commodities?

N.T. MITCHELL: This could be an important aspect of the transfer process, and may substantially modify both the concentration in the product and the dose received. The Windscale laverbread pathway again provides a good example and illustrates two important points. The first is that the exposed population may be some distance from the source where the radioactivity is introduced into the environment; the second is that the contaminated foodstuffs may get diluted with other identical material from areas remote from the source of radioactivity.

H. T. DAW: In calculating the man·rems to a population or sub-population group, it is possible to take a short cut and arrive at the answer without going in detail into the problem of food pathway. For example, if one knows the total amount of a certain food commodity consumed by the population, and its activity per gram, one can determine the total activity ingested by the population in question and hence the population dose. Of course, to arrive at the average individual dose for a critical group the model would be required.

N. T. MITCHELL: Yes, but it is essential that the concentration can be measured. Otherwise it is necessary to assess the level of contamination, for which data on the whole of the preceding transfer process may be required.

S. AMARANTOS: Do you not think that, in view of the accumulation of activity in sediments, we should pay more attention to the effect on species other than man?

N. T. MITCHELL: Another paper in this Seminar (paper IAEA-SM-184/1[illegible] goes into just this problem. While we find that — provided the stringent recommendations of the ICRP are followed — environmental resources are also safeguarded, it still behoves us all to ensure that the whole environment is protected, and for this reason we included studies of the effect of radiation on the resources in our research programme.

K.-J. VOGT: Could you comment on the possibility that, after becoming attached to the sediments, radionuclides can again enter the drinking water exposure pathway by means of resuspension?

N. T. MITCHELL: This is certainly a potential transfer process, but its importance will depend on many factors, for example the efficiency of the water filtration plant, and the behaviour of individual radionuclides. It might be a significant transfer process for nuclides of elements such as strontium, which are adsorbed to sediments by ion exchange.

TRANSFER OF ^{137}Cs AND ^{90}Sr FROM THE ENVIRONMENT TO THE JAPANESE POPULATION VIA MARINE ORGANISMS

T. UEDA, Y. SUZUKI, R. NAKAMURA
National Institute of Radiological Sciences,
Chiba,
Japan

Abstract

TRANSFER OF ^{137}Cs AND ^{90}Sr FROM THE ENVIRONMENT TO THE JAPANESE POPULATION VIA MARINE ORGANISMS.

During the period from 1963 to 1971 we investigated the contamination levels of marine organisms, total diet and individual foodstuffs due to ^{137}Cs and ^{90}Sr from fall-out. The results are summarized as follows. (1) The concentrations of ^{137}Cs and ^{90}Sr in marine organisms show a decreasing tendency since 1963, closely resembling that of seawater. (2) The concentration factors and observed ratios, calculated from radioactive elements, are in good agreement with those from stable elements. (3) The ratios of the specific activities of marine organisms to that of seawater (expressed in annual mean values) were found to be almost 1, except the case of Sr in fish bone. (4) The total mean ^{137}Cs contribution for each of nine foodstuffs was observed; the contribution of cereals was about 40% of the total ^{137}Cs intake, followed by beans, milk and its products, leafy vegetables and fish. (5) The ^{90}Sr contribution of leafy vegetables was the highest (34%), followed by beans (20%) and cereals (18%). (6) The internal radiation dose rate received by the Japanese population is estimated to be 7.3 mrad, of which the contribution of marine foods was less than 10% from ^{137}Cs in the total diet between 1963 and 1971.

INTRODUCTION

The radioactive contamination of marine organisms is an important problem to the Japanese who utilize seafood as one of their main protein sources. We have studied the contamination of marine organisms by radioactive nuclides originating from various sources, both by field studies and by tracer experiments. There are two causes of the contamination of marine organisms: (1) global contamination by fall-out; and (2) local contamination by release from various nuclear establishments.

This paper, based on the results of field studies, considers the accumulation pattern of Sr and Cs by marine organisms from their environmental water from three aspects: concentration factors, observed ratios and specific activities. In addition, the contribution of marine organisms to the total Japanese diet, with regard to radioactive contamination by ^{137}Cs and ^{90}Sr, has been estimated and the internal dose rate calculated.

MATERIALS

The marine organisms from several districts (Niigata, Fukui, Fukushima, Ibaraki, Chiba and Hiroshima) and the total diet from Hokkaido, Niigata, Tokyo, Ibaraki, Osaka and Fukushima were collected from 1963 onwards, while the individual foodstuffs from Hokkaido, Niigata and

TABLE I. SPECIFIC ACTIVITIES OF CAESIUM IN FISH MUSCLE AND SEAWATER

	1964	1965	1966	1967	1968	1969	1970	
Fish ($\times 10^3$) [A]	1.3(25)	1.4(36)	0.8(28)	0.7(32)	0.8(32)	0.4(40)	0.3(33)	
Seawater ($\times 10^3$) [B]	1.0	0.9	0.8	0.8	1.2	0.5	0.5	mean
[A]/[B]	1.3	1.6	1.0	0.9	0.7	0.8	0.6	1.3 ± 0.3 [a]

Figures in parentheses are the number of samples analysed.
[a] Standard deviation.

Kagoshima were collected from 1966 onwards. The individual foodstuffs consist of nine categories (cereals, beans, potatoes, milk, eggs, fish and shellfish, leafy vegetables and root vegetables).

METHODS

The collected samples were ignited at 450°C [1] in an electric muffle furnace. The ashes were radiochemically analysed according to the following procedures:

(i) ^{137}Cs was fixed as chloroplatinate after co-precipitation with ammonium phosphomolybdate and then the radioactivity was measured by a low-background gas-flow counter (Tracerlab Omni/Guard; background: 0.2 counts/min).

(ii) ^{90}Sr was analysed by the fuming nitric acid method [2] and the radioactivity of ^{90}Y, the daughter nuclide, was measured in the low-background counter mentioned above.

(iii) The amounts of stable Cs and Sr were determined by atomic absorption spectrophotometry [3, 4] (Perkin-Elmer Model 303, wavelengths: 8521 Å for Cs and 4608 Å for Sr, acetylene-air flame).

(iv) The amount of potassium was determined flame-photometrically (Beckman, 786 nm).

(v) The amount of Ca was determined by the titration method using ethylenediamine tetraacetic acid (0.01247M).

RESULTS

1. The amount of Cs in the muscles of marine fish

The annual variation in the total mean concentration of ^{137}Cs in the muscles of fish is shown in Fig. 1, together with that in seawater quoted from the data on the same regions and years that fish was collected by Ohmomo et al. [5, 6]. No regional variation in ^{137}Cs concentration in fish muscles was evident, even among the same species. The level of ^{137}Cs in fish muscle was 80 pCi/kg flesh in 1963 and 7 pCi/kg in 1970. The decreasing tendency was parallel to that for seawater. The mean concentration of ^{137}Cs in *Limanda irrdorum* (benthic fish) was 60 pCi/kg flesh in 1963 and then decreased to about 8 pCi/kg flesh in 1970. The levels of ^{137}Cs in *Trachurus japonicus* and *Scomber japonicus* (migratory fishes) were comparable with that in *Limanda irrdorum*. The correlation between level and habitat was not apparent. The amount of stable Cs in fish flesh ranged from 17 (*Ditrema temmincki*) to 26 μg/kg flesh (*Mylio macrocephalus*) [7].

2. Cs concentration factor

The concentration factors of ^{137}Cs in the muscles of 154 fish samples ranged from 11 to 81 and the average was 43 with a standard deviation of 12. The amount of stable Cs in fish muscle ranged from 17 to 26 μg/kg flesh and the concentration factor is calculated to be between 34 and 52, with an average of 42 [7], using the data of Goldberg [8] for stable Cs in seawater.

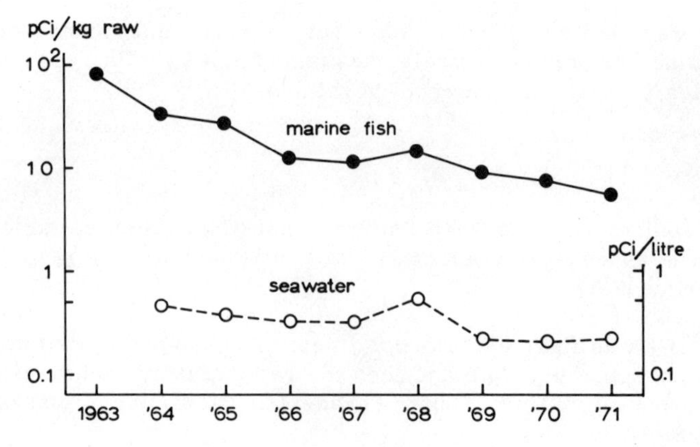

FIG. 1. Annual variation in the total mean concentration of ^{137}Cs in fish muscle and seawater.

3. Cs observed ratio in fish muscle

The Cs observed ratios (Cs/K in fish flesh/Cs/K in seawater) were calculated from both the radioactive and the stable elements and the average for the former was 5.7 with a standard deviation of 1.9. This figure is in good agreement with that based on stable Cs (5.9 ± 0.8) obtained by taking the amount of stable Cs and K to be 21.2 μg/kg and 2.78 g/kg in fish muscle [7], and 0.5 μg/litre and 0.387 g/litre [9] in seawater, respectively.

4. Cs specific activity in fish muscle

The specific activities (^{137}Cs/stable Cs) in fish muscle and seawater are shown in Table I. The averaged ratio of both values was 1.0 with a standard deviation of 0.3. This figure indicates that the fish organism is unable to distinguish between radioactive and stable Cs when this element is taken up from the environmental water, i.e. ^{137}Cs originating from radioactive fall-out shows the same physico-chemical behaviour in seawater as does stable Cs.

From the above results, the radiation dose ($\overline{D}$: mrad/a per caesium unit (CU) in the human body) to which the Japanese population is exposed from the intake of seafood (fish muscle) contaminated by ^{137}Cs is estimated, using the data (0.018 mrad/a per CU in the human body) of Spiers [10], as follows (OR = observed ratio):

$\overline{D}$ (mrad/a per CU in human body)

$$= CU_{seawater} \times OR_{fish:seawater} \times OR_{human:fish} \times 0.018$$

$$= \frac{^{137}Cs \text{ pCi/litre of seawater}}{0.387} \times 5.9 \times 2.3 \times 0.018$$

$$= 0.63 \times {}^{137}Cs \text{ pCi/litre of seawater}$$

where

$OR_{fish : seawater}$: 5.9 (calculated from stable Cs and K)

$OR_{human : fish}$: 2.3 (calculated with the data of Ref. [11] for the levels in the human body and the present data for the ^{137}Cs level in the total diet)

since the amount of K in seawater is 387 mg/litre.

5. The concentration of ^{90}Sr in marine organisms

The values of the SU (strontium unit = ^{90}Sr · pCi/g of Ca) in Scomber japonicus (migratory fish), Mylio macrocephalus (sedentary fish) and Limanda irrdorum (benthic fish) were similar to each other and no correlation between ^{90}Sr level and habitat was observed. The annual variation in the total mean concentration of ^{90}Sr in fish bone and seawater is shown in Fig. 2. The SU in molluscan shell was lower than that of fish bone, while the values in brown weed were much higher than those in both fish bone and molluscan shell (Fig. 3).

The ratios of the amounts of stable Sr in various marine organisms to that in fish muscle are shown in Table II. From the analysis of any one of these organisms the ^{90}Sr of the other organisms could be estimated.

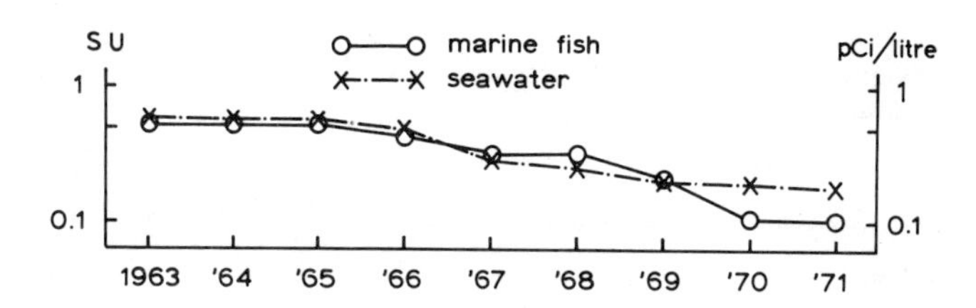

FIG. 2. Annual variation in the total mean concentration of ^{90}Sr in fish bone and seawater.

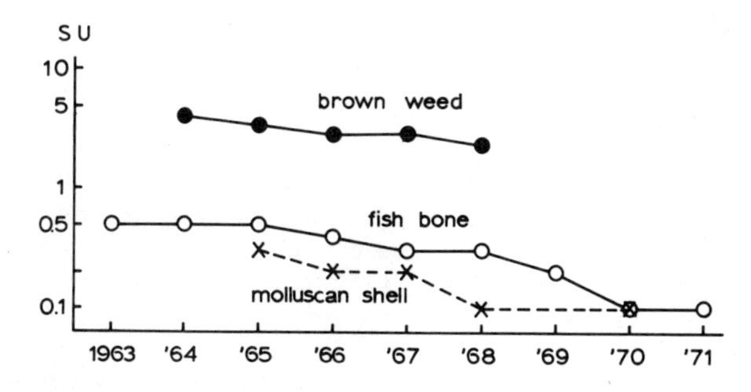

FIG. 3. Annual variation in the total mean concentration of ^{90}Sr in marine organisms.

TABLE II. THE AMOUNTS OF STABLE STRONTIUM, THE CONCENTRATION FACTORS IN MARINE ORGANISMS AND THEIR RATIOS TO THAT IN FISH MUSCLE

		Sr mg/kg raw	Ratio	CF	Ratio
Fish	Muscle	3	(1)	0.5	(1)
	Bone	200 ± 40	(70)	26 ± 7	(50)
Crustacea (exoskeleton)					
Decapoda					
	Anomura	1461	(490)	183	(370)
	Brachyura	1000 ± 400	(300)	120 ± 50	(200)
	Macrura	500 ± 300	(200)	70 ± 40	(100)
Stomatopoda		567	(190)	100 ± 50	(200)
Mollusca					
Bivalvia		1000 ± 200	(330)	130 ± 30	(260)
Gastropoda		1000 ± 200	(330)	130 ± 30	(260)
Brown weed		192	(64)	24	(48)

TABLE III. OBSERVED RATIOS OF STRONTIUM

	^{90}Sr	Stable Sr
Fish bone	0.3	0.27
Fish (whole body)	0.3	-
Molluscan shell	0.2	0.19
Brown weed	3.7	3.74

6. Sr concentration factor

The concentration factors in fish bone obtained by the measurement of ^{90}Sr ranged from 14 to 63 and were comparable with those obtained by stable Sr analysis (10 - 37) [12]. The concentration factors in molluscan shell obtained by the measurement of ^{90}Sr ranged from 40 to 205 in comparison with those obtained by stable Sr analysis (61-194) [12]. For brown weed the concentration factors obtained by ^{90}Sr measurement were 4-10 but the results from stable Sr analysis were 18-31. These values in fish bone and molluscan shell are in agreement with those of Bryan [13] and Mauchline [14]

7. Sr observed ratio in marine organisms

As seen in Table III, the observed ratio from the analyses of ^{90}Sr and stable Sr are in good agreement. The observed ratio in brown weed was one order of magnitude higher than those in fish bone and molluscan shell. This may be due to the affinity of the alginic acid in brown weed for Sr, as pointed out by Haug and Smidsrod [15, 16].

8. Sr specific activity

The ratios of the Sr specific activities in marine organisms to those in seawater were found to be almost 1.0, except for fish bone sampled between 1963 and 1967. These relatively high values in fish bone may be due to a slow Sr turnover rate in fish bone. From these figures it can be said that ^{90}Sr from fall-out, like ^{137}Cs, behaves in a manner physico-chemically similar to stable Sr in seawater. The radiation dose ($\bar{D}$) for the Japanese population, due to the intake of seafoods contaminated with ^{90}Sr, is estimated using the data of 1.4 and 0.7 mrad/a per SU for bone-lining cells and bone marrow, respectively [17]:

$\bar{D}$ (mrad/a per SU in human bone)

$$= SU_{seawater} \times OR_{fish:\,seawater} \times OR_{human\ bone:\,food} \times 1.4$$

$$= \frac{^{90}Sr\ pCi/litre\ seawater}{0.4} \times 0.3 \times 0.25 \times 1.4$$

$$= 0.26 \times {}^{90}Sr\ pCi/litre\ of\ seawater$$

where

$OR_{fish:\,seawater}$: 0.3 (from our investigations)

$OR_{human\ bone:\,fish}$: 0.25

since the amount of Ca in seawater is 400 mg/litre. Similarly, $\bar{D}$ in bone marrow is calculated as follows:

$\bar{D}$(mrad/a per SU in bone)

$$= SU_{seawater} \times 0.3 \times 0.25 \times 0.7$$

$$= 0.13 \times {}^{90}Sr\ pCi/litre$$

9. The level of ^{137}Cs in total diet and individual foodstuffs consumed by the Japanese population

Since 1964 the annual mean levels of ^{137}Cs in the Japanese total diet has decreased with time and the general tendency is similar to that for the ^{137}Cs fall-out rate [18], as shown in Fig. 4. The contribution (in per

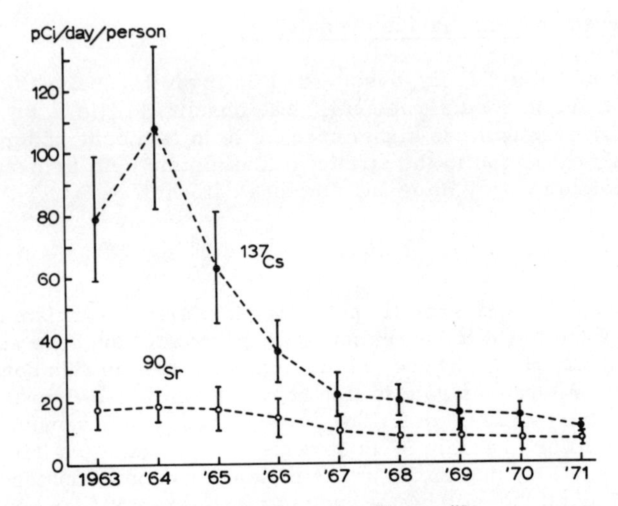

FIG. 4. Annual variations in the total mean concentration of ^{90}Sr and ^{137}Cs in the total diet of the Japanese population.

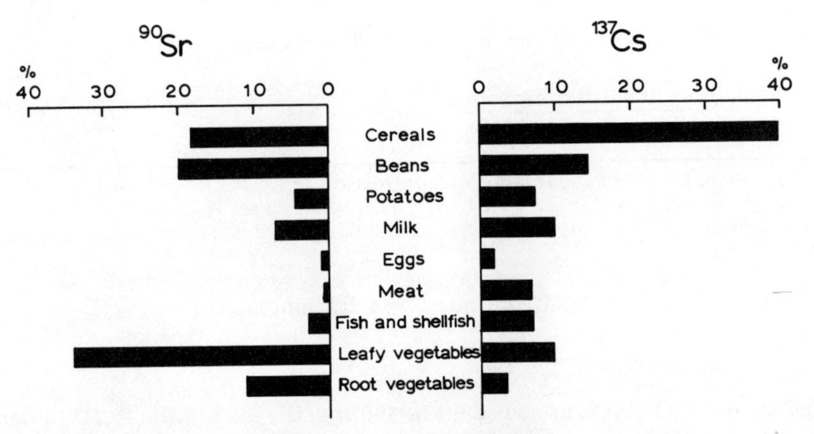

FIG. 5. The contribution of ^{90}Sr and ^{137}Cs in individual foodstuffs to the total diet of the Japanese population (1966-1971).

cent) of ^{137}Cs for each of nine foodstuffs is presented in Fig. 5. The contribution of ^{137}Cs from cereals was about 40% of the total ^{137}Cs intake, followed by beans, milk and milk products, leafy vegetables and fish. Consequently, the category of foodstuffs that contributes most to the ^{137}Cs intake by individual members of the Japanese population is not clear, unlike the case of milk and milk products for Western populations, even cereals contributing less than half of the total ^{137}Cs intake of the Japanese population.

Supposing that 1 CU (= ^{137}Cs pCi/g of K) in the human body gives an internal radiation dose of 18 μrad/a [9], the dose received by members

TABLE IV. THE ESTIMATED INTERNAL DOSES RECEIVED BY MEMBERS OF THE JAPANESE POPULATION BETWEEN 1963 AND 1971

	Total diet (CU)	Body burden (CU)	Internal dose rate (mrad)
1963	33.8	100.0	1.80
1964	41.0	98.4	1.77
1965	26.9	56.5	1.02
1966	18.8	38.7	0.77
1967	14.1	32.3	0.58
1968	12.8	24.2	0.44
1969	11.1	23.2	0.42
1970	6.9	(15.6)	0.28
1971	6.5	(14.7)	0.27
		Total (1963-1971)	7.28 mrad

Figures in parentheses were calculated from an OR (Body burden/Total diet) of 2.26.

of the Japanese population is estimated to have been 7.3 mrad between 1963 and 1971 (Table IV). During the period from 1966 to 1971 the ^{137}Cs internal radiation dose received by the Japanese population via marine fish amounted to about 10% of the ^{137}Cs radiation dose received via the total diet.

10. ^{90}Sr level in total diet and individual foodstuffs consumed by the Japanese population

The annual variation in the ^{90}Sr level in the total diet is shown in Fig. 3. The highest level occurred in 1964 and has since decreased. The contributions (in per cent) of each of the nine categories are shown in Fig. 4. The contribution of leafy vegetables was the highest (34%), followed by beans (20%) and cereals (18%).

DISCUSSION AND CONCLUSION

There have been many reports on the effects of the food chain upon the contamination of marine organisms by ^{137}Cs and ^{90}Sr. However, the results of this investigation strongly suggest that the ^{137}Cs and ^{90}Sr contamination of marine organisms is mainly defined by the level of these nuclides in seawater. However, this investigation was performed on the global contamination of sea by fall-out and therefore, further investigations should be carried out to determine whether this concept is also applicable in the situation where the Cs and Sr concentration levels vary in coastal seas as a result of release from various nuclear establishments, with a view to establishing a correlation with the biological turnover of Cs and Sr. Furthermore, the statistical average amounts of foods consumed by the Japanese was employed to estimate the internal radiation dose resulting from the Cs and Sr contamination of marine foods, although the pattern of consumption of marine foods might vary widely, i.e. fishermen eat considerable amounts of seafood and some of them eat marine organisms as their sole source of protein. An effort should be made to reduce their dose of internal radiation as far as possible by considering the amounts of food consumed.

ACKNOWLEDGEMENTS

The authors wish to express their gratitude to Dr. M. Izawa, Chief of our Division, and to Professor Y. Hashimoto, University of Tokyo, for their valuable advice and to Mrs. Kawachi for her excellent technical assistance.

REFERENCES

[1] NAKAMURA, R., SUZUKI, Y., UEDA, T., The loss of radionuclides in marine organisms during thermal decomposition, J. Radiat. Res. 13 3 (1972) 149.

[2] TANAKA, G., TOMIKAWA, A., KAWAMURA, H., OYAGI, Y., The determination of Rb and Cs by atomic absorption spectrophotometry, Rep. NIRS-AR 10 (1966) 108.

[3] BRYANT, F.J., MORGAN, A., SPICER, G.S., The Determination of Radiostrontium in Biological Materials, Rep. AERE-R 3030 (2959).

[4] TANAKA, G., TOMIKAWA, A., KAWAMURA, H., OYAGI, Y., Determination of strontium by atomic absorption spectrophotometry, Nippon Kagaku Zasshi 89 (1968) 175.

[5] OHMOMO, Y., YAMAGUCHI, H., SAIKI, M., The concentration of radionuclides in surface seawater collected from Japanese coast (1963-1966), Radioactivity Survey Data in Japan, Scientific Technical Agency 5.

[6] ARAI, M., OHMOMO, Y., SAIKI, M., "Radionuclides in the surface sea water collected from Japanese coast", Proc. 12th Conf. Radioactivity Survey in Japan, Scientific Technical Agency (1970) 51.

[7] SUZUKI, Y., NAKAMURA, R., UEDA, T., Cesium-137 contamination of marine fishes from the coasts of Japan, J. Radiat. Res. 14 4 (1973) 382.

[8] GOLDBERG, E.D., "The minor constituents of sea water", Chemical Oceanography (RILEY, J.P., SKIRROW, G., Eds), Academic Press, London and New York (1965) Chap. 5.

[9] CULKIN, F., "The major constituents of sea water", Chemical Oceanography (RILEY, J.P., SKIRROW, G., Eds) Academic Press, London and New York (1965) Chap. 4.

[10] SPIERS, F.W., Radioisotopes in the Human Body, Academic Press, New York (1968).

[11] SAIKI, M., IINUMA, T., UCHIYAMA, M., Cesium-137 content in human body, Radioactivity Survey Data in Japan 7 (1965).

[12] UEDA, T., SUZUKI, Y., NAKAMURA, R., Accumulation of Sr in marine organisms-I Strontium and calcium contents, CF and OR values in marine organisms, Bull. Japan Soc. Scient. Fish 39 12 (1973) 1253.

[13] BRYAN, G.W., PRESTON, A., TEMPLETON, W.L., "Accumulation of radionuclides by aquatic organisms of economic importance in the United Kingdom", Disposal of Radioactive Wastes into Seas, Oceans and Surface Waters (Proc. Symp. Vienna, 1966), IAEA, Vienna (1966) 623.

[14] MAUCHLINE, J., TEMPLETON, W.L., Strontium, calcium and barium in marine organisms from the Irish sea, J. Cons. Perm. Int. Explor. Mer. 30 (1966) 161.

[15] HAUG, A., SMIDSROD, O., Strontium and magnesium in brown algae, Nature (London) 215 (1967) 1167.

[16] HAUG, A., SMIDSROD, O., Strontium-calcium selectivity of alginate, Nature (London) 215 (1967) 757.

[17] UNITED NATIONS SCIENTIFIC COMMITTEE ON THE EFFECTS OF ATOMIC RADIATION Rep. 14 (A/6314) (1966) 69.

[18] MIYAKE, Y., SARUHASHI, K., KATSURAGI, Y., KANAZAWA, T., Monthly and cumulative deposition of strontium-90 and cesium-137, Radioactivity Survey Data in Japan 1 (1963); 6 (1965); 7 (1965); 9 (1965); 10 (1966); 14 (1967); 17 (1967); 22 (1969); 28 (1969); 31 (1971).

EVALUATION OF THE RESUSPENSION PATHWAY TOWARD PROTECTIVE GUIDELINES FOR SOIL CONTAMINATION WITH RADIOACTIVITY *

L.R. ANSPAUGH, J.H. SHINN, D.W. WILSON
Bio-Medical Division,
Lawrence Livermore Laboratory,
University of California,
Livermore, Calif.,
United States of America

Abstract

EVALUATION OF THE RESUSPENSION PATHWAY TOWARD PROTECTIVE GUIDELINES FOR SOIL CONTAMINATION WITH RADIOACTIVITY.

The resuspension and subsequent inhalation of surface-deposited radioactivity released to the environment can be a significant mode of exposure for a few radionuclides such as ^{239}Pu. Two simple, interim models, which may be used to predict the average concentration of resuspended aerosols, are developed on an empirical basis. One method uses the time-dependent resuspension factor approach, and differs from previous work in placing more emphasis on resuspension at late times. The second method is appropriate only for aged sources and uses a straight-forward mass-loading approach. The relative significance of the resuspension pathway is also modelled in comparison to the initial exposure resulting from a non-nuclear explosion that disperses radioactivity. Two hypothetical ^{239}Pu contamination situations are modelled. In the first case the 50-a dose commitment resulting from an initial deposition of 1 $\mu Ci/m^2$ is calculated as a function of time post deposition. Half of the total dose commitment is accumulated in the first 100 d. In the second situation the reoccupation of an area contaminated many years previously is considered. Protective guidelines for ^{239}Pu soil contamination are derived from these studies — 1 $\mu Ci/m^2$ for a freshly deposited source and 7 nCi/g in the top 10 mm of soil for a source that has aged several years. An estimate of the biological cost of not cleaning up contaminated areas is compared with the engineering and agricultural costs of soil removal.

1. INTRODUCTION

A need exists for a set of guidelines to define maximum acceptable levels of radionuclide contamination of soil with reference to the resuspension and consequent inhalation of the deposited material. Although most radionuclides, including ^{90}Sr, ^{131}I, and ^{137}Cs, present greater public health problems due to food-chain transport or external gamma exposure, the resuspension route is generally accepted as important for a few radionuclides which have insignificant gamma emission and which are highly discriminated against by living organisms. The most notable example of such a radionuclide is ^{239}Pu. Because of the use of plutonium in weapon systems and its projected use in the future energy economy, there is an important need for realistic secondary soil contamination standards.

Accidental release incidents have already demonstrated the need for such standards. The cost-benefit aspect of applying any such standard also poses difficulties. It is costly to remove or otherwise treat a contaminated soil surface, and the treatment process itself may greatly increase the airborne concentration of resuspended contaminant. The long-term ecological

* This work was performed under the auspices of the US Atomic Energy Commission.

consequences of soil removal or disturbance must also be balanced against the long-term potential for health consequences. Finally, some practical consideration should be given to the fact that the contaminating event itself may already have produced an unavoidable dose commitment to an exposed population. In such cases it may be inappropriate to mount a large effort which would not have a major impact on the total dose commitment.

The purpose of this paper is to develop a rationale for setting protective guidelines for soil contamination based upon an evaluation of the resuspension pathway, and to discuss some of cost-benefit aspects of the application of such guidelines.

2. EVALUATION OF THE RESUSPENSION PATHWAY

The wind-driven resuspension of contaminant material initially deposited in an open environment has been studied on numerous occasions [1-13]. Several general conclusions may be drawn from the results of these experiments: 1) The airborne concentration of radioactive material produced by explosions declines rapidly immediately after the detonation and is lower by a factor of 100 to 1000 by 100 h after the detonation [3, 8]; 2) After the initial rapid decline, a further decrease with time is noted with half-times of about 5 weeks observed during the first 6 to 20 weeks following release [1, 2, 4, 8], but a larger half-time of decrease of about 10 weeks was reported for one experiment conducted 12 to 40 weeks post release [8]; 3) Such slow decreases with time are not due to an appreciable net loss of contaminant from the area, but to a "weathering in" process whereby the contaminant becomes less erodible [2, 3, 5]; 4) Areas which were contaminated 10 to 17 yr previously are still significant, although greatly attenuated resuspension sources [7, 9]; 5) For explosion produced sources, the quotient of air concentration divided by ground deposition increases with distance from the source [2, 4, 10], this may be partly due to the decrease in the particle size of the contaminant which is deposited at further distances [3]; 5) Short-term, order of magnitude fluctuations of the airborne concentration of resuspended contaminant are frequently observed which are presumably due to changes in meteorological conditions such as wind velocity, stability, type of weather event, and soil moisture content although the functional relationships are not well defined [2-5, 7-9]; 6) Artificial disturbances such as vehicular traffic in a contaminated environment can also produce orders of magnitude increases in the airborne concentration of contaminant [11, 12]; and 7) Particle size measurements of resuspended plutonium aerosols indicate that the respirable fraction averages about 0.25 [4, 9, 13].

There is as yet no general model of the complex resuspension process which may be used to predict accurately either the correct magnitude of the average concentration of resuspension aerosols for all source conditions or the short-term fluctuations in concentration as a function of meteorological variables and artificial disturbances. There is enough information available, however, that simple, interim models can be constructed which include the general features listed above and which can be used to estimate the average concentration of resuspended aerosols. Because the factors which influence the rate of resuspension are not yet well-defined, it is prudent that such models be conservative.

2.1 Time-dependent resuspension factor approach

The earliest efforts to define the magnitude of the resuspension pathway used the empirical approach of defining a resuspension factor, K, which is defined as [1]

$$K = \frac{\text{Resuspended air concentration (activity/m}^3\text{)}}{\text{Surface deposition (activity/m}^2\text{)}}$$

Such an approach neglects many variables which undoubtedly are important in determining the rate of resuspension, but the fact remains that data expressed in such a manner provide almost the entire data base which may be used to predict the airborne concentration of resuspended contaminant aerosols. In addition, this approach does at least provide an important normalization to the potential source strength.

Stewart [6] and Mishima [14] have tabulated values of K from a variety of experiments. For open environments soon after the contaminating event, calculated values of the resuspension factor range from 10^{-7} to 10^{-3} m^{-1}, with only a few values greater than 10^{-4} m^{-1} which were associated with artificial disturbances. Many of the lower values are derived from measurements made in close proximity to explosive detonation points [4, 10] where the particle size of the contaminant would be expected to be larger.

A convenient way to model the airborne concentration of resuspended contaminant over long periods of time is to make the resuspension factor a function of time to account for the observed decrease in air concentration which has been noted to occur in the absence of a significant net loss of the deposited contaminant. Conceptionally, it would be more appropriate to define a time-dependent fraction of the total deposition which is available for resuspension. However, there is no realistic way in which such a fraction can be experimentally determined, so this approach will be avoided for the present purpose. With the time dependency inherent in the resuspension factor, it follows that the average airborne concentration, $\overline{\chi}$, of resuspended contaminant will be given by

$$\overline{\chi}\,(t) = K\,(t)\,S_A$$

where S_A is the total amount of contaminant deposited per unit area. S_A is therefore considered a constant although the actual distribution of the contaminant with soil depth will change with time.

Kathren [15] and Langham [16] have each formulated predictive resuspension models which, when expressed in the above format, give the following time dependency

$$K\,(t) = K_o \exp\,(-\lambda t)$$

with values of λ corresponding to half-times of 45 d and 35 d. Such a formulation appears to simulate reasonably well the available observations up to several weeks post deposition [4, 8]. After a few years, however, such a formulation underestimates by many orders of magnitude the airborne concentration of resuspended contaminants which have been measured over aged sources [5, 7, 9]. For example, the Kathren and Langham models would predict values for K (t) of 10^{-29} and 10^{-38} m^{-1} respectively 10 yr after a contaminating event whereas an average value determined from 236 individual air concentration measurements at a location contaminated with plutonium 17 yr previously was found to be 10^{-9} m^{-1} [9].

We have derived a different formulation of the time dependency of the resuspension factor which more accurately reflects the resuspension process as it is observed in the proximity of aged sources. This model was empirically derived to conform to the following constraints: 1) The apparent half-time of decrease during the first 10 weeks should approximate a value of 5 weeks and should approximately double over the next 30 weeks; 2) The initial resuspension factor should be 10^{-4} m^{-1}; and 3) The resuspension factor 17 yr after the contaminating event should approximate 10^{-9} m^{-1}.

A simple model which closely approximates these constraints is

$$K(t) = 10^{-4} \exp(-0.15\, d^{-\frac{1}{2}}\sqrt{t})\ m^{-1} + 10^{-9}\ m^{-1}$$

The second term was added based upon the assumption that there may be no further measurable decrease in the resuspension process after 17 yr which is the longest period post deposition for which measurements have been reported. This was deemed appropriate because such a "model" was derived empirically to simulate experimental measurements, and contains no fundamental understanding of the resuspension process. A graphical representation of this model, both with and without the second term is given in Fig. 1; the model equations used by Kathren and Langham are also shown for comparison.

This model is an attempt to provide protective guidance for the evaluation of the resuspension pathway over long time periods beginning with the initial contaminating event. It assumes that resuspension is a local phenomenon and that the concentration in air drops off rapidly downwind of the deposited source; this is consistent with experimental observations [3, 9]. The model is suitable only for the prediction of long-term averages of airborne concentrations; extreme short-term fluctuations are to be expected and suitable long-term averages of experimental data were used to define the model constrainsts. The initial value for the resuspension factor, however, was deliberately chosen to be sufficiently high to include the effects of artificial disturbance.

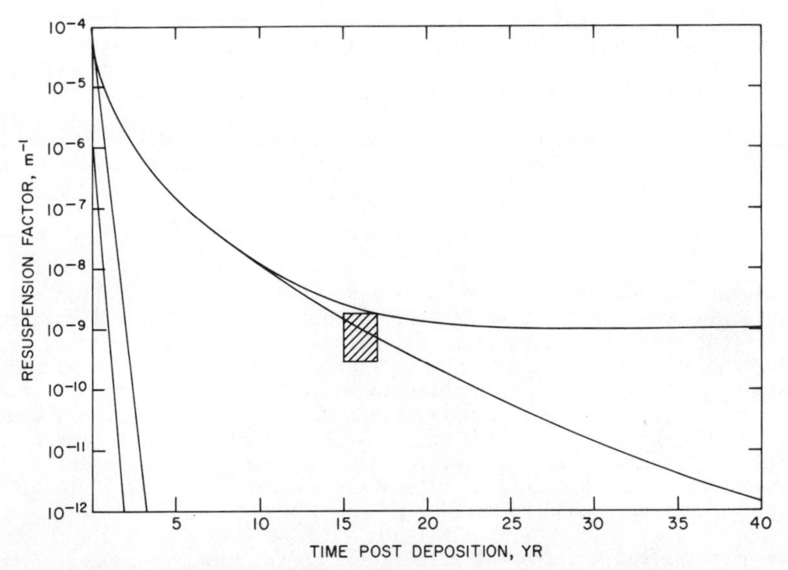

FIG. 1. A graphical representation of several time-dependent resuspension factor models. The two curves on the far left represent the models of Langham [16] and Kathren [15]. The remaining two curves represent models developed in the paper, both with and without a constant term of 10^{-9} m^{-1}. The hatched area represents measurements recently reported at an aged source [9].

2.2 Mass-loading approach

Nearly all of the experimental measurements of the concentration of resuspended contaminants have been conducted in the vicinity of freshly deposited sources. This is appropriate for most situations of practical concern such as accidental events. However, there are some situations such as the contemplated reoccupation of test areas contaminated many years previously where an alternate method of predicting the concentration of resuspended contaminant may be used to supplement the results derived from the resuspension factor model.

It has been observed by many authors [17-20] that radionuclides deposited on the earth's surface in either solution or particulate form penetrate within a few months to depths of more than 10 mm. Eventually their distribution with depth is well approximated by an exponential function characterized by a relaxation depth of 10 to 100 mm. Such a distribution with depth implies an intimate mixing of the contaminant with the host soil. Therefore, a method of predicting the average airborne concentration of the contaminant several years after the contaminating event is to simply multiply the measured activity of the contaminant per unit weight of soil taken from the top 10 mm by the concentration of particulate matter in the atmosphere.

For predictive purposes, an average atmospheric concentration of 100 $\mu g/m^3$ appears to be reasonable [9]. The choice of this value is partly based upon measurements of particulate concentration reported for 30 nonurban locations in the United States [21]. Annual arithmetic averages varied from 9 to 79 $\mu g/m^3$ with a mean for all 30 stations of 38 $\mu g/m^3$.

Several experimental results are available to check the accuracy of this simple prediction method [9, 22-26]. These values are tabulated in Table I; the agreement between the predicted and measured values is generally excellent.

2.3 Relative importance of resuspension in the inhalation pathway

Some accident situations may produce an initial contaminant aerosol cloud with a resulting dose commitment to an exposed population. In such cases, it is of practical interest to compare such an unavoidable dose commitment to that predicted via the resuspension pathway.

If the initial integrated air activity is A_o (activity-time/volume), the ground deposition may be calculated by multiplying by a deposition velocity, V, (length/time). The integrated air activity, A, due to resuspension is then given by

$$A = VA_o \int K(t)\,dt$$

which may then be compared conveniently with the initial integrated air activity, A_o. One problem in such a comparison, however, is the choice of an appropriate value for the deposition velocity. This varies as a function of particle size and wind speed, and in close proximity to an explosion the actual deposition would be strongly influenced by the ballistic effects of the explosion itself. At such close-in distances, however, the initial resuspension factor is also lower [2, 4, 10]. Because the initial resuspension factor chosen for use in the model is relatively high and is presumed to be appropriate for smaller particle size distributions of the contaminant aerosol, it is assumed that an appropriate value for V is 40 m/h. This is an average value which can be derived from several years of fallout data [27].

TABLE I. A COMPARISON OF OBSERVED AND PREDICTED AIR CONCENTRATIONS BASED UPON A SIMPLE MASS LOADING MODEL

Location, etc.	Radionuclide	Air Concentration	
		Predicted[a]	Measured[b]
GMX site, USAEC Nevada Test Site [9]			
NE, 1971-1972	^{239}Pu	7200 aCi/m^3	6600 aCi/m^3
GZ, 1972, 2 weeks	^{239}Pu	120 fCi/m^3	23 fCi/m^3
Lawrence Livermore Laboratory			
1971 [22]	^{238}U	150 pg/m^3	52 pg/m^3
1972 [23]	^{238}U	150 pg/m^3	100 pg/m^3
1973 [24]	^{238}U	150 pg/m^3	86 pg/m^3
1973 [24]	^{40}K	1000 aCi/m^3	980 aci/m^3
Argonne National Laboratory [25]			
1972	^{232}Th	320 pg/m^3	240 pg/m^3
1972	^{nat}U	215 pg/m^3	170 pg/m^3
Sutton, England [26]			
1967-1968	^{nat}U	110 pg/m^3	62 pg/m^3

[a]Predicted value is equal to the soil concentration (activity/g) x 10^{-4} g/m^3.

[b]Most values are annual averages.

2.4 Dose commitment due to the resuspension of ^{239}Pu

In order to derive a protective guideline for soil contamination, some reference must be made to a primary standard. For ^{239}Pu, which will be used as an example, it will be assumed that the desired primary reference standard is the accumulated 50 yr dose commitment. The dose commitment calculations were made with the additional assumptions that the plutonium is non-transporta the lung is the critical organ [28], 25% of the inhaled plutonium is deposited in the pulmonary region, and that the metabolic parameters listed by Morgan [2 for Class Y compounds were appropriate.

Representative calculations are presented in Tables II and III for two hypothetical situations. The first assumes a nonnuclear explosion which produces a ^{239}Pu deposition of 1 μCi/m^2. The 50 yr dose commitment due to the initial cloud passage and resuspension of deposited material as a function of time post deposition were calculated using the models given in sub-sections 2.1 and 2.3. The results in Table II indicate that any protective action must be undertaken fairly rapidly if a significant reduction in the 50 yr dose commitment is to be achieved.

TABLE II. CALCULATED DOSE COMMITMENT TO THE LUNG FROM A HYPOTHETICAL ACCIDENT DISPERSING ^{239}Pu IN THE ENVIRONMENT WHICH PRODUCES A GROUND DEPOSITION OF 1 $\mu Ci/m^2$

Time	Accumulated 50 yr dose commitment, rem	
	Resuspension	Total
Intial cloud passage		6.1
1 d	0.53	6.6
5 d	2.4	8.5
10 d	4.3	10.
50 d	15.	21.
100 d	23.	29.
1 yr	41.	47.
10 yr	52.	58.
50 yr	52.	58.

TABLE III. CALCULATED DOSE COMMITMENT TO THE LUNG FOR POPULATION REENTRY INTO AN AREA OF AGED ^{239}Pu CONTAMINATION WITH 1 $\mu Ci/g$ IN THE TOP 10 mm OF SOIL

Time	Accumulated 50 yr dose commitment, rem
1 yr	210
10 yr	2,100
30 yr	6,400
50 yr	10,000

The second situation assumes the reoccupation of landscape contaminated many years previously and where the ^{239}Pu concentration in the top 10 mm of soil is 1 $\mu Ci/g$. Calculations of the 50 yr dose commitment as a function of time post reentry were made using the model given in sub-section 2.2, and are presented in Table III.

If we assume that an acceptable 50 yr dose commitment in such situations is 50 x 1.5 rem, or 75 rem, then protective guidelines of 1 $\mu Ci/m^2$ and 7 nCi/g are derived for the two hypothetical situations. It is emphasized, however, that this analysis has considered only the inhalation of ambient air. Other pathways such as personal contamination and ingestion may be more restrictive in some situations.

TABLE IV. ESTIMATES OF THE COSTS AND BENEFITS RESULTING FROM THE CLEAN-UP OF AGRICULTURAL LAND CONTAMINATED WITH ^{239}Pu. All values are in US $.

Factor	System value	Cost	Benefit
Biological effect[a]			2.0×10^5
Soil pick-up		4.5×10^3	
Waste transportation[b]		1.0×10^6	
Agricultural land[c]	5.0×10^4		
Annual crop yield[c]	1.6×10^4		

[a] Population density of 10/km^2 and a 50 yr dose commitment of 75 rem.

[b] Transportation of the top 30 mm a distance of 500 km.

[c] US average.

3. Cost-benefit aspects of protective guideline application

Models developed in this study can be applied to a general analysis of costs and benefits associated with the clean-up of environmental plutonium. In this instance, costs are defined as all expenditures and monetary losses derived from clean-up action. Benefits are the reductions in potential health costs which may be derived from any proposed clean-up action.

Comparative costs have been derived for clean-up of a hypothetical km^2 of agricultural land containing one Ci of ^{239}Pu in the soil surface (Table IV). The objective of this exercise is to provide a semi-quantitative analysis of the economic trade-offs between clean-up and no clean-up.

Costs of removal of soil depend upon availability of equipment, costs of transportation, and the complexity of terrain. In 1974, U.S. public works projects involving soil surface pick-up cost U.S. \$0.15/m^3 of soil; associated transportation costs for long hauls were U.S. \$ 0.06/m^3 - km [30]. Clean-up costs for removing the top 30 mm of soil from a km^2 of land would be approximately U.S. \$4500. Transportation of contaminated soil to a storage site 500 km away would be approximately U.S. \$ 1×10^6/km^2 of surface removed.

Thus, clean-up costs are likely to be small compared to transportation costs for clean-up in impacted areas which are remote from a radioactive waste management area. Costs associated with long-term waste management will be ignored here.

Land values are highly variable and depend upon the specific land use. Currently, U.S. agricultural land is valued at approximately U.S. \$50,000/km^2 [31]. Restriction against crop production would result in this economic loss as well as losses of annual crop production, valued at U.S. \$ 2×10^4/km^2 [31]. Clean-up could result in costs from temporary decreased agricultural production due to interruptions in the crop cycle or decreased soil fertility. For locations with marginal food supply, restrictions in agricultural production could lead to more serious biological costs than costs from radiation exposure.

Potential health costs arising from radiation exposure in populations have been evaluated in monetary terms by others [32]. Such figures are useful for comparison with the monetary costs of remedial action against radiation exposure. Health cost estimates are developed in this study using a value of U.S. $250/man-rad. At a dose commitment level of 75 rem, the calculated cost per person exposed is U.S. $ 2×10^4. On typical agricultural land, one could reasonably expect a population density of one to ten persons per km^2. Potential biological costs could be approximately U.S. $ 2×10^5. The benefit-cost ratio for clean-up is therefore approximately 0.2. This ratio is very sensitive to the costs of transportation and the number of human receptors which are involved. Alternatives, such as restriction against crop production incur costs which are comparable to the benefits of reduced radiation exposure; for example, the economic crop loss of creating an exclusion area for ten years is approximately U.S. $ 2×10^5.

In summary, the deposition of a Ci per km^2 of ^{239}Pu on agricultural land can be evaluated as leading to potential health costs which are comparable in economic value to the costs of remedial actions which remove the potential exposure. The judgement regarding proper action involves consideration of qualitative features, such as economic policies and societal priorities for utilizing finite monetary resources to promote human welfare. Some consideration should also be given to an implied long-term commitment should a decision be made not to clean-up a contaminated area. If such an area is used for agricultural production, some monitoring program will be necessary to assess the level of ^{239}Pu in food crops. While the food-chain transport of ^{239}Pu is generally believed to be negligible based upon short-term studies, it cannot be stated with certainty that this will remain true over long periods of time [19].

The short time period available for making an effective decision is also apparent from the results in Table II. By 100 d after the contaminating event, one-half of the total 50 yr dose commitment has already been accumulated. An alternate course of action would be to evacuate the area and/or plow the affected ground surfaces. This should be effective in greatly reducing the biological costs, but also substantially increases the ultimate cost of clean-up should it eventually be deemed necessary.

Finally, some mention should be made of urban environments. Here, the potential for rapid spread of the material and the large population density appear to offer no choice but to contain and remove the contamination as rapidly as possible.

REFERENCES

[1] LANGHAM, W.H., Plutonium distribution as a problem in environmental science, Environmental Plutonium Symposium (Proc. Symp. Los Alamos, New Mexico, 1971) USAEC Rep. LA-4756 (1971) 3.

[2] LARSON, K.H., NEEL, J.W., HAWTHRONE, H.A., MORK, H.M., ROWLAND, R.H., BAURMASH, L., LINDBERG, R.G., OLAFSON, J.H., KOWALEWSKY, B.W., Distribution, Characteristics, and Biotic Availability of Fallout, Operation Plumbbob, USAEC Rep. WT-1488 (1966).

[3] SHREVE, J.D., JR., Summary Report, Test Group 57, USAEC Rep. ITR-1515 (DEL) (1958).

[4] WILSON, R.H., THOMAS, R.G., STANNARD, J.N., Biomedical and Aerosol Studies Associated with a Field Release of Plutonium, USAEC Rep. WT-1511 (1960).

[5] OLAFSON, J.H., LARSON, K.H., Plutonium, Its Biology and Environmental Persistence, USAEC Rep. UCLA-501 (1961).

[6] STEWART, K., The resuspension of particulate material from surfaces, Surface Contamination (Proc. Symp. Gatlinburg, Tenn., 1964) Pergamon, London (1964) 63.

[7] VOLCHOK, H.L., Resuspension of plutonium-239 in the vicinity of Rocky Flats, Environmental Plutonium Symposium (Proc. Symp. Los Alamos, New Mexico, 1971) USAEC Rep. LA-4756 (1971) 99.

[8] ANSPAUGH, L.R., PHELPS, P.L., KENNEDY, N.C., BOOTH, H.G., Wind-driven redistribution of surface-deposited radioactivity, Environmental Behaviour of Radionuclides Released in the Nuclear Industry (Proc. Symp. Aix-en-Provence, France, 1973) IAEA, Vienna (1973) 167.

[9] ANSPAUGH, L.R., PHELPS, P.L., KENNEDY, N.C., BOOTH, H.G., GOLUBA, R.W., REICHMAN, J.M., KOVAL, J.S., Resuspension element status report, The Dynamics of Plutonium in Desert Environments, USAEC Rep. NVO-142 (in press).

[10] ANSPAUGH, L.R., PHELPS, P.L., HOLLADAY, G., HAMBY, K.O., Distribution and redistribution of airborne particulates from the Schooner cratering event, Health Physics Aspects of Nuclear Facility Siting (Proc. Symp. Idaho Falls, Idaho, 1970) 2, Eastern Idaho Chapter, Health Physics Society, Idaho Falls (1971) 428.

[11] MORK, H.M., Redistribution of Plutonium in the Environs of the Nevada Test Site, USAEC Rep. UCLA-12-590 (1970).

[12] SEHMEL, G.A., Particle resuspension from an asphalt road caused by car and truck traffic, Atmos. Environ. 7 (1973) 291.

[13] VOLCHOK, H.L., KNUTH, R.H., The respirable fraction of plutonium at Rocky Flats, Health Phys. 23 (1972) 395.

[14] MISHIMA, J., A Review of Research on Plutonium Releases during Overheating and Fires, USAEC Rep. HW-83668 (1964).

[15] KATHREN, R.L., Towards interim acceptable surface contamination levels for environmental PuO_2, Radiological Protection of the Public in a Nuclear Mass Disaster (Proc. Symp. Interlaken, Swit., 1968) EDMZ, Bern (1968) 460.

[16] LANGHAM, W.H., Biological Considerations of Nonnuclear Incidents Involving Nuclear Warheads, USAEC Rep. UCRL-50639 (1969).

[17] BECK, H.L., Environmental gamma radiation from deposited fission products, 1960-1964, Health Phys. 12 (1966) 313.

[18] ROGOWSKI, A.S., TAMURA, T., Erosional behavior of cesium-137, Health Phys. 18 (1970) 467.

[19] ROMNEY, E.M., MORK, H.M., LARSON, K.H., Persistence of plutonium in soil, plants and small mammals, Health Phys. 19 (1970) 487.

[20] KREY, P.W., HARDY, E.P., Plutonium in Soil Around the Rocky Flats Plant, USAEC Rep. HASL-235 (1970).

[21] NATIONAL AIR POLLUTION CONTROL ADMINISTRATION, Air Quality Data from the National Air Sampling Networks and Contributing State and Local Networks, 1966 Edition, USHEW Rep. APTD 68-9 (1968).

[22] GUDIKSEN, P.H., LINDEKEN, C.L., GATROUSIS, C., ANSPAUGH, L.R., Environmental Levels of Radioactivity in the Vicinity of the Lawrence Livermore Laboratory, January through December 1971, USAEC Rep. UCRL-51242 (1972).

[23] GUDIKSEN, P.H., LINDEKEN, C.L., MEADOWS, J.W., HAMBY, K.O., Environmental Levels of Radioactivity in the Vicinity of the Lawrence Livermore Laboratory, 1972 Annual Report, USAEC Rep. UCRL-51333 (1973).

[24] SILVER, W.J., LINDEKEN, C.L., MEADOWS, J.W., HUTCHIN, W.H., MCINTYRE, D.R., Environmental Levels of Radioactivity in the Vicinity of the Lawrence Livermore Laboratory, 1973 Annual Report, USAEC Rep. UCRL-51547 (1974).

[25] SEDLET, J., GOLCHERT, N.W., DUFFY, T.L., Environmental Monitoring at Argonne National Laboratory, Annual Report for 1972, USAEC Rep. ANL-8007 (1973).

[26] HAMILTON, E.I., The concentration of uranium in air from contrasted natural environments, Health Phys. 19 (1970) 511.

[27] KLEINMAN, M.T., VOLCHOK. H.L., Radionuclide concentrations in surface air: Direct relationship to global fallout, Science 166 (1969) 376.

[28] ICRP COMMITTEE 4, The Assessment of Internal Contamination Resulting from Recurrent or Prolonged Uptakes, ICRP Publication 10A, Pergamon, Oxford (1971).

[29] MORGAN, K.Z., Proper use of information on organ and body burdens of radioactive material, Assessment of Radioactive Contamination in Man (Proc. Symp. Stockholm, 1971) IAEA, Vienna (1972) 3.

[30] MCMAHON, L.A., Dodge Estimating Guide for Public Works Construction, 1974 Annual Ed., No. 6, McGraw-Hill, New York (1974).

[31] US DEPT. OF AGRICULTURE, Agricultural Statistics, 1973, USGPO, Washington (1973).

[32] COHEN, J.J., HIGGINS, G.H., The socioeconomic impact of low-level tritium releases to the environment, Tritium (Proc. Symp. Las Vegas, 1971) Messenger Graphics, Las Vegas (1973) 14.

DISCUSSION

H. T. DAW: The final sentence of your paper prompts me to remark that if the cost of taking remedial action is equal to the cost of taking no action, it would surely be better, at the decision-making point, to opt for the procedure by which people are protected. In other words the cost-benefit analysis does not help here.

L.R. ANSPAUGH: It is true that the cost-benefit analysis did not provide us with a great deal of guidance. As is often the case, it depends how one selects one's priorities with only limited financial resources.

H.P. JAMMET: Your cost-benefit analysis does have the merit of allowing decisions to be based on numerical data. Nevertheless, it must be borne in mind that there is a qualitative difference in the two things being compared: on the one hand you have the cost, based on reality, and on the other you have the probability of detriment, which is difficult to assess.

F.C.J. TILDSLEY: A somewhat similar study, undertaken as a joint United Kingdom/French project, has been made of this problem and of the costs resulting from releasing a quantity of plutonium into the environment, and it is interesting to note that we arrived at conclusions similar to those of your very interesting paper, though by somewhat different routes. For instance, we concluded that the level of contamination for clean-up purposes is close to the practical limit of detection for plutonium alpha-emitters.

Work of this nature is important, and enables sound decisions to be made on such aspects as site storage quantities and conditions. However, I do not think that the decision whether or not to clean up will depend on calculations of this type. Public opinion would probably demand decontamination levels as low as possible in urban areas, while in agricultural areas it would probably prove difficult if not impossible to sell crops from contaminated farm land if no clean-up procedure had been followed — regardless of the time the land had been allowed to lie fallow in isolation.

L.R. ANSPAUGH: I would certainly agree that public opinion may be a dominating factor in clean-up decisions, and in some cases it may force action that is not justified. We have not attempted to calculate a monetary value for public opinion or 'good will'.

INCIDENCE DES RECONCENTRATIONS RADIOECOLOGIQUES SUR LA VALEUR DE LA CONCENTRATION MAXIMALE ADMISSIBLE DES EAUX DE RIVIERE

R. SCHAEFFER
Electricité de France,
Paris, France

Abstract-Résumé

EFFECTS OF RADIOECOLOGICAL RECONCENTRATION ON MAXIMUM PERMISSIBLE CONCENTRATIONS FOR RIVER WATER.

The MPC_w values recommended by ICRP take into account only one contamination pathway – the ingestion of water. In order to take into account the second contamination pathway – the food chain in the case of river water – it is therefore necessary to reduce the MPC value. The author describes the method for calculating the reduction coefficient, taking by way of example contamination by caesium-137 and strontium-99. Account is taken of two essential phenomena – the adsorption of radionuclides by river sediments and the movement of fish – which result in a higher concentration factor for fish with increasing distance from the point of discharge; this would largely explain the disparity of the concentration factor values given in the literature. However, the increase in the concentration factor value is less rapid than the decrease in the water concentration resulting from adsorption by sediments, so that the value of the coefficient by which the MPC value should be divided decreases as the distance from the discharge point at which the fish is caught increases. Bearing in mind also the concentration factors typical of plants, the coefficient barely exceeds 20 under the worst possible circumstances, if the population in question is near the discharge point, and declines to a much lower value in the case of more distant populations.

INCIDENCE DES RECONCENTRATIONS RADIOECOLOGIQUES SUR LA VALEUR DE LA CONCENTRATION MAXIMALE ADMISSIBLE DES EAUX DE RIVIERE.

Les CMA_{eau} recommandées par la CIPR ne considèrent qu'une seule voie de contamination: l'ingestion d'eau. Pour prendre en compte la deuxième voie de contamination que constitue la chaîne alimentaire quand il s'agit d'eau de rivière, il faut donc réduire la valeur de ces CMA. Cette étude expose la méthode de calcul de ce coefficient de réduction, en prenant comme exemple la contamination par le césium-137 et le strontium-90. Il est tenu compte de deux phénomènes essentiels: l'adsorption des radioéléments par les sédiments de la rivière et les déplacements des poissons. Ils ont pour effet d'augmenter le facteur de concentration chez les poissons avec la distance au point de rejet, ce qui expliquerait, en grande partie, la disparité des valeurs citées dans la littérature pour ce facteur. Mais cette augmentation est moins rapide que la décroissance de la concentration de l'eau résultant de l'adsorption par les sédiments, de sorte que la valeur du coefficient par lequel il convient de diviser la CMA diminue avec la distance où le poisson est pêché. Compte tenu également des facteurs de concentration propres aux végétaux, ce coefficient ne dépasse guère 20 dans les pires conditions si les populations sont rapprochées du point de rejet et s'abaisse bien en dessous de cette valeur dans le cas de populations éloignées.

1. INTRODUCTION

Les concentrations maximales admissibles (CMA), que la Commission Internationale de Protection Radiologique (CIPR) a recommandées en 1959 pour l'eau [1], ne considèrent qu'une seule voie de contamination : l'ingestion de l'eau elle-même, sous la forme d'eau de boisson ou d'eau imbibant les aliments. Cette $(CMA)_{eau}$ délivre, à elle seule, la dose maximale admissible (DMA) à l'organisme.

Or, les reconcentrations radioécologiques observées chez les animaux abreuvés et les végétaux arrosés avec l'eau de rivière conduisent à une voie supplémentaire de contamination des populations : celle de la chaîne alimentaire.

La CIPR précise [1] que dans ce cas il convient d'apporter aux valeurs de la $(CMA)_{eau}$ qu'elle a définie un facteur de correction tel que la somme des doses délivrées par les deux voies de contamination n'excède pas la DMA.

L'objet de notre étude est de déterminer la valeur de ce facteur de correction. A cet effet, nous envisagerons l'hypothèse la plus défavorable: celle d'un homme qui, à la fois, boirait de l'eau contaminée à la même concentration que l'eau de la rivière, se nourrirait de légumes arrosés avec cette eau et mangerait à chaque repas du poisson pêché dans la rivière.

Nous exposerons d'abord notre méthode de calcul, puis nous donnerons nos résultats que nous discuterons.

2. METHODE DE CALCUL

Pour caractériser l'importance des phénomènes de reconcentration, les écologistes ont défini un coefficient F, appelé facteur de concentration :

$$F = \frac{\text{activité contenue dans 1 kg de produit frais}}{\text{activité contenue dans 1 litre d'eau}}$$

étant précisé que :

a) l'activité dont il s'agit pour le produit frais est la valeur constante atteinte à l'équilibre,

b) l'activité dont il s'agit pour l'eau est celle qui est mesurée à l'endroit où le produit est récolté, c'est-à-dire, dans le cas des poissons, l'activité présentée par l'eau à l'endroit même où ceux-ci sont pêchés.

Dans ces conditions, le facteur de concentration représente l'équivalent en litres d'eau de la quantité de radioéléments contenus dans 1 kilogramme d'aliment .

Nous désignerons par m (kg) la masse d'aliments consommés par jour et nous affecterons de l'indice p les paramètres relatifs aux poissons et de l'indice l ceux relatifs aux légumes.

Les sédiments de la rivière ont pour propriété de fixer par adsorption une fraction des radioéléments, plus ou moins importante selon la nature de ces derniers, de sorte que l'activité volumique de l'eau décroît en fonction de la distance x au point de rejet selon une loi que l'on peut approximativement traduire par une exponentielle de la forme e^{-kx} (fig.1).

La concentration maximale C_o de l'eau de la rivière, immédiatement après dilution des effluents rejetés, doit être telle que l'activité journalière absorbée par notre individu soit égale à celle qui est permise par la CIPR. Etant donné que selon cette commission l'homme boit en moyenne

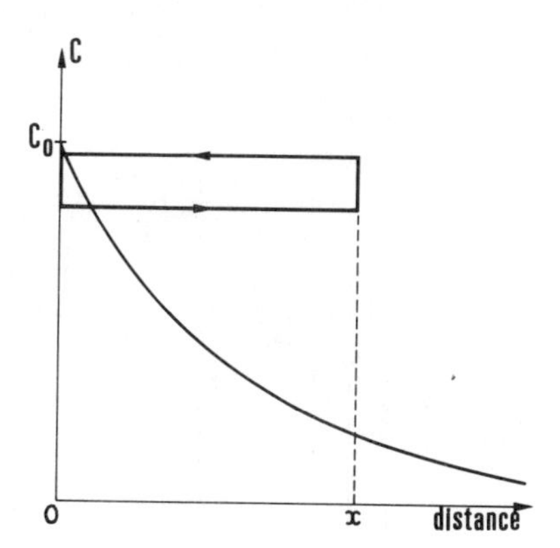

FIG.1. Variation de la concentration C de l'eau de la rivière en fonction de la distance au point de rejet et cycle des déplacements du poisson pêché en x.

1,2 litres et ingère par l'intermédiaire des aliments 1 litre d'eau par jour, soit un total de 2,2 litres d'eau par jour, on doit donc avoir, pour un individu habitant à une distance x du point de rejet, l'égalité :

$$C_o e^{-kx} (1,2 + m_p F_p + m_l F_l) = (CMA) \times 2,2$$

On en tire l'expression suivante du coefficient η par lequel il faut diviser la CMA :

$$\eta = \frac{1,2 + m_p F_p + m_l F_l}{2,2} e^{-kx} \qquad (1)$$

Il nous faut donc rechercher les valeurs des facteurs de concentration F et des quantités m d'aliments ingérés par jour, relatives aux poissons et aux légumes.

2.1. Valeurs du facteur de concentration relatif aux poissons

Les poissons ne mènent pas une vie sédentaire, mais se déplacent le long du cours d'eau en effectuant de nombreux va-et-vient. Mis à part les grandes migrations de frai de certaines espèces (anguille, saumon), ces déplacements ne semblent pas obéir à des lois spécifiques connues, mais seraient simplement soumis au jeu du hasard.

Considérons un poisson pêché à une distance x du point de rejet (fig.1). Puisque nous recherchons la valeur maximale du facteur de concentration, c'est l'activité (appelée encore charge) la plus grande que le poisson peut accumuler dans son corps qui nous intéresse ici. Cette charge sera obtenue quand, parmi tous les déplacements possibles, le poisson adopte l'un d'eux : celui dans la partie du cours de la rivière où l'eau présente la plus grande activité volumique, c'est-à-dire entre les points o et x.

La charge varie évidemment le long du trajet suivi par le poisson et augmente avec le temps de séjour du poisson dans l'eau contaminée, c'est-à-dire avec le nombre de cycles parcourus (fig.2).

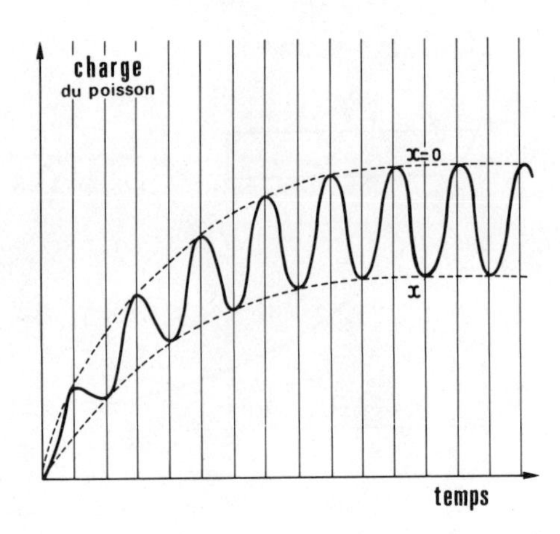

FIG.2. Variation de la charge du poisson en fonction du temps, entre le point de rejet (x = o) et le lieu de pêche (x).

Désignons par :

v la vitesse moyenne de déplacement du poisson,

f la fraction de l'activité volumique de l'eau qui est transférée à l'unité de masse du poisson par unité de temps,

T la période d'élimination effective du radionucléide chez le poisson, à laquelle correspond une constante d'élimination effective $\lambda = 0,693/T$.

Nous établissons en annexe l'expression mathématique de l'activité massique q que le poisson présente chaque fois qu'il repasse au point x et nous montrons qu'au bout d'un certain nombre de cycles cette charge atteint un équilibre.

En divisant cette expression par la valeur de l'activité volumique de l'eau au point x , nous obtenons l'expression du facteur de concentration en ce point :

$$F_p = f \frac{\dfrac{1 - e^{-(\lambda - kv)x/v}}{\lambda - kv} + \dfrac{e^{-(\lambda - kv)x/v} - e^{-2\lambda x/v}}{\lambda + kv}}{1 - e^{-2\lambda x/v}} \qquad (2)$$

Les valeurs numériques des constantes k, f et T sont rapportées dans le tableau I pour deux des plus importants nucléides : le césium-137 et le strontium-90, que nous avons choisis à titre d'exemple.

Les valeurs de k sont tirées d'observations effectuées sur des canaux du Rhône [2], qui montrent que le césium est beaucoup plus adsorbé par les sédiments que le strontium.

Les valeurs de f et de T sont déduites de résultats d'expériences de contamination en aquariums [3, 4] et concernent uniquement la partie comestible du poisson, c'est-à-dire les muscles, à l'exclusion du sang, des

TABLEAU I

Valeurs numériques des constantes

	k (km^{-1})	f (j^{-1})	T (j)
^{137}Cs	0,05	0,06	400
^{90}Sr	0,01	0,015	100

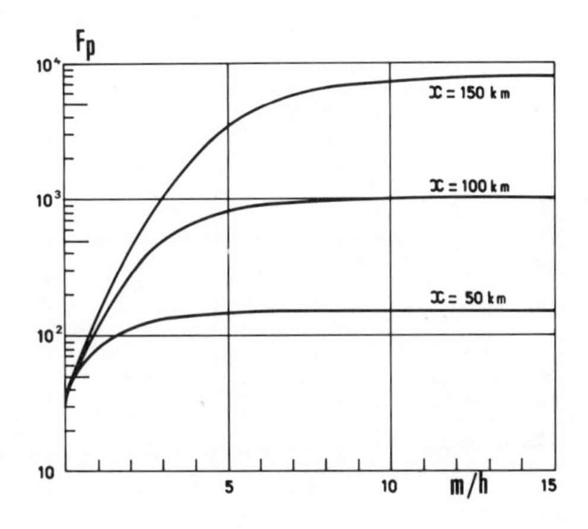

FIG.3. Variation, en fonction de la vitesse moyenne de déplacement, du facteur de concentration F_p présenté par le poisson à différentes distances x du point de rejet, dans le cas d'une contamination par le césium-137.

viscères et du squelette. Toutefois, les valeurs de f ainsi déduites ont été multipliées par deux pour tenir compte de ce que ces expériences ne comportaient pas de contamination du poisson par la nourriture; en effet, la plupart des chercheurs pensent que les apports d'activité se répartissent sensiblement à parts égales entre l'eau et la nourriture [5] .

L'application de ces valeurs numériques à l'équation (2) conduit aux courbes de la figure 3 représentant la variation de F_p en fonction de v et x dans le cas d'une contamination par le césium-137. Nous constatons que, pour des vitesses moyennes de déplacement du poisson supérieures à une dizaine ou une vingtaine de mètres par heure, F_p devient pratiquement indépendant de v . De telles vitesses paraissent fort plausibles si on les compare aux vitesses maximales de déplacement des poissons [6] rapportées dans le tableau II.

Nous pouvons donc considérer F_p comme ne dépendant que de x . L'expression (2) se simplifie et devient :

$$F_p = f \frac{e^{kx} - 1}{k\lambda x} \qquad (3)$$

TABLEAU II

Vitesses maximales (km/h) des poissons

Brème	2	Gardon	16
Epinoche	11	Perche	16
Anguille	12	Barbeau	18
Carpe	12	Brochet	33
Vairon	13	Truite	37
Vandoise	15	Saumon	40

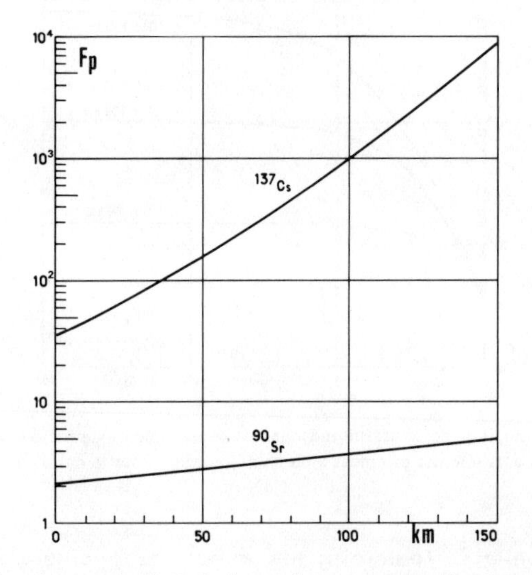

FIG.4. Variation de la valeur maximale F_p du facteur de concentration chez le poisson en fonction de la distance au point de rejet.

TABLEAU III

Valeurs du facteur de concentration dans la chair des poissons, selon la distance au point de rejet

km	0	25	50	100	150
^{137}Cs	35	70	150	1000	8000
^{90}Sr	2	2,5	2,8	3,7	5

TABLEAU IV

Valeurs du facteur de concentration pour quelques végétaux, selon la nature du terrain

légumes	Césium-137			Strontium-90		
	sable	calcaire	argile	sable	calcaire	argile
choux	4			78		20
haricots verts		12			5	
pommes de terre	50		2	15		2,5
salades		42	9	51	14	28
tomates	5		0,4	2,5		0,5

Cette fonction est représentée graphiquement par les courbes de la figure 4 et, pour préciser les idées, nous rapportons quelques-unes de ses valeurs dans le tableau III. Cette importante variation de F_p avec la distance x où le poisson est pêché pourrait expliquer, en grande partie, le large éventail des valeurs de F_p citées dans la littérature (de quelques dizaines à près de dix mille en ce qui concerne le césium-137).

2.2. Valeurs du facteur de concentration relatif aux légumes

Le tableau IV rapporte quelques-unes des valeurs à long terme de ce facteur, établies selon la nature du terrain par l'Institut National de la Recherche Agronomique en collaboration avec le Commissariat à l'Energie Atomique [7, 8].

Ces valeurs résultent d'expériences d'irrigation effectuées dans les conditions de la pratique sur sols en place dans des régions choisies en raison de leur culture intensive du végétal considéré. Tous les terrains contenaient de l'humus, ce qui avait pour effet d'augmenter la réversibilité de la fixation des radioéléments par le sol et par conséquent d'accroître la quantité de radioéléments disponible pour la plante.

Nous ne retiendrons du tableau IV que ses plus fortes valeurs :

- en terrain sablonneux, régime alimentaire composé à parts égales des deux légumes les plus contaminés (choux et pomme de terre),

- en terrain argileux ou calcaire, régime alimentaire composé uniquement de salade (ou d'un légume dont le facteur de concentration serait sensiblement analogue).

2.3. Valeurs des quantités d'aliments consommées par jour

La masse totale d'aliments consommés par jour se déduit de la quantité d'eau (1 litre) admise par la CIPR comme contenue dans l'alimentation quotidienne et de la teneur moyenne en eau (85 %) des aliments. On trouve :

$$m = 1/0,85 = 1,2 \text{ kg}$$

La valeur de m_p peut être déterminée à partir de la masse de protéines dont l'homme a besoin (70 grammes par jour) et de la teneur en protéines de la chair de poisson (17,5 %) [9]. On trouve :

$$m_p = 0,4 \text{ kg}$$

La valeur de m_l est égale à la différence entre m et m_p, soit :

$$m_l = 0,8 \text{ kg}$$

3. RESULTATS

Les valeurs du coefficient η (équation 1) sont données, en fonction de la distance x et selon la nature du terrain, par les courbes de la figure 5.

Malgré les hypothèses très défavorables que nous avons adoptées pour le régime alimentaire, ce coefficient ne dépasserait guère 20 si les populations étaient rapprochées du point de rejet et s'abaisserait bien en dessous de ces valeurs dans le cas de populations éloignées.

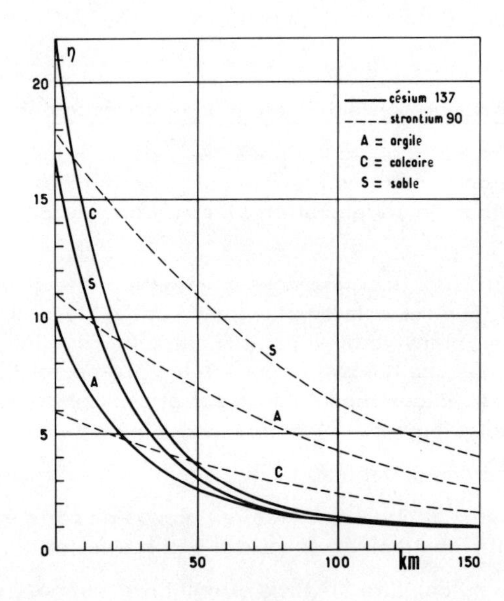

FIG. 5. Variation, en fonction de l'éloignement des populations par rapport au point de rejet, du coefficient η par lequel la CMA doit être divisée.

4. DISCUSSION

Notre calcul met en évidence deux variations que l'on serait tenté de croire, a priori, comme contradictoires : d'une part l'augmentation de F_p avec la distance x, d'autre part la diminution du coefficient η avec cette distance. Ceci mérite une explication.

Les parts relatives prises par les poissons et les végétaux dans la contamination des populations sont également intéressantes à connaître. Nous les évaluerons.

Enfin, la curiosité nous incite à comparer nos résultats avec les valeurs déjà proposées dans la littérature.

4.1. Augmentation de F_p et diminution de η avec la distance x

L'explication est donnée par le fait que la charge du poisson diminue avec x moins vite que la concentration de l'eau.

De cette variation, il résulte en effet :

- d'une part, que le dénominateur du rapport définissant le facteur de concentration devient de plus en plus petit par rapport au numérateur : F_p augmente (à la limite, pour un poisson qui migrerait dans une eau totalement épurée, on trouverait un facteur de concentration infini, puisque le dénominateur s'annulerait),

- d'autre part, que la quantité de radioéléments ingérée par l'homme en consommant le poisson pêché au point x diminue : η décroît.

4.2. Parts relatives des poissons et des végétaux dans la contamination des populations

La variation des parts relatives prises par les poissons et les végétaux dans la contamination des populations, en fonction de la distance x où ces dernières sont localisées, est représentée par les courbes de la figure 6 dans le cas d'un rejet de césium 137 et de terrains sablonneux : la contamination par les végétaux l'emporte sur celle par les poissons près du point de rejet; c'est l'inverse loin de ce point.

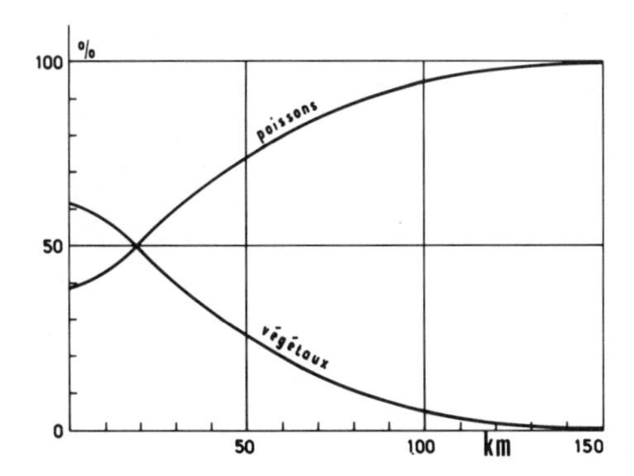

FIG.6. Variation, en fonction de la distance au point de rejet, des parts relatives prises par les végétaux et les poissons dans la contamination des populations par le césium-137 en terrain sablonneux.

TABLEAU V

Coefficient de division de la $(CMA)_{eau}$
dans le cas où les reconcentrations écologiques
sont limitées aux poissons

	ici (maximum)	A.I.E.A.
^{137}Cs	7	3000
^{90}Sr	1,25	100

En général, la littérature ne fait pas état de ces changements de prédominance, mais accorde toujours une part beaucoup plus importante aux poissons qu'aux végétaux [10, 11]. Ceci provient de ce que dans ces études il n'est tenu aucun compte de l'adsorption par les sédiments et qu'il est attribué au facteur de concentration des végétaux des valeurs à court terme qui ne prennent pas en considération les possibilités d'accumulation progressive des radioéléments dans le sol.

4.3. Comparaison avec les valeurs proposées dans la littérature

Une étude publiée par l'Agence Internationale de l'Energie Atomique (AIEA) [12] donne des valeurs du coefficient η par lequel il faudrait diviser les CMA dans le Danube pour tenir compte des reconcentrations radioécologiques limitées aux poissons. Le tableau V compare les valeurs que nous obtiendrions dans cette hypothèse à celles de l'étude en question. Les différences sont considérables.

5. CONCLUSION

En résumé, d'après l'exemple que nous avons pris d'une contamination par le césium 137 et le strontium 90, les reconcentrations radioécologiques dans la chair des poissons et les végétaux obligeraient à diviser les CMA par un coefficient qui dans les pires conditions n'excèderait guère 20.

L'adsorption des radioéléments sur les sédiments de la rivière et les déplacements du poisson nous paraissent constituer deux phénomènes essentiels.

Aussi les chercheurs devraient-ils indiquer les points kilométriques auxquels les poissons dont ils mesurent l'activité ont été pêchés. Traiter les poissons comme des animaux sédentaires n'est pas une attitude réaliste.

REFERENCES

[1] International Commission on Radiological Protection, ICRP publication 2, Report of Committee II on Permissible Dose for Internal Radiation, Pergamon Press, New York (1959).

[2] PICAT, P., GRAUBY, A., "Evolution de la radiocontamination des canaux d'irrigation alimentés par le Rhône", Actes du Symposium International de Radioécologie, C.E.A, C.E.N. Cadarache, Isère, France (1969).

[3] FOULQUIER, L. et GRAUBY, A., "Propositions pour un radioindicateur biologique international des milieux aquatiques à partir d'un bilan expérimental écologique et économique", Colloque sur la Surveillance de l'Environnement auprès des Installations Nucléaires, Varsovie (1973), A.I.E.A., Vienne, vol. 2, IAEA-SM-180/50 (à paraître).

[4] FOULQUIER, L., MARCOUX, G. et GRAUBY, A., "Etude de la fixation et de la désorption du radiostrontium par Anguilla (L) en eau douce et en eau salée et de son transfert à l'homme par l'alimentation", International Symposium on Radioecology applied to the protection of man and his environment, Rome (1971), EUR 4800 d.f.i.e.

[5] MAREY, A.N., SAUROV, M.M., "Material for assessment of the role played by food chains in 90 Sr migration from fresh-water reservoirs into the diet of man"; Radioecological concentration processes, proceed.of an intern. Symposium held in Stockholm, Pergamon Press (1966) 135-142.

[6] BURTON, M., "La vie des poissons", Editions Fernand NATHAN, Paris (1973) 49.

[7] I.N.R.A. et S.H.A.R.P., "Absorption des radioéléments du sol par divers légumes cultivés dans les conditions de la pratique", Rapport C.E.A. n° 1860, C.E.N. Saclay, France (1961).

[8] I.N.R.A. et C.E.A., "Etude expérimentale de la contamination radioactive par l'eau d'irrigation de certaines plantes cultivées, Rapport C.E.A. - R 2625, C.E.N. Fontenay-aux-Roses, France (1964).

[9] FREKE, A.M., Health Physics, 13 (1967).

[10] U.S. Atomic Energy Commission, Final environmental statement concerning proposed rule making action, 1 (1973).

[11] FLETCHER, J.F., DOTSON, W.L.(compilers), "Hermes, a digital computer code for estimating regional radiological effects from the nuclear power industry", USAEC Report HEDL-TME-71-168, Handford Engineering Development Laboratory, Richland, WA (1971).

[12] International Atomic Energy Agency, Disposal of Radioactive Wastes into Rivers, Lakes and Estuaries, Safety series, n°36, Vienne (1971) 62.

ANNEXE

Calcul de l'activité massique q du poisson

Nous utiliserons les notations déjà explicitées dans le texte (§ 2).

Nous considèrerons ici la distance x où le poisson est pêché comme fixe.

Désignons par "d" la distance variable, comprise entre o et x, où le poisson se trouve à l'instant t.

La variation de l'activité massique du poisson par unité de temps est égale à la différence entre les apports et les pertes, représentée par :

$$\frac{dq}{dt} = f\,C_o\,e^{-kd} - \lambda q \qquad (4)$$

Chaque cycle comprend un trajet aller de x à o et un trajet retour de o à x (fig. 1).

a) Trajet aller de x à o

Si l'origine du temps est prise au début de ce trajet, on a :

$$x - d = vt$$

L'équation (4) s'écrit :

$$\frac{dq}{dt} = f\,C_o\,e^{-k(x-vt)} - \lambda q$$

L'intégration donne :

$$q = \frac{f\,C_o\,e^{-kx}}{\lambda + kv}\left(e^{kvt} - e^{-\lambda t}\right) + q\,(t = o)\,e^{-\lambda t}$$

On a en fin de trajet :

$$t = x/v\ ,$$

et s'il s'agit du $n^{ième}$ cycle :

$$q = q_{n-0,5} \qquad\qquad q\,(t = o) = q_{n-1}$$

$$q_{n-0,5} = \frac{f\,C_o\,e^{-kx}}{\lambda + kv}\left(e^{kx} - e^{-\lambda x/v}\right) + q_{n-1}\,e^{-\lambda x/v} \qquad (5)$$

b) Trajet retour de o à x

Si l'origine du temps est prise au début de ce trajet, on a :

$$d = vt$$

et l'on aboutirait comme précédemment à l'expression :

$$q_n = \frac{f\,C_o}{\lambda - kv}\left(e^{-kx} - e^{-\lambda x/v}\right) + q_{n-0,5}\, e^{-\lambda x/v} \qquad (6)$$

Portons (5) dans (6). Il vient :

$$q_n = a + b + (1 - c)\, q_{n-1} \qquad (7)$$

où :

$$a = f\,C_o\, e^{-kx}\, \frac{1 - e^{-(\lambda - kv)x/v}}{\lambda - kv}$$

$$b = f\,C_o\, e^{-kx}\, \frac{e^{-(\lambda - kv)x/v} - e^{-2\lambda x/v}}{\lambda + kv}$$

$$c = 1 - e^{-2\lambda x/v}$$

L'équation (7) peut encore s'écrire sous la forme différentielle :

$$q_n - q_{n-1} = \frac{dq}{dn} = a + b - cq$$

L'intégration donne :

$$q = \frac{a + b}{c}\left(1 - e^{-cn}\right)$$

Au bout d'un certain nombre de cycles parcourus, q atteint la valeur d'équilibre $\frac{a + b}{c}$.

DISCUSSION

A. BAYER: Do the normal changes of the river flow rate in the course of a year have any influence?

R. SCHAEFFER: The important thing is not the flow rate but the constancy of the volumetric activity at each particular point. Whatever the seasonal variations in the flow, this constancy is assured as the radioactive effluents are released in quantities depending on the flow rate.

N. T. MITCHELL: Instead of attempting to modify MPCs for water, a particularly useful approach which we have adopted over many years with great success is to use them only as a means of calculating maximum permissible rates of intake of radionuclides. This is now becoming ICRP practice and is referred to in the forthcoming version of Publication 2.

I should also like to ask a question. Do you feel that the assumed equal subdivision of activity between water and food pathways is an adequate approximation? For example, with caesium we have found that the activity is predominantly associated with the food of the fish — to the extent of more than 90% in your conditions of hardness.

R. SCHAEFFER: There are certainly many exceptions to the rule I assumed for the apportionment of activity between water and food, but differences in this apportionment would have only a limited effect on the MPC division coefficient η. For example, if the apportionment factor for caesium-137 was 90% instead of 50%, the coefficient η would only be doubled. In radiological protection an error of one order of magnitude is acceptable.

IAEA-SM-184/41

SORPTION-DESORPTION OF RADIOACTIVE CAESIUM, STRONTIUM AND CERIUM ON EARTH COMPONENTS

M. PIRŠ
Jožef Stefan Institute,
University of Ljubljana,
Ljubljana,
Yugoslavia

Abstract

SORPTION-DESORPTION OF RADIOACTIVE CAESIUM, STRONTIUM AND CERIUM ON EARTH COMPONENTS.

The sorption capacity on humus material, calcite and dolomite for caesium, strontium and cerium has been determined by static and dynamic measurements. The dependence of the equilibrium concentration on the solution, pH, the presence of foreign salts and the stability of the fixation of the above-mentioned ions has been investigated. The elution curves for Cs, Sr and Ce from the humus material, sand, calcite and dolomite with water, salt solutions and acids have been calculated.

INTRODUCTION

Liquid effluents from a nuclear power plant contaminate the environment to a very small amount. A 600-MW(e) power plant discharges 10 - 12 Ci/a of various isotopes in the cooling water, which, at a flow of 100 m^3/s, is a negligible quantity. But in the event of an accident (damaged cooling system) great quantities of contaminated water can pour out into the near environment and soak into the ground.

When the radioactive water comes in contact with the soil the ions react with the soil constituents and bind on the soil particles. Generally the anions bind weakly and the cations strongly, and they are seldom detectable in the groundwater [1-3]. The ground acts as a mechanical filter and as an ion exchanger column. The processes that take place are: adsorption, absorption, ion exchange and precipitation. All these processes will be called here sorption. The degree to which the ions are removed from the solution depends on the sorption capacity, which is dependent on the concentration of the solution, on the composition of the soil and on the presence of foreign ions. Under the influence of precipitation the sorbed ions slowly migrate into deeper layers of the ground.

For our investigations of the retention properties of ground soils, we collected representative samples of humus soil, sand, calcite and dolomite from Brežiško Polje, the site of the future nuclear power plant. The samples were analysed and the capacity determined in the static and dynamic way. The desorption was investigated by elution experiments.

EXPERIMENTAL

Sampling

Samples of the humus soil, which is about 50 - 60 cm thick, were taken in eight places. By chemical analyses it was found that the composition of the humus samples and the sand was the same and therefore the samples were mixed and two average samples were made for the following investigations. Samples of sand were taken to 100 cm thickness. The humus soil and sand were air dried, homogenized, sieved and chemical (Table I) and granulometric analyses (Table II) were made.

TABLE I. CHEMICAL ANALYSIS OF HUMUS SOIL AND SAND

	Humus soil (%)	Sand (%)
SiO_2	52.4	47.0
R_2O_3	7.8	9.1
CaO	9.7	9.2
MgO	4.8	1.8
Insoluble	8.2	8.6
Moisture content 105°C	2.3	0.7
Weight loss on ignition 600°C	4.0	2.5
pH	8.3	8.9

TABLE II. GRANULOMETRIC ANALYSIS OF SAMPLES

mm	Sand (%)	Humus soil (%)
>1	29.5	
0.5 - 1	3	
0.3 - 5	5	16 > 0.3 mm
0.25 - 0.3	4	7
0.2 - 0.25	3	8
0.15 - 0.2	4	10
0.12 - 0.15	6	8
0.10 - 0.12	8	5
0.09 - 0.1	7	6
0.09	29.5	40

TABLE III. SORPTION CAPACITY OF CAESIUM, STRONTIUM AND CERIUM AS A FUNCTION OF EQUILIBRIUM SOLUTION CONCENTRATION

Material	Caesium		Strontium		Cerium	
	Capacity (meq/g $\times 10^{-2}$)	Equ. sol. ($\underline{N} \times 10^{-3}$)	Capacity (meq/g $\times 10^{-2}$)	Equ. sol. ($\underline{N} \times 10^{-3}$)	Capacity (meq/g)	Equ. sol. ($\underline{N} \times 10^{-3}$)
Humus soil	0.35	0.03	0.29	0.04	2.5×10^{-4}	-
	2.4	0.52	2.5	0.05	2.5×10^{-2}	-
	7.0	8.6	10.0	5.8	4.5×10^{-2}	9
	7.2	100.0	10.2	10.0	3.5×10^{-1}	93
Sand	0.25	0.05	0.28	0.04		
	1.0	0.8	2.1	0.58		
	3.0	9.4	4.1	9.1		
	3.1	100.0				
Calcite and dolomite	0.01	0.098	0.08	0.8	2.5×10^{-4}	10^{-4}
					2.5×10^{-2}	10^{-2}
					0.3	10^{2}

Determination of sorption capacity

Batch method: Standard solutions of Cs, Sr, Ce of various concentrations were made and spiked with ^{137}Cs, ^{85}Sr, ^{89}Sr and ^{144}Ce. One-gram portions of humus soil and sand were weighed into polyethylene bottles; 50 ml of each solution was added to separate samples and shaken for 24 hours. The suspensions were centrifuged, the centrifugates decanted, and the samples washed with a small portion of water to make up the amount of solution to 50 ml. Cs, Sr and Ce were then determined by counting the liquid on a NaI gamma-counter. From the results shown in Table III it is evident that the sorption capacity increases with the increasing concentration of equilibrium solution, after which it attains a limit and with increasing concentration remains constant.

The influence of pH on the capacity

The pH of the solution does not influence the sorption capacity of caesium. In contrast, the pH of the solution has a great influence on the sorption of strontium and cerium. These two elements sorb only in solutions with pH greater than 6-7. The results are shown in Table IV.

Influence of the presence of foreign salts in the solution

Addition of foreign salts diminishes the sorption capacity. A concentration of 0.01N solution has an insignificant effect on the sorption capacity, but a greater concentration of foreign salts diminishes the sorption capacity. The results for caesium are shown in Table V. The diminution in the sorption capacity of strontium and cerium was found to be identical.

TABLE IV. SORPTION OF CAESIUM, STRONTIUM AND CERIUM AT DIFFERENT pH ON HUMUS SOIL

pH	2	3	5.1	7.1	9.1	9.8
Sorption % Cs	83	84	83	96	94	95
% Sr	15	16	25	48	84	85
% Ce	5	7	6	25	95	98

TABLE V. SORPTION % OF CAESIUM ON HUMUS SOIL IN DEPENDENCE OF CONCENTRATION OF FOREIGN ELECTROLYTES

Normality of CsCl	Normality of NaCl				
	0	10^{-4}	10^{-3}	10^{-2}	10^{-1}
$\sim 10^{-9}$	95	96	89	82	79
$\sim 10^{-3}$	56	50	52	45	32
$\sim 10^{-2}$	18	16	15	11	6

TABLE VI. BREAKTHROUGH CAPACITY OF HUMUS SOIL BY COLUMN METHOD

Ion	Normality of influent	Capacity (meq/g)
Cs	10^{-4}	2×10^{-3}
Cs	10^{-3}	1.3×10^{-2}
Sr	10^{-4}	1.6×10^{-3}
Sr	10^{-3}	1.0×10^{-2}
Ce	10^{-3}	2.2

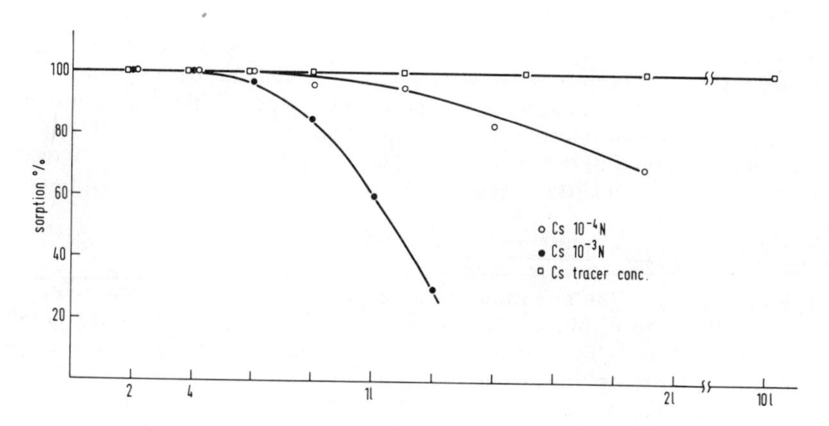

FIG. 1. Filtering of Cs ions through 1 cm layer of humus material.

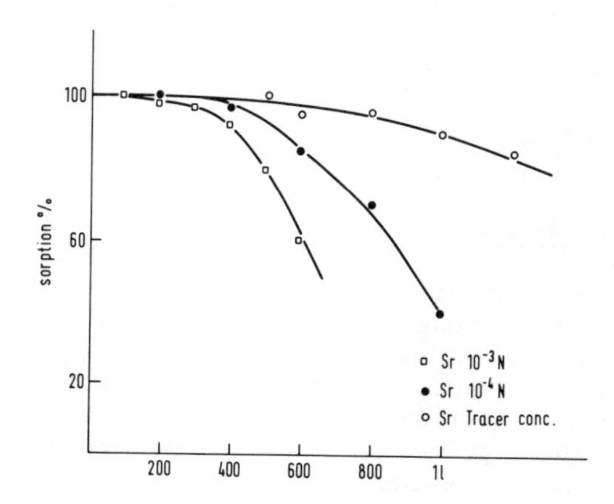

FIG. 2. Filtering of Sr ions through 1 cm layer of humus material.

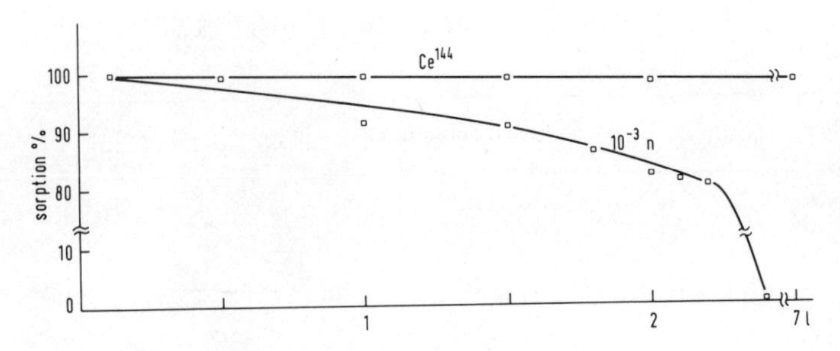

FIG. 3. Filtering of Ce ions through 1 cm layer of humus material.

Column method

We tried to determine the breakthrough capacity by column experiments. Weighted amounts of humus soil, which had been soaked for 48 hours in water, were placed in a tube with a perforated bottom so that the layer was 1-2 cm thick. The radioactive solution was passed through the column under slightly reduced pressure. The activity of the influent and effluent was measured. The results are evident from Table VI and Figs 1, 2 and 3.

Sorption rate of Cs, Sr and Ce

The efficiency of the sorption process depends on two parameters; capacity and the rate of binding. The velocity of the penetration of the contaminated water through the ground depends on the geological formation of the ground. If the material is compact, the contact time is great and equilibrium between the fixed ions and the ions in solution is established. Under a known water flow rate we can estimate what part of the total amount of radionuclides is sorbed in a given depth.

Measurement of the sorption rate

Two grams of material were put into a bottle, 200 ml of the tracer solution were added and mixed at 500 rev/min. Samples were taken at time intervals and measured (the samples were returned to the solution after measuring). The results are shown in Table VII.

The stability of the fixation of radionuclides on the soil particles

The stability of the fixation is the most important parameter for the migration of radionuclides into the ground. The bound radionuclides on the soil particles are washed by precipitation into deeper layers of the ground. The rate of migration depends on the quantity of the water, the ion and the process by which the ions are fixed on the soil particles. Many theories deal with the migration of radionuclides [3-8] but usually the parameters are difficult to evaluate.

We tried to investigate the stability of the fixation by both static and dynamic methods.

TABLE VII. SORPTION RATE OF CAESIUM, STRONTIUM AND CERIUM

Time	Percentage of activity sorbed				
	Humus soil			Calcite and dolomite	
	Cs	Sr	Ce	Cs	Ce
30 s	45	42	97		96
1 min	53	44	98	62	96
5 min	68	56	96	61	98
10 min	76	57	96	57	97
30 min	85	56	97		
1 h	89	67	98	71	96
3 h	90	78	98		98
6 h	88	80	98		98
1 d	90	79	97	55	97
2 d	91	80		57	
6 d	95	82	98		
30 d	93	80	97		97

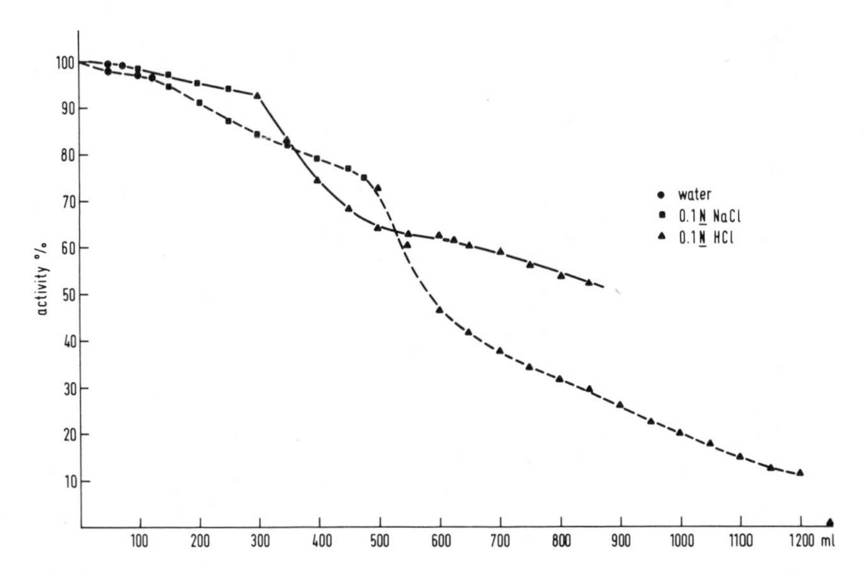

FIG.4. Leaching of Cs from humus soil —— and sand ----- saturated with tracer solution.

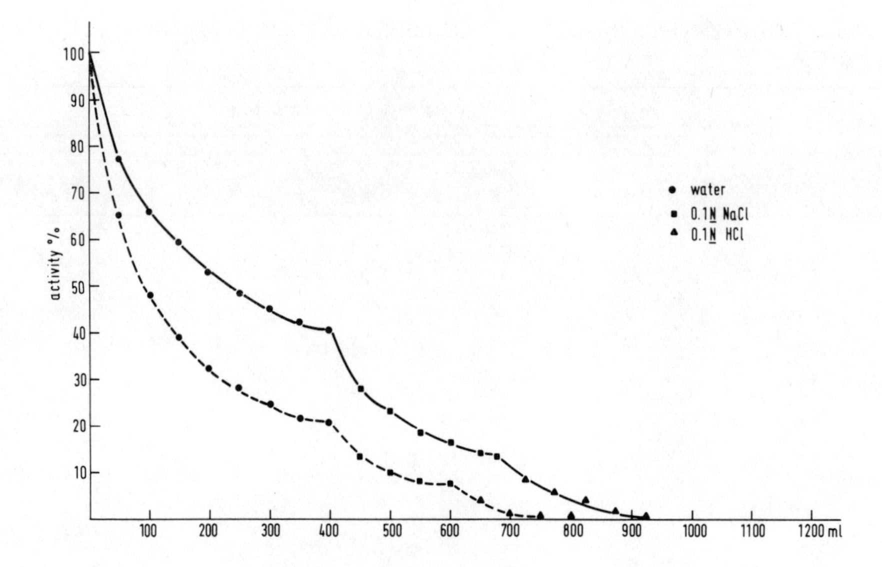

FIG. 5. Leaching of Cs from humus soil ——— and sand ---- saturated with 10^{-3}N solution.

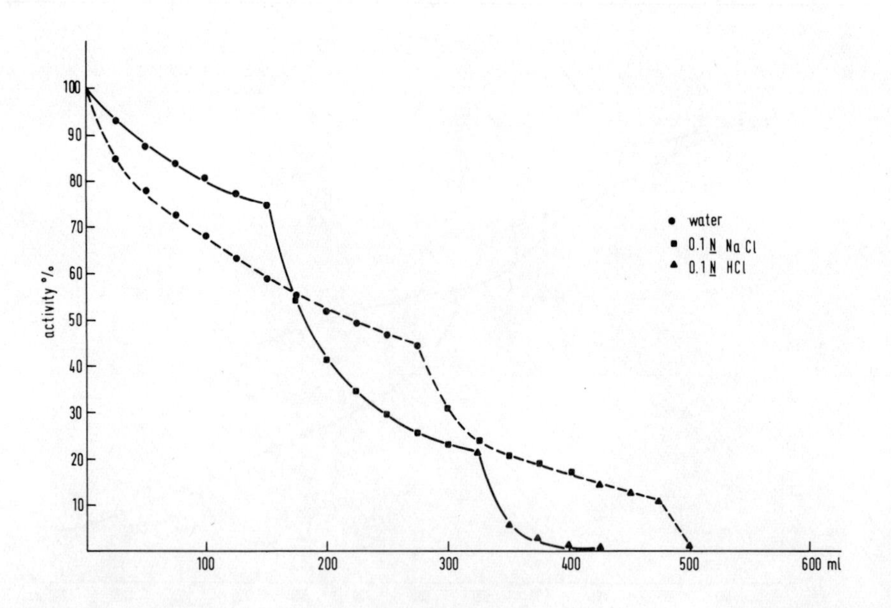

FIG. 6. Leaching of Sr from humus soil ——— and sand ---- saturated with tracer solution.

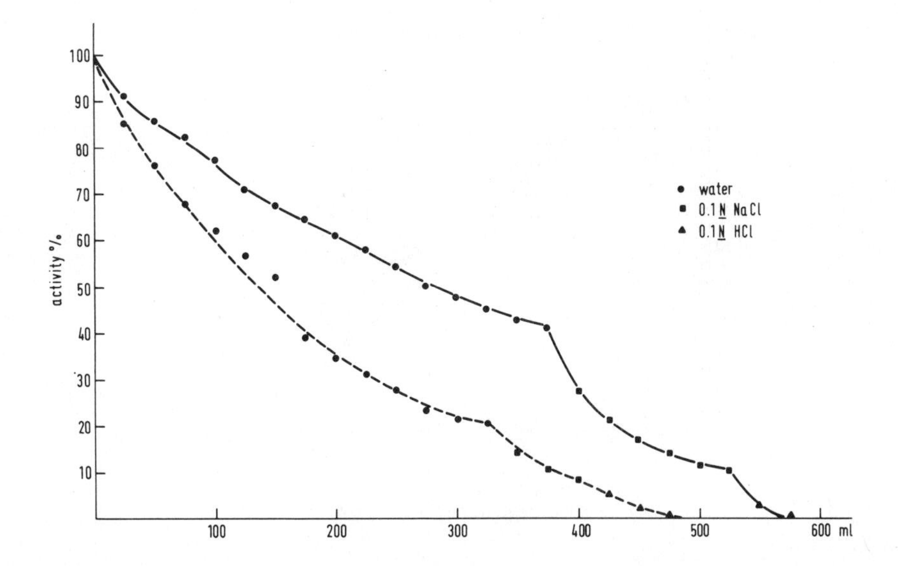

FIG. 7. Leaching of Sr from humus soil ——— and sand ----- saturated with 10^{-3} N solution.

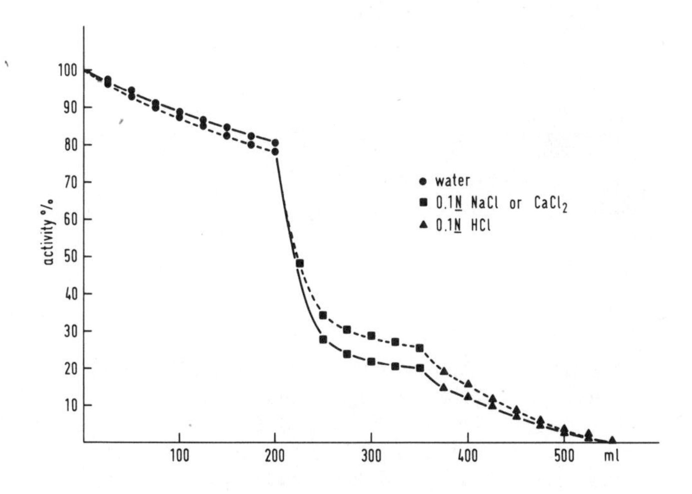

FIG. 8. Leaching of Cs from dolomite ——— and calcite ----- .

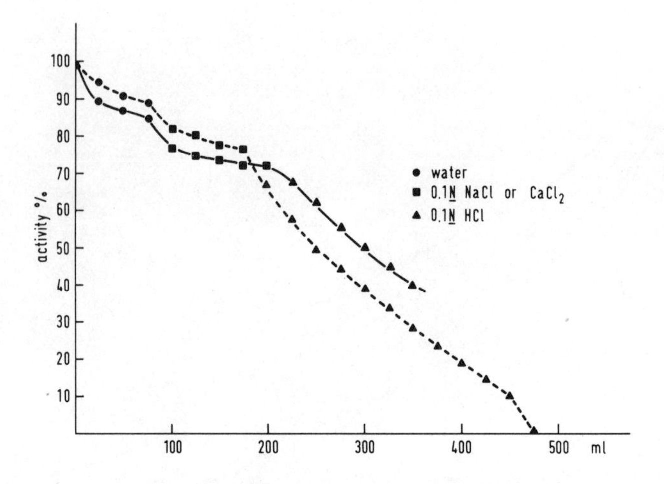

Fig. 9. Leaching of Sr from dolomite ——— and calcite -----.

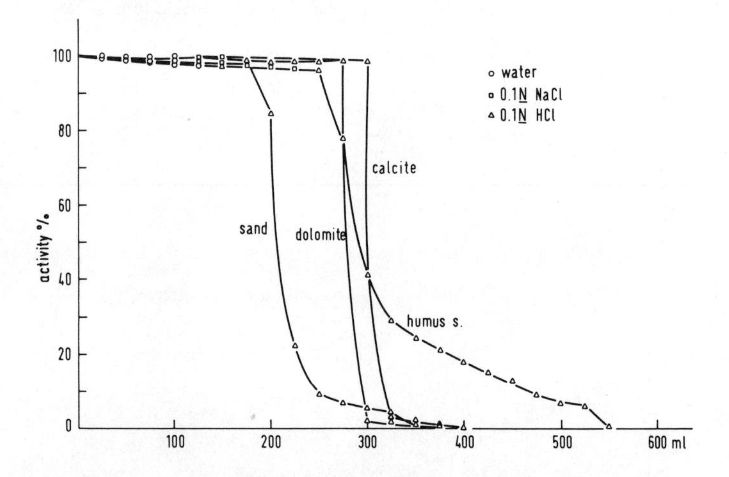

FIG. 10. Leaching from Ce from humus soil, sand, calcite and dolomite saturated with tracer solution.

Static method

One gram of material saturated with solutions of different concentrations of Cs, Sr and Ce was put into centrifuge tubes, 25 ml of water, salt solution or acid were added and mixed for 1 hour at 500 rev/min. The samples were then centrifuged, the activity measured, a new portion of extractant added and the elution repeated. This procedure was repeated until all the activity was extracted or the activity of the extract became a constant fraction of the whole sorbed activity. The results are shown in Figs 4-11.

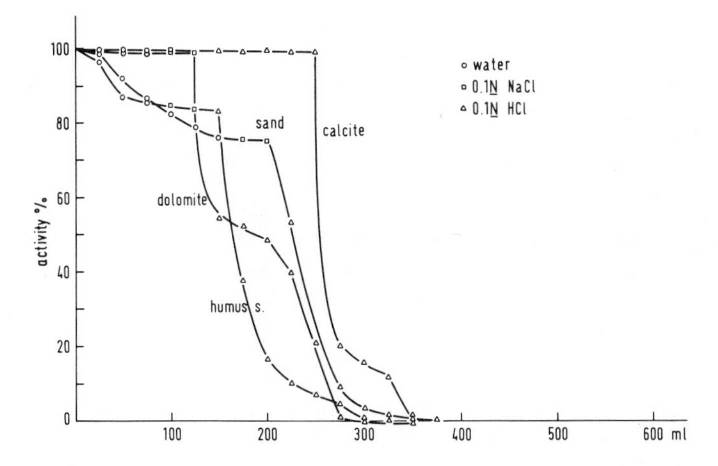

FIG. 11. Leaching from Ce from humus soil, sand, calcite and dolomite saturated with 10^{-3}N solution.

From the elution curves it is evident that water leached Cs from humus soil, sand, calcite and dolomite in small amounts. The percentage of eluted activity is greater if the amount of bound ions is greater. Salt solutions and acids are better leaching agents. Strontium behaves similarly, but the amount of activity leached is greater. The stability of the fixation of strontium ions is weaker and therefore the migration is faster. Cerium-144 does not elute with water and salt solutions but with a small amount of acid it elutes completely.

Column method

A column was filled with humus material and soaked in water for 24 hours. The diameter of the column was 2 cm and the height of the bed 15 cm. One to two μCi of ^{137}Cs and ^{85}Sr were put on the top of the column and the water allowed to pass through the column under reduced pressure at a flow rate of 1 ml/min. The distribution of Cs and Sr activity along the column was monitored after elution of 50 column volumes with a gamma scintillation counter set up behind a 1-cm lead shield with a 0.5-cm wide window. The migration of caesium and strontium was detected only qualitatively. After 150 bed volumes a movement of 2 cm ± 50% was detected.

CONCLUSIONS

Ground soil samples from Brežiško Polje showed very good retention properties for ^{137}Cs, ^{85}Sr and ^{144}Ce. In dynamic conditions the ground sorbs about 0.1-1 meq/100 g; under static conditions and at the same concentration of the equilibrium solutions the sorption capacity is greater. Should, in the case of an accident, the contaminated liquid come into contact with the ground, we can expect that the total radioactivity would be retained in the first 5-10 cm of the ground. This ground would of course become unusable for agricultural purposes.

The migration of radionuclides into the ground under the influence of water is very slow. The rate depends on the kind of binding of the radionuclides on the soil particles. From elution experiments and from the rate of sorption we can estimate that Ce is bound by a precipitation process. The reaction between soil particles and Ce ions is instantaneous and the cerium is practically unleached in neutral solutions, but very quickly leached in acid solutions.

Caesium is probably bound by ion exchange. At high concentrations, when the ion exchange capacity is saturated, it binds by absorption and adsorption. These parts of the caesium are more weakly bound and are more easily leached by water.

The bond between strontium and soil particles is very difficult to define. The sorption on humus soil and sand is similar to caesium but is easier to leach out with water. On calcite and dolomite it binds immediately but later comes again into solution. Probably strontium is bound by all four processes.

ACKNOWLEDGEMENT

This paper is partly based on work financed through the Boris Kidrič Foundation.

REFERENCES

[1] MERRIT, W.F., MAWSON, C.A., "Experiences with ground disposal at Chalk River", Disposal of Radioactive Wastes into the Ground (Proc. Symp. Vienna, 1967), IAEA, Vienna (1967) 79-93.

[2] MARTER, W.L., "Ground waste disposal practices at the Savannah River Plant", Disposal of Radioactive Wastes into the Ground (Proc. Symp. Vienna, IAEA 1967), IAEA, Vienna (1967) 95-107.

[3] BEARD, S.J., GODFREY, W.L., "Waste disposal into the ground at Hanford", Disposal of Radioactive Wastes into the Ground (Proc. Symp. Vienna, 1967), IAEA, Vienna (1967) 121-134.

[4] THORNTHWAITE, C.W., MATHER, J.R., NAKAMURA, J.K., C.W. Thornthwaite Associates Laboratory of Climatology, Tech. Rep. 1 (1961).

[5] SPITSYN, V.I., BALEKOVA, V.D., NAUMOVA, A.F., GROMOV, V.V., SPIRIDONOV, F.M., VETROR, E.M., GRAFOV, G.I., Int. Conf. peaceful Uses atom. Energy (Proc. Conf. Geneva, 1958) 18, UN, New York (1958) 439-48.

[6] MAYER, S.W., THOMPKINS, E.R., J. Amer. Chem. Soc. 69 (1947) 2866-73.

DISCUSSION

D. BENINSON: Did you measure the amounts of clay mineral, such as montmorillonite, in the soils studied? They would greatly influence the retention of radionuclides.

M. PIRŠ: No, we did not.

P.L. BOVARD: In reply to Mr. Beninson, perhaps I could say that the complexed ruthenium does not become fixed on the soil and rapidly returns to the water table.

I should also like to ask Mr. Pirš a question. Don't you think that if ^{90}Sr and ^{137}Cs spread over the soil in solution form, the soil structure can be an important consideration? Is your soil sufficiently sandy and homogeneous for its structure to be neglected in the risk evaluation?

M. PIRŠ: No studies were performed on soil structure.

R. BITTEL: Have any measurements been made on the quality of the organic matter in the soil samples? In France we have found that caesium retention is less for soils rich in fulvic acids than for those rich in humic acids.

M. PIRŠ: Only inorganic chemical analyses were performed.

P.L. BOVARD: Perhaps I can say also a few words to the question asked by Mr. Bittel. Fixation on the organic colloid differs from fixation on the clay colloid in that it is less stable. It is true that the binding differs, depending on whether the acids are fulvic or humic, the former being more labile.

Control of population doses
(Session IX)

Chairmen: L.-E. LARSSON (Sweden)
H.P. JAMMET (France)

THE ESTIMATION OF RADIATION DOSE RATES TO FISH IN CONTAMINATED ENVIRONMENTS, AND THE ASSESSMENT OF THE POSSIBLE CONSEQUENCES

D.S. WOODHEAD
Ministry of Agriculture, Fisheries and Food,
Fisheries Radiobiological Laboratory,
Lowestoft, Suffolk,
United Kingdom

Presented by N.T. Mitchell

Abstract

THE ESTIMATION OF RADIATION DOSE RATES TO FISH IN CONTAMINATED ENVIRONMENTS, AND THE ASSESSMENT OF THE POSSIBLE CONSEQUENCES.

The disposal of radioactive waste into coastal waters, rivers and lakes may produce localized conditions in which fish experience radiation dose rates in excess of that from the natural background. Reasonably realistic estimates of these dose rates may be obtained by combining data on the levels of radioactivity in the water, sediments and fish with simple dosimetry models. Where the dose rates are sufficiently high it has been possible to provide confirmation of the estimates, using in situ thermoluminescent dosimeters. A knowledge of the radiation regime, together with information on the effects of irradiation on fish, provides a basis for assessing the consequences for fish populations in contaminated environments.

1. INTRODUCTION

The control of the disposal of radioactive waste into the aquatic environment is currently based on the requirement that the resultant increment of radiation dose-rate received by human populations should be restricted within specified limits[1,2]. In safeguarding man in this way it is implicitly assumed that the living resources in aquatic ecosystems are also protected. Support for this position can only come from information relating to dose-rate regimes in both contaminated and uncontaminated (or background) areas combined with a knowledge of the responses of aquatic organisms to irradiation. For example, the assumption has been questioned in the case of marine fish eggs[3], but estimates of the dose-rate likely to be received by developing embryos in a contaminated area indicate that such concern is unwarranted[4]. Evidence is produced below which shows that a similar conclusion can be arrived at for fish inhabiting two areas in the British Isles which receive radioactive waste, viz. the brown trout in Lake Trawsfynydd, North Wales and the plaice in the north-east Irish Sea.

2. LEVELS OF NATURAL RADIOACTIVITY IN AQUATIC ENVIRONMENTS

2.1. Sea water and fresh water

The sources and levels of naturally-occurring radionuclides in sea water have been discussed previously [5] and the data are given in Table I. A similar range of radionuclides is also present in fresh water, although the levels of particular isotopes often show differences from those found in sea water (see Table I). For example, the increased level of tritium is a consequence of the higher proportion of recent rainwater in rivers and lakes as compared with surface ocean waters.

TABLE I. LEVELS OF NATURAL RADIOACTIVITY IN SEA WATER AND FRESH WATER

Radionuclide	Sea water		Fresh water	
	Concentration, pCi l^{-1}	Reference	Concentration pCi l^{-1}	Reference
^{3}H	0.6-3	6	5.4-16.5	24, 25
^{14}C	0.16-0.18	7, 8	-	-
^{40}K	320	-	0.1-6.6	26, 27, 28
^{87}Rb	2.9	-	2.4×10^{-2}	28
^{238}U	1.2	9, 10	5×10^{-3}-1.7	7, 28, 29
^{234}U	1.3	9, 10	10^{-2}-3.4	-
^{230}Th	$(0.6\text{-}14) \times 10^{-4}$	11, 12	-	-
^{226}Ra	$(4\text{-}4.5) \times 10^{-2}$	13, 14, 15	10^{-2}-3	6, 7, 24, 29
^{222}Rn	$\approx 2 \times 10^{-2}$	13	0.2-180	6, 7
^{210}Pb	$(1\text{-}6.8) \times 10^{-2}$	16, 17, 18, 19	$(0.25\text{-}3.6) \times 10^{-1}$	6, 18, 30
^{210}Po	$(0.6\text{-}4.2) \times 10^{-2}$	18, 19, 20, 21, 22	7×10^{-3}-0.23	18
^{232}Th	$(0.1\text{-}7.8) \times 10^{-4}$	11, 12, 23	$(0.1\text{-}1.1) \times 10^{-2}$	28, 29
^{228}Ra	$(0.1\text{-}10) \times 10^{-2}$	45	-	-
^{228}Th	$(0.2\text{-}3.1) \times 10^{-3}$	11, 12	-	-
^{235}U	5×10^{-2}	-	$(0.2\text{-}70) \times 10^{-3}$	-

The levels of the other natural radionuclides in fresh waters are dependent on the rock and mineral deposits with which the water has been in contact and will therefore reflect, to some extent, the geology of the watershed resulting in a range of activity levels. In Table I the concentrations of ^{234}U have been calculated from those of ^{238}U, on the basis that the activity ratio of two determined by Blanchard[31] is generally applicable. It has also been assumed that the relative abundances of ^{235}U and ^{238}U are the same as those found in terrestrial rocks.

2.2. Marine and freshwater sediments

In marine sediments in coastal waters and in the beds of lakes and rivers, it is reasonable to assume that the mineral fraction has uranium, thorium and potassium contents similar to those of terrestrial rocks, for which average values are given in Table II[32, 33].

TABLE II. TYPICAL URANIUM, THORIUM AND POTASSIUM LEVELS IN BEACH SANDS AND COMMON TERRESTRIAL ROCKS

Material	Uranium (ppm)	Thorium (ppm)	Potassium (%)
Beach sands	3.0	6.4	0.33
Granite	5.0	18	3.8
Shale	3.7	12	1.7
Limestone	1.3	1.1	0.2
Sandstone	0.45	1.7	0.6
Basalt	0.50	2.0	0.5

2.3. Marine and freshwater fish

Table III summarizes the available information on the levels of natural radioactivity in marine and freshwater fish. The values given for tritium, carbon-14 and potassium-40 have been derived from the stable element composition of the fish[34] and the known specific activity of the nuclides in the environment, on the assumption that there is no isotope fractionation during the accumulation process. The concentration of rubidium-87 in sea fish was calculated from the concentration factor given by Lowman _et al._[35] and the activity in sea water.

TABLE III. LEVELS OF NATURAL RADIOACTIVITY IN MARINE AND FRESHWATER FISH

Radionuclide	Marine fish			Freshwater fish		
		Concentration, pCi g^{-1}	Reference		Concentration, pCi g^{-1}	Reference
^{3}H		$(0.5-2.7) \times 10^{-3}$	-		$(5-14) \times 10^{-3}$	-
^{14}C		0.4	-		0.4	-
^{40}K		2.5	-		3.5	27
^{87}Rb		5×10^{-2}	-		-	-
^{226}Ra	Soft tissue	$(2-51) \times 10^{-4}$	36	Whole fish	1-3.5	38
				Flesh	$(2-20) \times 10^{-4}$	36, 39
				Bone	$6 \times 10^{-3}-2.1$	36, 39
^{210}Pb	Flesh	$(2-23) \times 10^{-4}$	18, 20, 36	Flesh	$(0.1-4.7) \times 10^{-3}$	18, 36
	Stomach	$(2-85) \times 10^{-2}$	20	Stomach	-	-
	Liver	$(11-24) \times 10^{-3}$	18, 20	Liver	$(0.9-24) \times 10^{-3}$	18
	Bone	$(9-130) \times 10^{-3}$	18, 20	Bone	$(8-84) \times 10^{-3}$	18, 36
^{210}Po	Flesh	$(4-1400) \times 10^{-4}$	18, 20, 22, 36	Flesh	$(3.7-110) \times 10^{-3}$	18, 36
	Stomach	$(2-26) \times 10^{-1}$	20	Stomach	-	-
	Liver	$(2-9) \times 10^{-1}$	18, 20, 37	Liver	$(7.5-50) \times 10^{-2}$	18
	Bone	$(2-22) \times 10^{-2}$	18, 20	Bone	$(5-130) \times 10^{-3}$	18, 36

TABLE IV. ANNUAL DISCHARGES OF RADIOACTIVITY (CURIES) TO THE AQUATIC ENVIRONMENT AT LAKE TRAWSFYNYDD AND WINDSCALE

Lake Trawsfynydd[a]				Windscale[b]					
Isotope	1970	1971	1972	Isotope	1968	1969	1970	1971	1972
^{3}H	34.3	27.5	29.0	^{3}H	20 300	24 800	32 000	31 500	33 600
^{60}Co	3.0×10^{-3}	15×10^{-3}	14×10^{-3}	^{89}Sr	40	230	470	390	1 080
^{90}Sr	0.58	0.32	1.55	^{90}Sr	1 370	2 950	6 000	12 300	15 200
^{125}Sb	0.23	3.34	16.6	^{95}Zr	28 100	31 600	9 100	18 000	25 600
^{134}Cs	0.23	0.81	1.06	^{95}Nb	37 100	30 000	9 900	17 300	23 500
^{137}Cs	0.93	3.41	4.16	^{103}Ru	1 800	1 400	890	830	1 160
^{144}Ce	4.7×10^{-2}	0.11	0.52	^{106}Ru	24 200	22 900	27 600	36 400	30 500
				^{134}Cs	-	630	7 010	6 370	5 820
				^{137}Cs	1 500	12 000	30 100	35 800	34 800
				^{144}Ce	10 000	13 500	12 400	17 200	13 700
				^{91}Y + R.E.[c]	-	-	3 000	4 400	14 200

(a) These values are based on data supplied by the CEGB and effluent analyses carried out at the Fisheries Radiobiological Laboratory.

(b) These values are based on data supplied to the Fisheries Radiobiological Laboratory by the United Kingdom Atomic Energy Authority and by British Nuclear Fuels Limited.

(c) R.E. ≡ rare earths.

TABLE V. CONCENTRATIONS OF RADIOACTIVITY IN THE ENVIRONMENT AT LAKE TRAWSFYNYDD DURING 1971 AND AT WINDSCALE DURING 1968

Material	^{60}Co	^{90}Sr		^{125}Sb		^{134}Cs		^{137}Cs		^{144}Ce
		Average	Range	Average	Range	Average	Range	Average	Range	
LAKE TRAWSFYNYDD										
Water, pCi l^{-1}	-	5.3	3.2-6.9	-	-	3.9	0.9-7.2	18	3.0-35	-
Trout flesh, pCi g^{-1}	-	0.13	0.12-0.15	-	-	2.0	1.2-4.0	12	7.7-20	-
Perch flesh, pCi g^{-1}	-	0.27	0.22-0.34	-	-	4.3	1.5-8.9	26	14-45	-
Mud, pCi g^{-1} wet	0.02	-	-	0.32	0-0.98	0.12	0.04-0.30	2.4	1.4-5.2	-
Peat, pCi g^{-1} wet	0.004	-	-	0.22	0.07-0.45	0.07	0.02-0.11	0.47	0.1-0.96	0.04

Material	$^{95}Zr/^{95}Nb$		^{106}Ru		^{134}Cs		^{137}Cs		^{144}Ce	
	Average	Range	Average	Range	Average	Range	Average	Range	Average	Range
WINDSCALE										
Water, pCi l^{-1}	350	49-1250	87	23-230	-	-	58	14-120	27	15-57
Plaice flesh, pCi g^{-1}	-	-	-	-	0.4	0.2-0.7	1.2	0.7-2.0	-	-
Sediment, pCi g^{-1} wet	2340	78-9510	700	86-1990	-	-	14	4-29	440	59-1220

3. LEVELS OF ARTIFICIAL RADIOACTIVITY IN LAKE TRAWSFYNYDD AND THE NORTH-EAST IRISH SEA

Lake Trawsfynydd receives the low-level waste produced during the operation of a Magnox nuclear power station by the Central Electricity Generating Board. Apart from tritium, the primary source of radioactivity is the water from the cooling ponds used to store the irradiated fuel elements prior to their transport to Windscale. From the environmental viewpoint the important nuclides in the effluent are the fission-products caesium-134 and -137 and, to a lesser extent, strontium-90, antimony-125 and cerium-144. Other nuclides present in the effluent include a relatively large quantity of tritium and some sulphur-35 and cobalt-60; of these only cobalt-60 is detected in the environment. The annual discharges of radionuclides, based on analyses of effluent samples carried out in this laboratory, are given in Table IV for the years 1970-72. Lake Trawsfynydd is relatively small (approximately 3 km across) and the radioactivity discharged becomes fairly uniformly distributed throughout the water mass. Table V gives the data on the levels of radioactivity measured in samples of water, fish flesh and lake bed collected during 1971[40].

At Windscale the irradiated fuel arising from the British nuclear power programme is reprocessed, resulting in a considerable volume of low-level liquid waste. This effluent is discharged to the north-east Irish Sea via a pipeline extending 2.5 km beyond high-water mark, and as can be seen from the data given in Table IV it contains mainly fission-product radionuclides. Tidal action disperses and dilutes the radioactivity, and the levels of contamination decrease with increasing distance from the outfall. In Table V data are given for the year 1968 on the levels of radioactivity in the environment in the vicinity of the discharge point, these being representative of the highest levels of contamination[5, 41, 42].

4. DOSE-RATES TO FISH FROM ENVIRONMENTAL RADIOACTIVITY

It would be difficult and probably unnecessary to devise a dosimetry model which matched the geometry of a particular fish exactly and thus allowed precise estimation of the dose-rate received from internal and external radioactivity. In the case of internal natural radioactivity, the data have been obtained from many different species of fish, of varying geometries, and since a complete set of data is not available for a particular species it must be assumed that the information given in Table III is representative of the levels of activity which exist in any marine or freshwater fish species. Thus simplified idealized models must be used to obtain estimates of dose-rates, despite their obvious limitations as descriptions of the true situation.

The models adopted to represent the fish being considered are as follows:

plaice : a flat cylinder 1.8 cm high and 16 cm diameter,
trout : a cylinder 20 cm long and 6 cm diameter,

and it has been assumed that the radioactivity is uniformly distributed in unit density tissue throughout the volume. Although differential distribution of radioactivity between tissues in fish is to be expected, it is only rarely that the available information justifies the development of more refined models for dosimetry purposes[42].

These models are of such dimensions that it can reasonably be assumed that the dose-rates from internal α- and β- radiation are equal to $D_\alpha(\infty)$ and $D_\beta(\infty)$, where $D_\alpha(\infty) = 2.13\ \overline{E}_\alpha C\ \mu rad\ h^{-1}$, and similarly $D_\beta(\infty) = 2.13\ \overline{E}_\beta C\ \mu rad\ h^{-1}$, these being the dose-rates in effectively infinite volumes (i.e. linear dimensions greater than the maximum α- and β-particle ranges) uniformly contaminated with C pCi g^{-1} of radionuclides emitting α-particles (or β-particles) of average energy $\overline{E}_\alpha$ (or $\overline{E}_\beta$) MeV per disintegration.

The calculation of the γ-ray dose-rate within a relatively small volume containing a distributed source requires the consideration of geometrical factors[43], since a substantial proportion of the emitted energy is dissipated outside the volume. The average dose-rate is given by:

$$\overline{D}_\gamma = \Gamma C p_a \overline{g} \times 10^{-3}\ \mu rad\ h^{-1},$$

where Γ = the specific γ-ray constant, in cm^2 rad h^{-1} mCi^{-1},

C = the activity, in pCi g^{-1},

p_a = the density of material, in g cm^{-3}, and is taken to be unity for fish,

$\overline{g}$ = the mean geometrical factor, in cm. For the geometries of the models adopted for plaice and trout $\overline{g}$ has a value of 26 cm. In view of the approximations inherent in the derivation of values for $\overline{g}$, the value of 26 cm has also been assumed to be appropriate for perch.

For irradiation from external sources the geometry is again important. The ranges of α-particles are such that α-activity in either the water or the sediment can only irradiate the superficial cell layers of the body, contributing negligibly to the average dose-rate throughout the fish.

The dose-rate at the surface of the fish from external sources of β-radiation is $0.5\ D_\beta(\infty)$ (evaluated for the levels of activity in the water, or the sediment when the fish is in contact with it); it falls rapidly with depth in tissue, becoming negligible at approximately 1 g cm^{-2} from the surface, even for a high energy β-ray spectrum such as that from ^{106}Ru-^{106}Rh ($E_{\beta max}$ = 3.53 MeV). For most purposes it is sufficient to give the magnitude of the surface dose to indicate the radiation regime. However, in the particular case of the plaice at Windscale, where the most important radiation source is the underlying sediment, it is of interest to know the contribution to the dose-rate to the gonads from the β-radiation from this source, and thus allowance must be made for the attenuation of the β-ray flux in the tissue between the body surface and the gonads. The simple geometry given in Fig. 1 has been used as a model for dose-rate estimation. It is assumed that the sea bed is a plane-uniform β-ray source of effectively infinite thickness and extent, and the expression given by Loevinger et al. [46, Equ. 24, p. 722] has been used to calculate the appropriate attenuation factors.

For external sources of γ-radiation the dose-rates throughout the volumes representing the fish can be taken to be uniform. The natural radioactivity in the sediment and in the water is, in each case, assumed to constitute a uniform, infinitely thick, plane source, and thus the dose-rate at the sediment-water boundary

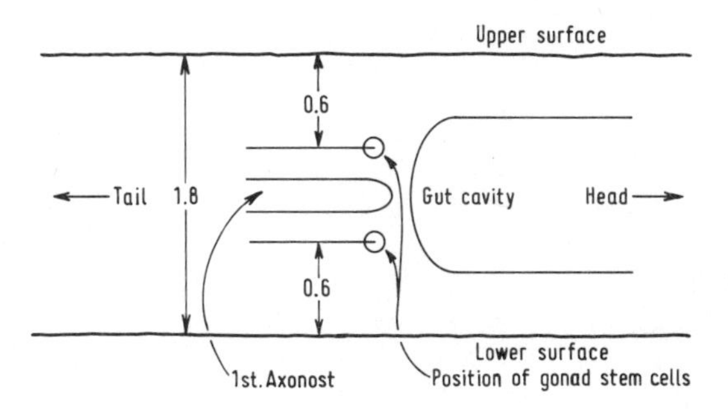

FIG.1. Simple geometry used to estimate the radiation dose rate to the plaice gonads (dimensions in cm).

from each source is $0.5D_\gamma(\infty)$ evaluated for the activity levels in sediment and water. This would be the dose-rate for a fish such as the plaice, which spends much of its time on or very close to the sea bed; for a trout in midwater and more than one metre from either the sediment or the surface, the γ-ray dose-rate from the sediment becomes negligible and the γ-ray dose-rate from the water approaches $D_\gamma(\infty)$ for water. In calculating the γ-ray dose-rate from the sediment at Windscale, account has been taken of the finite thickness of the source and the variation of activity with depth in the sediment [44]. Although there is no equivalent information available for the bed of Lake Trawsfynydd, it has been assumed that similar factors apply. The dose-rates to fish at these two sites, calculated on the basis of these simple models, are given in Tables VI to X.

5. DISCUSSION

For both marine and freshwater fish the internal activity contributes the greater proportion of the total background dose-rate, with the α-emitting nuclides ^{210}Po and ^{226}Ra respectively being the main sources and ^{40}K making up most of the balance. In the case of the α-activity, the assumption of uniform distribution within the fish or even within discrete organs can result in underestimates of the local dose-rates if differential accumulation occurs. Of the natural radionuclides present in sea water only ^{40}K contributes significantly to the exposure, either of the whole body from γ-radiation or the body surface from β-radiation. In fresh-water environments it is the occasional presence of relatively high levels of ^{222}Rn and its daughters (assumed to be in equilibrium) in the water which can result in significant exposure. The underlying sediment is an important source of irradiation of fish which live on, or spend much time very close to, the bed of the sea, lake or river. Cosmic radiation is also a significant source of exposure for fish occupying surface or shallow waters, the dose-rate being 4 μrad h^{-1} at the surface, falling to 0.5 μrad h^{-1} at a depth of 20 m.

In both Lake Trawsfynydd and the north-east Irish Sea the dose-rate from the waste radioactivity in the water is unimportant compared with that from internal contamination or from activity adsorbed on to sediment.

TABLE VI. DOSE RATES ($\mu rad \cdot h^{-1}$) TO MARINE AND FRESHWATER FISH FROM NATURAL RADIOACTIVITY IN THE WATER

Isotope	Marine fish			Freshwater fish	
	γ-ray dose-rate	β-ray dose-rate		γ-ray dose-rate	β-ray dose-rate at the surface
		at the surface	at 0.6 g cm^{-2}		
^{3}H	-	$(0.4-1.9) \times 10^{-5}$	-	-	$(0.4-1.1) \times 10^{-4}$
^{14}C	-	$(0.9-1.0) \times 10^{-5}$	-	-	-
^{40}K	5.3×10^{-2}	0.16	-	$(0.3-22) \times 10^{-4}$	$(0.5-32) \times 10^{-4}$
^{87}Rb	-	2.8×10^{-4}	-	-	2.4×10^{-6}
^{238}U + daughters	3×10^{-5}	1.0×10^{-3}	1.7×10^{-5}	$(0.3-85) \times 10^{-6}$	$(0.4-150) \times 10^{-5}$
^{234}U	-	-	-	-	-
^{230}Th	-	-	-	-	-
^{226}Ra	-	-	-	-	-
^{222}Rn + daughters	3.9×10^{-5}	1.6×10^{-5}	5.4×10^{-7}	$(0.8-710) \times 10^{-3}$	$(0.2-150) \times 10^{-3}$
^{210}Pb + ^{210}Bi	-	$(0.4-2.9) \times 10^{-5}$	-	-	$(0.1-1.5) \times 10^{-4}$
^{210}Po	-	-	-	-	-
^{232}Th	-	-	-	-	-
^{228}Ra-^{228}Ac	$(0.13-13) \times 10^{-5}$	$(0.43-43) \times 10^{-6}$	$(0.12-12) \times 10^{-8}$	-	-
^{228}Th + daughters	$(0.4-5.9) \times 10^{-6}$	$(1.5-23) \times 10^{-7}$	$(0.1-2.0) \times 10^{-8}$	-	-
^{235}U	1×10^{-5}	-	-	$(0.1-30) \times 10^{-6}$	-

TABLE VII. DOSE RATES ($\mu rad \cdot h^{-1}$) FROM NATURAL RADIOACTIVITY IN SEDIMENTS

Isotope	γ-ray dose-rate at sediment surface	β-ray dose-rate at sediment surface	β-ray dose-rate at 0.6 g cm^{-2} above sediment surface (i.e. at plaice gonad)
^{238}U + daughters in equilibrium	0.3-4	0.3-4	(0.6-7) x 10^{-2}
^{232}Th + daughters in equilibrium	0.7-6	0.3-2	(0.2-1.5) x 10^{-2}
^{40}K	0.5-6	1-15	-

TABLE VIII. DOSE RATES ($\mu rad \cdot h^{-1}$) TO MARINE AND FRESHWATER FISH FROM INTERNAL NATURAL RADIOACTIVITY

Isotope	Marine fish		Freshwater fish	
^{3}H		(0.6-3.5) x 10^{-5}		(0.6-1.8) x 10^{-4}
^{14}C		4.5 x 10^{-2}		4.5 x 10^{-2}
^{40}K		2.5		3.5
^{87}Rb		9.8 x 10^{-3}		-
^{226}Ra		(0.2-5.2) x 10^{-2}		10-36
^{210}Pb-^{210}Bi		(1.7-20) x 10^{-4}		(0.8-40) x 10^{-4}
^{210}Po	Flesh	(0.5-160) x 10^{-2}	Flesh	(0.4-12) x 10^{-1}
	Stomach	2.3-29	Stomach	-
	Liver	2.3-10	Liver	0.8-5.7
	Bone	0.2-2.5	Bone	(0.6-15) x 10^{-1}

TABLE IX. DOSE RATES ($\mu rad \cdot h^{-1}$) IN THE ENVIRONMENT AT LAKE TRAWSFYNYDD

Isotope	Water		Sediment				Internal	
	γ-ray dose-rate	β-ray dose-rate at fish surface	Mud		Peat		Trout	Perch
			γ-ray dose-rate at sediment surface	β-ray dose-rate at sediment surface	γ-ray dose-rate at sediment surface	β-ray dose-rate at sediment surface		
^{60}Co	-	-	2.2×10^{-2}	2.1×10^{-3}	4.4×10^{-3}	4.3×10^{-4}	-	-
$^{90}Sr-^{90}Y$	-	$(3.8-8.1) \times 10^{-3}$	-	-	-	-	0.28-0.35	0.52-0.80
$^{125}Sb-^{125m}Te$	-	-	0-0.19	0-0.15	$(1.4-8.7) \times 10^{-2}$	$(1.1-7.0) \times 10^{-2}$	-	-
^{134}Cs	$(0.3-2.5) \times 10^{-2}$	$(0.2-1.3) \times 10^{-3}$	$(0.3-2.2) \times 10^{-1}$	$(0.7-5.3) \times 10^{-2}$	$(1.5-8.1) \times 10^{-2}$	$(0.4-2.0) \times 10^{-2}$	0.7-2.8	0.9-5.1
^{137}Cs	$(0.4-4.5) \times 10^{-2}$	$(0.6-6.8) \times 10^{-3}$	0.4-1.4	0.3-1.0	$(0.3-2.6) \times 10^{-1}$	$(0.2-1.9) \times 10^{-1}$	3.6-9.4	6.6-21
$^{144}Ce-^{144}Pr$	-	-	-	-	8.1×10^{-4}	5.4×10^{-2}	-	-

TABLE X. DOSE RATES ($\mu rad \cdot h^{-1}$) IN THE ENVIRONMENT AT WINDSCALE

Isotope	Water		Sediment			Internal
	γ-ray dose-rate	β-ray dose-rate at fish surface	γ-ray dose-rate at sediment surface	β-ray dose-rate at sediment surface	β-ray dose-rate of 0.6 g cm^{-2} above sediment surface (i.e. at plaice gonad)	
^{95}Zr-^{95}Nb	0.04-1.0	$(0.04-1.0) \times 10^{-1}$	25-3100	6.3-770	-	-
^{106}Ru-^{106}Rh	$(0.5-5.3) \times 10$	$(0.3-3.3) \times 10$	8.7-200	120-2900	19-430	-
^{134}Cs	-	-	-	-	-	0.1-0.4
^{137}Cs	$(0.9-7.8) \times 10^{-2}$	$(0.3-2.3) \times 10^{-2}$	1.4-10	0.8-5.6	-	0.3-0.9
^{144}Ce-^{144}Pr	$(0.7-2.7) \times 10^{-3}$	$(2.0-7.7) \times 10^{-2}$	1.2-25	80-1700	5.6-120	-

TABLE XI. SUMMARY OF DOSE RATES ($\mu rad \cdot h^{-1}$) TO FISH FROM ENVIRONMENTAL RADIOACTIVITY

Source		Marine fish (Windscale)		Freshwater fish (Lake Trawsfynydd, trout)	
		1 m depth, more than 1 m from the sea bed	20 m depth, on the sea bed	1 m depth, more than 1 m from lake- or river-bed	3 m depth, on lake- or river-bed
NATURAL BACKGROUND					
Cosmic radiation		2.7	0.5	2.7	1.9
Internal activity		2.6-4.2	2.6-4.2	14-41	14-41
Water activity	Whole body γ	5.3×10^{-2}	2.7×10^{-2}	$(0.8-710) \times 10^{-3}$	$(0.4-360) \times 10^{-3}$
	Body surface β	0.16	0.16	$(0.3-150) \times 10^{-3}$	$(0.3-150) \times 10^{-3}$
Sediment activity	Whole body γ	-	1.5-16	-	1.5-16
	Body surface β	-	1.6-21	-	1.6-21
	Gonad β	-	$(0.8-8.5) \times 10^{-2}$	-	$(0.8-8.5) \times 10^{-2}$
Total		5.4-7.0	4.6-21	17-44	17-59
RADIOACTIVE WASTE					
Internal activity		0.4-1.3	0.4-1.3	4.6-12	4.6-12
Water activity	Whole body γ	0.06-1.1	0.03-0.55	$(0.7-7.0) \times 10^{-2}$	$(0.4-3.5) \times 10^{-2}$
	Body surface β	0.06-0.53	0.06-0.53	$(4.6-16) \times 10^{-3}$	$(4.6-16) \times 10^{-3}$
Sediment activity	Whole body γ	-	36-3300	-	0.6-2.3
	Body surface β	-	210-5400	-	0.09-1.2
	Gonad β	-	25-550	-	-
Total		0.5-2.4	36-3300	4.6-12	5.2-14

Table XI gives a summary of the dose-rates received by fish from different sources in various environments. The totals do not include a contribution from external β-radiation because this varies with the geometry; it is clear, however, that this contribution can be significant, as in the case of the dose-rate to the plaice gonad from β-radiation from the sediment at Windscale [47].

6. In situ DOSIMETRY

At Windscale, the quantities of activity discharged result in a large area of the sea bed becoming contaminated. The values given in Table X are indicative of the highest dose-rates in the vicinity of the outfall. Fig. 2 shows the dose-rate contours at the seabed-seawater surface, based on the measured activity of seabed samples collected from a grid of 26 stations during September 1968. It can be seen that the dose-rate due to the contamination exceeds that from the natural background over an area of some 2 500 km^2. In this situation, in situ measurements of the radiation regime experienced by the plaice were possible, using lithium fluoride thermoluminescent dosimeters. The distribution of dose-rates indicated by the dosimeters on the underside of the plaice is given in Fig. 3. The results broadly confirm the estimates obtained from dose calculations, but show that the natural behaviour of the fish reduces the average exposure [47]. After account had been taken of the effects of fish behaviour and the attenuation of the average β-ray spectrum in the tissues between the gonads and the body surfaces, it was concluded that the average dose-rate to the gonads was 250 μrad h^{-1}.

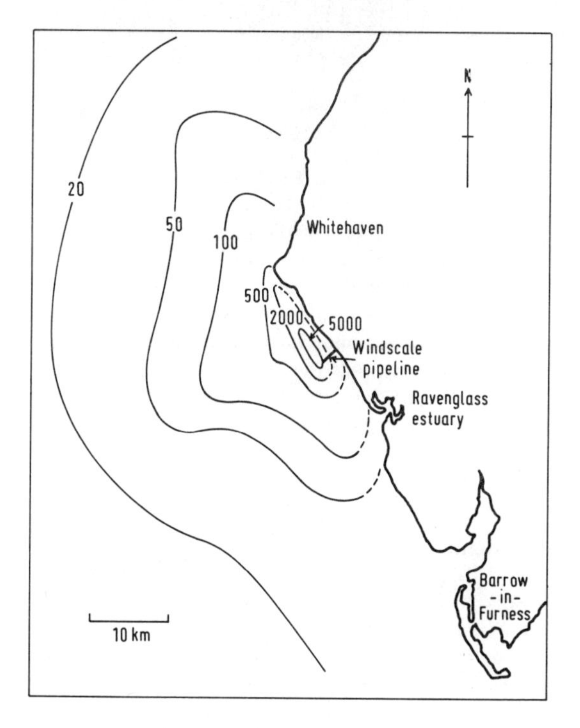

FIG. 2. Dose-rate contours on the sea bed in the vicinity of Windscale (dose rates in μrad·h^{-1}).

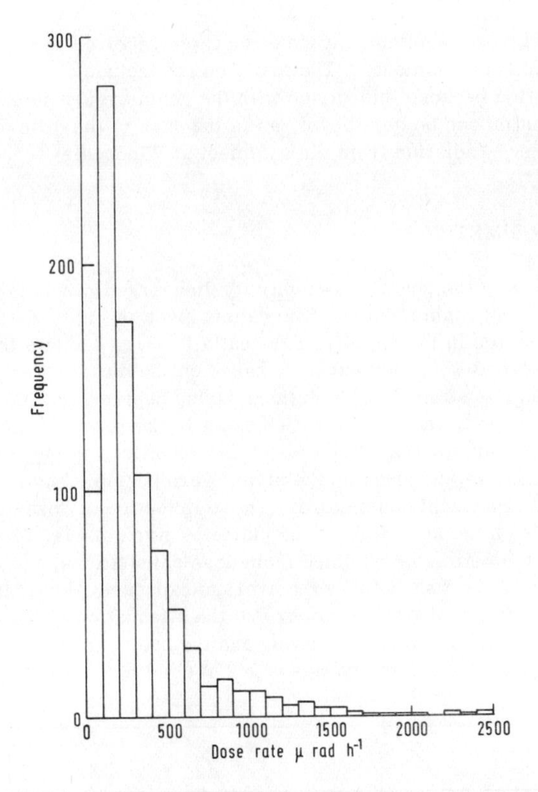

FIG. 3. Distribution of dose rates indicated by dosimeters on the underside of plaice in the north-east Irish Sea.

7. ASSESSMENT OF RADIATION EFFECTS

Fish have been shown to be one of the more radiosensitive components of aquatic ecosystems[48, 3, 49], and the mechanisms whereby chronic irradiation might result in resource damage include reduced lifespan, reduced fertility and increased mutation rate. All of these effects can act to give a reduction in fecundity at the population level and influence its long-term survival. The majority of experimental work has involved doses and dose-rates far in excess of those in contaminated environments, and any attempt to assess the possible effects in these situations requires considerable extrapolation.

7.1. Reduced lifespan

The acute irradiation $LD_{50/30}$ for fish appears to be in the range 1.1-5.6 krad[48, 3, 49], although it has been shown that other environmental parameters such as salinity[50] and particularly temperature[50, 51, 52] can interact with radiation to ameliorate or accentuate the response of a given species. It has also been found that, as the period of observation after acute irradiation is increased, a marked reduction in the median lethal dose occurs[49, 53]. These data do not, however, cover intervals equivalent to the normal lifespan of the fish and therefore cannot indicate what marginal effect might be expected from a small dose.

Fractionation or protraction of a given total dose over period of time produces less effect than the same dose given as an acute exposure, although here again temperature can exert a modifying effect[54]. Guppies (Poecilia reticulata) have been shown to survive integrated doses of chronic irradiation (at dose-rates of 1.3, 0.40, 0.17 rad h^{-1}) in excess of those which are lethal as acute exposures[55].[1]

From these results it would appear that the reduction in lifespan induced by irradiation is likely to be very small for fish in either of the environments being considered. The effect of any increased mortality on the number of fish available for exploitation can be demonstrated with a simple model. In conventional fish population dynamics the total mortality of the population is represented by the expression:

$$N = N_o \exp(-[F+M]t) ,$$

where N_o is the population size at t = o,

N is the number surviving at time t,

F is the instantaneous fishing mortality coefficient,

M is the instantaneous natural mortality coefficient.

From this it can be shown that the mean lifespan of a fish is 1/(F+M), and that, of the number of fish lost from population in any time interval, a fraction F/(F+M) were captured[56]. Thus a decrease in the mean lifespan following from an increase in the value of M due to irradiation would result in a proportionate decrease in the fish catch if F remains constant.

The reduction in lifespan would, if it occurred, also bring about a decrease in the fecundity of the population, since the reproductive potential of the individuals dying earlier, due to irradiation, will have been lost.

7.2. Reduced fertility

Large acute doses of irradiation in the range 4-8 krad have been shown to produce complete sterility in fish of both sexes[57, 58]. Doses of 500 and 1 000 rad to male guppies reduced the number of germ cells at all stages, with the radiosensitivity decreasing as development progressed from spermatogonia through to spermatozoa. After some temporary sterility, recovery took place as undamaged spermatogonia proliferated and repopulated the testes[59]. Irradiation of female loach (Misgurnus anguillicaudatus), with whole or partial body doses less than 1 krad, produced no marked histological effect within two months; 2 krad to the whole body, or to the anterior part of the body, produced degeneration of some large oocytes whether they had been irradiated or not, and it was concluded that the effect was a result of damage to the pituitary[60]. Chronic irradiation from internally-deposited radionuclides also produces sterility, although no estimates

---

[1] The dose-rates and integrated doses given in ref. 55 were based on measurements of the absorbed dose-rate in air at the beginning of the experiment. The values given above are derived from measurements using small thermoluminescent dosimeters placed in the aquaria and, by including the effects of attenuation in the water, source decay and average daily exposure time, provide more accurate estimates of the average radiation dose-rates experienced by the fish.

of the absorbed doses to the gonads are available[61, 62]. At a dose-rate of 1.3 rad h^{-1}, guppies ceased to produce young after an integrated dose of approximately 4 100 rad, and histological examination after an accumulated dose of 7 200 rad showed the ovaries and testes to be devoid of germ cells. At dose-rates of 0.40 and 0.17 rad h^{-1}, brood production ceased before that of the controls, but in every case the gross histological appearance of the gonads was normal even though the maximum accumulated doses were of the order of 7 400 and 4 100 rad respectively[55 and unpublished results]. Thus effects other than destruction of germ cell populations were responsible for the loss of fecundity, and it is possible that the induction of dominant lethal mutations was the cause. A population of mosquito fish (Gambusia affinis affinis) which had inhabited a contaminated lake for many generations under a dose-rate regime estimated to be of the order of 0.5 rad h^{-1} has been found to have a fecundity greater than that of a control population from a nearby uncontaminated lake. In this instance, therefore, the increased radiation level cannot have had any damaging effect on the fertility of the fish[63].

On the basis of these results it is certain that any reduction in fertility as a consequence of irradiation from the contamination at Windscale or Lake Trawsfynydd, if it exists at all, must be extremely small.

7.3. Increased mutation rate

Although it is known that radiation is mutagenic in fish, there are very few numerical data on induced mutation rates which could serve as a basis for estimating the magnitude of mutagenic effects in contaminated environments[64]. For the guppy, values of (0.4-11) x 10^{-5}/rad/gamete and 2.5 x 10^{-7}/rad/locus for the radiation-induced mutation rates have been obtained[65]. Similarly, Purdom has indicated that the specific locus mutation rate in the guppy is probably not greater than 2 x 10^{-7}/rad/locus[55]. It has recently been demonstrated that the radiation-induced specific locus mutation rate is directly proportional to the nuclear DNA content in organisms from widely different phyla[66]. Measurements of the nuclear DNA content of a wide variety of fish species (not including the guppy, for which no information appears to be available) give values in the range 0.48 to 4.4 pg/haploid genome, with a modal value of 1 pg[67]. This, with the data from reference 66, implies a specific locus mutation rate in the region of 7 x 10^{-8}/rad/locus; that is, somewhat lower than the experimental values which have been obtained for the guppy. If it is assumed that the number of loci coding for functional genes is 10^4 (Purdom, pers. comm.), then the mutation rate would be of the order of 7 x 10^{-4}/rad/zygote. Conservatively it can be assumed that all the mutants are unconditionally harmful and result in non-viable zygotes or embryonic death, giving an effective reduction in fecundity.

Under the dose-rate regimes being considered, the loss in fecundity due to an increased mutation rate of this magnitude will be negligible.

8. CONCLUSIONS

It is clear that the increased dose-rates experienced by trout in Lake Trawsfynydd and plaice in the north-east Irish Sea cannot lead to any significant or even detectable damage to the populations. In the case of the plaice, which receive measurable doses, the reduction in fecundity of the population through the mechanisms discussed above, if it exists at all, can only be marginal, and would easily be accommodated within the reserve reproductive capacity which has been demonstrated to exist[68].

REFERENCES

[1] Recommendations of the International Commission on Radiological Protection (Adopted 17 September 1965), ICRP Publication 9, Pergamon Press, London (1966).

[2] PRESTON, A., MITCHELL, N. T., Radioactive Contamination of the Marine Environment (Proc. Symp. Seattle, 1972), IAEA, Vienna (1973) 575.

[3] POLIKARPOV, G. G., Radioecology of Aquatic Organisms, North Holland Publishing Co., Amsterdam (1966).

[4] WOODHEAD, D. S., The assessment of the radiation dose rate to developing fish embryos due to the accumulation of radioactivity by the egg, Radiat. Res. 43 (1970) 582.

[5] WOODHEAD, D. S., Levels of radioactivity in the marine environment and the dose commitment to marine organisms, Radioactive Contamination of the Marine Environment (Proc. Symp. Seattle, 1972), IAEA, Vienna (1973) 499.

[6] ANON., Report of the United Nations Scientific Committee on the Effects of Atomic Radiation, General Assembly Official Records, 21st Session, Suppl. No. 14 (A/6314), United Nations, New York (1966).

[7] ANON., Report of the United Nations Scientific Committee on the Effects of Atomic Radiation, General Assembly Official Records, 17th Session, Suppl. No. 16 (A/5216), United Nations, New York (1962).

[8] BROECKER, W. S., GERARD, R., EWING, M., HEEZEN, B. C., J. Geophys. Res. 65 (1960) 2903.

[9] MIYAKE, Y., SUGIMURA, Y., UCHIDA, T., J. Geophys. Res. 71 (1966) 3083.

[10] MIYAKE, Y., SUGIMURA, Y., MAYEDA, M., J. Oceanographical Soc. Japan 26 (1970) 123.

[11] MIYAKE, Y., SUGIMURA, Y., YASUJIMA, T., J. Oceanographical Soc. Japan 26 (1970) 130.

[12] MOORE, W. S., SACKETT, W. M., J. Geophys. Res. 69 (1964) 5401.

[13] BROECKER, W. S., LI, Y. H., CROMWELL, J., Science 158 (1967) 1307.

[14] SZABO, B. J., Geochim. Cosmochim. Acta 31 (1967) 1321.

[15] KU, T. L., LI, Y. H., MATHIEU, G. G., WONG, H. K., J. Geophys. Res. 75 (1970) 5286.

[16] GOLDBERG, E. D., Radioactive Dating (Proc. Symp. Athens, 1962), IAEA, Vienna (1963) 121.

[17] RAMA, KOIDE, M., GOLDBERG, E. D., Science 134 (1961) 98.

[18] KAURANEN, P., MIETTINEN, J. K., Helsinki Univ. Dept Radiochemistry Annual Report, August 1969-August 1970, NYO-3446-14, paper 27 (1970).

[19] SHANNON, L. V., CHERRY, R. D., OWEN, M. J., Geochim. Cosmochim. Acta 34 (1970) 701.

[20] BEASLEY, T. M., Lead-210 in Selected Marine Organisms, Ph.D. Thesis, Oregon State University, Corvallis, Oregon, USA (1968).

[21] KAUFMAN, A., USAEC Rep. CU-3139-1 App. A (1967).

[22] FOLSOM, T. R., BEASLEY, T. M., Manuscript submitted to the committee on "Effect of Atomic Radiation on Oceanography and Fisheries" of the National Academy of Sciences (1968) (unpublished).

[23] SOMAYAJULU, B. L. K., GOLDBERG, E. D., Earth Planet. Sci. Letters 1 (1966) 102.

[24] ANON., Report of the United Nations Scientific Committee on the Effects of Atomic Radiation, General Assembly Official Records, 13th Session, Suppl. No. 17 (A/3838), United Nations, New York (1958).

[25] LIBBY, W. F., Tritium in the Physical and Biological Sciences (Proc. Symp. Vienna, 1961), IAEA, Vienna (1962) 1 5.

[26] FRANTZ, A., Limnologie der Donau 2 (1966) 84.

[27] PRESTON, A., JEFFERIES, D. F., DUTTON, J. W. R., Water Res. 1 (1967) 475.

[28] GOLDBERG, E. D., BROECKER, W. S., GROSS, M. G., TUREKIAN, K. K., "Marine chemistry", Ch. 5, Radioactivity in the Marine Environment, US National Academy of Sciences:National Research Council, Washington DC (1971).

[29] MIYAKE, Y., SUGIMURA, Y., TSUBOTA, H., "Content of uranium, radium and thorium in river waters in Japan", Ch. 11, The Natural Radiation Environment (ADAMS, J. A. S., LOWDER, M., Eds), University of Chicago Press, Chicago (1964).

[30] HOLTZMAN, R. B., "Lead-210 (RaD) and polonium-210 (RaF) in potable waters in Illinois", Ch. 12, The Natural Radiation Environment (ADAMS, J. A. S., LOWDER, M., Eds), University of Chicago Press, Chicago (1964).

[31] BLANCHARD, R. L., J. Geophys. Res. 70 (1965) 4055.

[32] MOXHAM, R. M., "Some aerial observations on the terrestrial component of environmental γ-radiation", Ch. 44, The Natural Radiation Environment (ADAMS, J. A. S., LOWDER, M., Eds), University of Chicago Press, Chicago (1964).

[33] MAHDAVI, A., "The thorium, uranium and potassium content of Atlantic and Gulf coast beach sands", Ch. 5, The Natural Radiation Environment (ADAMS, J. A. S., LOWDER, M., Eds), University of Chicago Press, Chicago (1964).

[34] VINOGRADOV, A. P., The Elementary Chemical Composition of Marine Organisms, Memoir 2, Sears Foundation for Marine Research, New Haven (1953).

[35] LOWMAN, F. G., RICE, T. R., RICHARDS, F. A., "Accumulation and redistribution of radionuclides by marine organisms", Ch. 7, Radioactivity in the Marine Environment, US National Academy of Sciences:National Research Council, Washington DC (1971).

[36] HOLTZMAN, R. B., Proc. 2nd National Symposium on Radioecology (Ann Arbor, 1967) (NELSON, D. J., EVANS, F. C., Eds), USAEC, Oak Ridge (1969) 535.

[37] CHERRY, R. D., SHAY, M. M., SHANNON, L. V., Nature, Lond. 228 (1970) 1002.

[38] DeBORTOLI, M., GAGLIONE, P., Health Phys. 22 (1972) 43.

[39] ANDERSON, J. B., TSIVGLOU, E. C., SHEARER, J. D., Radioecology (Proc. Symp. Colorado, 1961) (SCHULTZ, V., KLEMENT, A. W., Eds), Reinhold, New York (1963) 373.

[40] MITCHELL, N. T., Radioactivity in Surface and Coastal Waters of the British Isles 1971, Ministry of Agriculture, Fisheries and Food Tech. Rep. FRL 9 (1973).

[41] MITCHELL, N. T., Radioactivity in Surface and Coastal Waters of the British Isles 1968, Ministry of Agriculture, Fisheries and Food Tech. Rep. FRL 5 (1969).

[42] PENTREATH, R. J., WOODHEAD, D. S., JEFFERIES, D. F., Radionuclides in Ecosystems (Proc. 3rd National Symposium on Radioecology, Oak Ridge, 1971) (NELSON, D. J., Ed), USAEC, Oak Ridge (1973) 731.

[43] LOEVINGER, R., HOLT, J. G., HINE, G. J., "Internally administered isotopes", Ch. 17, Radiation Dosimetry, 1st ed. (HINE, G. J., BROWNELL, G. L., Eds), Academic Press, New York (1956).

[44] JEFFERIES, D. F., Environmental Surveillance in the Vicinity of Nuclear Facilities (Proc. Symp. Augusta, 1968) (REINIG, W. C., Ed), Charles C. Thomas, Illinois (1970) 205.

[45] KAUFMAN, A., TRIER, R. M., BROECKER, W. S., FEELY, H. W., J. Geophys. Res. 78 (1973) 8827.

[46] LOEVINGER, R., JAPHA, E. M., BROWNELL, G. L., "Discrete radioisotope sources", Ch. 16, Radiation Dosimetry, 1st ed. (HINE, G. J., BROWNELL, G. L., Eds), Academic Press, New York (1956).

[47] WOODHEAD, D. S., Health Physics 25 (1973) 115.

[48] DONALDSON, L. R., FOSTER, R. E., "Effects of radiation on aquatic organisms", Ch. 10, The Effects of Atomic Radiation on Oceanography and Fisheries, US National Academy of Sciences:National Research Council, Washington DC (1957).

[49] WHITE, J. C., ANGELOVIC, J. W., Chesapeake Sci. 7 (1966) 36.

[50] ANGELOVIC, J. W., WHITE, J. C., DAVIS, E. M., Annual Report of the Bureau of Commercial Fisheries Radiobiological Laboratory, Beaufort, N.C. for the Fiscal Year ending June 30 1967, US Dept Int., US Fish and Wildl. Serv. Bur. Comm. Fish. Circ. No. 289 (1968) 35.

[51] BLAYLOCK, B. G., MITCHELL, T. J., Radiat. Res. 40 (1969) 503.

[52] EGAMI, N., ETOH, H., Radiat. Res. 27 (1966) 630.

[53] SHECHMEISTER, I. L., WATSON, L. J., COLE, V. W., JACKSON, L. L., Radiat. Res. 16 (1962) 89.

[54] ETOH, H., EGAMI, N., Radiat. Res. 32 (1967) 884.

[55] PURDOM, C. E., WOODHEAD, D. S., Genetics and Mutagenesis of Fish (Proc. Symp. Munich, 1972) (SCHRÖDER, J. H., Ed), Springer-Verlag, Berlin (1973) 67.

[56] GULLAND, J. A., Manual of Methods for Fish Stock Assessment, Part 1, Fish population analysis, FAO, Rome (1969).

[57] KONNO, K., EGAMI, N., Annot. Zool. Japan 39 (1966) 63.

[58] EGAMI, N., HYODO, Y., Annot. Zool. Japan 38 (1965) 171.

[59] KOBAYASHI, J., YAMAMOTO, T., Tokushima J. Exp. Med. 18 (1971) 21.

[60] EGAMI, N., AOKI, K., Annot. Zool. Japan 39 (1966) 7.

[61] SRIVASTAVA, P. N., RATHI, S. K., Experientia 23 (1967) 229.

[62] YOSHIMURA, N., ETOH, H., EGAMI, N., ASAMI, K., YAMADA, T., Annot. Zool. Japan 42 (1969) 75.

[63] BLAYLOCK, B. G., Radiat. Res. 37 (1969) 108.

[64] SCHRÖDER, J. H., Genetics and Mutagenesis of Fish (Proc. Symp. Munich, 1972) (SCHRÖDER, J. H., Ed), Springer-Verlag, Berlin (1973) 91.

[65] SCHRÖDER, J. H., Mutation Res. 7 (1969) 75.

[66] ABRAHAMSON, S., BENDER, M. A., CONGER, A. D., WOLFF, S., Nature, Lond. 245 (1973) 460.

[67] HINEGARDNER, R., ROSEN, D. E., Amer. Naturalist 106 (1972) 621.

[68] SIMPSON, A. C., Ministry of Agriculture, Fisheries and Food, Fishery Investigations, Series II 22 (8) (1959) 30.

MISCELLANEOUS SOURCES OF IONIZING RADIATIONS IN THE UNITED KINGDOM: THE BASIS OF SAFETY ASSESSMENTS AND THE CALCULATION OF POPULATION DOSE

A.D. WRIXON, G.A.M. WEBB
National Radiological Protection Board,
Harwell, Didcot, Berks,
United Kingdom

Abstract

MISCELLANEOUS SOURCES OF IONIZING RADIATIONS IN THE UNITED KINGDOM: THE BASIS OF SAFETY ASSESSMENTS AND THE CALCULATION OF POPULATION DOSE.

In the report miscellaneous sources are defined as material sources that might involve the general public. This definition includes such diverse sources as isotopic batteries for cardiac pacemakers, by-product gypsum used as a building material and luminous devices. The most common miscellaneous sources in the UK are reviewed and compared. The current UK legislation affecting the control of miscellaneous sources is described, together with relevant international recommendations. Since to evaluate the desirability of a given use the complete detriment from a source is required, a general method for calculating the population dose from a gamma-emitting source is described. It is concluded that even for a source emitting high energy gamma radiation under extreme conditions of population distribution, only the doses to persons in the immediate vicinity of the source need be included to obtain the collective dose with sufficient accuracy. On this basis the current average UK gonad dose from miscellaneous sources is estimated to be in the range 0.2 - 1.0 $mrad \cdot yr^{-1}$.

INTRODUCTION

The main sources of population exposure to ionising radiations are:

(a) Natural background radiation

(b) Medical applications of ionising radiations

(c) Fallout from nuclear weapons tests

(d) Effluent discharges from the nuclear power industry

(e) Occupational exposure

(f) Miscellaneous sources of radiation

In its broadest sense the term "miscellaneous sources" includes all forms of population exposure to ionising radiations not included within the other five categories. Most of these 'sources' are consumer products containing radioactive material. In the United Kingdom the term was first used in a report by the Medical Research Council /1/ in 1956 and is assumed in accordance with that report to mean material sources which might involve the general public.

It is the purpose of this paper to discuss the radiological safety assessments used in evaluating miscellaneous sources in the United Kingdom and the present contributions to individual and collective doses from these sources. Transport of radioactive material as a contributor to public exposure may be considered a miscellaneous source but is not included in this paper.

THE BASIS OF SAFETY ASSESSMENTS

(a) International Recommendations

In recent years the approach to miscellaneous sources has been formalised by a number of publications from the International Commission on Radiological Protection (ICRP) /2,3/ and the Nuclear Energy Agency (NEA) of OECD /4/.

Several assessment criteria can be extracted from the ICRP publications:

1) Individual and population doses should be within the appropriate ICRP dose limits

2) Doses from unnecessary exposures should be avoided

3) Doses should be as low as is reasonably achievable, economic and social considerations being taken into account

4) Doses should be justifiable by the expected benefits of the procedures.

The last three criteria are an enunciation of the principles of cost-benefit analysis applied to radiation protection. In general the application of these principles is a two-stage procedure:

1) a basic policy decision has to be made whether a particular application is justifiable, weighing doses against expected benefits,

2) the manner in which approved activities are carried out must be optimised so that resultant exposures to radiation are as low as is reasonably achievable.

This second stage is the process of differential cost-benefit analysis described in ICRP Publication 22 /3/. For miscellaneous sources the basic policy decision is generally the more important stage. Once this decision has been made it is not difficult to keep doses as low as is reasonably achievable through selecting the least hazardous radionuclide in terms of type, activity, chemical and physical form and half life consistent with the useful life of the product.

In keeping with this two-stage procedure the NEA document /4/ provides a useful guide defining the general radiation protection and safety considerations to be taken into account in authorising the distribution, use and disposal of products containing radionuclides and intended for the general public. In particular it concerns all products exempt from the normal regulations governing radionuclides in which the presence of radionuclides is intentional. It specifically excludes isotopic batteries and products in which the presence of radionuclides is unintentional and is therefore somewhat limited in its application to miscellaneous sources. The major prohibition is of the intentional addition of radionuclides to products intended for ingestion, inhalation or application to man, i.e. foodstuffs, beverages and cosmetics. This prohibition is frequently extended to toys, articles for personal adornment and substances for domestic use. Consideration of other proposals is governed by two policy considerations:

1) Approval of a product should be contingent upon an adequate demonstration that the radioactive product performs a

function which can be fulfilled only by a radioactive method or so fulfilled that the radioactive method has clear advantages over any other practical method and upon a justification for the use of the specific radionuclide selected;

2) Generally the radiation dose to the average individual user and to the population from all exempt products should not exceed a small fraction (a few per cent) of applicable ICRP limits.

As a guide to basic policy decision making each level of benefit has been tentatively allocated a certain fraction of the individual and population dose limits (Table I). This is a direct application of the first stage of cost-benefit analysis (the term 'risk/benefit' in the NEA guide is used with the same meaning as 'cost-benefit' by ICRP). Where the benefit is easily recognised and the risk (cost) small, decisions present little problem but where both the risk and benefit are small, which is the more typical situation, decision making is more difficult.

To obtain a complete evaluation of the risk from a single product the Guide states that the external radiation dose or dose commitment to individuals resulting from the following three categories of exposure must be taken into account:

Category 1. Exposure resulting from normal use and disposal of the product.

Category 2. Exposure resulting from normal handling of quantities of the product while engaged in activities related to marketing, distribution, installation or servicing of the product.

Category 3. Exposure due to credible abuse, accidents or fire.

The Guide also recommends calculating the possible doses to individual members of the public from multiple products and the genetic and somatic population doses.

The NEA has been considering several specific applications of radionuclides in consumer products based on this guide. The Radiation Protection Standards for Gaseous Tritium Light Devices have recently been published /5/. The NEA has also been drafting radiation protection standards for the design, construction, testing and control of radioisotopic cardiac pacemakers. The first international recommendations, however, on consumer products containing radionuclides, in this case radioluminous timepieces, were formulated by a joint NEA/IAEA group of experts before the publication of the NEA Guide and published in the IAEA Safety Series /6/.

(b) UK Legislation and Control

Just as there is no single general statute concerned with radiological protection, there is no general Act concerned with miscellaneous sources. Certain Acts are nonetheless concerned with certain aspects.

Control over consumer products in general is exercised by the Home Office under the Consumer Protection Act 1961 /7/. This Act does not refer specifically to radioactive materials and only empowers the Secretary of State to impose requirements 'as are in his opinion expedient to prevent

TABLE I. DOSE APPORTIONMENT FOR EXEMPT PRODUCTS BASED ON RISK/BENEFIT CONSIDERATIONS[a]

Order of Benefit	Individual[b] Dose	Population[c] Dose
Outstanding benefit (such as life-saving devices)	< 0.1 ICRP dose limits	$< 10^{-4}$ ICRP dose limits
Safety and security devices Improve reliability or dependability of technical devices Special technical devices	< 0.01 ICRP dose limits[d]	$< 10^{-4}$ ICRP dose limits[d]
Lower order of benefit	< 0.001 ICRP dose limits[d]	$< 10^{-5}$ ICRP dose limits[d]

[a] The dose limits established in this chart were not derived from precise technical information. The figures have been estimated on the basis of normal use of the product. This table should be regarded as provisional only and is likely to undergo changes depending on developments in the nuclear industry, and on social and economic needs.

[b] Refers to a single article.

[c] Refers to the total distribution of the article under consideration.

[d] It is recommended that the total exposure from all exempt products (except for those with outstanding benefit) should not exceed 10% of the dose limit recommended by ICRP for individual members of the public and 1% of the population dose limit.

or reduce risk of death or personal injury'. The only occasion when the Home Office has published requirements relating to radioactivity in consumer products was for shoe-fitting fluoroscopes specifying radiation levels and the need for regular inspection /8/.

Some powers are available under the Radioactive Substances Act 1948 /9/ to control the sale and supply of radioactive substances; these powers have not yet been used.

The Radioactive Substances Act 1960 /10/ is primarily intended to regulate the accumulation and disposal of radioactive waste. The definition of "radioactive material" includes all artificial radionuclides and natural radionuclides in concentrations greater than those specified. Since this definition includes a number of minor uses and commonplace materials where the detailed control requirements of the Act would not be justified a number of exemption orders have been made /11/. Many of these exemption orders relate to miscellaneous sources in consumer products. Although it looks superficially possible to effectively prohibit a particular radioactive consumer product by refusing to either register it or grant an exemption order, the power to refuse is limited to reasons connected with the final disposal as radioactive waste; if this can be shown to be safe and in accordance with accepted practice registration cannot, in principle, be refused under the Act, however undesirable the proposed use may be. Nevertheless in the preparation of the exemption orders it was found possible in many cases to include restrictions on waste disposal grounds that were also deemed desirable on public health protection grounds.

Despite the seeming lack of statutory control the importance of exercising some form of voluntary control over miscellaneous sources was recognised very early in the UK and resulted in the setting up of the Miscellaneous Sources Panel (M.S.P.) of the Radioactive Substances Advisory Committee, appointed under the Radioactive Substances Act 1948. The functions of this panel together with its parent committee were transferred to the National Radiological Protection Board by the Radiological Protection Act 1970 /12/. In particular under the first section of the Act one of the Board's functions is "to provide information and advice to persons (including government departments) with responsibilities in the United Kingdom in relation to the protection from radiation hazards either of the community as a whole or of particular sections of the community".

The effect of this is that if the Board gives advice on a particular miscellaneous source the advice is taken very seriously by both manufacturers and government departments. In practice the Board's considered opinion has invariably been accepted. This advice is not given by the Board in isolation but after consultation with government departments, manufacturers and other interested parties and follows very closely the guidelines laid down by ICRP and NEA. In addition advice is also drawn from a number of non-statutory standards, where available /13,14,15/; that for radioluminous time measurement instruments /14/ follows closely the IAEA radiation protection standards /6/.

A major change in the present voluntary approval scheme for miscellaneous sources in the UK may be needed in view of the European Community (Euratom) Directives /16/ which require a statutory system of prior approval.

A SURVEY OF MISCELLANEOUS SOURCES IN THE UNITED KINGDOM

(a) Sources in which the radioactivity is a necessary and useful part of the product

Luminous devices

The production of luminosity was one of the first widespread applications of radioactivity and it still represents the largest single application of radionuclides in products available to the general public. The traditional radium-226 paints have been replaced by tritium and promethium-147 paints and even these are being superseded by the intrinsically safer gaseous tritium light sources.

The most common use of radioluminising is in timepieces. A survey carried out in the UK /17/ when the use of radium-226 was predominant indicated that the average radiation dose to the gonads from luminous clocks and watches was about 0.5 mrad yr^{-1}. Even with the use of the less toxic nuclides, particularly tritium, doses to individuals from normal usage have not been altogether eliminated. Tritiated water or simple tritiated organic molecules evolve slowly from the tritium paints leading to annual gonad doses to wearers estimated at 1 mrad /18/.

Doses to members of the public from other applications of luminising are more difficult to determine and are probably insignificant by comparison with watches. Instances of promethium-147 luminised compasses in which the paint has been inadequately sealed giving high levels of removable contamination and in some cases high surface dose rates have come to the attention of the NRPB. The Board has made recommendations for improved sealing although a better solution, also recommended, would be the replacement of the luminous compound with gaseous tritium light sources. Other instances involving old luminous instrument dials and accidents in which the luminous compound from watches and clocks is liberated are also difficult to assess. Assuming ingestion or inhalation of radioactivity equal to one-tenth of that of a watch containing the average permissible quantity it has been shown /6/ that annual doses would still be below the dose limit of 0.5 rem applicable to the whole body, the gonads and the blood-forming organs. Doses from the normal procedures of disposal of luminous articles would no doubt be negligible.

Although they contain higher activities gaseous tritium light sources (GTLS's) consisting of a sealed glass tube coated internally with phosphor and filled with tritium gas would only be expected to contribute significantly to individual and population exposure in the event of breakage through accident or disposal. Any external radiation is solely due to bremsstrahlung emission, which is strongly attenuated by the device in which the GTLS is incorporated. In the great majority of existing applications in the United Kingdom the dose rate at 1 cm, measured in air, from any point on the surface of a device incorporating a GTLS has not been found to exceed 0.1 mrad h^{-1}. In the event of a GTLS being broken the principal hazard to persons in the vicinity is from inhalation of the tritiated water which might be present. Assuming a tritiated water content of 2% and that a person in the vicinity inhales one tenth of the released vapour, it has been calculated that the smallest tritium content of a GTLS which could lead to a maximum permissible annual intake of tritiated water is 2.5 Ci /5/. For this reason 2 Ci per device has been adopted by NEA as the limit above which the use of GTLDs should be controlled. Few accident statistics are available but indications are that breakages are extremely rare.

Isotopic batteries

The British design of isotopic batteries for cardiac pacemakers uses plutonium-238 (about 200 mg) as the heat source for thermoelectric conversion. Assuming safe containment of the fuel the principal hazard is the gamma ray and neutron emission. A particularly important contributor to the gamma ray emission is the thallium-208 (2.62 MeV gamma) daughter of plutonium-236 present as an impurity. The neutron emission arises from spontaneous fission of plutonium-238, alpha particle reactions with impurity elements of low atomic number and gamma interactions with impurity beryllium. Organ doses to bearers from such batteries have been determined both by measurement and calculation /19/.

Smoke detectors

The present situation with smoke detectors is somewhat uncertain in view of currently changing practices. Originally detectors containing 15 μCi radium-226 were used but these have been replaced by detectors containing 60 μCi americium-241 sources. More recently lower activities of radium-226 (0.5 μCi - 5 μCi) and americium-241 (15 μCi) have been used. Under normal conditions of use external irradiation (X and γ) from these sources represents the principal hazard; evolution of radon gas is also of concern with radium but is not limiting with normal ventilation conditions.

Anti-Static devices

Anti-static devices for both industrial and public use usually contain polonium-210 sources; devices containing americium-241 sources are also manufactured but are not currently distributed in the UK. Since polonium-210 is essentially a pure α-emitter the potential hazard is inhalation and ingestion of the radioactive material resulting from leakage due to unsatisfactory sealing or in the event of an accident /20/. Anti-static brushes available to the general public are currently under investigation in the Board's laboratories.

(b) Sources which emit ionising radiation incidental to their function

A very wide range of miscellaneous sources are included in this category; only a few of the more important ones are mentioned here.

By-Product Gypsum as a building material

Of particular concern has been a proposal to use by-product gypsum produced as a waste product in the manufacture of phosphoric acid as a replacement for natural gypsum in building materials. Much of the radium contained in the parent rock is carried over into the by-product gypsum. An assessment of individual exposure in a typical house built from this material indicated the potential doses given in table II /21/. In terms of currently accepted exposures for members of the public and normal variations in radon content of air, the exposure due to β-radiation and radon are acceptable. The γ-radiation exposure is also well below ICRP recommended dose limits and is broadly comparable with the extent of the regional variations in dose due to natural background. The Board has approved the use of by-product gypsum subject to manufacturers keeping the Board informed of its usage so that the collective dose can be assessed from time to time.

Uses of Uranium and Thorium

The main uses of uranium are either as a pigment or in applications making use of its high density properties. Thorium is used in incandescent mantles and in certain optical lenses.

The principal hazard from the uses of uranium and thorium under normal conditions is the somatic dose from the beta-emitting daughters. In general doses received will be small due to sufficient attenuation distance between the device and the exposed person. In particular cases this may not apply, however. Some optical lenses containing

TABLE II. RADIATION EXPOSURE FROM BY-PRODUCT GYPSUM[a]

Type of exposure	Critical Organ	Magnitude
β -radiation	Skin and lens of the eye	<20 mrad yr^{-1}
γ -radiation	Gonads and bone-marrow	30 mrads yr^{-1}
Radon	Lungs	0.15 pCi l^{-1} or 0.04 WLM yr^{-1}[b]

[a] Radium concentration taken as 25 pCi g^{-1}

[b] A working level month (WLM) is the product of the length of exposure in working months and the concentration of radon daughters in working levels. A working level is any combination of radon daughters in one litre of air that will result in the ultimate emission of 1.3 x 10^{5} MeV of alpha energy from the decay of these daughters to RaD.

up to 30% by weight natural thorium may deliver doses to the lens of the eye. Another example which has been of particular interest to the Board is the practice of incorporating uranium oxide in dental porcelain to give it an ivory hue. Although uranium concentrations are low, ranging up to 0.2% by weight, the dose is delivered continuously to the surrounding tissue. If a suitable alternative to uranium could be found it would obviously be in keeping with the NEA Guide to discontinue the practice since the benefit is marginal.

Television receivers

Television, particularly colour receivers, are a source of population exposure involving large numbers of people. A recent assessment carried out in the UK estimates that the mean gonad dose is 0.4 μrads yr^{-1} assuming normal viewing habits /22/.

POPULATION DOSES FROM MISCELLANEOUS SOURCES

With the increasingly widespread distribution of miscellaneous sources and emphasis /3/ on the total detriment from a given source for purposes of cost-benefit analysis, it is necessary to consider the collective and population doses from these sources in addition to the individual doses as criteria in their assessment.

(a) Population Dose from Gamma Sources

One example of the general type of formulation which might be used in the calculation of collective dose from a distribution of gamma sources is:

$$S = \iint \frac{CqBe^{-\mu|\underline{r}'-\underline{r}|}\rho(r)\rho'(r')dAdA'}{|r' - r|^2}$$

Where S = collective dose to the population group considered

$$C = \frac{3.7 \times 10^{10} f}{4\pi K(E)}$$

f	=	branching ratio
$K(E)$	=	ratio of flux density to dose rate
q	=	activity in curies
B	=	γ-ray multiple scattering build-up factor
μ	=	photon attenuation coefficient
r	=	vector position of person
r'	=	" " " source
$\rho(r)$	=	population density distribution function
$\rho'(r')$	=	source " " "
A	=	area covered by population
A'	=	" " " source distribution

This general formulation permits an evaluation of collective doses from sources both within a given population or exterior to that population.

To obtain an indication of the way in which collective dose might change with size of population or distance of the source from a large population an unshielded 1 Ci source emitting one 1 MeV gamma ray per disintegration was considered in two situations:

1) with the source placed at the centre of a circular city of uniform population density. A figure of 5,000 people km^{-2} was taken as typical /23/;

2) with the source placed at various distances from a city with a population of one million.

The results for these two hypothetical cases are shown in tables III and IV. Two main conclusions may be drawn. The effect of the source on even a very large population in a distant town or city is negligible. More than half the collective dose is received by individuals within a radius of 20 metres. ICRP /3/ states that there is no need to pursue the summation beyond the point where it becomes clear that the further contribution to the sum will not change the estimate of population dose by more than a factor of about 2 or 3. It therefore is reasonable with a cardiac pacemaker, for example, to consider only the collective dose to those living and working in close proximity (i.e. a few metres) to the bearer, particularly the spouse. For miscellaneous sources a critical group approach is still acceptable for external exposure under normal conditions and is used here to estimate collective doses from some of the miscellaneous sources surveyed above.

For cardiac pacemakers the γ-dose rate at 2 cms distance from the source used in the UK battery has been found to be 2.9 $mrads^{-1}$ /19/. Assuming an inverse square relationship holds and the spouse spends 8 h d^{-1} at a distance of 50 cms from the bearer, this gives an annual gonad dose of about 10 mrads. If there were eventually 10,000 bearers of plutonium-238 fuelled pacemakers in the UK the population dose would then be about 100 man rads yr^{-1}.

TABLE III. COLLECTIVE DOSE FROM A POINT SOURCE AT THE CENTRE OF A CIRCULAR CITY SUMMED TO DIFFERENT RADII

Radius of summation (metres)	Number of people within that radius	Dose Rate at circumference (rads h^{-1})	Collective Dose (man rads yr^{-1})
5	0.4	2.0×10^{-2}	224
10	1.6	5.1×10^{-3}	320
20	6.3	1.3×10^{-3}	417
50	39	2.0×10^{-4}	543
100	157	4.4×10^{-5}	633
250	982	3.9×10^{-6}	722
500	3930	2.5×10^{-7}	750
1000	15,700	2.8×10^{-9}	754
2000	62,800	7.7×10^{-13}	754
3000	141,000	3.0×10^{-16}	754

Source Activity: 1 Ci emitting a 1 MeV gamma ray per disintegration (no allowance is made for attenuation by materials other than air).

Population Density : 5,000 people per square kilometre.

Integration Error : 1%.

The collective dose from smoke detectors is more difficult to assess. Assuming that the majority of detectors in use contain 60 μCi americium-241 sources, the gonad dose from the 60 keV gamma emission of this nuclide to a person at an average distance of 2 m from the source for 40 h/week would be about 0.3 mrads yr^{-1}. If 100,000 people were involved then the collective dose would be 30 man rads yr^{-1}.

(b) Population Doses from Tritium

Gaseous tritium light devices, GTLDs, which contain tritium above the exemption limit of 2 Ci are subject to recovery or disposal requirements /5/, others of lower activity will be discarded after the useful life of the device and the tritium remaining will be released as waste. This tritium will rapidly mix with the atmosphere and will be slowly converted

TABLE IV. COLLECTIVE DOSE FROM A POINT SOURCE AT A DISTANCE FROM A LARGE CENTRE OF POPULATION

Distance of Population Centre from Source (kms)	Collective Dose (man rads yr^{-1})
1	24.4
2	6.8×10^{-3}
3	2.6×10^{-6}
5	5.2×10^{-13}

Source Activity: 1 Ci emitting a 1 MeV gamma ray per disintegration (no allowance is made for attenuation by materials other than air).

Population : 10^6 people, assumed to be all at the same point.

TABLE V. COLLECTIVE AND POPULATION DOSES FROM THE ANNUAL DISPOSAL OF 10^6 Ci TRITIUM

	U.K.[a] Collective Dose (man rad yr^{-1})	N. Hemisphere[b] Population Dose (man rad yr^{-1})
Initially	0.5	32
At equilibrium	9.4	600

[a] U.K. Population taken as 5×10^7

[b] N. Hemisphere Population taken as 3.2×10^9

TABLE VI. SUMMARY OF COLLECTIVE DOSES TO THE U.K. POPULATION FROM VARIOUS MISCELLANEOUS SOURCES

Source	Collective Dose to U.K. Population (man rads yr^{-1})
Gaseous Tritium Light Sources[a]	10
Luminised Clocks and Watches[b]	10,000-50,000
Cardiac Pacemakers[a]	100
Smoke Detectors	30
Television Receivers	20
By-Product Gypsum[c]	-

a Potential doses. See text.

b Depending on what proportion of the population wear luminous watches.

c Collective dose depends on extent of use.

to tritiated water. In this assessment it is assumed that 10^6 Ci tritium are released annually. This will eventually mix with the circulating waters of the Northern Hemisphere to a depth of 75 metres to give an average tritium concentration of 2 x 10^{-3} pCi g^{-1} at equilibrium. Calculations have shown that tritiated water in body fluids at an equilibrium concentration of 1 pCi g^{-1} delivers 10^{-4} rad yr^{-1} to the whole body including the gonads /18/. Collective doses to the UK population and population doses to the inhabitants of the Northern Hemisphere from this rate of disposal of tritium are given in table V.

A summary of potential or actual collective doses to the UK population from the miscellaneous sources surveyed is given in table VI.

CONCLUSION

Early assessments /1,17/ of individual and collective doses from miscellaneous sources have revealed that the levels are only a very small per cent of relevant ICRP dose limits and often trivial. Although the use of radioactive materials in products available to the general public is increasing, it is probable that the average annual gonad dose is no greater now than when these early assessments were carried out and is less than 1 mrad. The reasons for this are the gradual phasing out of radium-226 and its replacement by nuclides such as promethium-147 and tritium and the continual improvement of control. In the future an even greater influence of international controls can be expected. This should result in an easing of trade restrictions and a harmonisation of safety standards.

Since doses are usually low most assessments of the safety and desirability of a particular miscellaneous source have concentrated on the doses to individuals in the immediate vicinity of the source. It has been shown that this approach is justified for all sources when used normally and that collective and population doses calculated on this basis will be well within the accuracy limits suggested by ICRP.

REFERENCES

/1/ Medical Research Council, The Hazards to Man of Nuclear and Allied Radiations, Cmnd. 9780, HMSO London (1956).

/2/ Recommendations of the International Commission on Radiological Protection, ICRP Publication 9, Pergamon Press, Oxford (1966).

/3/ Recommendations of the International Commission on Radiological Protection, Implications of Commission Recommendations that Doses be kept as low as Readily Achievable, ICRP Publication 22, Pergamon Press, Oxford (1973).

/4/ European Nuclear Energy Agency, Basic approach for safety analysis and control of products containing radionuclides and available to the general public. Paris (1970).

/5/ Nuclear Energy Agency, Radiation Protection Standards for Gaseous Tritium Light Devices, NEA, Paris (1973).

/6/ International Atomic Energy Agency, Radiation Protection Standards for Radioluminous Timepieces, Safety Series No.23, IAEA, Vienna (1967).

/7/ The Consumer Protection Act 1961, HMSO London.

/8/ Home Office Requirements for the Control of Shoe-Fitting Fluoroscopes, 30 April 1963.

/9/ The Radioactive Substances Act 1948, HMSO London.

/10/ The Radioactive Substances Act 1960, HMSO London.

/11/ The Radioactive Substances, Exemption Orders England and Wales (Corresponding Orders exist for Scotland and N. Ireland), HMSO London.

/12/ The Radiological Protection Act 1970, HMSO London.

/13/ Specification for Self-Luminous Exit Signs, British Standard Institution, BS 4218: 1967.

/14/ Specification for Radioluminous Time Measurement Instruments Part 1. Instruments bearing radioactive luminous materials, British Standard Institution, BS 4333: Part 1: 1968.

/15/ The Protection of Structures Against Lightning, British Standard Institution, CP 326: 1965.

/16/ Basic Safety Standards for Protection against Ionizing Radiation, Commission of the European Communities, proposed amendment November (1973).

/17/ Medical Research Council, The Hazards to Man of Nuclear and Allied Radiations, A Second Report to the Medical Research Council, Cmnd.1225, HMSO London (1960).

/18/ United Nations Scientific Committee on the Effects of Atomic Radiation, Ionizing Radiations: Levels and Effects, Vol.1: Levels, U.N. New York (1972).

/19/ GIBSON, J.A.B., MARSHALL, M., DOCKERTY, J., In-Phantom Dosimetry for a Nuclear Heart Pacemaker, Proceedings of Second International Symposium on Power from Radioisotopes, Madrid, p.647, NEA (1972).

/20/ ROBERTSON, M.K. and RANDLE, M.W., Hazards from the Industrial Use of Radioactive Static Eliminators, Health Physics 26 (1974), 245.

/21/ O'RIORDAN, M.C., DUGGAN, M.J., ROSE, W.B. and BRADFORD, G.F., The Radiological Implications of Using By-Product Gypsum as a Building Material, NRPB-R7 (1972).

/22/ O'RIORDAN, M.C. and CASBOLT, P.N., X-rays from Domestic Colour Television Receivers in Britain. Nature 228 (1970), 420-421.

/23/ Britain 1973, An Official Handbook, Prepared by Central Office of Information p.16, London, Sept.1972.

DISCUSSION

R.F. BARKER: I just want to comment that, in the United States of America, the National Council of Radiation Protection and Measurements is preparing a report on population doses due to radioactive materials in consumer products. The report should be issued later in 1974. Also, the US Atomic Energy Commission is developing a method of calculating collective and individual doses using a computer code, into which the parameters are fed. Since the parameters are not well known in most cases, the initial results, which are also expected later this year, will be rather imprecise.

K.-J. VOGT: There appears to be a discrepancy in the judgement of doses. On the one hand we have nuclear plants such as power reactors, with rather strict limits, or fuel processing plants, where an α-emitter release resulting in 10^{-7} of the natural exposure is going to be reduced to zero release; on the other hand we have radiation exposure due to building materials such as by-product gypsum, for which 30 mrad/a are regarded as tolerable in the United Kingdom. Could you comment on this discrepancy in the light of the particular benefits of the different sources?

A.D. WRIXON: In the case of by-product gypsum, there is some justification for comparing the gonad dose with that from natural background. For instance, we have already heard (IAEA-SM-184/2) of measurements of doses ranging from 4.2 to 35.7 μR/h in houses in Germany, depending on the building materials; the highest values were in brick-built houses. In order to keep doses as low as possible should we therefore favour wooden houses or prefabricated houses and forbid brick ones? In the same way should we encourage people living in areas of higher background to move to areas of lower background?

One further point. The value of 30 mrad/a is the individual's gonad dose; the average gonad dose to the population or the population dose will depend on the extent to which by-product gypsum is used. It is probable, and the NRPB will be watching this closely, that the average gonad dose will increase only very slowly, say, over decades, and if eventually one home in ten is constructed of the material, the average gonad dose will only have increased by 3 mrad/a.

D. BENINSON: I think the case of building materials is a good example of the possibility of assessing the collective dose without knowing the actual distribution of individual doses.

As regards radon levels in houses, they depend very strongly on ventilation, rather than on the concentration of ^{226}Ra in the materials. What assumptions were made in the calculations shown in your slide?

A. D. WRIXON: The radon emanation from panels made of by-product gypsum was measured and for the assessment a ventilation rate of one air change per hour was assumed. This is probably a conservative assumption since in British homes the ventilation rate may be appreciably higher and the radon concentrations correspondingly lower.

H. P. JAMMET: You mentioned the contribution of the habitat to the collective dose to the population. It must be realized that this contribution should be included with the controllable artificial irradiation sources, and not with natural irradiation. In the future, cost-benefit analyses should be performed for the habitat.

A. D. WRIXON: We have not actually carried out a rigorous cost-benefit analysis with by-product gypsum, although the building industry has given us assurances of substantial benefit from its use. However, it is probable that such an analysis would be feasible, unlike the case with many other miscellaneous sources. One could balance the costs of disposal of the by-product gypsum, the mining of natural gypsum, and so on, against the possible detriment associated with the use of this otherwise waste material.

K. J. KOREN: You mentioned shoe-fitting by means of X-rays. Have you any idea of the risk-benefit weighting of this kind of irradiation? In Norway we considered the benefit so slight that we have prohibited shoe-fitting by X-rays except under medical supervision.

A. D. WRIXON: The problem of shoe-fitting fluoroscopes was considered in the United Kingdom many years ago and in 1963 the Home Office, a body with statutory responsibility for consumer products, published requirements which, among other things, limit the dose rates in the principal beam (see Ref. [8]). Since then the numbers of machines in use have been substantially reduced.

ICRP Publication 15 recommends that the use of X-rays for shoe-fitting purposes should be restricted to medical examinations and that they should be prohibited for commercial shoe-fitting purposes. It is my personal feeling also that there is little or no benefit to be gained from them for normal shoe-fitting.

S. O. W. BERGSTRÖM: The ventilation rate no doubt does reduce the lung dose from gypsum material in buildings, but not the genetically significant dose.

A. D. WRIXON: Ventilation in the assessment of doses from by-product gypsum was only considered in the case of radon, when the lung is the critical organ. The gonad doses were calculated for γ-emission from the radium-226 in the walls, and ventilation would have no effect on this.

M. B. BILES: Perhaps I could mention the situation in Grand Junction, Colorado, which is an example of a community dose resulting from building materials, in this case uranium tailings used as sub-base for concrete floors and foundations. As a result of concern over gamma and radon levels in buildings, Congress appropriated $5 million and the State of Colorado about $1.7 million for remedial action; this action primarily consists of removing the tailings and carrying out sealing against radon. The remedial programme, based on special guidelines prepared by the US Surgeon General, is well under way and should be completed in two years. Of some 18 000 buildings in the city, about 3500 have had tailings used in their construction.

P. CANDÈS: In Table V you give an equilibrium dose of 600 man·rads for the population of the northern hemisphere from the release of 10^6 Ci of tritium. I recall that Mr. Beninson gave much lower figures with a much bigger tritium release for the world population. Could you comment on this?

D. BENINSON: Perhaps I could answer Mr. Candès. The model used in my paper assumed total dispersion in the circulating waters of the hemisphere, and the results are in fact consistent with those of the paper just presented. The equilibrium values of Mr. Wrixon's Table V are equivalent to dose commitment per 10^6 Ci of discharge. Therefore 1 MW(e)·a, producing about 20 Ci, implies a collective dose commitment of the order of 10^{-2} man·rems.

S. AMARANTOS: What is your experience with the use of lightning conductors containing radioactive materials?

A. D. WRIXON: Lightning conductors with radioactive attachments are manufactured in the United Kingdom and in various other countries. Under the Radioactive Substances Act (1960) an Exemption Order exists which in effect allows lightning conductors to be installed on public buildings, etc., without the need for registration. Exemption is subject to strict conditions, however: limitation of activity, display of notices, control of disposal, notification of the competent authority.

The concept of lightning conductors with radioactive attachments was proposed early in this century by a scientist of repute and has since been accepted in certain countries. In the UK, however, there has been some doubt as to the value of these devices, and the drafting committee for a British Code of Practice concerned with lightning protection (Ref. [15]) included a statement to the effect that, on the basis of the evidence considered, the committee was unable to recommend the use of means of artificially increasing the lightning attraction of protection systems. A recent publication, "Lightning Protection" by R. H. Golde, states that "a radioactive lightning conductor is no better than a conventional conductor installed in the same position".

In the United Kingdom scientific community there is therefore some doubt as to the effectiveness of these devices, and it is my opinion that a full scientific study would be useful finally to resolve the problem.

F. D. SOWBY: Since there seems to be considerable doubt as to whether radioactive lightning conductors are more efficacious than the non-radioactive type, is there any justification for their use?

A. D. WRIXON: The approach to a new proposal differs in some respects from the approach to an established tradition. Present-day manufacture of radioactive lightning conductors is based upon scientific concepts evolved many years ago; nowadays, there is a body of expert opinion in the United Kingdom which feels that there is no advantage to be gained from this use of radioactive material. It is my own personal feeling, therefore, that if this application were to be proposed for the first time today, it would not be considered justified until it had been conclusively demonstrated that there are real advantages to be gained from it.

T. NIEWIADOMSKI: The only building material you considered in your paper is by-product gypsum. What about other industrial wastes, such as boiling slag and fly ash?

A. D. WRIXON: I regret that I have no information on these materials.

P. GRANDE: Are beta lights fabricated in the United Kingdom for use as floats?

A. D. WRIXON: Perhaps I should first point out that the term 'beta light' is a trade name; the correct term is 'gaseous tritium light source'.

Gaseous tritium light sources have been accepted for use in fishing floats provided the conditions given in the NEA standards (Ref. [5]) are met. Among other things these conditions stipulate that the sources shall always be contained in a device and shall not be directly accessible.

P. GRANDE: I think the NEA Health and Safety Committee agreed that this is a good example of what should not be permitted.

A. D. WRIXON: The NEA standards, at all events, (Ref. [5]) make no mention of this particular application; the standards only forbid the use of gaseous tritium light devices in toys, for personal adornment, or for frivolous purposes. If the NEA Health and Safety Committee has agreed that the float application should not be permitted, this view is certainly not reflected in the standards.

N. T. MITCHELL: I should just like to add, in this connection, that the use of radioisotope power-batteries has been considered in the United Kingdom, particularly for application in remote areas where they have a clear advantage over conventional power sources. One of the most obvious uses is as an aid to navigation — for warning lights in remote positions marking the entrance to navigable channels. The resulting population exposure will clearly be very low indeed.

A. MARTIN: To revert to the question of the dose commitment from tritium, there seems to be a large discrepancy between the estimates in your paper and those in my own paper (IAEA-SM-184/9).

A. D. WRIXON: The model used in my assessment was based on the assumption that the tritium gas is released to atmosphere following disposal and is converted into tritiated water which then mixes with the circulating waters of the Northern Hemisphere, including the water in body fluids, and the method of dose assessment is similar to that adopted by Mr. Beninson in his paper (IAEA-SM-184/102). Our values and Mr. Beninson's should be in fairly good agreement if it is assumed that the tritium is disposed of directly to the sea as tritiated water. Your values for release to atmosphere do differ from mine by a factor of about 20, but this appears to be due to the contribution to the population dose from the more highly exposed local critical group assumed in your assessment. This contribution might be important where tritiated water is being discharged from a stack, since there may be some 'wash-out' into local drinking waters. With discharges of tritium gas, whether from a stack or following the disposal of gaseous tritium light sources, there is unlikely to be much local removal into drinking waters since tritium gas is only very slowly converted into tritiated water in the atmosphere. This justifies an assessment on a more general basis.

RADIATION DOSE TO POPULATION (CREW AND PASSENGERS) RESULTING FROM THE TRANSPORTATION OF RADIOACTIVE MATERIAL BY PASSENGER AIRCRAFT IN THE UNITED STATES OF AMERICA

R.F. BARKER, D.R. HOPKINS, A.N. TSE
United States Atomic Energy Commission,
Washington, D.C.,
United States of America

Abstract

RADIATION DOSE TO POPULATION (CREW AND PASSENGERS) RESULTING FROM THE TRANSPORTATION OF RADIOACTIVE MATERIAL BY PASSENGER AIRCRAFT IN THE UNITED STATES OF AMERICA.

Radiation exposures to passengers and crew members in passenger aircraft carrying packages of radioactive material are controlled by regulations that limit the radiation dose rate outside each package and the number and positioning of such packages as loaded in one type of aircraft. The actual radiation level in mrem/h at 3 feet from the external surface of the package is defined in the USA as the package transport index (TI). The regulation specifies that the TI for any package shall be no greater than 10. A table in the regulations specifies a maximum value for the sum of TI for the packages that can be loaded into a single aircraft, depending on the available separation distances between the surface of the packages to the partition of the passenger's compartment. Observation of TI and separation distance ensures that radiation levels in passenger and crew compartments are within predetermined levels.

Measurements were made of radiation dose rates in passenger seats and at crew stations after routine shipments of packages of radioactive material had been loaded normally by carrier personnel into the cargo compartments of regular flights. By analysing these dose rates and the corresponding TI of the packages loaded, a relationship between average radiation dose rate and total TI per flight was established. Statistics on the total TI per flight, number of passengers and crew, and duration of flights were gathered by 9 airlines for all their passenger flights from 33 airports across the country, and these data were extrapolated to account for all flights nationwide. The total TI on each flight was translated into average radiation dose rates in the passenger and crew compartments, and the data were used to estimate the annual cumulative dose from the transportation of radioactive material by passenger aircraft.

The methods of analysis are described, and the calculation used in arriving at the estimated annual cumulative dose is presented. The annual cumulative dose due to cosmic radiation at the elevation at which commercial aircraft fly is estimated and compared.

1. Introduction and Summary

Many packages containing radioactive material are moving through the world by various modes of transportation. It is estimated that close to a million such packages are transported in the United States alone each year. Most of the packages contain radiopharmaceuticals which, because of their short radioactive half-lives, require rapid shipment to hospitals and clinics for diagnostic and therapeutic treatment of patients. Almost all radiopharmaceutical packages in the U.S. are shipped by passenger aircraft because there is no other equally satisfactory mode of transport available.

This paper reports on a study to estimate the annual radiation exposures to passengers and crew in passenger aircraft carrying packages of radioactive material in the United States. The annual cumulative dose and other parameters were used to consider ways in which the regulations under which the radioactive materials are transported in passenger aircraft might be modified to reduce the exposure to a level as low as practicable.

Let us first examine why controls are necessary on stowage of packages of radioactive materials in the transportation of such packages by passenger aircraft. Assuming 800,000 packages are transported by passenger aircraft, and each package having dose rates up to maximum allowed under the packaging regulations, that is, 200 mrem/hr at surface and 10 mrem/hr at 3 feet from the surface. In the absence of carrier control, a package carried in the hold of an aircraft can be as close as about 2 feet from the gonads of a seated passenger directly above the package and about 3 feet, 5 feet, and 7 feet from the next rows of passengers. As a rough estimate, this could result in a maximum exposure rate of 16 mrem/hr to the passenger directly above the package and an average exposure rate of about 5 mrem/hr to 35 passengers in nearby seats. On a typical 2 hour flight then, a single package could impose a cumulative radiation dose exceeding 300 mrem. 800,000 such packages might have the potential to impose a population radiation dose of 240,000 man-rem per year. While only a small percentage of packages being transported exhibit maximum allowable radiation levels, and all packages would not be stowed in the worst available position even in the absence of regulatory controls, a potential exposure of this magnitude demands an adequate system of control.

It is possible but not practical to provide radiation shielding in a package so that little or no radiation is emitted from the package. The U.S. and IAEA regulations permit up to 200 mrem/hr at the surface of a package and 10 mrem/hr at 3 feet or 1 meter from the package. The radiation levels permitted outside of packages in the transportation of radioactive materials require that controls be exercised to limit radiation exposure of persons who are associated with the transportation.

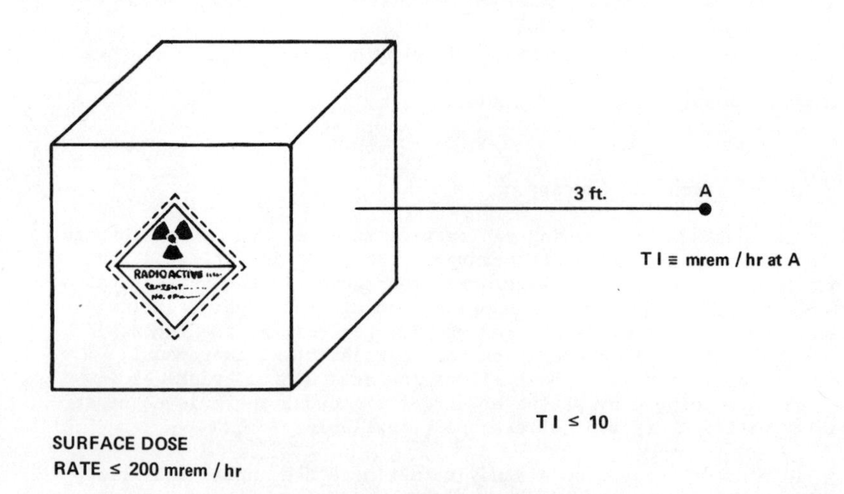

FIG.1. Limitations on package radiation levels.

These controls are necessary to ensure that no person receives a radiation dose in excess of the permissible dose limits.[1,2] A comprehensive discussion of the problems associated with the transportation of radioactive materials is given by R. D. Evans.[3] More recently, general reviews of IAEA regulations and Euratom basic safety standards involved in transportation are given by Gibson[4] and Failla.[5] Gibson's book discusses technical aspects of the IAEA's 1964 transport regulations and the means in which these regulations are incorporated into national transport regulations. Failla's report contains detailed analysis of the application to the transport of radioactive materials of the classification of workers and population groups, definitions of controlled and protected areas, and types of radiation exposure controls recommended in the basic radiation safety standards of Euratom and international standards issued by ICRP, IAEA, and OCDE-ENEA.

For passenger-carrying aircraft, allowable radiation levels are not specified in the regulations or measured by the carriers, but are controlled by regulatory limits on packages and conditions of loading. The actual radiation level in mrem/hr at 3 feet from the external surface of each package is called the package transport index (TI), and that number is specified on the package label for purposes of carrier control.[6] Figure 1 shows a typical package and the limitations of package external radiation levels. Because metric units are used by most nations, one meter is used by IAEA in defining transport index. The U.S. is expected to convert to the one meter distance to determine TI within the next year.

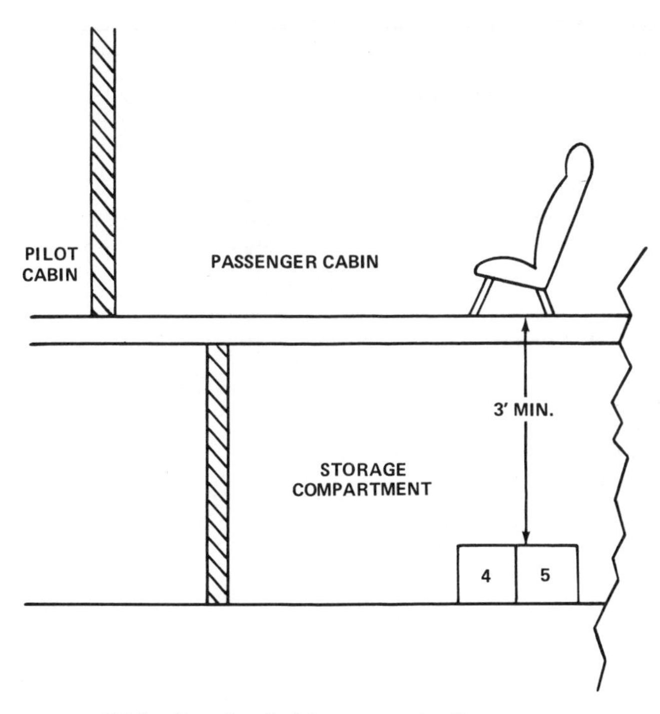

FIG. 2. Examples of minimum separation distances.

Regulations applicable to the air carrier designate minimum separation distances between the surface of the package and the partition of the passenger compartment, varying with the total cumulative TI of the packages to be placed in the cargo compartment with a maximum total TI of 50.[7] Observation of these separation distances ensures that radiation levels in passenger compartments are within predetermined levels. Figure 2 shows examples of separation distances.

There are approximately five million aircraft departures annually in the United States. For assessing the transportation picture for the entire U.S., data on flights and packages from a large number of airports and over extended periods of time would be necessary to obtain accurate results. Because of constraints in time and sampling resources, methods of estimating the annual cumulative doses to passengers and crew members were developed to simplify the collection of data. Resources were allocated in areas with a higher-than-average concentration of radioactive traffic. The results obtained in this work are considered to be the best estimates possible under the limitations of time and resources.

The gathering of data consisted of two separate parts. Measurement data which relates actual radiation levels on board passenger aircraft to the total TI stored in the baggage compartments was collected at Boston Logan and Chicago O'Hare Airports by health physics graduate students from Harvard and Northwestern Universities. This effort produced data on some 142 flights carrying radioactive material. Part two consisted of collection of data on the numbers of flights carrying packages of radioactive material, the total TI and passengers aboard those flights, and the flight times by nine major U.S. airlines through the cooperation and coordination of the Air Transport Association. The cooperation of all persons involved in the data-gathering effort, including many services by Federal Aviation Administration field and headquarters personnel, is attributed to the general desire for factual information in this area.

The analysis of the collected data show annual cumulative doses of 1400, 1, and 70 man-rem for passengers, pilots, and stewardesses, respectively, who are exposed to radiation emitted by packages of radioactive material carried in the passenger aircraft. By comparison, the annual cumulative doses to these same population groups due to cosmic radiation at the altitudes at which commercial aircraft fly are 2000, 3000, and 4000 man-rem, respectively; the corresponding doses to all air passengers, pilots, and stewardesses are 70000, 6000, and 8000 man-rem, respectively.

2. Method

Assuming the number of people receiving doses between H and H+dH is N(H), the cumulative dose, D, is given by

$$D = \int N(H)\, dH \qquad (1)$$

where the integration is carried over the dose distribution for all exposed occupants in passenger aircraft, including passengers, pilots, and stewardesses.

It is difficult to apply equation (1) to calculate cumulative dose because one does not know the function N(H). However, one might rewrite equation (1) into a summation form:

$$D = \sum_j D_j \quad (2)$$

where D_j is the cumulative dose for a single flight j. By summing up flights in the U.S. for one year, one obtains the annual cumulative dose as expressed by equation (1).

The cumulative dose for a single flight, D_j, can be calculated by the following equation:

$$D_j = \left(\sum_{i=1}^{3} R_i N_i \right)_j t_j \quad (3)$$

where R = average dose rate to gonads, in mrem/hour, at seat level for passengers or pilots or at 90 cm above floor for stewardesses in a single flight.

N = number of passengers (or pilots or stewardesses) in the flight

t = exposed time, in hours, in the flight.

subscript i denotes the three subgroups: passengers, pilots, and stewardesses.

Since the passengers are seated almost randomly over the entire cabin, one might assume that the average dose rate for passengers is the sum of dose rates of all passenger seats divided by the total number of seats in the cabin, i.e.,

$$R_{passenger} = \frac{\sum_k H_k}{k} \quad (4)$$

where H = dose rate at each passenger seat

k = total number of passenger seats

The dose rate at each passenger seat is dependent on the transport index and physical size of the package, the location in the aircraft at which the package is loaded, the shielding effect due to the aircraft structure, and the shielding effect of other cargo loaded between the package and the dose point. If several packages of radioactive materials are carried in a single aircraft, the dose rate, H, at each passenger seat is the sum of the dose rates contributed by each package, i.e.,

$$H = \sum_m (TI)_m \frac{C_m^2}{\vec{x}_m^2} (TF)_m \quad (5)$$

where TI = transport index of each package

$\vec{x}$ = distance between the source and the seat

C = distance between the source and a point which is 3 feet away from the surface of the package

TF = transmission factor due to the shielding effect of aircraft floor structure and other cargo loaded between the package and the dose point

subscript m denotes individual packages containing radioactive materials loaded in the aircraft.

The shielding effect of the aircraft floor structure has been investigated in a separate study carried out at the facilities of the Federal Aviation Administration at Oklahoma City. The study involved setting up packages of radioactive materials in cargo compartments of two types of passenger-carrying aircraft, a Douglas DC-9 and a Boeing 727, and the measurement of radiation levels produced in the aircraft to which passengers and crew might be subjected during the flight. The average transmission factor of the aircraft floor structure was determined to be 0.7 for the seat directly above the package. The total shielding effect should also include the effect of other cargo loaded between the passenger seat and the packages containing radioactive materials.

For pilots and copilots, the average dose rate, R_{pilots}, can be calculated by equations (4) and (5) except that the seats under consideration are in the cockpit. Since stewardesses spend most of their time walking up and down the aisle to serve passengers, the dose rate to which they are exposed is assumed to be the dose rate at 90 cm above the floor (approximately gonad height) averaged over the entire aisle.

Equations (2) to (5) could be used to calculate the annual cumulative dose if all parameters identified in these equations were known. However, it is difficult to collect information on the two parameters $\vec{x}$ and TF for regular flights because of the short aircraft turnaround time and other operational problems at the airport.

To avoid the need for determining these two quantities, direct radiation measurements were made for all seats in the passenger and crew compartments after routine shipments of packages had been loaded normally by carrier personnel into the cargo compartment of regular flights. Average dose rates for each flight were calculated. Limitations on resources and time, however, make it almost impossible to perform such a survey for all aircraft departures at every significant airport. As an alternative, effort was concentrated on radiation surveys at two major airports for an extended period of time during which data were collected on total TI per flight, radiation levels on seats at seat level and at 90 cm in the aisles. Since the measurements were performed under the normal loading conditions, we have already included the variation in loading patterns and the transmission factors for aircraft structures and, in some cases, the intervening cargo. Assuming the loading patterns for all other flights are similar to those observed in the radiation survey at the two major airports, we may express the average radiation dose rates in terms of only one parameter--total transport index per flight.

A regression line with proper confidence intervals can be established to represent the functional relationship between average dose rates and

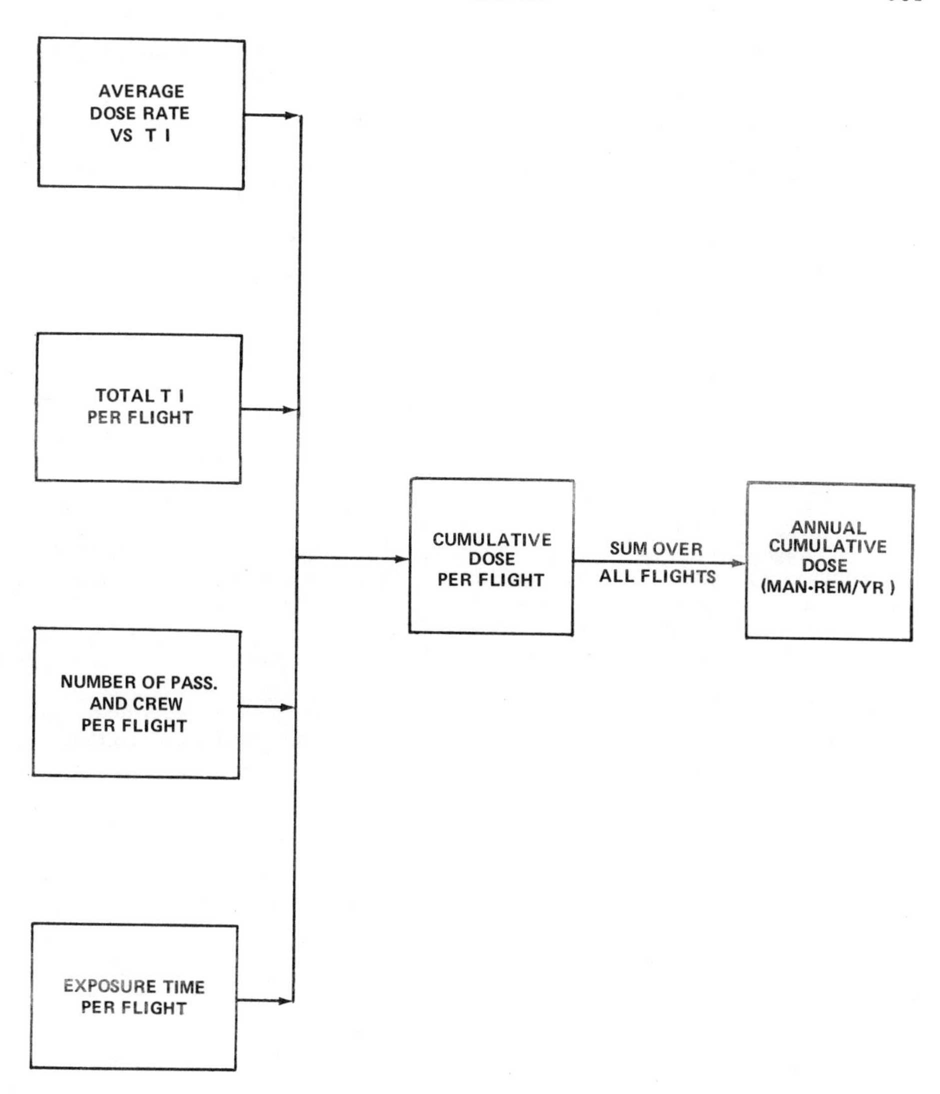

FIG.3. Parameters required to evaluate annual cumulative exposure.

the total transport index carried on each flight. Thus, equations (4) and (5) can be replaced by

$$R = f\ (TI) \tag{6}$$

The next step was to collect information on total transport index, number of people, and exposure time for each flight in the U.S., in order that the annual cumulative dose could be calculated from equations (6), (3), and (2).

Figure 3 is a block diagram to illustrate the parameters required to estimate the annual cumulative dose. In the next section, the radiation surveys on board regular flights will be discussed. Statistical data collection will be discussed in Section 4.

3. Radiation Surveys

To establish the functional relationship between the average dose rates and total transport index per flight as described in equation (6), radiation surveys were conducted for a period of four weeks in Chicago O'Hare and Boston Logan airports. O'Hare and Logan were selected for the survey because of the large numbers of air shipments of radioactive material which originate at these airports.

The survey teams measured radiation dose rates at passenger seat levels, at 90 cm above the aisle, and at seat levels in the cockpit of regular scheduled passenger aircraft after the radioactive shipments had been stowed in the aircraft cargo compartments. The aircraft measured were selected on the basis that they were known to be carrying packages of radioactive material. A total of 142 flights were measured. To prevent delay of flights and to avoid creating curiosity in passengers, the measurements were limited to those flights originating at the airport for which measurements could be made prior to loading passengers in the cabin. Since cargo was generally scheduled to be placed on board between 20 and 30 minutes prior to flight time, while passengers began boarding the aircraft 20 to 15 minutes before departure, the amount of

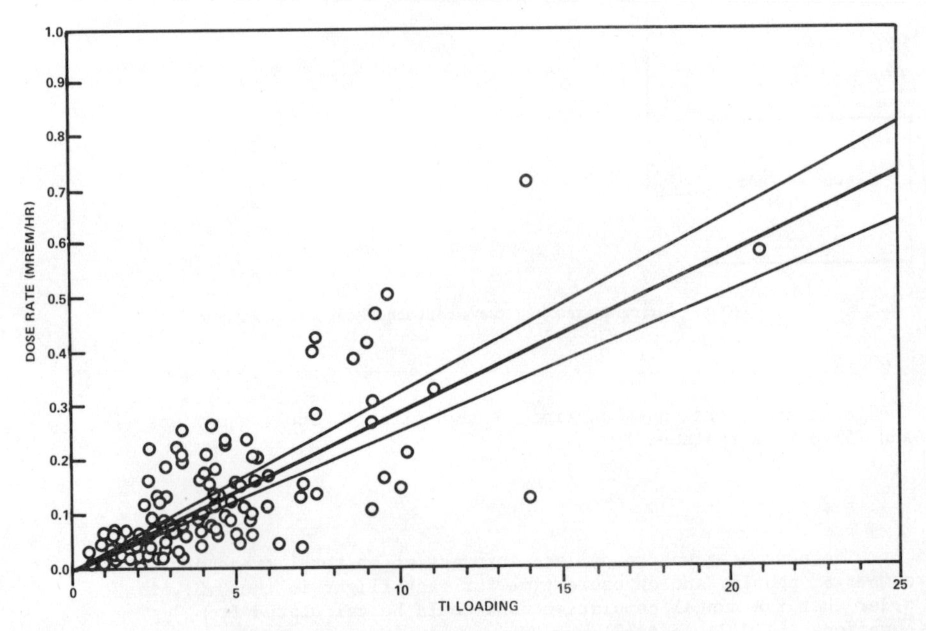

FIG.4. Passenger average dose rate plotted against TI.

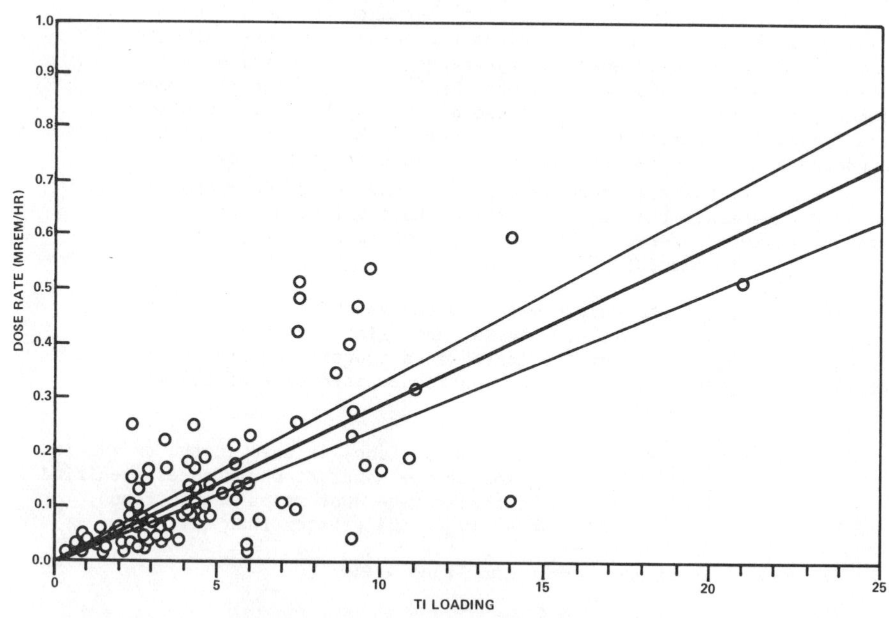

FIG. 5. Stewardess average dose rate plotted against TI.

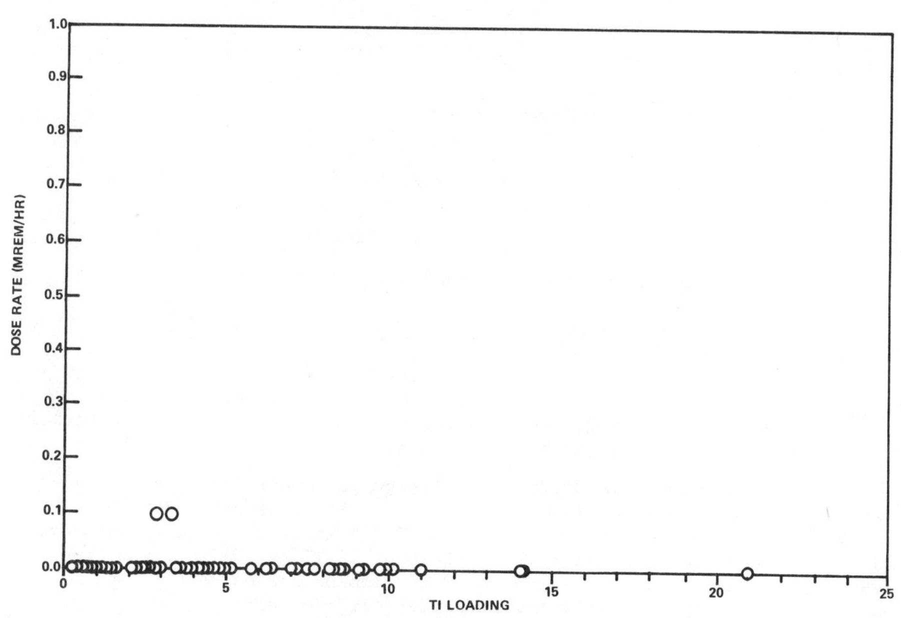

FIG 6. Pilot average dose rate plotted against TI.

time available for making measurements was limited. An average of 10 dose rates were measured at passenger seat level per flight. Interpolation and extrapolation were used to estimate the dose rate for the unmeasured seats. Average dose rates for passengers were calculated based on equation (4). Average dose rates for stewardesses were estimated from the dose rate measured at 90 cm above the floor and averaged over the entire aisle of the aircraft. Most of the shipments were loaded in the rear cargo compartment of the aircraft; therefore, the dose rate in the cockpit was undetectable in most cases.

Almost all the measurements were performed with the Victoreen Model 440 survey meters. They are ionization chamber instruments, incorporating a vibrating-reed electrometer. The specifications give an accuracy of $\pm$ 10 percent of full scale.

The results of the survey were plotted as shown in Figures 4, 5, and 6. Average passenger, stewardess, and pilot radiation dose rates for each of 142 flights were indicated as a function of transport index per flight. A regression line and confidence intervals of the line were computed[8] as shown in each plot.

If the metric system (1 meter from surface) is used in the definition of a given transport index $(TI)_m$ and if the separation distances specified in the U.S. regulations are used, the average dose rates can be found from the curves by using a TI about 10 to 15% greater than $(TI)_m$.

The equations of the regression lines are:

for passengers $Y = 0.029\ x + 0.005$ (7)

where x denotes TI and Y denotes average radiation dose rate in mrem/hr.

correlation coeff = 0.82

$\sigma = 0.07$

for stewardesses $Y = 0.028\ x + 0.003$ (8)

correlation coeff = 0.8

$\sigma = 0.07$

for pilots $Y = 0.001$ (9)

correlation coeff = 0.001

$\sigma = 0.01$

The above values obtained for pilots are not significant because, in the 142 flights surveyed, the dose rates at the cockpit were undetectable for 140 flights and 0.1 mrem/hr for 2 flights.

Based on the above analysis, the average passenger dose rate in all aircraft measured is about 0.029 mrem/hr per unit transport index while the average stewardess dose rate is about 0.028 mrem/hr per unit transport index. We have 95% confidence that, at about 10 TI per flight, the true average dose rate is between 0.34 and 0.25 mrem/hr for passengers and between 0.33 and 0.24 mrem/hr for stewardesses. For pilots, the average dose rate is zero for all practical purposes.

We believe the data collected in the Boston-Chicago surveys can be realistically applied nationwide for the following reasons:

1. The data correspond very closely to a more limited (in terms of number of aircraft surveyed) but more widespread survey by the authors at airports in San Francisco, California, St. Louis, Missouri, and in JFK and Laguardia Airports in New York City.

2. Operating guidelines for loading techniques and restrictions are issued by the airline carriers, not the airports. This means, for example, that United Airlines loading personnel in Newark Airport in New Jersey load hazardous materials under the same operating guidelines as United Airlines loading personnel in O'Hare Airport in Chicago, Ill. By sampling a wide cross section of carriers in Chicago and Boston, we believe we have a good sampling of the different carrier's operations nationwide. There could very well be a difference in emphasis on carriage of radioactive material by the same carriers in different cities, but this effect is small based on the statistical data collected nationwide (as discussed in the following section).

4. Statistical Data Collection

There are approximately five million aircraft departures from over 300 airports in the United States each year. It is necessary to obtain a significant sampling of the number of passengers, exposure time, and transport index for the flights in the U.S. for 1973 to calculate annual cumulative dose based on equations (2) and (3). The Air Transport Association of America and nine of its member airlines cooperated to record the above information for every flight carrying packages of radioactive material departing from 33 airports for a period of seven consecutive days during the last quarter of 1973. A one-week period was selected because the radiopharmaceutical manufacturers believed that their shipments are relatively constant from week to week throughout the year.

A stratified sampling of airports was selected based on the national ranking of the airports according to the number of air carrier departures.[9] Since most radiopharmaceutical manufacturers are located in or near large cities, and many packages were distributed to hospitals in larger cities or transferred through major airports to hospitals in smaller cities, we anticipated that we would account for most radiopharmaceutical air shipments by surveying departure flights from the 20 major airports which represent about half of all flight departures in the United States. In addition, two airports were selected as special airports--Knoxville, Tennessee, and Syracuse, New York--because Knoxville is close to Oak Ridge National Laboratory which produces radioisotopes and Syracuse is believed to be the major transfer point for shipments imported from Canada. The remaining airports were divided into two groups: medium (from rank 21 to 80) and small (from rank 81 to about 300). Five medium and six small airports were randomly selected to represent the two groups. Figure 7 is a block diagram to show the scheme of stratified sampling. To obtain the annual cumulative dose over the entire United States, the following equations were used.

$$D_y = \sum_{\substack{\text{all airports} \\ \text{in Y group}}} f_{\text{airport}} \left[\sum_{\substack{\text{flights surveyed} \\ \text{from one airport}}} \sum_{i=1}^{3} (RN)_i t \right] \quad (10)$$

where Y denotes large and special, medium, or small airport group;
i denotes passenger, pilot, and stewardess group;

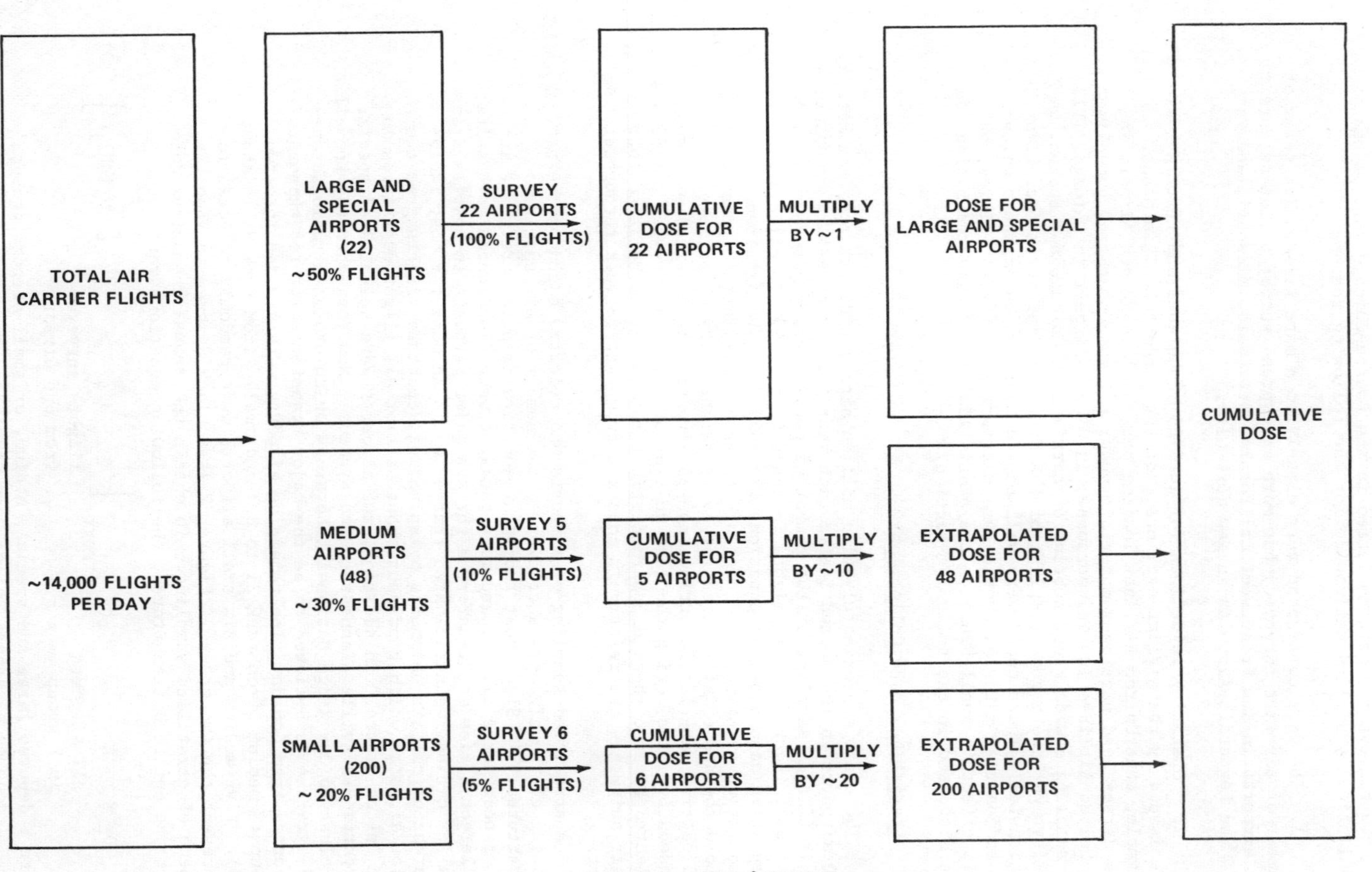

FIG. 7. Sampling of airports.

$$\text{and } f_{\text{airport}} = \frac{\text{total air carrier departures from the airport}}{\text{departures from 9 airlines from the airport}}$$

Since the flights from the nine airlines included in the survey constitute only a portion of the total flights from each of the 33 airports sampled, it was necessary to extrapolate, based on number of departures for each air carrier, the sampled data from each airport to include all air carrier flights. The use of parameter f in equation (10) is based on the assumption that the airlines not included in the survey had the same distribution of parameters involved in the calculation as the nine surveyed airlines. Since the selection of airlines was based on the information that they carry more packages of radioactive material than other airlines, that assumption is a conservative one.

The annual cumulative dose can be obtained by extrapolating over all airports and over the entire year, i.e.,

$$D = \left(\sum_{Y=1}^{3} D_Y \, g_Y\right) \times 52 \tag{11}$$

$$\text{where } g = \frac{\text{departures from all airports in one group}}{\text{departures from surveyed airports in this group}}$$

and Y denotes large and special, medium, or small airport group.

5. Annual Cumulative Dose

A computer program was set up to read transport index, number of passengers and crew, and exposure time (scheduled time plus 15 minutes for ground delay time) for every sampled flight. After converting transport index to average radiation dose rate based on the regression line, the cumulative doses for each flight were calculated. Finally, the doses were summed in each airport and extrapolated according to equations (10) and (11).

The final results showed average annual cumulative doses for passengers of 1700 man-rem; for stewardesses, 80 man-rem; and less than 1 man-rem for pilots.

The above results were obtained based on no differentiation for types of aircraft involved because of limited data for aircraft other than DC-9 and B-727 aircraft. If separate regression lines and the subsequent calculations were made for specific types of aircraft, the results would show a reduction in doses to 1400 man-rem, for passengers, 70 man-rem for stewardesses, and less than 1 man-rem for pilots.

The authors believe the second set of results may be more realistic than the first set because each type of aircraft has different sizes of cargo compartments, a different total number and arrangement of seats, and a different floor structure.

Finally, we may compare the annual cumulative dose estimated in this study to the dose to passengers and crew due to cosmic radiation at the elevations at which commercial aircraft fly. Direct measurements of dose equivalent in aircraft conducted by Brookhaven National Laboratory[10] indicated the radiation dose rate due to cosmic radiation varies from

0.3 mrem/hr to 0.6 mrem/hr at elevations from 30,000 ft (for short flights) to 40,000 ft (for long flights). The annual cumulative doses to passengers and crew members can be estimated as follows.

Based on 1971 FAA-CAB data,[9] the total number of passengers enplaned was approximately 175 x 10^6 with the average flying time about two hours. If 2/3 of the flying time was at an altitude of 30,000 ft, the annual cumulative dose for all passengers due to cosmic radiation would be approximately 70,000 man-rem. The number of passengers who are exposed to radiation from the carrying of packages of radioactive material can be estimated by multiplying the average number of passengers per flight (60 passengers) by the annual number of flights (100,000 flights) which carry such packages, a total of 6,000,000 passengers. The annual cumulative dose due to cosmic radiation for the same group of passengers is approximately 2,000 man-rem (about 1/30 of the dose received by all aircraft passengers).

There are about 30,000 flight crew members and 40,000 cabin crew members in the United States. The flying time for each member is about 1000 hours annually. Again, if 2/3 of the flying time was at an altitude of 30,000 feet, the annual cumulative dose for the flight crew and the cabin crew would be approximately 6,000 and 8,000 man-rem respectively. It is estimated that one-half of the crew members fly at least once per year in a flight carrying packages of radioactive material; the annual cumulative doses due to cosmic radiation for these pilots and stewardesses are approximately 3,000 and 4,000 man-rem, respectively.

Annual cumulative doses received by different population groups and from different sources were tabulated in Table I. The average annual doses for individuals of each group were estimated by dividing the cumulative dose by the number of persons in that group.

It should be recognized that, if limiting the total number of packages or total TI that can be carried in one aircraft causes spreading of the packages into several aircraft for transport, the cumulative dose will not be reduced although the individual doses will be decreased.

TABLE I. ANNUAL CUMULATIVE DOSES

Population groups	No. of persons	Annual cumulative doses, man-rem/yr	
		Radioactive packages	Cosmic radiation
Persons Exposed to Radiation from Packages			
Passenger	6×10^6	1 400	2 000
Pilot	1.5×10^4	1	3 000
Stewardess	2×10^4	70	4 000
All Persons			
Passenger	175×10^6	1 400	70 000
Pilot	3×10^4	1	6 000
Stewardess	4×10^4	70	8 000

The cumulative dose can be reduced, for example, if the total transport index of the packages which must be shipped by passenger aircraft is reduced either by increasing the shielding of packages or by finding alternative modes of transport.

Although annual doses to individuals of the most exposed groups are not discussed in this paper, such doses are important parameters and are now under evaluation. Studies have been initiated to investigate various alternatives for reducing the cumulative and individual doses from transporting packages of radioactive material in passenger-carrying aircraft.

6. Conclusion

The method described in this paper offers a means of obtaining a suitable sample for deriving a reasonable and sufficient estimate of the population dose from transporting packages of radioactive material on passenger aircraft.

The annual cumulative population dose to a total of 6 million passengers and crew members from the transportation of packages of radioactive material by passenger aircraft is less than 1500 man-rem. That dose is small and represents only a minor addition to the annual cumulative radiation dose from natural background radiation of 600,000 man-rem received by the same 6 million people who are exposed to the radiation from packages being transported. For purposes of comparison, the same groups of passengers, pilots, and stewardesses receive an annual cumulative radiation dose of 9000 man-rem from cosmic radiation as a result of their flying time at altitudes of from 30,000 to 40,000 feet, which are common for commercial flights.

The 1500 man-rem annual radiation dose from packages corresponds to an individual average annual dose of 0.25 millirem per year. This is a very small fraction of the 170 millirem per year which is recommended by the National Committee on Radiation Protection in the United States and by the International Committee for Radiological Protection as the annual average radiation dose not to be exceeded by population groups from all sources of radiation other than natural background.

The radiological risks associated with the routine transportation of those packages, mostly radiopharmaceuticals, by passenger aircraft are small.

REFERENCES

[1] "Background Material for the Development of Radiation Protection Standards" Federal Radiation Council Report No. 1 (1960)

[2] "Basic Safety Standards for Radiation Protection" 1967 edition, Safety Series No. 9, IAEA, Vienna (1967) STI/PUB/147

[3] Evans, R.D., "Problems Associated with the Transportation of Radioactive Substances" NAS-NRC Publication 205, (1951)

[4] Gibson, R. "The Safe Transport of Radioactive Materials" Pergamon Press, London (1966)

[5] Failla, L., Faloci, C., Susana, A., EUR 4884e, Commission of the European Communities (1972)

[6] 49 CFR 173, USDOT Regulations

[7] 49 CFR 103 USDOT-FAA Regulations

[8] Natrella, M. G., "Experimental Statistics" NBS Handbook 91 (1963)

[9] "FAA Air Traffic Activity - calendar year 1971" USDOT-FAA, (1972)

[10] Cowan, F. P., Chester, J. D., Kuehner, A. V., Phillips, L. F., "Direct Measurements of Dose Equivalent in Aircraft" BNL 17060, Presented in "Natural Radiation Environment Symposium" Houston, Texas, August 7-11, (1972)

DISCUSSION

P. CANDÈS: The study you have described is very interesting and useful, since it shows that the true dose is lower than the possible theoretical dose by a factor of at least 100. However, the study is limited to aircrew and passengers. Other persons are also involved, such as those responsible for handling following departure and preceding arrival. There are not very many of these people, but theoretically they can be exposed to dose rates exceeding 100 mrad/h. Could you comment on this?

R.F. BARKER: You are right. The doses to cargo handlers, sorters and loaders have been looked at and we believe they are within ICRP limits. The collective doses from such exposures will be the subject of further studies about to begin. We expect that the doses in question will be appreciable, but probably not as great as those to passengers and crew.

F.D. SOWBY: What maximum dose rate to passengers was found in your surveys?

R.F. BARKER: In one of the surveys conducted in early 1973 a dose rate as high as 20 mrem/h was measured. After a change in the regulations and appropriate instruction of the carriers, the maximum dose rate found later in 1973 was 4.5 mrem/h.

M. ČOPIČ: With such large-scale operations would it not be appropriate for the analysis to include the various incidents that can arise? A parallel can be drawn with the analysis for nuclear power plants, where we consider normal operation and accident situations.

R.F. BARKER: We have looked at the probability of such incidents in transport. Based on experience over 25 years, during which millions of packages have been safely transported, the radiation risk from such incidents seems extremely small.

S. IGNJATOVIĆ: Does the pilot of an aircraft have to know that he is carrying a radioactive source?

R.F. BARKER: Yes, the regulations in the United States of America require that the pilot be notified if his aircraft is carrying hazardous goods, such as radioactive materials.

P. CANDÈS: I should like to make two further comments of a more general nature. First of all, I think that a study of this kind will be followed by a cost-benefit study. The figures you have given seem to show that the mean dose from transport is not very different from that due to nuclear power stations, and this situation is not likely to change in the near future. I don't know whether the benefits of radioisotopes are greater than those of electricity, but if a cost-benefit study unfortunately showed that they are less, I think there would be considerable pressure for amendments in the transport regulations, which at present are quite satisfactory.

My second comment is in a way the converse of what I have just said. It is clear that the collective doses involved in the transport of fuel elements, which is performed with extremely great care, are low considering all the transport operations undertaken. It would seem, therefore, that:

(1) Apart from reasons of nuclear safety there is no need to stop the transport of fuel elements by introducing the concept of nuclear parks;

(2) Dose levels could be increased for this kind of transport, and the experts concerned would be very pleased at this.

R. F. BARKER: We are planning a cost-benefit analysis in connection with changes which are being considered in the air traffic regulations. Such a study will be necessary before any changes are made affecting the standards applicable to air transport of small sources or of irradiated fuel.

R. LE QUINIO: Did you extrapolate your values for, say, the next 20 years?

R. F. BARKER: No, but we are planning to do this with help from the radiopharmaceutical people.

ETUDE D'UN ECOSYSTEME AQUATIQUE NATUREL CONTAMINE IN SITU PAR DES EFFLUENTS LIQUIDES TRITIES, EN VUE DE L'EVALUATION DE LA SENSIBILITE DES PARAMETRES DES NIVEAUX D'EXPOSITION DU PUBLIC

R. BITTEL
Association Euratom-CEA «Niveaux de pollution du milieu ambiant»,
Centre d'études nucléaires de Fontenay-aux-Roses,
Département de protection,
Fontenay-aux-Roses, France

R. KIRCHMANN*, G. VAN GELDER-BONNIJNS*, G. KOCH**
Centre d'étude de l'énergie nucléaire,
Mol, Belgique

Abstract–Résumé

IN SITU STUDY OF A NATURAL AQUATIC ECOSYSTEM CONTAMINATED BY TRITIATED LIQUID EFFLUENTS WITH A VIEW TO EVALUATING THE SENSITIVITY OF THE PARAMETERS INVOLVED IN ASSESSING POPULATION EXPOSURE LEVELS.

The ecosystem in question is a small watercourse that receives liquid effluents from complex nuclear facilities and intermittently feeds a series of ponds used for fish-rearing. The system was investigated for about a year, the tritium concentrations in water and sediment samples and in plant and animal organisms being determined. The accent was placed on measurement of the tritium incorporated in organic matter, the specific activity of the combustion water being determined in each case. It is higher than that of the stream and pond water by a factor of as much as 100. This conflicts with observations, made in situ and in laboratory aquaria, regarding the transfer of tritium in the form of tritiated water so that one must look for biologically available tritiated compounds in the effluents. It would appear, therefore, that the physico-chemical state of the tritium is a sensitive parameter of particular importance for predicting and evaluating population exposures.

ETUDE D'UN ECOSYSTEME AQUATIQUE NATUREL CONTAMINE IN SITU PAR DES EFFLUENTS LIQUIDES TRITIES, EN VUE DE L'EVALUATION DE LA SENSIBILITE DES PARAMETRES DES NIVEAUX D'EXPOSITION DU PUBLIC.

L'écosystème envisagé comporte un petit cours d'eau recevant les effluents liquides d'installations nucléaires complexes et alimentant de façon intermittente une série d'étangs de pisciculture. Cet ensemble a été prospecté pendant un an environ et divers échantillons d'eau, de sédiments, d'organismes végétaux et animaux ont été prélevés et analysés pour leur teneur en tritium. On a mis l'accent sur la mesure du tritium incorporé à la matière organique, en évaluant dans chaque cas l'activité spécifique de l'eau de combustion. Cette activité spécifique est supérieure à celle de l'eau de la rivière et des étangs d'un facteur qui peut atteindre 100. Cette constatation est en désaccord avec les observations faites in situ et au laboratoire en aquarium relativement au transfert du tritium sous forme d'eau tritiée, ce qui conduit à la recherche, dans les effluents, de composés tritiés biologiquement disponibles. Il apparaît donc que l'état physico-chimique du tritium est un paramètre sensible particulièrement important à considérer pour prévoir et évaluer l'exposition des populations.

---

* Département de radiobiologie.

** Section «Mesures bas-niveau».

1. INTRODUCTION

Le tritium est actuellement l'un des radionucléides qui peut apparaître comme critique, dans le cadre notamment de la radioprotection de la population dans son ensemble ou de groupes très larges de population, en raison de la multiplicité des causes de production et des sites d'utilisation, des quantités mises en jeu, du développement prévisible de l'énergie nucléaire et des options concernant les types de réacteurs [1]. Cependant on hésite parfois encore à admettre le tritium parmi les polluants radioactifs dangereux, en raison de la discrimination qui existe en faveur de l'isotope léger de l'hydrogène, d'une part, et des paramètres radiochimiques du radionucléide, d'autre part.

La caractéristique essentielle du tritium est d'être un radioisotope de l'hydrogène, élément vraiment fondamental des tissus. Des expérimentations in vitro ont montré que le tritium de l'eau peut s'échanger avec l'hydrogène de l'eau légère et l'hydrogène de diverses molécules organiques.

Dans certains cas, le tritium reste échangeable et on peut parler de tritium mobile, comme on parle, en chimie organique, d'hydrogène mobile. Mais le tritium de certaines molécules organiques se trouve parfois dans un état difficilement échangeable avec l'hydrogène léger.

Des expérimentations in vivo ont clairement mis en évidence l'incorporation du tritium à partir d'eau tritiée dans des molécules organiques. Ceci a été montré dans le cas de chaînes alimentaires terrestres par Kirchmann et al. [5-8] et par Koranda et Martin [9], et, dans celui des chaînes aquatiques, dans diverses expérimentations [10-14]. En particulier, Kirchmann et Bonotto [12] ont montré que l'algue unicellulaire *Acetabularia mediterranea* incorporait le tritium dans différentes fractions organiques. Il en est de même pour les poissons [15-16]. Mais, on observe toujours une discrimination plus ou moins nette en faveur de l'hydrogène léger. Des recherches effectuées à Mol sur des poissons contaminés par de l'eau tritiée l'ont récemment confirmé.

Mais le tritium des effluents liquides peut avoir des origines multiples et revêtir des formes physico-chimiques diverses. La variété des combinaisons organiques tritiées, préparées par les Centres, commercialisées et utilisées dans maintes applications, laisse prévoir l'existence dans les effluents d'un nombre important de composés tritiés très variés, en particulier de composés chimiquement et biochimiquement voisins des molécules de grand intérêt biologique [2, 3]. Il paraît donc insuffisant de baser une évaluation des risques sur des études de contamination par l'eau tritiée. Diverses expérimentations ont montré d'ailleurs que le tritium des molécules tritiées présentes dans le milieu s'incorporait dans le compartiment organique des tissus plus intensément que le tritium de l'eau, qui tendait à aller vers le compartiment «eau libre» [6, 17].

Il faut encore remarquer que les limons, notamment les limons riches en argile et surtout en matières organiques, sorbent les composés tritiés présents dans l'eau avec une très grande intensité [18], notamment dans leur fraction organique ou organominérale. Il en résulte que les animaux aquatiques prélevant leurs aliments par filtration peuvent trouver dans le milieu des formes tritiées très diverses dont l'activité spécifique peut être relativement élevée.

En résumé, les organismes aquatiques, entrant dans les chaînes alimentaires et vivant dans des eaux recevant des effluents radioactifs

TABLEAU I. TRITIUM DANS DES ECHANTILLONS D'EAU

Endroit de prélèvement	Date de prélèvement	pCi/ml	pCi/g H
Etang 1	12 mai 1972	42,2	380
Etang 5		1,3	12
Etang 2	8 juin 1972	29	261
Etang 3	28 juin 1972	0,8	7
Etang 8	29 août 1972	13,4	120
Essenboom		19,0	171
Etang 6		14,1	127
Netevallei	28 août 1972	0,5	5
Essenboom	Février 1973	4,4	40
Essenboom	17 mai 1973	56,2	506

tritiés, se trouvent en présence de tritium à l'état d'eau tritiée, de formes organiques solubles, de formes organiques ou organominérales colloïdales ou particulaires. L'objet de la présente communication est de montrer quelles conséquences pour l'homme peuvent résulter d'une contamination chronique d'un écosystème naturel recevant des effluents, où, précisément le tritium se trouve sous ces différents états.

2. ETUDE EXPERIMENTALE

2.1. Site de la Molse-Nete

Les effluents liquides des installations du Centre d'étude de l'énergie nucléaire de Mol (Belgique) sont collectés dans des citernes individuelles pour chaque bâtiment, d'où ils sont pompés vers une station de traitement, qui reçoit également les rejets liquides d'activité faible ou moyenne de l'Usine de retraitement de combustibles irradiés, Eurochemic. Après épuration, ces effluents sont déversés dans un petit cours d'eau, la Molse-Nete, dont le débit moyen est de 50 000 m^3 par jour. Un petit chapelet d'étangs s'égrène le long de cette rivière et certains s'alimentent en eau dans celle-ci par pompage périodique ou lui sont même reliés directement.

Les teneurs en tritium des eaux de la Molse-Nete sont très variables dans le temps, en fonction des rejets périodiques d'effluents. Pour la période du 1er janvier 1970 au 15 décembre 1972, les extrêmes sont: 500 pCi/l et 350 000 pCi/l, soit en pCi par gramme d'hydrogène, respectivement 4,5 et 3150. Dans les étangs, les niveaux du tritium sont également très variables en fonction de l'alimentation par les eaux de la rivière et de leur situation dans le site. Les teneurs observées en 1972 et au début de 1973 s'échelonnent de 400 à 56 000 pCi/l, soit 3,6 à 500 pCi par gramme d'hydrogène (tableau I).

TABLEAU II. TRITIUM DANS DES ECHANTILLONS VEGETAUX (exemples)

Nature	Date de prélèvement	Eau de combustion (pCi ^{3}H/g H)
Myriophyllum	8 juin 1972	12 843
Valisneria		35 775
Potamogeton natans		27 216
Lemna		19 017
Lemna		792
Valisneria	28 juin 1972	8 703
Lemna		4 437
Elodea		315
Potamogeton natans		8 757
Elodea		3 294
Valisneria		23 265
Lemna		288
Myriophyllum	10 août 1972	4 959
Elodea		2 826
Valisneria		4 950

2.2. Analyse des échantillons

Sur ce site, on a prélevé des échantillons d'eau, de végétaux et d'animaux aquatiques dont l'espèce a fait l'objet de déterminations soigneuses.

La mesure du tritium dans l'eau ne présente pas de difficultés particulières. On a opéré par scintillation liquide. La solution scintillante est composée de xylène, de 2,5 diphényl-oxazole (PPO), de 2, p-phénylène-bis (5 phényloxazole) (POPOP) et de tensioactifs.

En ce qui concerne les échantillons solides, on a déterminé les teneurs en tritium de leur eau libre, d'une part, de leurs matières organiques, d'autre part. La détermination du tritium organique est réalisée de la manière suivante: on brûle les échantillons après dessiccation dans un four tubulaire en quartz, à 700°C et sous un courant d'oxygène, en présence de nickel qui s'oxyde en NiO qui sert de catalyseur. L'eau de combustion est recueillie et son activité est mesurée par scintillation liquide.

2.3. Résultats

Un grand nombre d'organismes ont été analysés pour leur teneur en tritium organique. Certains des résultats obtenus figurent dans les tableaux II (végétaux aquatiques) et III (poissons). Le tableau IV concerne les mesures sur des œufs d'oiseaux qui vivent sur l'écosystème étudié. On a exprimé les résultats en termes d'activité spécifique de l'hydrogène

TABLEAU III. TRITIUM DANS DES ECHANTILLONS DE POISSONS (exemples)

Nature	Date de prélèvement	Eau d'hydratation (libre) (pCi ^{3}H/g H)	Eau de combustion (^{3}H organique) (pCi ^{3}H/g H)
Poisson	8 juin 1972		37 386
Filet de carpe	24 septembre 1972		724
Filet de carpe	24 septembre 1972		1 628
Brèmes	17 mai 1973		
Branchies, Ecailles, Opercules, Nageoires			2 835
Reins, App. génital, Vessie natatoire		468	2 556
Muscles, Cœur		621	5 730
Os		666	2 880
Tête		495	2 529
Filet		477	1 224
Filet		207	1 782

et on a confronté les valeurs ainsi obtenues aux activités spécifiques de l'hydrogène de l'eau des étangs, en déterminant le rapport

$$R = \frac{\text{A.S. de H organique dans les tissus}}{\text{A.S. de H de l'eau du milieu}}$$

R ayant donc la signification d'un «rapport observé».

La valeur maximale de l'activité spécifique de l'eau de la rivière entre avril et août 1972 a été 350 pCi ^{3}H par cm^3, soit 3,15 nCi ^{3}H par gramme d'hydrogène. L'activité spécifique de l'hydrogène organique des végétaux récoltés durant cette période est généralement supérieure à cette valeur d'un facteur allant jusqu'à 11,5 (R = 11,5). Si on se rapporte aux valeurs médianes de l'activité spécifique de l'hydrogène de l'eau, on obtient, pour les végétaux aquatiques, des valeurs de R échelonnées de 0,7 à 32. En ce qui concerne les animaux aquatiques, on constate également que R est en général supérieur à l'unité et peut, pour des poissons notamment, atteindre des valeurs de l'ordre de 100. Les mêmes faits sont observés dans le cas des œufs de canes vivant sur l'écosystème de la Molse-Nete et de ses étangs, le rapport R étant de l'ordre de 15 pour le blanc et de 50 pour le jaune des œufs. On remarquera que, pour les poissons tout au moins, l'activité spécifique de l'hydrogène de l'eau libre est très inférieure à l'activité spécifique de l'hydrogène organique (tableau III).

TABLEAU IV. TRITIUM DANS DES ECHANTILLONS D'ŒUFS DE CANES

Nature	Date de prélèvement	Eau de combustion (^3H organique) (pCi ^3H/g H)
Jaune	27 avril 1972	60 633
Blanc	27 avril 1972	15 165
Œuf entier	7 novembre 1972	4 752
Œuf entier	5 février 1973	6 102
Jaune	18 avril 1973	1 530
Blanc	18 avril 1973	792

L'ensemble des constatations précédentes est en contradiction avec les résultats des expérimentations poursuivies par ailleurs, dans lesquelles le tritium est introduit dans le milieu sous forme d'eau tritiée uniquement. Des recherches réalisées au CEN de Mol simultanément aux présentes études confirment que, dans le cas d'une contamination par l'eau tritiée, le rapport R évalué précédemment est en général inférieur à l'unité. Il semble donc que les résultats obtenus in situ avec des effluents réels soient à mettre en relation avec la forme chimique du tritium présent dans ces effluents. La Section «Mesures bas-niveau» du CEN de Mol effectue actuellement des analyses de ces effluents en vue d'y déterminer le tritium présent à l'état d'eau tritiée, le tritium organique volatil et le tritium organique non distillable, et de fractionner ce dernier afin d'identifier si possible quelques-unes des combinaisons tritiées présentes. Quoi qu'il en soit pour le moment, il est intéressant de quantifier les conséquences des faits observés in situ en ce qui concerne la dose délivrée à l'homme, consommateur final des chaînes aquatiques.

3. INCIDENCE SUR LA DOSE DELIVREE A L'HOMME DU FAIT DE L'INGESTION

Le débit de la dose délivrée à un organe de référence de l'homme du fait de l'ingestion d'un radionucléide peut s'exprimer par la relation

$$D = \frac{k\,Q\,F\,f_e \epsilon\,x}{m\,\lambda}$$

où:

Q est la masse du vecteur alimentaire ingéré par unité de temps,

F, le facteur de transfert milieu-aliment, c'est-à-dire le rapport de l'activité spécifique de l'aliment de l'homme à l'activité spécifique du milieu,

f_e, la fraction du radionucléide ingéré par l'homme qui parvient à l'organe de référence,

ϵ, l'énergie par désintégration du radionucléide,

x, l'activité spécifique du milieu,

m, la masse de l'organe de référence,

λ, la constante de décroissance effective du radionucléide dans l'organe de référence de l'homme,

k, un facteur de proportionnalité dépendant des unités adoptées.

La Commission internationale de protection radiologique (CIPR) a admis que le tritium se répartissait dans le corps entier de l'homme comme dans l'eau libre de l'organisme et a proposé pour f_e la valeur 1; pour m, 70 kg, masse de l'homme standard; pour ϵ, 10^{-2} MeV; pour λ, 21 $(\text{ans})^{-1}$ [19]. Si on admet que le tritium se répartit d'une manière uniforme dans l'ensemble de l'eau de l'organisme aquatique ingéré par l'homme, l'activité spécifique de l'hydrogène de l'organisme étant, à l'équilibre, égale à celle du milieu, on est conduit à une valeur de F de l'ordre de 0,7. Il apparaît cependant indispensable de revoir ces conceptions, en fonction des apports expérimentaux récents, montrant l'incorporation du tritium du milieu dans les composants organiques des tissus. Une méthode d'approche consiste à évaluer la variabilité de la dose D en fonction de la variabilité des paramètres intervenant dans son expression mathématique, en se plaçant dans des situations opposées.

Situation A

Le tritium est rejeté dans les eaux à l'état d'eau tritiée. Il est incorporé uniquement dans le compartiment «eau libre» des aliments «aquatiques» de l'homme et, après ingestion par celui-ci, il reste dans le compartiment «eau libre» de l'organisme.

Situation B

Le tritium est rejeté dans les eaux à l'état de molécules organiques tritiées. Il est incorporé uniquement dans le compartiment organique des aliments de l'homme et, après ingestion, dans le compartiment organique de l'homme.

Situation A

On admettra R = 1. Comme

$$R = F \frac{\text{H eau/milieu}}{\text{H eau/tissus}} \qquad F = 0{,}7$$

Pour l'homme, la période effective est prise égale à celle proposée par la CIPR, T = 12 jours, d'où $\lambda = 21\ (\text{ans})^{-1}$. L'organe critique de l'homme est le compartiment «eau libre» de masse m égale à 43 kg. Par suite, $QF/m\lambda$, terme de variabilité de D, vaut $8 \cdot 10^{-4}$ an pour une consommation d'aliment Q unitaire.

Situation B

On admettra R = 100. R s'exprime en fonction de F par l'expression

$$R = F \frac{\text{H eau/milieu}}{\text{H organique/tissus}}$$

L'hydrogène de l'eau représente 0,11 du poids de l'eau, celui de la matière organique des tissus, environ 0,02 du poids des tissus. Par suite

$$R = F \frac{0,11}{0,02} \qquad F \approx 20$$

On ne doit considérer que la masse des matières organiques de l'homme comme organe critique. Pour l'homme standard de la CIPR, m ≈ 20 kg. L'analyse des modèles conduit à admettre pour le tritium organique des tissus de l'homme une période de l'ordre d'un an [20, 21]; donc $\lambda \approx 0,7\ (\text{an})^{-1}$. Il en résulte que $QF/m\lambda$ prend dans cette seconde éventualité une valeur de l'ordre de 1,5 an, pour une valeur unitaire de Q.

Dans l'évaluation qui précède, on n'a pas tenu compte d'une variabilité de ϵ. On admettra ici que l'énergie effective par désintégration ne dépend pas de la forme du tritium présent dans l'organisme humain, bien qu'il semble probable que l'efficacité biologique relative du tritium incorporé dans les combinaisons organiques soit plus grande que celle du tritium présent dans l'organisme à l'état d'eau tritiée [1, 2, 22].

L'étude de variabilité précédente montre donc que l'incidence de la forme physico-chimique du tritium dans les effluents, dans l'hypothèse où elle retentit directement sur celle du tritium dans les différents échelons du transfert jusqu'à l'homme, est très grande. Tous autres facteurs égaux, les doses peuvent varier dans le rapport de $8 \cdot 10^{-4}$ à 1,5 c'est-à-dire dans le rapport de 1 à 2000, suivant que le tritium chemine à l'état d'eau tritiée ou sous forme de molécules organiques tritiées. Bien entendu, la situation B est trop sévère et trop exclusive pour refléter la réalité, même dans les cas concrets les plus défavorables: il est en effet peu pensable que, quelle que soit la forme rejetée dans le milieu, le tritium ne passe pas en partie dans le compartiment eau des organismes. L'analyse précédente attire seulement l'attention sur les variabilités extrêmes possibles. Même si cette variabilité est dans la réalité plus faible, elle n'en est pas moins probablement importante.

4. CONCLUSION

Les résultats expérimentaux et la discussion qui les ont suivis montrent combien il serait intéressant de préciser la forme physico-chimique du tritium des effluents liquides et le comportement des molécules organiques tritiées dans le milieu et chez l'homme. C'est précisément sur ces points qu'on va se concentrer maintenant, de manière à obtenir une évaluation fidèle des risques résultant de rejets de tritium par des installations nucléaires complexes.

REFERENCES

[1] WOODARD, W.Q., The biological effects of tritium, USAEC Rep. HASL-229 (1970).

[2] EVANS, E.A., Tritium and its Compounds, Butterwords, Londres (1966).

[3] FEINENDEGEN, L.E., Tritium-Labeled Molecules in Biology and Medicine, Academic Press, New York (1967).

[4] SMITH, T.E., TAYLOR, R.T., Incorporation of Tritium from Tritiated Water into Carbohydrates, Lipids and Nucleic Acids, Rep. UCRL-50-781 (1969).

[5] KIRCHMANN, R., LAFONTAINE, A., VAN DEN HOEK, J., KOCH, G., Transferts et répartition du tritium dans les constituants principaux du lait de vaches alimentées avec de l'eau contaminée, C.R. Soc. Biol. 163 6 (1969) 1459-63.

[6] KIRCHMANN, R., VAN DEN HOEK, J., LAFONTAINE, A., Transfert et incorporation du tritium dans les constituants de l'herbe et du lait en conditions naturelles, Health Phys. 21 1 (1971) 61-66.

[7] KIRCHMANN, R., VAN DEN HOEK, J., KOCH, G., ADAM, V., «Studies on the foodchain contamination by tritium», Tritium (Moghissi, A.A., Carter, M.W., Eds), Messenger Graphics, Phoenix (1973) 341-48.

[8] VAN DEN HOEK, J., KIRCHMANN, R., «Tritium secretion into cow's milk after administration of organically bound tritium and of tritiated water», La Radioécologie appliquée à la protection de l'homme et de son environnement, Rapport Euratom EUR 4800 dfie (1972) 1121-33.

[9] KORANDA, J.J., MARTIN, J.R., «The movement of tritium in ecological systems», Tritium (Moghissi, A.A., Carter, M.W., Eds), Messenger Graphics, Phoenix (1973) 430-55.

[10] WEINBURGER, D., PORTER, J.W., Metabolism of hydrogen isotopes by rapidly growing Chlorella pyrenoïdosa cells, Arch. Biochem. Biophys. 50 (1954) 160-68.

[11] KANAZAWA, T., KANAZAWA, K., BASSHAM, J.A., Tritium incorporation in the metabolism of Chlorella pyrenoïdosa, Environ. Sci. Technol. 6 7 (1972) 638-42.

[12] KIRCHMANN, R., BONOTTO, S., «Pénétration et distribution du tritium dans l'algue marine Acetabularia mediterranea», Actes du 5e Colloque international d'océanographie médicale, Libraria Bonanzinga, Messine (1973) 325-33.

[13] ROSENTHAL, G.M., STEWART, M.L., «Tritium incorporation in algae and transfer in simple aquatic food chains», Symposium on Radioecology, USAEC CONF-7105-01-24.

[14] BRUNER, H.D., «Distribution of tritium between the hydrosphere and invertebrates», Tritium (Moghissi, A.A., Carter, M.W., Eds), Messenger Graphics, Phoenix (1973) 303-13.

[15] HARRISSON, F.L., KORANDA, J.J., TUCKER, J.S., «Tritiation of aquatic animals in an experimental marine pool», Tritium (Moghissi, A.A., Carter, M.W., Eds), Messenger Graphics, Phoenix (1973) 363-78.

[16] PATZER, R.G., MOGHISSI, A.A., McNELIS, D.N., «Accumulation of tritium in various species of fish reared in tritiated water», Environmental Behaviour of Radionuclides Released in the Nuclear Industry (C.R. Coll. Aix-en-Provence, 1973), AIEA, Vienne (1973) 403-12.

[17] HATCH, F.T., MAZRIMAS, J.A., Tritiation of animals from tritiated water, Radiat. Res. 50 (1972) 339-57.

[18] COHEN, L.K., KNEIP, T.J., «Environmental tritium studies at PWR power plant», Tritium (Moghissi, A.A., Carter, M.W., Eds), Messenger Graphics, Phoenix (1973) 623-38.

[19] COMMISSION INTERNATIONALE DE PROTECTION RADIOLOGIQUE, Rapport du Comité 11 sur la dose admissible en cas d'irradiation interne, Pergamon Press, Oxford (1959); Trad. française, Gauthier-Villars, Paris (1963).

[20] MOGHISSI, A.A., CARTER, M.W., LIEBERMANN, R., Long-term evaluation of the biological half-time of tritium, Health Phys. 21 (1971) 57-60.

[21] ROBERTSON, J.S., «Tritium turnover rates in mammals», Tritium (Moghissi, A.A., Carter, M.W., Eds), Messenger Graphics, Phoenix (1973) 322-26.

[22] JOHNSON, H.A., «The quality factor for tritium radiation», Tritium (Moghissi, A.A., Carter, M.W., Eds), Messenger Graphics, Phoenix (1973) 231-39.

DISCUSSION

Suzanne VIGNES: In dealing with the important question of tritium transfer to man via different radioactive effluents, you consider the case of transfer in a purely organic form through the consumption of fish. Actually, this type of transfer should be reversible since these organic molecules participate in metabolism, and during catabolism are capable of yielding tritiated water. The risk of concentration is therefore reduced.

R. BITTEL: As you say, part of the organic tritium forms tritiated water during catabolism, but during anabolism fractions of organic molecules take part in the synthesis of new organic molecules. Some of the organic tritium therefore remains in the organic compartment during the transfer chains.

RAPPORT GENERAL: Synthèse de la situation actuelle et remarques finales

H. JAMMET
Département de protection,
CEA, Centre d'études nucléaires de Fontenay-aux-Roses,
Fontenay-aux-Roses, France

Il est bien difficile de résumer et de commenter, comme on me l'a demandé, ce qui a été dit au cours de cette semaine très dense et, à partir de tout ce que nous avons entendu, de présenter en un exposé succinct la situation actuelle et les perspectives d'avenir. Je ne reprendrai pas les sessions une par une, bien entendu, mais au contraire je tâcherai de résumer la situation par domaines particuliers, et j'en ai retenu cinq: l'évolution des concepts, les modalités d'irradiation des populations à partir des sources, l'estimation des doses, l'estimation du détriment, et enfin les analyses décisionnelles.

L'évolution des concepts

L'historique de l'évolution des recommandations de la CIPR, thème d'une des communications présentées à ce Colloque, est un sujet des plus intéressants. Cette évolution, en effet, permet d'expliquer certains aspects de la situation actuelle. On est parti de doses qui initialement visaient à protéger les travailleurs, à l'époque très peu nombreux. Ensuite, la Commission internationale de protection radiologique s'est intéressée à la protection des personnes du public à partir du moment où on a pris conscience de l'irradiation croissante des populations. Il est évident que ses premières recommandations avaient pour but de limiter les irradiations des travailleurs et du public et étaient essentiellement basées sur la notion de dose individuelle. Peu à peu on s'est aperçu que cette limitation des doses ne suffisait pas et que l'on pouvait faire mieux; par conséquent on a été amené à introduire des recommandations générales portant sur la limitation des doses globales, et ceci sous forme d'abord de doses moyennes qui ont en partie été développées par l'UNSCAR et également sous forme de doses collectives, ou de doses population quand il s'agit de la population mondiale tout entière. Par ailleurs et parallèlement, depuis un certain nombre d'années l'UNSCAR avait commencé à s'occuper non seulement, comme le faisait la CIPR, du risque, mais aussi du risque pondéré — le détriment — de façon à avoir une idée suffisamment correcte des dommages éventuellement subis par des groupes de la population. C'est ainsi que dans sa publication 22 la CIPR a été amenée à introduire en plus du concept de risque radiologique le concept de détriment radiologique. Par ailleurs, cette publication a attiré l'attention sur l'importance de deux recommandations qui étaient très condensées dans les recommandations générales de la CIPR puisqu'elles se trouvaient toutes les deux dans une seule phrase du paragraphe 52 de la publication 9; ces recommandations portaient, d'une part, sur le fait que l'on devait éviter toute exposition inutile et, d'autre part, sur le fait que l'on devait réduire l'exposition à des niveaux aussi bas que cela est pratiquement réalisable, compte tenu de considérations économiques et sociales. Donc, on a tenu ces derniers temps à porter l'accent sur le fait que les recommandations de la Commission reposent sur trois concepts: le premier est la justification des activités, le second est l'optimisation de la radioprotection et le troisième est, bien entendu, que de toute façon il y a une limitation des doses qui, du point de vue sanitaire, ne pourrait être qu'absolument impérative. Voilà donc quelle a été l'évolution des concepts. Ce que nous avons vu au cours de cette semaine, c'est comment ces concepts étaient appliqués ou commençaient à l'être, et je voudrais mettre en évidence les facilités ou les difficultés que l'on rencontre dans l'exécution de ces tâches.

Les modalités d'irradiation

Il faut bien être conscient de ce que les modalités d'irradiation sont extrêmement diverses. Nous savons qu'il y a les deux grands modes d'exposition et de contamination radioactives, mais ce qu'il faut bien saisir, c'est que l'exposition par exemple à des sources externes se fait de façons fort différentes: il y a des expositions qui se font à bas débit, par exemple les expositions à l'irradiation naturelle, les expositions à l'irradiation après les rejets d'effluents, les expositions après les retombées; et il y a des expositions qui se font à haut débit, par exemple la plupart des expositions médicales ou les expositions accidentelles. Il y a d'autre part des différences dans la répartition des champs d'irradiation, et il y a également des différences selon la nature des rayonnements en cause. Donc, quand on utilise ces concepts, on a souvent tendance à traiter de la même façon des conditions d'exposition assez différentes et à utiliser les mêmes hypothèses, en particulier pour l'évaluation du détriment, comme nous le verrons plus loin. Après tout, on peut le faire puisqu'on fait des estimations qui donnent des ordres de grandeur, mais il ne faut pas leur attribuer une précision qui n'y est pas. Pour la contamination, j'ai été frappé de constater ici qu'on continue à évoluer, mais d'une façon assez lente, en ce qui concerne les différents facteurs qui interviennent dans le processus de contamination à partir des sources jusqu'à l'homme. Finalement, on étudie l'émission proprement dite, les transferts et l'incorporation. Or, on doit reconnaître que dans les études de transferts on a attaché beaucoup d'importance dans le passé aux transferts initiaux, c'est-à-dire aux rejets dans les deux milieux ou vecteurs initiaux que sont l'air et l'eau, et on a établi des modèles de transferts très nombreux, variés, qui correspondent plus ou moins bien à la réalité, et souvent même assez bien. Nous venons de voir qu'on s'intéressait beaucoup également aux transferts que je qualifierais d'intermédiaires, c'est-à-dire à partir de ces vecteurs initiaux la reprise par le milieu des différents polluants et leur cheminement écologique. Ces facteurs écologiques interviennent de façons fort diverses, en constituant des barrages, des filtres, des zones de rétention, des zones au contraire d'évacuation accélérée, et tout ceci est assez complexe. Il y a par contre un domaine qui est toujours assez peu étudié et présenté, c'est le troisième facteur: la dispersion terminale. Il ne faut pas oublier que l'air, l'eau et les aliments sont tous de la nourriture pour l'homme: on vit d'oxygène, on vit d'eau et on vit d'aliments divers. Or peu de communications ont souligné l'importance de ces facteurs. On a attiré l'attention au cours d'une discussion sur le fait que la distribution des produits alimentaires jouait un rôle important parce que l'autoconsommation est rare. Il y a autoconsommation pour l'air, il y a autoconsommation très limitée pour l'eau, et finalement ce sont les aliments qui comptent. La production, la distribution et la consommation alimentaires sont des facteurs de dispersion qui sont peut-être aussi importants que les facteurs liés à l'émission, et je pense qu'il y aurait intérêt à ce que ces questions soient développées.

L'estimation des doses

Lorsqu'on a parlé des problèmes relatifs aux estimations de doses, il m'a paru que l'on n'a pas insisté suffisamment sur les imprécisions en matière de dosimétrie qui sont courantes. Il est évident que la dosimétrie sanitaire, c'est-à-dire celle qui a pour but de déterminer les doses effectivement reçues à l'intérieur de l'organisme par les différents organes, les tissus ou même les cellules, a fait des progrès au cours des dernières années; mais elle doit en faire encore. Il ne faut pas se faire d'illusions: souvent on fait des estimations de doses moyennes à des organes qui sont assez osées. Quelques communications ont eu le mérite de montrer que ces estimations moyennes à des organes devaient être pondérées par le fait qu'il y avait des surexpositions et des sous-expositions de ces organes, qui pouvaient avoir une grande influence sur le détriment et, par conséquent, l'estimation des conséquences des doses individuelles ou des doses collectives. En ce qui concerne l'estimation des doses individuelles, je crois que l'on peut dire à juste titre, comme on l'a fait remarquer, qu'elle continue d'être une des actions essentielles. La CIPR, et en particulier son Comité 4, a au cours des dernières années publié un certain nombre de documents relatifs à la méthodologie de l'estimation des doses individuelles, avec toute la philosophie des nucléides critiques, des voies critiques de transfert, des organes critiques pour les personnes et des groupes critiques pour la population. Par conséquent tout ceci reste valable et c'est une des façons de traiter des problèmes auxquels nous sommes confrontés, il ne faut pas l'oublier. Ces estimations de doses individuelles, même pour la population, ne sont pas en compétition avec l'estimation des doses collectives, elles sont complémentaires: on doit faire les deux démarches et on doit faire les deux estimations. Il est bon d'une part d'estimer la dose collective, mais il est bon aussi de vérifier que dans aucun cas les groupes critiques ne risquent d'atteindre ou surtout de dépasser les doses individuelles qui ont été prescrites. Les doses collectives évidemment sont à l'ordre du jour: c'est une conception nouvelle et par conséquent il est normal que tout le monde cherche à œuvrer dans ce domaine. Pourtant cette conception n'est pas aussi nouvelle qu'elle le paraît car au fond les doses moyennes, en particulier les doses moyennes génétiques par exemple, étaient en fait des doses collectives. Mais évidemment les doses collectives ont connu un regain d'intérêt lorsqu'elles ont été considérées comme la base indispensable de l'estimation correcte du détriment. La notion de dose collective, il ne faut pas l'oublier, est assez complexe. Certaines communications ont évoqué des doses population, c'est-à-dire finalement des doses collectives étendues à l'ensemble de la population du monde. Mais le problème complexe auquel on aura à faire face dans l'avenir sera la nécessité de traiter ces problèmes à partir d'une source donnée pour un groupe de population ou pour l'ensemble de la population, et également de les traiter à partir d'un certain nombre de sources — qui ne seront pas vraiment en compétition mais qui se surajouteront — dont il faudra connaître la résultante, soit pour un groupe déterminé qui serait le groupe critique par rapport à l'ensemble des sources, soit pour l'ensemble de la population. Donc, il s'agit là de travaux pour l'avenir qui seront importants et où l'on sera obligé, soit pour les bassins fluviaux soit pour des territoires météorologiquement relativement homogènes, d'envisager les interférences et les résultantes des différentes sources.

L'estimation du détriment

En ce qui concerne le détriment, je crois qu'il faut éviter les querelles byzantines sur les hypothèses qui sont à la base de son estimation. Il est évident qu'initialement les recommandations de la CIPR visaient à mettre les travailleurs à l'abri des effets somatiques, et des effets somatiques immédiats, parce qu'à cette époque-là, il y a quelques décennies, des travailleurs présentaient des manifestations précoces d'irradiations importantes. Puis on s'est aperçu que, en fait, ce qui était le plus limitatif c'était soit les effets somatiques de type stochastique, soit les effets génétiques, qui sont eux-mêmes liés à la dose par des relations stochastiques. Et de ce fait, ce sont les effets stochastiques qui ont pris le devant de la scène. Il est évident que pour les effets stochastiques, l'état actuel de la radiobiologie et de l'épidémiologie ne nous permet pas de décider si, à des doses très faibles, on a ou n'a pas un seuil et si on a ou non une relation linéaire, mais la CIPR a choisi la voie de la prudence en admettant l'hypothèse pessimiste, c'est-à-dire l'absence de seuil et la linéarité. On ne peut pas le lui reprocher. Mais du moment qu'on choisit une hypothèse de travail, il n'y a pas lieu de remettre continuellement en question les résultats qui sont liés à cette hypothèse. On sait très bien que l'évaluation du détriment signifie simplement que si l'on parle de cancers induits, c'est que le nombre est compris entre le nombre calculé dans le cadre de cette hypothèse et zéro, et ceci doit être dit d'une façon très claire. Pour cette raison je ne pense pas que le plus judicieux, quand on parle du détriment, soit de faire des estimations absolues mais d'essayer toujours, comme d'ailleurs cela a été le cas dans plusieurs communications, de faire des comparaisons. En particulier, si on parle d'induction de cancer — pour prendre un exemple — il serait bon, quand on fait des estimations, de ne pas présenter un nombre de leucémies ou de cancers du poumon en tant que tels, mais de montrer quelle est l'estimation du détriment dû à l'activité considérée, de comparer ce détriment à celui d'autres activités impliquant des irradiations, de comparer ces irradiations artificielles avec le détriment dû à l'irradiation naturelle, et puis de comparer tous ces cancers hypothétiques avec les cancers réels et connus, qui sont ceux donnés par les statistiques des fréquences de cancers dans l'espèce humaine. A ce moment-là les hypothèses qui sont à la base de l'estimation du détriment reprennent leur place et on n'a plus besoin de se quereller pour savoir si elles sont plus ou moins bien fondées.

Les analyses décisionnelles

Ce dernier point présente un intérêt tout particulier. Il faut bien admettre que jusqu'à présent on prenait des décisions — on en a toujours pris, chacun de nous dès sa naissance prend des décisions. Il est évident que quand on sort dans la rue et qu'on traverse, on prend la décision de traverser et on sait qu'il y a une probabilité de se faire écraser, mais comme cette probabilité est suffisamment faible, on traverse quand même. Or, il faut reconnaître que jusqu'à présent les décisions en matière de protection étaient prises sur des bases qui étaient souvent floues ou inconnues, même pour ceux qui prenaient les décisions. Le mérite des analyses décisionnelles n'est pas de permettre de prendre des décisions différentes, c'est de permettre à ceux qui ont la responsabilité de prendre des décisions de les prendre d'une façon rationnelle. Je crois qu'il faut distinguer ici entre les deux recommandations qui portent, d'une part sur la justification des activités, d'autre part sur l'optimisation de la radioprotection. La justification des activités est une analyse décisionnelle du type coût/avantage, c'est-à-dire que normalement on compare les avantages de l'activité au coût du détriment sur le plan individuel, social, etc. Ceci est une entreprise délicate; elle est délicate parce que souvent il est difficile d'estimer exactement les avantages et que, d'autre part, les coûts, qu'il s'agisse de ceux du détriment ou des coûts opérationnels, sont souvent difficiles à évaluer dès que l'on ne veut pas seulement tenir compte des cancers induits mais aussi de toutes sortes de considérations économiques et sociales. Il est évident que, quand on traite ces questions d'une façon absolue, on arrive à des difficultés d'interprétation de l'analyse par ceux qui vont prendre la décision, et on en a vu des exemples au cours de cette semaine. Par contre, on a vu que ces analyses, quand elles sont faites par une méthode comparative et qu'elles tiennent compte, par exemple, de la production d'énergie nucléaire et de la production d'énergie thermique par le charbon ou par le pétrole, permettent de comparer dans l'analyse décisionnelle. Parce que l'avantage est le même on n'a plus qu'à comparer les détriments entre eux sur la base d'hypothèses qui sont aussi les mêmes, et on arrive ainsi à des conclusions qui sont tout à fait claires pour ceux qui ont à prendre des décisions. Si donc on n'utilise pas cette méthode comparative mais des méthodes absolues, on laisse la responsabilité à celui qui va prendre la décision, et en général sa décision sera une décision politique, souvent même une décision psychologique, ou sociologique, mais ce ne sera pas forcément une décision sanitaire.

Par contre, quand on aborde l'autre aspect de la question, celui de l'optimisation de la radioprotection, c'est-à-dire la réduction des doses au-dessous des limites à un niveau aussi bas que cela est pratiquement réalisable, on a affaire non plus à une analyse coût/avantage mais à une analyse coût/efficacité. On compare le coût de l'augmentation de la protection à l'avantage de la diminution du détriment, et on s'aperçoit qu'à un certain moment la diminution du détriment obtenue ne se justifie plus par rapport au coût de la protection. Le but d'une telle analyse, c'est d'essayer de déterminer à quel moment on est arrivé à l'optimum, c'est-à-dire à la meilleure solution. La décision n'est pas alors une décision politique, c'est une décision technique. D'ailleurs c'est en général à ceux qui ont la responsabilité de la protection ou la responsabilité des opérations qu'il appartient de la prendre, et ils peuvent le faire sans remonter, en quelque sorte, aux autorités politiques pour traiter ce problème. Mais là aussi il faut reconnaître que la méthode comparative est la meilleure. En fait, ce que l'on compare ce sont différents procédés de protection et leur efficacité effective à diminuer le détriment. Le présent Colloque nous a permis de constater que ces méthodes commençaient à être appliquées dans de nombreux pays et par de nombreux organismes. J'ai été agréablement surpris, étant donné la difficulté de traiter ces problèmes, de voir qu'ils l'étaient d'une façon assez remarquable, c'est-à-dire qu'on prenait bien en considération les différents domaines, les différents facteurs à retenir. Je dois dire que j'en suis d'autant plus heureux que, lorsque nous avons préparé la publication 22 au Comité 4 de la CIPR, nous avons ressenti une certaine anxiété, et je crois que la Commission de la CIPR elle-même était un peu inquiète d'ouvrir, dans une certaine mesure, la voie dans ce domaine. Mais on peut penser au terme de ce Colloque que les confusions qui existaient il y a quelques années se dissipent peu à peu et qu'on commence à traiter ces problèmes convenablement.

En terminant sur cette note optimiste je voudrais ajouter que les travaux qui sont actuellement faits permettront certainement de compléter les informations dont disposait le Comité scientifique des Nations Unies pour établir le bilan comparé des irradiations des populations, et on se rendra compte sans doute, comme on a pu le voir cette semaine, qu'il y a des contributions auxquelles on pensait moins et qui, au point de vue des doses collectives ou des doses population, sont loin d'être négligeables et peuvent même devenir importantes. On a parlé de l'habitat, des transports aériens; ce sont des choses que jusqu'à présent on avait, sinon négligées, du moins estimées de façon trop imprécise pour mettre en évidence leur contribution effective à l'irradiation des populations. Les travaux effectués présenteront sans doute un grand intérêt parce qu'ils permettront à l'opinion publique d'y voir clair et de savoir ce que représente l'impact du développement de l'énergie nucléaire en termes de considérations sur l'habitat, sur les transports aériens, sur l'irradiation médicale, et également par rapport à l'irradiation naturelle. On aura donc un classement, par la méthode comparative, quelles que soient les hypothèses de base, quelles que soient les unités employées. On pourra même se passer d'unités. Il sera possible, en prenant par exemple l'irradiation naturelle comme référence 100, d'établir les autres simplement par comparaison et d'avoir ainsi un instrument qui permettra d'une part à l'opinion publique d'être informée et d'autre part aux gouvernants d'agir pour le bien-être de l'humanité.

DISCUSSION

Suzanne VIGNES: Although, as you say, the proportionality hypothesis may have made it possible for ICRP to fix prudent exposure standards, the extended use of this hypothesis to calculate detriment to the population from infinitesimal additive doses creates a wrong impression and is indeed likely to terrify the uninitiated. According to this procedure, a population dose of 1 mrem in the United States of America leads to cancer deaths and genetic deaths. Actually, of course, an additional dose of 1 mrem is insignificant in the wide range of exposures to which people are subjected in everyday life.

When you speak of optimizing radiological protection, therefore, it is difficult to understand where the optimum should be fixed, since standards are already below the harmful dose level. Why, then, should they be reduced still further? And where will this process stop if one adopts the linear law?

H. P. JAMMET: In my opinion the ICRP adopted the linear dose-effect relationship not because it is exact but because it is conservative. By the very nature of its protective function, ICRP is more or less obliged to follow this conservative course and to predict the detriment resulting under the most adverse conditions. As you say, however, figures can be misleading, especially when they are not presented on a comparative basis. Considered alone a dose of 1 mrem may, under adverse circumstances, be serious, but compared with the natural background dose, which people don't bother about, it is insignificant. The fact that people may be terrified by certain predictions results mainly from their misinterpretation of the figures presented.

H. T. DAW: I should like to comment on another problem which needs attention, namely the contribution of occupational exposure to population dose. An important point in that connection is the contribution of lung cancer in uranium miners to the total detriment due to the production of nuclear electrical energy.

H. P. JAMMET: I agree that these matters should be properly considered.

M. COPPOLA: I am pleased to note that Mr. Jammet has touched on at least two subjects which are basic to the formulation of recommendations on radiation protection. Firstly, there is the need for better dosimetry in relation to the various organs: the effect of a certain dose varies depending on whether the dose is delivered to a particular structure of an organ or uniformly to the entire organ. This subject, which could be called 'Organ-dosimetry', should therefore receive adequate attention. The second important point is the need for better radiobiological and microdosimetric studies with a view to settling the question of the dose-effect relationship. In the dose region for which it is valid, Rossi's hypothesis is that the relation between dose and effect is in general not just linear, but polynomial, with a second-order term. In some cases, therefore, a linear relation may not necessarily be very conservative.

H. P. JAMMET: According to all the information available, the linear relationship is certainly a prudent one for the purposes of protection. Of course, in evaluating detriment we are of necessity dealing with hypotheses. The important thing is that we should all agree on the same hypotheses, so as to facilitate comparisons.

A. MARTIN: I do not believe that the best interests of safety are served by a simple and apparently cautious approach. In some ranges, for instance over 100 rads, the present basis of assessments may be optimistic. A different dose-risk hypothesis would have a profound effect on reactor siting and on emergency procedures.

Like the previous speakers I would oppose any playing-down of the importance of the dose-risk relationship.

H. P. JAMMET: I did say during my talk that these hypotheses had been adopted with reference to small doses. Accidents naturally bring up a different set of problems. At doses of 100 rads we are no longer dealing just with stochastic effects but with definite, predictable ones, and we therefore do not have to rely on hypotheses in considering these high doses.

E. SHALMON (Scientific Secretary, WHO): I would like to say a few concluding words on behalf of the international organizations involved in this work. The present Seminar is a good example of the close collaboration between IAEA (International Atomic Energy Agency), WHO (World Health Organization) and UNEP (United Nations Environment Programme) concerning the health aspects of medical radiation exposure and the evaluation of population doses. We do not view the Seminar as an end in itself, but also anticipate follow-up activities, including the promotion of standard terminology and methodology, for which there is such an urgent need. We should also like to organize a combined workshop and training station, where workers from various countries could discuss methods of measuring and assessing medical radiation as a first step to defining the health problems involved and proposing solutions. We should also like to continue with the work of collecting and evaluating comparable information on population exposure to radiation from all sources.

CHAIRMEN OF SESSIONS

Session I	M. ČOPIČ	Yugoslavia
Session II	D. BENINSON	Argentina
Session III	O. ILARI	Italy
Session IV	N.T. MITCHELL	United Kingdom
Session V	W.J. BAIR	United States of America
Session VI	M.M. SAUROV	Union of Soviet Socialist Republics
Session VII	J.C. VILLFORTH	United States of America
Session VIII/IX	L.-E. LARSSON H.P. JAMMET	Sweden France

SECRETARIAT

Scientific Secretaries	H.T. DAW	Division of Nuclear Safety and Environmental Protection, IAEA
	E. SHALMON	Radiation Health Unit, WHO
Administrative Secretary	Caroline DE MOL VAN OTTERLOO	Division of External Relations, IAEA
Editor	C.N. WELSH	Division of Publications, IAEA
Records Officer	D.J. MITCHELL	Division of Languages and Policy-making Organs, IAEA
Liaison Officer	B. CVIKL	Jožef Stefan Institute, Ljubljana

LIST OF PARTICIPANTS

ARGENTINA

Beninson, D.	Comisión Nacional de Energía Atómica, Gerencia de Protección Radiológica y Seguridad, Av. Libertador 8250, Buenos Aires

AUSTRALIA

Watson, G.M.	Australian Atomic Energy Commission, Sutherland, New South Wales 2232

AUSTRIA

Krejsa, P.P.	Österreichische Studiengesellschaft für Atomenergie, Lenaugasse 10, 1080 Wien
Pusch, W.	Bundesministerium für Gesundheit und Umweltschutz, Stubenring 1, 1010 Wien

BELGIUM

Fieuw, G.H.M.	Centre d'Etude de l'Energie Nucléaire, SCK/CEN, Boeretang 233, 2400 Mol
Hublet, P.F.	Ministère du Travail, 53 rue Belliard, 1040 Brussels
Stallaert, P.F.	Ministère du Travail, 53 rue Belliard, 1040 Brussels

BOLIVIA

Martinez Pacheco, J.	Comisión Boliviana de Energía Nuclear, Av.6 de Agosto 2905, Casilla 4821, La Paz

BULGARIA

Getschev, G.H.	Bulgarian Atomic Energy Commission, P.O. Box 102, Sofia

CANADA

Das Gupta, A.K.	Radiation Protection Bureau, Department National Health and Welfare, Ottawa
Shah, J.	Environmental Protection Service, Environment Canada, Ottawa

CSSR

Chorvát, D.	Institute of Indistrial Hygiene and Occupational Diseases, Duklianska 20, Bratislava

Hladký, E.	Power Research Institute, Jaslovské Bohunice
Kunz, E.	Institute of Hygiene and Epidemiology, Šrobárova 48, 10042 Prague 10
Neuman, F.	Czechoslovak Atomic Energy Commission, Slezská 9, Prague-Vinohrady

DENMARK

Grande, P.	State Institute of Radiation Hygiene, 378, Frederikssundsvej, 2700 Copenhagen BRH
Øhlenschlager, N.	National Health Service, 1 St. Kongensgade, 1264 Copenhagen
Schultz-Larsen, J.	Institute of Medical Genetics, 71, Rådmandsgade, 2200 Copenhagen N

FINLAND

Salo, Anneli L.	Institute of Radiation Physics, P.O. Box 268, 00101 Helsinki 10
Toivola, A.A.	Institute of Radiation Physics, P.O. Box 268, 00101 Helsinki 10

FRANCE

Ausset, R.	Centre de la Hague, B.P. 209, 50107 Cherbourg
Bahu, R.-J.	FRAMATOME, 77-81, rue du Mans, 92400 Courbevoie
Bataller, G.N.	Centre d'Etudes Nucléaires de Cadarache, B.P. No.1, 13115 St. Paul-lez-Durance
Bittel, R.	CEA, Département de Protection SPS, B.P. No.6, 92260 Fontenay-aux-Roses
Bovard, P.L.	CEA, Département de Protection SPS, B.P. No.6, 92260 Fontenay-aux-Roses
Bresson, G.	CEA, B.P. No.6, 92260 Fontenay-aux-Roses
Candès, P.	Centre d'Etudes Nucléaires de Saclay, B.P. No.2, 91190 Gif-sur-Yvette
de Choudens, H.A.	Centre d'Etudes Nucléaires, B.P. No.85, Centre de Tri, 38041 Grenoble
Delpla, M.J.A.	Electricité de France, 3, rue de Messine, 75008 Paris

Estournel, R.	CEA, Centre de Marcoule, B.P. No. 206, 30200 Bagnols-sur-Cèze
Fagnani, F.	CEA, B.P. No.6, 92260 Fontenay-aux-Roses
Garnier, Arlette	CEA, B.P. No.6, 92260 Fontenay-aux-Roses
Hébert, J.G.	Electricité de France, 3, rue de Messine, 75008 Paris
Jammet, H.P.	CEA, B.P. No. 6, 92260 Fontenay-aux-Roses
Lacourly, G.E.	CEA, Département de Protection, B.P. No.6, 92260 Fontenay-aux-Roses
Lasseur, C.G.	CEA, B.P. No.6, 92260 Fontenay-aux-Roses
Le Quinio, R.	CEA, B.P. No.2, 91190 Gif-sur-Yvette
Lulin, R.	Electricité de France, Etudes et Recherches, 17, av. du Gén. de Gaulle, 92 Clamart
Mattei, Janine	CEA, B.P. No.6, 92260 Fontenay-aux-Roses
Mazaury, E.C.	CEA, 33, rue de la Fédération, 75015 Paris
Mechali, D.	CEA, B.P. No.6, 92260 Fontenay-aux-Roses
Morlat, G.	Electricité de France et CEA, 3, rue de Messine, 75008 Paris
Nicolai, R.A.	CEA, 33, rue de la Fédération, 75015 Paris
Planet, J.	CEA, Département de Protection SPS, B.P. No.6, 92260 Fontenay-aux-Roses
Schaeffer, R.	Electricité de France, Comité de Radioprotection, 3, rue de Messine, 75008 Paris
Slizewicz, P.	CEA, B.P. No.2, 91190 Gif-sur-Yvette

Stolz, J.M. Electricité de France,
3, rue de Messine,
75008 Paris

Vignes, Suzanne Electricité de France,
Comité de Radioprotection,
3, rue de Messine,
75008 Paris

GERMAN DEMOCRATIC REPUBLIC

Schuettmann, W.O. Staatliches Amt für Atomsicherheit und Strahlenschutz,
Waldowallee 117,
1157 Berlin

GERMANY, FEDERAL REPUBLIC OF

Bayer, A. Kernforschungszentrum Karlsruhe,
Postfach 3640, 7500 Karlsruhe

Edelhäuser, H. Bundesministerium des Inneren,
Rheindorferstr. 198, 53 Bonn

Koelzer, W. Kernforschungszentrum Karlsruhe,
Postface 3640, 7500 Karlsruhe

Kowalewsky, H.K. Bundesanstalt für Materialprüfung,
Unter den Eichen 87, 1000 Berlin 45

Narrog, J.J. Ministerium für Arbeit, Gesundheit und Sozialordnung Baden-Württemberg,
Postfach 1250, 7000 Stuttgart 1

Riedel, H.H. Bundesgesundheitsamt,
Ingolstädter Landstrasse 1,
8042 Neuherberg

Roedler, H.D. Klinikum Steglitz der Freien Universität Berlin,
Hindenburgdamm 30, 1000 Berlin 45

Rosenbaum, O.E.R. Ministerium für Arbeit, Gesundheit und Soziales Nordrhein-Westfalen,
Horionpl. 1, 4000 Düsseldorf

Scheider, R.H.W. Institut für Reaktorsicherheit,
Glockeng. 2, 5000 Cologne 1

Stäblein, G. Kernforschungszentrum Karlsruhe,
Postfach 3640, 7500 Karlsruhe

Vogt, K.-J. Kernforschungsanlage Jülich,
Postfach 365, 517 Jülich

GREECE

Amarantos, S. Greek Atomic Energy Commission,
Democritos Nuclear Research Centre,
Aghia Paraskevi Attikis,
Athens

Karaoulani, Elisabeth	Democritos Nuclear Research Centre, Health Physics Division, Aghia Paraskevi Attikis, Athens

HUNGARY

Bíró, T.	HAS Institute of Isotopes, P.O. Box 77, 1525 Budapest
Deme, S.	Central Research Institute for Physics, P.O. Box 49, 1525 Budapest
Predmerszky, T.	Institute for Radiobiology and Radiation Hygiene, Pentz K. u. 5, P.O. Box 101, 1221 Budapest

INDIA

Supe, S.J.	Bhabha Atomic Research Centre, Trombay, Bombay 85

ITALY

Boeri, G.	CNEN, Viale Regina Margherita 125, 00198 Rome
Brofferio, Carla	CNEN, Viale Regina Margherita 125, 00198 Rome
Ilari, O.	CNEN, Divisione Protezione Sanitarie e Controlli, Viale Regina Margherita 125, 00198 Rome
Maiani, L.F.	Laboratori di Fisica, Istituto Superiore di Sanità, Viale Regina Elena 299, 00161 Rome
Muscolino, D.A.	Università di Roma, Città Universitaria, Rome
Salvadori, P.	Laboratori di Fisica, Istituto Superiore di Sanità, Viale Regina Elena 299, 00161 Rome

JAPAN

Ueda, T.	National Institute of Radiological Sciences, 4-9-1, Anagawa, Chiba shi, 280

KUWAIT

Al-Mudaires, J.M.	Ministry of Public Health, Radiotherapy Centre, Al-Sabah Hospital, Kuwait

NETHERLANDS

Baas, J.L.	Ministry of Public Health and Environmental Affairs, Dr. Reyersstraat 8-12, Leidschendam
van Daatselaar, C.J.	Ministry of Social Affairs, Balen van Andelplein 2, Voorburg
van der Heijde, H.B.	Shell International Research, Royal Dutch/Shell Laboratory, Badhuisweg 3, Amsterdam
Idema, H.W.J.	Scientific Council for Fission Energy, The Hague
Strackee, L.	National Institute of Public Health, P.O. Box 1, Bilthoven

NORWAY

Devik, F.	State Institute of Radiation Hygiene, Medical Section, Institute of Pathology, Rikshospital, Oslo 1
Koren, K.J.	State Institute of Radiation Hygiene, Oslo 3
Løken, P.C.	NVE, Middeltungsgt. 29, Oslo 3
Tveten, U.	Institutt for Atomenergi, 2007 Kjeller

PAKISTAN

Orfi, S.D.	Pakistan Institute of Nuclear Science and Technology, P.O. Nilore, Rawalpindi

PERU

Guzman Acevedo, C.	Junta de Control de Energía Atómica, Luis Aldana 120, Lima

POLAND

Kowalski, M.	Polish Atomic Energy Authority, Pałac Kultury i Nauki, Warsaw
Majle, T.	Ministry of Health and Social Welfare, National Institute of Hygiene, 24, Chocimska Str., Warsaw
Niewiadomski, T.	Institute of Nuclear Physics, Radzikowskiego 152, 31-342 Krakow

Peńsko, J.	Institute of Nuclear Research, Radiation Protection Department, 05-400 Świerk/Otwock
Szumski, W.	Atomic Energy Authority, Pałac Kultury i Nauki, Warsaw

SOUTH AFRICA

Simpson, D.M.	Atomic Energy Board, Private bag x256, Pretoria

SPAIN

Díaz de la Cruz, F.	Junta de Energía Nuclear, Avenida Complutense 22, Madrid 3
Pleite Sánchez, J.	Subdirección General de Sanidad Ambiental y Medicina Preventiva, Dirección General de Sanidad, Ministerio de la Gobernación, Plaza de Espania 13, Madrid 3

SUDAN

El Hadi, Z.El H.A.	Department of Physics, University of Khartoum

SWEDEN

Bergström, S.O.W.	AB Atomenergi, Fack, 61101 Nyköping
Cederqvist, H.G.	ASEA-ATOM, 72104 Västerås
Cervin, T.G.	Sydkraft, Fack, 20070 Malmö 5
Gyllander, Christina	AB Atomenergi, Studsvik, 61101 Nyköping
Larsson, L.-E.	National Institute of Radiation Protection, Fack, 10401 Stockholm 60
Mandahl, B.	Atomkraftkonsortiet, Birger Jarlsgatan 41A, Stockholm

SWITZERLAND

Michaud, B.M.	Federal Commission for Surveillance of Radioactivity, Physics Institute of the University, 1700 Fribourg
Wagner, G.	Federal Commission for Radiological Protection, 3045 Meikirch

TURKEY

Alkan, H.	Çekmece Nuclear Research Center, P.K. 1, Havaalani, Istanbul
Yülek, Gürcü G.	Ankara Nuclear Research Center, Faculty of Science, Beşeuler, Ankara

USSR

Povaliaev, A.P.	Biophysical Institute of the Health Ministry, Zhivopisnaja 46, Moscow D 182
Saurov, M.M.	Institute of Agriculture and Radiology, St. Prijnishni 6, block 17, Moscow

UNITED KINGDOM

Goddard, A.J.H.	Imperial College of Science and Technology, London SW7 2BX
Hookway, B.R.	Department of Environment, Whitehall, London S.W.1
Martin, A.	Associated Nuclear Services, 14-16 Regent St., London S.W.1
Mitchell, N.T.	Ministry of Agriculture, Fisheries and Food, Fisheries Radiobiological Laboratory, Hamilton Dock, Lowestoft, Suffolk
Shaw, J.	Nuclear Engineering Department, Queen Mary College, London El 4NS
Tildsley, F.C.J.	Nuclear Installations Inspectorate, Department of Energy, Thames House South, London SW1P 4QJ
Wrixon, A.D.	National Radiological Protection Board, Harwell, Didcot

UNITED STATES OF AMERICA

Anspaugh, L.R.	Lawrence Livermore Laboratory, University of California, P.O. Box 808, Livermore, Calif. 94550
Bair, W.J.	Battelle Pacific Northwest Laboratory, Richland, Wash. 99352
Barker, R.F.	Directorate of Regulatory Standards, Washington, D.C. 20545
Biles, M.B.	US Atomic Energy Commission, Washington, D.C. 20545

Hub, K.A. — Argonne National Laboratory, 9700 South Cass, Argonne, Ill.60439

Meinhold, C.B. — Brookhaven National Laboratory, Upton, New York 11973

Rowe, W.D. — Environmental Protection Agency, Washington, D.C. 20460

Siegel, J.R. — Gibbs &Hill, Inc., 393 7th Ave., New York, N.Y. 10001

Villforth, J.C. — US Department of Health, Education and Welfare, Bureau of Radiological Health, Rockville, Maryland 20852

VENEZUELA

Lopez Barrios, A. — Instituto Venezolano de Investigaciones Científicas (IVIC), Apartado 1827, Caracas

YUGOSLAVIA

Arsov, L.S. — Elektrostopanstvo, Tiranska 2, Skopje

Cvikl, B. — Jožef Stefan Institute, Jamova 39, 61000 Ljubljana

Čopič, M. — Jožef Stefan Institute, Jamova 39, 61000 Ljubljana

Djukić, Z. — Health Protection Department, Boris Kidrič Institute of Nuclear Sciences, Vinča, P.O. Box 522, 11001 Belgrade

Draganović, B. — Veterinary Faculty, Bul JNA 18, 11000 Belgrade

Gabrovšek, Z. — Nuklearna Elektrarna Krško, ul.4 julija 38, 62870 Krško,

Hrušovar, G. — Rep.Sanitary Inspectorate, Parmova 33, 61000 Ljubljana

Ignjatović, S. — Institute of Occupational and Radiological Health, Belgrade

Janković, Olga — Boris Kidrič Institute of Nuclear Sciences, Vinča, P.O. Box 522, 11001 Belgrade

Jovanović, P.	Federal Committee for Health and Social Welfare, Bulevar Avnoj 104, Belgrade
Karuza, J.	Nuklearna Elektrarna Krško, ul.4 julija 38, 62870 Krško
Kocić, A.M.	Boris Kidrič Institute of Nuclear Sciences, Vinča, P.O. Box 522, 11001 Belgrade
Kristan, J.	Jožef Stefan Institute, Jamova 39, 61000 Ljubljana
Kubelka, V.	Rudjer Bošković Institute, P.O. Box 1016, 41001 Zagreb
Legat, F.	Geološki Zavod, Dimičeva 16, 61000 Ljubljana
Lulić, S.	Rudjer Bošković Institute, P.O. Box 1016, 41001 Zagreb
Maksić, R.	Federal Office for Scientific, Cultural, Educational and Technical Co-operation, Kosančićev venac 29, Belgrade
Marković, P.	Boris Kidrič Institute of Nuclear Sciences, Vinča, P.O. Box 522, 11001 Belgrade
Mitrović, S.M.	Boris Kidrič Institute of Nuclear Sciences, Vinča, P.O. Box 522, 11001 Belgrade
Ninković, M.M.	Boris Kidrič Institute of Nuclear Sciences, Vinča, P.O. Box 522, 11001 Belgrade
Ozretić, B.I.	Center for Marine Research, Rudjer Bošković Institute, Rovinj
Pirš, M.	Jožef Stefan Institute, Jamova 39, 61000 Ljubljana
Popović, V.	Institute for Medical Research, M. Pijade 758, 41000 Zagreb
Radosavljević, Ž.	Boris Kidrič Institute of Nuclear Sciences, Vinča, P.O. Box 522, 11001 Belgrade

Radovanović, R.G.	Institute of Occupational and Radiological Health, Belgrade
Ristić, V.D.	Boris Kidrič Institute of Nuclear Sciences, Vinča, P.O. Box 522, 11001 Belgrade
Simić, B.	Institute of Hygiene, Faculty of Medicine, M. Pijade 6, Sarajevo
Smiljanić, G.	Rudjer Bošković Institute, P.O. Box 101, 41001 Zagreb
Sterle, M.	Institute for Work Safety, Radiation Protection Department, Korytkova 3/A, 61000 Ljubljana
Stojanović, D.B.	Boris Kidrič Institute of Nuclear Sciences, Vinča, P.O. Box 522, 11001 Belgrade
Strohal, P.	Rudjer Bošković Institute, P.O. Box 101, 41001 Zagreb
Sušnik, J.	Jožef Stefan Institute, Jamova 39, 61000 Ljubljana
Tasovac, T.	Boris Kidrič Institute of Nuclear Sciences, Vinča, P.O. Box 522, 11001 Belgrade
Terček, V.	Nuklearna Elektrarna Krško, ul. 4 julija 38, 62870 Krško
Tomaš, P.	Rudjer Bošković Institute, P.O. Box 101, 41001 Zagreb
Vidmar, M.	Boris Kidrič Institute of Nuclear Sciences, Vinča, P.O. Box 522, 11001 Belgrade
Vidmar, M.	Rep.Sekret. za Gospodarstvo, Gregorčičeva 25, Ljubljana

ORGANIZATIONS

CEC (Commission of European Communities)

Coppola, M.	CCR/Euratom, Biology Service, 21020 Ispra, Italy

Recht, P. | CEC,
rue Aldringen,
Luxemburg

ICRP (International Commission on Radiological Protection)

Sowby, F.D. | Clifton Ave., Sutton,
Surrey SM2 5PU,
England

UNITED NATIONS

Edvarson, K.E. | Research Institute of National Defence,
10450 Stockholm, Sweden

AUTHOR INDEX

Arabic numerals underlined refer to the first page of a paper by the author concerned.
Further Arabic numerals denote contributions to discussions.
Literature references are not indexed.

CONVERSION TABLE:
FACTORS FOR CONVERTING UNITS TO SI SYSTEM EQUIVALENTS*

SI base units are the metre (m), kilogram (kg), second (s), ampere (A), kelvin (K), candela (cd) and mole (mol). [For further information, see International Standards ISO 1000 (1973), and ISO 31/0 (1974) and its several parts]

Multiply		*by*	*to obtain*
Mass			
pound mass (avoirdupois)	1 lbm	= 4.536×10^{-1}	kg
ounce mass (avoirdupois)	1 ozm	= 2.835×10^{1}	g
ton (long) (= 2240 lbm)	1 ton	= 1.016×10^{3}	kg
ton (short) (= 2000 lbm)	1 short ton	= 9.072×10^{2}	kg
tonne (= metric ton)	1 t	= 1.00×10^{3}	kg
Length			
statute mile	1 mile	= 1.609×10^{0}	km
yard	1 yd	= 9.144×10^{-1}	m
foot	1 ft	= 3.048×10^{-1}	m
inch	1 in	= 2.54×10^{-2}	m
mil (= 10^{-3} in)	1 mil	= 2.54×10^{-2}	mm
Area			
hectare	1 ha	= 1.00×10^{4}	m^2
(statute mile)^2	1 mile^2	= 2.590×10^{0}	km^2
acre	1 acre	= 4.047×10^{3}	m^2
yard^2	1 yd^2	= 8.361×10^{-1}	m^2
foot^2	1 ft^2	= 9.290×10^{-2}	m^2
inch^2	1 in^2	= 6.452×10^{2}	mm^2
Volume			
yard^3	1 yd^3	= 7.646×10^{-1}	m^3
foot^3	1 ft^3	= 2.832×10^{-2}	m^3
inch^3	1 in^3	= 1.639×10^{4}	mm^3
gallon (Brit. or Imp.)	1 gal (Brit)	= 4.546×10^{-3}	m^3
gallon (US liquid)	1 gal (US)	= 3.785×10^{-3}	m^3
litre	1 l	= 1.00×10^{-3}	m^3
Force			
dyne	1 dyn	= 1.00×10^{-5}	N
kilogram force	1 kgf	= 9.807×10^{0}	N
poundal	1 pdl	= 1.383×10^{-1}	N
pound force (avoirdupois)	1 lbf	= 4.448×10^{0}	N
ounce force (avoirdupois)	1 ozf	= 2.780×10^{-1}	N
Power			
British thermal unit/second	1 Btu/s	= 1.054×10^{3}	W
calorie/second	1 cal/s	= 4.184×10^{0}	W
foot-pound force/second	1 ft·lbf/s	= 1.356×10^{0}	W
horsepower (electric)	1 hp	= 7.46×10^{2}	W
horsepower (metric) (= ps)	1 ps	= 7.355×10^{2}	W
horsepower (550 ft·lbf/s)	1 hp	= 7.457×10^{2}	W

* Factors are given exactly or to a maximum of 4 significant figures

Multiply			by	to obtain
Density				
pound mass/inch^3	1 lbm/in^3	=	2.768×10^4	kg/m^3
pound mass/foot^3	1 lbm/ft^3	=	1.602×10^1	kg/m^3
Energy				
British thermal unit	1 Btu	=	1.054×10^3	J
calorie	1 cal	=	4.184×10^0	J
electron-volt	1 eV	$\simeq$	1.602×10^{-19}	J
erg	1 erg	=	1.00×10^{-7}	J
foot-pound force	1 ft·lbf	=	1.356×10^0	J
kilowatt-hour	1 kW·h	=	3.60×10^6	J
Pressure				
newtons/metre^2	1 N/m^2	=	1.00	Pa
atmosphere[a]	1 atm	=	1.013×10^5	Pa
bar	1 bar	=	1.00×10^5	Pa
centimetres of mercury (0°C)	1 cmHg	=	1.333×10^3	Pa
dyne/centimetre^2	1 dyn/cm^2	=	1.00×10^{-1}	Pa
feet of water (4°C)	1 ftH_2O	=	2.989×10^3	Pa
inches of mercury (0°C)	1 inHg	=	3.386×10^3	Pa
inches of water (4°C)	1 inH_2O	=	2.491×10^2	Pa
kilogram force/centimetre^2	1 kgf/cm^2	=	9.807×10^4	Pa
pound force/foot^2	1 lbf/ft^2	=	4.788×10^1	Pa
pound force/inch^2 (= psi)[b]	1 lbf/in^2	=	6.895×10^3	Pa
torr (0°C) (= mmHg)	1 torr	=	1.333×10^2	Pa
Velocity, acceleration				
inch/second	1 in/s	=	2.54×10^1	mm/s
foot/second (= fps)	1 ft/s	=	3.048×10^{-1}	m/s
foot/minute	1 ft/min	=	5.08×10^{-3}	m/s
mile/hour (= mph)	1 mile/h	=	4.470×10^{-1} 1.609×10^0	m/s km/h
knot	1 knot	=	1.852×10^0	km/h
free fall, standard (= g)		=	9.807×10^0	m/s^2
foot/second^2	1 ft/s^2	=	3.048×10^{-1}	m/s^2
Temperature, thermal conductivity, energy/area·time				
Fahrenheit, degrees −32	°F − 32		$\frac{5}{9}$	°C
Rankine	°R		$\frac{5}{9}$	K
1 Btu·in/ft^2·s·°F		=	5.189×10^2	W/m·K
1 Btu/ft·s·°F		=	6.226×10^1	W/m·K
1 cal/cm·s·°C		=	4.184×10^2	W/m·K
1 Btu/ft^2·s		=	1.135×10^4	W/m^2
1 cal/cm^2·min		=	6.973×10^2	W/m^2
Miscellaneous				
foot^3/second	1 ft^3/s	=	2.832×10^{-2}	m^3/s
foot^3/minute	1 ft^3/min	=	4.719×10^{-4}	m^3/s
rad	rad	=	1.00×10^{-2}	J/kg
roentgen	R	=	2.580×10^{-4}	C/kg
curie	Ci	=	3.70×10^{10}	disintegration/s

[a] atm abs: atmospheres absolute;
atm (g): atmospheres gauge.

[b] lbf/in^2 (g) (= psig): gauge pressure;
lbf/in^2 abs (= psia): absolute pressure.